Multiscale Biomechanics

Multiscale Biomechanics

Theory and Applications

Soheil Mohammadi
School of Civil Engineering
University of Tehran
Tehran, Iran

WILEY

This edition first published 2023

Registered Office(s)
John Wiley & Sons, Inc., 111 River Street, Hoboken, NJ 07030, USA
John Wiley & Sons Ltd, The Atrium, Southern Gate, Chichester, West Sussex, PO19 8SQ, UK

For details of our global editorial offices, customer services, and more information about Wiley products visit us at www.wiley.com.

Wiley also publishes its books in a variety of electronic formats and by print-on-demand. Some content that appears in standard print versions of this book may not be available in other formats.

Library of Congress Cataloging-in-Publication Data
Name: Mohammadi, S. (Soheil), author.
Title: Multiscale biomechanics : theory and applications / Soheil Mohammadi
Description: Hoboken, NJ : John Wiley & Sons Ltd., 2023. | Includes bibliographical references and index.
Identifiers: LCCN 2022054751 (print) | LCCN 2022054752 (ebook) | ISBN 9781119033691 (hardback) | ISBN 9781119033738 (pdf) | ISBN 9781119033721 (epub) | ISBN 9781119033714 (ebook)
Subjects: LCSH: Biomechanics.
Classification: LCC QH513 .M642 2023 (print) | LCC QH513 (ebook) | DDC 571.4/3--dc23/eng/20230418
LC record available at https://lccn.loc.gov/2022054751
LC ebook record available at https://lccn.loc.gov/2022054752

Cover image: © SCIEPRO/Getty Images, Courtesy of Soheil Mohammadi
Cover design by Wiley

Set in 9.5/12.5pt STIXTwoText by Integra Software Services Pvt. Ltd, Pondicherry, India
C9781119033691_010623

“Love is the unique mysterious solution to the complicated question of Life”

To: Mansoureh, Sogol & Soroush

Contents

Preface

Life and space have been the most fascinating scientific concepts that I used to cogitate from my childhood; watching the "Star Trek" series or thinking on the way all living organisms have evolved from non-living atoms and molecules.

As a civil engineer with numerical skills, however, it may seem quite unusual to get involved with biological problems. Nevertheless, the computational mechanics has bridged over the different pillars of science, from buildings to aerospace structures, and from spectacular suspension bridges to intelligent nano additives in biological systems.

It all started about a decade ago, when my former student, Shahrokh Shahi, began his thesis endeavour on multiscale biomechanics. After his graduation, we planned for a book on the subject, although he had to separate from the project to follow his future quests.

Biomechanics is primarily used to study the wide range of mechanical responses of biosystems: from biomolecules scales up to the organ and body levels, and from routine medical procedures to synthesized tissues. The wealth of well-developed mathematical and numerical methods of solving general engineering and physical problems is now available to assist clinical staff and medical industries in assessing the existing procedures and products or to propose new engineered designs and concepts for future research and development.

I have tried to provide the theoretical and computational bases of biomechanics in this textbook and to present my small contribution to encourage young talents to further advance the numerical capabilities for analysis of biomechanical applications.

The book, which can be regarded as an introduction to multiscale biomechanics, is composed of three parts. The preliminary part is meant to provide an introduction and insight into the general concept of biomechanics and the wide variety of biological problems that can be solved numerically through single and multiscale methods.

Part II is dedicated to analytical and numerical bases. In its first set of chapters, the general concepts of continuum mechanics associated with solid materials, fluid flow and diffusion problems are provided. The second set of chapters covers the basics of numerical analysis methods, including the finite element method, the extended finite element method, the isogeometeric analysis, the principles of meshless methods, and the variable node element. The final chapter of Part II discusses the multiscale methods. It begins by examining the homogenization technique. Then the atomistic/molecular dynamics and statistical mechanics are discussed in detail. The sequential multiscale method is thoroughly discussed by a sample case, which spans the extremely wide range of atomistic simulations, nanoscale analysis, microscale computations, mesoscale investigation and macroscale study. The multiscale chapter is concluded by a comprehensive review of concurrent multiscale schemes.

Part III is dedicated to discussions on single and multiscale biomechanical simulations. Its first chapter is devoted to the modelling of soft tissues. It begins with explaining the composition and physiology of soft tissues and their macro and micro hyperelastic constitutive laws. This chapter includes several single and multiscale simulations of soft tissue applications, including damaged tissues, the aortic heart valve, skin damage, arterial wall degradation, wound healing and the viscoelastic response of the brain.

The next chapter of this part is dedicated to hard tissues, and briefly explains the composition of bones and the necessary mechanical models. A number of single and multiscale simulations are presented to provide more insight into the way hard tissue biomechanical studies are performed. The chapter closes with a discussion on the healing processes of hard tissues.

Part III is concluded by a brief complementary chapter on a number of supplementary topics, covering the principles of stenting simulations, multiscale modelling of the eye, the concept of a shape memory polymer drug delivery system and an introduction to artificial intelligence and deep learning in biomechanical applications.

This book is a result of intense research works of the computational community for many years. I have learned from them and tried to pass this package of knowledge to others. I am always obliged to the outstanding works of researchers all over the world and I have done my best to explicitly acknowledge their achievements within the text, figures, tables and formulae, and I sincerely apologize for any unintended failing in appropriately acknowledging them.

I would like to express my acknowledgement to the professional team of John Wiley & Sons, Ltd., for their excellent work that facilitated the whole process of publication of the book: in particular to M. Hammond, E. Willner, A. Hunt, T.P. Koh, S. Lemore, V. Saminathan, D. Shanmugasundram, L. Poplawski and I. Proietti, and to P. Bateson for excellent professional editing of the manuscript.

My sincere thanks to S. Shahi, who jointly began this venture about a decade ago. A number of my colleagues and former students helped me in preparing this book, which I am indebted to acknowledge and sincerely appreciate: Dr. H. Bayesteh reviewed and commented on the homogenization chapter in detail and on time, Dr. O. Alizadeh examined the discussion on concurrent multiscale methods, H. Moslemzadeh prepared the data for a number of sections of the multiscale chapter and the sequential multiscale chapter is credited to Dr. M. Eftekhari. Contributions of M. Vokhshouri and P. Fatemi, and the help of E. Baniasad are acknowledged and the outstanding assistance and contribution of S. Hatefi in different parts are gratefully appreciated.

All numerical simulations of Part III are from the excellent works of my former and present students in the School of Civil Engineering, University of Tehran, beginning with S. Shahi, who developed our basic multiscale solution across multiple levels of organ, tissue and cell, and analysed the heart valve leaflet. Dr. F. Fathi and S. Hatefi then extended the work to damaged soft tissues, and M. Janfada and S. Zolghadr developed the hyperelastic and viscoelastic models of the human brain. In parallel, K. Khaksar solved the complicated coupled system of equations of the healing process in soft tissues, followed by the same approach for hard tissues by M. Zamani. A.R. Torabizadeh eagerly simulated the problems of hard tissues across multiple scales and A. Foyouzat proposed the idea of a nano scale shape memory drug delivery system.

The power for completing the book has been the love, trust and understanding of my beloved family, as they had to comply with all my commitments and to adapt to my long absences. This work is proudly dedicated to noble Iranian students who accomplish academic achievements while challenging for prosperous life and freedom. A living entity may be damaged in harsh environments, but it heals itself, evolves and rises to a flourishing future.

Soheil Mohammadi
University of Tehran
Summer–Fall 2022
Tehran, Iran

List of Abbreviations

1D	One Dimensional
2D	Two Dimensional
3D	Three Dimensional
AI	Artificial Intelligence
AIS	Abbreviated Injury Scale
ALE	Arbitrary Lagrangian Eulerian
ANN	Artificial Neural Networks
ASME	American Society of Mechanical Engineers
AtC	Atomistic to Continuum Method
AV	Aortic Valve
BCC	Body-Centred Cubic Crystal
BDM	Bridging Domain Method
bFDM	Backward Finite Difference Method
BSM	Bridging Scale Method
CADD	Coupled Atomistics and Discrete Dislocation Mechanics
CAR	Cell Aspect Ratio
CDM	Continuum Damage Mechanics
CEMHYD3D	A Three-Dimensional Cement Hydration and Microstructure Development Modeling Package
CFD	Computational Fluid Dynamics
cFDM	Central Finite Difference Method
CHARMM	Chemistry at Harvard Macromolecular Mechanics (MD Potential)
CLE	Coupled Lagrangian Eulerian
CLL	Coupled Lagrangian-Lagrangian
CNN	Convolutional Neural Networks
CNT	Carbon Nano Tube
CQC	Cluster Quasicontinuum
CSH	Silicate-Cement-Hydrate
CSPM	Corrected SPH Method
CT	Computer Tomography
CTOA	Crack Tip Opening Angle
CTOD	Crack Tip Opening Displacement

DCMM	Disordered Concurrent Multiscale Method
DL	Deep Learning
DNA	Deoxyribonucleic Acid
DOF	Degree of Freedom
DWCNT	Double Wall Carbon Nano Tube
EAM	Embedded Atom Model
ECM	Extracellular Matrix
EDI	Equivalent Domain Integral
EFG	Element Free Galerkin
EMM	Enriched Multiscale Method
EMT	Effective Medium Theory
EPFM	Elastoplastic Fracture Mechanics
FDM	Finite Difference Method
Feat	Finite Element and Atomistic Model
FEM	Finite Element Method
fFDM	Forward Finite Difference Method
FGM	Functionally Graded Material
FPM	Finite Point Method
FS	Finnis–Sinclair
FSI	Fluid-Structure Interaction
FVM	Finite Volume Method
GA	Genetic Algorithm
GAN	Generative Adversarial Networks
HIC	Head Injury Criterion
HPC-Lab	High Performance Computing Laboratory
IC	Interstitial Cell
IGA	Isogeometric Analysis
IKN	Irving–Kirkwood–Noll Stress
IOP	Intraocular Pressure
ITZ	Interface Transition Zone
LAMMPS	Large Scale Atomic/Molecular Massively Parallel Simulator (Software)
LEFM	Linear Elastic Fracture Mechanics
LJ	Lennard–Jones
LSM	Level Set Method
MAAD	Macroscopic Atomistic ab Initio Dynamics
ML	Machine Learning
MD	Molecular Dynamics
MEAM	Modified Embedded Atom Model
MEMS	Micro Electromechanical System
MLPG	Meshless Local Petrov Galerkin
MLS	Moving Least Square
MMP	Matrix Metalloproteinase
MPS	Maximum Principal Strain
MS	Molecular Statics
MSC	Mesenchymal Stem Cell

MSD	Mean Square Displacement Function
MSE	Mean-Squared Error
MWCNT	Multi Wall Carbon Nano Tube
NASGRO	A computer program for analysis of fracture and fatigue crack growth.
NCOP	Non-Collagenous Organic Proteins
NIST	National Institute of Standards and Technology
NiTi	Nickle-Titanium Alloy
NPT	Isothermal-Isobaric Ensemble (Constant Number of Atoms, Pressure and Temperature)
NTG	Normal Tension Glaucoma
NURBS	Non-Uniform Rational B-Spline
NVE	Microcanonical Ensemble (Constant Number of Atoms, Volume and Energy)
NVT	Canonical Ensemble (Constant Number of Atoms, Volume and Temperature)
ONH	Optical Nerve Head
OVITO	Open Visualization Tool
PDE	Partial Differential Equations
PIM	Point Interpolation Method
PINN	Physics-Informed Neural Networks
POAG	Primary Open Angle Glaucoma
PPIM	Polynomial Point Interpolation Method
PTCA	Percutaneous Transluminal Coronary Angioplasty
PU	Partition of Unity
QC	Quasi-continuum
RBF	Radial Basis Function
RDF	Radial Distribution Function
REP	Recovery by Equilibrium on Patch
RKPM	Reproducing Kernel Particle Method
RL	Reinforcement Learning
RNA	Ribonucleic acid
RNN	Recursive (Recurrent) Neural Networks
RPIM	Radial Point Interpolation Method
RPPIM	Radial-Polynomial Point Interpolation Method
RVE	Representative Volume Element
SIF	Stress Intensity Factor
SMA	Shape Memory Alloy
SME	Shape Memory Effect
SMP	Shape Memory Polymer
SPH	Smoothed Particle Hydrodynamics
SPR	Superconvergent Patch Recovery
SW	Stone–Wales Defect
SWCNT	Single Wall Carbon Nano Tube
TBI	Traumatic Brain Injury
TPS	Thin Plate Spline
UV	Ultra Violet
VAF	Velocity Acceleration Function

VNE	Variable Node Element
VNME	Variable Node Multiscale Element
VNMM	Variable Node Multiscale Method
VV	Velocity Verlet
WLS	Weighted Least Square
XFEM	Extended Finite Element Method
XIGA	Extended Isogeometric Analysis
μ-CT	Micro-Computed Tomography
μic	Particle Kinetics Chemical Hydration Model (Software)

Part I

Introduction

1

Introduction

1.1 Introduction to Biomechanics

One of the astonishing secrets of life is probably the fact that any living organism is composed of non-living objects: atoms and molecules. Nevertheless, they form biomolecules and eventually constitute the living tissues and organs.

Biology is a comprehensive set of science that is related to living bodies, organs, tissues, cells, molecules and even smaller constituents, and their potential interactions. While biochemistry is part of the biology that is dedicated to what is going on within a biomolecule and their interactions, biomechanics is primarily used to study the wide range of mechanical responses of biosystems; from the biomolecule level up to the organ and body levels.

In this chapter, an introduction is provided to present an insight into the general concept of biomechanics and the wide variety of biological problems that can be solved by conventional or multiscale numerical methods. The numerical approach is widely used in research and development activities for medical-related industries and can be assumed as a reliable tool to assist the medical staff in assessing the existing damages or failures of living tissues or to resemble a specific procedure/operation based on the realistic conditions of a patient and to predict the level of success or potential consequences, before actually performing it.

1.2 Biology and Biomechanics

The word 'biology' gradually replaced the phrase of 'natural history' well over the centuries to denote the branches of science which deal with living animals or plants (Fung 1993). In fact, it was originally applied to all world-related scientific contributions and not to any individual living object. Nowadays, biology is adopted to designate any science related to the living bodies, organs, tissues, cells, molecules and even smaller constituents, and their potential interactions. It has also spanned to synthetic tissues and includes the technologies of making tailored and engineered materials to replace the existing malfunctioning tissues/organs. The important issue of the drug delivery mechanisms can also be considered within the general concept of biology.

Biomechanics represents an engineering approach to biology, where the general concepts of mechanics are adopted to formulate the biological phenomena and to study them

Multiscale Biomechanics: Theory and Applications, First Edition. Soheil Mohammadi.

through the wealth of well-developed mathematical and numerical methods of solving general mechanical problems. While some basic mechanical concepts, such as stress or strain, constitutive equations, strength, fatigue and fracture, fluid flow, diffusion, composite action, etc., may not be well defined or used in biology, they are efficiently adopted in biomechanics to define and study the behaviour of living tissues and the existing interactions of different living objects in an organ or in a more complex set of tissues.

Biomechanics spans virtually all aspects of biology, ranging from molecular scales to surgery procedures and from micromodelling of aneurysm disease to the design of large orthopaedic items and haemodialysis machines.

While molecular and cell biology may seem to be only involved with biological and/or chemical phenomena in very small scales, nevertheless, the biomechanics has been adopted to simulate complex organs down to their cell microstructure, such as the brain and its nerve cells, as presented in Figure 1.1, and the heart and its leaflets, the corresponding microscale layers and even an interstitial cell, as depicted in Figure 1.2.

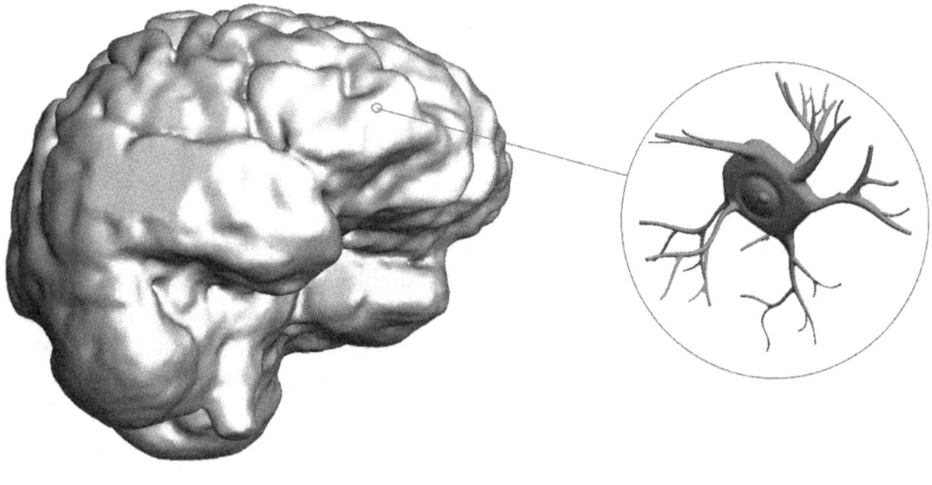

Figure 1.1 The brain organ and its nerve cell.

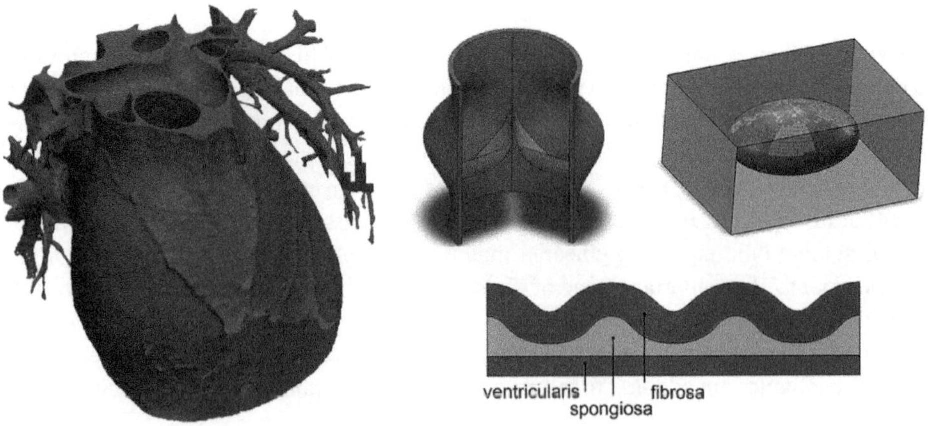

Figure 1.2 Sample biomechanical simulation of an organ, its tissue layers and the structural molecular/cell level.

It is not always necessary to examine a biomechanical problem down to the cell scale. For instance, in the study of the behaviour of a bone, it is sufficient to take into account its heterogeneous or porous microstructure, as typically presented in Figure 1.3.

In surgery-related topics, the whole process of healing of scars or damages of a soft tissue, as typically illustrated in Figure 1.4, can now be fully investigated by biomechanics. These predictions can also be quite useful in clinical orthopaedic or cosmetic surgeries. Similarly, modelling the way a tumour is evolved by the analysis of growth and mass transfer across a membrane/interface to predict how the tumour is contained or expanded by time may help the specialist to obtain a better diagnosis and more effective medical treatment.

Biomechanics has been involved in the study of various aspects of the cardiovascular system. For instance, it includes modelling of heart valves, blood flow, hemodynamic disorders and aneurysms, etc. More recently, modelling of the complete procedure of stenting

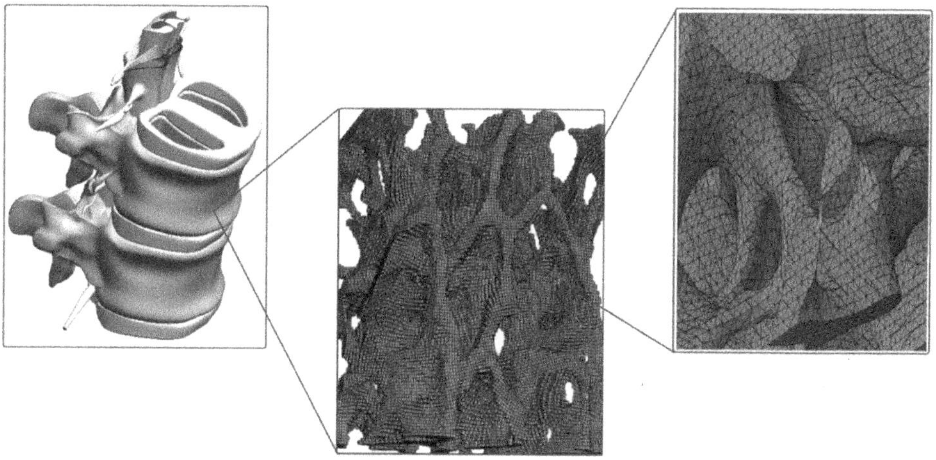

Figure 1.3 A typical bone with a porous microstructure.

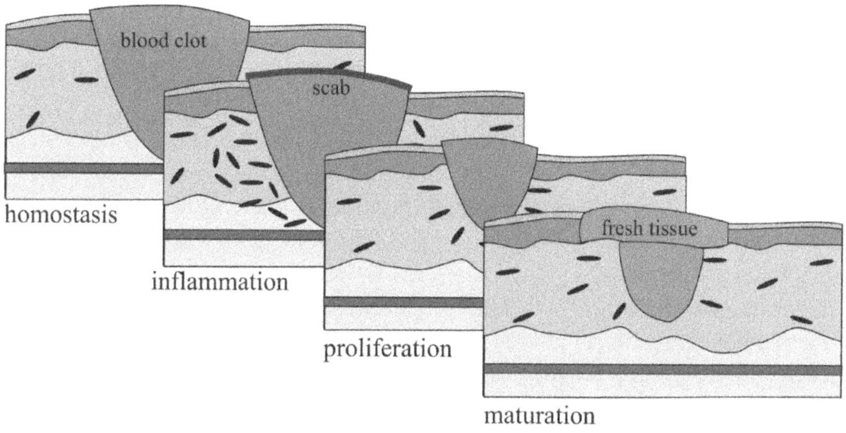

Figure 1.4 An illustration of the process of healing in soft tissues.

by balloon or through the shape memory effects by detailed macro/microstructural simulation of the blood vessel helps the cardiovascular specialist to decide on a better way for stenting to proceed for any specific patient.

On the other hand, a large number of annual car and industrial accidents and military and terrorist activities have created a growing mass of traumatized and injured people who need orthopaedic replacements in order to get back to a normal life. The corresponding economic impact has justified governments and industries to provide sufficient funding to support the research and development of related biomechanical studies on synthetic biosystems and artificial tissues.

In general, the following steps are involved in any biomechanical computations:

1) Geometrical macro- and microstructures of the organ/tissue should be determined.
2) The living organ/tissue performs in a specific environment that should be resembled by biomechanical computations.
3) The mechanical properties of the constituents must be known. This is probably one of the most challenging parts of a biomechanical analysis due to the size and conditions of a living tissue, which does not readily allow for accurate mechanical tests.
4) The multiphase and complicated composite nature of the tissue requires a sufficiently accurate model with known constitutive equations.
5) The results should be verified and calibrated by experiments on the real tissue/organ.

1.3 Types of Biological Systems

Looking at the governing equations and the way the biomechanical systems can be simulated, the following types can be distinguished:

1.3.1 Biosolids

The word biosolid is used to denote solid biological systems. It is noted that biosolid terminology is also used in some public health-related industries to discuss the solid wastes and even toxic chemicals, which are not related to this work.

The biosolid organs and tissues are usually of a composite nature, both in macro- and microscales. While they are solid in nature and are governed by the conventional rules of solid mechanics, they are usually in direct interaction with other types of biomaterials, even with different scales of their biological constituents.

For instance, a soft skin tissue is composed of several distinguishing layers. Some of them have a complex microstructure of collagen fibres within the grounds of a gel-like matrix. The collagen fibre itself is composed of a composite microstructure of fibrils and matrix. The fibres may form a straight or helical shape in different tissues, which constitute the major functionality of the tissue.

As another example, a fracture is a relatively common mode of failure for a bone (Figure 1.5). The bone, as a biosolid, is composed of several layers and a heterogeneous microstructure with different constituents and holes (representing the blood vessels, etc.). Both two- and three-dimensional simulations have been performed to study the fracture resistance of the bone as a single-scale porous solid medium or by advanced

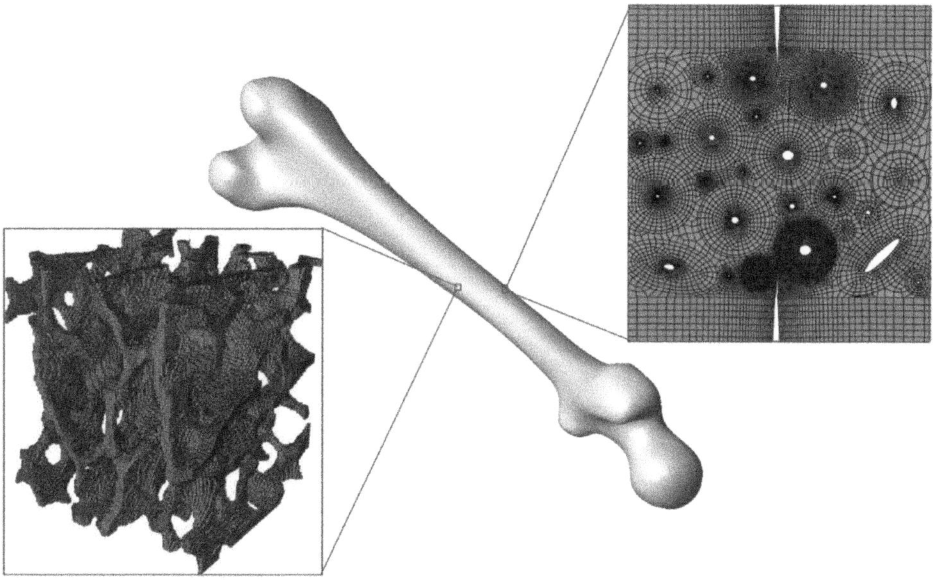

Figure 1.5 Simulations of bones on different scales.

multiscale methodology. Such analyses are in practice based on the conventional continuum or fracture mechanics of solids or employ the multiscale analysis of solid materials. Recently, the study of healing of fractured bone has gained a lot of attention and in addition to conventional solid equilibrium equations, requires several coupled equations of the mass transfer and diffusion nature, to be solved simultaneously.

1.3.2 Biofluids

While the earliest written document on blood circulation may be dated back to centuries ago, the breaking work can be attributed to one in the 17th century, where the motion of the heart and blood in animals was discussed for the first time (Harvey 1628), developing the so-called Poiseuille flow and pressure gradient theory in long tubes. It was followed by another major work based on the theory of circulation. Nowadays, massive research studies are being performed on various biofluid applications.

Biofluids (biological flows) cover a wide range of biological applications from organs to cells. Cardiovascular systems, lung and breathing mechanisms, urinary and reproductive systems and even some neurological systems may partially be studied using the biofluid mechanics/dynamics method (Figure 1.6).

While the fundamentals of fluid dynamics govern the mechanical responses of biofluids, the well-developed computational fluid dynamics (CFD) can practically and efficiently be used to analyse such systems and to study their interactions with other biological mechanisms. The complexity of the biofluid simulations arise from the fact that in addition to their basic physiological characteristics, their motion and subsequent interactions should also be considered. This is the case when drug delivery conditions are investigated or DNA/RNA and some proteins are sustained in the biofluid.

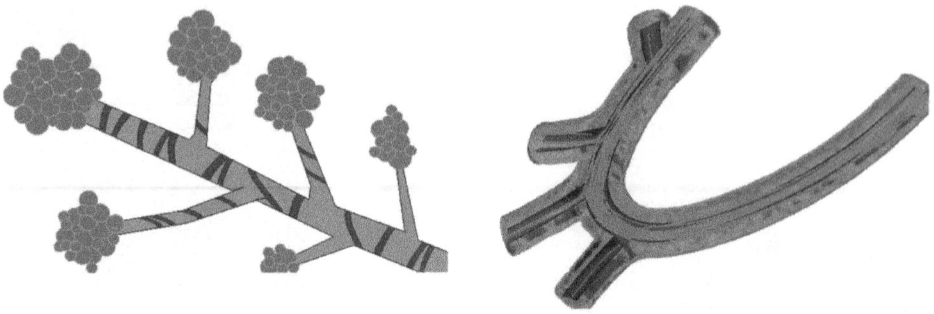

Figure 1.6 Typical bronchiole and biofluid (blood flow) simulations.

Moreover, CFD can be combined with conventional solid mechanics and advanced multiscale solutions to study the biomechanical response of different types of diseases such as aneurysms, coronary heart failure and pulmonary problems, leading to better understanding of the corresponding implications, diagnostic and even treatment procedures.

The cardiovascular system is composed of the heart and a blood-carrying network of arteries and veins. The blood flow through the pulmonary and systematic circulation system into the lung arterioles and capillaries allow for the exchange of oxygen and carbon dioxide. The oxygenated blood is then transferred to various organs and cells through systematic circulation. Biofluid mechanics is expected to contribute in the handling of all such complicated mechanisms.

The blood itself is a complex viscous biological fluid, composed of different cells suspended in a plasma, which exchanges oxygen and carbon dioxide within the lungs and other substances within other organs such as kidneys, etc. The plasma contains proteins and other constituents. In addition to the main role of flow mechanics of blood, it performs other roles, such as transfer and diffusion mechanisms of various components that affect the healing process of soft and hard tissues. These effects can be seen in the form of propagation and diffusion processes (Waite and Fine 2007), as will be described in dedicated sections.

1.3.3 Biomolecules

Biomolecules are organic combinations of molecular structures that exist in bio-organisms and contribute to various processes of the bio-object (Figure 1.7). These processes are vital to the existence and performance of living organisms, such as growth factors, differentiation factors, etc.

The following biomolecular topics are among the important subjects of research in recent decades:

- Proteins are among the primary and essential bases of any biosystem, which affect the life-related functioning of the system.
- Lipids, which are mainly fatty acids, are the basic constituent of biological membranes.
- Vitamins are composed of a biomolecule or set of chemically related biomolecules and are essential for an organism to continue its specific functioning. Many types of vitamins may not be synthesized sufficiently within a bio-organism and therefore should be obtained by drug or diet prescriptions.

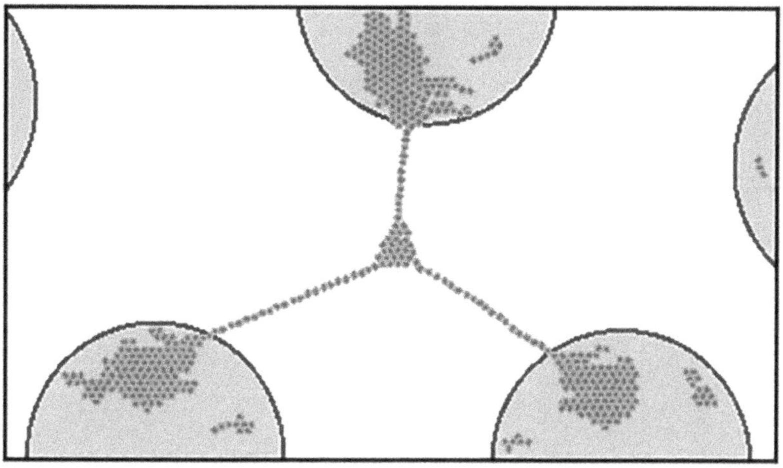

Figure 1.7 An illustration of a biomolecular structure.

- Carbohydrates generally consist of carbon, hydrogen and oxygen atoms. They may perform as a structural constituent, even for RNA, and a biostorage of energy. Carbohydrates are an essential part of the food and agriculture industries, and are important ingredients in pharmacological technologies and various clinical procedures.
- Nucleosides are molecules of more complex formations. For instance, DNA is dominated by a double helix structure, which constitutes a highly stable and reliable basis for genetic information storage.
- Amino acids are observed in proteins and influence their structure and interactions between proteins. They are also used in different industrial activities, such as cosmetics and drugs in the pharmaceutical industry, agriculture fertilizers and animal feeds, and the food industry.
- ...

In biomechanics, the formation, design and function of the biomolecules can be studied. Simulation of biomolecules may require fundamentals of cell mechanics and molecular simulations, which should sufficiently employ the effects of both biochemistry and biomechanics.

1.3.4 Synthesized Biosystems

Tissue engineering is an important interdisciplinary engineering and biological subject, with a variety of goals from function improvement of existing biosystems to repair and even replacement of existing biosystems with engineered artificial tissues/organs (Langer and Vacanti 1993; Galletti and Mora 1995). Human or animal tissues and organs and processed or synthesized materials can be used to design various biochemical processes and artificial biosystems. Extensive research and development reports on tailored engineering and artificial forms of skin, bone and blood vessels, etc. illustrate the extensive needs of the public health and the attention of corresponding industries towards this important issue. Such tailored or functional tissue engineering may allow the development of more realistic

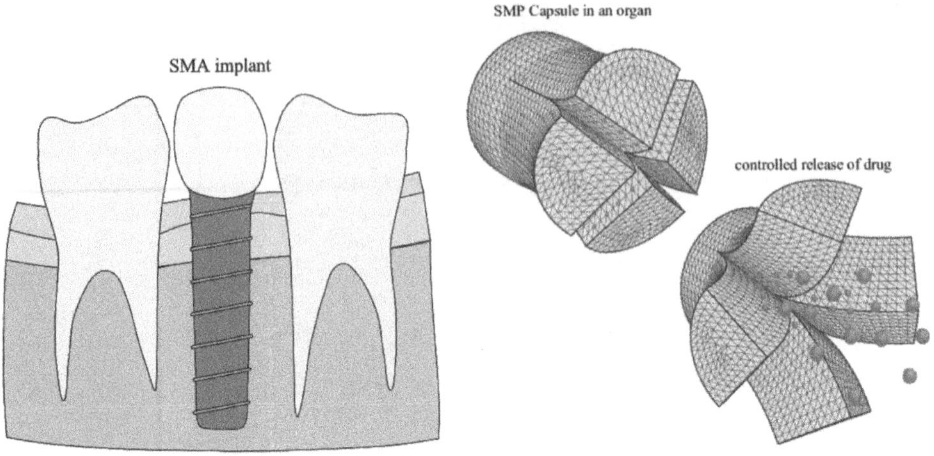

Figure 1.8 Synthesized systems: a shape memory alloy (SMA) implant and a nano shape memory polymer (SMP) drug delivery capsule.

biological substitutes that match the important properties of biological systems better, such as anisotropy, constitutive properties, inhomogeneity and microstructure, permeability, hardness and wear, multiphase nature, rate and viscous characteristics and physico-chemo-mechanical effects.

Nevertheless, there are many important issues that should be taken into consideration. For instance, while many engineered tissues should perform according to specific mechanical functioning, their micromechanical properties may not be explicitly similar to the corresponding biological tissue and the way they control the overall biological response. There is also a lack of sufficient and integrated collection of data for some tissues, especially when they are related to complex microstructural characteristics, inhomogeneity, and rate-dependent properties. Moreover, the issue is further complicated by the fact that the synthesized tissues may be developed and implanted in environments far different from the actual biological system.

Another important concern is the lack of unified standards in many synthesized applications, as they may be adopted for a very wide and different range of applications. On the other hand, the long-term effects, such as mechanical fatigue (for example in implants) and biological reactions against the constituents (for instance in a drug-delivery system), should also be considered (Figure 1.8). They may also have biomechanical side-effects, which should be considered in modelling and design of the synthesized systems.

1.4 Biomechanical Hierarchy

1.4.1 Organ Level

Biomechanical analysis of organs usually involves the study of several types of tissues and may require analysis of different phases (such as solids, fluids, etc.) in order to provide a realistically verifiable prediction of the way an organ model performs.

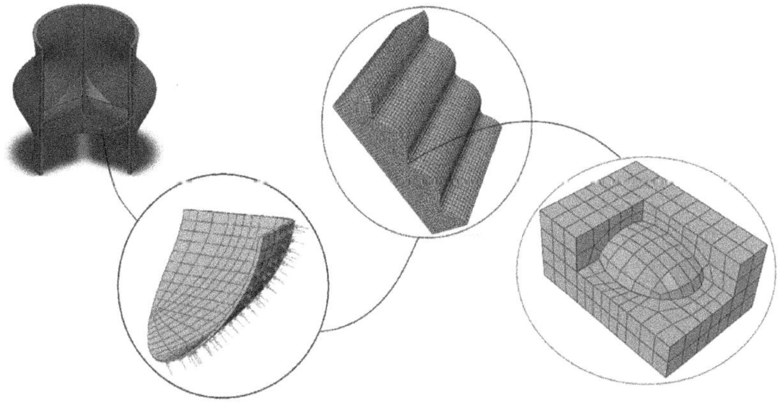

Figure 1.9 Simulation of an aortic heart valve leaflet.

As an example, the aortic heart valve is a one-way valve located between the left ventricle and the aorta, preventing the blood from flowing backwards into the heart. Its tissue is a composite one similar to other soft tissues, and the available hyperelastic formulations may well define the mechanical behaviour of the tissue and its constituents. The heart valve is mainly subjected to a pre-set blood flow pressure in most conventional and advanced multiscale simulations. Nevertheless, there may be cases that require a full fluid–solid interaction (FSI) analysis for the heart/valve/blood simulation (Shahi 2013). The model may need to be further refined to finer scales, even to the cell level, to provide a physiologically related outcome that can be used clinically (Figure 1.9).

1.4.2 Tissue Level

Accurate simulation of soft biological tissues, which may involve complex nonlinear analyses, has been active in recent decades. From a microscopic point of view, there exists a non-uniform distribution of collagen fibres embedded in a ground of gel like matrix. The collagen fibres are made from the fibrils, and the fibrils are composed of layers of collagen monomers, defining the whole microstructure in multiple scales (Figure 1.10).

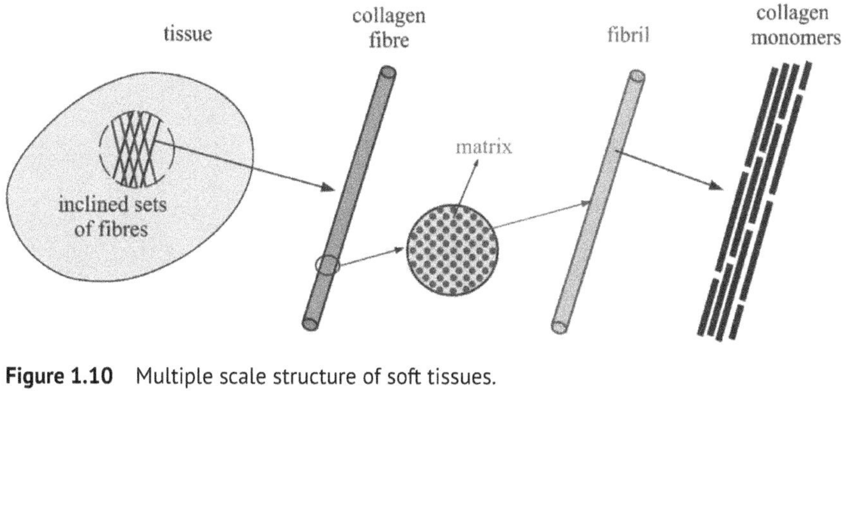

Figure 1.10 Multiple scale structure of soft tissues.

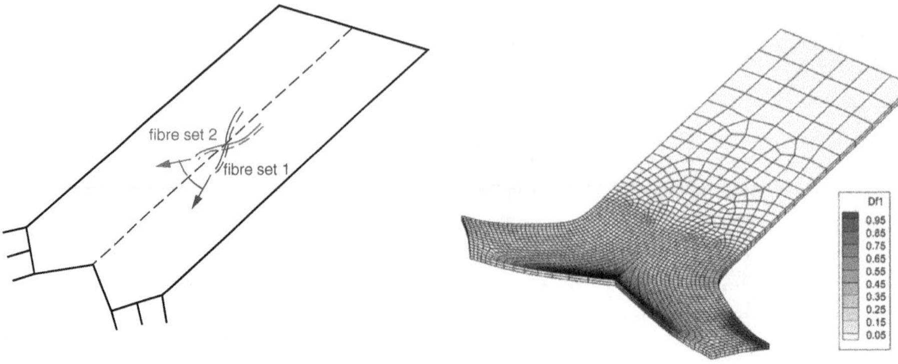

Figure 1.11 Sample modelling of the dissection problem.

Such a complex multiscale characteristic makes the analysis and capture of the highly nonlinear mechanical response of soft tissue problems very complex (Figure 1.11), which requires a set of advanced coupled numerical formulations to obtain feasible results (Fathi et al. 2017).

1.4.3 Cellular and Lower Levels

Complex individual and interactive responses of molecular and cellular structures of a tissue determine its physiological characteristics. Therefore, it is important to study the behaviour of biosystems at a cellular or molecular level (Figure 1.12).

For instance, assemblies of biomolecules or clusters of subcellular structures and any change in their functioning may indicate potential disordered phenomena that may eventually lead to a pathological disease. Such biological disorders, for instance in the extracellular matrix (ECM), may somehow be related to the biomechanical characteristics, such as porosity, viscosity, stiffness, etc., and then linked to higher scale properties by some sort of

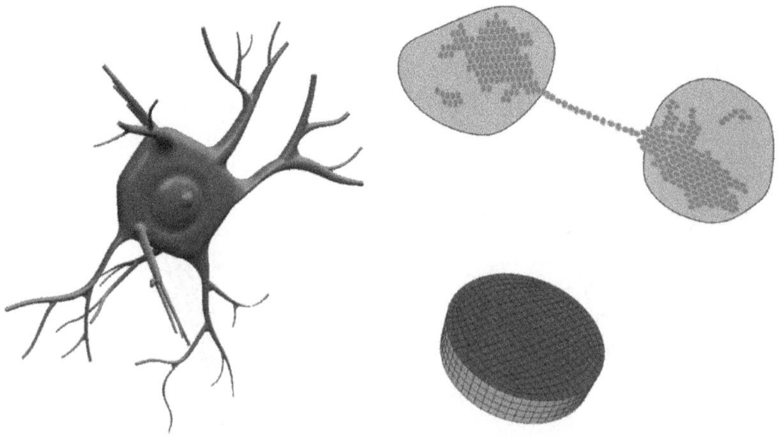

Figure 1.12 A variety of cells, from a complex nerve cell to a simplified model.

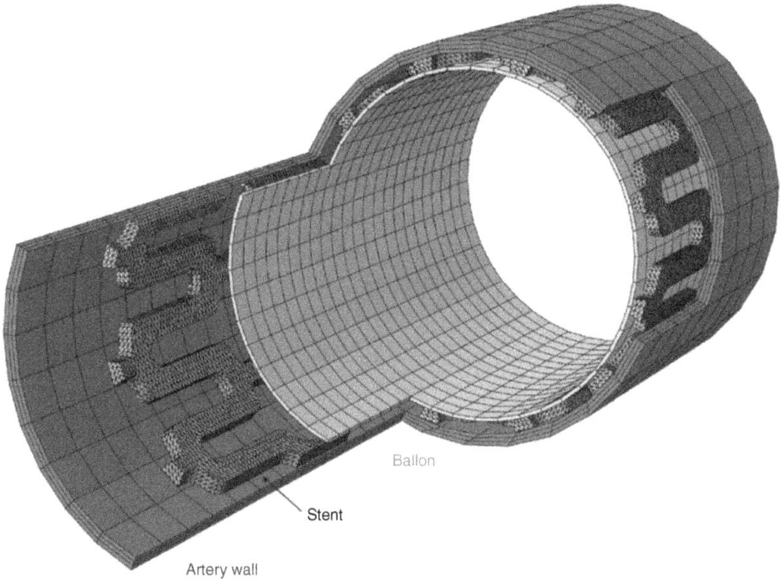

Figure 1.13 Modelling of the shape memory stenting procedure.

multiscale approach. The same concept can be extended to examine any potential contamination with hostile particles and the corresponding biological effects.

1.4.4 Complex Medical Procedures

Advanced technologies have revolutionized the way medical treatments are performed nowadays. They have simplified the procedures, provided more accurate solutions and reduced the risk of conventional operations. Biomechanics has been involved in the study of various aspects of such high-tech procedures and their ultimate effects on the human body. For instance, modelling of heart valves, blood flow, hemodynamic disorders and aneurysms in the cardiovascular system are frequently performed nowadays. More recently, modelling of the complete procedure of stenting by balloon or through shape memory effects by detailed macro/microstructural simulation of a blood vessel, helps the cardiovascular specialist to make a better decision on the way stenting should proceed for any specific patient (see Figure 1.13).

1.5 Multiscale/Multiphysics Analysis

Multiscale and multiphysics methods involve the set of advanced computational techniques that allow for accurate, yet computationally affordable, analysis of highly complicated engineering problems and physical phenomena, which cover a vast wide range of problems from cosmology to subatomic physics and from conventional civil engineering projects to complex biological applications.

Due to the large range of applications, several categories of multiscale solutions have been developed over the years. Some of them provide tailored solutions for specific problems, whereas many others provide general procedures for a larger set of applications.

Conceptually, atomistic and molecular analysis should provide an accurate solution for any physical or engineering problem. However, it is limited to very small-scale models due to the extremely large cost of computations, required to handle billions of atoms necessary for a very tiny solid, even on advanced cloud-based cluster supercomputers. A logical remedy with affordable computational costs is to use the multiscale concept, where the micromodelling is performed only where needed and the rest of the problem is modelled by one of the conventional single-scale solutions. A proper link between the scales should also be designed and implemented to ensure the consistency of formulations and responses of different scales.

Among the multiscale methods, the sequential approaches are more straightforward and may span across a wider length scale. They avoid fully coupled solutions between the scales and perform in a one-way scheme, either from the large scales to the small ones or vice versa. For instance, in simulating a reinforced concrete with additive carbon nanotube (CNT) fibres, one may adopt a sequential multiscale solution by atomistic modelling of the CNT to determine its nanoscale properties. Then, the obtained nano properties can be used in a silicate–cement–hydrate (CSH) paste model to determine interactions of CNT and cement with hydrate reactions to assess a CSH equivalent response, which may include microscale damages and defects. The next stage is to use the microscale homogenized characteristics in a mesoscale model that includes aggregates and the cement (CNT-reinforced cement). This mesoscale solution leads to characteristics that can be used in the actual reinforced concrete specimen to evaluate some engineering goals, such as resiliency, durability, toughness, seismic response or impact resistance (Figure 1.14).

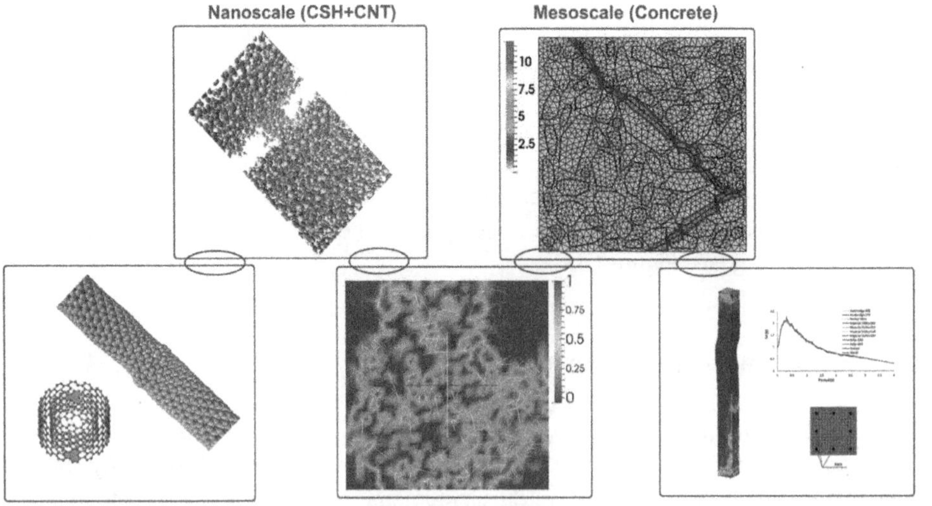

Figure 1.14 A sequential micro- to macroscale modelling.

Another important multiscale solution can be designed quite in an opposite direction: from macro- to microscale. For instance, to study heart valve functioning, one may begin with a fluid–structure interaction to evaluate the effect of blood flow pressure on the valve leaflet tissue. Then, for the most critical point, a lower scale model may be constructed to determine the induced stress, strain and other state variables within the microstructures of the tissue, which includes the collagen fibres, etc. Another step can be followed to analyse a further lower scale cell model of the biological constituent to determine if a biological criterion for cell activity is met (Figure 1.9). The same principles can be followed to study the excessive vitreous pressure on the optical nerve head (an indication of potential glaucoma), where the simulation begins from the full eye model and reaches to the cells, as typically presented in Figure 1.15.

A major category of multiscale solutions is the homogenization technique. While a simple one-way homogenization is usually used in micro to macro sequential techniques, the main concept of homogenization is defined in a coupled fashion to relate the macroscale response with the microstructure characteristics (Figure 1.16). Mathematical and computational homogenizations have been well developed over the years and applied to many engineering and physical problems. They have also been adopted in specific biomechanical applications.

Some biomechanical problems may require a further insight well into the cell and molecular levels. Such simulations may be performed by one of the concurrent multiscale models, which consider two or more scales at the same time. In fact, they use different scales to formulate different parts of the domain or to describe different responses of a domain at the same time.

Turning back to biomechanics, there has been a large amount of literature on the multiscale solutions in recent decades. They include almost every organ or tissue or even every physiological phenomena. To gain some idea of the extent of multiscale simulations of the main human organs, Figure 1.17 shows a typical illustration of the potential multiscale

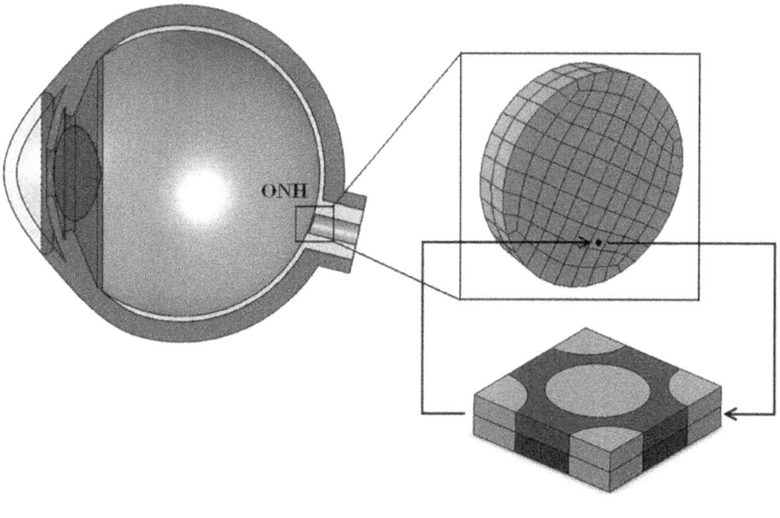

Figure 1.15 From macro- to microscale modelling of an eye.

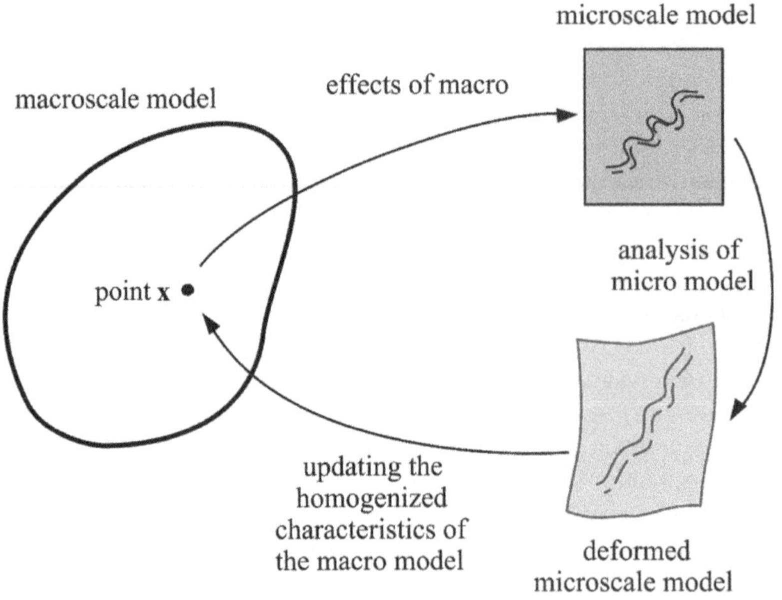

Figure 1.16 Fully coupled macro-micro multiscale homogenization.

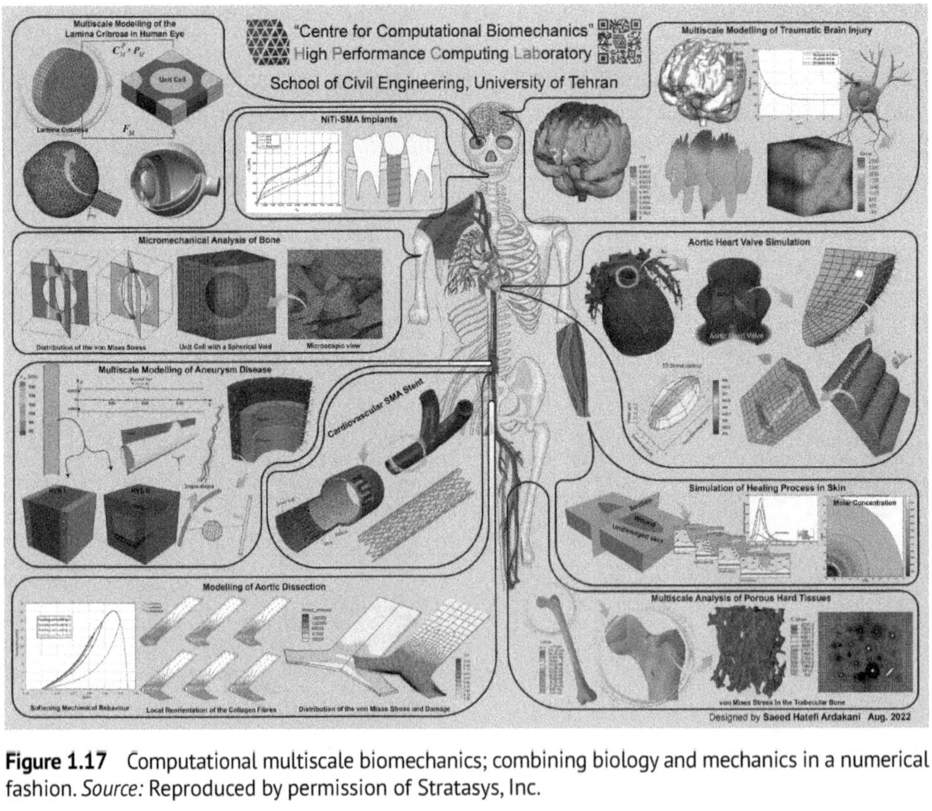

Figure 1.17 Computational multiscale biomechanics; combining biology and mechanics in a numerical fashion. *Source:* Reproduced by permission of Stratasys, Inc.

biomechanics models, as originally presented by the HPC-Lab, University of Tehran (2020) (the central part design is by GrabCAD).

Several types of multiscale solutions may be performed on the biomechanical issues related to human organs or diseases. Some of the important topics of biomechanical applications can be presented as:

- Lamina cribrosa in the human eye
- NiTi SMA implants
- Micromechanics of bones
- Modelling of aneurysm disease
- Cardiovascular SMA stenting
- Modelling of aortic dissection
- Healing process of damaged soft tissues
- Healing in hard tissues
- Aortic heart valve simulation
- Traumatic brain injury
- Growth of cancerous tumours
- Drug delivery
- ...

Some of the mentioned topics will be dealt with and discussed in detail in Part III (based on the theoretical formulations and numerical methods presented in Part II).

1.6 Scope of the Book

The book is composed of three parts. This preliminary part is meant to provide an introduction and insight on the general concept of the biomechanics and the wide variety of biological problems that can be solved numerically, to assist the clinical staff and medical-related industries in assessing the existing procedures and products or to propose new engineered designs and concepts for future research and development.

Part II is dedicated to analytical and numerical bases in a systematic way, beginning with the very basics and approaching advanced computational techniques.

This part first covers the general concepts of continuum mechanics. For solid mechanics problems, a review is provided for elasticity, followed by principles of plasticity and damage mechanics. Then, fundamentals of fracture mechanics are presented. Viscoelasticity and rate-dependent behaviours are discussed. Poroelasticity is briefly reviewed and the concepts of large deformation are discussed in more detail, as it is the basis for many subsequent theories.

For fluid flow and field problems, the governing equations of fluid flow, convection and diffusion problems are explained and the fluid–structure interaction (FSI) is examined by discussing the coupled Lagrangian–Eulerian (CLE) formulation and the Arbitrary Lagrangian–Eulerian (ALE) technique.

The next chapters cover the conventional and advanced numerical methods, which are essential for explaining the multiscale concepts and techniques. It begins by briefly reviewing the finite difference method (FDM), followed by an introduction to the finite volume method (FVM) and the finite element method (FEM). Then, the extended finite element method (XFEM) is described as an extension to the classical FEM for accurate and efficient analysis of a wide variety of discontinuity and singularity problems. Then, an introduction is provided for the extended isogeometeric analysis (IGA and XIGA). The principles of the powerful meshless methods are described and some of its main approaches, the element free Galerkin (EFG), the meshless local Petrov-Galerkin (MLPG) and the smoothed particle hydrodynamics (SPH), are further discussed. That chapter concludes by examining the variable node element method, which is a combination of finite element and meshless principles to provide a very flexible new element for multiscale purposes.

The final chapter of Part II is dedicated to the comprehensive discussion on the multiscale methods. It begins by an introduction to provide an overview of the subject. The next section discusses the concepts of the homogenization technique and covers both mathematical and computational homogenization techniques and presents the state-of-the-art concepts of the enriched multiscale homogenization, with sample results on microscale cracking problems.

Then, the atomistic and molecular modellings are discussed in detail. This covers the concepts of statistical mechanics and the governing equations of motion along with a review on major available potential functions. The section concludes by a sample modelling of a set of polymeric chains.

Then, the important subject of the sequential multiscale method is thoroughly discussed by providing details of a practical simulation, which spans the extremely wide range of molecular dynamics (MD) simulations, nanoscale analysis, microscale computations, mesoscale investigation and macroscale study of concrete specimen with additive CNT fibres.

The multiscale chapter is concluded by a comprehensive review of the concurrent schemes. It begins by briefly examining the main concepts of the concurrent techniques. Then the quasi-continuum (QC) approach is discussed and its main assumptions are presented, followed by the fundamentals of the bridging domain (BDM) and the bridging scale (BSM) methods. The so-called disordered multiscale method is presented for amorphous and polymeric material structures, which is expected to perform well for polymer and biomolecule applications. Finally, the variable node multiscale method (VNMM) and its extension to the enriched multiscale method (EMM) are presented.

Part III is dedicated to single and multiscale biomechanical simulations. It adopts the previously discussed theoretical concepts and numerical methods to discuss the way various biomechanical problems can be analysed. The chapter begins by reviewing the physiology of the problem and moves towards the governing equations and the way they can be numerically simulated and solved. They include several numerical examples to explain the adopted procedure for efficient simulation of the complex phenomena involved in various biomechanics problems.

The first chapter of Part III is dedicated to the modelling of soft tissues. Again, it begins with explaining the composition and physiology of soft tissues, with special attention given to its fibrous microstructure by collagen fibres, fibrils, etc. Then, the macro- and microstructure mechanical properties are discussed in detail and the corresponding material models and constitutive laws in the form of hyperelastic models are presented. This chapter includes several single- and multiscale simulations of a number of soft tissue applications, including tendon and ligament, aortic heart valve, skin damage, arterial wall degradation, wound healing and the viscoelastic response of the brain.

The next chapter of this part is dedicated to hard tissues. It begins by explaining the composition and architecture of hard tissues and looks at their macro- and microstructures. Then, the necessary mechanical models are discussed. A number of single- and multiscale simulations are presented to provide more insight into the way hard tissue biomechanical studies are performed. Moreover, a discussion is provided on the healing processes of hard tissues by examining the governing equations and the numerical simulations.

Part III is concluded by a complementary chapter on brief reviews of supplementary topics, covering principles of artery stenting simulations, an initial assessment of the multiscale modelling of the eye, a look at the pulsatile blood flow in the aorta, analysis of the concept of a shape memory polymer drug delivery system and an introduction to the emerging computational technology of artificial intelligence and deep learning in biomechanical applications.

The contents of the book are completed by the list of references.

Part II

Analytical and Numerical Bases

2

Theoretical Bases of Continuum Mechanics

2.1 Introduction

In this chapter, the theoretical bases of continuum mechanics are reviewed. The aim is to provide a review of the basic concepts and the necessary formulations, as required by the upcoming chapters of the book. The discussions are presented without mathematical proofs and readers are expected to refer to the references for detailed discussions.

After this brief introduction, the chapter begins by examining the solid mechanics problems. It includes a review of elasticity-based stress–strain relations, followed by discussions on plastic behaviour of materials and related nonlinear stress update procedures. The next section is dedicated to damage mechanics, as it provides the theoretical bases for material degradation in many physical, engineering and biomechanical problems. Then the chapter covers a review on the concepts of fracture mechanics and fatigue phenomena. Viscoelastic and poroelastic theories will be examined and the corresponding formulations are presented. The chapter continues with a review of the large deformation formulation, which is the basis for many practical multiscale problems, including biomechanics applications. The final section is dedicated to brief discussions on fluid mechanics, convection and diffusion problems.

2.2 Solid Mechanics

In this section, the basic definitions and governing equations for solid mechanics problems are briefly reviewed. Beginning with linear elastic stress–strain constitutive relations, the fundamental concepts of plasticity are described. The section then summarizes damage mechanics formulations for both scalar (isotropic) and tensorial (anisotropic) damage problems. The next part examines the fundamentals of computational fracture mechanics for cracked media. Viscoelasticity is required in a number of studies of biomechanical problems that involve the time and rate effects and the theory of poroelasticity governs the behaviour of a porous medium with fluid flow inside the pores. The section concludes with the important subsection on presenting the basics of the theory of large deformation. All preliminary concepts can be reformulated to account for large deformation and large strain regimes, as will be partially discussed in a number of soft and hard biomechanical problems discussed in Chapters 5 and 6.

Multiscale Biomechanics: Theory and Applications, First Edition. Soheil Mohammadi.

2.2.1 Elasticity

Description of the basic equations of linear elasticity begins with the definition of stress and strain tensors, $\boldsymbol{\sigma}$ and $\boldsymbol{\varepsilon}$, respectively:

$$\boldsymbol{\sigma} = \begin{bmatrix} \sigma_{11} & \sigma_{12} & \sigma_{13} \\ \sigma_{12} & \sigma_{22} & \sigma_{23} \\ \sigma_{13} & \sigma_{23} & \sigma_{33} \end{bmatrix} \tag{2.2.1}$$

$$\boldsymbol{\varepsilon} = \begin{bmatrix} \varepsilon_{11} & \varepsilon_{12} & \varepsilon_{13} \\ \varepsilon_{12} & \varepsilon_{22} & \varepsilon_{23} \\ \varepsilon_{13} & \varepsilon_{23} & \varepsilon_{33} \end{bmatrix} \tag{2.2.2}$$

or in a vector notation,

$$\boldsymbol{\sigma} = \begin{Bmatrix} \sigma_1 \\ \sigma_2 \\ \sigma_3 \\ \sigma_4 \\ \sigma_5 \\ \sigma_6 \end{Bmatrix} = \begin{Bmatrix} \sigma_{11} \\ \sigma_{22} \\ \sigma_{33} \\ \sigma_{23} \\ \sigma_{13} \\ \sigma_{12} \end{Bmatrix} \tag{2.2.3}$$

$$\boldsymbol{\varepsilon} = \begin{Bmatrix} \varepsilon_1 \\ \varepsilon_2 \\ \varepsilon_3 \\ \varepsilon_4 \\ \varepsilon_5 \\ \varepsilon_6 \end{Bmatrix} = \begin{bmatrix} \varepsilon_{11} \\ \varepsilon_{22} \\ \varepsilon_{33} \\ \gamma_{12} = 2\varepsilon_{12} \\ \gamma_{23} = 2\varepsilon_{23} \\ \gamma_{13} = 2\varepsilon_{13} \end{bmatrix} \tag{2.2.4}$$

The generalized linear elastic constitutive law can be defined as

$$\boldsymbol{\sigma} = \mathbf{D}\boldsymbol{\varepsilon} \tag{2.2.5}$$

$$\boldsymbol{\varepsilon} = \mathbf{C}\boldsymbol{\sigma} \tag{2.2.6}$$

where $\mathbf{D}$ and $\mathbf{C}$ are the fourth-order elasticity and compliance tensors, respectively. Equations (2.2.5) and (2.2.6) can be written in a component form as

$$\sigma_{ij} = D_{ijkl}\varepsilon_{kl}, \qquad i, j = 1,2,3 \tag{2.2.7}$$

$$\varepsilon_{ij} = C_{ijkl}\sigma_{kl}, \qquad i, j = 1,2,3 \tag{2.2.8}$$

These equations can also be defined in terms of the stress and strain vectors and the two-dimensional matrix forms of **D** and **C**:

$$\sigma_i = d_{ij}\varepsilon_j, \qquad i,j = 1,2,3,\ldots,6 \tag{2.2.9}$$

$$\varepsilon_i = c_{ij}\sigma_j, \qquad i,j = 1,2,3,\ldots,6 \tag{2.2.10}$$

For an orthotropic material with three arbitrarily orthogonal planes of elastic symmetry, d_{ij} and c_{ij} are reduced to

$$\begin{bmatrix} d_{11} & d_{12} & d_{13} & 0 & 0 & 0 \\ d_{12} & d_{22} & d_{23} & 0 & 0 & 0 \\ d_{13} & d_{23} & d_{33} & 0 & 0 & 0 \\ 0 & 0 & 0 & d_{44} & 0 & 0 \\ 0 & 0 & 0 & 0 & d_{55} & 0 \\ 0 & 0 & 0 & 0 & 0 & d_{66} \end{bmatrix} \tag{2.2.11}$$

$$\begin{bmatrix} c_{11} & c_{12} & c_{13} & 0 & 0 & 0 \\ c_{12} & c_{22} & c_{23} & 0 & 0 & 0 \\ c_{13} & c_{23} & c_{33} & 0 & 0 & 0 \\ 0 & 0 & 0 & c_{44} & 0 & 0 \\ 0 & 0 & 0 & 0 & c_{55} & 0 \\ 0 & 0 & 0 & 0 & 0 & c_{66} \end{bmatrix} \tag{2.2.12}$$

where d_{ij} for the principal directions of orthotropy can be explicitly defined as (Bower 2012):

$$\begin{bmatrix} E_1(1-\nu_{23}\nu_{32})\bar{\nu} & E_1(\nu_{12}-\nu_{31}\nu_{23})\bar{\nu} & E_1(\nu_{31}-\nu_{21}\nu_{31})\bar{\nu} & 0 & 0 & 0 \\ E_1(\nu_{12}-\nu_{31}\nu_{23})\bar{\nu} & E_2(1-\nu_{13}\nu_{31})\bar{\nu} & E_2(\nu_{32}-\nu_{12}\nu_{31})\bar{\nu} & 0 & 0 & 0 \\ E_1(\nu_{31}-\nu_{21}\nu_{31})\bar{\nu} & E_2(\nu_{32}-\nu_{12}\nu_{31})\bar{\nu} & E_3(1-\nu_{12}\nu_{21})\bar{\nu} & 0 & 0 & 0 \\ 0 & 0 & 0 & \mu_{23} & 0 & 0 \\ 0 & 0 & 0 & 0 & \mu_{13} & 0 \\ 0 & 0 & 0 & 0 & 0 & \mu_{12} \end{bmatrix} \tag{2.2.13}$$

where

$$\frac{1}{\bar{\nu}} = 1 - \nu_{12}\nu_{21} - \nu_{23}\nu_{32} - \nu_{31}\nu_{13} - 2\nu_{21}\nu_{32}\nu_{13} \tag{2.2.14}$$

and the corresponding c_{ij} can be written as

$$\begin{bmatrix} \frac{1}{E_1} & -\frac{\nu_{21}}{E_2} & -\frac{\nu_{31}}{E_3} & 0 & 0 & 0 \\ -\frac{\nu_{12}}{E_1} & \frac{1}{E_2} & -\frac{\nu_{32}}{E_3} & 0 & 0 & 0 \\ -\frac{\nu_{13}}{E_1} & -\frac{\nu_{23}}{E_2} & \frac{1}{E_3} & 0 & 0 & 0 \\ 0 & 0 & 0 & \frac{1}{\mu_{23}} & 0 & 0 \\ 0 & 0 & 0 & 0 & \frac{1}{\mu_{31}} & 0 \\ 0 & 0 & 0 & 0 & 0 & \frac{1}{\mu_{12}} \end{bmatrix} \tag{2.2.15}$$

Note that the following additional relations must be respected:

$$E_1\nu_{21} = E_2\nu_{12},\ E_2\nu_{32} = E_3\nu_{23},\ E_3\nu_{13} = E_1\nu_{31} \tag{2.2.16}$$

d_{ij} can be further simplified for the generalized plane stress problems:

$$\begin{bmatrix} d_{11} & d_{12} & 0 \\ d_{12} & d_{22} & 0 \\ 0 & 0 & d_{66} \end{bmatrix} = \begin{bmatrix} \bar{\nu}E_1 & \bar{\nu}\nu_{12}E_2 & 0 \\ \bar{\nu}\nu_{12}E_2 & \bar{\nu}E_2 & 0 \\ 0 & 0 & \mu_{12} \end{bmatrix} \tag{2.2.17}$$

and for the generalized plane strains (Garcia-sanchez et al. 2008),

$$\begin{bmatrix} d_{11} & d_{12} & 0 \\ d_{12} & d_{22} & 0 \\ 0 & 0 & d_{66} \end{bmatrix} = \begin{bmatrix} \bar{\nu}E_1(1-\nu_{12}\nu_{21}) & \bar{\nu}E_1\left(\nu_{21}+\frac{E_2}{E_1}\nu_{13}\nu_{32}\right) & 0 \\ \bar{\nu}E_1\left(\nu_{21}+\frac{E_2}{E_1}\nu_{13}\nu_{32}\right) & \bar{\nu}E_2(1-\nu_{13}\nu_{31}) & 0 \\ 0 & 0 & \mu_{12} \end{bmatrix} \tag{2.2.18}$$

The components of the compliance matrix c_{ij} for plane stress and plain strain problems can be obtained by the following simple rule (Ghorashi et al. 2011a):

$$c_{ij} = \begin{cases} c_{ij}, & i,j = 1,2,6 \quad \text{plane stress} \\ c_{ij} - \frac{c_{i3}c_{j3}}{c_{33}}, & i,j = 1,2,6 \quad \text{plane strain} \end{cases} \tag{2.2.19}$$

For isotropic materials, the elasticity equations are substantially simplified. Components d_{ij} for three-dimensional isotropic problems are reduced to (Neto et al. 2008)

$$\begin{bmatrix} \lambda+2\mu & \lambda & \lambda & 0 & 0 & 0 \\ \lambda & \lambda+2\mu & \lambda & 0 & 0 & 0 \\ \lambda & \lambda & \lambda+2\mu & 0 & 0 & 0 \\ 0 & 0 & 0 & \mu & 0 & 0 \\ 0 & 0 & 0 & 0 & \mu & 0 \\ 0 & 0 & 0 & 0 & 0 & \mu \end{bmatrix} \tag{2.2.20}$$

where λ and μ are the Lame and shear modules

$$\lambda = \frac{\nu E}{(1+\nu)(1-2\nu)} \tag{2.2.21}$$

$$\mu = \frac{E}{2(1+\nu)} = G \tag{2.2.22}$$

For the isotropic plane stress case, the stress–strain relation is reduced to

$$\begin{Bmatrix} \sigma_{11} \\ \sigma_{22} \\ \sigma_{12} \end{Bmatrix} = \frac{E}{1-\nu^2} \begin{bmatrix} 1 & \nu & 0 \\ \nu & 1 & 0 \\ 0 & 0 & \frac{1-\nu}{2} \end{bmatrix} \begin{Bmatrix} \varepsilon_{11} \\ \varepsilon_{22} \\ \gamma_{12} \end{Bmatrix} \tag{2.2.23}$$

with an out-of-plane component of strain, ε_{33}:

$$\varepsilon_{33} = -\frac{\nu}{1-\nu}(\varepsilon_{11}+\varepsilon_{22}) \tag{2.2.24}$$

Similarly, for the isotropic plane strain state,

$$\begin{Bmatrix} \sigma_{11} \\ \sigma_{22} \\ \sigma_{33} \\ \sigma_{12} \end{Bmatrix} = \frac{E}{(1+\nu)(1-2\nu)} \begin{bmatrix} 1-\nu & \nu & 0 \\ \nu & 1-\nu & 0 \\ \nu & \nu & 0 \\ 0 & 0 & \frac{1-2\nu}{2} \end{bmatrix} \begin{Bmatrix} \varepsilon_{11} \\ \varepsilon_{22} \\ \gamma_{12} \end{Bmatrix} \tag{2.2.25}$$

It is sometimes useful to define the effective Young's modulus, E':

$$E' = \begin{cases} E & \text{plane stress} \\ \dfrac{E}{1-\nu^2} & \text{plane strain} \end{cases} \tag{2.2.26}$$

2.2.2 Plasticity

Linear elasticity is a very idealized description of the material response. Most materials behave inelastically and/or nonlinearly. The source of such behaviour differs based on the microstructure of material and its constituents, the rate of loading and deformation, thermal effects and various sources of failure and damage. These phenomena may be analysed by macroscopic plasticity, as briefly reviewed in this section, or can be studied by the concepts of micromechanics.

Figure 2.1 illustrates typical unidirectional stress–strain responses of an alloy (for example a tensile steel bar) and a cementitious material (for instance, a compressive concrete specimen).

The source of plasticity in metals can be attributed to their polycrystalline microstructures, which may generate plastic deformations through the relative slips in crystalline planes. In fact, in addition to elastic deformations in microstructures, potential breakage in interatomic bonds may occur. Therefore, the deformation process is no longer fully reversible.

Most biomechanical problems show little plasticity and are usually governed by other macro-based theories, such as hyperelasticity, fracture mechanics and damage, combined with viscosity and time-dependent effects. Therefore, this section is kept in a summarized form.

Plasticity in metals is usually discussed in crystal plasticity. Some of the main observations are (Dunne and Petrinic 2004):

1) The plastic deformation is usually governed by the shearing phenomena, which are pressure independent.
2) The plastic deformation is an incompressible process.
3) The plastic process follows an isotropic rule.

Typical uniaxial stress–strain responses are depicted in Figure 2.2. The first part of the response is the linear elastic path up to the yielding stress limit σ_{yld}. The slope of the linear part is the Young's modulus of elasticity E. Any unloading would follow back on the same path. Afterwards, the inelastic part begins, which may be in the form of perfectly plastic, linear strain hardening or nonlinear strain hardening paths. Any unloading after the yield

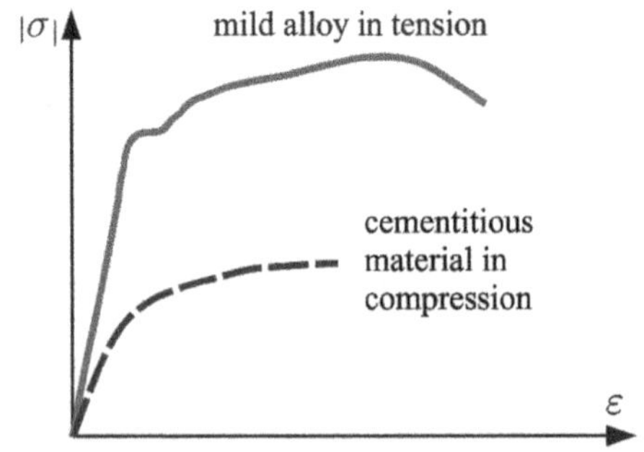

Figure 2.1 Typical stress–strain responses of a mild alloy in tension and a cementitious specimen in compression.

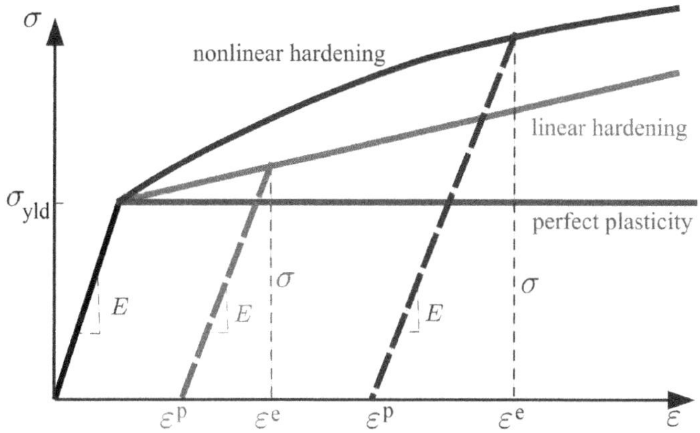

Figure 2.2 Elastoplastic response with perfect, linear and nonlinear hardening plasticity.

stress σ_{yld} is assumed to be along a linear path parallel to the original linear elastic response. As a result, while any full stress unloading from a plastic response will be linear elastic in nature, the overall response is accompanied by a remaining plastic strain ε^{p}.

In a perfect plastic behaviour, the stress level of the linear elastic branch remains unaltered by the level of experienced plastic strain. In contrast, the yield stress increases in strain hardening regimes. In other words, on a reloading after an unloading from an elastoplastic state, the material may endure a higher limit of linear elastic response (yield stress σ_{yld}) if it follows a strain-hardening rule.

The simple uniaxial case can be similarly generalized to multiaxial states in terms of stress and strain tensors, $\boldsymbol{\sigma}$ and $\boldsymbol{\varepsilon}$, respectively. Moreover, a yield criterion f should be adopted to determine the onset of plasticity:

$$f = \sigma_{\text{eff}} - \sigma_{\text{yld}}(\kappa_{\text{p}}) \tag{2.2.27}$$

where σ_{eff} is a scalar effective stress that describes the whole multiaxial state of stress tensor and κ_{p} is the hardening parameter. Any definition of the effective stress depends on the way the components of a multiaxial state can be combined for a specific elastoplastic constitutive law. For instance, one of the most frequently used effective stress definitions in the absence of hardening effects is the von Mises effective stress:

$$(\sigma^{\text{M}})^2 = (\sigma_{\text{eff}})^2 = \frac{3}{2}\text{tr}\left[(\boldsymbol{\sigma}_{\text{dev}})^{\text{T}}(\boldsymbol{\sigma}_{\text{dev}})\right] \tag{2.2.28}$$

where $\boldsymbol{\sigma}_{dev}$ is the deviatoric part of the stress tensor:

$$\boldsymbol{\sigma}_{\text{dev}} = \boldsymbol{\sigma} - \mathbf{p} = \boldsymbol{\sigma} - \frac{1}{3}\text{tr}(\boldsymbol{\sigma})\mathbf{I} \tag{2.2.29}$$

Figure 2.3 illustrates a two-dimensional representation of the yield surface and the plastic strain.

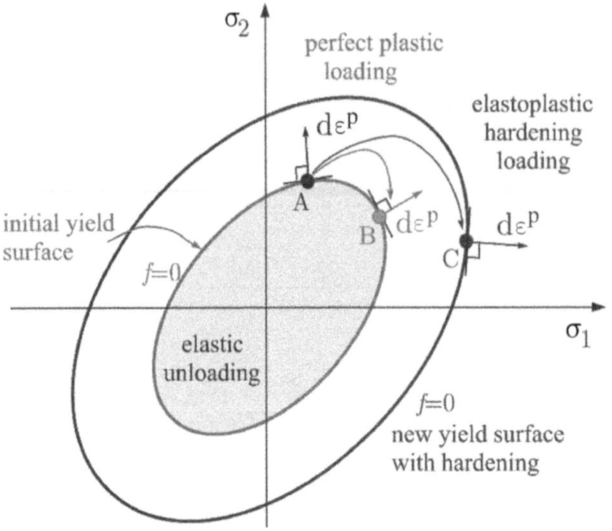

Figure 2.3 Yield surface and plastic loading.

The classical theory of plasticity is based on the additive decomposition of strains (or strain increments):

$$d\varepsilon = d\varepsilon^{e} + d\varepsilon^{p} \tag{2.2.30}$$

where

$$d\boldsymbol{\sigma} = \mathbf{D}d\varepsilon^{e} = \mathbf{D}(d\varepsilon - d\varepsilon^{p}) \tag{2.2.31}$$

The normality condition of the associated flow rule is defined based on the magnitude of the plastic strain increment ($d\lambda$) and its direction ($\partial f / \partial \boldsymbol{\sigma}$):

$$d\varepsilon^{p} = d\lambda \frac{\partial f}{\partial \boldsymbol{\sigma}} \tag{2.2.32}$$

The incremental form of the elastoplastic constitutive law can be written as (Dunne and Petrinic 2004)

$$d\boldsymbol{\sigma} = \mathbf{D}_{ep} d\varepsilon \tag{2.2.33}$$

where $\mathbf{D}_{ep}$ is the tangential material modulus:

$$\mathbf{D}_{ep} = \mathbf{D} - \mathbf{D}\frac{\partial f}{\partial \boldsymbol{\sigma}} \frac{\frac{\partial f}{\partial \boldsymbol{\sigma}}\mathbf{D}}{\left[\frac{\partial f}{\partial \boldsymbol{\sigma}}\mathbf{D}\frac{\partial f}{\partial \boldsymbol{\sigma}} - \frac{2}{3}\frac{\partial f}{\partial \kappa_p}\frac{\partial f}{\partial \boldsymbol{\sigma}}\left(\frac{\partial f}{\partial \boldsymbol{\sigma}}\right)^{1/2}\right]} \tag{2.2.34}$$

and κ_p is the effective plastic strain, which is expressed similarly to the definition of the effective stress:

$$(\mathrm{d}\kappa_p)^2 = \frac{2}{3}\mathrm{tr}\left[(\mathrm{d}\boldsymbol{\varepsilon}^{\mathrm{p}})^{\mathrm{T}}(\mathrm{d}\boldsymbol{\varepsilon}^{\mathrm{p}})\right] \tag{2.2.35}$$

A variety of implicit and explicit algorithms are available to integrate the incremental forms of multiaxial elastoplastic constitutive laws, leading to the derivation of the consistent elastoplastic tangent material modulus. For comprehensive details, refer to Owen and Hinton (1980), Neto et al. (2008) and Dunne and Petrinic (2004).

2.2.3 Damage Mechanics

Initiation and propagation of various microdefects, such as microcracks, microvoids, and other forms of microdiscontinuities in microstructures, constitute the occurrence of damage, as depicted in Figure 2.4. While such phenomena are primarily discontinuous in nature, the continuous damage mechanics (CDM), as originally proposed in concept by Kachanov (1958), offers a continuous formulation in terms of an average density of microdefects to describe the behaviour of such complex phenomena. Accordingly, the theory of ductile damage (Lemaitre 1984) is adopted to describe the necessary constitutive equations for determining the evolution of damage variables (Voyiadjis and Kattan 1996).

There are similarities between the formulations of continuum damage mechanics and the well-developed theory of plasticity. Figure 2.5 illustrates a typical uniaxial stress–strain response for general elastoplastic damage problems. One of the main differences of continuum damage and plasticity models may be attributed to the fact that while the slope of an unloading response remains parallel to the initial linear elastic loading part in elastoplastic problems, the stiffness (slope) is reduced on the occurrence of damage in elastoplastic damage problems. The reduced stiffness may be assumed to be an effective modulus of elasticity. Such an effective modulus, or other effective state variables, may either be related directly to the physical damage phenomena, such as porosity and size of defects, or can be formulated indirectly in terms of an equivalent damage stress or the corresponding elastic energy.

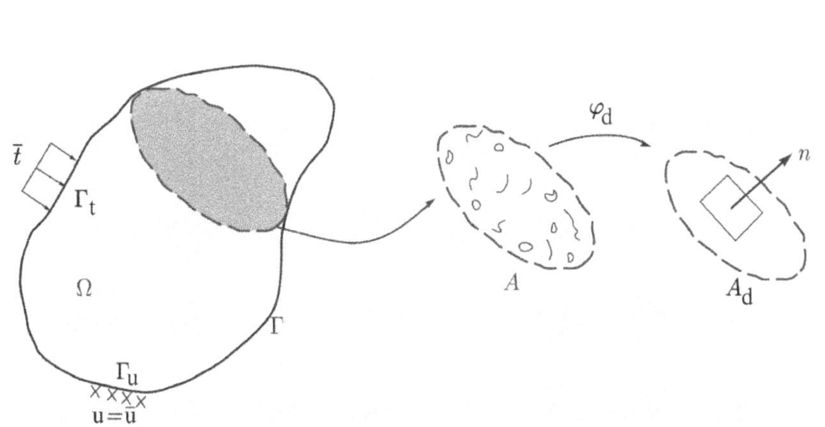

Figure 2.4 Description of a damaged body and the existing microdefects.

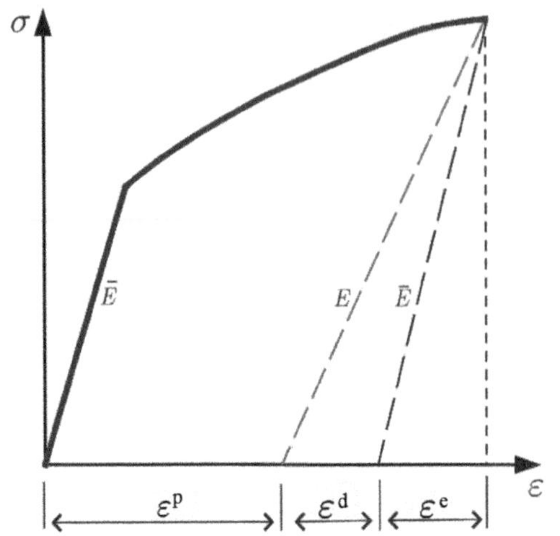

Figure 2.5 A typical elastoplastic damage behaviour.

It is observed that, on unloading, the permanent strain includes two parts of plastic and damage strains, ε^{p} and ε^{d}, respectively. The postulate of additive decomposition of the strain tensor can then be adopted for the elastoplastic damage problem:

$$\boldsymbol{\varepsilon} = \boldsymbol{\varepsilon}^{\mathrm{e}} + \boldsymbol{\varepsilon}^{\mathrm{p}} + \boldsymbol{\varepsilon}^{\mathrm{d}} \tag{2.2.36}$$

The corresponding constitutive equation is usually derived based on the incremental form of the additive decomposition, as depicted in Figure 2.5:

$$\mathrm{d}\boldsymbol{\varepsilon} = \mathrm{d}\boldsymbol{\varepsilon}^{\mathrm{e}} + \boldsymbol{\varepsilon}^{\mathrm{p}} + \mathrm{d}\boldsymbol{\varepsilon}^{\mathrm{d}} \tag{2.2.37}$$

The solution procedure requires proper formulation of the damage state and the corresponding evolution mechanisms.

2.2.3.1 Isotropic Damage

Consider a body with internal microdefects, typically shown in Figure 2.4. Evolution of the microdefects is affected by the deformation of the body. For a typical representative surface section with area δA (and normal $\boldsymbol{n}$), the effective load-bearing area δA_{eff} is clearly smaller than δA, due to the existence of defects with the total area of δA_{d}. This concept can be better illustrated in Figure 2.6.

The scalar damage variable φ_{d} is defined as

$$\varphi_{\mathrm{d}} = \frac{\delta A_{\mathrm{d}}}{\delta A}, \forall \boldsymbol{n} \tag{2.2.38}$$

which can be interpreted as the surface density of microdefects. $\varphi_{\mathrm{d}} = 0$ represents an undamaged element and $\varphi_{\mathrm{d}} = 1$ shows that the element is fully damaged. In an isotropic damage model, the scalar damage φ_{d} is not related to the surface normal $\boldsymbol{n}$.

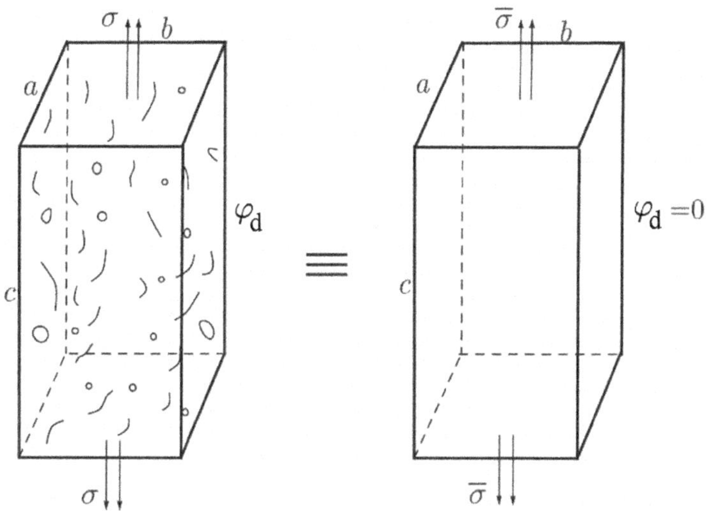

Figure 2.6 Equivalent stress and the scalar damage variable.

The effective area A_{eff} can now be written as

$$\delta A_{eff} = \delta A - \delta A_d = \delta A(1 - \varphi_d) \tag{2.2.39}$$

and the effective stress tensor can then be defined as

$$\boldsymbol{\sigma}_{eff} = \frac{\boldsymbol{\sigma}}{1 - \varphi_d} \tag{2.2.40}$$

There are different hypotheses for deriving the damage mechanics formulation based on the damage variable φ_d. The concept of strain equivalence assumes that the conventional forms of undamaged constitutive laws can be similarly adopted in terms of the effective stress $\boldsymbol{\sigma}_{eff}$. On the other hand, the principle of equivalence of elastic energy assumes that the strain energy of the damaged material can be written in the form of an undamaged material but with equivalent properties (effective stress $\boldsymbol{\sigma}_{eff}$, etc.).

For a simple case of uniaxial stress, the principle of strain equivalence leads to

$$\sigma_{eff} = E\ \varepsilon_{eff} \tag{2.2.41}$$

$$\varepsilon_{eff} = \frac{\varepsilon}{(1 - \varphi_d)} = \frac{\sigma}{(1 - \varphi_d)E} \tag{2.2.42}$$

and the principle of elastic energy equivalence (Voyajilis and Kattan 2005)

$$\frac{1}{2}\sigma_{eff}\ \varepsilon_{eff} = \frac{1}{2}\sigma\varepsilon \tag{2.2.43}$$

with $\sigma = E\varepsilon$ and $\sigma_{\text{eff}} = E_{\text{eff}}\ \varepsilon_{\text{eff}}$, leads to

$$\varepsilon_{\text{eff}} = (1 - \varphi_{\text{d}})\varepsilon \tag{2.2.44}$$

$$E_{\text{eff}} = \frac{E}{(1 - \varphi_{\text{d}})^2} \tag{2.2.45}$$

2.2.3.2 Damage Potential

In a conventional computational approach, the simple concept of one-dimensional damage can be generalized to three-dimensional cases by defining an energy potential for the damage problem. Such a definition allows the damage-based constitutive equation and state variables to be derived.

Beginning with the Helmholtz free energy ψ_{H} in the form of elastic strain energy for a damage problem (Lemaitre 1996; Voyjadjis and Kattan 1996),

$$\psi_{\text{d}} = \psi_{\text{H}}(\varepsilon^e, T, h_{\text{d}}, \varphi_{\text{d}}) \tag{2.2.46}$$

where h_{d} is the isotropic hardening variable, the second principle of thermodynamics and the Clausius–Duham inequality allow for determination of stress σ and entropy s:

$$\sigma = \rho \frac{\partial \psi_{\text{d}}}{\partial \varepsilon^e} \tag{2.2.47}$$

$$s = -\frac{\partial \psi_{\text{d}}}{\partial T} \tag{2.2.48}$$

and

$$H_{\text{d}} = \rho \frac{\partial \psi_{\text{d}}}{\partial h_{\text{d}}} \tag{2.2.49}$$

$$Y_{\text{d}} = \rho \frac{\partial \psi_{\text{d}}}{\partial \varphi_{\text{d}}} \tag{2.2.50}$$

where H_{d} is the hardening stress variable associated with the hardening variable h_{d} and Y_{d} represents a generalized thermodynamic force associated with the damage variable φ_{d}.

Experimental data or micromechanical considerations determine the way the analytical form of ψ_{d} is selected. For an isothermal problem with hardening behaviour, ψ_{d} can be explicitly defined as

$$\psi_{\text{d}} = \frac{1}{\rho}\left[\frac{1}{2}(1 - \varphi_{\text{d}})\mathbf{D}\varepsilon^e\varepsilon^e\right] + \psi_{\text{h}} \tag{2.2.51}$$

where **D** is the fourth-order tensor of elastic modulus and the hardening part ψ_h is defined in terms of the scalar hardening variables h_d and H_d^∞ (Kattan and Voyiadjis 2001; Lemaitre and Desmorat 2005),

$$\psi_h = H_d^\infty \left[1 + \frac{1}{b} \exp(-bh_d) \right] \tag{2.2.52}$$

where b is an isotropic hardening constant.

The associated damage rules define the evolution of state variables in terms of the damage potential $g_d \equiv 0$ (similar to the associated flow rule and plastic potential):

$$d\varepsilon^p = d\lambda \frac{\partial g_d}{\partial \sigma} \tag{2.2.53}$$

$$dh_d = -d\lambda \frac{\partial g_d}{\partial H_d} \tag{2.2.54}$$

and

$$d\varphi_d = -d\lambda \frac{\partial g_d}{\partial Y_d} \tag{2.2.55}$$

The damage potential g_d, also known as the potential of dissipation, is defined in terms of the state variables and the available experimental data for any specific material (Lemaitre 1996; Voyjadjis and Kattan 1996):

$$g_d \equiv 0 \tag{2.2.56}$$

The damage loading/unloading conditions can then be categorized by

$$g_d < 0 : \text{elastic unloading} \tag{2.2.57}$$

$$g_d = 0, \frac{\partial g_d}{\partial \dot{Y}_d} \dot{Y}_d \begin{cases} < 0 & \text{elastic unloading} \\ = 0 & \text{neutral loading} \\ > 0 & \text{loading} \end{cases} \tag{2.2.58}$$

2.2.3.3 Anisotropic Damage

In an anisotropic damage formulation, the microdefects may also change the normal direction of the representative surface. As a result, the simplest form of description of an anisotropic damage formulation is by a second-order damage tensor φ_d:

$$(\boldsymbol{I} - \varphi_d)\boldsymbol{n}\delta A = \boldsymbol{n}_{\text{eff}} \delta A_{\text{eff}} \tag{2.2.59}$$

Accordingly, the effective stress tensor $\boldsymbol{\sigma}_{\text{eff}}$ is related to the stress tensor $\boldsymbol{\sigma}$ by

$$\boldsymbol{\sigma}_{\text{eff}} = \boldsymbol{\sigma}(\boldsymbol{I} - \varphi_{\text{d}})^{-1} \tag{2.2.60}$$

Furthermore, the anisotropic damage formulation can be described by the fourth-order damage tensor ϕ_d, which also accounts for the surface distortions,

$$\phi_{\text{d}}^{-1}(\boldsymbol{vn}\delta A) = (\boldsymbol{vn}_{eff}\delta A_{eff}) \tag{2.2.61}$$

Generally, the effective stress tensor $\boldsymbol{\sigma}_{\text{eff}}$ is related to the stress tensor $\boldsymbol{\sigma}$ by

$$\boldsymbol{\sigma}_{\text{eff}} = \boldsymbol{\sigma}\left(\phi_{\text{d}}\right) \tag{2.2.62}$$

which results in a non-symmetric σ_{eff}, complicating the governing formulation and the numerical implementation.

The effective elasticity tensor $\mathbf{D}_{\text{eff}}$ can be defined in terms of the elasticity tensor $\mathbf{D}$ and the damage tensor ϕ_d as

$$\mathbf{D}_{\text{eff}} = \left(\boldsymbol{I} - \phi_{\text{d}}\right)\mathbf{D} \tag{2.2.63}$$

and the corresponding free energy of the damage state

$$\psi_{\text{d}} = \frac{1}{\rho}\left[\frac{1}{2}\left(\boldsymbol{I} - \phi_{\text{d}}\right)\mathbf{D}\varepsilon^{\text{e}}\varepsilon^{\text{e}}\right] + \psi_{\text{h}} \tag{2.2.64}$$

where the hardening part ψ_{h} is defined in the tensorial forms of $\boldsymbol{h}$ and $\boldsymbol{H}_{\text{d}}^{\infty}$ for a kinematic hardening process (Lemaitre and Desmorat 2005; Voyiadis and Woelke 2008):

$$\psi_{\text{h}} = \frac{1}{2}\gamma \boldsymbol{H}_{\text{d}}^{\infty}\left[\boldsymbol{hh}\right] \tag{2.2.65}$$

where γ is a hardening constant.

For further details on the integration of elastoplastic damage constitutive equations, refer to the reference textbooks on damage (Lemaitre and Desmorat 2005; Voyajilis and Kattan 2005).

2.2.4 Fracture Mechanics

Crackings may substantially reduce the strength of a material/specimen even if they are not apparent and readily detectable. In addition to substantial financial costs, fractures and defects may lead to major safety problems and even catastrophes, with huge social consequences.

The modern analytical approach to fracture can be attributed to Kircsh (1898) and Kolosov (1909), who were the first to analyse the effect of a circular flaw in an infinite tensile plate and discussed the biaxial stress state around the flaw with a stress concentration factor of 3. Then, in 1913 Inglis published a stress analysis for an elliptical hole in an infinite linear elastic tensile plate and discussed a sharp crack by degeneration of the elliptical hole into a line (Inglis 1913). Other major related analytical studies were published by Westergaard (1939) and Williams (1952) for stress and displacement solutions near a crack tip.

The major quantitative connection between the fracture state and flaw size came from the work of Griffith in 1920, who studied the effects of scratches and similar flaws on industrial components (Griffith 1921, 1924).

Later, Irwin (1948) and Orowan (1948) independently extended the Griffith approach to metals by including energy dissipation by local plastic flow. Irwin (1957) correlated the concept of the energy release rate G in the Griffith theory to the stress intensity factor (SIF) as a means for expressing the stress and displacement fields near a crack tip.

In 1955, Wells used the fracture mechanics to show that the fuselage failures resulted from fatigue cracks reaching a critical size (Wells 1955). A major study on general fatigue analysis was later presented by Paris et al. (1961), who provided experimental and theoretical arguments for their approach on applying fracture mechanics principles to fatigue crack growth.

The finite element method (FEM) was originally used as a simple analytical tool for obtaining the continuum-based displacement and stress fields. Later, sophisticated singular elements were proposed and efficiently implemented by Owen and Fawkes (1983) to simulate the singularity condition at crack tips. Major numerical progresses, however, include the development of the meshless methods such as the element-free Galerkin method (EFG) (Belytschko et al. 1994), the meshless local Petrov–Galerkin method (MLPG) (Atluri and Shen 2002), the smoothed particle hydrodynamics method (SPH) (Belytschko et al. 1996), the extended finite element method (XFEM) (Mohammadi 2008, 2012a) and the extended isogeometric analysis method (XIGA) (Ghorashi et al. 2012, 2015).

2.2.4.1 Linear Elastic Fracture Mechanics (LEFM)

Any defects such as a hole, crack and even inclusion are sources of inhomogeneity and concentration in solids. They are also responsible for various modes of failure and reduction in material strength.

Historically, about a century ago, analytical solutions for an infinite plate with circular and elliptical holes contributed largely in understanding the concept of stress concentration and change of stress state due to internal defects. For instance, in a tensile plate with a circular hole (Kircsh 1898), the stress concentration factor of 3 is obtained and the stress state becomes biaxial and non-uniform with even compressive components (Figure 2.7). Later, a similar study on a tensile plate with an elliptical hole (Inglis 1913) concluded the same patterns for stress contours, but with a stress concentration factor that was inversely proportional to the square root of the radius of curvature of the hole boundary ρ:

$$\left(\sigma_{\beta\beta}\right)_{\substack{\alpha=\alpha_0\\ \beta=0,\pi}} = \sigma_0\left(1+2\sqrt{\frac{a}{\rho}}\right) \overset{\rho\ll a}{\longrightarrow} \left(\sigma_{\beta\beta}\right)_{\substack{\alpha=\alpha_0\\ \beta=0,\pi}} \approx 2\sigma_0\sqrt{\frac{a}{\rho}} \qquad (2.2.66)$$

An immediate conclusion is that for a line crack (a degenerated ellipse: $b=0$), ρ becomes zero and an infinite stress in generated.

Consider a unidirectional tensile plate, shown in Figure 2.8, which includes a number of microdefects and inhomogeneities. In an intact specimen, the uniaxial stress component

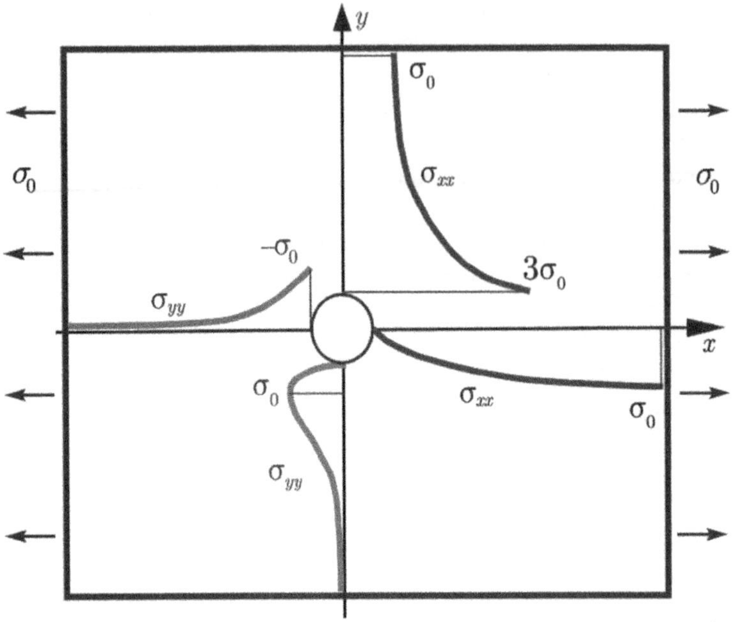

Figure 2.7 Distribution of stress components in an infinite tensile plate with a circular hole.

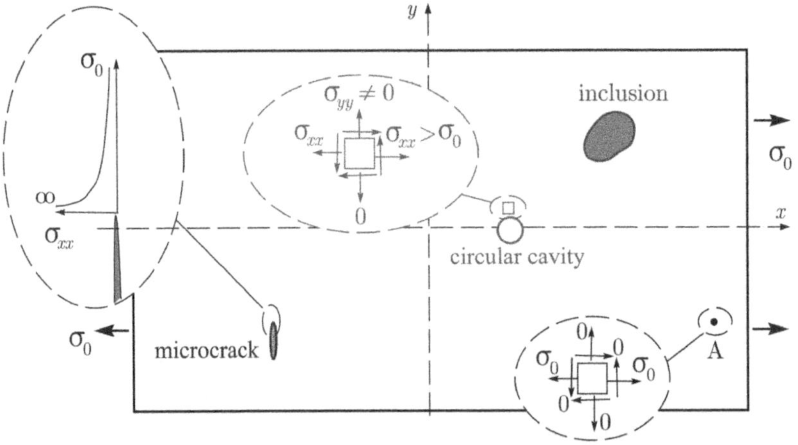

Figure 2.8 A typical tensile plate with various microinhomogeneities and defects.

$\sigma_{xx} = \sigma_0$ is uniformly distributed throughout the specimen. This is also the case everywhere away from the microholes and microcracks. Around the microhole, however, the stress state changes considerably into a biaxial non-uniform stress distribution with a stress concentration factor of 3. The situation becomes extremely more severe around the tip of a microcrack, as the theoretical solution predicts an infinite stress concentration.

This simple illustration shows that if a criterion based on macro or average stress is adopted for design purposes, it will fail due to ignoring the stress concentration effects, especially for the infinite stresses at the crack tip. On the other hand, while most engineering materials contain microdefects with even infinite stresses, they remain stable for a wide range of loadings. A remedy is to develop and adopt non-local theories. Among them, the theory of fracture mechanics adopts the concepts of the stress intensity factor and fracture energy release rate to examine the stability of cracked materials and the potential propagation responses.

2.2.4.2 Asymptotic Solution Around a Crack Tip

Westergaard (1939) adopted a biharmonic stress function to derive the stress field around a crack tip in an infinite tensile plate with a central traction-free crack. Accordingly, the final near crack tip ($r \ll a$) polar stress solutions can be obtained from (Meguid 1989)

$$\sigma_{xx} = \sigma_0 \sqrt{\frac{a}{2r}} \cos\frac{\theta}{2}\left(1 - \sin\frac{\theta}{2}\sin\frac{3\theta}{2}\right) \tag{2.2.67}$$

$$\sigma_{yy} = \sigma_0 \sqrt{\frac{a}{2r}} \cos\frac{\theta}{2}\left(1 + \sin\frac{\theta}{2}\sin\frac{3\theta}{2}\right) \tag{2.2.68}$$

$$\sigma_{xy} = \sigma_0 \sqrt{\frac{a}{2r}} \sin\frac{\theta}{2}\cos\frac{\theta}{2}\cos\frac{3\theta}{2} \tag{2.2.69}$$

where $2a$ is the crack length and σ_0 is the uniform biaxial stress, as depicted in Figure 2.9. Note that Equations (2.2.67) to (2.2.69) are only valid for relatively small values of r/a.

Alternatively, Williams (1952) adopted a different solution strategy to solve for stress distribution of a wedge problem, as depicted in Figure 2.9. The wedge can be degenerated into a crack problem. The methodology of the Williams solution was later extended to various asymptotic solutions of complicated elasticity problems such as a crack at the interface of two dissimilar materials (Esna Ashari and Mohammadi 2011a), and even to the general problem of a stick/slip crack between dissimilar materials (Ebrahimi et al. 2013).

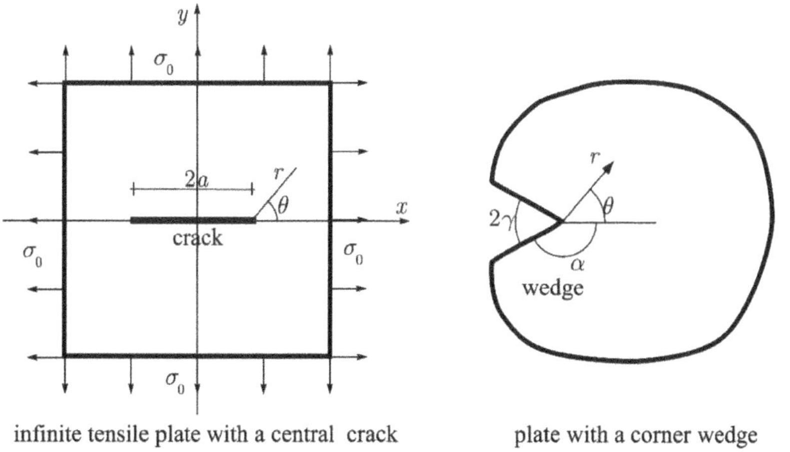

Figure 2.9 Definitions of the Westergaard and Williams problems. *Source:* Reproduced from Mohammadi, 2008/John Wiley & Sons Ltd.

The Williams solution for components of the stress field can be written as

$$\sigma_{rr} = \frac{1}{\sqrt{r}} \sum_{n} \left\{ b_n \left[\cos\frac{\theta}{2}\left(1+\sin^2\frac{\theta}{2}\right)\right] + a_n \left[-\frac{5}{4}\sin\frac{\theta}{2} + \frac{3}{4}\sin\frac{3\theta}{2}\right]\right\} \tag{2.2.70}$$

$$\sigma_{\theta} = \frac{1}{\sqrt{r}} \sum_{n} \left\{ b_n \left[\cos\frac{\theta}{2}\left(1-\sin^2\frac{\theta}{2}\right)\right] + a_n \left[-\frac{3}{4}\sin\frac{\theta}{2} - \frac{3}{4}\sin\frac{3\theta}{2}\right]\right\} \tag{2.2.71}$$

$$\sigma_{r\theta} = \frac{1}{\sqrt{r}} \sum_{n} \left\{ b_n \left[\sin\frac{\theta}{2}\cos^2\frac{\theta}{2}\right] + a_n \left[\frac{1}{4}\cos\frac{\theta}{2} + \frac{3}{4}\cos\frac{3\theta}{2}\right]\right\} \tag{2.2.72}$$

where a_n and b_n are

$$a_n = \frac{B_n}{A_n} = -\frac{\cos(\lambda_n - 1)\pi}{\cos(\lambda_n + 1)\pi}, \lambda_n = \frac{n}{2}, n = 1,3,4,\ldots \tag{2.2.73}$$

$$b_n = \frac{C_n}{D_n} = -\frac{\sin(\lambda_n - 1)\pi}{\sin(\lambda_n + 1)\pi}, \lambda_n = \frac{n}{2}, n = 1,3,4,\ldots \tag{2.2.74}$$

and λ is the solution of the characteristic equation (Saouma 2000)

$$\sin 2\lambda_n \alpha + \lambda_n \sin 2\alpha = 0 \tag{2.2.75}$$

2.2.4.3 Stress Intensity Factor, *K*

A major breakthrough in fracture mechanics was achieved by Irwin (1957) with the genius definition of the stress intensity factor (SIF), which quantifies the strength of singularity of a crack problem. Accordingly, all elastic stress fields around a crack tip can be formulated in terms of the stress intensity factor *K*.

Considering the basic problem of an infinite tensile plate with a central crack, the stress intensity factor is defined as

$$K_0 = \lim_{\substack{r\to 0\\ \theta=0}} \sigma_{yy}\sqrt{2\pi r} \tag{2.2.76}$$

which can be simplified using (2.2.68):

$$K_0 = \lim_{\substack{r\to 0\\ \theta=0}} \sqrt{2\pi r}\,\sigma_0 \sqrt{\frac{a}{2r}} \cos\frac{\theta}{2}\left(1+\sin\frac{\theta}{2}\sin\frac{3\theta}{2}\right) = \sigma_0\sqrt{\pi a} \tag{2.2.77}$$

In fact, K_0 represents all the geometric, loading and boundary characteristics of the fracture problem.

For a general form of crack problem, the stress state can be written in terms of the radial singularity, three independent stress intensity factors K_i and the corresponding angular distribution functions $f^i(\theta)$:

$$\sigma = \frac{1}{\sqrt{r}}\left\{K_{\mathrm{I}} f^{\mathrm{I}}(\theta) + K_{\mathrm{II}} f^{\mathrm{II}}(\theta) + K_{\mathrm{III}} f^{\mathrm{III}}(\theta)\right\} \tag{2.2.78}$$

The most general form of the stress state (Equation 2.2.78) categorizes three independent fracture modes (see Figure 2.10):

1) The opening mode I: crack surfaces are only pulled apart in the normal direction (y).
2) The shearing mode II: crack surfaces slide in the shearing (x) direction.
3) The tearing mode III: crack surfaces slide in the out-of-plane direction (z).

The corresponding stress intensity factors $K_{\text{I}}, K_{\text{II}}$ and K_{III} are defined as (Figure 2.10)

$$\begin{Bmatrix} K_{\text{I}} \\ K_{\text{II}} \\ K_{\text{III}} \end{Bmatrix} = \lim_{\substack{r \to 0 \\ \theta = 0}} \begin{Bmatrix} \sigma_{yy} \\ \sigma_{xy} \\ \sigma_{yz} \end{Bmatrix} \sqrt{2\pi r} \tag{2.2.79}$$

In fracture mechanics, the computed stress intensity factors for any specific loading condition are compared with their corresponding critical values or toughness K_C, which is a material property, to determine the state of crack stability. For general problems where more than one pure fracture mode governs the problem simultaneously, a mixed mode criterion is required to assess the stability of the crack.

The final polar stress components for the opening mode I can be written in terms of the stress intensity factor K_{I}:

$$\sigma_{rr} = \frac{K_{\text{I}}}{\sqrt{2\pi r}} \cos\frac{\theta}{2}\left(1 + \sin^2\frac{\theta}{2}\right) = \frac{K_{\text{I}}}{\sqrt{2\pi r}}\left(\frac{5}{4}\cos\frac{\theta}{2} - \frac{1}{4}\cos\frac{3\theta}{2}\right) \tag{2.2.80}$$

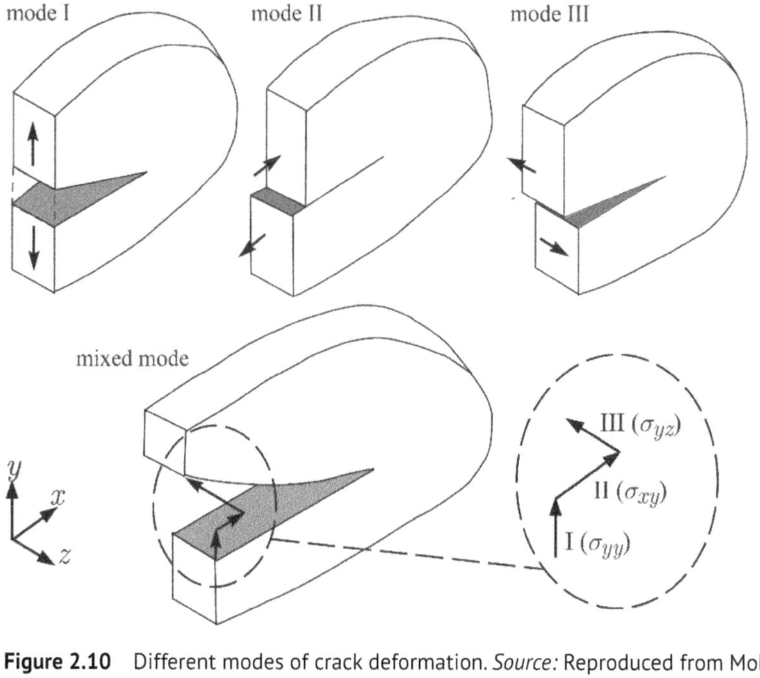

Figure 2.10 Different modes of crack deformation. *Source:* Reproduced from Mohammadi, 2008/ John Wiley & Sons Ltd.

$$\sigma_{\theta\theta} = \frac{K_I}{\sqrt{2\pi r}} \cos^3\frac{\theta}{2} = \frac{K_I}{\sqrt{2\pi r}}\left(\frac{3}{4}\cos\frac{\theta}{2} + \frac{1}{4}\cos\frac{3\theta}{2}\right) \tag{2.2.81}$$

$$\sigma_{r\theta} = \frac{K_I}{\sqrt{2\pi r}} \sin\frac{\theta}{2}\cos^2\frac{\theta}{2} = \frac{K_I}{\sqrt{2\pi r}}\left(\frac{1}{4}\sin\frac{\theta}{2} + \frac{1}{4}\sin\frac{3\theta}{2}\right) \tag{2.2.82}$$

and for the shear mode II:

$$\sigma_{rr} = \frac{K_{II}}{\sqrt{2\pi r}} \sin\frac{\theta}{2}\left(1 - 3\sin^2\frac{\theta}{2}\right) = \frac{K_{II}}{\sqrt{2\pi r}}\left(-\frac{5}{4}\sin\frac{\theta}{2} + \frac{3}{4}\sin\frac{3\theta}{2}\right) \tag{2.2.83}$$

$$\sigma_{\theta\theta} = \frac{-3K_{II}}{2\sqrt{2\pi r}} \sin\frac{\theta}{2}\cos\frac{\theta}{2} = \frac{K_{II}}{\sqrt{2\pi r}}\left(-\frac{3}{4}\sin\frac{\theta}{2} - \frac{3}{4}\sin\frac{3\theta}{2}\right) \tag{2.2.84}$$

$$\sigma_{r\theta} = \frac{K_{II}}{\sqrt{2\pi r}} \cos\frac{\theta}{2}\left(1 - 3\sin^2\frac{\theta}{2}\right) = \frac{K_{II}}{\sqrt{2\pi r}}\left(\frac{1}{4}\cos\frac{\theta}{2} + \frac{3}{4}\cos\frac{3\theta}{2}\right) \tag{2.2.85}$$

The tearing mode III has only two non-zero stress components:

$$\sigma_{xz} = -\frac{K_{III}}{\sqrt{2\pi r}} \sin\frac{\theta}{2} \tag{2.2.86}$$

$$\sigma_{yz} = \frac{K_{III}}{\sqrt{2\pi r}} \cos\frac{\theta}{2} \tag{2.2.87}$$

Having known, or computed, the values of the stress intensity factors for any fracture problem, the full stress states around the crack tip are known.

2.2.4.4 Fracture Energy and Crack Growth

Griffith (1921) demonstrated that the tensile strength of brittle specimens decreases with the increase of their size. He adopted the Inglis solution for an internal flaw to postulate his thermodynamic criterion based on the total change in energy balance of a cracked body in terms of changes of the crack length.

For a general deformable cracked body, Griffith re-wrote the first law of thermodynamics and assumed that for a virtual crack extension of a fractured body, the rate of external supplies is equal to the sum of the rate of kinetic and internal strain energies, and the rate of surface energy dissipated during the crack propagation:

$$\frac{\partial W}{\partial a} + \frac{\partial Q}{\partial a} = \frac{\partial U^k}{\partial a} + \frac{\partial U^s}{\partial a} + \frac{\partial U^\Gamma}{\partial a} \tag{2.2.88}$$

where W represents the external work, Q is the heat source, U^k and U^s are the kinetic and internal strain energies, respectively, and U^Γ characterizes the so-called surface energy. Equation (2.2.88) constitutes a criterion for the crack growth based on the comparison of the external energy available for crack growth with the total resistance of the deformable material.

Decomposing the strain energy U^{s} into elastic $U_{\mathrm{e}}^{\mathrm{s}}$ and plastic $U_{\mathrm{p}}^{\mathrm{s}}$ parts for quasi-static materials and in the absence of thermal effects, Equation (2.2.88) can be simplified to

$$\frac{\partial W}{\partial a}=\frac{\partial U_{\mathrm{e}}^{\mathrm{s}}}{\partial a}+\frac{\partial U_{\mathrm{p}}^{\mathrm{s}}}{\partial a}+\frac{\partial U_{\Gamma}}{\partial a} \tag{2.2.89}$$

or in terms of the potential energy $\Pi=U_{\mathrm{e}}^{\mathrm{s}}-W$:

$$-\frac{\partial \Pi}{\partial a}=-\frac{\partial U_{\mathrm{e}}^{\mathrm{s}}}{\partial a}+\frac{\partial W}{\partial a}=\frac{\partial U_{\mathrm{p}}^{\mathrm{s}}}{\partial a}+\frac{\partial U^{\Gamma}}{\partial a} \tag{2.2.90}$$

For a perfectly brittle material, U_p^s vanishes and Equation (2.2.90) reduces to the classical Griffith crack growth energy release rate per unit crack extension, G:

$$G=-\frac{\partial \Pi}{\partial a}=-\frac{\partial U_{\mathrm{e}}^{\mathrm{s}}}{\partial a}+\frac{\partial W}{\partial a}=\frac{\partial U^{\Gamma}}{\partial a}=2\gamma_s \tag{2.2.91}$$

where γ_s is the surface energy, which should be determined experimentally. $2\gamma_s$ can be interpreted as the critical fracture energy release rate G_c and the crack growth criterion is written as

$$\frac{\mathrm{d}\Pi}{\mathrm{d}a}\geq 2\gamma_s \rightarrow G\geq G_c \tag{2.2.92}$$

This is somehow comparable to the criterion for fracture growth in terms of the stress intensity factor:

$$K=K_c \tag{2.2.93}$$

Equation (2.2.94) defines the relation between K and G for mode I problems:

$$G=\frac{K_{\mathrm{I}}^2}{E'} \tag{2.2.94}$$

where E' is the effective Young's modulus, defined in Equation (2.2.26).

It should be noted that, contrary to the stress intensity factor, only a unique G is defined for a fracture problem and it cannot be directly decomposed into modes I and II components. Therefore, the values of two stress intensity factors cannot be obtained from a single value of the fracture energy release rate:

$$G=\frac{K_{\mathrm{I}}^2+K_{\mathrm{II}}^2}{E'} \tag{2.2.95}$$

The stability of a brittle crack growth can be examined by the second derivative of $(\Pi+U^{\Gamma})$ (Gtoudos 1993):

$$\frac{\partial^2\left(\Pi+U^{\Gamma}\right)}{\partial a^2}\begin{cases}<0 & \text{unstable fracture}\\ >0 & \text{stable fracture}\\ =0 & \text{neutral equilibrium}\end{cases} \tag{2.2.96}$$

2.2.4.5 Mixed Mode Crack Propagation

Engineering fracture problems are usually in the mixed mode state subjected to multiaxial loadings and/or inclined cracks (Figure 2.11). The most general approach is to compare the energy release rate, G, of the actual problem with a material property, G_c (critical energy release rate). Unfortunately, computation of G remains a difficult task for many engineering problems.

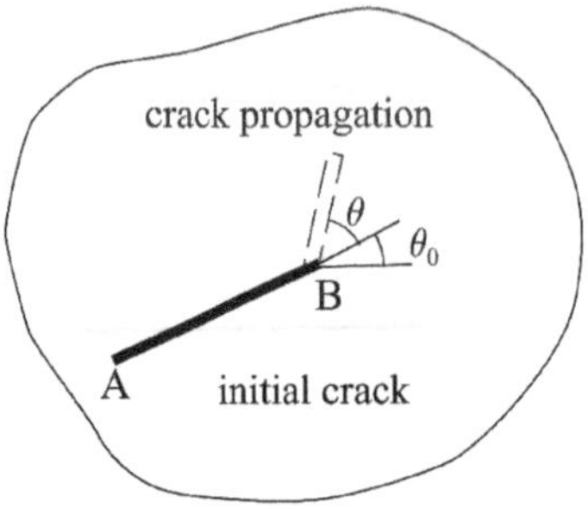

Figure 2.11 Mixed mode crack propagation.

An efficient numerical replacement is the maximum circumferential/hoop tensile stress criterion, which is the most frequently used approach to assess the stability state of a mixed mode crack problem and to determine its potential propagation angle. According to this criterion, the crack propagates in a radial direction normal to the direction of maximum circumferential tensile stress $(\sigma_\theta)_{\max}$ (Erdogan and Sih 1963). The crack propagation criterion for a general θ_0-inclined crack with a potential propagation angle θ can be expressed as

$$K_{\theta\theta} = \sigma_{\theta\theta}\sqrt{2\pi r} = K_{\mathrm{I}}\cos^3\frac{\theta}{2} - \frac{3}{2}K_{\mathrm{II}}\cos\frac{\theta}{2}\sin\theta = K_{\mathrm{Ic}} \tag{2.2.97}$$

Alternative criteria include the minimum strain energy density criterion and the maximum energy release rate, mainly for research-oriented problems, and a variety of experimental and empirical formulae for engineering applications.

For orthotropic problems, the fracture toughness also depends on the orientation. Therefore, the ratio of $K_{\theta\theta}\,/\,K_{Ic}^{\theta}$ has to be maximized to assess the state of crack growth and to compute the crack growth angle θ:

$$\max\left\{\frac{K_{\theta\theta}}{K_{\mathrm{Ic}}^{\theta}}\right\} = \max\left\{\frac{\sqrt{2\pi r}\sigma_{\theta\theta}}{K_{\mathrm{Ic}}^{\theta}}\right\} \tag{2.2.98}$$

Having known the two values of K_{Ic}^1 and K_{Ic}^2 along the principal planes of elastic symmetry (Figure 2.12), Equation (2.2.99) can be adopted to determine the toughness K_{Ic}^{θ} at any crack propagation angle θ:

$$K_{\mathrm{Ic}}^{\theta} = K_{\mathrm{Ic}}^1\cos^2\theta + K_{\mathrm{Ic}}^2\sin^2\theta \tag{2.2.99}$$

Furthermore, the following assumption is widely accepted to avoid performing two separate fracture toughness tests (Saouma et al. 1987):

$$K_{\mathrm{Ic}}^2 = K_{\mathrm{Ic}}^1\frac{E_1}{E_2} \tag{2.2.100}$$

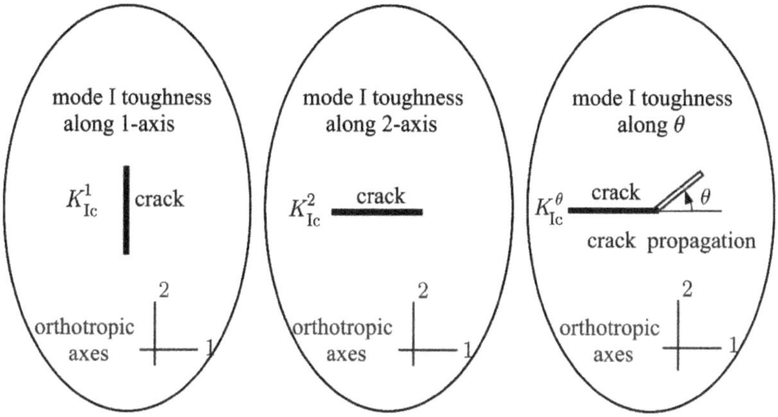

Figure 2.12 Fracture toughness for an orthotropic medium. *Source:* Adapted from Saouma et al. 1987.

2.2.4.6 Quarter Point Singular Elements

Singular finite elements provide an efficient method for analysis of crack problems with a singular stress state of the order of analytical asymptotic solutions of fracture mechanics. They also allow for fast and accurate evaluation of stress intensity factors (Shih et al. 1976). Nevertheless, the singular elements become inefficient in crack propagation problems and cannot be extended to non-isotropic and other complex fracture problems.

The concept of the singular finite element is simple. It is well known that by moving a mid-side node of a six-node or eight-node finite element into the quarter point position (Figure 2.13), the singular characteristic of LEFM can be reproduced at the corresponding corner node (Owen and Fawkes 1983).

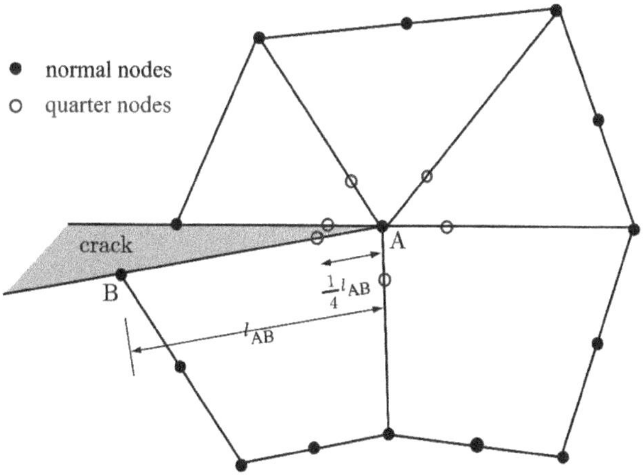

Figure 2.13 Singular (quarter point) finite elements around a crack tip. *Source:* Reproduced from Mohammadi, 2008/John Wiley & Sons Ltd.

For instance, if the mid-side node of the element in Figure 2.13 is located at the quarter point position, the determinant of the Jacobian matrix $\mathbf{J}$ becomes zero at the nearby corner node A (crack tip), leading to the singular $\mathbf{J}^{-1}$:

$$\mathbf{J}^{-1} = \frac{1}{\det \mathbf{J}} \begin{bmatrix} \dfrac{\partial y}{\partial \eta} & -\dfrac{\partial y}{\partial \xi} \\ -\dfrac{\partial x}{\partial \eta} & \dfrac{\partial x}{\partial \xi} \end{bmatrix} \tag{2.2.101}$$

Accordingly, the $\mathbf{B}$ matrix (derivatives of shape functions) becomes singular, leading to the singular stress tensor at the corner node A:

$$\sigma = \mathbf{DBU} \tag{2.2.102}$$

The method is very fast and efficient for stationary crack problems with no change in the finite element formulation and only requires a local re-positioning of the mid-side nodes around a crack tip. The concept can be similarly extended to three-dimensional crack problems.

2.2.4.7 *J* Integral

Rice and Rosengren (1968) modified the original work of Eshelby (1956, 1974) on path independent contour J integrals for general elasticity problems to develop a fracture energy related concept for linear elastic and monotonically nonlinear elastic crack problems (Anderson 1995).

In the absence of body force and crack tractions ($\boldsymbol{f}^b = \boldsymbol{f}^c = \mathbf{0}$), the path independent J integral on a crack-closed counter-clockwise contour Γ is defined as (Figure 2.14a)

$$J = \int_\Gamma \left(w_s dy - \boldsymbol{t}\frac{\partial \mathbf{u}}{\partial x} d\Gamma \right) = \int_\Gamma \left(\frac{1}{2}\sigma_{ij}\varepsilon_{ij} dy - \boldsymbol{t}\frac{\partial \mathbf{u}}{\partial x} d\Gamma \right) \tag{2.2.103}$$

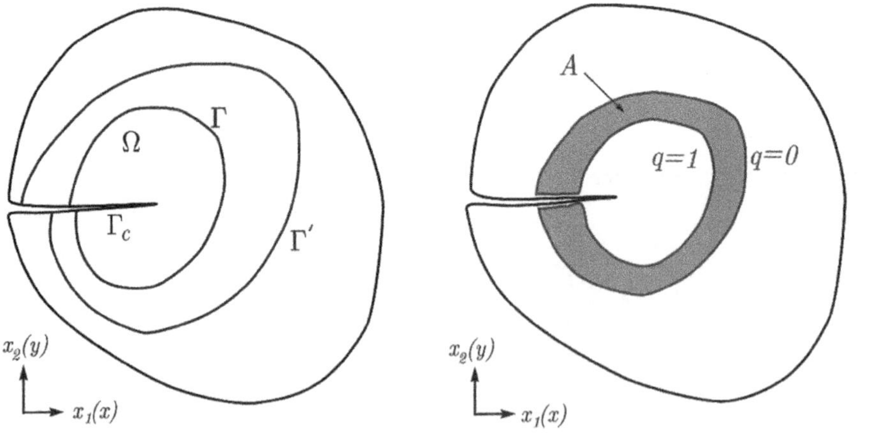

Figure 2.14 Definitions of the contour and equivalent domain integrals. *Source:* Reproduced from Mohammadi, 2008/John Wiley & Sons Ltd.

or in the component form,

$$J=\int_{\Gamma}\left(w_s\delta_{1j}-\sigma_{ij}\frac{\partial u_i}{\partial x}\right)n_j d\Gamma=\int_{\Gamma}\left(\frac{1}{2}\sigma_{ij}\varepsilon_{ij}\delta_{1j}-\sigma_{ij}\frac{\partial u_i}{\partial x}\right)n_j d\Gamma \tag{2.2.104}$$

where $\boldsymbol{t}-\boldsymbol{\sigma n}$ represents the traction vector on a plane defined by the outward normal $\boldsymbol{n}$.

The J integral can be readily extended to account for the effect of crack surface tractions $\boldsymbol{f}^c$ (Karlsson and Backlund 1978):

$$J=\int_{\Gamma}\left\{\frac{1}{2}\sigma_{ij}\varepsilon_{ij}dy-\boldsymbol{t}\frac{\partial \mathbf{u}}{\partial x}d\Gamma\right\}-\int_{\Gamma_c}\boldsymbol{f}^c\frac{\partial \mathbf{u}}{\partial x}d\Gamma \tag{2.2.105}$$

where Γ_c is the portion of the crack surfaces along the contour Γ. Similarly, the presence of body force $\mathbf{f}^{\mathrm{b}}$ leads to (Atluri 1982)

$$J=\int_{\Gamma}\left\{\frac{1}{2}\sigma_{ij}\varepsilon_{ij}dy-\boldsymbol{t}\frac{\partial \mathbf{u}}{\partial x}d\Gamma\right\}-\int_{\Omega}\boldsymbol{f}^{\mathrm{b}}\frac{\partial \mathbf{u}}{\partial x}d\Omega \tag{2.2.106}$$

The J integral can also be defined in the form of the equivalent domain integral (EDI) on an area A bound by the two contours, as illustrated in Figure 2.14b (Babuska and Miller 1984; Li et al. 1985):

$$J=\int_{A}\left[\sigma_{ij}\frac{\partial u_i}{\partial x_1}-w_s\delta_{1j}\right]\frac{\partial q}{\partial x_j}dA \tag{2.2.107}$$

where q is an arbitrary smoothing function,

$$q=\begin{cases}0 & \text{on outer contour}\\ 1 & \text{on inner contour}\end{cases} \tag{2.2.108}$$

In practice, the contours are selected in a way to conform with the finite element edges.

It is important to note that the J integral can be related to the fracture energy release rate G, which allows for its application to more general fracture problems (Mohammadi 2008):

$$J=-\frac{d\Pi}{da}=G \tag{2.2.109}$$

2.2.4.8 Interaction Integral Method

The interaction integral method is a numerical approach to compute both stress intensity factors of a mixed mode problem. It is derived from the superposition of two actual and auxiliary states (Sih et al. 1965). The auxiliary state is adopted from the well-known Westergaard solution for the displacement and stress fields in the vicinity of the crack tip.

The EDI form of the M integral for a combined set of actual and auxiliary states can be defined as

$$M=\int_{A}\left[\sigma_{ij}\frac{\partial u_i^{\mathrm{aux}}}{\partial x_1}+\sigma_{ij}^{\mathrm{aux}}\frac{\partial u_i}{\partial x_1}-\frac{1}{2}\left(\sigma_{ij}\varepsilon_{ij}^{\mathrm{aux}}+\sigma_{ij}^{\mathrm{aux}}\varepsilon_{ij}\right)\delta_{1j}\right]\frac{\partial q}{\partial x_j}dA \tag{2.2.110}$$

M is related to the stress intensity factors by

$$M = \frac{2}{E'}\left(K_I K_I^{aux} + K_{II} K_{II}^{aux}\right) \tag{2.2.111}$$

Equations (2.2.112) allow for the computation of mode I and II stress intensity factors by two independent evaluations of M from Equation (2.2.110):

$$\begin{aligned} K_I &= \frac{E'}{2} M\left(K_I^{aux} = 1, K_{II}^{aux} = 0\right) \\ K_{II} &= \frac{E'}{2} M\left(K_I^{aux} = 0, K_{II}^{aux} = 1\right) \end{aligned} \tag{2.2.112}$$

2.2.4.9 Elastoplastic Fracture Mechanics (EPFM)

LEFM assumes an ideal material that may withstand infinite stress levels. This is clearly unacceptable from a physical point of view. Moreover, it leads to conservatively expensive solutions for many engineering ductile materials by ignoring their energy-absorbing plastic phenomena at and around the crack tip. Elastoplastic fracture mechanics (EPFM) is a simplified approach, designed to modify the LEFM formulations, in order to account for the crack tip plasticity.

A simple strategy is to determine the size of the plastic zone r_p in one-dimensional uniaxial problems. Three different formulations are available to estimate r_p:

$$\begin{aligned} \text{1st order}: \; r_p &= \frac{1}{2\pi} \frac{K_I^2}{\sigma_{yld}^2} \\ \text{2nd order}: \; r_p &= \frac{1}{\pi} \frac{K_I^2}{\sigma_{yld}^2} \\ \text{Dugdale}: r_p &= a\left[\cos\left(\frac{\pi}{2}\frac{\sigma_0}{\sigma_{yld}}\right)\right]^{-1} \approx \frac{\pi}{8}\left(\frac{K_I}{\sigma_{yld}}\right)^2 \end{aligned} \tag{2.2.113}$$

The first-order approximation, based on simply ignoring all stresses above σ_{yld} (Figure 2.15), clearly violates the 1D stress equilibrium. The second-order solution improves the approximation by assuming a redistribution of stress to satisfy both the equilibrium and the yield criterion (Irwin 1961). The Dugdale assumption replaces the actual crack length with an effective one to ensure that the overall stress intensity factor (from the applied load and the closing plastic effect σ_{yld}) vanishes at the crack tip.

Alternatively, Cottrell (1961) and Wells (1963) proposed a different local criterion to indirectly determine the effects of crack tip plasticity. Dissimilar to LEFM sharp cracks, a blunted crack tip is generated in an elastoplastic problem. Accordingly, the so-called crack tip opening displacement (CTOD) is not negligible and can be adopted to indirectly determine the state of ductile fracture (Figure 2.16). Alternative measures, such as the crack tip opening angle (CTOA) can also be adopted (Bayesteh and Mohammadi 2011).

The crack tip opening displacement can be analytically related to the energy release rate:

$$\text{CTOD} \approx \frac{G}{\sigma_{yld}} \tag{2.2.114}$$

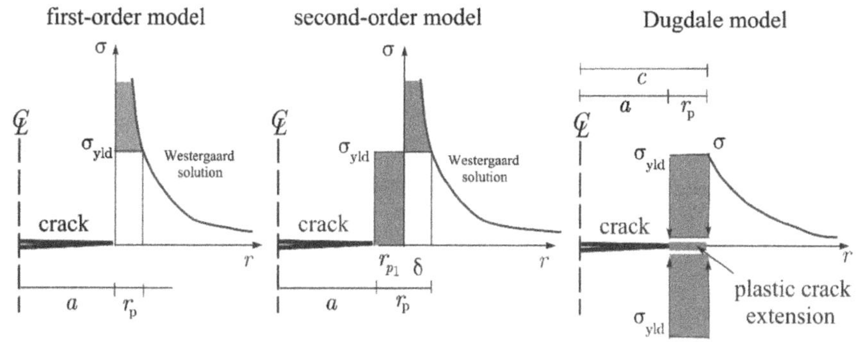

Figure 2.15 Different models for approximation of the size of the crack tip plastic zone. *Source:* Reproduced from Mohammadi, 2008/John Wiley & Sons Ltd.

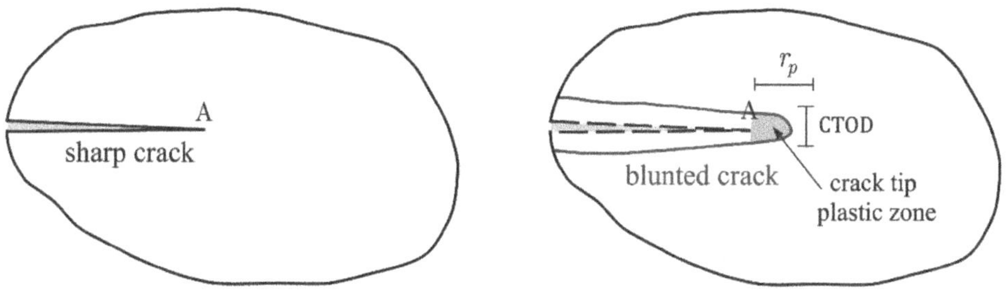

Figure 2.16 Definition of the crack opening displacement.

Estimation of CTOD can be performed by the first- and second-order approximations, based on the Irwin and Dugdale EPFM solutions, respectively (Kanninen 1984):

$$
\begin{aligned}
&\text{1st order: CTOD} = \frac{4}{\pi}\frac{K_I^2}{E\sigma_{\text{yld}}} \\
&\text{2nd order: CTOD} = \frac{8}{\pi}\frac{a\sigma_{\text{yld}}}{E}\ln\left[\sec\frac{\pi}{2}\frac{\sigma}{\sigma_{\text{yld}}}\right] = \frac{K_I^2}{E\sigma_{\text{yld}}}\left[1+\frac{\pi^2}{24}\frac{\sigma^2}{\sigma_{\text{yld}}^2}+\cdots\right]
\end{aligned}
\tag{2.2.115}
$$

2.2.4.10 Fatigue

It is well known that the strength of material is reduced if it is subjected to repeated loading conditions. Type and brittleness of the material, number of loading cycles, size of the specimen, temperature and environmental effects, and residuals and stress measures, among others, affect the magnitude of strength reduction. It is also observed that the fatigue failure in both brittle and ductile materials is mainly a brittle phenomenon (Pokluda and Sandera 2010). Analysis of fatigue is of paramount importance in the design and manufacture of biomedical objects such as artificial organs, which should endure a sufficiently high number of performance cycles in their pre-set service life.

The concept of fatigue failure may be studied by the stress- or strain-based criteria or can be examined by the fatigue crack propagation criterion.

Figure 2.17 defines the stress measures that may affect the fatigue strength of material subjected to cyclic loading. They are the maximum stress $\sigma_{\max}$, the minimum stress $\sigma_{\min}$, the stress range $\Delta\sigma = \sigma_{\max} - \sigma_{\min}$, the mean stress $\sigma_m = 0.5(\sigma_{\max} + \sigma_{\min})$, the alternating stress $\sigma_a = 0.5\Delta\sigma$, the stress ratio $R = \sigma_{\min} / \sigma_{\max}$ and the alternating stress ratio $R_a = \sigma_a / \sigma_m$.

In a stress-life approach ($\sigma_a - N$), variations of stress with time (in the logarithmic scales) follows the typical trend of Figure 2.18 with three distinguished parts. The low cycle fatigue region represents the cases where the fatigue failure occurs in a low number of cycles with an initial strength of S_u. In the high cycle fatigue region, a linear approximation may be adopted to relate the failure stress to the number of cycles. The third region of the figure is a plateau line for cases with a very high number of cycles, representing the fatigue limit stress S_f, which is assumed to be independent of the additional number of cycles. An estimation of $S_f = aN^b$ can be used to determine S_f based on experimentally curve-fitted constants a and b.

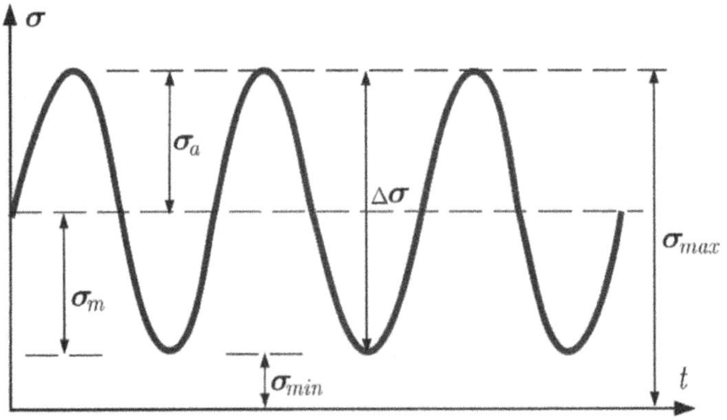

Figure 2.17 Stress measures affecting a fatigue problem.

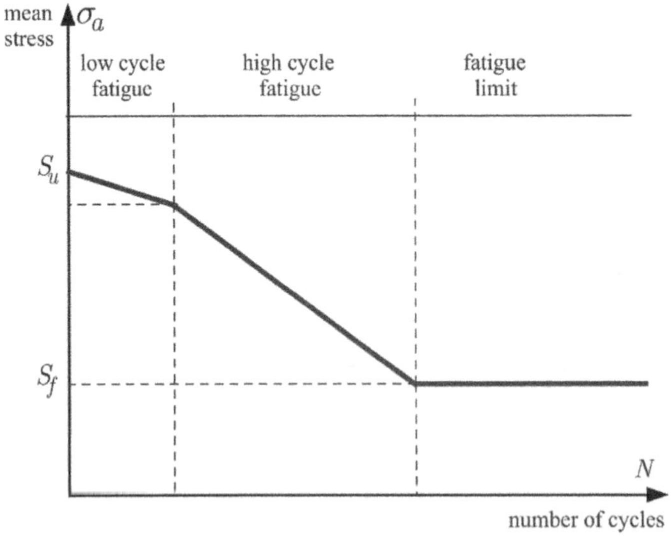

Figure 2.18 Stress-life fatigue description.

Some of the available experimental-based stress measures are:

$$\begin{cases} \text{Goodman} & \dfrac{\sigma_a}{S_f}+\dfrac{\sigma_m}{S_u}=1 \\ \text{Greber} & \dfrac{\sigma_a}{S_f}+\left(\dfrac{\sigma_m}{S_u}\right)^2-1 \\ \text{ASME} & \left(\dfrac{\sigma_a}{S_u}\right)^2+\left(\dfrac{\sigma_m}{S_y}\right)^2=1 \end{cases} \tag{2.2.116}$$

Alternatively, the strain-life approach ($\varepsilon_a - N$) adopts the variation of strain with time (in logarithmic scales), as typically depicted in Figure 2.19, to examine the fatigue process. The strain-life approach may provide a basis for determining the elastic and plastic dominant phenomena in a fatigue analysis.

Some of the available experimental-based 1D ($\varepsilon_a - N$) measures are:

$$\begin{cases} \text{Marrow} & \varepsilon_a=\dfrac{\sigma_f^e-\sigma_m}{E}\left(2N_f\right)^b+\varepsilon_f^p\left(2N_f\right)^c \\ \text{Generalized marrow} & \varepsilon_a=\dfrac{\sigma_f^e-\sigma_m}{E}\left(2N_f\right)^b+\varepsilon_f^p\left(\dfrac{\sigma_f^e-\sigma_m}{\sigma'_f}\right)^{\frac{c}{b}}\left(2N_f\right)^c \\ \text{SWT} & \varepsilon_a=\dfrac{1}{\sigma_{\max}E}\left(\sigma_f^e\right)^2\left(2N_f\right)^{2b}+\dfrac{1}{\sigma_{\max}}\sigma_f^e\varepsilon_f^p\left(2N_f\right)^{b+c} \end{cases} \tag{2.2.117}$$

where $\sigma_f^e = E\varepsilon_f^e$, ε_f^e and ε_f^p represent the elastic and plastic limits of the strain components, respectively, as shown in Figure 2.19, b and c are their corresponding slopes, and N_f is the expected life of the specimen.

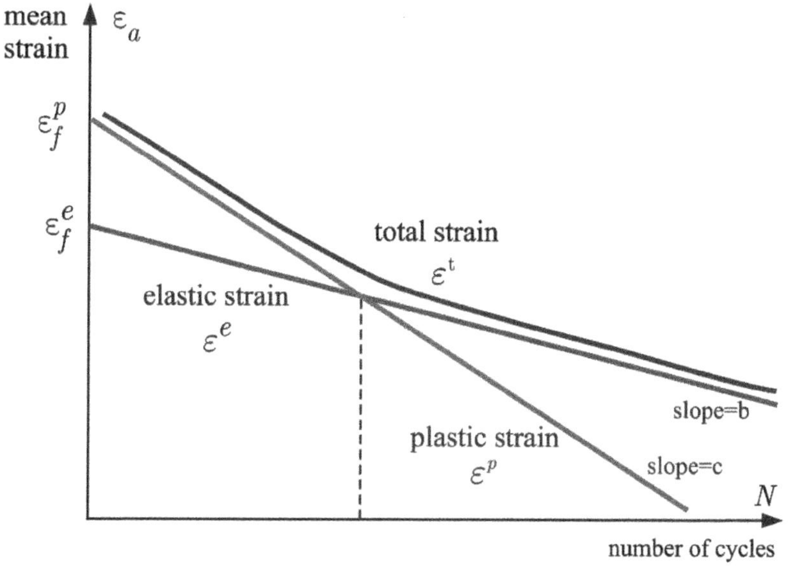

Figure 2.19 Strain-life fatigue description.

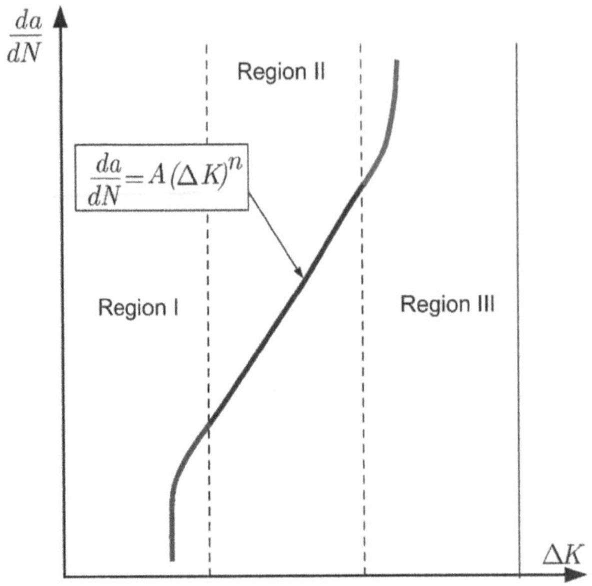

Figure 2.20 Fatigue crack growth criterion.

The fatigue crack propagation criterion is frequently used to study the fatigue strength and the potential crack growth simultaneously. Figure 2.20 shows a typical variation of crack growth rate $\mathrm{d}a/\mathrm{d}N$ in terms of the variations of the stress intensity factor ΔK. Accordingly, region I represents the conditions where no crack growth occurs. In region II, the Paris law governs the response of stable macroscopic crack growth (Paris et al. 1961; Paris and Erdogan 1963):

$$\frac{\mathrm{d}a}{\mathrm{d}N} = A(\Delta K)^n \tag{2.2.118}$$

where n and A are constants. Other modified relations based on a number of experimentally measured constants are also available (Ritchie and Knott 1973; Bolotin 2019):

$$\begin{cases} \text{Forman} & \dfrac{\mathrm{d}a}{\mathrm{d}N} = \dfrac{B(\Delta K)^n}{(1-\mathrm{R})K_\mathrm{c} - \Delta K} \\ \text{Walker} & \dfrac{\mathrm{d}a}{\mathrm{d}N} = \dfrac{C(\Delta K)^n}{(1-\mathrm{R})^{n(1-\lambda)}} \\ \text{NASGRO} & \dfrac{\mathrm{d}a}{\mathrm{d}N} = D\left(\dfrac{1-f}{1-R}\Delta K\right)^n \dfrac{\left(1-\dfrac{\Delta K_{th}}{\Delta K}\right)^p}{\left(1-\dfrac{K_{\max}}{K_c}\right)^q} \end{cases} \tag{2.2.119}$$

where $R = \sigma_{\min} / \sigma_{max}$.

Finally, region III is governed by the fracture instability, which is characterized by the fracture toughness.

For further details on fatigue analysis refer to Suresh (1998).

2.2.5 Viscoelasticity

It is well known that time plays an important role in changing behaviour of many materials, from man-made engineering substances to natural biological tissues. Such effects are usually observed in the form of creep, relaxation, aging, etc., and may largely influence the load-bearing capacity of structures. In this section, a brief review is presented on basic definitions and governing equations of the viscoelastic response. For further details, see Marques and Creus (2012).

2.2.5.1 Viscoelastic Models

Simple descriptions of creep and relaxation behaviours, which are appropriate for defining necessary experimental tests, are illustrated in Figure 2.21. In a creep test, the strain (and deformation) is increased in time under a constant stress condition, whereas in a relaxation test, the stress is reduced in time under a constant strain. On a sudden unloading (dotted curves in Figure 2.21), the creep response shows a sudden drop in strain, followed by a decreasing trend. Such a sudden drop is similarly observed in the relaxation stress response, which may even lead to a change in the sign of unloaded stress. The response continues to decrease in magnitude (absolute value) in time (Marques and Creus 2012).

Viscoelastic characteristics may also be affected by the chemical processes in the constituents of material and the environmental effects such as radiation, temperature, humidity, etc. This is particularly important in studying the viscoelastic response of many biological tissues.

The viscoelastic models are constructed from various combinations of springs and dashpots, representing the elastic (solid) and viscous (fluid) effects, respectively.

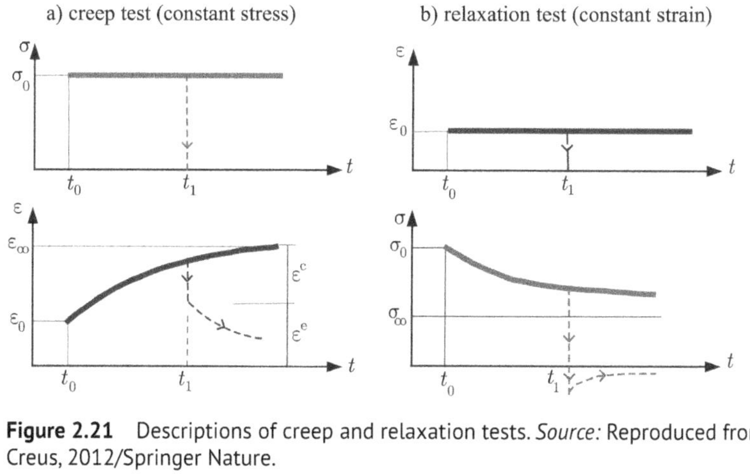

Figure 2.21 Descriptions of creep and relaxation tests. *Source:* Reproduced from Marques and Creus, 2012/Springer Nature.

Beginning with a simple model of linear elastic spring, the stress–strain relation can be readily defined as

$$\sigma(t) = E\varepsilon(t) \tag{2.2.120}$$

The viscous response is modelled by an ideal dashpot, which relates the rate of strain $\dot{\varepsilon}$ to the applied stress σ with the viscosity coefficient η:

$$\dot{\varepsilon}(t) = \frac{1}{\eta}\sigma(t) \tag{2.2.121}$$

These two very basic definitions allow for the description of the main viscoelastic models of Maxwell and Kelvin. The Maxwell model is based on a serial combination of spring and dashpot, as depicted in Figure 2.22.

The differential equation for the Maxwell model can be written as

$$\dot{\varepsilon}(t) = \frac{\dot{\sigma}(t)}{E} + \frac{\sigma(t)}{\eta} \tag{2.2.122}$$

Solution of Equation (2.2.122) for a typical creep condition can be obtained from

$$\varepsilon(t) = \int_{\tau=\tau_0}^{t} \left(\frac{1}{E} + \frac{t-\tau}{\eta}\right)\dot{\sigma}(\tau)d\tau \tag{2.2.123}$$

Alternatively, in a typical relaxation condition, solution of Equation (2.2.122) is obtained from

$$\sigma(t) = \int_{\tau=\tau_0}^{t} Ee^{-\frac{E}{\eta}(t-\tau)}\varepsilon(\tau)d\tau \tag{2.2.124}$$

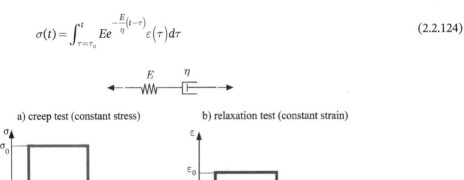

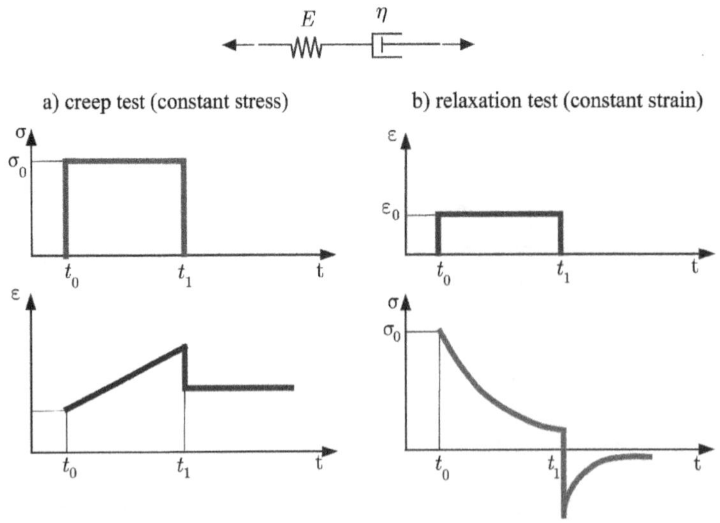

Figure 2.22 Creep and relaxation responses of the Maxwell model. *Source:* Reproduced from Marques and Creus, 2012/Springer Nature.

Accordingly, the creep compliance C^c and the relaxation modulus D^r can be defined as

$$C^c\ (t-\tau)=\frac{1}{E}+\frac{t-\tau}{\eta} \tag{2.2.125}$$

$$D^r(t-\tau_0)=Ee^{-\frac{E}{\eta}(t-\tau_0)} \tag{2.2.126}$$

The Kelvin model uses a parallel combination of spring and dashpot, as depicted in Figure 2.23.

The differential equation for the Kelvin model can be written as

$$\sigma(t)=E\varepsilon(t)+\eta\dot{\varepsilon}(t) \tag{2.2.127}$$

It should be noted that the Kelvin model cannot reproduce a relaxation response, as it requires a non-physical infinite stress. Solution of the Kelvin differential equation for a creep condition can then be obtained from

$$\varepsilon(t)=\int_{\tau=\tau_0}^{t}\frac{1}{\eta}e^{-\frac{E}{\eta}(t-\tau)}\sigma(\tau)d\tau \tag{2.2.128}$$

and the creep compliance C^c for the Kelvin model can be written as

$$C^c(t-\tau_0)=\frac{1}{E}\left[1-e^{-\frac{E}{\eta}(t-\tau_0)}\right] \tag{2.2.129}$$

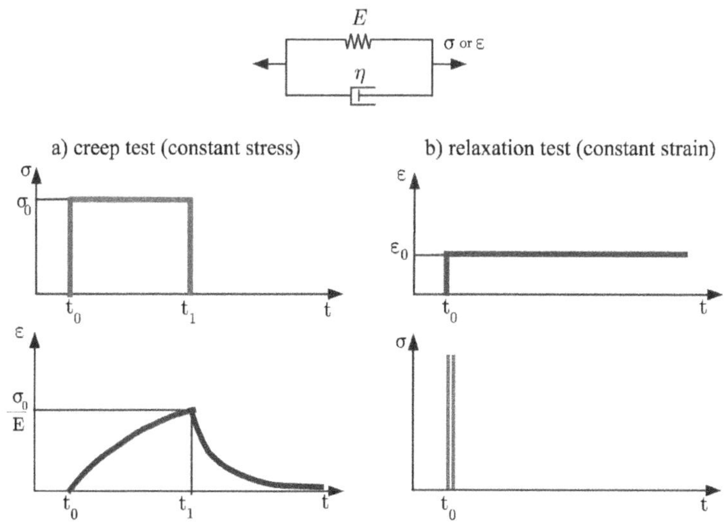

Figure 2.23 Creep and relaxation responses of the Kelvin model. *Source:* Reproduced from Marques and Creus, 2012/Springer Nature.

The generalized Maxwell and Kelvin models can be adopted to enhance the applicability of the original Maxwell and Kelvin assumptions for practical viscoelastic materials and to better fit the available viscoelastic experimental data.

2.2.5.2 The Generalized Maxwell Model

Figure 2.24 illustrates a typical generalized Maxwell model, consisting of n parallel Maxwell models (with springs E_p and dashpots η_p) and an individual spring E_∞. This model is usually adopted for relaxation problems. All constituents bear the same strain in this model and the total stress can be computed from

$$\sigma(t) = D^r \varepsilon(t) \tag{2.2.130}$$

$$D^r(t-\tau) = E_\infty + \sum_{p=1}^{n} E_p e^{-\frac{E_p}{\eta_p}(t-\tau)} \tag{2.2.131}$$

The integral form of the stress–strain equation can then be written as

$$\sigma(t) = \int_{\tau=\tau_0}^{t} D^r(t-\tau) \mathrm{d}\varepsilon(\tau) = \int_{\tau=\tau_0}^{t} D^r(t-\tau)\dot{\varepsilon}(\tau) d\tau \tag{2.2.132}$$

or in terms of the Prony exponential series,

$$\sigma(t) = \underbrace{E_\infty \varepsilon(t)}_{\sigma_0(t)} + \sum_{p=1}^{n} E_p q_p^r(t) \tag{2.2.133}$$

$$q_p^r(t) = \int_{\tau=\tau_0}^{t} e^{-\frac{E_p}{\eta_p}(t-\tau)} \dot{\varepsilon}(\tau) d\tau \tag{2.2.134}$$

where q_p^r represents the strain in the Maxwell spring p. Instead of the integral form of Equation (2.2.134), the following differential equation can be adopted for each Maxwell chain p:

$$\dot{q}_p^r(t) + \frac{E_p}{\eta_p} q_p^r(t) = \dot{\varepsilon}(t) \tag{2.2.135}$$

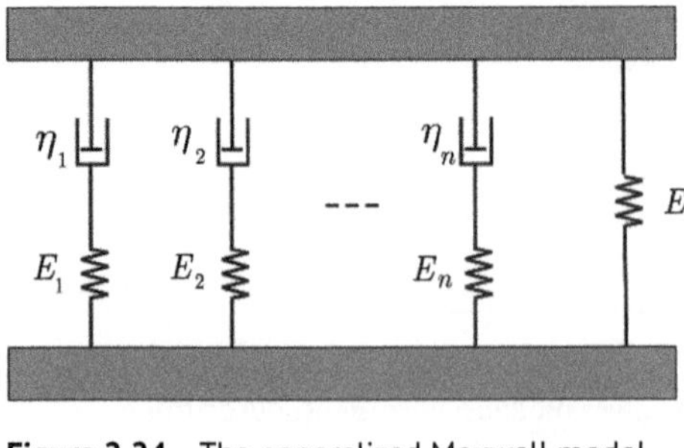

Figure 2.24 The generalized Maxwell model.

The rate form of Equation (2.2.135) can be directly solved to compute $q_p^r(t+\Delta t)$:

$$q_p^r(t+\Delta t) = q_p^r(t) + \Delta q_p^r(t) \tag{2.2.136}$$

$$\Delta q_p^r(t) = -\left[\frac{E_p}{\eta_p} q_p^r(t) - \dot{\varepsilon}(t)\right]\Delta t \tag{2.2.137}$$

A recursive solution allows for a more accurate solution (Goh et al. 2004; Marques and Creus 2012):

$$q_p^r(t+\Delta t) = e^{-\frac{E_p}{\eta_p}\Delta t} q_p^r(t) + \frac{1}{E_\infty \frac{E_p}{\eta_p}\Delta t}\left[1 - e^{-\frac{E_p}{\eta_p}\Delta t}\right]\left[\sigma_0(t+\Delta t) - \sigma_0(t)\right] \tag{2.2.138}$$

The elastic modulus D_e^r is defined from

$$\sigma_0(t+\Delta t) = D_e^r \varepsilon(t+\Delta t) \tag{2.2.139}$$

and the viscos modulus D_v^r is updated from (Kaliske and Rothert 1997)

$$D_v^r(t+\Delta t) = \frac{\partial\sigma(t+\Delta t)}{\partial\varepsilon(t+\Delta t)} = \left[1 + \sum_{p=1}^{n} \frac{1}{\frac{E_p}{\eta_p}\Delta t}\left[1 - e^{-\frac{E_p}{\eta_p}\Delta t}\right]\right] \tag{2.2.140}$$

Equation (2.2.133) can be reformulated as (Abolfathi et al. 2009)

$$\sigma(t) = g_\infty \sigma_0(t) + \sum_{p=1}^{n} g_p w_p^r(t) \tag{2.2.141}$$

$$w_p^r(t) = \int_{\tau=\tau_0}^{t} e^{-\frac{E_p}{\eta_p}(t-\tau)} \dot{\sigma}(\tau) d\tau \tag{2.2.142}$$

where the dimensionless g_p and the time interval τ_p are the equivalent viscoelastic coefficients defined by

$$g_p = \frac{E_p}{E_0} \tag{2.2.143}$$

$$\tau_p = \frac{\eta_p}{E_p} \tag{2.2.144}$$

and

$$E_0 = E_\infty + \sum_{p=1}^{n} E_p \tag{2.2.145}$$

A similar procedure can be followed to derive the tensorial forms of the solution for three-dimensional viscoelastic problems. For details see Marques and Creus (2012).

2.2.5.3 The Generalized Kelvin Model

The generalized Kelvin model, depicted in Figure 2.25, consists of n series of Kelvin models (with springs E_p and dashpots η_p) and an individual spring E_0. This model is usually adopted for creep modelling problems. In this model, all constituents bear the same stress, and the total strain can be computed from

$$\varepsilon(t) = C^c \sigma(t) \tag{2.2.146}$$

$$C^c(t-\tau) = \frac{1}{E_0} + \sum_{p=1}^{n} \frac{1}{E_p}\left[1 - e^{-\frac{E_p}{\eta_p}(t-\tau)}\right] \tag{2.2.147}$$

It should be noted that it is always possible to find a generalized Kelvin model equivalent to the generalized Maxwell model, and vice versa (Marques and Creus 2012). As a result, relations (2.2.131) and (2.2.147) to compute modules D^r and C^c become important in solving the practical generalized models.

The integral form of the strain–stress can be written as

$$\varepsilon(t) = \int_{\tau=\tau_0}^{t} C^c(t-\tau)\mathrm{d}\sigma(\tau) = \int_{\tau=\tau_0}^{t} C^c(t-\tau)\dot{\sigma}(\tau)d\tau \tag{2.2.148}$$

or in terms of the Prony exponential series,

$$\varepsilon(t) = \frac{\sigma(t)}{E_0} + \sum_{p=1}^{n} q_p^c(t) \tag{2.2.149}$$

$$q_p^c(t) = \int_{\tau=\tau_0}^{t} \frac{1}{\eta_p} \mathrm{e}^{-\frac{E_p}{\eta_p}(t-\tau)} \sigma(\tau)d\tau \tag{2.2.150}$$

Similar to the generalized Maxwell formulation, the following differential equation can be adopted for each Kelvin chain p:

$$\dot{q}_p^c(t) + \frac{E_p}{\eta_p} q_p^c(t) = \frac{1}{\eta_p}\sigma(t) \tag{2.2.151}$$

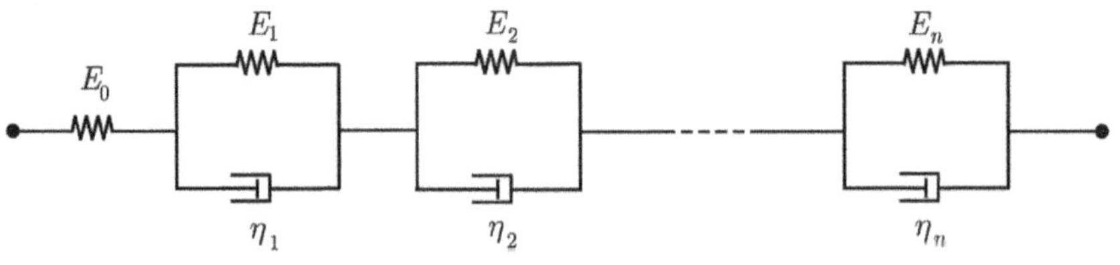

Figure 2.25 The generalized Kelvin model.

The rate form (2.2.151) can be directly solved to compute $q_p^c(t+\Delta t)$:

$$q_p^c(t+\Delta t)=q_p^c(t)+\Delta q_p^c(t) \tag{2.2.152}$$

$$\Delta q_p^c(t)=-\left[\frac{E_p}{\eta_p}q_p^c(t)-\frac{1}{\eta_p}\sigma(t)\right]\Delta t \tag{2.2.153}$$

Moreover, Equation (2.2.151) can be solved in time using a number of recursive formulations. If the stress $\sigma(t)$ is assumed constant over the time interval $[t,t+\Delta t]$, the following recursive procedure can be adopted (Hu and Suo 2012):

$$q_p^c(t+\Delta t)=\mathrm{e}^{-\frac{E_p}{\eta_p}\Delta t}q_p^c(t)+\frac{1}{\eta_p}\left[1-\mathrm{e}^{-\frac{E_p}{\eta_p}\Delta t}\right]\sigma(t) \tag{2.2.154}$$

and based on the assumption of a linear variation of stress $\sigma(t)$ over the time interval $[t,t+\Delta t]$:

$$q_p^c(t+\Delta t)=\mathrm{e}^{-\frac{E_p}{\eta_p}\Delta t}q_p^c(t)+\frac{1}{\eta_p}\left[1-\mathrm{e}^{-\frac{E_p}{\eta_p}\Delta t}\right]\sigma(t)+\Delta q^l \tag{2.2.155}$$

$$\Delta q^l=\frac{\sigma(t)-\sigma(t-\Delta t)}{E_p\Delta t}\left[\Delta t-\frac{\eta_p}{E_p}\left(1-\mathrm{e}^{-\frac{E_p}{\eta_p}\Delta t}\right)\right] \tag{2.2.156}$$

A similar procedure may be followed to reformulate the generalized Kelvin model in terms of non-dimensional viscoelastic parameters g_p and τ_p.

2.2.6 Poroelasticity

The theory of poroelasticity is briefly reviewed in this section due to the fact that a number of biological tissues and organs are made from poroelastic constituents (Berger 2015; Sowinski et al. 2021). The characteristics of poroelastic materials are related to the coupled phenomena raised from the properties of the porous solid and the fluid flow.

The theory of poroelasticity can be developed either by adopting a micromechanical concept or is based on a macrodescription with some ideas of average properties of the mixed problem of porous solid and fluid flow. The micromechanical approach requires a detailed description of the porous geometry of a solid and the complex fluid flow within the porous medium. It is useful in solving specific poroelasticity problems, but cannot be adopted for deriving the general governing equations on a macro level. In this section, only the basic governing equations of the macro approach are briefly presented.

Consider an isothermal porous solid body Ω that also includes fluid flow, as typically depicted in Figure 2.26. The voids may be interconnected to form natural paths for the fluid flow or may be formed in individually isolated patterns that do not allow any fluid flow.

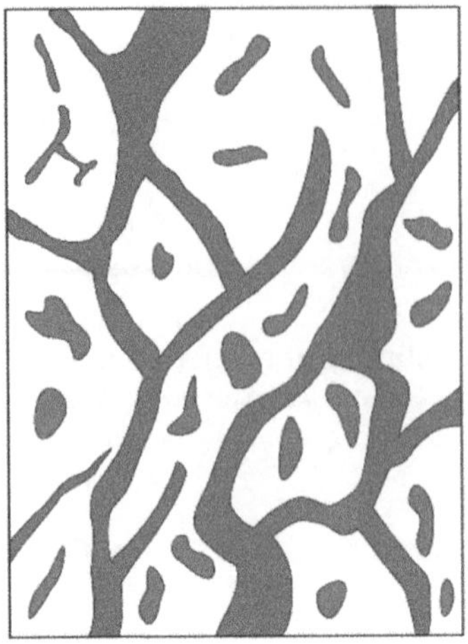

Figure 2.26 A typical porous medium.

2.2.6.1 Governing Equations

There are different descriptions for deriving the governing equations of poroelastic problems. Among the available solutions based on two, three and even four independent field variables, the frequently used $p-\boldsymbol{u}$ formulation is briefly reviewed (Wang 2000; Harper et al. 2018).

The governing poroelastic equations are written in terms of the displacements $\boldsymbol{u}$ and the pore pressure p. The conventional equilibrium equations for the solid part can be written as (Merxhani 2016)

$$\nabla \cdot \boldsymbol{\sigma} + \boldsymbol{f}^{\mathrm{b}} = 0 \tag{2.2.157}$$

with the following solid boundary conditions:

$$\boldsymbol{u} = \bar{\boldsymbol{u}} \text{ on } \Gamma_u \tag{2.2.158}$$

$$\boldsymbol{t} = \boldsymbol{\sigma} \cdot \boldsymbol{n} = \boldsymbol{f}^t \text{ on } \Gamma_t \tag{2.2.159}$$

In loading a saturated porous medium, only part of the stress is carried by the solid and the pore pressure compensates for the rest. As a result, the constitutive equation of the solid should be defined on the so-called effective stress σ' acting on the solid skeleton,

$$\boldsymbol{\sigma}' = \mathbf{D}\varepsilon = \mathbf{D}\nabla_s \boldsymbol{u} \tag{2.2.160}$$

where

$$\boldsymbol{\sigma} = \boldsymbol{\sigma}' - \alpha \boldsymbol{m} p \tag{2.2.161}$$

$$\boldsymbol{m}^{\mathrm{T}} = \left[1,1,1,0,0,0\right] \tag{2.2.162}$$

and α is Biot's coefficient.

The general form of Darcy's law defines the fluid flow in a porous media:

$$\boldsymbol{q} = -\frac{k_{\mathrm{c}}}{\mu_{\mathrm{f}}} \nabla p \tag{2.2.163}$$

where k_c is the intrinsic permeability of the solid, μ_{f} is the viscosity of the pore fluid (Kumpel 2004; Sowinski et al. 2021), and $\boldsymbol{q}$ is the specific discharge, defined in terms of the relative velocity of the fluid $\boldsymbol{v}_{\mathrm{f}}$ with respect to the velocity of the solid $\boldsymbol{v}_{\mathrm{s}}$

$$\boldsymbol{q} = n_{\mathrm{p}}(\boldsymbol{v}_{\mathrm{f}} - \boldsymbol{v}_{\mathrm{s}}) \tag{2.2.164}$$

and n_p is the porosity of the medium. The following fluid boundary conditions should be satisfied:

$$p = \bar{p} \text{ on } \Gamma_p \tag{2.2.165}$$

$$\boldsymbol{q} \cdot \boldsymbol{n} = \bar{q} \text{ on } \Gamma_q \tag{2.2.166}$$

The fluid mass balance can be obtained in the form of

$$\alpha\frac{\partial\epsilon^{\nu}}{\partial t}+S_{\epsilon}\frac{\partial p}{\partial t}=-\boldsymbol{\nabla}\cdot\boldsymbol{q}=\boldsymbol{\nabla}\cdot\left(\frac{k_{\mathrm{c}}}{\mu_{\mathrm{f}}}\boldsymbol{\nabla}p\right) \tag{2.2.167}$$

where ϵ^{ν} is the volumetric strain

$$\epsilon^{v}=\varepsilon_{xx}+\varepsilon_{yy}+\varepsilon_{zz}\,(=\varepsilon_{kk}) \tag{2.2.168}$$

and S_{ϵ} is the constrained storage capacity

$$S_{\epsilon}=\alpha C_{\mathrm{s}}+n_{\mathrm{p}}(\beta_{\mathrm{f}}-C_{\mathrm{p}}) \tag{2.2.169}$$

C_{s} and C_{p} are the bulk and pore compressibility coefficients, respectively, and β_{f} is the fluid compressibility coefficient.

2.2.6.2 Governing Equations in Terms of the Fluid Content

The fluid content ζ is defined by

$$\zeta=\alpha\varepsilon^{v}+S_{\varepsilon}p \tag{2.2.170}$$

Theoretically, while stress $\boldsymbol{\sigma}$ and strain $\boldsymbol{\varepsilon}$ are work conjugates in a mechanical state, pressure p and fluid content ζ are work conjugates for the fluid flow description (Gaspar et al. 2018). The increment of the fluid content ζ in time can be related to the specific discharge $\boldsymbol{q}$ by the fluid mass balance equation (ignoring the fluid source term)

$$\frac{\partial\zeta}{\partial t}=-\boldsymbol{\nabla}\cdot\boldsymbol{q} \tag{2.2.171}$$

The governing equations of fluid in the absence of body forces $\boldsymbol{f}^{\mathrm{b}}$ can be written in the form of the diffusion equation:

$$\frac{\partial\zeta}{\partial t}=d_{\mathrm{h}}\nabla^{2}\zeta \tag{2.2.172}$$

and with the body force $\boldsymbol{f}^{\mathrm{b}}$,

$$\frac{\partial\zeta}{\partial t}=d_{\mathrm{h}}\nabla^{2}\zeta+\frac{\alpha}{\lambda+2\mu}c\boldsymbol{\nabla}\cdot\boldsymbol{f}^{\mathrm{b}} \tag{2.2.173}$$

where (λ,μ) are Lame's coefficients of the porous medium and d_{h} is the coefficient of hydraulic diffusivity (Kumpel 2004):

$$d_{\mathrm{h}}=\frac{k_{\mathrm{c}}}{\mu_{f}}\frac{\lambda+2\mu}{\alpha^{2}+(\lambda+2\mu)S_{\varepsilon}} \tag{2.2.174}$$

The undrained stress–strain constitutive equation can be described by

$$\boldsymbol{\sigma} = 2\mu\boldsymbol{\varepsilon} + \left[\left(\lambda + \alpha B_s K_u\right)\varepsilon^v - \frac{\alpha\zeta}{S_\varepsilon}\right]\boldsymbol{I} \tag{2.2.175}$$

where B_s is the Skempton pore pressure coefficient,

$$B_s = -\frac{\Delta p}{\Delta p_c}|_{\zeta=0} \tag{2.2.176}$$

p_c is the confining pressure

$$p_c = \frac{1}{3}(\sigma_{xx} + \sigma_{yy} + \sigma_{zz})\left(= \frac{1}{3}\sigma_{kk}\right) \tag{2.2.177}$$

and K_u is the undrained bulk modulus,

$$p_c = K_u\epsilon^v \tag{2.2.178}$$

The governing equilibrium equations can be written as (Wang and Kumpel 2003; Harper et al. 2018)

$$\mu\nabla^2\boldsymbol{u} + (\lambda + \mu)\boldsymbol{\nabla}\varepsilon^v = \alpha\boldsymbol{\nabla}p - \boldsymbol{f}^b = \frac{\alpha}{S_\varepsilon}\boldsymbol{\nabla}\zeta - \boldsymbol{f}^b \tag{2.2.179}$$

Note that for a drained porous material, the governing Equation (2.2.179) takes the form of

$$\mu'\nabla^2\boldsymbol{u} + (\lambda' + \mu')\boldsymbol{\nabla}\varepsilon^v = -\boldsymbol{f}^b \tag{2.2.180}$$

where (λ', μ') are Lame's coefficients of the solid material.

2.2.6.3 Discretized Governing Equations

The finite element shape functions $\mathbf{N}_u$ and $\mathbf{N}_p$ are adopted for discretization of displacement $\boldsymbol{u}$ and pressure $\boldsymbol{p}$ fields, respectively,

$$\boldsymbol{u} = \mathbf{N}_u\boldsymbol{u}^n \tag{2.2.181}$$

$$\boldsymbol{p} = \mathbf{N}_p\boldsymbol{p}^n \tag{2.2.182}$$

where $\boldsymbol{u}^n$ and $\boldsymbol{p}^n$ are the vectors of nodal displacement and pressure variables, respectively. The discretized matrix form of the Galerkin weighted residual weak form of the governing equations can then be written as (Detournay and Cheng 1993; Li and Borja 2005; Merxhani 2016)

$$\begin{bmatrix} \mathbf{K} & -\mathbf{Q} \\ \mathbf{0} & \mathbf{H} \end{bmatrix}\begin{bmatrix} \boldsymbol{u}^n \\ \boldsymbol{p}^n \end{bmatrix} + \begin{bmatrix} \mathbf{0} & \mathbf{0} \\ \mathbf{Q}^T & \mathbf{S} \end{bmatrix}\begin{bmatrix} \dot{\boldsymbol{u}}^n \\ \dot{\boldsymbol{p}}^n \end{bmatrix} = \begin{bmatrix} \mathbf{f} \\ \mathbf{q} \end{bmatrix} \tag{2.2.183}$$

where

$$\mathbf{K} = \int_{\Omega} \mathbf{B}_u^{\mathrm{T}} \mathbf{D} \mathbf{B}_u d\Omega = \int_{\Omega} (\nabla_s \mathbf{N}_u)^{\mathrm{T}} \mathbf{D} (\nabla_s \mathbf{N}_u) \mathrm{d}\Omega \tag{2.2.184}$$

$$\mathbf{Q} = \int_{\Omega} \mathbf{B}_u^{\mathrm{T}} \alpha \boldsymbol{m} \mathbf{N}_p \mathrm{d}\Omega = \int_{\Omega} (\nabla_s \mathbf{N}_u)^{\mathrm{T}} \alpha \boldsymbol{m} \mathbf{N}_p \mathrm{d}\Omega \tag{2.2.185}$$

$$\mathbf{S} = \int_{\Omega} \mathbf{N}_p^{\mathrm{T}} S_{\epsilon} \mathbf{N}_p^{\mathrm{T}} \mathrm{d}\Omega \tag{2.2.186}$$

$$\mathbf{H} = \int_{\Omega} (\nabla \mathbf{N}_p)^{\mathrm{T}} \frac{k_c}{\mu_f} (\nabla \mathbf{N}_p) \mathrm{d}\Omega \tag{2.2.187}$$

and

$$\mathbf{f} = \int_{\Gamma_t} \mathbf{N}_u^{\mathrm{T}} \boldsymbol{f}^t \mathrm{d}\Gamma + \int_{\Omega} \mathbf{N}_u^{\mathrm{T}} \boldsymbol{f}^{\mathrm{b}} \mathrm{d}\Omega \tag{2.2.188}$$

$$\mathbf{q} = \int_{\Gamma_q} \mathbf{N}_p^{\mathrm{T}} \overline{q} \mathrm{d}\Gamma \tag{2.2.189}$$

2.2.7 Large Deformation

Many biological systems may undergo large deformations, and even large strains, during their performance or subject to environmental effects. Such large deformations are usually well within the elastic regime; the system may return to its initial state without any permanent plastic strains. This is usually studied within the concept of hyperelastic formulation, as will be discussed in detail for soft biological tissues in Chapter 5. Here, the general concepts of large deformation formulation are briefly reviewed.

2.2.7.1 Deformation Gradient

Consider a body in its initial or reference undeformed configuration, as depicted in Figure 2.27. The body is subjected to a loading and deforms into its current configuration. The current configuration is assumed to be totally different from the initial configuration of the body due to large deformations. Upper- and lower-case letters are usually adopted to denote the variables in initial and current configurations, respectively.

The current position $\mathbf{x}$ of a material point can be defined in terms of its initial position $\mathbf{X}$ and the deformation $\boldsymbol{u}$:

$$\mathbf{x} = \mathbf{X} + \boldsymbol{u} \tag{2.2.190}$$

The deformation gradient $\boldsymbol{F}$ relates the infinitesimal/differential length vectors of the initial and current configurations:

$$d\mathbf{x} = \boldsymbol{F}\ \mathrm{d}\mathbf{X} \tag{2.2.191}$$

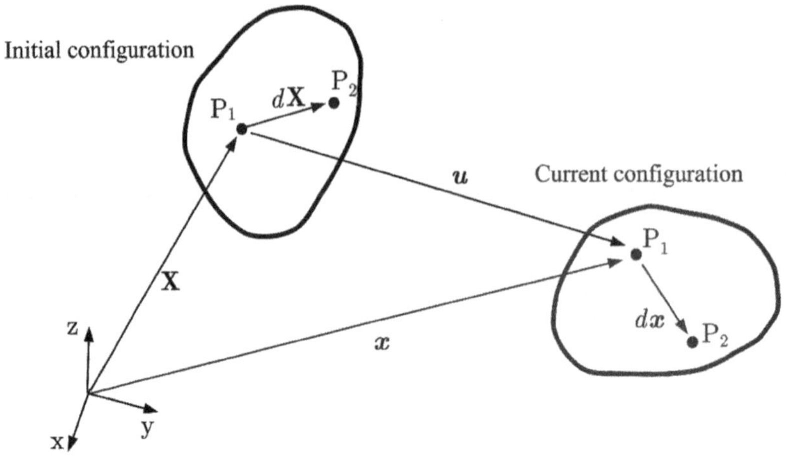

Figure 2.27 Initial and current configurations of a body in a large deformation process.

The deformation gradient $\boldsymbol{F}$ fully describes both the stretch and rotational parts of a deformation and can be decomposed into the rigid body rotation and stretch parts using the polar decomposition

$$\boldsymbol{F} = \boldsymbol{RU} = \boldsymbol{VR} \tag{2.2.192}$$

where $\boldsymbol{U}$ and $\boldsymbol{V}$ are symmetric stretch tensors and $\boldsymbol{R}$ is an orthogonal rotation tensor

$$\boldsymbol{U} = \boldsymbol{U}^{\mathrm{T}}, \boldsymbol{V} = \boldsymbol{V}^{\mathrm{T}} \tag{2.2.193}$$

$$\boldsymbol{RR}^{\mathrm{T}} = \boldsymbol{I} \tag{2.2.194}$$

The changes in volume (from dV^{Ω_0} to dv^{Ω}) and area vector (from $\boldsymbol{N}\,\mathrm{d}A$ to $\boldsymbol{n}\,\mathrm{d}a$) in a deformation process can be obtained from $\boldsymbol{F}$:

$$\frac{dv^{\Omega}}{dV^{\Omega_0}} = \det \boldsymbol{F} = J \tag{2.2.195}$$

$$\left(\boldsymbol{n}\,da\right) = J\boldsymbol{F}^{-\mathrm{T}}(NdA) \tag{2.2.196}$$

where the Jacobian J represents a measure of change in volume.

$\boldsymbol{F}$ can also be used to define measures of stretches in the forms of the right and left Cauchy–Green tensors, $\boldsymbol{B}$ and $\boldsymbol{C}$, respectively,

$$\boldsymbol{B} = \boldsymbol{FF}^{\mathrm{T}} = \boldsymbol{V}^2 \tag{2.2.197}$$

$$\boldsymbol{C} = \boldsymbol{F}^{\mathrm{T}}\boldsymbol{F} = \boldsymbol{U}^2 \tag{2.2.198}$$

2.2.7.2 Strain Measures

Unlike the conventional theory of small deformation, definitions of strain, stress and their rates are not unique in the large deformation theory. Such measures should also meet a number of conditions, such as objectivity (or frame invariance). Some of the important strain measures are:

1) The Cauchy–Green tensor $\boldsymbol{B}$ is a measure of stretch. Its major drawback is that it includes non-zero components even for pure rotation, which makes it less popular.
2) The Almansi (also Eulerian or spatial) strain tensor $\boldsymbol{e}$ is associated with the reference configuration and is defined directly from $\boldsymbol{B}$. Dissimilar to $\boldsymbol{B}$, it becomes null for a pure rotation:

$$\boldsymbol{e} = \frac{1}{2}\left(\boldsymbol{I} - \boldsymbol{B}^{-1}\right) \tag{2.2.199}$$

3) The Green–Lagrange (or Lagrangian) strain tensor $\boldsymbol{E}$,

$$\boldsymbol{E} = \frac{1}{2}(\boldsymbol{C} - \boldsymbol{I}) \tag{2.2.200}$$

can also be written as

$$\boldsymbol{E} = \frac{1}{2}\left(\frac{\partial \boldsymbol{u}}{\partial \mathbf{X}}\right) + \frac{1}{2}\left(\frac{\partial \boldsymbol{u}}{\partial \mathbf{X}}\right)^{\mathrm{T}} + \frac{1}{2}\left(\frac{\partial \boldsymbol{u}}{\partial \mathbf{X}}\right)^{\mathrm{T}}\left(\frac{\partial \boldsymbol{u}}{\partial \mathbf{X}}\right) \tag{2.2.201}$$

The first two terms represent the conventional small strain definition and the third term (second order) accounts for the effects of large deformations.
4) The logarithmic (true) strain ε:

$$\varepsilon = -\frac{1}{2}\ln\left(\boldsymbol{B}^{-1}\right) = \ln(\boldsymbol{V}) \tag{2.2.202}$$

Note that while $\boldsymbol{F}$ may not be symmetric, the strain measures $\boldsymbol{E}$ and ε always remain symmetric.

2.2.7.3 The Effects of Time/Rate

The common approach for solving nonlinear problems is to adopt an incremental formulation, where the increments are written in a pseudo rate (of loading) formulation. Moreover, the dynamic and viscoelastic problems are naturally described in the time/rate forms.

The velocity gradient $\boldsymbol{l}$ is defined as

$$\boldsymbol{l} = \frac{\partial \boldsymbol{v}}{\partial \boldsymbol{x}} = \dot{\boldsymbol{F}}\boldsymbol{F}^{-1} \tag{2.2.203}$$

The rate of deformation tensor $\boldsymbol{d}$ is the symmetric part of $\boldsymbol{l}$,

$$\boldsymbol{d} = \frac{1}{2}\left(\boldsymbol{l} + \boldsymbol{l}^{\mathrm{T}}\right) \tag{2.2.204}$$

and the spin tensor $\boldsymbol{w}$ is the anti-symmetric part of $\boldsymbol{l}$,

$$\boldsymbol{w} = \frac{1}{2}\left(\boldsymbol{l} - \boldsymbol{l}^{\mathrm{T}}\right) \tag{2.2.205}$$

Note that the spin tensor cannot be regarded as a rate of pure rigid body rotation. In general, it can be written in terms of the polar components $\boldsymbol{R}$ and $\boldsymbol{U}$ as

$$\boldsymbol{w} = \dot{\boldsymbol{R}}\boldsymbol{R}^{\mathrm{T}} + \frac{1}{2}\boldsymbol{R}\left[\dot{\boldsymbol{U}}\boldsymbol{U}^{-1} - \left(\dot{\boldsymbol{U}}\boldsymbol{U}^{-1}\right)^{\mathrm{T}}\right]\boldsymbol{R}^{\mathrm{T}} \tag{2.2.206}$$

The first term depends only on the rigid body rotation, whereas the remaining terms depend on both the stretch and the rotation. This is important in defining the objective stress rates.

2.2.7.4 Stress Measures

The Cauchy stress tensor $\boldsymbol{\sigma}$ represents the true stress, relating the traction $\boldsymbol{t}$ and surface normal $\boldsymbol{n}$ at any internal point of a body to the current configuration,

$$\boldsymbol{t}(\boldsymbol{n}) = \boldsymbol{\sigma}\boldsymbol{n} \tag{2.2.207}$$

$\boldsymbol{\sigma}$ can be simply realised as the ratio of uniaxial loading to area, both in the current configuration. The Cauchy stress tensor $\boldsymbol{\sigma}$ is adopted to define the governing equilibrium equation at the spatial current configuration $\mathbf{x} \in \Omega$ of a deformable body Ω with the density ρ and subject to the body force $\boldsymbol{f}^b$ based on the local form of the balance of linear momentum,

$$\nabla \cdot \boldsymbol{\sigma} + \rho \boldsymbol{f}^b = \mathbf{0} \text{ on } \mathbf{x} \in \Omega \tag{2.2.208}$$

It is sometimes more convenient to transform Equation (2.2.208) into a Lagrangian initial description. Beginning with the definition of the Kirchhoff stress tensor,

$$\boldsymbol{\tau} = J\boldsymbol{\sigma} \tag{2.2.209}$$

Alternatively, the nominal stress or the first Piola–Kirchhoff stress tensor $\boldsymbol{P}$ is defined mathematically on the initial configuration $\mathbf{X} \in \Omega_0$, relating the corresponding traction $\boldsymbol{T}$ and surface normal $\boldsymbol{N}$,

$$\boldsymbol{T} = \boldsymbol{P}\boldsymbol{N} \tag{2.2.210}$$

or in terms of the Cauchy stress,

$$\boldsymbol{P} = \boldsymbol{\tau}\boldsymbol{F}^{-\mathrm{T}} = J\boldsymbol{\sigma}\boldsymbol{F}^{-\mathrm{T}} \tag{2.2.211}$$

$\boldsymbol{P}$ does not represent an independent physical meaning and is not symmetric. It is based on both the initial (material) and the current (spatial) configurations. The corresponding equilibrium equation based on the density ρ_0 and the body force $\boldsymbol{f}_0^{\mathrm{b}}$ at the initial configuration can then be written as

$$\nabla_0 \cdot \boldsymbol{P} + \rho_0 \boldsymbol{f}_0^{\mathrm{b}} = \mathbf{0} \text{ on } \mathbf{X} \in \Omega_0 \tag{2.2.212}$$

The symmetric second Piola–Kirchhoff stress tensor $\boldsymbol{S}$ is defined entirely in the initial configuration

$$\boldsymbol{S} = \boldsymbol{F}^{-1}\boldsymbol{P} = J\boldsymbol{F}^{-1}\boldsymbol{\sigma}\boldsymbol{F}^{-\mathrm{T}} \tag{2.2.213}$$

and the corresponding equilibrium equation can be written as

$$\boldsymbol{\nabla}_0 \cdot (\boldsymbol{F}\boldsymbol{S}) + \rho_0 \boldsymbol{f}_0^{\mathrm{b}} = \mathbf{0} \quad \text{on}\, \mathbf{X} \in \Omega_0 \tag{2.2.214}$$

Note that the second Piola–Kirchhoff stress tensor $\boldsymbol{S}$ provides no physical meaning.

Alternative stress measures include the symmetric Green–Naghdi co-rotated Cauchy stress tensor $\boldsymbol{\sigma}^{\mathrm{GN}}$

$$\boldsymbol{\sigma}^{\mathrm{GN}} = \boldsymbol{R}^{\mathrm{T}}\boldsymbol{\sigma}\boldsymbol{R} \tag{2.2.215}$$

and the non-symmetric Mandel stress tensor $\boldsymbol{\sigma}^{\mathrm{M}}$, usually for plastic applications,

$$\boldsymbol{\sigma}^{\mathrm{M}} = \boldsymbol{C}\boldsymbol{S} \tag{2.2.216}$$

2.2.7.5 Objective Stress Rates

The material objectivity is a necessary condition for stress rate measures in order to be used in constitutive equations. It is important that while a state measure may be objective, its decomposed parts or its rate may not remain objective. For instance, the Cauchy stress $\boldsymbol{\sigma}$ is an objective stress, whereas its rate $\dot{\boldsymbol{\sigma}}$ does not satisfy the objectivity conditions. Similarly, the deformation gradient $\boldsymbol{F}$ and the rate of deformation $\boldsymbol{d}$ are objective, but the velocity gradient $\boldsymbol{l}$ and the spin tensor $\boldsymbol{w}$ are non-objective.

The rate of Cauchy stress $\dot{\boldsymbol{\sigma}}$ from a known state of $\boldsymbol{\sigma}$ can generally be related to an objective stress rate measure $\boldsymbol{\sigma}^{\mathrm{obj}}$ by (Belytschko et al. 2014)

$$\boldsymbol{\sigma}^{\mathrm{obj}} = \dot{\boldsymbol{\sigma}} + \boldsymbol{w}\boldsymbol{\sigma} - \boldsymbol{\sigma}\boldsymbol{w} \tag{2.2.217}$$

There are various definitions for the stress measure $\boldsymbol{\sigma}^{\mathrm{obj}}$:

1) Truesdell stress rate

$$\dot{\boldsymbol{\sigma}}^{\mathrm{Tr}} = \boldsymbol{\sigma}^{\mathrm{obj}} = \dot{\boldsymbol{\sigma}} + (\mathrm{tr}\, \boldsymbol{l})\boldsymbol{\sigma} - \boldsymbol{l}\boldsymbol{\sigma} - \boldsymbol{\sigma}\boldsymbol{l}^{\mathrm{T}} \tag{2.2.218}$$

2) Oldroyd stress rate

$$\dot{\boldsymbol{\sigma}}^{\mathrm{O}} = \boldsymbol{\sigma}^{\mathrm{obj}} = \dot{\boldsymbol{\sigma}} - \boldsymbol{l}\boldsymbol{\sigma} - \boldsymbol{\sigma}\boldsymbol{l}^{\mathrm{T}} \tag{2.2.219}$$

3) Convective stress rate

$$\dot{\boldsymbol{\sigma}}^{\mathrm{con}} = \boldsymbol{\sigma}^{\mathrm{obj}} = \dot{\boldsymbol{\sigma}} + \boldsymbol{l}^{\mathrm{T}}\boldsymbol{\sigma} + \boldsymbol{\sigma}\boldsymbol{l} \tag{2.2.220}$$

4) Green–Naghdi stress rate

$$\dot{\boldsymbol{\sigma}}^{\mathrm{GN}} = \boldsymbol{\sigma}^{\mathrm{obj}} = \dot{\boldsymbol{\sigma}} - \dot{\boldsymbol{R}}\boldsymbol{R}^{\mathrm{T}}\boldsymbol{\sigma} + \boldsymbol{\sigma}\dot{\boldsymbol{R}}\boldsymbol{R}^{\mathrm{T}} \tag{2.2.221}$$

5) Jaumann stress rate

$$\dot{\boldsymbol{\sigma}}^{\mathrm{J}} = \boldsymbol{\sigma}^{\mathrm{obj}} = \dot{\boldsymbol{\sigma}} + \boldsymbol{\sigma}\boldsymbol{w} - \boldsymbol{w}\boldsymbol{\sigma} \tag{2.2.222}$$

2.2.7.6 Large Deformation and Elastoplasticity

The additive decomposition (2.2.30) of strain into elastic and plastic parts in small deformation plasticity is replaced by the multiplicative decomposition of deformation gradient into elastic and plastic parts, $\boldsymbol{F}^{\mathrm{e}}$ and $\boldsymbol{F}^{\mathrm{p}}$ respectively, in large deformations,

$$\boldsymbol{F} = \boldsymbol{F}^{\mathrm{e}}\boldsymbol{F}^{\mathrm{p}} \tag{2.2.223}$$

Accordingly, the velocity gradient $\boldsymbol{l}$, the rate of deformation $\boldsymbol{d}$ and the spin $\boldsymbol{w}$ tensors can be decomposed into elastic (e) and plastic (p) parts.

Moreover, the Jaumann stress rate and the elastic part of the rate of deformation are objective and can be used in the rate-form of the elastoplastic constitutive equation. For further details, refer to Neto et al. (2008).

2.2.7.7 Hyperelastic Constitutive Equations

Constitutive equations describe the fundamental stress–strain response of a material. Hyperelastic formulation is one of the efficient constitutive equations for the class of materials whose states can be expressed in terms of the total strain in a large deformation regime. As a result, the hyperelastic material response remains independent of the displacement history.

Hyperelastic formulation is based on the definition of the strain energy density function w^{s}. Due to the objectivity of the constitutive relation, the strain energy density w^{s} should be defined in terms of the invariants of the strain tensor ε. Alternatively, it is more appropriate to define w^{s} in terms of the invariants of the deformation gradient $\boldsymbol{F}$ or the right and left Cauchy–Green tensors, $\boldsymbol{B}$ and $\boldsymbol{C}$, respectively.

The general form of the hyperelastic model in the reference configuration can be defined as

$$w^{\mathrm{s}}(\boldsymbol{F}(\mathbf{X}),\mathbf{X}) \tag{2.2.224}$$

and the corresponding stress state is obtained by differentiation of w^s with respect to the strain,

$$\boldsymbol{\sigma} = \frac{\partial w^{\mathrm{s}}}{\partial \varepsilon} \tag{2.2.225}$$

or

$$\boldsymbol{P} = \frac{\partial w^{\mathrm{s}}}{\partial \boldsymbol{F}} \tag{2.2.226}$$

$$\boldsymbol{S} = \frac{\partial w^{\mathrm{s}}}{\partial \mathbf{E}} \tag{2.2.227}$$

The general form of the isotropic hyperelastic model in terms of the invariants I_1, I_2 and I_3 of $\boldsymbol{C}$ can be written as

$$w^{s}(\boldsymbol{F}(\mathbf{X}),\mathbf{X}) = w^{s}(\boldsymbol{C}(\mathbf{X}),\mathbf{X}) = w^{s}(I_1,I_2,I_3,\mathbf{X}) \tag{2.2.228}$$

where

$$I_1 = \mathrm{tr}(\boldsymbol{C}) = \boldsymbol{C}:\mathbf{I} \tag{2.2.229}$$

$$I_2 = tr(\boldsymbol{CC}) = \boldsymbol{C}:\boldsymbol{C} \tag{2.2.230}$$

$$I_3 = \det(\boldsymbol{C}) = J^2 \tag{2.2.231}$$

where $\mathbf{I}$ is the second-order tensor (Kronecker delta).

The second Piola–Kirchhoff stress is obtained by integration of w^{s} with respect to the Green–Lagrange strain tensor $\boldsymbol{E}$ of Equation (2.2.200):

$$\boldsymbol{S} = \frac{\partial w^{s}}{\partial \boldsymbol{E}} = 2\frac{\partial w^{s}}{\partial \boldsymbol{C}} = 2\frac{\partial w^{s}}{\partial I_1}\frac{\partial I_1}{\partial \boldsymbol{C}} + 2\frac{\partial w^{s}}{\partial I_2}\frac{\partial I_2}{\partial \boldsymbol{C}} + 2\frac{\partial w^{s}}{\partial I_3}\frac{\partial I_3}{\partial \boldsymbol{C}} \tag{2.2.232}$$

or

$$\boldsymbol{S} = 2\frac{\partial w^{s}}{\partial I_1}\mathbf{I} + 4\frac{\partial w^{s}}{\partial I_2}\boldsymbol{C} + 2J^2\frac{\partial w^{s}}{\partial I_3}\boldsymbol{C}^{-1} \tag{2.2.233}$$

Therefore, the derivatives of the invariants have to be determined to compute the stress state, which may lead to complex computations for general hyperelastic models.

Most hyperelasic problems follow an incompressible or nearly incompressible behaviour. The modified hyperelastic solution for these materials can be written as (Bonet and Wood 1997; Holzapfel 2002)

$$\boldsymbol{S} = 2\frac{\partial w^{s}}{\partial \boldsymbol{C}} + \gamma J\boldsymbol{C}^{-1} \tag{2.2.234}$$

where the constant γ becomes equivalent to pressure p for incompressible conditions of a homogeneous strain energy density function (Holzapfel 2002). The strain energy density function can be written in terms of the deviatoric form of $\boldsymbol{C}$:

$$\boldsymbol{C}_{\mathrm{d}} = (I_3)^{-\frac{1}{3}}\boldsymbol{C} \tag{2.2.235}$$

$$w_{\mathrm{d}}^{s}(\boldsymbol{C}) = w^{s}(\boldsymbol{C}_{\mathrm{d}}) \tag{2.2.236}$$

and the deviatoric component of S is obtained from

$$\boldsymbol{S}_{\mathrm{d}} = 2\frac{\partial w_{\mathrm{d}}^{s}}{\partial \boldsymbol{C}} \tag{2.2.237}$$

For nearly incompressible materials, the strain energy density is further modified by a volumetric term $w_{\mathrm{v}}^{\mathrm{s}}$,

$$w^{\mathrm{s}}(\boldsymbol{C}) = w_{\mathrm{d}}^{\mathrm{s}}(\boldsymbol{C}) + w_{\mathrm{v}}^{\mathrm{s}}(J) \tag{2.2.238}$$

with

$$w_{\mathrm{v}}^{\mathrm{s}}(J) = \frac{1}{2}\alpha(J-1)^2 \tag{2.2.239}$$

where α is a penalty number to enforce the incompressibility constraint.

The second Piola–Kirchhoff stress $\boldsymbol{S}$ is then computed from

$$\boldsymbol{S}_{\mathrm{d}} = 2\frac{\partial w^{\mathrm{s}}}{\partial \boldsymbol{C}} = \cdots = 2\frac{\partial w_{\mathrm{d}}^{\mathrm{s}}}{\partial \boldsymbol{C}} + \alpha J(J-1)\boldsymbol{C}^{-1} \tag{2.2.240}$$

In the following, a number of major hyperelastic models are briefly presented. Sophisticated hyperelastic models can be used for modelling biomechanical problems, as will be discussed in Chapter 5. The hyperelastic formulation may also be extended to viscoelastic problems and to contain various terms of damage to better capture the visco-hyperelastic physical responses of biomechanical applications (Holzapfel and Ogden 2009, 2010; Marques and Creus 2012; Wang and Hong 2012; Holzapfel and Fereidoonnezhad 2017).

2.2.7.7.1 Neo-Hookean Model

The neo-Hookean model is a very simple hyperelastic model, defined as

$$w^{\mathrm{s}} = \frac{1}{2}\mu(I_1 - 3) - \mu \ln J + \frac{1}{2}\lambda(\ln J)^2 \tag{2.2.241}$$

with

$$J^2 = I_3 \tag{2.2.242}$$

where μ and λ are the elastic coefficients of the material.

The second Piola–Kirchhoff stress is then obtained by

$$\boldsymbol{S} = \frac{\partial w^{\mathrm{s}}}{\partial \boldsymbol{E}} = \frac{\partial w^{\mathrm{s}}}{\partial I_1}\frac{\partial I_1}{\partial \boldsymbol{E}} + \frac{\partial w^{\mathrm{s}}}{\partial J}\frac{\partial J}{\partial \boldsymbol{E}} \tag{2.2.243}$$

which finally leads to

$$\boldsymbol{S} = \mu\left(\mathbf{I} - \boldsymbol{C}^{-1}\right) + \lambda(\ln J)\boldsymbol{C}^{-1} \tag{2.2.244}$$

and the Cauchy stress tensor,

$$\boldsymbol{\sigma} = \frac{\mu}{J}(\boldsymbol{B} - \mathbf{I}) + \frac{\lambda}{J}(\ln J)\mathbf{I} \tag{2.2.245}$$

The corresponding elasticity tensor $\mathbf{D}$ is obtained by further differentiation,

$$\mathbf{D} = \frac{\partial \boldsymbol{S}}{\partial \boldsymbol{E}} = 2\frac{\partial \boldsymbol{S}}{\partial \boldsymbol{C}} = 4\frac{\partial^2 w^s}{\partial \boldsymbol{C} \partial \boldsymbol{C}} = \cdots = \lambda \boldsymbol{C}^{-1} \otimes \boldsymbol{C}^{-1} - 2(\mu - \lambda \ln J)\mathbf{Y} \tag{2.2.246}$$

where $\mathbf{Y}$ is the fourth-order symmetric unity tensor,

$$\mathbf{Y}_{ijkl} = -\frac{\partial \left(\boldsymbol{C}^{-1}\right)_{ij}}{\partial \left(\boldsymbol{C}\right)_{kl}} \tag{2.2.247}$$

It is noted that $J = 1$ represents the incompressible state of the material, reducing the strain energy density function to

$$w^s = \frac{1}{2}\mu(I_1 - 3) \tag{2.2.248}$$

2.2.7.7.2 Mooney–Rivlin Model

The Mooney–Rivlin model is an extension of the neo-Hookean hyperelastic model,

$$w^s = A_1(J_1 - 3) + A_2(J_2 - 3) + A_3(J_3 - 1)^2 \tag{2.2.249}$$

where

$$J_1 = I_1(I_3)^{\frac{1}{3}} \tag{2.2.250}$$

$$J_2 = I_2(I_3)^{-\frac{2}{3}} \tag{2.2.251}$$

$$J_3 = (I_3)^{\frac{1}{2}} \tag{2.2.252}$$

The second Piola–Kirchhoff stress can be obtained by integration of Equation (2.2.249),

$$\boldsymbol{S} = \frac{\partial w^s}{\partial \boldsymbol{E}} = \frac{\partial w^s}{\partial J_1}\frac{\partial J_1}{\partial \boldsymbol{E}} + \frac{\partial w^s}{\partial J_2}\frac{\partial J_2}{\partial \boldsymbol{E}} + \frac{\partial w^s}{\partial J_3}\frac{\partial J_3}{\partial \boldsymbol{E}} \tag{2.2.253}$$

$$\boldsymbol{S} = A_1 J_{1,\boldsymbol{E}} + A_2 J_{2,\boldsymbol{E}} + 2A_3 J_{3,\boldsymbol{E}}(J_3 - 1) \tag{2.2.254}$$

where

$$J_{1,E} = 2\mathbf{I}(I_3)^{-\frac{1}{3}} - \frac{2}{3}I_1(I_3)^{-\frac{1}{3}}\boldsymbol{C}^{-1} \tag{2.2.255}$$

$$J_{2,E} = 2(I_1\mathbf{I} - \boldsymbol{C})(I_3)^{-\frac{2}{3}} - \frac{4}{3}I_2(I_3)^{-\frac{2}{3}}\boldsymbol{C}^{-1} \tag{2.2.256}$$

$$J_{3,E} = (I_3)^{\frac{1}{2}} I_3 \boldsymbol{C}^{-1} \tag{2.2.257}$$

The corresponding elasticity tensor $\mathbf{D}$ is then obtained as

$$\mathbf{D} = \frac{\partial \boldsymbol{S}}{\partial \mathbf{E}} = A_1 J_{1,\boldsymbol{EE}} + A_2 J_{2,\boldsymbol{EE}} + 2A_3 J_{3,\boldsymbol{EE}}(J_3 - 1) + 2J_{3,\boldsymbol{E}} \otimes J_{3,\boldsymbol{E}} \tag{2.2.258}$$

where a further set of differentiation is required:

$$J_{1,\boldsymbol{EE}} = \frac{\partial J_{1,\boldsymbol{E}}}{\partial \mathbf{E}} \tag{2.2.259}$$

$$J_{2,\boldsymbol{EE}} = \frac{\partial J_{2,\boldsymbol{E}}}{\partial \mathbf{E}} \tag{2.2.260}$$

$$J_{3,\boldsymbol{EE}} = \frac{\partial J_{3,\boldsymbol{E}}}{\partial \mathbf{E}} \tag{2.2.261}$$

Clearly, more complex hyperelastic models require more complicated and lengthy differentiation computations. For a comprehensive discussion refer to Bonet and Wood (1997) and Holzapfel (2000).

2.3 Flow, Convection and Diffusion

This section is dedicated to provide the basic formulation for a variety of field problems including fluid mechanics, gas dynamics, diffusion and convection. Again, only the necessary concepts and formulations are provided and the reader should refer to the reference texts for proofs or more details. The section begins with a review of the thermodynamics basic concepts and then covers the formulations of temperature-independent fluid flow. Concepts of compressible gas dynamics are reviewed and the governing equations are briefly presented. The section concludes by reviewing the convection/diffusion processes and the corresponding formulations.

2.3.1 Thermodynamics

Principles of thermodynamics govern the physical and engineering phenomena. In this section only a brief review of the necessary definitions of thermodynamics concepts are provided. They are required in defining the governing equations of fluid motion, gas dynamics and diffusion problems.

Thermodynamic characteristics are either extensive or intensive. The extensive parameters are related to the mass of system, such as E, S and H. Most of intensive parameters, such as temperature T, pressure p, internal energy e, entropy s and enthalpy $h = e + p\nu_\rho$ are independent of mass. For simple systems, each intensive parameter can only be expressed as a function of two other intensive properties.

The canonical forms of the set of thermodynamic equations of state can be written as

$$de = Tds - pd\nu_\rho \tag{2.3.1}$$

$$dh = Tds + \nu_s dp \tag{2.3.2}$$

$$\nu_\rho = \frac{1}{\rho} \tag{2.3.3}$$

From the definition of the internal energy $e = e(s, \nu_\rho)$, one may conclude that

$$T = \left.\frac{\partial e}{\partial s}\right|_{\nu_\rho} \tag{2.3.4}$$

$$p = -\left.\frac{\partial e}{\partial \nu_\rho}\right|_s \tag{2.3.5}$$

The enthalpy h, the Helmholtz free energy ψ_{H} and the Gibbs free energy ψ_{G} can be defined as

$$h = e + p\nu_\rho \tag{2.3.6}$$

$$\psi_{\mathrm{H}} = e - Ts \tag{2.3.7}$$

$$\psi_{\mathrm{G}} = h - Ts \tag{2.3.8}$$

Moreover,

$$T = \left.\frac{\partial h}{\partial s}\right|_p \tag{2.3.9}$$

$$p = -\left.\frac{\partial \psi_{\mathrm{H}}}{\partial \nu_\rho}\right|_T \tag{2.3.10}$$

$$\nu_\rho = \left.\frac{\partial h}{\partial p}\right|_s = \left.\frac{\partial \psi_{\mathrm{G}}}{\partial p}\right|_T \tag{2.3.11}$$

$$s = -\left.\frac{\partial \psi_{\mathrm{H}}}{\partial T}\right|_{\nu_\rho} = -\left.\frac{\partial \psi_{\mathrm{G}}}{\partial T}\right|_p \tag{2.3.12}$$

These equations can be used to derive the appropriate equations of state in the forms of $p(\nu_\rho, T)$ or $e(\nu_\rho, T)$.

The sound speed c can also be obtained from

$$c^2 = \left.\frac{\partial p}{\partial \rho}\right|_s = \left.\frac{\partial p}{\partial \rho}\right|_T + \frac{T}{c_v \rho^2}\left(\left.\frac{\partial p}{\partial T}\right|_\rho\right)^2 \tag{2.3.13}$$

Equation (2.3.13) is further simplified for the prefect gas:

$$c^2 = \gamma RT = \frac{c_p}{c_v} RT \tag{2.3.14}$$

2.3.2 Fluid Mechanics

Fluid flows are categorized into incompressible and compressible types. An incompressible fluid, usually liquids, takes the shape of a container while filling it, with a free surface determined according to its preserved volume (with respect to the volume of the container). An incompressible flow may well resist in compression, while it has no resistance in tensile conditions. In very high pressures, liquids may also show some levels of compressibility (Nakayama and Boucher 1999).

A viscous fluid may resist against sliding the layers of flow over each other, providing a shear resistance τ_s against the velocity strain rate $\dot{\varepsilon}$:

$$\dot{\varepsilon} = \frac{1}{\eta_v} \tau_s \tag{2.3.15}$$

where η_v is the coefficient of viscosity. A viscous flow allows for a shear/sliding flow, where the corresponding component of the velocity changes in the other direction. The coefficient of kinematic viscosity η_{vk} is defined as

$$\eta_{vk} = \frac{\eta_v}{\rho_f} \tag{2.3.16}$$

In contrast, a compressible fluid, such as a gas, does not preserve its volume and therefore it compresses or expands to completely fill the container, with no free surface. The compressibility coefficient β_c is defined from the bulk modulus K_f of the compressible flow as

$$\beta_c = \frac{1}{K_f} = \left(-V \frac{dp}{dV} \right)^{-1} \tag{2.3.17}$$

where V is the volume of the fluid element.

Moreover, the specific volume ν_ρ is defined as

$$\nu_\rho = \frac{1}{\rho_f} \tag{2.3.18}$$

In this section, only fluids (liquids) without temperature effects are reviewed. The effects of temperature will be included in the next section, which deals with gas dynamics.

2.3.2.1 Static Pressure

For a fluid at rest or when the dynamic effects of fluid flow are neglected, the fluid is in a static condition and the corresponding fluid pressure $p_f(z)$ at any point with the height z from the free surface is determined by the weight of the fluid column above it and the atmospheric pressure p_0 on the free surface (Figure 2.28):

$$p_f(z) = \rho_f g z + p_0 \tag{2.3.19}$$

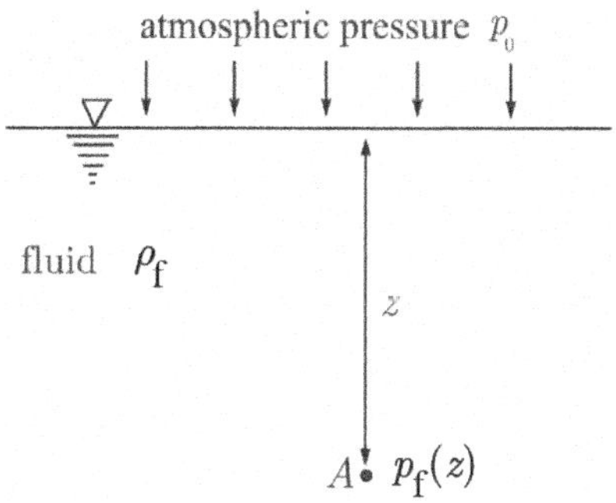

Figure 2.28 Static fluid pressure.

2.3.2.2 Laminar and Turbulent Flow

Another classification of fluid flow is based on its smooth laminar or turbulent nature, as typically illustrated in Figure 2.29. In low velocities without any external disturbances, the flow is assumed to be laminar. In higher velocities and subject to external effects, the fluid flow becomes turbulent.

A simple criterion to distinguish between laminar and turbulent flows is based on the Reynolds number R_e:

$$R_e = \frac{v_{ave} L_c}{\eta_{vk}} \tag{2.3.20}$$

where L_c is a characteristic length scale and v_{ave} is the average fluid velocity. Reynolds experimentally examined the critical value for R_e, which is associated with a transition from laminar to turbulent flow (Nakayama 1998). It somehow accounts for the effect of viscosity of a fluid.

2.3.2.3 General Compressible Viscous Flow

The Navier–Stokes equations govern the general forms of the fluid flow. The following subsections provide these equations for various types of fluid flows.

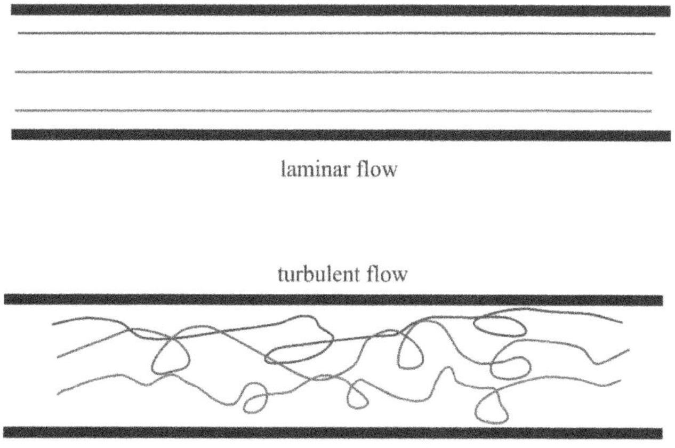

Figure 2.29 Schematic representation of laminar and turbulent flows.

The set of Navier–Stokes equations for a compressible Newtonian viscous fluid include the equation of motion, the continuity equation and the equation of state. Ignoring variations of temperature, and with constant viscosities in time and space, the equation of motion can be written as (Violeau 2012)

$$\underbrace{\frac{\partial \mathbf{v}_{\mathrm{f}}}{\partial t}}_{\text{inertia}}+\underbrace{\nabla \mathbf{v}_{\mathrm{f}} \cdot \mathbf{v}_{\mathrm{f}}}_{\text{convection}}=\underbrace{-\frac{1}{\rho_{\mathrm{f}}} \nabla p_{\mathrm{f}}}_{\text{pressure}}+\underbrace{\eta_{vk} \nabla^{2}\, \mathbf{v}_{\mathrm{f}}}_{\text{viscosity}}+\underbrace{\frac{\lambda_{\mathrm{f}}+\mu_{\mathrm{f}}}{\rho_{\mathrm{f}}} \nabla(\nabla \cdot \mathbf{v}_{\mathrm{f}})}_{\text{compressibility}}+\underbrace{\mathbf{g}}_{\text{body force}} \tag{2.3.21}$$

and the continuity equation

$$\frac{\partial \rho_{\mathrm{f}}}{\partial t}+\nabla \cdot(\rho_{\mathrm{f}} \mathbf{v}_{\mathrm{f}})=0 \tag{2.3.22}$$

An equation of state (EOS) is required to complete the set of equations for computing the unknowns $\mathbf{v}_{\mathrm{f}}$, p_{f} and ρ_{f} at each point of the fluid. For instance, the following EOS may be adopted to obtain the pressure p_{f} associated with the density ρ_{f},

$$p_{\mathrm{f}}=\frac{\rho_{\mathrm{f}0} c^{2}}{\gamma}\left[\left(\frac{\rho_{\mathrm{f}}}{\rho_{\mathrm{f}0}}\right)^{\gamma}-1\right] \tag{2.3.23}$$

where γ and c are the fluid constants and $\rho_{\mathrm{f}0}$ is the initial density of fluid.

2.3.2.4 Weakly Compressible Viscous Fluid

If the temperature variations are negligible and fluid is uniform, the equation of motion of the weakly compressible flow can be written as

$$\frac{\partial \mathbf{v}_{\mathrm{f}}}{\partial t}+\nabla \mathbf{v}_{\mathrm{f}} \cdot \mathbf{v}_{\mathrm{f}}=-\frac{1}{\rho_{\mathrm{f}}} \nabla p_{\mathrm{f}}+\eta_{vk} \nabla^{2}\, \mathbf{v}_{\mathrm{f}}+\mathbf{g} \tag{2.3.24}$$

$$\frac{\partial \rho_{\mathrm{f}}}{\partial t}+\nabla \cdot(\rho_{\mathrm{f}} \mathbf{v}_{\mathrm{f}})=0 \tag{2.3.25}$$

with an equation of state similar to Equation (2.3.23).

2.3.2.5 Incompressible Viscous Flow

For an incompressible fluid with spatially variable viscosity, the equation of motion is transformed to

$$\frac{\partial \mathbf{v}_{\mathrm{f}}}{\partial t}+\nabla \mathbf{v}_{\mathrm{f}} \cdot \mathbf{v}_{\mathrm{f}}=-\frac{1}{\rho_{\mathrm{f}}} \nabla p_{\mathrm{f}}+\frac{2}{\rho_{\mathrm{f}}}\, \nabla \cdot(\mu_{\mathrm{f}} \mathbf{s})+\mathbf{g} \tag{2.3.26}$$

$$\nabla \cdot \mathbf{v}_{\mathrm{f}}=0 \tag{2.3.27}$$

where $\mathbf{s}$ is the strain rate tensor,

$$\mathbf{s}=(\nabla \mathbf{v}_{\mathrm{f}})^{\mathrm{sym}} \tag{2.3.28}$$

2.3.2.6 Steady State Laminar Viscous Flow

Frictional resistance of a solid boundary may influence the flow of fluid in a thin layer around the solid. The Navier–Stokes boundary layer equations of motion for a two-dimensional viscous flow (along the x-direction) are written as

$$v_{fx}\frac{\partial v_{fx}}{\partial x}+v_{fy}\frac{\partial v_{fx}}{\partial y}=-\frac{1}{\rho_f}\frac{\partial p}{\partial x}+\eta_{vk}\frac{\partial^2 v_{fx}}{\partial y^2} \tag{2.3.29}$$

$$\frac{\partial p}{\partial y}=0 \tag{2.3.30}$$

Ignoring the variations of ρ_f for the thin boundary layer, the corresponding continuity equation is written as

$$\nabla\cdot\boldsymbol{v}_f=0 \tag{2.3.31}$$

2.3.2.7 Mean Steady Flow

The Navier–Stokes boundary layer equations of motion of a two-dimensional steady state flow (along the x-direction) can be expressed in terms of the mean velocity vector $\bar{\boldsymbol{v}}_f=(\bar{v}_{fx},\bar{v}_{fy})$ and the mean pressure $\bar{p}$ of the boundary layer:

$$\bar{v}_{fx}\frac{\partial \bar{v}_{fx}}{\partial x}+\frac{\partial \bar{v}_{fx}}{\partial y}=-\frac{1}{\rho_f}\frac{\partial \bar{p}}{\partial x}+\frac{1}{\rho_f}\frac{\partial \tau_b}{\partial y} \tag{2.3.32}$$

$$\frac{\partial \bar{p}}{\partial y}=0 \tag{2.3.33}$$

The shear stress τ_b of the boundary layer is defined as

$$\tau_b=\tau_{lam}+\tau_{tus}=\eta_{vk}\frac{\partial \bar{v}_{fx}}{\partial y}-\rho_f v'_{fx}v'_{fy} \tag{2.3.34}$$

where $\boldsymbol{v}'_f=(v'_{fx},v'_{fy})$ is the vector of fluctuating velocity of the boundary layer:

$$\boldsymbol{v}_f=\bar{\boldsymbol{v}}_f+\boldsymbol{v}'_f \tag{2.3.35}$$

The continuity equation is

$$\nabla\cdot\bar{\boldsymbol{v}}_f=0 \tag{2.3.36}$$

2.3.2.8 Transient Flow of an Ideal Inviscid Fluid

For a fluid with a large Reynolds number, the fluid can be treated as an ideal fluid without viscosity, governed by Euler's equation:

$$\frac{\partial \boldsymbol{v}_f}{\partial t}+\nabla\boldsymbol{v}_f\cdot\boldsymbol{v}_f=-\frac{1}{\rho_f}\nabla p_f+\mathbf{g} \tag{2.3.37}$$

along with the continuity equation

$$\frac{\partial \rho_f}{\partial t} + \nabla \cdot (\rho_f \boldsymbol{v}_f) = 0 \tag{2.3.38}$$

and an equation of state, similar to Equation (2.3.23)

2.3.2.9 Steady State Ideal Flow with No Body Force

The equation of motion of the steady state inviscid flow in the absence of the body force $\mathbf{g}$ can then be written as

$$\nabla \boldsymbol{v}_f \cdot \boldsymbol{v}_f = -\frac{1}{\rho_f} \nabla p_f \tag{2.3.39}$$

and the continuity equation as

$$\nabla \cdot (\rho_f \boldsymbol{v}_f) = 0 \tag{2.3.40}$$

2.3.3 Gas Dynamics

In gas thermodynamics, in addition to compressibility of fluid, the thermal effects and energy issues are vital in developing the equations of state (Powers 2019).

2.3.3.1 General Compressible Viscous Flow

The set of Navier–Stokes equations govern the behaviour of a compressible Newtonian viscous fluid flow. The conservation equations of mass, linear momentum and internal energy e_g can be written in terms of the main variables of fluid velocity $\boldsymbol{v}_g$, density ρ_g, pressure p_g and temperature T_g as

$$\frac{\partial \rho_g}{\partial t} + \nabla \cdot (\rho_g \boldsymbol{v}_g) = 0 \tag{2.3.41}$$

$$\frac{\partial (\rho_g \boldsymbol{v}_g)}{\partial t} + \nabla \cdot (\rho_g \boldsymbol{v}_g \boldsymbol{v}_g) = -\nabla p_g + \nabla \cdot \boldsymbol{\tau}_g + \rho_g \mathbf{g} \tag{2.3.42}$$

$$\frac{\partial (\rho_g E_g)}{\partial t} + \nabla \cdot (\rho_g \boldsymbol{v}_g E_g) = -\nabla \cdot \mathbf{q} - \nabla \cdot (p_g \boldsymbol{v}_g) + \nabla \cdot (\boldsymbol{\tau}_g \cdot \boldsymbol{v}_g) + \rho_g \boldsymbol{v}_g \cdot \mathbf{g} \tag{2.3.43}$$

or in the non-conservative form of internal energy e_g:

$$\rho_f \frac{de_g}{dt} = -\nabla \cdot \mathbf{q} - p_g \nabla \cdot \boldsymbol{v}_g + \boldsymbol{\tau}_g : \nabla \boldsymbol{v}_g \tag{2.3.44}$$

where the total energy E_g is defined as

$$E_g = e_g + \frac{1}{2} \boldsymbol{v}_g \cdot \boldsymbol{v}_g \tag{2.3.45}$$

and the heat flux $\mathbf{q}$ is related to the temperature T_g by

$$\mathbf{q} = -k_g \nabla T_g \tag{2.3.46}$$

where $k_g(\rho_g, T_g)$ is the thermal conductivity coefficient.

In a Newtonian fluid, the viscosity is defined as the ratio of the applied shear stress to the velocity strain rate:

$$\tau_g = \eta_v \dot{\varepsilon} \tag{2.3.47}$$

or in its general form in terms of the coefficients of viscosity $\eta_v(\rho_g, T_g)$ and $\xi_v(\rho_g, T_g)$,

$$\boldsymbol{\tau}_g = \eta_v(\nabla \boldsymbol{v}_g + \nabla \boldsymbol{v}_g \mathrm{T}) + \xi_v(\nabla \cdot \boldsymbol{v}_g)\,\mathbf{I} \tag{2.3.48}$$

where

$$\eta_v \geq 0 \tag{2.3.49}$$

$$\xi_v \geq -\frac{2}{3}\eta_v, \text{ usually}: \ \xi_v = -\frac{2}{3}\eta_v \tag{2.3.50}$$

The second viscosity coefficient ξ_v only contributes in compressible viscous flows, usually in high-speed conditions. As a result, assumption (2.3.50) is valid for low-speed conditions, which are usually encountered in biomechanical applications.

2.3.3.2 Constitutive Equation/Equations of State

The equation of state describes the way the gas pressure p_g is determined from the density ρ_g and the temperature T_g:

$$p_g = p_g(\rho_g, T_g) \tag{2.3.51}$$

It is important to note that the general laws of thermodynamics should always be respected in using the equations of state along with the set of governing Navier–Stokes equations.

The equation of state can also be written in the form of the internal energy

$$e_g = e_g(\rho_g, T_g) \tag{2.3.52}$$

with the thermodynamic condition of

$$\mathrm{d}e_g = c_v(T_g)\mathrm{d}T_g - \frac{1}{\rho_g^2}\left[T_g \left.\frac{\partial p_g}{\partial T_g}\right|_{\rho_g} - p_g\right]\mathrm{d}\rho_g \tag{2.3.53}$$

where c_v is the specific heat at constant volume,

$$c_v = \left.\frac{\partial e_g}{\partial T_g}\right|_{V_g} \tag{2.3.54}$$

A number of simple forms of thermal EOS are (Blazek 2001; Powers 2019)

$$\text{Perfect / ideal gas:} \begin{cases} p_g(\rho_g, T_g) = \rho_g R T_g \\ e_g(\rho_g, T_g) = c_{v0}(T_g - T_{g0}) + e_{g0} \end{cases} \tag{2.3.55}$$

$$\text{Virial gas:} \begin{cases} p_g(\rho_g, T_g) = \rho_g R T_g (1 + b\rho_g) \\ e_g(\rho_g, T_g) = \int_{T_{g0}}^{T_g} c_{v0}(T_g) \mathrm{d}T_g + e_{g0} \end{cases} \tag{2.3.56}$$

$$\text{Van der Waals:} \begin{cases} p_g(\rho_g, T_g) = \dfrac{RT_g}{\rho_g - b} - a\rho_g^2 \\ e_g(\rho_g, T_g) = \int_{T_{g0}}^{T_g} c_{v0}(T_g) \mathrm{d}T_g - a(\rho_g - \rho_{g0}) + e_{g0} \end{cases} \tag{2.3.57}$$

2.3.3.3 Incompressible Flow

The governing equations can be simplified for an incompressible inviscid fluid with constants ρ_g, k_g, η_v and c_p:

$$\nabla \cdot \boldsymbol{v}_f = 0 \tag{2.3.58}$$

$$\rho_g \frac{\partial \boldsymbol{v}_g}{\partial t} = -\nabla p_g + \eta_v \nabla^2 \boldsymbol{v}_g \tag{2.3.59}$$

$$\rho_g c_p \frac{\partial T_g}{\partial t} = k_g \nabla^2 T_g \tag{2.3.60}$$

where c_p is the specific heat at constant pressure,

$$c_p = \left. \frac{\partial h}{\partial T_g} \right|_{p_g} \tag{2.3.61}$$

2.3.3.4 Acoustic Problems

Acoustic problems are usually based on the Euler equations for an isentropic flow. An isentropic process is defined as a process that is both adiabatic ($\delta q = 0, \mathrm{d}s = 0$) and reversible ($\delta q = T\mathrm{d}s$). As a result, in an isentropic process, the change in energy is only caused by the reversible work of pressure–volume (Stynes and Stynes 2010). While many physical shock-wave problems may be studied in an isentropic condition, it is rarely encountered in biomechanical applications.

The governing equations of motion, continuity and EOS can be written as

$$\frac{\partial \rho_a}{\partial t} + \rho_0 \nabla \cdot \boldsymbol{v}_a = 0 \tag{2.3.62}$$

$$\rho_0 \frac{\partial v_a}{\partial t} + \nabla p_a = \mathbf{0} \tag{2.3.63}$$

$$p_a = c_0^2 \rho_a \tag{2.3.64}$$

where the subscripts '0' and 'a' denote the ambient and perturbed conditions, respectively.

Analytically, definition of the velocity potential ϕ_a

$$v_a = \nabla \phi_a \tag{2.3.65}$$

leads to a conventional wave equation

$$\frac{\partial^2 \phi}{\partial t^2} = c_0^2 \nabla^2 \phi_a \tag{2.3.66}$$

Having known the potential function ϕ_a, the acoustic variables p_a and ρ_a can then be obtained from:

$$p_a = -\rho_0 \frac{\partial \phi_a}{\partial t} \tag{2.3.67}$$

$$\rho_a = -\rho_0 c_0^2 \frac{\partial \phi_a}{\partial t} \tag{2.3.68}$$

2.3.4 Diffusion and Convection

Physical problems include quantities that are not always uniformly distributed within a domain. As a result, the physical quantity may diffuse towards the low/high concentration region (Figure 2.30). In a convection problem, an existing flow affects the concentration of an element at a specific position. A wide range of physical quantities may be considered: from mass particles to energy, from heat to electrical current and from financial concepts to information theory.

Both phenomena are frequently observed in biomechanical applications. For instance, in a healing process, both diffusion and convection may be involved in the biophysical processes around a wound to allow for creation of new tissue, as will be comprehensively discussed in Section 5.8. Osmosis effects, membrane simulations and other similar biomechanical phenomena can be studied by diffusion and convection formulations.

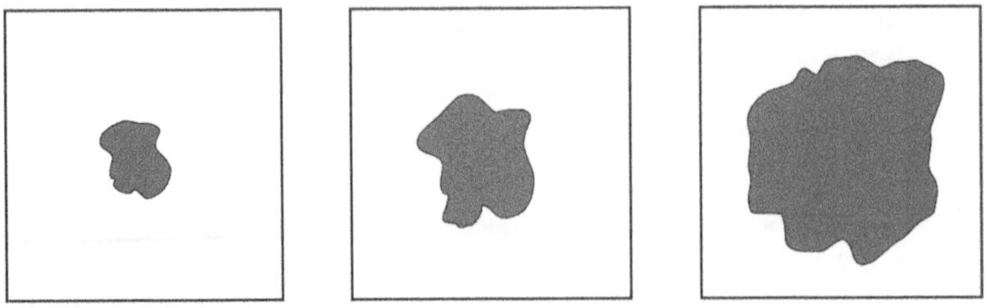

Figure 2.30 A typical illustration of the diffusion process.

The diffusion equation governs seemingly random physical and mathematical processes, as depicted in Figure 2.30. They may become deterministic processes in lower scales by considering the microscale effects or contributing components in detail. However, it may be computationally very expensive or may require inclusion of additional (unknown) physical or chemical microprocesses.

2.3.4.1 Governing Differential Equation

Due to similarities of diffusion and convection problems, they can be described in a unique formulation. The general form of the diffusion/convection differential equation for a physical quantity $c(\mathbf{x},t)$ can be expressed as (Nepf 2008; Stynes and Stynes 2010; Violeau 2012)

$$\frac{\partial c(\mathbf{x},t)}{\partial t} = \underbrace{\boldsymbol{\nabla}\cdot\left[D(\mathbf{x},t)\boldsymbol{\nabla}c(\mathbf{x},t)\right]}_{\text{diffusion}} - \underbrace{\boldsymbol{\nabla}\cdot(\boldsymbol{v}(\mathbf{x},t)c(\mathbf{x},t))}_{\text{convection}} + \underbrace{S(\mathbf{x},t)}_{\text{source/sink}} \tag{2.3.69}$$

where D represents the diffusion coefficient, $\boldsymbol{v}$ is the velocity of flow that affects c and S describes the source or sink of a physical quantity. The diffusion/convection Equation (2.3.69) is clearly nonlinear in time and space.

For many practical problems, however, D, $\boldsymbol{v}$ and S can be assumed to be constants, leading to

$$\frac{\partial c(\mathbf{x},t)}{\partial t} = D\nabla^2 c(\mathbf{x},t) - \boldsymbol{v}\cdot\boldsymbol{\nabla}c(\mathbf{x},t) + S \tag{2.3.70}$$

Various simplified versions of the diffusion/convection equation can then be considered:

1) Steady-state/stationary equation:

$$D\nabla^2 c(\mathbf{x},t) - \boldsymbol{v}\cdot\boldsymbol{\nabla}c(\mathbf{x},t) + S = 0 \tag{2.3.71}$$

2) Transport equation (with no source/sink):

$$\frac{\partial c(\mathbf{x},t)}{\partial t} + \boldsymbol{v}\cdot\boldsymbol{\nabla}c(\mathbf{x},t) = 0 \tag{2.3.72}$$

3) Diffusion equation (constant D with no source/sink):

$$\frac{\partial c(\mathbf{x},t)}{\partial t} = D\nabla^2 c(\mathbf{x},t) \tag{2.3.73}$$

Analytical solutions can be obtained for very simple 1D diffusion or convection problems. Such solutions represent the distribution function of the concentration of the physical quantity.

2.3.4.2 Boundary Conditions

The boundary conditions are usually expressed as a prescribed field $\bar{c}$ on the boundary Γ_c,

$$c = \bar{c} \quad \text{on } \Gamma_c \tag{2.3.74}$$

or as a prescribed flux $\bar{q}_n$,

$$D_n \frac{\partial c}{\partial n} - c v_n = \bar{q}_n \quad \text{on } \Gamma_q \tag{2.3.75}$$

where n is the normal vector on the flux boundary Γ_q and v_n is the normal velocity of flow. In a perfectly absorbing boundary, $\bar{c} = 0$, and a no-flux boundary is represented by $\bar{q}_n = 0$.

2.3.4.3 Transport of Particles

Equation (2.3.69) can be extended to include the effect of velocity of independent particles $\boldsymbol{v}_p$ in the convection term

$$\frac{\partial c(\mathbf{x},t)}{\partial t} = \boldsymbol{\nabla} \cdot \left[D(\mathbf{x},t) \boldsymbol{\nabla} c(\mathbf{x},t) \right] - \boldsymbol{\nabla} \cdot \left[(\boldsymbol{v}(\mathbf{x},t) + \boldsymbol{v}_p(\mathbf{x},t)) c \right] + S \tag{2.3.76}$$

Usually, $\boldsymbol{v}_p$ only includes the vertical component due to the difference in density of particles with respect to the fluid (Nepf 2008).

2.4 Fluid–Structure Interaction

Fluid–structure interactions (FSIs) govern many physical phenomena and engineering problems that contain both solid and fluid parts. For instance, in dam engineering, FSI plays an important role in obtaining a reliable solution for seismic effects on the structure of the dam as well as the water reservoir. Similarly, the process of blood flow in an artery can be considered by FSI algorithms that couple the incompressible blood (fluid) flow and the complex deformable soft tissue of the artery (solid). The study of a breathing mechanism requires modelling of complicated porous geometry of the lung and the flow of compressible air. In this section, the general concepts of the FSI are briefly reviewed. It begins with a look at the basic concepts of the Lagrangian and Eulerian descriptions and covers the fundamentals of the coupled Lagrangian–Eulerian (CLE) and the arbitrary Lagrangian–Eulerian (ALE) formulations.

2.4.1 Lagrangian and Eulerian Descriptions

In addition to the formulations of fluid flow and solid mechanics, FSI is expected to determine the profile of the evolving free surface of the fluid, to properly define and enforce the boundary conditions and to formulate the interaction of fluid and solid phases.

The profile of the free surface of a fluid can be determined by a number of techniques. Lagrangian methods automatically follow the movement of the boundaries and do not require any additional requirement. Moreover, history-dependent constitutive laws can be efficiently implemented. Unfortunately, they suffer from element distortions in large deformations. In particle methods such as the smoothed particle hydrodynamics (SPH), the free surface is automatically handled by the Lagrangian nature of the particle method. In Eulerian formulations, a so-called marker and cell method can be adopted to identify or mark the position of free surface by a number of particles moving with the fluid flow. Alternatively, the powerful ALE approach somehow combines the Eulerian and Lagrangian configurations to capture the free surface in a systematic way (Donea et al. 2004).

2.4.2 Fluid–Solid Interface Boundary Conditions

Different fluid–solid boundary conditions can be considered based on the type of fluid and boundary. The simplest form is a fixed wall, where the boundary is assumed rigid and fixed. In this case, the non-slip condition between the fluid and boundary interface Γ_i can be specified in terms of the velocity of fluid $\boldsymbol{v}_\mathrm{f}$ as

$$\boldsymbol{v}_\mathrm{f} = \mathbf{0} \text{ on } \Gamma_\mathrm{i} \tag{2.4.1}$$

For a slip condition, usually for non-viscous fluids, only the normal component of the relative velocity becomes zero:

$$\boldsymbol{n}_\mathrm{f}.\boldsymbol{v}_\mathrm{f} = 0 \text{ on } \Gamma_\mathrm{i} \tag{2.4.2}$$

where $\boldsymbol{n}_\mathrm{f}$ is the normal to the boundary. In high-speed compressible FSI problems, another boundary condition in the form of angular velocity can be defined (ADINA 2010). The tangential component should then be computed as an independent variable.

The second category of boundary conditions is the moving boundary for deformable/moving solids. Both non-slip and slip conditions can be assumed. In a non-slip moving boundary, the fluid velocity follows the velocity of the solid boundary $\boldsymbol{v}_\mathrm{s}(t)$,

$$\boldsymbol{v}_\mathrm{f} = \boldsymbol{v}_\mathrm{s}(t) \text{ on } \Gamma_\mathrm{i} \tag{2.4.3}$$

and for the slip moving wall,

$$\boldsymbol{n}_\mathrm{f}.\boldsymbol{v}_\mathrm{f} - \boldsymbol{n}_\mathrm{s}.\boldsymbol{v}_\mathrm{s}(t) = 0 \text{ on } \Gamma_\mathrm{i} \tag{2.4.4}$$

Both prescribed temperature ($\theta_\mathrm{b} = \overline{\theta}_\mathrm{b}$) or adiabatic boundary conditions ($\boldsymbol{n}.\nabla\theta_\mathrm{b} = 0$) can be assumed for fixed or moving wall boundary conditions (Figure 2.31).

Conditions for a fluid–structure interface are best described by an equation similar to Equation (2.4.3), with the velocity of a moving wall being an independent unknown. This velocity is computed for the Lagrangian solid in an FSI solution, while the fluid pressure is applied on the corresponding boundary of the solid:

$$p\boldsymbol{n}_\mathrm{f} = \boldsymbol{n}_\mathrm{s}.\boldsymbol{\sigma} \tag{2.4.5}$$

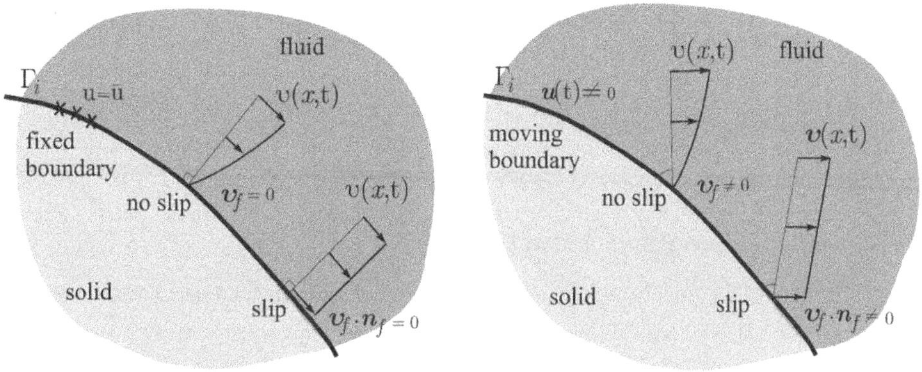

Figure 2.31 Different FSI boundary conditions. *Source:* Adapted from ADINA 2010.

For general shapes of fluid–solid interfaces, the level set technique may be efficiently adopted. This technique allows for accurate and efficient tracking of evolution of internal boundaries, as widely used in XFEM propagation analysis of discontinuities (Mohammadi 2012a).

The level set function ϕ for an FSI problem can be defined as:

$$\phi(\mathbf{x},t) = \begin{cases} <0 & \mathbf{x}\in\Omega_\mathrm{s} \\ =0 & \mathbf{x}\in\Gamma_\mathrm{i} \\ >0 & \mathbf{x}\in\Omega_\mathrm{f} \end{cases} \tag{2.4.6}$$

and the normal to the interface Γ_i can be obtained from the level set function

$$\boldsymbol{n}_\mathrm{s}(\mathbf{x},t) = -\boldsymbol{n}_\mathrm{f}(\mathbf{x},t) = \nabla\phi(\mathbf{x},t) \tag{2.4.7}$$

where

$$\phi(\mathbf{x},t) = \min_{\bar{\mathbf{x}}(t)\in\Gamma_\mathrm{i}} \| \mathbf{x}-\bar{\mathbf{x}}(t) \| \, sign\left[n_\mathrm{s}.(\mathbf{x}-\bar{\mathbf{x}}(t))\right] \tag{2.4.8}$$

2.4.3 Governing Equations in the Eulerian Description

The strong forms of the governing mass, momentum and energy conservation equations in the Eulerian configuration for both fluid and solid phases can be expressed as (Rugonyi and Bathe 2001):

$$\left.\frac{\partial\rho}{\partial t}\right|_x + \boldsymbol{v}\cdot\nabla\rho = -\rho\nabla\cdot\boldsymbol{v} \tag{2.4.9}$$

$$\rho\left[\left.\frac{\partial v}{\partial t}\right|_x + (\boldsymbol{v}\cdot\nabla)\boldsymbol{v}\right] = \nabla\cdot\boldsymbol{\sigma} + \rho\boldsymbol{f}^\mathrm{b} \tag{2.4.10}$$

$$\rho\left[\left.\frac{\partial E}{\partial t}\right|_x + \boldsymbol{v}\cdot\nabla E\right] = \nabla\cdot(\boldsymbol{\sigma}\cdot\mathbf{v}) + \boldsymbol{v}\cdot\rho\boldsymbol{f}^\mathrm{b} \tag{2.4.11}$$

where E is the total energy. Equation (2.4.11) can also be written in terms of the internal energy e,

$$\rho\left[\left.\frac{\partial e}{\partial t}\right|_{x}+\boldsymbol{v}\cdot\boldsymbol{\nabla}e\right]=\boldsymbol{\sigma}:\boldsymbol{\nabla}^{s}\boldsymbol{v} \tag{2.4.12}$$

2.4.4 Coupled Lagrangian–Eulerian (CLE)

The strong form of the governing equations based on the CLE description of a non-slip FSI problem can be written as (Rugonyi and Bathe 2001)

$$\rho_f\dot{\boldsymbol{v}}_f=\boldsymbol{\nabla}\cdot\boldsymbol{\sigma}_f+\rho_f\boldsymbol{f}_f^{b}\quad\text{in }\Omega_f \tag{2.4.13}$$

$$\rho_s\dot{\boldsymbol{v}}_s=\boldsymbol{\nabla}\cdot\boldsymbol{\sigma}_s+\rho_s\boldsymbol{f}_s^{b}\quad\text{in }\Omega_s \tag{2.4.14}$$

$$\boldsymbol{\sigma}_f\cdot\boldsymbol{n}_f-\boldsymbol{f}_f^{t}=\mathbf{0}\quad\text{on }\Gamma_{ft} \tag{2.4.15}$$

$$\boldsymbol{\sigma}_f\cdot\boldsymbol{n}_f-\boldsymbol{\lambda}=\mathbf{0}\quad\text{on }\Gamma_i \tag{2.4.16}$$

$$\boldsymbol{\sigma}_s\cdot\boldsymbol{n}_s-\boldsymbol{\lambda}=\mathbf{0}\quad\text{on }\Gamma_i \tag{2.4.17}$$

$$\boldsymbol{v}_f-\boldsymbol{v}_s=\mathbf{0}\quad\text{on }\Gamma_i \tag{2.4.18}$$

$$\dot{\rho}_f+\nabla(\rho_f\boldsymbol{v}_f)=0\quad\text{in }\Omega_f \tag{2.4.19}$$

Combining Equations (2.4.16) and (2.4.17) results in

$$\boldsymbol{\sigma}_s\cdot\boldsymbol{n}_s+\boldsymbol{\sigma}_f\cdot\boldsymbol{n}_f=0\quad\text{on }\Gamma_i \tag{2.4.20}$$

For a slip FSI problem, Equations (2.4.18) and (2.4.20) are replaced by

$$\boldsymbol{v}_f\cdot\boldsymbol{n}_f-\boldsymbol{v}_s\cdot\boldsymbol{n}_s=0\quad\text{on }\Gamma_i \tag{2.4.21}$$

$$\boldsymbol{\sigma}_s\boldsymbol{n}_s\cdot\boldsymbol{n}_s+\boldsymbol{\sigma}_f\boldsymbol{n}_f\cdot\boldsymbol{n}_f=0\quad\text{on }\Gamma_i \tag{2.4.22}$$

The final discretized system of simultaneous equations can be written as (Legay et al. 2006):

$$\overline{\overline{\mathbf{M}}}\dot{\mathbf{V}}+\overline{\mathbf{M}}\mathbf{V}=\mathbf{f} \tag{2.4.23}$$

or

$$\begin{bmatrix}\mathbf{M}_f & \mathbf{0} & \mathbf{0} & \mathbf{0}\\ \mathbf{0} & \mathbf{M}_s & \mathbf{0} & \mathbf{0}\\ \mathbf{0} & \mathbf{0} & \mathbf{M}_\rho & \mathbf{0}\\ \mathbf{0} & \mathbf{0} & \mathbf{0} & \mathbf{0}\end{bmatrix}\begin{bmatrix}\dot{\mathbf{V}}_f\\ \dot{\mathbf{V}}_s\\ \dot{\mathbf{R}}\\ \mathbf{0}\end{bmatrix}+\begin{bmatrix}\mathbf{0} & \mathbf{0} & \mathbf{0} & \mathbf{M}_{f\lambda}\\ \mathbf{0} & \mathbf{0} & \mathbf{0} & \mathbf{M}_{s\lambda}\\ \mathbf{0} & \mathbf{0} & \mathbf{0} & \mathbf{0}\\ \left(\mathbf{M}_{f\lambda}\right)^{T} & \left(\mathbf{M}_{s\lambda}\right)^{T} & \mathbf{0} & \mathbf{0}\end{bmatrix}\begin{bmatrix}\mathbf{V}_f\\ \mathbf{V}_s\\ \mathbf{R}\\ \boldsymbol{\Lambda}\end{bmatrix}=\begin{bmatrix}\mathbf{f}_f^{int}\\ \mathbf{f}_s^{int}\\ \mathbf{f}_\rho\\ \mathbf{0}\end{bmatrix}+\begin{bmatrix}\mathbf{f}_f^{ext}\\ \mathbf{0}\\ \mathbf{0}\\ \mathbf{0}\end{bmatrix} \tag{2.4.24}$$

where the vectors of nodal unknowns are

$$\mathbf{V}_\mathrm{f} = \begin{bmatrix} \mathbf{V}_\mathrm{f}^1 & \cdots & \mathbf{V}_\mathrm{f}^{n_\mathrm{f}} \end{bmatrix}^\mathrm{T} \tag{2.4.25}$$

$$\mathbf{V}_\mathrm{s} = \begin{bmatrix} \mathbf{V}_\mathrm{s}^1 & \cdots & \mathbf{V}_\mathrm{s}^{n_\mathrm{s}} \end{bmatrix}^\mathrm{T} \tag{2.4.26}$$

$$\mathbf{R} = \begin{bmatrix} \rho^1 & \cdots & \rho^{n_\mathrm{f}} \end{bmatrix}^\mathrm{T} \tag{2.4.27}$$

$$\mathbf{\Lambda} = \begin{bmatrix} \Lambda^1 & \cdots & \Lambda^{n_\mathrm{f}} \end{bmatrix}^\mathrm{T} \tag{2.4.28}$$

with the nodal matrices

$$\mathbf{M}_\mathrm{f} = \int_{\Omega_\mathrm{f}} \rho_\mathrm{f} \mathbf{N}_\mathrm{f}^\mathrm{T} \mathbf{N}_\mathrm{f} H(\phi) d\Omega \tag{2.4.29}$$

$$\mathbf{M}_\mathrm{s} = \int_{\Omega_\mathrm{s0}} \rho_\mathrm{s} \mathbf{N}_\mathrm{s}^\mathrm{T} \mathbf{N}_\mathrm{s} d\Omega \tag{2.4.30}$$

$$\mathbf{M}_\rho = \int_{\Omega_\mathrm{f}} \mathbf{N}_\mathrm{f}^\mathrm{T} \mathbf{N}_\mathrm{f} H(\phi) d\Omega \tag{2.4.31}$$

$$\mathbf{M}_{\mathrm{f}\lambda} = -\int_{\Gamma_\mathrm{i}} \mathbf{N}_\mathrm{f}^\mathrm{T} \mathbf{N}_\lambda d\Gamma \tag{2.4.32}$$

$$\mathbf{M}_{s\lambda} = \int_{\Gamma_\mathrm{i}} \mathbf{N}_s^\mathrm{T} \mathbf{N}_\lambda d\Gamma \tag{2.4.33}$$

and nodal forces

$$\mathbf{f}_\mathrm{f}^\mathrm{int} = -\int_{\Omega_\mathrm{f}} \left(\mathbf{B}_\mathrm{f}^\mathrm{T} \boldsymbol{\sigma}_\mathrm{f} + \rho_\mathrm{f} \mathbf{N}_\mathrm{f}^\mathrm{T} \mathbf{v}_\mathrm{f} \boldsymbol{\nabla} . \mathbf{v}_\mathrm{f} \right) H(\phi) d\Omega \tag{2.4.34}$$

$$\mathbf{f}_\mathrm{s}^\mathrm{int} = -\int_{\Omega_\mathrm{s0}} \mathbf{B}_\mathrm{s}^\mathrm{T} \boldsymbol{\sigma}_\mathrm{s} d\Omega \tag{2.4.35}$$

$$\mathbf{f}_\rho = -\int_{\Omega_\mathrm{f}} \mathbf{N}_\mathrm{f}^\mathrm{T} \boldsymbol{\nabla} (\rho_\mathrm{f} \mathbf{v}_\mathrm{f}) H(\phi) d\Omega \tag{2.4.36}$$

$$\mathbf{f}_\mathrm{f}^\mathrm{ext} = \int_{\Omega_\mathrm{f}} \mathbf{N}_\mathrm{f} \boldsymbol{f}_\mathrm{f}^\mathrm{b} d\Omega + \int_{\Gamma_\mathrm{ft}} \mathbf{N}_\mathrm{f} \boldsymbol{f}_\mathrm{f}^\mathrm{t} d\Gamma \tag{2.4.37}$$

2.4.5 Coupled Lagrangian–Lagrangian (CLL)

In this case, the Lagrangian description is adopted for both fluid and solid phases. The mass conservation equation is automatically satisfied and the discretized solution is substantially simplified. For an updated Lagrangian formulation, the governing equation can be written as

$$\mathbf{M}\dot{\mathbf{V}} = \mathbf{f}^\mathrm{ext} - \mathbf{f}^\mathrm{int} \tag{2.4.38}$$

with

$$\mathbf{M} = \int_{\Omega} \rho \mathbf{N}^{\mathrm{T}} \mathbf{N} d\Omega \left(= \int_{\Omega_0} \rho_0 \mathbf{N}^{\mathrm{T}} \mathbf{N} d\Omega \right) \tag{2.4.39}$$

$$\mathbf{f}^{\mathrm{int}} = -\int_{\Omega} \mathbf{B}^{\mathrm{T}} \sigma d\Omega \left(= -\int_{\Omega_0} \mathbf{B}^{\mathrm{T}} P d\Omega \right) \tag{2.4.40}$$

$$\mathbf{f}^{\mathrm{ext}} = \int_{\Omega} \mathbf{N}^{\mathrm{T}} \boldsymbol{f}^{b} d\Omega + \int_{\Gamma_t} \mathbf{N}^{\mathrm{T}} \boldsymbol{f}^{t} d\Gamma \left(= \int_{\Omega_0} \mathbf{N}^{\mathrm{T}} \boldsymbol{f}_0^{b} d\Omega + \int_{\Gamma_{t0}} \mathbf{N}^{\mathrm{T}} \boldsymbol{f}_0^{t} d\Gamma \right) \tag{2.4.41}$$

The terms in parenthesis are related to the total Lagrangian formulation.

2.4.6 Arbitrary Lagrangian–Eulerian (ALE)

The Lagrangian description of finite element modelling of FSI problems may lead to excessive deformation and distortion of elements. An adaptive finite element solution is a powerful remedy to avoid distortions by remeshing algorithms (Emamzadeh et al. 2015). Nevertheless, it is computationally expensive for large-scale problems. The arbitrary Lagrangian–Eulerian formulation (ALE) is an alternative approach that combines the benefits of Lagrangian and Eulerian formulations and adopts a local node movement technique based on a reference configuration to define the governing formulations. It avoids the element distortion of the Lagrangian description and does not suffer from the complicated definition of free surfaces in the Eulerian formulation. Figure 2.32 illustrates a typical simple comparison for the Lagrangian, Eulerian and ALE descriptions.

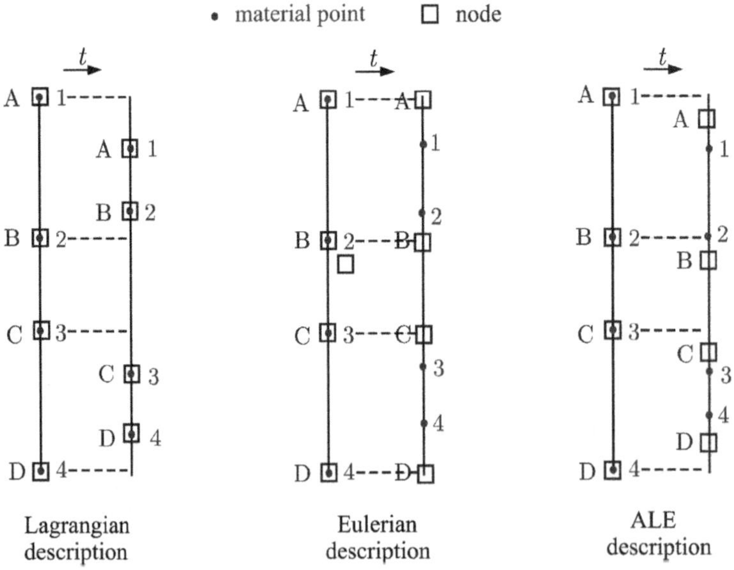

Figure 2.32 Lagrangian, Eulerian and ALE Descriptions.

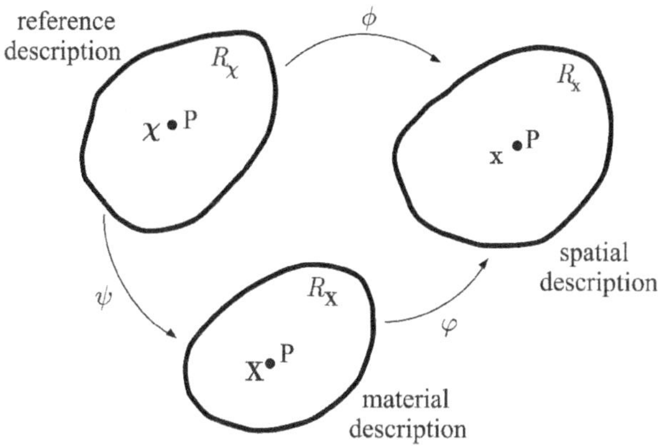

Figure 2.33 Material, spatial and reference configurations. *Source:* Reproduced from Stein et al. 2007/John Wiley & Sons Ltd.

The ALE formulation correlates the material and spatial configurations through a reference configuration, as depicted in Figure 2.33 (Stein et al. 2007).

A kinematic description of motion in ALE is not based on the Lagrangian movement of material particles, nor on the Eulerian fixed mesh, that the material moves through. Instead, ALE is based on a reference configuration, which moves with the continuum domain independently of the motion of the material point. As a result, it is somehow similar to a Eulerian approach where the material moves through the refence mesh. On the other hand, it resembles a Lagrangian approach by explicit definition of boundaries and avoiding element distortions.

Consider $\mathbf{X}$ and $\mathbf{x}$ as the position vectors in two consecutive material configurations, respectively. Mapping φ allows for the computation of the displacement vector $\mathbf{u}$ in the material description $\mathbf{X}$:

$$\mathbf{u} = \mathbf{x} - \mathbf{X} = \varphi(\mathbf{X}, t) - \mathbf{X}(t) \tag{2.4.42}$$

or in the spatial description $\mathbf{x}$

$$\mathbf{u} = \mathbf{x} - \mathbf{X} = \mathbf{x}(t) - \varphi^{-1}(\mathbf{x}, t) \tag{2.4.43}$$

The mapping φ is re-defined as $\mathbf{\Phi}$ in the reference configuration χ of the ALE description:

$$\mathbf{x} = \varphi(\mathbf{X}, t) = \mathbf{\Phi}(\chi, t) \tag{2.4.44}$$

The nodal displacement and velocity can then be determined in the material configurations as

$$\mathbf{u} = \varphi(\mathbf{X}, t) - \mathbf{X}(t) \tag{2.4.45}$$

$$\boldsymbol{v} = \left.\frac{\partial \mathbf{u}(\mathbf{X}, t)}{\partial \mathrm{t}}\right|_{x} \tag{2.4.46}$$

and with respect to the reference configuration,

$$\mathbf{u}^{\mathrm{r}} = \Phi(\chi,t) - \chi(t) \tag{2.4.47}$$

$$\boldsymbol{v}^{\mathrm{r}} = \left.\frac{\partial \mathbf{u}^{\mathrm{r}}(\chi,t)}{\partial t}\right|_{\chi} \tag{2.4.48}$$

Velocity of the reference configuration with respect to the material configuration, $\boldsymbol{w}_{\chi}$, can be obtained from

$$\boldsymbol{w}_{\chi} = \left.\frac{\partial \chi}{\partial t}\right|_{\chi} = \frac{\partial \chi}{\partial \mathbf{x}} \cdot \boldsymbol{c} \tag{2.4.49}$$

$$\mathbf{c} = \boldsymbol{v} - \boldsymbol{v}^{r} \tag{2.4.50}$$

The final strong forms of the governing mass, momentum and energy conservation equations in the reference configuration can now be expressed as (Donea et al. 2004)

$$\left.\frac{\partial \rho}{\partial t}\right|_{\chi} + \mathbf{c} \cdot \nabla_{\mathbf{x}} \rho = -\rho \nabla_{\mathbf{x}} \cdot \boldsymbol{v} \tag{2.4.51}$$

$$\rho \left[\left.\frac{\partial \boldsymbol{v}}{\partial t}\right|_{\chi} + (\mathbf{c} \cdot \nabla_{\mathbf{x}}) \boldsymbol{v} \right] = \nabla_{\mathbf{x}} \cdot \boldsymbol{\sigma} + \rho \boldsymbol{f}^{b} \tag{2.4.52}$$

$$\rho \left[\left.\frac{\partial E}{\partial t}\right|_{\chi} + \mathbf{c} \cdot \nabla_{\mathbf{x}} E \right] = \nabla_{\mathbf{x}} \cdot (\boldsymbol{\sigma} \cdot \boldsymbol{v}) + \boldsymbol{v} \cdot \rho \boldsymbol{f}^{b} \tag{2.4.53}$$

It is important to note that implementation of the constitutive law for the solid phase is not straightforward as in the Lagrangian description. The reason can be attributed to the fact that in ALE (as in the Eulerian description), the solid material inside each element does not remain the same in subsequent times. Therefore, the relative movement between the reference and material configurations should be considered when deriving the constitutive law. The rate forms of the constitutive laws should then be based on

$$\frac{D\sigma}{Dt} = \left.\frac{D\sigma}{Dt}\right|_{\chi} + (\mathbf{c} \cdot \nabla_{\mathbf{x}}) \sigma \tag{2.4.54}$$

The final discretized form of the governing equations can be written for both phases as

$$\mathbf{M}\dot{\mathbf{V}} - \mathbf{f}\mathbf{V} = \mathbf{f}^{\mathrm{int}} + \mathbf{f}^{\mathrm{ext}} \tag{2.4.55}$$

where

$$\mathbf{M} = \int_{\Omega(t)} \rho \mathbf{N}^{\mathrm{T}} \mathbf{N} \, d\Omega \tag{2.4.56}$$

$$\mathbf{f} = \int_{\Omega(t)} \mathbf{N}_{\mathrm{f}}^{\mathrm{T}} (\rho_{\mathrm{f}} \mathbf{c} \cdot \boldsymbol{\nabla}_{\mathbf{x}}) \mathbf{N}_{\mathrm{f}} d\Omega \tag{2.4.57}$$

$$\mathbf{f}^{\mathrm{int}} = -\int_{\Omega(t)} p \mathbf{B}_{\mathrm{f}} \Omega \text{ for fluid} \tag{2.4.58}$$

$$\mathbf{f}^{\mathrm{int}} = -\int_{\Omega(t)} \mathbf{B}_{\mathrm{s}}^{\mathrm{T}} \boldsymbol{\sigma}_{\mathrm{s}} d\Omega \text{ for solid} \tag{2.4.59}$$

$$\mathbf{f}^{\mathrm{ext}} = \int_{\Omega(t)} \mathbf{N}_{\mathrm{s}}^{\mathrm{T}} \rho_{\mathrm{s}} \boldsymbol{f}_{\mathrm{s}}^{b} d\Omega + \int_{\Gamma_{t}(t)} \mathbf{N}_{\mathrm{s}}^{\mathrm{T}} \boldsymbol{f}_{\mathrm{s}}^{t} d\Gamma \tag{2.4.60}$$

with a rate form of the constitutive equation ($\mathbf{q}$) for the solid

$$\mathbf{M}_{\mathrm{s}} \frac{d\boldsymbol{\sigma}}{dt} + \mathbf{L}_{\mathrm{s}} \boldsymbol{\sigma} = \mathbf{q} \tag{2.4.61}$$

where

$$\mathbf{M}_{\mathrm{s}} = \int_{\Omega_{\mathrm{s}}(t)} \mathbf{N}_{\mathrm{s}}^{\mathrm{T}} \mathbf{N}_{\mathrm{s}} d\Omega \tag{2.4.62}$$

$$\mathbf{L}_{\mathrm{s}} = \int_{\Omega_{\mathrm{s}}(t)} \mathbf{N}_{\mathrm{s}}^{\mathrm{T}} \mathbf{c} \boldsymbol{\nabla} \mathbf{N}_{\mathrm{s}} d\Omega \tag{2.4.63}$$

An important aspect of ALE analysis is to set accurate procedures for movement of the arbitrary mesh. This is achieved either by the general concepts of mesh adaptivity to reduce the overall error of the computational solution or through the geometry-based mesh regularization techniques to limit the level of element distortions.

Different mesh regularization techniques are available, which include, among the others, the transfinite mapping method, the Laplacian approach, the area-based approach and a combination of them (Gadalaa et al. 2002). All regularization methods lead to a new set of spatial nodal coordinates $\mathbf{x}^{\mathrm{reg}}$, and the corresponding displacement and velocity fields can be obtained from

$$\mathbf{u}^{\mathrm{reg}} = \mathbf{x}^{\mathrm{reg}} - \mathbf{x} \tag{2.4.64}$$

$$\mathbf{v}^{\mathrm{reg}} = \frac{1}{\Delta t} (\mathbf{x}^{\mathrm{reg}} - \mathbf{x}) \tag{2.4.65}$$

The transformation velocity c is then obtained from the material velocity $\boldsymbol{v}$ and the regularized velocity of the reference configuration $\boldsymbol{v}^{\mathrm{reg}}$

$$\mathbf{c} = \boldsymbol{v} - \boldsymbol{v}^{\mathrm{reg}} \tag{2.4.66}$$

The ALE solution can be performed in three stages. First, a pure Lagrangian solution without the transformation terms is performed to obtain the material state. In this stage, the material configuration is assumed to coincide with the reference configuration. A possible element distortion can then be regularized in the second stage to improve the nodal positions in the reference configuration (and not in the physical domain). The Godunov scheme (Stein et al. 2007) is a simple and straightforward technique to transfer element-based variables. Alternative adaptivity solutions such as the superconvergent patch recovery technique (SPR) or recovery by equilibrium on patch (REP) can also be adopted (Boroomand and Zienkiewicz 1997). The third stage is dedicated to account for the effects of transformation in the reference configuration with respect to a pure Lagrangian description.

3

Numerical Methods

3.1 Introduction

In this chapter, a number of main numerical methods for solving engineering and physical problems are reviewed. The methods include the finite difference method, the finite volume method, the finite element method, the extended finite element method, the extended isogeometric analysis, the meshless methods and the variable node element method.

The aim is to provide basics of the methods, as they may be adopted for single and multiscale analysis of various problems in general, and biomechanical applications in particular. For details on each method, the readers should refer to the corresponding textbooks and reference publications.

3.2 Finite Difference Method (FDM)

The finite difference method (FDM) is one of the simplest numerical methods used to solve the governing equations of a system by transforming the partial differential equations of a variable to algebraic relations in terms of the values of the variable at a structured grid of nodes. It is a strong form of solution and can be similarly adopted for spatial and temporal differential equations.

The basic idea of the FDM is to approximate the derivative of a variable with its change in a finite interval:

$$u'(x) = \frac{du}{dx} \approx \frac{\Delta u}{\Delta x} \tag{3.2.1}$$

Any differential equations can be transformed to an algebraic relation by applying the set of equations similar to Equation (3.2.1) for various orders of derivatives.

The finite difference formulation may be presented in the forms of forward FDM (fFDM), central FDM (cFDM) and backward FDM (bFDM) schemes, which are briefly reviewed in this section.

Multiscale Biomechanics: Theory and Applications, First Edition. Soheil Mohammadi.

3.2.1 One-Dimensional FDM

Consider a one-dimensional domain, which is discretized by uniform Δx intervals. The value of function u is assumed to be u_i at each position x_i, as depicted in Figure 3.1. The fFDM, bFDM and cFDM approximations are presented for derivatives (up to order 3) of a typical function u.

3.2.1.1 Forward Finite Difference Method (fFDM)

The forward finite difference formulation for the first derivative of u with respect to x at position x_i can be written as

$$u'_i = u^{(1)}(x_i) \approx \frac{1}{\Delta x}(u_{i+1} - u_i) \tag{3.2.2}$$

or simply

$$u_i^{(1)} \approx \frac{1}{\Delta x}(-u_i + u_{i+1}) \tag{3.2.3}$$

Similarly, the fFDM approximation for the second and third derivatives can be written as (Thomas 1995)

$$u_i^{(2)} \approx \frac{1}{(\Delta x)^2}(u_i - 2u_{i+1} + u_{i+2}) \tag{3.2.4}$$

$$u_i^{(3)} \approx \frac{1}{(\Delta x)^3}(-u_i + 3u_{i+1} - 3u_{i+2} + u_{i+3}) \tag{3.2.5}$$

3.2.1.2 Backward Finite Difference Method (bFDM)

In the backward finite difference, the derivatives at each point x_i are approximated by the values of the function at predecessor points:

$$u_i^{(1)} \approx \frac{1}{\Delta x}(-u_{i-1} + u_i) \tag{3.2.6}$$

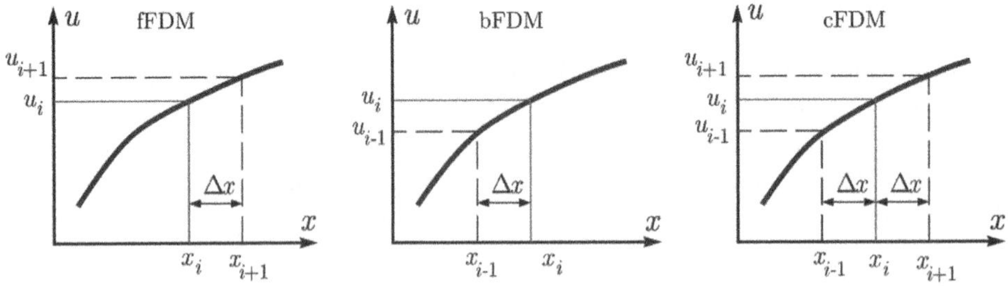

Figure 3.1 One-dimensional representation of forward, backward and central finite difference schemes.

and (Thomas 1995)

$$u_i^{(2)} \approx \frac{1}{(\Delta x)^2}(u_{i-2} - 2u_{i-1} + u_i) \tag{3.2.7}$$

$$u_i^{(3)} \approx \frac{1}{(\Delta x)^3}(-u_{i-3} + 3u_{i-2} - 3u_{i-1} + u_i) \tag{3.2.8}$$

3.2.1.3 Central Finite Difference Method (cFDM)

The central finite difference uses the values of a number of points distributed symmetrically around the point x_i to approximate the derivatives at that point:

$$u_i^{(1)} \approx \frac{1}{2\Delta x}(-u_{i-1} - u_{i+1}) \tag{3.2.9}$$

and (Thomas 1995)

$$u_i^{(2)} \approx \frac{1}{(2\Delta x)^2}(u_{i+2} - 2u_i + u_{i-2}) \tag{3.2.10}$$

or in terms of the half-interval values $\Delta x / 2$,

$$u_i^{(2)} \approx \frac{1}{(\Delta x)^2}(u_{i-1} - 2u_i + u_{i+1}) \tag{3.2.11}$$

and

$$u_i^{(3)} \approx \frac{1}{(\Delta x)^3}(-\frac{1}{2}u_{i-2} + u_{i-1} - u_{i+1} + \frac{1}{2}u_{i+2}) \tag{3.2.12}$$

The central finite difference method is frequently used in the explicit time integration schemes.

3.2.2 Higher Order One-Dimensional FDM

It is also possible to increase the accuracy of finite difference solutions by using additional terms in the Taylor series and additional points for values of the variable (Thomas 1995).

3.2.2.1 bFDM

$$u_i^{(1)} \approx \frac{1}{(2\Delta x)^1}(u_{i-2} - 4u_{i-1} + 3u_i) \tag{3.2.13}$$

$$u_i^{(2)} \approx \frac{1}{(\Delta x)^2}(-u_{i-3} + 4u_{i-2} - 5u_{i-1} + 2u_i) \tag{3.2.14}$$

$$u_i^{(3)} \approx \frac{1}{2(\Delta x)^3}(3u_{i-4} - 14u_{i-3} + 24u_{i-2} - 18u_{i-1} + 5u_i) \tag{3.2.15}$$

3.2.2.2 fFDM

$$u_i^{(1)} \approx \frac{1}{(2\Delta x)^1}(-3u_i + 4u_{i+1} - u_{i+2}) \tag{3.2.16}$$

$$u_i^{(2)} \approx \frac{1}{(\Delta x)^2}(2u_i - 5u_{i+1} + 4u_{i+2} - u_{i+3}) \tag{3.2.17}$$

$$u_i^{(3)} \approx \frac{1}{2(\Delta x)^3}(-5u_i + 18u_{i+1} - 24u_{i+2} + 14u_{i+3} - 3u_{i+4}) \tag{3.2.18}$$

3.2.2.3 cFDM

$$u_i^{(1)} \approx \frac{1}{(12\Delta x)^1}(-u_{i-2} - 8u_{i-1} + 8u_{i+1} - u_{i+2}) \tag{3.2.19}$$

$$u_i^{(2)} \approx \frac{1}{12(\Delta x)^2}(-u_{i-2} + 16u_{i-1} - 30u_i + 16u_{i+1} - u_{i+2}) \tag{3.2.20}$$

$$u_i^{(3)} \approx \frac{1}{8(\Delta x)^3}(u_{i-3} - 8u_{i-2} + 13u_{i-1} - 13u_{i+1} + 8u_{i+2} - u_{i+3}) \tag{3.2.21}$$

3.2.2.4 Imposition of Boundary Conditions

One of the main problems with FDM is its inefficiency in handling the boundary conditions. For instance, in modelling of a beam with a simple hinge support at point x_0, the boundary conditions require

$$u_0 = 0 \tag{3.2.22}$$

$$M_0 = -EI\left.\frac{d^2u}{dx^2}\right|_{x_0} = 0 \tag{3.2.23}$$

Imposition of the moment free condition by FDM requires an extra (ghost) node u_{-1} outside the domain, as depicted in Figure 3.2:

$$M_0 = -EI\left.\frac{d^2u}{dx^2}\right|_{x_0} = 0 \rightarrow u_0^{(2)} \approx \frac{1}{(\Delta x)^2}(u_{-1} - 2u_0 + u_1) = 0 \tag{3.2.24}$$

which results in

$$u_{-1} = -u_1 \tag{3.2.25}$$

A similar procedure for a fixed/clamped boundary (Figure 3.2) needs a zero slope at x_0 (clamped end):

$$u_{-1} = u_1 \tag{3.2.26}$$

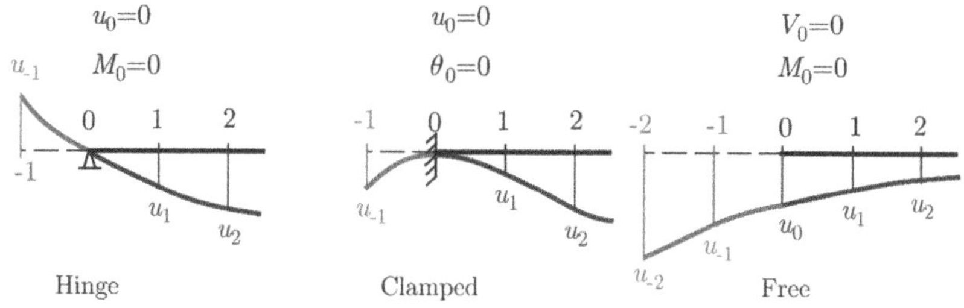

Figure 3.2 Imposition of hinge, clamped and free boundary conditions for a one-dimensional flexural beam.

and a free end at x_0 with zero shear force and bending moment (free end in Figure 3.2),

$$\begin{cases} u_{-1} = 2u_0 - u_1 \\ u_{-2} = 4u_0 - 4u_1 + u_2 \end{cases} \tag{3.2.27}$$

Clearly, imposition of boundary conditions is not straightforward in FDM and may become quite complicated for general geometries and various forms of boundary conditions.

Having imposed the boundary conditions, FDM approximations are directly applied on the strong form of the governing differential equations at all nodes x_i individually, resulting in a set of nonlinear algebraic equations in terms of the vector of nodal values $\bar{\mathbf{U}}$,

$$\mathbf{K}\bar{\mathbf{U}} = \mathbf{f} \tag{3.2.28}$$

The coefficient matrix $\mathbf{K}$ is always banded, and most of the rows contain similar coefficients, except for the nodes near the boundaries, which are affected by the ghost nodes for imposition of the boundary conditions. For instance, a sample coefficient matrix $\mathbf{K}$ can be in the following form:

$$\begin{bmatrix} 1 & 0.5 & 0.7 & & & & & & & & \\ 4 & 3 & 5 & 1 & & & & & & & \\ 1 & 7 & 8 & 3 & 2 & & & & & & \\ & 1 & 7 & 8 & 3 & 2 & & & & & \\ & & 1 & 7 & 8 & 3 & 2 & & & & \\ & & & \ddots & \ddots & \ddots & \ddots & \ddots & & & \\ & & & & 1 & 7 & 8 & 3 & 2 & & \\ & & & & & 1 & 7 & 8 & 3 & 2 & \\ & & & & & & 1 & 7 & 8 & 3 & 2 \\ & & & & & & & 1 & 5 & 3 & 4 \\ & & & & & & & & 0.7 & 0.5 & 1 \end{bmatrix} \tag{3.2.29}$$

3.2.2.5 Richardson Approximation

Assume two FDM solutions u_1 and u_2 (of orders 2 or 4) are available based on n_1 and n_2 number of nodes, respectively. A better solution u can then be obtained by the Richardson approximation (Thomas 1995),

$$u = \frac{n_1^2 u_1 - n_2^2 u_2}{n_1^2 - n_2^2} \tag{3.2.30}$$

3.2.3 FDM for Solving Partial Differential Equations

FDM can be extended to solve partial differential equations. Here, only the central finite difference formulation is briefly presented for a number of important partial differential operators. In two-dimensional problems, a structured grid of points is adopted to represent the domain, as depicted in Figure 3.3.

Beginning with the cFDM of first derivatives of a function $u(x, y)$ with respect to x and y at node (i, j) (note that ',' does not mean a derivative in this section),

$$\left(\frac{\partial u}{\partial x}\right)_{i,j} \approx \left(\frac{\Delta u}{\Delta x}\right)_{i,j} = \frac{1}{2h_x}(u_{i+1,j} - u_{i-1,j}) \tag{3.2.31}$$

$$\left(\frac{\partial u}{\partial y}\right)_{i,j} \approx \left(\frac{\Delta u}{\Delta y}\right)_{i,j} = \frac{1}{2h_y}(u_{i,j+1} - u_{i,j-1}) \tag{3.2.32}$$

and the following cFDM approximations can be derived:

$$\left(\frac{\partial u}{\partial x} + \frac{\partial u}{\partial y}\right)_{i,j} \approx \frac{1}{2h_x}(-u_{i-1,j} + u_{i+1,j}) + \frac{1}{2h_y}(-u_{i,j-1} + u_{i,j+1}) \tag{3.2.33}$$

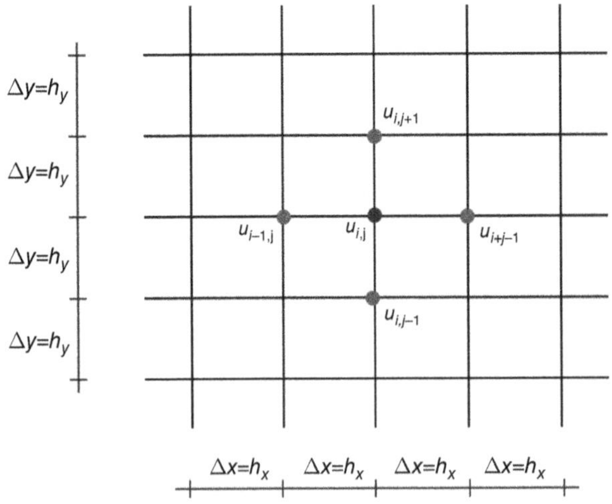

Figure 3.3 A sample structured grid of points for the FDM solution of 2D partial differential equations.

$$\left(\frac{\partial^2 u}{\partial x^2}\right)_{i,j} \approx \frac{1}{h_x^2}(u_{i-1,j} - 2u_{i,j} + u_{i+1,j}) \tag{3.2.34}$$

$$\left(\frac{\partial^2 u}{\partial y^2}\right)_{i,j} \approx \frac{1}{h_y^2}(u_{i,j-1} - 2u_{i,j} + u_{i,j+1}) \tag{3.2.35}$$

$$\left(\frac{\partial^2 u}{\partial x \partial y}\right)_{i,j} \approx \frac{1}{4h_x h_y}(u_{i-1,j-1} - u_{i-1,j+1} - u_{i+1,j-1} + u_{i+1,j+1}) \tag{3.2.36}$$

and for the ∇^2 operator,

$$(\nabla^2 u)_{i,j} \approx \frac{1}{h_x^2}(u_{i-1,j} - 2u_{i,j} + u_{i+1,j}) + \frac{1}{h_y^2}(u_{i,j-1} - 2u_{i,j} + u_{i,j+1}) \tag{3.2.37}$$

which is simplified for equal grid sizes $h_x = h_y = h$,

$$(\nabla^2 u)_{i,j} \approx \frac{1}{h^2}(u_{i-1,j} + u_{i,j-1} + u_{i+1,j} + u_{i,j+1} - 4u_{i,j}) \tag{3.2.38}$$

Finally, for the ∇^4 operator with equal grid sizes h,

$$\begin{aligned}\left(\nabla^4 u\right)_{i,j} \approx \frac{1}{h^4}\Big[&\left(u_{i-2,j} + u_{i,j-2} + u_{i+2,j} + u_{i,j+2}\right) \\ &+2\left(u_{i-1,j-1} + u_{i+1,j-1} + u_{i+1,j+1} + u_{i-1,j+1}\right) \\ &-8\left(u_{i-1,j} + u_{i,j-1} + u_{i+1,j} + u_{i,j+1}\right) + 20u_{i,j}\Big]\end{aligned} \tag{3.2.39}$$

$$\begin{aligned}\left(\frac{\partial^4 u}{\partial x^2 \partial y^2}\right)_{i,j} \approx \frac{1}{h^4}\Big[&(u_{i-1,j-1} + u_{i+1,j-1} + u_{i+1,j+1} + u_{i-1,j+1}) \\ &-2(u_{i-1,j} + u_{i,j-1} + u_{i+1,j} + u_{i,j+1}) + 4u_{i,j}\Big]\end{aligned} \tag{3.2.40}$$

The method can be generalized to non-uniform grids, polar systems and inclined (non-orthogonal) coordinates, which are complicated and far less efficient, especially for imposition of boundary conditions in general geometries (LeVeque 2007). Instead, the meshless methods, which are far more efficient, flexible and accurate in handling complex problems, are recommended (see Section 3.7).

3.3 Finite Volume Method (FVM)

The finite volume method (FVM) is a numerical approach for solving the governing equations of a system by transforming the terms of volume integrals into the surface integrals. The concept of using the divergence theorem for such a transformation is frequently adopted in various numerical methods. In FVM, however, it is employed in problems with the divergence terms in the governing equations, to allow for their evaluation as a surface flux on

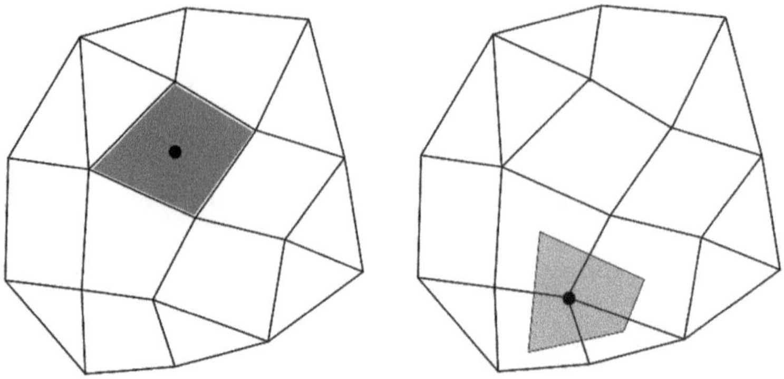

Figure 3.4 Cell and vertex centred FVM algorithms. *Source:* Reproduced from Moukalled et al. 2015/Springer Nature.

each small finite volume that represents the domain geometry around a node (Figure 3.4). FVM acts accurately on an average sense over the finite volumes (Fallah et al. 2000).

The finite volume method can be readily implemented on unstructured meshes, which are frequently employed in computational field problems and fluid dynamics. Various forms of FVM are available. Here, a brief review on the basic concepts of FVM is presented according to Moukalled et al. (2015).

Beginning with the general form of the diffusion/convection differential equation for a physical quantity $f(\mathbf{x},t)$ (Eymard et al., 2019),

$$\frac{\partial f(\mathbf{x},t)}{\partial t} = \underbrace{\boldsymbol{\nabla}\cdot\left[D(\mathbf{x},t)\boldsymbol{\nabla}f(\mathbf{x},t)\right]}_{\text{diffusion}} - \underbrace{\boldsymbol{\nabla}\cdot(\rho(\mathbf{x},t)\boldsymbol{v}(\mathbf{x},t)f(\mathbf{x},t))}_{\text{convection}} + \underbrace{S(\mathbf{x},t)}_{\text{source/sink}} \tag{3.3.1}$$

where D represents the diffusion coefficient, $\boldsymbol{v}$ is the velocity that affects f and S describes the source or sink of the physical quantity. The steady state form of Equation (3.3.1) can be written as

$$\boldsymbol{\nabla}\cdot\left[D\boldsymbol{\nabla}f\right] - \boldsymbol{\nabla}\cdot(\rho\boldsymbol{v}f) + S = 0 \tag{3.3.2}$$

and its integral form over the control volume V_C (associated with the finite volume C) is expressed as (Figure 3.5)

$$\int_{V_C} \boldsymbol{\nabla}\cdot(D\boldsymbol{\nabla}f)\mathrm{d}V - \int_{V_C} \boldsymbol{\nabla}\cdot(\rho\boldsymbol{v}f)\mathrm{d}V + \int_{V_C} S\mathrm{d}V = 0 \tag{3.3.3}$$

The first two terms of Equation (3.3.3) can be transformed by the divergence theorem,

$$\int_{V_C} \boldsymbol{\nabla}\cdot(D\boldsymbol{\nabla}f)\mathrm{d}V = \oint_{\Gamma_C} (D\boldsymbol{\nabla}f)\cdot\mathrm{d}\boldsymbol{A} \tag{3.3.4}$$

$$\int_{V_C} \boldsymbol{\nabla}\cdot(\rho\boldsymbol{v}f)\mathrm{d}V = \oint_{\Gamma_C} (\rho\boldsymbol{v}f)\cdot\mathrm{d}\boldsymbol{A} \tag{3.3.5}$$

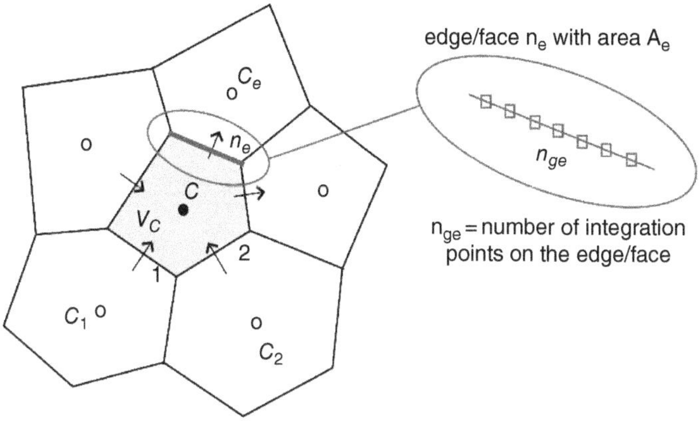

Figure 3.5 Description of the balance form over a finite volume and fluxes across volume edges/faces. *Source:* Reproduced from Moukalled et al. 2015/Springer Nature.

leading to

$$\oint_{\Gamma_C} (D\nabla f)\cdot d\boldsymbol{A} - \oint_{\Gamma_C} (\rho\boldsymbol{v}f)\cdot d\boldsymbol{A} + \int_{V_C} S dV = 0 \tag{3.3.6}$$

where Γ_C is the boundary of the finite volume V_C.

The convection and diffusion fluxes, $\boldsymbol{q}_c$ and $\boldsymbol{q}_d$, respectively, can be defined by

$$\boldsymbol{q}_c = \rho\boldsymbol{v}f \tag{3.3.7}$$

$$\boldsymbol{q}_d = -D\nabla f \tag{3.3.8}$$

and the total flux $\boldsymbol{q}_t$ is

$$\boldsymbol{q}_t = \boldsymbol{q}_c + \boldsymbol{q}_d \tag{3.3.9}$$

Equation (3.3.6) can then be re-written as

$$\oint_{\Gamma_C} \boldsymbol{q}_d \cdot d\boldsymbol{A} + \oint_{\Gamma_C} \boldsymbol{q}_c \cdot d\boldsymbol{A} = \int_{V_C} S dV \tag{3.3.10}$$

or

$$\oint_{\Gamma_C} \boldsymbol{q}_t \cdot d\boldsymbol{A} = \int_{V_C} S dV \tag{3.3.11}$$

which can be interpreted as the equivalence of the total flux $\boldsymbol{q}_t$ and the source flux S.

Numerical computation of Equation (3.3.10) over n_e edges/faces of the control volume (Figure 3.5) can be performed by

$$\sum_{n_e}\left[\int_{A_e} \boldsymbol{q}_d \cdot d\boldsymbol{A}\right] + \sum_{n_e}\left[\int_{A_e} \boldsymbol{q}_c \cdot d\boldsymbol{A}\right] = \int_{V_C} S dV \tag{3.3.12}$$

Integration can then be performed by a conventional integration rule,

$$\int_{A_e} \boldsymbol{q}_c \cdot d\boldsymbol{A} = \int_{A_e} \boldsymbol{q}_c \cdot \boldsymbol{n} dA = \int_{A_e} (\rho \boldsymbol{v} f \cdot \boldsymbol{n}) dA = \sum_{g=1}^{n_{ge}} \left[w_g (\rho \boldsymbol{v} f \cdot \boldsymbol{n})_g A_g \right] \tag{3.3.13}$$

$$\int_{A_e} \boldsymbol{q}_d \cdot d\boldsymbol{A} = \int_{A_e} \boldsymbol{q}_d \cdot \boldsymbol{n} dA = \int_{A_e} (-D\nabla f \cdot \boldsymbol{n}) dA = \sum_{g=1}^{n_{ge}} \left[w_g (-D\nabla f \cdot \boldsymbol{n})_g A_g \right] \tag{3.3.14}$$

$$\int_{V_C} S dV = \sum_{g=1}^{n_{gc}} \left[w_{gc} S_{gc} \right] V_{Cg} \tag{3.3.15}$$

where $\boldsymbol{n}$ is the normal vector of an edge/face, n_{ge} is the number of integration points on all edges/faces of the control volume and n_{gc} is the number of integration points on the control volume V_C, usually assumed to be one.

Computation of the mean value of flux over the cell is of the order of accuracy $\mathcal{O}(|\mathbf{x} - \mathbf{x}_e|^2)$, Similarly, it can be discussed that the corresponding mean values of the convection and diffusion fluxes over an edge e of a cell C are of order of accuracy $\mathcal{O}(|\mathbf{x} - \mathbf{x}_e|^2)$ (Moukalled et al. 2015).

For a full discussion on the finite volume method, refer to Moukalled et al. (2015).

3.4 Finite Element Method (FEM)

The finite element method is undoubtably the most frequently used numerical approach for analysis of a wide range of engineering problems, from linear elastic applications to very complicated nonlinear elastic-viscoplastic-damage problems in large strain regimes. It covers fluid structure interactions and hydrothermal porous interactions and is frequently adopted for various modellings of biomechanical applications. FEM can be adopted for simulations of physical problems in macroscales and even in micro- and nanoscales and is one of the main methods in multiscale simulations.

There is no need to review the development history of FEM, as it has been fully developed in various disciplines and has become a mature numerical technique available as commercial general purpose finite element packages. For a complete discussion on FEM, refer to the text books by Zienkiewicz et al. (2005) and Bathe (1982).

3.4.1 Basics of FEM Interpolation

In the finite element method, the physical domain of an engineering problem is discretized by a mesh of finite elements, as typically shown in Figure 3.6 for a biomechanical application. The finite element mesh is generated inside the domain of the problem and should conform with the boundaries of the domain. The elements are connected by common nodal points (nodes) and the number and type of the finite elements are selected based on the required accuracy and the computational software and hardware resources.

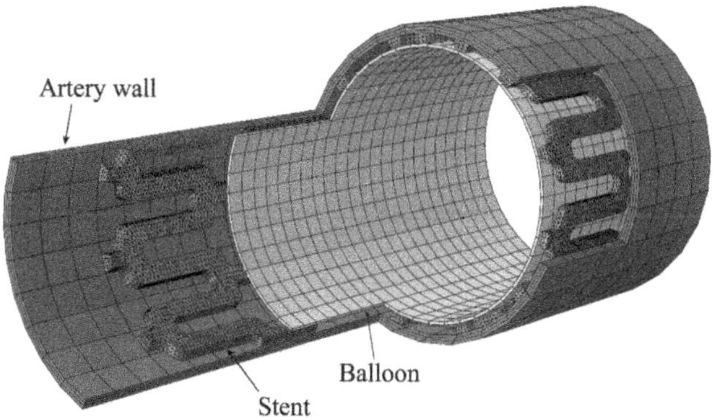

Figure 3.6 A typical finite element modelling of the biomechanical balloon-stent-artery simulation. *Source:* Reproduced from Torabizadeh and Mohammadi 2022a/the University of Tehran.

In an isoparametric finite element formulation, the coordinates $\mathbf{x}^{\mathrm{T}} = (x, y, z)$ are interpolated from the nodal values $\bar{\mathbf{x}}^{\mathrm{T}} = (\bar{x}, \bar{y}, \bar{z})$:

$$\mathbf{x} = \sum_{j=1}^{n} N_j(\mathbf{x})\bar{\mathbf{x}}_j \tag{3.4.1}$$

where n is the number of nodes for each finite element. Similarly, the displacement field $\boldsymbol{u}^{\mathrm{T}} = (u_x, u_y, u_z)$ can be written in terms of the nodal displacements $\bar{\boldsymbol{u}}^{\mathrm{T}} = (\bar{u}_x, \bar{u}_y, \bar{u}_z)$:

$$\boldsymbol{u}(\mathbf{x}) = \sum_{j=1}^{n} N_j(\mathbf{x})\bar{\boldsymbol{u}}_j \tag{3.4.2}$$

where N_j are the nodal finite element shape functions, as discussed in Section 3.4.2.

The concept of finite element method is to assume a predefined approximation of the main field variable (for example, the displacement $\boldsymbol{u}$) inside each element based on the nodal values ($\bar{\boldsymbol{u}}_i$). The overall solution is then obtained by discretization of the weak form solution of the governing equations with the corresponding essential and traction boundary conditions.

3.4.2 FEM Basis Functions/Shape Functions

Interpolation inside a finite element is usually performed by the shape functions N_j, which are known functions for various conventional finite elements:

$$\boldsymbol{u}^h(\mathbf{x}) = \mathbf{N}(\mathbf{x})\bar{\boldsymbol{u}} = \sum_{j=1}^{n} N_j(\mathbf{x})\bar{\boldsymbol{u}}_j \tag{3.4.3}$$

where $\bar{\boldsymbol{u}}_i$ are the known nodal values.

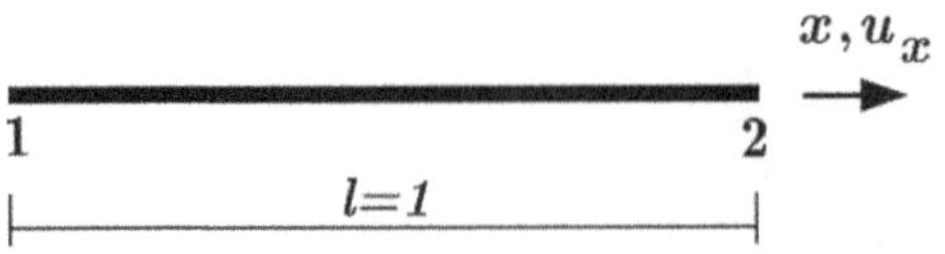

Figure 3.7 One-dimensional linear element.

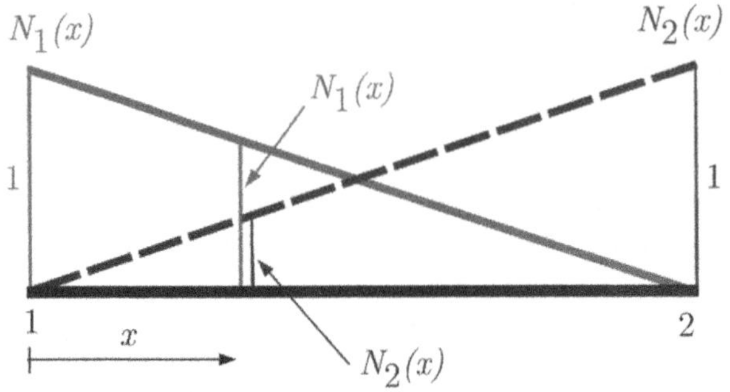

Figure 3.8 Shape functions of the one-dimensional linear finite element.

Alternatively, the element interpolation may be defined in terms of the basis functions p_j:

$$\boldsymbol{u}^h(\mathbf{x}) = \mathbf{p}^{\mathrm{T}}(\mathbf{x})\boldsymbol{a} = \sum_{j=1}^{n} p_j(\mathbf{x})a_j \tag{3.4.4}$$

where a_j are the unknown constants.

To discuss the subject in more detail, the one-dimensional linear element of Figure 3.7 is considered.

The shape functions are well defined in this simple problem (see Figure 3.8):

$$\begin{cases} N_1(x) = 1 - x \\ N_2(x) = x \end{cases} \tag{3.4.5}$$

and the finite element approximation at any point x inside the element becomes

$$u^h(x) = \sum_{j=1}^{2} N_j(x)\bar{u}_j = N_1(x)\bar{u}_1 + N_2(x)\bar{u}_2 \tag{3.4.6}$$

Now, if the basis function formulation is adopted, the element requires two basis function components,

$$\mathbf{p}^T = \{1, x\} \tag{3.4.7}$$

and the FEM approximation u^h at point x becomes

$$u^h(x) = \sum_{j=1}^{2} p_j(x)a_j = p_1(x)a_1 + p_2(x)a_2 = (1)a_1 + (x)a_2 = a_1 + a_2 x \tag{3.4.8}$$

The values of interpolation must match the nodal values:

$$\left.\begin{aligned} \bar{u}_1 &= u^h(x=0) = a_1 \\ \bar{u}_2 &= u^h(x=1) = a_1 + a_2 \end{aligned}\right\} \rightarrow \left\{\begin{aligned} a_1 &= \bar{u}_1 \\ a_2 &= \bar{u}_2 - \bar{u}_1 \end{aligned}\right. \tag{3.4.9}$$

and the finite element formulation:

$$u^h(x) = a_1 + a_2 x = \bar{u}_1 + (\bar{u}_2 - \bar{u}_1)x = \underbrace{(1-x)}_{N_1(x)}\,\bar{u}_1 + \underbrace{(x)}_{N_2(x)}\,\bar{u}_2 \tag{3.4.10}$$

$$u^h(x) = N_1(x)\bar{u}_1 + N_2(x)\bar{u}_2 \tag{3.4.11}$$

which is similar to the conventional finite element approximation derived directly from the shape functions.

Generally, the shape functions can be derived from the basis functions $\mathbf{p}$ by

$$N_j(\mathbf{x}) = \mathbf{p}^{\mathrm{T}}(\mathrm{x})\mathbf{P}_j^{-1} \tag{3.4.12}$$

where

$$\mathbf{P}^{\mathrm{T}} = \left[\mathbf{p}^{\mathrm{T}}(\mathbf{x}_1) \quad \mathbf{p}^{\mathrm{T}}(\mathbf{x}_2) \quad \cdots \quad \mathbf{p}^{\mathrm{T}}(\mathbf{x}_n)\right] \tag{3.4.13}$$

Many primary characteristics of the finite element formulations, including both deficiencies or advantages, can be studied by examining the basis function formulation.

3.4.3 Properties of the Finite Element Interpolation

The finite element shape functions ensure the following important characteristics.

3.4.3.1 Continuity

The continuity of the displacement filed inside each element is trivial. The displacement field continuity in between the adjacent elements sharing a common edge/face is guaranteed, leading to a consistent and continuous displacement field over the whole model. This is not necessarily the case for the derivatives of the displacement field, which may result in discontinuous strain or stress fields across the element edges/faces.

3.4.3.2 Partition of Unity (PU)

The finite element shape functions satisfy the partition of unity (PU) condition

$$\sum_{j=1}^{n} N_j(\mathbf{x}) = 1 \tag{3.4.14}$$

This is important in generalization of FEM to more advanced solutions such as the extended finite element method, which will be discussed in Section 3.5. As a result of PU property, the finite element shape functions allow for any arbitrary function $\psi(\mathbf{x})$ to be reproduced,

$$\sum_{j=1}^{n} N_j(\mathbf{x})\psi(\mathbf{x}) = \psi(\mathbf{x}) \tag{3.4.15}$$

3.4.3.3 Kronecker Delta

The Kronecker delta property ensures that the finite element approximation is an interpolation, meaning that the finite element formulation reproduces the nodal values at $\mathbf{x}_j$:

$$N_j(\mathbf{x}_j) = \delta_{ij} = \begin{cases} 1, & i = j \\ 0, & i \neq j \end{cases} \tag{3.4.16}$$

$$\boldsymbol{u}(\mathbf{x}_j) = \bar{\boldsymbol{u}}_j \tag{3.4.17}$$

As a result, imposition of the essential boundary conditions in FEM is straightforward and does not require any additional constraint enforcement technique.

3.4.4 Physical and Parametric Coordinate Systems

While descriptions of the basis or shape functions can be performed in general physical (x, y, z) space, it is preferable to define them in the parametric space (ξ, η, ζ), which is uniquely defined independent of the physical coordinates. Mapping in between the physical and parametric spaces is required to transform the necessary variables of the finite element formulation (Figure 3.9).

3.4.5 Main Types of Finite Elements

3.4.5.1 Linear and Quadratic Triangular Elements

Figure 3.10 illustrates typical triangular elements that are frequently used in practical finite element simulations.

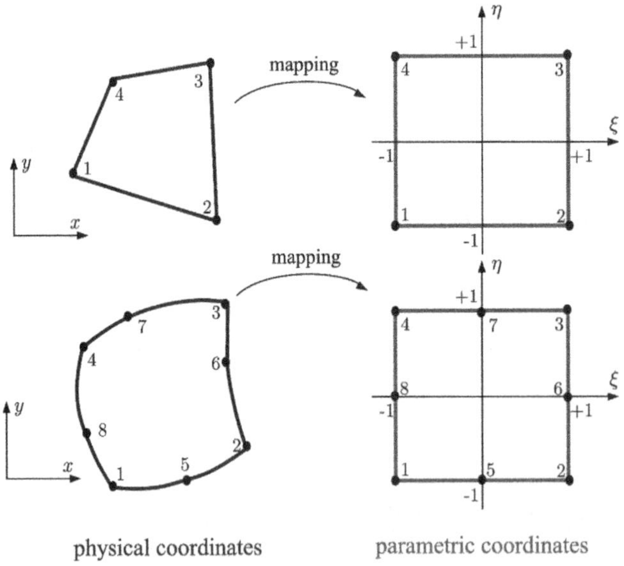

Figure 3.9 Mapping of quadrilateral finite elements from the physical space to the parametric coordinates.

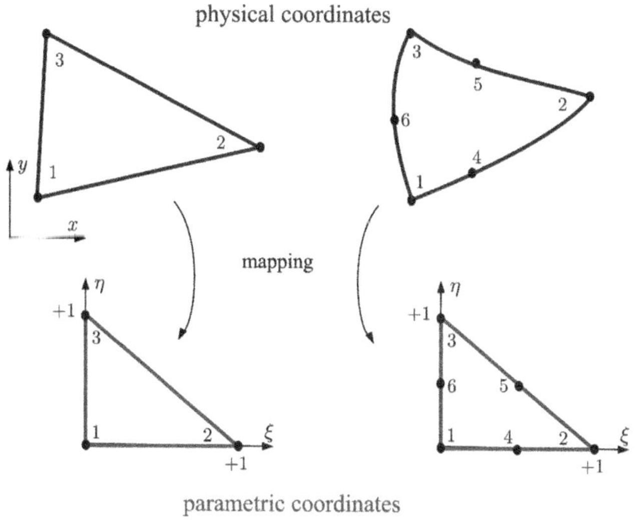

Figure 3.10 Triangular finite elements in physical and parametric coordinates.

The basis function **p** for linear (three-node) and quadratic (six-node) triangular elements, depicted in Figure 3.10, is defined as

$$\begin{cases} \text{Linear element:} & \mathbf{p}^{\mathrm{T}} = \{1, \xi, \eta\} \\ \text{Quadratic element:} & \mathbf{p}^{\mathrm{T}} = \{1, \xi, \eta, \xi^2, \xi\eta, \eta^2\} \end{cases} \tag{3.4.18}$$

The shape functions for the three-node element can be defined in terms of the area coordinates. However, it is preferable to define them in the local coordinate system (Figure 3.10):

$$\begin{cases} N_1 = \xi \\ N_2 = \eta \\ N_3 = 1 - \xi - \eta \end{cases} \tag{3.4.19}$$

and for the six-node quadratic triangular element as

$$\begin{cases} N_1 = 2(1-\xi-\eta)\left(\dfrac{1}{2}-\xi-\eta\right) \\ N_2 = 2\xi\left(\xi-\dfrac{1}{2}\right) \\ N_3 = 2\eta\left(\eta-\dfrac{1}{2}\right) \\ N_4 = 4\xi(1-\xi-\eta) \\ N_5 = 4\xi\eta \\ N_6 = 4\eta(1-\xi-\eta) \end{cases} \tag{3.4.20}$$

3.4.5.2 Bilinear Quadrilateral Lagrangian Element

A typical four-node quadrilateral element is illustrated in the physical and parametric coordinate systems in Figure 3.9.

The basis function $\mathbf{p}$ for the four-node quadrilateral element is defined as

$$\mathbf{p}^{\mathrm{T}} = \{1, \xi, \eta, \xi\eta\} \tag{3.4.21}$$

and the corresponding shape functions are

$$N_j = \frac{1}{4}(1+\xi_j\xi)(1+\eta_j\eta),\ j = 1-4 \tag{3.4.22}$$

3.4.5.3 Quadrilateral Serendipity and Lagrange Elements

Higher order quadrilateral elements are adopted in the finite element simulations, where higher orders of continuity and approximation are required. They are usually adopted in the forms of the Serendipity and Lagrange elements, as depicted in Figure 3.11.

The basis function $\mathbf{p}$ for the eight-node quadrilateral serendipity element (Figure 3.11) is defined as

$$\mathbf{p}^{\mathrm{T}} = \left\{1, \xi, \eta, \xi^2, \xi\eta, \eta^2, \xi^2\eta, \xi\eta^2\right\} \tag{3.4.23}$$

and the corresponding shape functions are

$$\begin{cases} N_j = \frac{1}{4}\left(1+\xi_j\xi\right)\left(1+\eta_j\eta\right)\left(\xi_j\xi+\eta_j\eta-1\right), & \text{corner nodes} \\ N_j = \frac{1}{2}\left(1-\xi^2\right)\left(1+\eta_j\eta\right), & \xi_j = 0, j = 5, 7 \\ N_j = \frac{1}{2}\left(1+\xi_j\xi\right)\left(1-\eta^2\right), & \eta_j = 0, j = 6, 8 \end{cases} \tag{3.4.24}$$

For the nine-node Lagrangian quadrilateral element, depicted in Figure 3.11, the basis function $\mathbf{p}$ is defined as

$$\mathbf{p}^{\mathrm{T}} = \left\{1, \xi, \eta, \xi^2, \xi\eta, \eta^2, \xi^2\eta, \xi\eta^2, \xi^2\eta^2\right\} \tag{3.4.25}$$

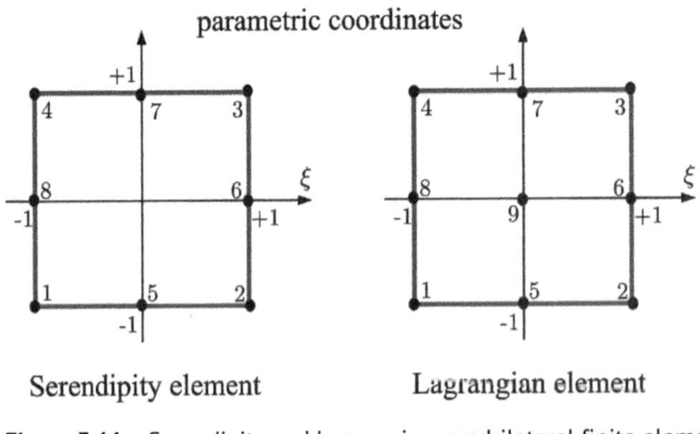

Figure 3.11 Serendipity and Lagrangian quadrilateral finite elements in the parametric coordinates.

and the shape functions are

$$\begin{cases} N_j = \frac{1}{4}\xi\eta\left(\xi+\xi_j\right)\left(\eta+\eta_j\right), & \text{corner nodes} \\ N_j = \left(1-\xi^2\right)\left(1-\eta^2\right), & \text{centre node} \\ N_j = \frac{1}{2}\eta\left(1-\xi^2\right)\left(\eta+\eta_j\right), & \xi_j=0,\ j=5,7 \\ N_j = \frac{1}{2}\xi\left(\xi+\xi_j\right)\left(1-\eta^2\right), & \eta_j=0,\ j=6,8 \end{cases} \tag{3.4.26}$$

3.4.5.4 3D Eight-Node Brick Element

The basis function **p** for the eight-node 3D brick element (Figure 3.12) is defined as

$$\mathbf{p}^{\mathrm{T}} = \{1, \xi, \eta, \zeta, \xi\eta, \eta\zeta, \zeta\xi, \xi\eta\zeta\} \tag{3.4.27}$$

and the corresponding shape functions are

$$N_j = \frac{1}{8}(1+\xi_j\xi)(1+\eta_j\eta)(1+\zeta_j\zeta) \tag{3.4.28}$$

3.4.6 Governing Equations of the Boundary Value Problem

Figure 3.13 illustrates a body Ω that is in the static state of equilibrium with the applied body force $\boldsymbol{f}^b$, the external traction $\boldsymbol{f}^t$ on Γ_t and the prescribed displacement $\bar{\boldsymbol{u}}$ on Γ_u.

The strong form of the equilibrium equation, in terms of the stress tensor $\boldsymbol{\sigma}$, can be written as

$$\nabla.\boldsymbol{\sigma} + \boldsymbol{f}^b = \mathbf{0} \quad \text{in } \Omega \tag{3.4.29}$$

with the following displacement and traction boundary conditions:

$$\boldsymbol{u} = \bar{\boldsymbol{u}} \quad \text{on } \Gamma_u \tag{3.4.30}$$

$$\boldsymbol{\sigma} \cdot \mathbf{n} = \boldsymbol{f}^t \quad \text{on } \Gamma_t \tag{3.4.31}$$

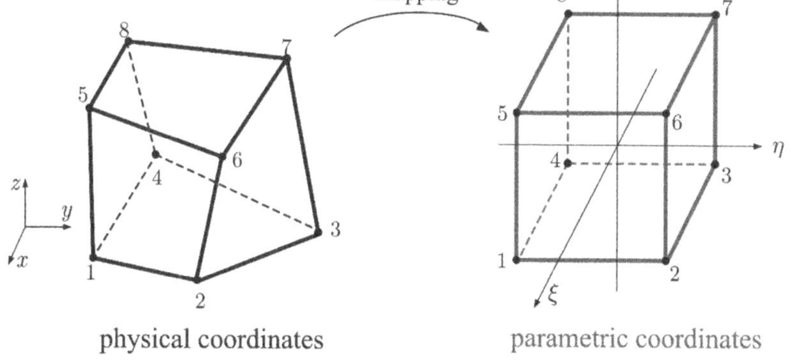

Figure 3.12 3D brick element in the physical and parametric coordinates.

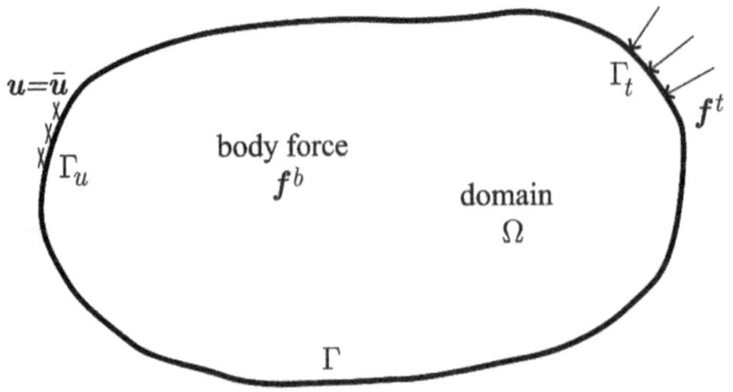

Figure 3.13 A body subjected to body force, tractions and displacement boundary conditions.

The governing weak form of the boundary value problem can then be defined as

$$\int_{\Omega} \boldsymbol{\sigma} \cdot \delta\boldsymbol{\varepsilon}\, \mathrm{d}\Omega = \int_{\Omega} \boldsymbol{f}^{b} \cdot \delta\boldsymbol{u}\, \mathrm{d}\Omega + \int_{\Gamma_t} \boldsymbol{f}^{t} \cdot \delta\boldsymbol{u}\, \mathrm{d}\Gamma \tag{3.4.32}$$

Discretization of Equation (3.4.32) can be performed by the FEM approximation (3.4.2) to define the displacement approximation $\boldsymbol{u}$ for a point $\mathbf{x}$ inside the finite element as

$$\boldsymbol{u} = \boldsymbol{u}^{h}(\mathbf{x}) = \sum_{j=1}^{n} N_j(\mathbf{x})\boldsymbol{u}_j \tag{3.4.33}$$

or

$$\boldsymbol{u} = \mathbf{N}\bar{\mathbf{U}} \tag{3.4.34}$$

where $\bar{\mathbf{U}}$ is the vector of classical FEM nodal degrees of freedom.

The strain field is computed directly from Equation (3.4.33):

$$\boldsymbol{\varepsilon} = \sum_{j=1}^{n} \mathbf{B}_j \mathbf{u}_j \tag{3.4.35}$$

or in the compact form of

$$\boldsymbol{\varepsilon} = \mathbf{B}\bar{\mathbf{U}} \tag{3.4.36}$$

where $\mathbf{B} = \nabla\mathbf{N}$ is the matrix of derivatives of the shape functions $\mathbf{N}$,

$$\mathbf{B}_j = \begin{bmatrix} N_{j,x} & 0 & 0 \\ 0 & N_{j,y} & 0 \\ 0 & 0 & N_{j,z} \\ 0 & N_{j,z} & N_{j,y} \\ N_{j,z} & 0 & N_{j,x} \\ N_{j,y} & N_{j,x} & 0 \end{bmatrix} \tag{3.4.37}$$

and the chain rule is invoked to determine the coefficients of $\mathbf{B}_j$:

$$\left\{\begin{array}{c}\frac{\partial N}{\partial x}\\ \frac{\partial N}{\partial y}\\ \frac{\partial N}{\partial z}\end{array}\right\} = \mathbf{J}^{-1}\left\{\begin{array}{c}\frac{\partial N}{\partial \xi}\\ \frac{\partial N}{\partial \eta}\\ \frac{\partial N}{\partial \zeta}\end{array}\right\} \tag{3.4.38}$$

where $\mathbf{J}$ is the Jacobian matrix

$$\mathbf{J} = \begin{bmatrix}\frac{\partial x}{\partial \xi} & \frac{\partial y}{\partial \xi} & \frac{\partial z}{\partial \xi}\\ \frac{\partial x}{\partial \eta} & \frac{\partial y}{\partial \eta} & \frac{\partial z}{\partial \eta}\\ \frac{\partial x}{\partial \zeta} & \frac{\partial y}{\partial \zeta} & \frac{\partial z}{\partial \zeta}\end{bmatrix} \tag{3.4.39}$$

The stress tensor can be obtained from the strain tensor using a constitutive law. For a simple linear elastic constitutive law,

$$\sigma = \mathbf{D}\varepsilon \tag{3.4.40}$$

and

$$\sigma = \mathbf{D}\mathbf{B}\bar{\mathbf{U}} \tag{3.4.41}$$

where $\mathbf{D}$ is the material modulus (2.2.5).

Discretization of Equation (3.4.32) results in the following set of equations:

$$\mathbf{K}\bar{\mathbf{U}} = \mathbf{f} \tag{3.4.42}$$

$\bar{\mathbf{U}}$ is the vector of nodal degrees of freedom,

$$\bar{\mathbf{U}}^{\mathrm{T}} = \{\boldsymbol{u}_1 \quad \cdots \quad \boldsymbol{u}_n\} \tag{3.4.43}$$

and the global stiffness matrix $\mathbf{K}$ and force vector $\mathbf{f}$ are calculated by the assembling operator $\cap$ on the corresponding stiffness matrix $\mathbf{K}^e$ and force vector $\mathbf{f}^e$ of each element e,

$$\mathbf{K} = \bigcap_{e=1}^{m} \mathbf{K}^e \tag{3.4.44}$$

$$\mathbf{f} = \bigcap_{e=1}^{m} \mathbf{f}^e \tag{3.4.45}$$

where

$$\mathbf{K}_{ij}^e = \int_{\Omega^e} \mathbf{B}_i^{\mathrm{T}} \mathbf{D} \mathbf{B}_j \mathrm{d}\Omega \tag{3.4.46}$$

$$\mathbf{f}_i^e = \int_{\Gamma_t} N_i^{\mathrm{T}} \boldsymbol{f}^t \mathrm{d}\Gamma + \int_{\Omega^e} N_i^{\mathrm{T}} \boldsymbol{f}^b \mathrm{d}\Omega \tag{3.4.47}$$

Integral forms Equations (3.4.46) and (3.4.47) are defined in the physical coordinates (x, y, z). Therefore, they have to be transformed to the parametric coordinates (ξ, η, ζ). For instance, the stiffness matrix **K** is computed from

$$\mathbf{K}_{ij}^e = \int_x \int_y \int_z \mathbf{B}_i(x, y, z)^{\mathrm{T}} \mathbf{D} \mathbf{B}_j(x, y, z) \,\mathrm{d}x \,\mathrm{d}y \,\mathrm{d}z \tag{3.4.48}$$

$$\mathbf{K}_{ij}^e = \int_{\xi=-1}^{1} \int_{\eta=-1}^{1} \int_{\zeta=-1}^{1} (\det \mathbf{J}) \, \mathbf{B}_i(\xi, \eta, \zeta)^{\mathrm{T}} \mathbf{D} \mathbf{B}_j \, (\xi, \eta, \zeta) \, \mathrm{d}\xi \, \mathrm{d}\eta \, \mathrm{d}\zeta \tag{3.4.49}$$

which should then be numerically evaluated by a standard integration scheme.

3.4.7 Numerical Integration

Equations (3.4.46) to (3.4.47) must be accurately integrated in order to provide sufficient accuracy for solving the discretized FEM Equation (3.4.42). The well-developed Gauss quadrature rule is adopted for numerical evaluation of those integrals in the parametric coordinate system,

$$I = \int_{-1}^{+1} f(\xi) d\xi = \sum_{g=1}^{ng} w_g f(\xi_g) \tag{3.4.50}$$

where the positions of Gauss points ξ_g and the corresponding weights w_g are well defined for different Gauss integration orders, as defined in Table 3.1 for one-dimensional integrations. The order of Gauss integration is chosen based on the type and order of the integrand $f(\xi)$ to provide a balance between the most accuracy of numerical integration and the least computational costs.

Similarly, for two- and three-dimensional quadrilateral elements:

$$I = \int_{\xi=-1}^{+1} \int_{\eta=-1}^{+1} f(\xi, \eta) \mathrm{d}\xi \, \mathrm{d}\eta = \sum_{g1=1}^{ng1} \sum_{g2=1}^{ng2} w_{g1} w_{g2} f(\xi_{g1}, \eta_{g2}) \tag{3.4.51}$$

$$I = \int_{\xi=-1}^{+1} \int_{\eta=-1}^{+1} \int_{\zeta=-1}^{+1} f(\xi, \eta, \zeta) \mathrm{d}\xi \, \mathrm{d}\eta \, \mathrm{d}\eta = \sum_{g1=1}^{ng1} \sum_{g2=1}^{ng2} \sum_{g3=1}^{ng3} w_{g1} w_{g2} w_{g3} f(\xi_{g1}, \eta_{g2}, \zeta_{g2}) \tag{3.4.52}$$

Figure 3.14 shows the positions of sample Gauss quadrature points for a number of finite elements.

Table 3.1 Basic specifications of the Gauss integration rule on a one-dimensional [−1,1] domain.

Order	ng	ξ_g	w_g
1	1	0	2
3	2	−0.577350	1
		0.577350	1
5	3	−0.774597	0.555556
		0	0.888889
		0.774597	0.555556
7	4	−0.861136	0.347855
		−0.3399981	0.652145
		0.3399981	0.652145
		0.861136	0.347855

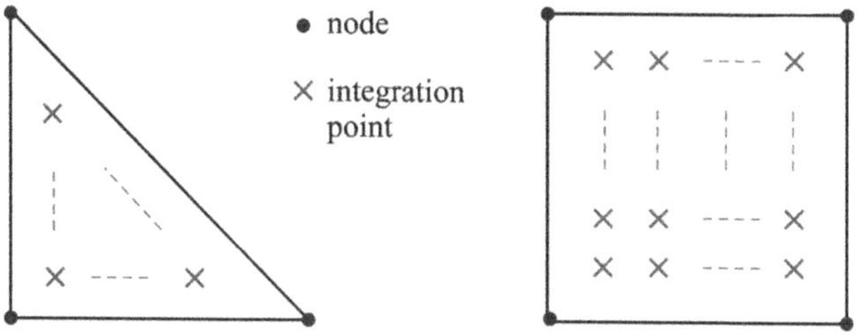

Figure 3.14 Positions of quadrature points for different finite elements.

3.5 Extended Finite Element Method (XFEM)

This section, which is a compact extraction from a similar chapter in Mohammadi (2012a), is presented to briefly review various theoretical and computational bases of the extended finite element method (XFEM).

3.5.1 A Review of XFEM Development

After the initial idea of a minimal remeshing finite element method for crack growth problems by Belytschko and Black (1999), Moës et al. (1999) improved the so-called extended finite element method (XFEM) and adopted the concept of enriched approximation to represent the entire crack. The work was further advanced to 2D elasticity and plate problems by Dolbow (1999) and Dolbow et al. (2000a, 2000b, 2000c) using the partition of unity concept and the asymptotic near tip fields to represent the singular stress state at the crack tip and the Heaviside function to take into account the displacement discontinuity across the crack edges.

Following the initial success of the method, XFEM has been extensively applied in various engineering applications in the past two decades, and has become the standard technique for analysis of different discontinuity or singular problems. For instance, some of the following major developments, among others, can be distinguished:

- Cohesive cracks by Moës and Belytschko (2002a).
- Crack analysis in composites by Asadpoure et al. (2006, 2007) and Asadpoure and Mohammadi (2007) by introducing new orthotropic enrichment functions.
- Localisation problems by Daneshyar and Mohammadi (2011, 2012a, 2012b) who developed a strong tangential discontinuity model for XFEM simulation of shear band initiation and propagation.
- Sliding contacts by Ebrahimi (2012) proposing a numerical approach for determining the order of singularity of a sliding contact corner.
- Large deformation regimes by Rashetnia and Mohammadi (2015), presenting various aspects of geometrically nonlinear XFEM analyses.
- Damage mechanics by Hatefi and Mohammadi (2012).
- Dynamic fracture analysis by Motamedi and Mohammadi (2010a; 2010b, 2012) who developed new orthotropic enrichment functions for moving cracks and Parchei et al. (2011) who used the dynamic XFEM combined with a contact approach to simulate near field solution for various induced fault rupture problems.
- Time domain discontinuity by introduction of a locally enriched space–time extended finite element method by Chessa and Belytschko (2004, 2006) and Afshar et al. (2018) for simulating various dynamic fracture problems.
- Thermomechanical analysis of orthotropic FGMs by Bayesteh and Mohammadi (2013).
- Orthotropic bimaterial interfaces by Esna Ashari and Mohammadi (2009; 2010, 2011a, 2011b, 2012) and Afshar et al. (2016).
- Phase change in cracked shape memory alloys (SMA) by Ahmadian et al. (2015), Hatefi et al. (2015), Hashemi et al. (2015) and Afshar et al. (2015b).
- SMA fibre bridging and healing of cracked media (Karimi et al. 2019).
- Dynamic bimaterial interface cracks (Afshar et al. 2018).

For a complete discussion on XFEM refer to the textbooks by Mohammadi (2008, 2012a).

3.5.2 Partition of Unity

The concept of partition of unity (PU) can be defined as a set of m functions $f_k(\mathbf{x})$ within a domain Ω_{pu} such that (Melenk and Babuska 1996)

$$\sum_{k=1}^{m} f_k(\mathbf{x}) = 1 \tag{3.5.1}$$

Consequently, any arbitrary function $\psi(\mathbf{x})$ can be reproduced by the partition of unity functions $f_k(\mathbf{x})$,

$$\sum_{k=1}^{m} f_k(\mathbf{x})\psi(\mathbf{x}) = \psi(\mathbf{x}) \tag{3.5.2}$$

The most commonly used PU functions are the standard finite element shape functions, N_j,

$$\sum_{j=1}^{n} N_j(\mathbf{x}) = 1 \tag{3.5.3}$$

where n is the number of nodes for each finite element.

3.5.3 Enrichments

Enrichment is a numerical way to enhance the reproduction property of an approximation based on an *a priori* solution of a problem. In fracture mechanics, enrichment is aimed at increasing the accuracy of approximation near a crack by using the analytical near crack tip solutions to derive the enrichment functions.

3.5.4 Signed Distance Function

The signed distance function $\xi(\mathbf{x})$ is defined as

$$\xi(\mathbf{x}) = \underbrace{\min \| \mathbf{x} - \mathbf{x}_\Gamma \|}_{\mathbf{x}_\Gamma \in \Gamma} \operatorname{sign}(\boldsymbol{n} \cdot (\mathbf{x} - \mathbf{x}_\Gamma)) \tag{3.5.4}$$

where $\boldsymbol{n}$ is the unit normal vector, $\mathbf{x}_\Gamma$ is the normal projection of $\mathbf{x}$ on Γ (Figure 3.15) and $d = \| \mathbf{x} - \mathbf{x}_\Gamma \|$ is the distance from a point $\mathbf{x}$ to an interface Γ.

3.5.5 XFEM Approximation for Cracked Elements

In the extended finite element modelling, the conventional finite element mesh is generated regardless of the presence of crack or discontinuity. Then, necessary additional degrees of freedom are added to the main finite element model in a selected set of nodes near the crack.

The XFEM displacement approximation for a point $\mathbf{x}$ inside a cracked finite element can be defined in terms of the classical and enriched approximations:

$$\boldsymbol{u}^{\mathrm{h}}(\mathbf{x}) = \boldsymbol{u}^{\mathrm{FE}} + \boldsymbol{u}^{\mathrm{enr}} = \sum_{j=1}^{n} N_j(\mathbf{x}) \boldsymbol{u}_j + \sum_{k=1}^{m} N_k(\mathbf{x}) \psi(\mathbf{x}) \boldsymbol{a}_k \tag{3.5.5}$$

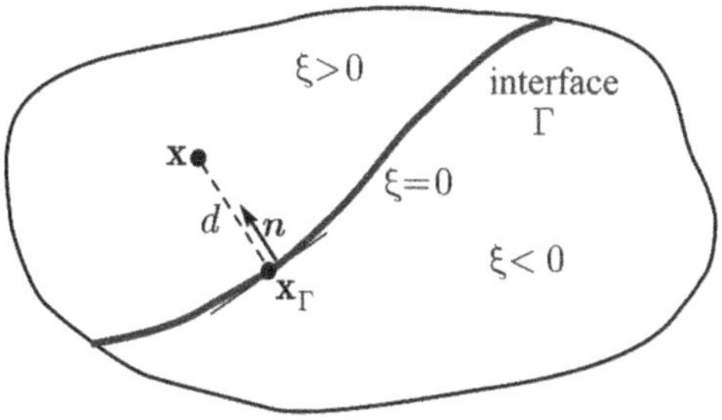

Figure 3.15 Definition of the signed distance function.

where $\boldsymbol{u}_j$ is the vector of classical FEM nodal degrees of freedom, $\boldsymbol{a}_k$ is the set of degrees of freedom added to the standard finite element model and $\psi(\mathbf{x})$ is the set of enrichment functions associated with the set of nodes that directly affect the crack. The enrichment function and its associated degrees of freedom allow for accurate modelling of the response of a crack without explicitly modelling it in the FE mesh.

Equation (3.5.5) for can be further expanded to account for the displacement discontinuity across the crack surfaces ($\boldsymbol{u}^{\mathrm{H}}$), stress singularity near the crack tips ($\boldsymbol{u}^{\mathrm{tip}}$) and material interface ($\boldsymbol{u}^{\mathrm{mat}}$)

$$\boldsymbol{u}^{\mathrm{h}}(\mathbf{x}) = \boldsymbol{u}(\mathbf{x}) + \boldsymbol{u}^{\mathrm{H}}(\mathbf{x}) + \boldsymbol{u}^{\mathrm{tip}}(\mathbf{x}) + \boldsymbol{u}^{\mathrm{mat}}(\mathbf{x}) \tag{3.5.6}$$

or, more explicitly,

$$\begin{aligned} \boldsymbol{u}^{\mathrm{h}}(\mathbf{x}) = &\left[\sum_{j=1}^{\mathrm{n}} N_j(\mathbf{x})\mathbf{u}_j\right] + \left[\sum_{h=1}^{\mathrm{mh}} N_h(\mathbf{x})\left(H(\mathbf{x}) - H(\mathbf{x}_h)\right)\boldsymbol{a}_h\right] \\ &+ \left[\sum_{k=1}^{\mathrm{mt}} N_k(\mathbf{x})\left(\sum_{l=1}^{\mathrm{mf}=4}\left[F_l(\mathbf{x}) - F_l(\mathbf{x}_k)\right]\boldsymbol{b}_k^l\right)\right] + \left[\sum_{m=1}^{mm} N_m(\mathbf{x})\chi_m(\mathbf{x})\boldsymbol{c}_m\right] \end{aligned} \tag{3.5.7}$$

where the enrichment functions include the Heaviside function $H(\mathbf{x})$, the crack tip enrichment functions $F_l(\mathbf{x})$, $(l = 1-4)$ and the weak/material discontinuity enrichment function $\chi_m(\mathbf{x})$, with the corresponding additional degrees of freedom $\boldsymbol{a}_h$, $\boldsymbol{b}_k^l(l = 1-4)$ and $\boldsymbol{c}_m$ for modelling crack faces, crack tip and material interface, respectively. mh, mt and mm represent the number of nodes associated with crack face, crack tip and weak enrichments, respectively.

A simplified form of Equation (3.5.7) can also be adopted (without the interpolation property of the enriched solution):

$$\begin{aligned} \boldsymbol{u}^{\mathrm{h}}(\mathbf{x}) = &\left[\sum_{j=1}^{\mathrm{n}} N_j(\mathbf{x})\boldsymbol{u}_j\right] + \left[\sum_{h=1}^{\mathrm{mh}} N_h(\mathbf{x})H(\mathbf{x})\boldsymbol{a}_h\right] + \left[\sum_{k=1}^{\mathrm{mt}} N_k(\mathbf{x})\left(\sum_{l=1}^{\mathrm{mf}} F_l(\mathbf{x})\boldsymbol{b}_k^l\right)\right] \\ &+ \left[\sum_{m=1}^{\mathrm{mm}} N_m(\mathbf{x})\chi_m(\mathbf{x})\boldsymbol{c}_m\right] \end{aligned} \tag{3.5.8}$$

The discontinuity enrichment function H can be readily expressed in terms of the Heaviside or step function,

$$H(\xi) = \begin{cases} 1, & \forall \xi > 0 \\ 0, & \forall \xi < 0 \end{cases} \tag{3.5.9}$$

or alternatively in the form of the sign function,

$$H(\xi) = \mathrm{sign}(\xi) = \begin{cases} 1, & \forall \xi > 0 \\ -1, & \forall \xi < 0 \end{cases} \tag{3.5.10}$$

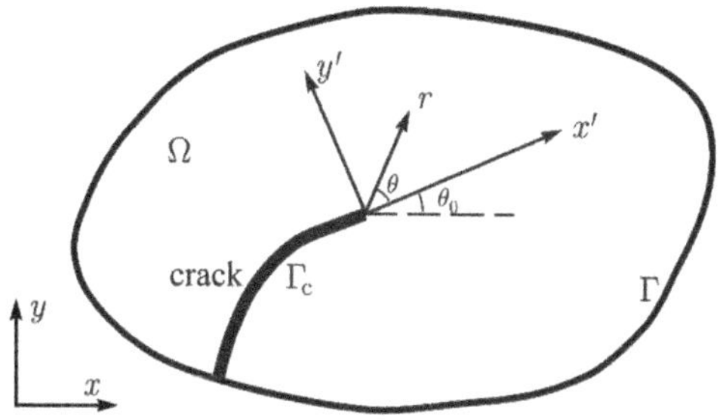

Figure 3.16 Global, local and polar coordinates at the crack tip.

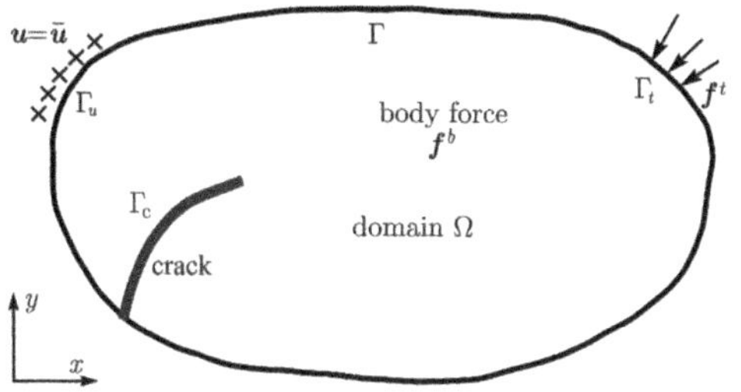

Figure 3.17 A cracked body subjected to body force, tractions and displacement boundary conditions.

Crack tip enrichments F_α are derived from the asymptotic analytical solution near a crack tip in the local (r,θ) polar coordinate system (Figure 3.16),

$$F_\alpha(r,\theta) = \left\{\sqrt{r}\sin\frac{\theta}{2}, \sqrt{r}\cos\frac{\theta}{2}, \sqrt{r}\sin\theta\theta\sin\frac{\theta}{2}, \sqrt{r}\sin\theta\cos\frac{\theta}{2}\right\} \tag{3.5.11}$$

A comprehensive review of different tip enrichment functions is presented in Section 3.5.12.

Finally, the weak discontinuity function $\chi_m(\boldsymbol{x})$, which represents a material interface, is defined as

$$\chi_m(\mathbf{x}) = |\xi(\mathbf{x})| - |\xi(\mathbf{x}_m)| \tag{3.5.12}$$

3.5.6 Boundary Value Problem for a Cracked Body

Figure 3.17 illustrates a body Ω with a traction free crack Γ_c, which is in the static state of equilibrium with the applied body force $\boldsymbol{f}^b$, external traction $\boldsymbol{f}^t$ on Γ_t and displacement $\bar{\boldsymbol{u}}$ on Γ_u.

The strong form of the equilibrium equation, in terms of the stress tensor $\boldsymbol{\sigma}$, can be written as

$$\nabla.\boldsymbol{\sigma} + \boldsymbol{f}^b = 0 \ \text{in}\Omega \tag{3.5.13}$$

with the following displacement and traction boundary conditions:

$$\boldsymbol{u} = \bar{\boldsymbol{u}} \qquad \text{on } \Gamma_u \tag{3.5.14}$$

$$\boldsymbol{\sigma} \cdot \mathbf{n} = \boldsymbol{f}^t \quad \text{on } \Gamma_t \tag{3.5.15}$$

$$\boldsymbol{\sigma} \cdot \mathbf{n} = \mathbf{0} \qquad \text{on } \Gamma_c \tag{3.5.16}$$

The governing weak form of the boundary value problem can then be defined as

$$\int_\Omega \boldsymbol{\sigma} \cdot \delta\boldsymbol{\varepsilon} \, \mathrm{d}\Omega = \int_\Omega \boldsymbol{f}^b \cdot \delta\boldsymbol{u} \, \mathrm{d}\Omega + \int_{\Gamma_t} \boldsymbol{f}^t \cdot \delta\boldsymbol{u} \, \mathrm{d}\Gamma \tag{3.5.17}$$

3.5.7 XFEM Discretisation of the Governing Equation

Similar to the conventional finite element method, discretization of Equation (3.5.17) can be performed by using the XFEM approximation of Equation (3.5.7), which results in the following linear set of equations:

$$\mathbf{K}\mathbf{U}^{\mathrm{h}} = \mathbf{f} \tag{3.5.18}$$

where $\mathbf{U}^{\mathrm{h}}$ is the global vector of nodal degrees of freedom, including the classical and enriched degrees of freedom:

$$\mathbf{U}^{\mathrm{h}} = \{\boldsymbol{u} \quad \boldsymbol{a} \quad \boldsymbol{b}_1 \quad \boldsymbol{b}_2 \quad \boldsymbol{b}_3 \quad \boldsymbol{b}_4 \quad \mathbf{c}\}^{\mathrm{T}} \tag{3.5.19}$$

and the global stiffness matrix $\mathbf{K}$ and force vector $\mathbf{f}$ are calculated by assembling the corresponding stiffness matrix $\mathbf{K}_{ij}^e$ and force vector $\mathbf{f}_i^e$ of each element, respectively. $\mathbf{f}_i^e$ is the vector of external force for a typical element e:

$$\mathbf{f}_i^e = \left\{ \mathbf{f}_i^u \quad \mathbf{f}_i^a \quad \mathbf{f}_i^{b1} \quad \mathbf{f}_i^{b2} \quad \mathbf{f}_i^{b3} \quad \mathbf{f}_i^{b4} \quad \mathbf{f}_i^c \right\}^{\mathrm{T}} \tag{3.5.20}$$

with

$$\mathbf{f}_i^u = \int_{\Gamma_t} N_i^{\mathrm{T}} \boldsymbol{f}^t \mathrm{d}\Gamma + \int_{\Omega^e} N_i^{\mathrm{T}} \boldsymbol{f}^b \mathrm{d}\Omega \tag{3.5.21}$$

$$\mathbf{f}_i^a = \int_{\Gamma_t} N_i^{\mathrm{T}} \left[H(\xi) - H(\xi_i)\right] \boldsymbol{f}^t \mathrm{d}\Gamma + \int_{\Omega^e} N_i^{\mathrm{T}} \left[H(\xi) - H(\xi_i)\right] \boldsymbol{f}^b \mathrm{d}\Omega \tag{3.5.22}$$

$$\mathbf{f}_i^{b\alpha} = \int_{\Gamma_t} N_i^{\mathrm{T}} (F_\alpha - F_{\alpha i}) \boldsymbol{f}^t \mathrm{d}\Gamma + \int_{\Omega^e} N_i^{\mathrm{T}} (F_\alpha - F_{\alpha i}) \boldsymbol{f}^b \mathrm{d}\Omega (\alpha = 1,2,3 \text{ and } 4) \tag{3.5.23}$$

$$\mathbf{f}_i^c = \int_{\Gamma_t} N_i^{\mathrm{T}} \chi \boldsymbol{f}^t \mathrm{d}\Gamma + \int_{\Omega^e} N_i^{\mathrm{T}} \chi \boldsymbol{f}^b \mathrm{d}\Omega \tag{3.5.24}$$

$\mathbf{K}_{ij}^{e}$ is the stiffness matrix for a typical element e,

$$\mathbf{K}_{ij}^{\mathrm{e}} = \begin{bmatrix} \mathbf{K}_{ij}^{uu} & \mathbf{K}_{ij}^{ua} & \mathbf{K}_{ij}^{ub} & \mathbf{K}_{ij}^{uc} \\ \mathbf{K}_{ij}^{au} & \mathbf{K}_{ij}^{aa} & \mathbf{K}_{ij}^{ab} & \mathbf{K}_{ij}^{ac} \\ \mathbf{K}_{ij}^{bu} & \mathbf{K}_{ij}^{ba} & \mathbf{K}_{ij}^{bb} & \mathbf{K}_{ij}^{bc} \\ \mathbf{K}_{ij}^{cu} & \mathbf{K}_{ij}^{ca} & \mathbf{K}_{ij}^{cb} & \mathbf{K}_{ij}^{cc} \end{bmatrix} \tag{3.5.25}$$

with

$$\mathbf{K}_{ij}^{rs} = \int_{\Omega^e} (\mathbf{B}_i^r)^{\mathrm{T}} \mathbf{D} \mathbf{B}_i^s \, \mathrm{d}\Omega \tag{3.5.26}$$

$$\mathbf{K}_{ij}^{ac} = \mathbf{K}_{ij}^{ca} = 0 \tag{3.5.27}$$

$\mathbf{B}$ is the matrix of shape function derivatives, defined for different classical and enrichment parts as

$$\mathbf{B}_i^{u} = \begin{bmatrix} N_{i,x} & 0 \\ 0 & N_{i,y} \\ N_{i,y} & N_{i,x} \end{bmatrix} \tag{3.5.28}$$

$$\mathbf{B}_i^{a} = \begin{bmatrix} \left(N_i\left[H(\xi) - H(\xi_i)\right]\right)_{,x} & 0 \\ 0 & \left(N_i\left[H(\xi) - H(\xi_i)\right]\right)_{,y} \\ \left(N_i\left[H(\xi) - H(\xi_i)\right]\right)_{,y} & \left(N_i\left[H(\xi) - H(\xi_i)\right]\right)_{,x} \end{bmatrix} \tag{3.5.29}$$

$$\mathbf{B}_i^{b} = \begin{bmatrix} \mathbf{B}_i^{b1} & \mathbf{B}_i^{b2} & \mathbf{B}_i^{b3} & \mathbf{B}_i^{b4} \end{bmatrix} \tag{3.5.30}$$

$$\mathbf{B}_i^{b\alpha} = \begin{bmatrix} \left[N_i(F_\alpha - F_{\alpha i})\right]_{,x} & 0 \\ 0 & \left[N_i(F_\alpha - F_{\alpha i})\right]_{,y} \\ \left[N_i(F_\alpha - F_{\alpha i})\right]_{,y} & \left[N_i(F_\alpha - F_{\alpha i})\right]_{,x} \end{bmatrix} \quad (\alpha = 1,2,3 \text{ and } 4) \tag{3.5.31}$$

$$\mathbf{B}_i^{c} = \begin{bmatrix} (N_i\chi)_{,x} & 0 \\ 0 & (N_i\chi)_{,y} \\ (N_i\chi)_{,y} & (N_i\chi)_{,x} \end{bmatrix} \tag{3.5.32}$$

3.5.8 Numerical Integration

Equations (3.5.21) to (3.5.24) and (3.5.26) must be accurately integrated in order to provide sufficient accuracy for solving the discretized XFEM Equation (3.5.18). While the Gauss quadrature rule is sufficiently accurate for polynomial integrands, it may not provide sufficient accuracy and may lead to substantial numerical error for some non-polynomial integrands, especially if discontinuity or singularity exists. The reason can be attributed to the fact that the existence of singularity within a finite element highly complicates the displacement and stress fields, which cannot be accurately integrated even by higher order Gauss quadrature rules.

Three efficient procedures are usually adopted for numerical integration on enriched elements, as presented in Figure 3.18. In addition to being capable of handling different types of the enrichments, the procedures should be consistent with the geometry of straight or curved crack or discontinuity.

The first approach, the sub-quad technique, is the simplest method, which subdivides the enriched element into predefined sub-quads (Figure 3.18). Then, the conventional Gauss quadrature rule is adopted to integrate each sub-quad (Dolbow 1999). A fine grid of sub-quads should be used for elements that include a crack tip, due to its high gradient stress state. It should be noted that the sub-quads are only used for integration purposes and do not possess any independent degrees of freedom.

Alternatively, the sub-triangle approach, which is practically the most efficient integration technique in XFEM crack analysis, can be adopted. It subdivides the enriched element into sub-triangles, as depicted in Figure 3.18 (Dolbow 1999). Again, a finer set of sub-triangulations is required for elements with a crack tip.

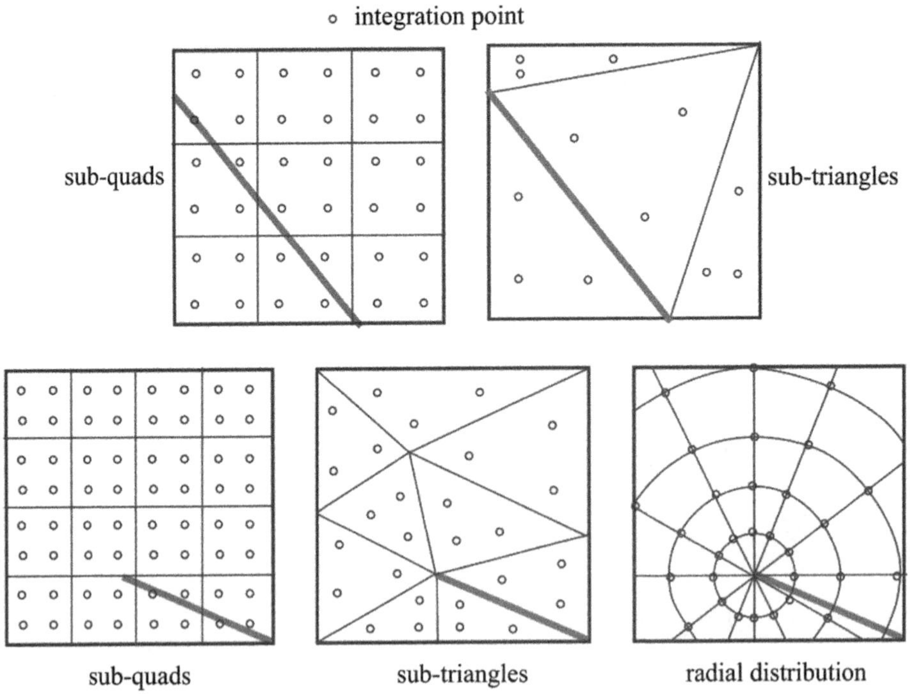

Figure 3.18 Three methods for numerical integration of a cracked element.

Finally, in the polar distribution technique, distribution of the integration points in a crack tip element is defined in a polar format, which is more consistent with the polar stress distribution around the crack tip. Accordingly, the method considers closer integration points in the vicinity of the crack tip, and gradually changes the pattern into far integration points towards the edges of the element (Figure 3.18). The method has been successfully implemented in XIGA fracture analysis (Ghorashi et al. 2012).

It should be noted that if a crack passes through a node, the standard XFEM enrichment fails and must be avoided. Instead, a split-node technique can be simply adopted. This is also the case if the crack locates very close to a node, which may lead to numerical ill-conditioning. A simple geometric criterion, originally proposed by Dolbow (1999), may be adopted to avoid such numerical consequences:

$$A^a / A < \varepsilon \text{ or } A^b / A < \varepsilon \tag{3.5.33}$$

where A^a and A^b denote the area of the element at the two sides of the crack, A is the area of the element, and ε is an allowable small tolerance value. Alternatively, a finite element node is not enriched if the corresponding area does not include any integration point.

3.5.9 Selection of Enrichment Nodes for Crack Propagation

One of the most efficient aspects of XFEM is the way it deals with propagating discontinuities such as cracks, as depicted in Figure 3.19. The nodes of the elements that need to be enriched by the Heaviside and crack tip enrichments change by the progress of crack propagation. The corresponding enrichment degrees of freedom should also be modified accordingly. For instance, at the first stage of Figure 3.19, elements A, B, C and D are enriched by the Heaviside function and element E is enriched by the crack tip enrichment. Other elements remain as the standard finite elements. As the crack propagates to the second stage of Figure 3.19, the enrichment of element E is changed from the crack tip to the Heaviside enrichment, and the classical element F is now enriched by the crack tip

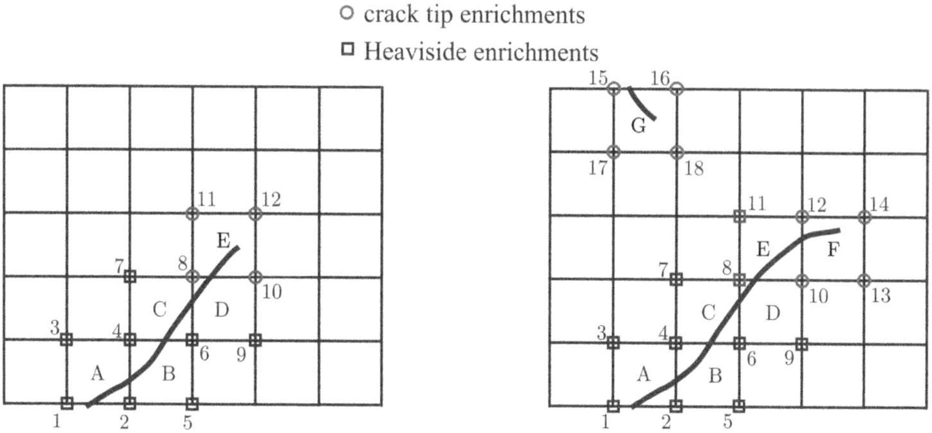

Figure 3.19 Typical nodal enrichment for different stages of a propagating crack.

enrichments. No independent Heaviside enrichment is required in the crack tip element because the crack tip enrichment functions can automatically and accurately represent the discontinuity of the displacement field across the crack faces.

This simple local on and off procedure for enrichment functions and their corresponding DOFs allows for an efficient numerical procedure of XFEM modelling of propagating cracks.

3.5.10 Incompatible Modes of XFEM Enrichments

Incompatible solutions may occur for partially enriched elements, as depicted in Figure 3.20. The highlighted elements are partially enriched and may have one, two or three enriched nodes. Clearly, XFEM approximation of the displacement field for the partially enriched elements can no longer satisfy the partition of unity condition:

$$\boldsymbol{u}^h(\mathbf{x}) = \sum_{j=1}^{4} N_j(\mathbf{x})\boldsymbol{u}_j + \sum_{k=1}^{3} N_k(\mathbf{x})\psi(\mathbf{x})\boldsymbol{a}_k \tag{3.5.34}$$

Nevertheless, it has a very limited effect on the overall approximation, as such an element is not supposed to show any discontinuity at all. If the shifted XFEM formulation with a nodal interpolation characteristic is adopted, the displacement consistency across the XFEM–FEM edges is ensured, and these elements may simply be considered as the standard finite elements without any full or partial enrichments.

Different orders of crack tip and Heaviside enrichments for two neighbouring elements may also lead to incompatibility of solution or its derivative across the common edge of

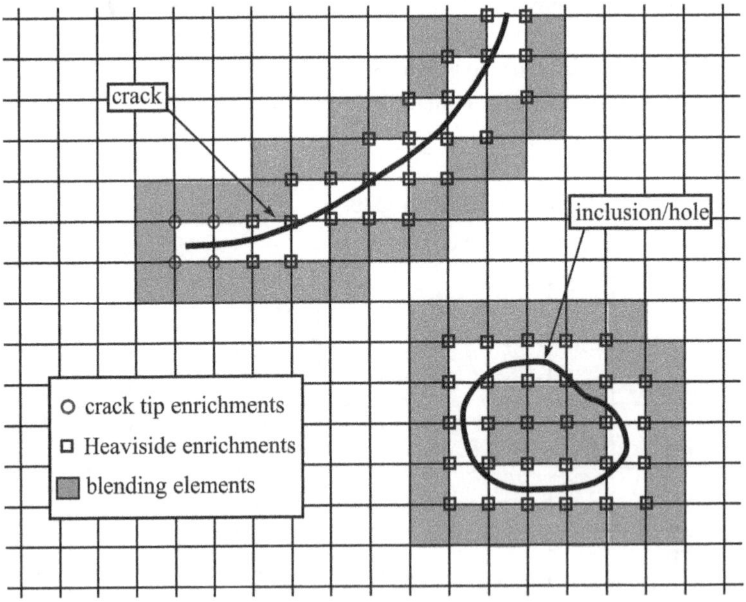

Figure 3.20 Standard, enriched and blending elements. *Source:* Reproduced from Mohammadi 2008/John Wiley & Sons Ltd.

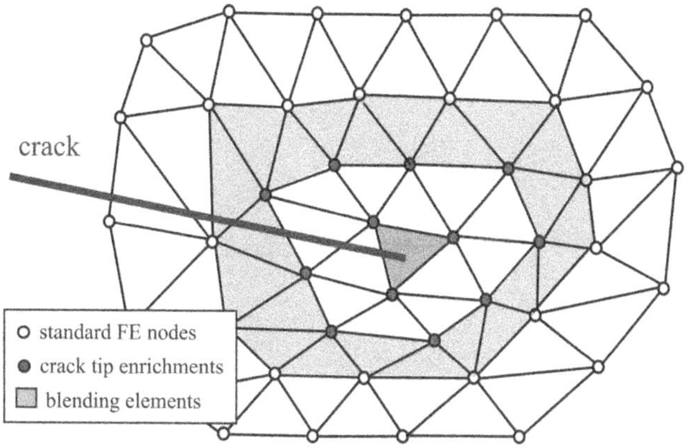

Figure 3.21 Blending elements for a smooth transition between the enriched (XFEM) and standard (FEM) approximations.

elements. Similar incompatibility exists between the enriched and classical finite elements. A computational remedy to remove or reduce the enrichment incompatibility is to adopt a smooth transition by the use of blending elements, which bridge the elements with and without enrichments (or with different enrichment strategies) (see Figure 3.21):

$$\boldsymbol{u}^h(\mathbf{x}) = (1 - R)\boldsymbol{u}(\mathbf{x}) + R\boldsymbol{u}^{\text{enr}}(\mathbf{x}) \tag{3.5.35}$$

where R is a smooth blending ramp function, set to 0 on the edge of the standard domain and 1 on the enriched edge. The order of ramp function R is selected based on the preferred order of continuity between the two domains. For instance, a linear R only ensures the continuity of the displacement field, whereas higher order blending functions are required to ensure a continuous stress/strain field.

3.5.11 The Level Set Method for Tracking Moving Boundaries

The level set method (Osher and Sethian 1988) is frequently used in many geometrical representations of evolving curves/surfaces. On its first unusual look, the level set method (LSM) generates a cone-shaped surface (the level set function) from the original curve in order to compute its intersection with the xy plane, which is the curve itself (Mohammadi 2012a). Despite its apparently complicated concept, LSM computes the motion of an interface on a fixed Eulerian mesh.

Consider two non-overlapping subdomains, Ω_1 and Ω_2, with the common interface Γ, as illustrated in Figure 3.22. The level set function $\varphi(x)$ is defined as

$$\varphi(\mathbf{x}) = \begin{cases} > 0, & \mathbf{x} \in \Omega_1 \\ = 0, & \mathbf{x} \in \Gamma \\ < 0, & \mathbf{x} \in \Omega_2 \end{cases} \tag{3.5.36}$$

The interface Γ can be interpreted as the zero-level contour of the level set function $\varphi(\boldsymbol{x})$.

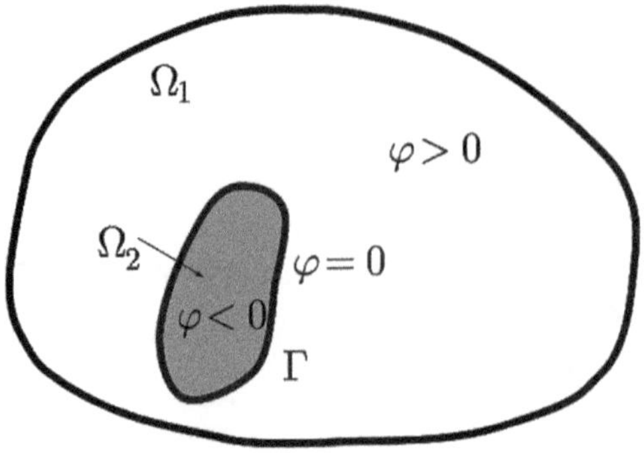

Figure 3.22 Definition of the level set function. *Source:* Reproduced from Mohammadi 2008/John Wiley & Sons Ltd.

The level set function, $\varphi(\boldsymbol{x})$, is usually defined in terms of the signed distance function (3.5.4) (see Figure 3.15). Then, the domains Ω_1 and Ω_2 can be defined as

$$\begin{cases} \Omega_1 = \{\mathbf{x} \in \Omega, H(\varphi(\mathbf{x})) = 1\} \\ \Omega_2 = \{\mathbf{x} \in \Omega, H(-\varphi(\mathbf{x})) = 1\} \end{cases} \tag{3.5.37}$$

The vector $\mathbf{n}$ (normal to the interface Γ at a point $\mathbf{x} \in \Gamma$) is defined as

$$\mathbf{n} = \frac{\nabla\varphi(\mathbf{x})}{\|\nabla\varphi(\mathbf{x})\|} \tag{3.5.38}$$

It should be noted that the position of a moving surface is not known *a priori*. An incremental form can be considered based on the first-order time integration scheme and the fact that the level set is zero on the surface (Osher and Sethian, 1988):

$$\frac{D\varphi(\mathbf{x},t)}{Dt} = 0 \rightarrow \frac{\partial\varphi(\mathbf{x},t)}{\partial t} + \nabla\varphi(\mathbf{x},t)\cdot\mathbf{v}(\mathbf{x},t) = 0 \tag{3.5.39}$$

The original definition of the level set should be modified for open-curve crack problems by defining an extra level set $\psi(\mathbf{x},t)$ at the crack tip, which is generally assumed to be orthogonal to φ. The crack evolution can then be expressed as the intersection of the two zero level sets $\varphi(\mathbf{x},t)$ and $\psi(\mathbf{x},t)$.

3.5.12 XFEM Tip Enrichments

This section briefly presents a similar discussion by Mohammadi (2012a) to review different available tip enrichment functions for various physical problems.

3.5.12.1 Isotropic Enrichment

The most frequently used crack tip enrichment functions are

$$\{F_l(r,\theta)\}_{l=1}^{4} = \left\{\sqrt{r}\cos\frac{\theta}{2}, \sqrt{r}\sin\frac{\theta}{2}, \sqrt{r}\sin\theta\cos\frac{\theta}{2}, \sqrt{r}\sin\theta\sin\frac{\theta}{2}\right\} \tag{3.5.40}$$

which are the basic tip enrichment functions, obtained in ideal isotropic, homogeneous, static and traction-free crack conditions. Nevertheless, these basic functions have been widely adopted, at least as a first estimate for better XFEM approximation, in several complex problems, such as cracking in orthotropic materials, functionally graded materials, bimaterials, dynamic problems, elastoplastic solutions, hydraulic fracture, electromechanical problems, etc.

3.5.12.2 Orthotropic Crack Tip Enrichment Functions

General orthotropic crack tip enrichment functions can be written as (Asadpoure et al. 2006, 2007; Asadpoure and Mohammadi 2007)

$$\{F_l(r,\theta)\}_{l=1}^{4}=\left\{\sqrt{r}\cos\frac{\theta_1}{2}\sqrt{g_1(\theta)},\sqrt{r}\cos\frac{\theta_2}{2}\sqrt{g_2(\theta)},\sqrt{r}\sin\frac{\theta_1}{2}\sqrt{g_1(\theta)},\right.$$
$$\left.\sqrt{r}\sin\frac{\theta_2}{2}\sqrt{g_2(\theta)}\right\} \tag{3.5.41}$$

where functions $g_k(\theta)$ and θ_k are defined as

$$g_k(\theta)=\sqrt{(\cos\theta+s_{kx}\sin\theta)^2+(s_{ky}\sin\theta)^2} \tag{3.5.42}$$

$$\theta_k=\arctan\left(\frac{s_{ky}\sin\theta}{\cos\theta+s_{kx}\sin\theta}\right) \tag{3.5.43}$$

s_{kx} and s_{ky} are the complex roots of the following characteristic equation (Lekhnitskii 1968; Sadd, 2005):

$$c_{11}s^4-2c_{16}s^3+(2c_{12}+c_{66})s^2-2c_{26}s+c_{22}=0 \tag{3.5.44}$$

and c_{ij} are the components of the material compliance modulus (Equation (2.2.12)):

$$\varepsilon_i=c_{ij}\sigma_j,\ i,j=1,2,3,\ldots,6 \tag{3.5.44}$$

3.5.12.3 Isotropic Bimaterial Crack Enrichments

Bimaterial enrichment functions for an interface crack tip in isotropic bimaterials can be expressed as (Sukumar et al. 2004)

$$\{F_l(r,\theta)\}_{l=1}^{12}=\left\{\sqrt{r}\cos(\varepsilon\ln r)e^{-\varepsilon\theta}\sin\frac{\theta}{2},\sqrt{r}\cos(\varepsilon\ln r)e^{-\varepsilon\theta}\cos\frac{\theta}{2},\right.$$
$$\sqrt{r}\cos(\varepsilon\ln r)e^{\varepsilon\theta}\sin\frac{\theta}{2},\sqrt{r}\cos(\varepsilon\ln r)e^{\varepsilon\theta}\cos\frac{\theta}{2},$$
$$\sqrt{r}\sin(\varepsilon\ln r)e^{-\varepsilon\theta}\sin\frac{\theta}{2},\sqrt{r}\sin(\varepsilon\ln r)e^{-\varepsilon\theta}\cos\frac{\theta}{2},$$
$$\sqrt{r}\sin(\varepsilon\ln r)e^{\varepsilon\theta}\sin\frac{\theta}{2},\sqrt{r}\sin(\varepsilon\ln r)e^{\varepsilon\theta}\cos\frac{\theta}{2}, \tag{3.5.45}$$
$$\sqrt{r}\cos(\varepsilon\ln r)e^{\varepsilon\theta}\sin\frac{\theta}{2}\sin\theta,\sqrt{r}\cos(\varepsilon\ln r)e^{\varepsilon\theta}\cos\frac{\theta}{2}\sin\theta,$$
$$\left.\sqrt{r}\sin(\varepsilon\ln r)e^{\varepsilon\theta}\sin\frac{\theta}{2}\sin\theta,\sqrt{r}\sin(\varepsilon\ln r)e^{\varepsilon\theta}\cos\frac{\theta}{2}\sin\theta\right\}$$

where ε is the index of oscillation,

$$\varepsilon = \frac{1}{2\pi}\ln\left(\frac{1-\beta}{1+\beta}\right) \tag{3.5.46}$$

and β for isotropic bimaterials is defined by

$$\beta = \frac{\mu_1(\kappa_2-1)-\mu_2(\kappa_1-1)}{\mu_1(\kappa_2+1)+\mu_2(\kappa_1+1)} \tag{3.5.47}$$

3.5.12.4 Orthotropic Bimaterial Crack Enrichments

Orthotropic bimaterial interface crack tip enrichment functions can be defined as (Esna Ashari and Mohammadi 2011a, 2012)

$$\begin{aligned}\{F_l(r,\theta)\}_{l=1}^{8} = \Bigg[& e^{-\varepsilon\theta_l}\cos\left(\varepsilon\ln(r_l)+\frac{\theta_l}{2}\right)\sqrt{r_l}, e^{-\varepsilon\theta_l}\sin\left(\varepsilon\ln(r_l)+\frac{\theta_l}{2}\right)\sqrt{r_l}, \\ & e^{\varepsilon\theta_l}\cos\left(\varepsilon\ln(r_l)-\frac{\theta_l}{2}\right)\sqrt{r_l}, e^{\varepsilon\theta_l}\sin\left(\varepsilon\ln(r_l)-\frac{\theta_l}{2}\right)\sqrt{r_l}, \\ & e^{-\varepsilon\theta_s}\cos\left(\varepsilon\ln(r_s)+\frac{\theta_s}{2}\right)\sqrt{r_s}, e^{-\varepsilon\theta_s}\sin\left(\varepsilon\ln(r_s)+\frac{\theta_s}{2}\right)\sqrt{r_s}, \\ & e^{\varepsilon\theta_s}\cos\left(\varepsilon\ln(r_s)-\frac{\theta_s}{2}\right)\sqrt{r_s}, e^{\varepsilon\theta_s}\sin\left(\varepsilon\ln(r_s)-\frac{\theta_s}{2}\right)\sqrt{r_s}\Bigg]\end{aligned} \tag{3.5.48}$$

with

$$r_j = r\sqrt{\cos^2\theta + Z_j^2\sin^2\theta}, j = l,s, \begin{cases} Z_l = p \\ Z_s = q \end{cases} \tag{3.5.49}$$

$$\theta_j = \tan^{-1}(Z_j\tan\theta), j = l,s, \begin{cases} Z_l = p \\ Z_s = q \end{cases} \tag{3.5.50}$$

where ε is the index of oscillation and β for orthotropic bimaterials is defined by

$$\beta = \frac{h_{11}}{\sqrt{h_{12}h_{21}}} \tag{3.5.51}$$

with

$$\begin{cases} h_{11} = (l_{11})_1 - (l_{11})_2 \\ h_{21} = (l_{21})_1 + (l_{21})_2 \\ h_{12} = (l_{12})_1 + (l_{12})_2 \end{cases}, \begin{cases} (l_{11})_k = \left\{\dfrac{p_s p - p_l q}{q-p}\right\}_k = \left\{\dfrac{q_s - q_l}{q-p}\right\}_k \\ (l_{21})_k = \left\{\dfrac{q q_l - p q_s}{q-p}\right\}_k \\ (l_{12})_k = \left\{\dfrac{p_l - p_s}{q-p}\right\}_k \end{cases}, k = 1,2 \tag{3.5.52}$$

Subscripts $k = 1,2$ denote the upper and lower materials, respectively. Also,

$$p,q = \sqrt{B_{12} \mp \sqrt{B_{12}^2 - B_{66}}}, \quad \begin{cases} B_{12} = \dfrac{1}{2}\dfrac{(2c_{12} + c_{66})}{c_{11}} \\ B_{66} = \dfrac{c_{22}}{c_{11}} \end{cases} \tag{3.5.53}$$

$$p_l = -p^2 c_{11} + c_{12}, p_s = -q^2 c_{11} + c_{12}, q_l = \frac{-p^2 c_{12} + c_{22}}{p}, q_s = \frac{-q^2 c_{12} + c_{22}}{q} \tag{3.5.54}$$

Note that $\varepsilon\pi$ should be replaced by $-\varepsilon\pi$ for the material under the interface ($k = 2$).

3.5.12.5 Dynamic Crack Tip Enrichment

A simplified set of enrichment functions for a propagating crack in an isotropic medium can be written as (Belytschko and Chen 2004)

$$\{F_l(r,\theta)\}_{l=1}^4 = \left\{ r_d \cos\frac{\theta_d}{2}, r_s \cos\frac{\theta_s}{2}, r_d \sin\frac{\theta_d}{2}, r_d \sin\frac{\theta_s}{2} \right\} \tag{3.5.55}$$

where

$$r_d^2 = x^2 + \beta_d^2 y^2, r_s^2 = x^2 + \beta_s^2 y^2 \tag{3.5.56}$$

$$\tan\theta_d = \beta_d \tan\theta, \tan\theta_s = \beta_s \tan\theta \tag{3.5.57}$$

with

$$\beta_d^2 = 1 - \frac{v_c^2}{c_d^2}, \beta_s^2 = 1 - \frac{v_c^2}{c_s^2} \tag{3.5.58}$$

where v_c is the crack tip velocity and c_d and c_s are the dilatational and shear wave speeds, respectively, defined in terms of the Lame coefficients, μ and λ, and the material density ρ,

$$c_d = \sqrt{\frac{\lambda + 2\mu}{\rho}} \tag{3.5.59}$$

$$c_s = \sqrt{\frac{\mu}{\rho}} \tag{3.5.60}$$

3.5.12.6 Dynamic Enrichments for Moving Cracks in Orthotropic Media

Dynamic enrichment functions for moving cracks in orthotropic media can be defined as (Motamedi and Mohammadi 2010a, 2010b, 2012)

$$\begin{aligned} \{F_l(r,\theta)\}_{l=1}^4 = \Big\{ &\sqrt{r}\cos\frac{\theta_1}{2}\sqrt{g_1(\theta)}, \sqrt{r}\cos\frac{\theta_2}{2}\sqrt{g_2(\theta)}, \\ &\sqrt{r}\sin\frac{\theta_1}{2}\sqrt{g_1(\theta)}, \sqrt{r}\sin\frac{\theta_2}{2}\sqrt{g_2(\theta)} \Big\} \end{aligned} \tag{3.5.61}$$

where $g_k(\theta)$ and θ_k, $(k=1,2)$ are defined as

$$g_k(\theta)=\sqrt{(\cos\theta+m_{kx}\sin\theta)^2+(m_{ky}\sin\theta)^2},\ k=1,2 \tag{3.5.62}$$

$$\theta_k=\mathrm{arctg}\left(\frac{m_{ky}\sin\theta}{\cos\theta+m_{kx}\sin\theta}\right),\ k=1,2 \tag{3.5.63}$$

and $m_k=m_{kx}+im_{ky}$ are the roots of the following characteristic equation:

$$m^4+2B_{12}m^2+B_{66}=0 \tag{3.5.64}$$

$$2B_{12}=\frac{1}{c_{11}}\left[2c_{12}+c_{66}+\rho v_c^2\left(c_{12}^2-c_{11}c_{66}-c_{11}c_{22}\right)\right] \tag{3.5.65}$$

$$B_{66}=\frac{1}{c_{11}}\left\{c_{22}+\rho v_c^2\left[c_{12}^2-c_{22}c_{66}-c_{11}c_{22}+\rho v_c^2 c_{66}\left(c_{11}c_{22}-c_{12}^2\right)\right]\right\} \tag{3.5.66}$$

3.5.12.7 Electro-mechanical Enrichment

In an electro-mechanical plane strain problem (Figure 3.23), the following constitutive equation relates the set of mechanical strain tensor ε_{ij} and the electric field E_i to the set of Cauchy stress tensor σ_{ij} and the electric displacement vector D_i (Sharma et al. 2012):

$$\begin{Bmatrix}\varepsilon_{xx}\\ \varepsilon_{yy}\\ \gamma_{xy}\\ E_x\\ E_y\end{Bmatrix}=\begin{bmatrix}c_{11}&c_{12}&0&0&b_{21}\\ c_{12}&c_{22}&0&0&b_{22}\\ 0&0&c_{33}&b_{13}&0\\ 0&0&-b_{13}&\delta_{11}&0\\ -b_{21}&-b_{22}&0&0&\delta_{22}\end{bmatrix}\begin{Bmatrix}\sigma_{xx}\\ \sigma_{yy}\\ \tau_{xy}\\ D_x\\ D_y\end{Bmatrix} \tag{3.5.67}$$

where c_{ij}, b_{ij} and δ_{ij} are mechanical and electrical material constants.

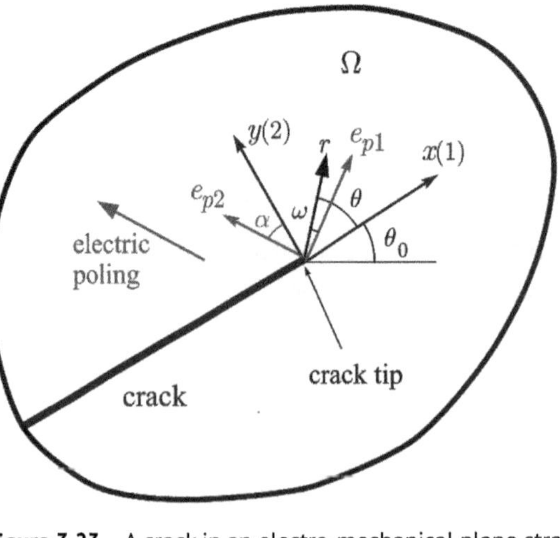

Figure 3.23 A crack in an electro-mechanical plane strain problem.

Crack tip enrichment functions for the mechanical displacement $\boldsymbol{u}$ and the electric potential ϕ can be written as (Bechet et al. 2009; Sharma et al. 2012)

$$\{F_l(r,\theta,\mu_k)\}_{l=1}^{6}=\{\sqrt{r}g_1(\theta),\sqrt{r}g_2(\theta),\sqrt{r}g_3(\theta),\sqrt{r}g_4(\theta),\sqrt{r}g_5(\theta),\sqrt{r}g_6(\theta)\} \tag{3.5.68}$$

where $\omega=\theta-\alpha$ and

$$g_m(\theta)=\begin{cases}\rho_m\left(\omega,s_m\right)\cos\left(\dfrac{\psi\left(\omega,s_m\right)}{2}\right), & s_{m2}>0\\ \rho_m\left(\omega,s_m\right)\sin\left(\dfrac{\psi\left(\omega,s_m\right)}{2}\right), & s_{m2}<0\end{cases} \tag{3.5.69}$$

$$\psi(\omega,s_m)=\frac{\pi}{2}+\pi\,\mathrm{int}\left(\frac{\omega}{\pi}\right)-\tan^{-1}\left(\frac{\cos\left[\omega-\pi\,\mathrm{int}\left(\frac{\omega}{\pi}\right)\right]+s_{m1}\sin\left(\omega-\pi\,\mathrm{int}\left(\frac{\omega}{\pi}\right)\right)}{\left|s_{m2}\right|\sin\left(\omega-\pi\,\mathrm{int}\left(\frac{\omega}{\pi}\right)\right)}\right) \tag{3.5.70}$$

$$\rho_m(\omega,s_m)=\frac{1}{\sqrt{2}}\left\{\left|s_{m2}\right|^2+s_{m1}\sin(2\omega)-\left[\left|s_m\right|^2-1\right]\cos(2\omega)\right\}^{\frac{1}{4}} \tag{3.5.71}$$

$s_m=s_{m1}+is_{m2}$ are the roots of the following electro-mechanical characteristic equation (Sharma et al. 2012)

$$\begin{aligned}&c_{11}\delta_{11}s^6+\left\{c_{11}\delta_{22}+(2c_{12}+c_{33})\delta_{11}+b_{12}\left(b_{12}+2b_{13}\right)+b_{13}^2\right\}s^4\\&+\left\{c_{22}\delta_{11}+(2c_{12}+c_{33})\delta_{22}+2b_{22}\left(b_{12}+b_{13}\right)\right\}s^2+\left\{c_{22}\delta_{22}+b_{22}^2\right\}=0\end{aligned} \tag{3.5.72}$$

3.5.12.8 Dislocation Enrichment

In a dislocation problem, the model is solved for the known prescribed sliding $\boldsymbol{b}=b\boldsymbol{e}_t$ along the sliding plane (Mohammadi and Malekafzali 2011; Keyhani et al. 2015). As a result, the XFEM solution is enriched without any additional degrees of freedom.

Referring to Figure 3.24 for the glide plane of an edge dislocation, one of the available forms for the dislocation enrichment is the so-called tapered Somigliana function (Gracie et al. 2007)

$$\boldsymbol{\Psi}_H=\boldsymbol{b}\sum_{h=1}^{mh}N_h(\mathbf{x})\left[H(f(\mathbf{x}))-H(f(\mathbf{x}_I))\right]\psi_H(g(\mathbf{x})) \tag{3.5.73}$$

$$\psi_H(g(\mathbf{x}))=\begin{cases}1, & g(\mathbf{x})>0\\ \dfrac{(a^m-g(\mathbf{x})^m)^n}{a^{mn}}, & -a<g(\mathbf{x})<0\\ 0, & g(\mathbf{x})<-a\end{cases} \tag{3.5.74}$$

where n, m and a are the material properties.

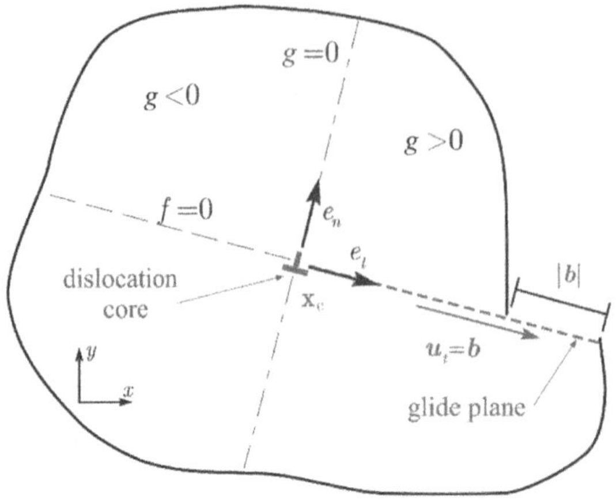

Figure 3.24 An edge dislocation and its glide plane.

3.5.12.9 Hydraulic Fracture Enrichment

In hydraulic fracture problems, the crack tip enrichment functions can be defined in terms of the power of singularity λ (Lecampion 2009; Goodarzi et al. 2011),

$$\{F_l(r,\theta)\}_{l=1}^{4} = \{r^{\lambda}\cos(\lambda\theta), r^{\lambda}\sin(\lambda\theta), r^{\lambda}\sin\theta\sin(\lambda\theta), r^{\lambda}\sin\theta\cos(\lambda\theta)\} \tag{3.5.75}$$

where λ is determined from

$$K_m = \frac{K_c}{\sqrt{12\mu E' v_c}} \tag{3.5.76}$$

and E' is the modulus of solid material, K_c is the fracture toughness, μ is the fluid viscosity and v_c is the crack propagation velocity.

According to Lecampion (2009) and Goodarzi et al. (2015), the hydraulic fracture problem is viscosity controlled for low values of K_m, and Equation (3.5.76) is simplified into

$$\lambda = \frac{2}{3} \rightarrow \{F_l(r,\theta)\}_{l=1}^{4} = \left\{0, (\mathbf{x}-\mathbf{x}_{tip})^{-\frac{1}{3}} H(\mathbf{x}-\mathbf{x}_{tip}), 0, 0\right\} \tag{3.5.77}$$

In contrast, the problem becomes toughness dominated for large values of K_m:

$$\lambda = \frac{1}{2} \rightarrow \{F_l(r,\theta)\}_{l=1}^{4} = \{0, H(\mathbf{x}-\mathbf{x}_{tip}), 0, 0\} \tag{3.5.78}$$

and the fluid pressure p inside the crack can be approximated by

$$p^h(\mathbf{x}) = \sum_{j=1}^{n} N_j(\boldsymbol{x}) p_j + \sum_{k=1}^{\mathrm{mt}} N_k(\mathbf{x}) \left(\sum_{l=1}^{\mathrm{mf}} F_l(x) b_k^l \right) \tag{3.5.79}$$

3.5.12.10 Plastic Enrichment

There is very limited literature available on analytical solutions for plastic problems that can be used for extraction of plastic enrichments. For the Hutchinson–Rice–Rosengren power law hardening material model (Hutchinson 1968),

$$\frac{\varepsilon}{\varepsilon_{\text{yld}}} = \frac{\sigma}{\sigma_{\text{yld}}} + k_0 \left(\frac{\sigma}{\sigma_{\text{yld}}} \right)^n \tag{3.5.80}$$

the following plastic crack tip enrichments are derived (Elguedj et al. 2006):

$$r^{\frac{1}{n+1}} \left\{ \left(\cos\frac{k\theta}{2}, \sin\frac{k\theta}{2} \right); k \in [1,3,5,7] \right\} \tag{3.5.81}$$

In practice, the following simplified forms can be cautiously adopted (Elguedj et al. 2006):

$$r^{\frac{1}{n+1}} \left\{ \sin\frac{\theta}{2}, \cos\frac{\theta}{2}, \sin\frac{\theta}{2}\sin\theta, \cos\frac{\theta}{2}\sin\theta, \sin\frac{\theta}{2}\sin 2\theta, \cos\frac{\theta}{2}\sin 2\theta \right\} \tag{3.5.82}$$

$$r^{\frac{1}{n+1}} \left\{ \sin\frac{\theta}{2}, \cos\frac{\theta}{2}, \sin\frac{\theta}{2}\sin\theta, \cos\frac{\theta}{2}\sin\theta, \sin\frac{\theta}{2}\sin 3\theta, \cos\frac{\theta}{2}\sin 3\theta \right\} \tag{3.5.83}$$

$$r^{\frac{1}{n+1}} \left\{ \sin\frac{\theta}{2}, \cos\frac{\theta}{2}, \sin\frac{\theta}{2}\sin\theta, \cos\frac{\theta}{2}\sin\theta, \sin\frac{\theta}{2}\sin 2\theta, \cos\frac{\theta}{2}\sin 2\theta, \sin\frac{\theta}{2}\sin 3\theta, \cos\frac{\theta}{2}\sin 3\theta \right\} \tag{3.5.84}$$

Due to the lack of reliable analytical solution for general elastoplastic problems, they are usually analysed by conventional enrichments of XFEM (Hatefi et al. 2019).

3.5.12.11 Contact Corner Enrichment

The enrichment functions associated with a contact wedge/corner at various free/stick/slip contact boundary conditions (Figure 3.25) can be written in terms of the dominant singularity power λ (Ebrahimi 2012):

$$\{F_l(r,\theta)\}_{l=1}^{4} = \left\{ r^\lambda \cos(\lambda\theta), r^\lambda \sin(\lambda\theta), r^\lambda \sin\theta \sin(\lambda\theta), r^\lambda \sin\theta \cos(\lambda\theta) \right\} \tag{3.5.85}$$

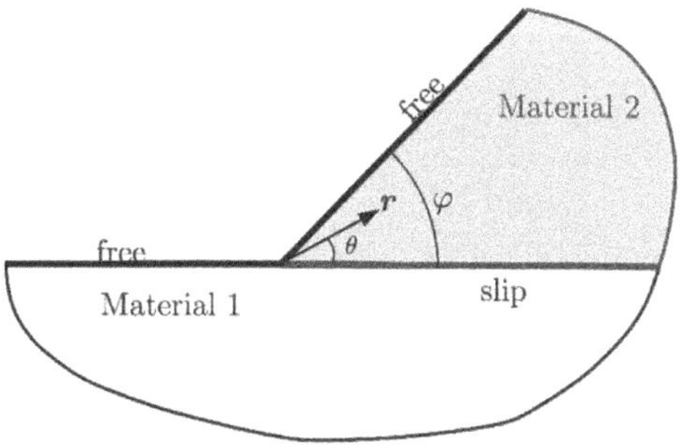

Figure 3.25 Configuration of a wedge sliding contact.

For a complete sliding punch on a dissimilar plane, λ is computed from the following characteristic equation (Ebrahimi 2012; Ebrihimi et al. 2013):

$$\begin{aligned} 8(1+\lambda)\sin\lambda\pi\Big\{(1+\alpha)\cos\lambda\pi\left(\sin^2\lambda\varphi-\lambda^2\sin^2\varphi\right) \\ +\frac{1}{2}(1-\alpha)\sin\lambda\pi\left(\sin 2\lambda\varphi+\lambda\sin 2\varphi\right) \\ +\mu\sin\lambda\pi\left[(1-\alpha)\lambda(1+\lambda)\sin^2\varphi-2\beta\left(\sin^2\lambda\varphi-\lambda^2\sin^2\varphi\right)\right]\Big\}=0 \end{aligned} \tag{3.5.86}$$

where φ is the wedge angle and α and β are defined as

$$\alpha=\frac{(\mu_2/\mu_1)(\kappa_1+1)-(\kappa_2+1)}{(\mu_2/\mu_1)(\kappa_1+1)+(\kappa_2+1)} \tag{3.5.87}$$

$$\beta=\frac{(\mu_2/\mu_1)(\kappa_1-1)-(\kappa_2-1)}{(\mu_2/\mu_1)(\kappa_1+1)+(\kappa_2+1)} \tag{3.5.88}$$

3.5.13 XFEM Enrichment Formulation for Large Deformation Problems

There is no universal solution for the wide range of large deformation fracture mechanics problems. As a result, most XFEM enrichment strategies have focused only on improved solutions. For instance, Bouchbinder et al. (2009, 2010) presented a weakly nonlinear quasi-static solution for the neo-Hookean materials and concluded that, in addition to the conventional radial term $\sqrt{r}$, a new logarithmic term $\ln r$ should be included in the enrichment function.

Apart from the potential changes in the form of the enrichment functions, the basic XFEM formulation should also be modified for large deformation problems. For instance, in a total Lagrangian formulation, the stiffness matrix is computed from (Rashetnia and Mohammadi 2015)

$$\mathbf{K}_{ij}^{rs}=\int_{\Omega^e}(\mathbf{B}_{0a}^{r})^{\mathrm{T}}\mathbf{D}\mathbf{B}_{0b}^{s}\,\mathrm{d}\Omega \tag{3.5.89}$$

The deformation gradient $\boldsymbol{F}$ and the derivatives of the shape functions with respect to the initial configuration $\mathbf{X}=(X,Y)$ contribute in defining the standard, H-enriched and crack tip enriched components of $\mathbf{B}_0$ (Rashetnia and Mohammadi 2015):

$$\mathbf{B}_{0i}^{\boldsymbol{u}}=\begin{bmatrix} N_{i,X}\boldsymbol{F}_{11} & N_{i,X}\boldsymbol{F}_{21} \\ N_{i,Y}\boldsymbol{F}_{11} & N_{i,Y}\boldsymbol{F}_{22} \\ N_{i,X}\boldsymbol{F}_{12}+N_{i,Y}\boldsymbol{F}_{11} & N_{i,X}\boldsymbol{F}_{22}+N_{i,Y}\boldsymbol{F}_{21} \end{bmatrix} \tag{3.5.90}$$

$$\mathbf{B}_{i}^{\boldsymbol{a}}=\begin{bmatrix} N_{i,X}\boldsymbol{F}_{11}\left[H(\xi)-H(\xi_i)\right] & N_{i,X}\boldsymbol{F}_{21}\left[H(\xi)-H(\xi_i)\right] \\ N_{i,Y}\boldsymbol{F}_{11}\left[H(\xi)-H(\xi_i)\right] & N_{i,Y}\boldsymbol{F}_{22}\left[H(\xi)-H(\xi_i)\right] \\ \left[N_{i,X}\boldsymbol{F}_{12}+N_{i,Y}\boldsymbol{F}_{11}\right]\left[H(\xi)-H(\xi_i)\right] & \left[N_{i,X}\boldsymbol{F}_{22}+N_{i,Y}\boldsymbol{F}_{21}\right]\left[H(\xi)-H(\xi_i)\right] \end{bmatrix} \tag{3.5.91}$$

$$\mathbf{B}_{i}^{\boldsymbol{b}}=\begin{bmatrix}\mathbf{B}_{i}^{b1} & \mathbf{B}_{i}^{b2} & \mathbf{B}_{i}^{b3} & \mathbf{B}_{i}^{b4}\end{bmatrix} \tag{3.5.92}$$

$$\mathbf{B}_i^{b\alpha} = \begin{bmatrix} \left(N_i\bar{F}_\alpha\right)_{,X} \boldsymbol{F}_{11} & \left(N_i\bar{F}_\alpha\right)_{,X} \boldsymbol{F}_{21} \\ \left(N_i\bar{F}_\alpha\right)_{,Y} \boldsymbol{F}_{12} & \left(N_i\bar{F}_\alpha\right)_{,Y} \boldsymbol{F}_{22} \\ \left(N_i\bar{F}_\alpha\right)_{,X} \boldsymbol{F}_{12} + \left(N_i\bar{F}_\alpha\right)_{,Y} \boldsymbol{F}_{11} & \left(N_i\bar{F}_\alpha\right)_{,X} \boldsymbol{F}_{22} + \left(N_i\bar{F}_\alpha\right)_{,Y} \boldsymbol{F}_{21} \end{bmatrix} \tag{3.5.93}$$

$$\bar{F}_\alpha = F_\alpha - F_{\alpha i} \ , \alpha = 1-4 \tag{3.5.94}$$

where the components of the deformation gradient $\boldsymbol{F}_{iI}$ are defined as

$$\boldsymbol{F}_{iI} = \frac{\partial \mathbf{x}_i}{\partial \mathbf{X}_I} = \begin{bmatrix} \boldsymbol{F}_{11} & \boldsymbol{F}_{12} \\ \boldsymbol{F}_{21} & \boldsymbol{F}_{22} \end{bmatrix} \tag{3.5.95}$$

3.6 Extended Isogeometric Analysis (XIGA)

3.6.1 Introduction

The isogeometric analysis (IGA) is based on the concept of non-uniform rational B-splines (NURBS) basis functions for a description of the geometry of the problem domain and approximation of the solution field (Hughes et al. 2005). It has been successfully employed for solution of a wide range of engineering problems and physical applications (Cottrell et al. 2009; Verhoosel et al. 2011a, 2011b). The original idea of IGA has also been expanded for crack propagation simulations by the extended isogemetric analysis (XIGA) (Benson et al. 2010; Ghorashi et al. 2012) and the T-spine XIGA (Ghorashi et al. 2015).

3.6.2 Isogeometric Analysis

3.6.2.1 The Basis Function

The B-spline and NURBS functions are briefly discussed in this section. A NURBS curve of order p for a set of n control points $\boldsymbol{T}_i$ is defined as

$$C(\xi) = \sum_{i=1}^{n} R_i^p(\xi)\boldsymbol{T}_i, \quad 0 \le \xi \le 1 \tag{3.6.1}$$

where

$$R_i^p(\xi) = \frac{Z_{i,p}(\xi) w_i}{\sum_{\hat{i}=1}^{n} Z_{\hat{i},p}(\xi) w_{\hat{i}}} \tag{3.6.2}$$

where R_i^p represents the NURBS functions, $\boldsymbol{T}_i = (x_{i_1}, x_{i_2})$ are the coordinate positions of the control points and w_i are the weights associated with each control point

$$\boldsymbol{w} = \left\{ w_1, w_2, \ldots, w_n \right\} \tag{3.6.3}$$

The B-spline $Z_{i,p}$ basis functions of order p are defined in the parametric space of the so-called knot vector Ξ,

$$\Xi = \left\{ \xi_1, \xi_2, \ldots, \xi_{n+p+1} \right\}, \ \xi_i \le \xi_{i+1}, i = 1, 2, \ldots, n+p \tag{3.6.4}$$

where the knots $0<\xi_i<1$ represent the coordinates in the parametric space. An open knot vector (with $p+1$ repeated end knots) is preferred in IGA to satisfy the Kronecker delta property at boundary points (Roh and Cho 2004).

The B-spline basis functions $Z_{i,p}(\xi)$ are defined in the recursive form of (Piegl and Tiller 1997)

$$Z_{i,0}(\xi)=\begin{cases}1, & \text{if } \xi_i\le\xi\le\xi_{i+1}\\ 0, & \text{otherwise}\end{cases} \tag{3.6.5}$$

$$Z_{i,p}(\xi)=\frac{\xi-\xi_i}{\xi_{i+p}-\xi_i}Z_{i,p-1}(\xi)+\frac{\xi_{i+p+1}-\xi}{\xi_{i+p+1}-\xi_{i+1}}Z_{i+1,p-1}(\xi)\;\; for\ p=1,2,3,\ldots \tag{3.6.6}$$

and its derivative can be computed from

$$\frac{dZ_{i,p}(\xi)}{d\xi}=\frac{p}{\xi_{i+p}-\xi_i}Z_{i,p-1}(\xi)+\frac{p}{\xi_{i+p+1}-\xi_{i+1}}Z_{i+1,p-1}(\xi) \tag{3.6.7}$$

The NURBS basis functions satisfy the partition of unity (PU) property. They are continuous of order C^{p-k} if a knot has multiplicity k. The support domain of each NURBS function $R_i^p(\xi)$ is within the interval $[\xi_i,\xi_{i+p+1}]$.

An extension to the NURBS surface of orders p and q in the ξ_1 and ξ_2 directions, respectively, is defined in terms of the control grid points $\boldsymbol{T}_{i,j}$ and the corresponding weights $w_{i,j}$:

$$S(\xi_1,\xi_2)=\sum_{i=1}^{n}\sum_{j=1}^{m}\frac{Z_{i,p}(\xi_1)Y_{j,q}(\xi_2)w_{i,j}}{\sum_{\hat{i}=1}^{n}\sum_{\hat{j}=1}^{m}Z_{\hat{i},p}(\xi_1)Y_{\hat{j},q}(\xi_2)w_{\hat{i},\hat{j}}}T_{i,j},\ 0\le\xi_1,\xi_2\le1 \tag{3.6.8}$$

where $\{Z_{i,p}\}$ and $\{Y_{j,q}\}$ are the B-spline basis functions defined on the knot vectors Ξ_1 and Ξ_2, respectively. The total number of control points per each knot span (element) is $n_{en}=(p+1)\times(q+1)$.

For a better illustration, a simple example of a disk of radius 10 is considered, as depicted in Figure 3.26. In order to generate the IGA model of the disk, nine control points $\mathbf{T}_i$ and their corresponding weights w_i are considered:

$$\mathbf{T}^{\mathrm{T}}=\left\{\begin{pmatrix}0\\-10a\end{pmatrix},\begin{pmatrix}5a\\-5a\end{pmatrix},\begin{pmatrix}10a\\0\end{pmatrix},\begin{pmatrix}5a\\5a\end{pmatrix},\begin{pmatrix}0\\10a\end{pmatrix},\begin{pmatrix}-5a\\5a\end{pmatrix},\begin{pmatrix}-10a\\0\end{pmatrix},\begin{pmatrix}-5a\\-5a\end{pmatrix},\begin{pmatrix}0\\0\end{pmatrix}\right\},a=\sqrt{2} \tag{3.6.9}$$

$$\boldsymbol{w}=\left\{\frac{1}{\sqrt{2}},1,\frac{1}{\sqrt{2}},1,\frac{1}{\sqrt{2}},1,\frac{1}{\sqrt{2}},1,1\right\} \tag{3.6.10}$$

Note that the control points are not positioned on the boundary of the disk. The knot vectors $\Xi_1=\Xi_2=\{0,0,0,1,1,1\}$ and NURBS function of order 2 are applied. Only one IGA element is sufficient to represent the exact geometry.

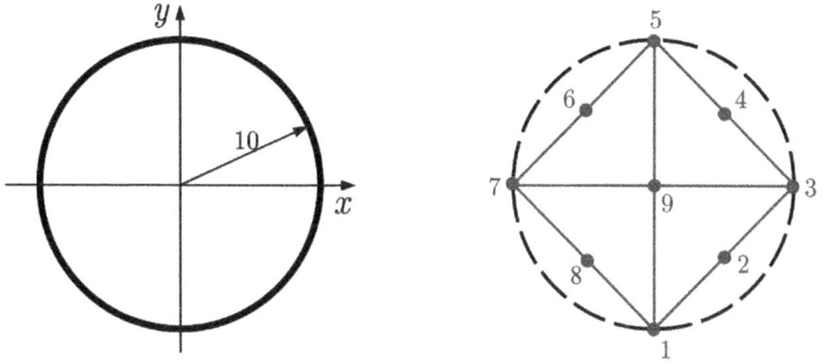

Figure 3.26 Mesh and control net for a disk of radius 10.

For more details on the B-spline and NURBS approximations, refer to Piegl and Tiller (1997).

3.6.2.2 Isogeometric Discretization

Similar to an isoparametric finite element formulation, IGA uses the NURBS basis for parameterization of both the geometry and the displacement field.

Approximation of the physical position $\mathbf{x} = (x_1, x_2)$ and the displacement $\boldsymbol{u}^h$ for a point $\xi = (\xi_1, \xi_2)$ in the parametric coordinates can be derived from

$$\mathbf{x}(\boldsymbol{\xi}) = \mathbf{R}(\boldsymbol{\xi})\mathbf{T} = \sum_{i=1}^{n_{en}} R_i(\boldsymbol{\xi})\boldsymbol{T}_i \tag{3.6.11}$$

$$\boldsymbol{u}^h(\boldsymbol{\xi}) = \mathbf{R}(\boldsymbol{\xi})\mathbf{U} = \sum_{i=1}^{n_{en}} R_i(\boldsymbol{\xi})\boldsymbol{u}_i \tag{3.6.12}$$

where $\mathbf{R}(\boldsymbol{\xi})$ is the vector of NURBS basis functions $R_i(i = 1, 2, \ldots, n_{en})$ in the parametric space of $\boldsymbol{\xi}$.

The strain and stress fields can be computed from

$$\varepsilon(\boldsymbol{\xi}) = \mathbf{B}(\boldsymbol{\xi})\mathbf{U}_s \tag{3.6.13}$$

$$\sigma(\boldsymbol{\xi}) = \mathbf{D}\varepsilon(\boldsymbol{\xi}) \tag{3.6.14}$$

where $\mathbf{B}$ is the matrix of basis function derivatives,

$$\mathbf{B} = \begin{bmatrix} \dfrac{\partial R_1}{\partial x_1} & 0 & \cdots & \dfrac{\partial R_{n_{en}}}{\partial x_1} & 0 \\ 0 & \dfrac{\partial R_1}{\partial x_2} & \cdots & 0 & \dfrac{\partial R_{n_{en}}}{\partial x_2} \\ \dfrac{\partial R_1}{\partial x_2} & \dfrac{\partial R_1}{\partial x_1} & \cdots & \dfrac{\partial R_{n_{en}}}{\partial x_2} & \dfrac{\partial R_{n_{en}}}{\partial x_1} \end{bmatrix} \tag{3.6.15}$$

and $n_{en} = (p+1)\times(q+1)$ is the number of non-zero basis functions for a given knot span (element).

The derivative of the basis functions with respect to the physical coordinates $\mathbf{x}$ in Equation (3.6.15) can be calculated from

$$\begin{Bmatrix} \dfrac{\partial R_i}{\partial x_1} \\ \dfrac{\partial R_i}{\partial x_2} \end{Bmatrix} = \mathbf{J}^{-1} \begin{Bmatrix} \dfrac{\partial R_i}{\partial \xi_1} \\ \dfrac{\partial R_i}{\partial \xi_2} \end{Bmatrix} \tag{3.6.16}$$

where the Jacobian matrix $\mathbf{J}$ for the transformation between physical and parametric spaces is defined as

$$\mathbf{J} = \begin{bmatrix} \dfrac{\partial x_1}{\partial \xi_1} & \dfrac{\partial x_2}{\partial \xi_1} \\ \dfrac{\partial x_1}{\partial \xi_2} & \dfrac{\partial x_2}{\partial \xi_2} \end{bmatrix} \tag{3.6.17}$$

3.6.3 Extended Isogeometric Analysis (XIGA)

In order to enhance the capabilities of IGA to efficiently analyse discontinuous problems such as cracks, the concepts of the XFEM (Mohammadi 2012a) can be included. In the extended isogeometric analysis, the isogeometric approximation is locally enriched to simulate discontinuities and singular fields. According to the location of any crack, a few degrees of freedom are added to the selected control points of the original IGA model near the crack. They contribute to the overall approximation through the use of enrichment functions.

In order to model crack edges and tips in XIGA, Equation (3.6.12) can be generalized by an extrinsic enrichment approximation of displacement for a typical point $\boldsymbol{\xi}$ (Ghorashi et al. 2012):

$$\boldsymbol{u}^h(\boldsymbol{\xi}) = \boldsymbol{u}^{IGA} + \boldsymbol{u}^{H} + \boldsymbol{u}^{tip} \tag{3.6.18}$$

$$\boldsymbol{u}^h(\boldsymbol{\xi}) = \underbrace{\sum_{i=1}^{n_{en}} R_i(\boldsymbol{\xi})\boldsymbol{u}}_{u^{IGA}} + \underbrace{\sum_{j=1}^{n_{cf}} R_j(\boldsymbol{\xi})H(\boldsymbol{\xi})\boldsymbol{d}_j}_{u^{H}} + \underbrace{\sum_{k=1}^{n_{ct}} R_k(\boldsymbol{\xi})(\sum_{l=1}^{4} F_l(\boldsymbol{\xi})\mathbf{c}_k^l)}_{u^{tip}} \tag{3.6.19}$$

where $H(\boldsymbol{\xi})$ and $F_l(\boldsymbol{\xi})$ $(l = 1,2,3,4)$ are the Heaviside and crack tip enrichment functions, respectively, and $\boldsymbol{d}_j$ and $\mathbf{c}_k^l$ are vectors of additional degrees of freedom associated with H and F_l, respectively, while n_{en} is the number of basis functions and n_{cf} and n_{ct} are the number of basis functions that have crack face and crack tip in their support domains, respectively.

In Equation (3.6.19), $H(\boldsymbol{\xi}(\mathbf{X}))$ is the generalized Heaviside function, defined in terms of the closest point $\mathbf{x}_\Gamma$ of a crack to the point $\mathbf{x}$ (see Figure 3.15):

$$H(\boldsymbol{\xi}(\mathbf{X})) = \begin{cases} +1, & \text{if}\left(\mathbf{x}-\mathbf{x}_\Gamma\right)\cdot\boldsymbol{n} > 0 \\ -1, & \text{otherwise} \end{cases} \tag{3.6.20}$$

where $\boldsymbol{n}$ is the unit normal vector in point $\mathbf{x}_\Gamma$ on the crack.

The crack tip enrichment functions F_l allow for reproduction of the singular stress fields near the crack tip,

$$F_l(r,\theta) = \left\{\sqrt{r}\sin\frac{\theta}{2}, \sqrt{r}\cos\frac{\theta}{2}, \sqrt{r}\sin\theta\sin\frac{\theta}{2}, \sqrt{r}\sin\theta\cos\frac{\theta}{2}\right\} \tag{3.6.21}$$

where the local polar coordinates (r,θ) in the physical space are defined as (see Figure 3.27)

$$\begin{cases} r = \sqrt{x_1^2 + x_2^2} \\ \theta = \arctan\left(\dfrac{x_2}{x_1}\right) \end{cases} \tag{3.6.22}$$

where (x_1, x_2) are the local Cartesian coordinates at the crack tip $\mathbf{x}_{tip} = (x_{1_{tip}}, x_{2_{tip}})$ in terms of the crack inclination angle θ_0 (see Figure 3.27),

$$\begin{Bmatrix} x_1 \\ x_2 \end{Bmatrix} = \begin{bmatrix} \cos\theta_0 & \sin\theta_0 \\ -\sin\theta_0 & \cos\theta_0 \end{bmatrix} \begin{Bmatrix} x_1 - x_{1_{tip}} \\ x_2 - x_{2_{tip}} \end{Bmatrix} \tag{3.6.23}$$

3.6.3.1 Selection of the Enriched Control Points

An equivalent number of basis functions and control points exist in an IGA analysis. As a result, a unique control point and its influence domain can be assigned to each basis function. However, the control points may not be located in the support domain of basis functions, because the control points may not be located in the physical space.

Figure 3.28 illustrates the procedure for selection of control points $\mathbf{T}_h$ and $\mathbf{T}_t$ for H and F_l enrichments. The concept is that the control points for H and F_l are selected for enrichments if the support domain of corresponding basis functions include the crack face and the crack tip, respectively.

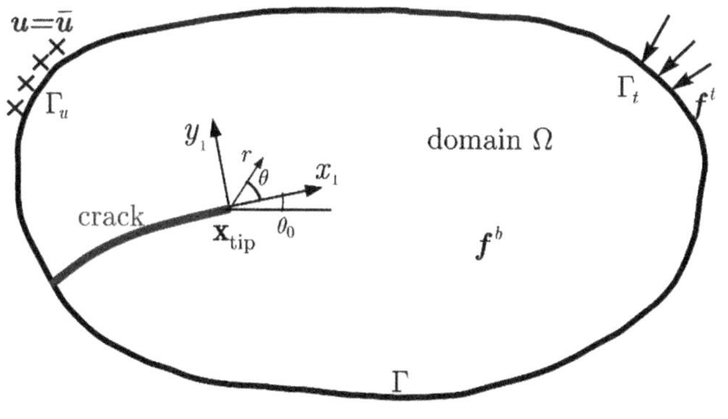

Figure 3.27 A two-dimensional cracked medium.

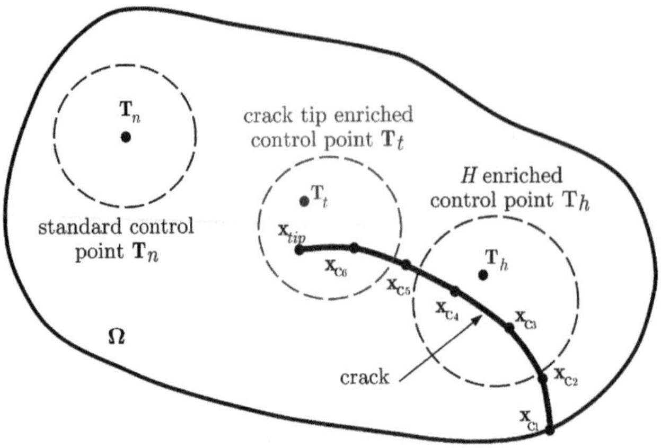

Figure 3.28 Procedure for selection of enriched control points: $\mathbf{T}_h$ and $\mathbf{T}_t$ are the Heaviside and crack tip enriched control points, respectively, and $\mathbf{T}_n$ represents the standard (un-enriched) control points. *Source:* Adapted from Ghorashi et al. 2012.

The first step is to compute the parametric coordinates of the crack tip (ξ_{tip}) and the corresponding NURBS functions. The crack tip enriched control points are specified by the non-zero NURBS values,

$$R_i(\xi_{tip}) \neq 0, i = 1, 2, \ldots, n_{cp} \tag{3.6.24}$$

Selection of control points for H enrichments is similarly performed based on some points on the crack face ($\mathbf{x}_{c1}, \mathbf{x}_{c2}, \ldots, \mathbf{x}_{c6}$ instead of $\mathbf{x}_{tip}$ in Figure 3.28):

$$R_j(\xi_{ci}) \neq 0 \tag{3.6.25}$$

For an ordinary control point (such as $\mathbf{T}_n$ in Figure 3.28),

$$R_l(\xi_{ci}) = 0, i = 1, 2, \ldots, n_{cp} \tag{3.6.26}$$

$$R_l(\xi_{tip}) = 0 \tag{3.6.27}$$

If a control point is selected for both the Heaviside and crack tip enrichments, it is only considered as a control point for crack tip enrichment (for instance, $\mathbf{T}_t$ in Figure 3.28).

3.6.4 XIGA Governing Equations

A typical 2D linear elastic domain that includes a traction-free crack is considered, as depicted in Figure 3.27.

The strong form and boundary conditions for this problem can be written as

$$\nabla \cdot \boldsymbol{\sigma} + \boldsymbol{f}^b = \mathbf{0} \quad \text{in } \Omega \tag{3.6.28}$$

$$\boldsymbol{\sigma} \cdot \boldsymbol{n} = \boldsymbol{f}^t \quad \text{on } \Gamma_t \tag{3.6.29}$$

$$\boldsymbol{u} = \bar{\boldsymbol{u}} \quad \text{on}\, \Gamma_u \tag{3.6.30}$$

where $\boldsymbol{\sigma}$ and $\boldsymbol{u}$ are the stress and displacement vectors, respectively, $\boldsymbol{f}^{\mathrm{b}}$ is the body force and $\boldsymbol{f}^{\mathrm{t}}$ is the prescribed traction on the traction boundary with the unit outward normal vector $\boldsymbol{n}$, and $\bar{\boldsymbol{u}}$ represents the prescribed displacement on the essential boundary (see Figure 3.27).

The lack of the Kronecker delta property of the NURBS basis functions necessitates a constraint enforcement method, such as the Lagrange method, to impose the essential boundary conditions (Wang and Xuan 2010). The governing weak form of Equation (3.6.28) with the Lagrange constraint method can be written as

$$\int_{\Omega} \delta \boldsymbol{\varepsilon}^{\mathrm{T}} \boldsymbol{\sigma} \mathrm{d}\Omega - \int_{\Omega} \delta \boldsymbol{u}^{\mathrm{T}} \boldsymbol{f}^{b} \mathrm{d}\Omega - \int_{\Gamma_t} \delta \boldsymbol{u}^{\mathrm{T}} \boldsymbol{f}^{t} \mathrm{d}\Gamma - \int_{\Gamma_t} \delta \boldsymbol{\lambda}^{\mathrm{T}} (\boldsymbol{u} - \bar{\boldsymbol{u}}) \mathrm{d}\Gamma - \int_{\Gamma_t} \delta \boldsymbol{u}^{\mathrm{T}} \lambda \mathrm{d}\Gamma = 0 \tag{3.6.31}$$

The new unknown vector field $\boldsymbol{\lambda}$ is defined for the set of n_λ control points on or close to the essential boundary. It can be discretized by the Lagrange interpolation shape functions $\mathbf{N}_\lambda(s)$ defined on the arc length s of the essential boundary,

$$\boldsymbol{\Lambda}^h = \mathbf{N}_\lambda(s) \boldsymbol{\lambda} \tag{3.6.32}$$

The final discretized form of the equation is derived as (Belytschko et al. 1994; Ghorashi et al. 2012)

$$\begin{bmatrix} \mathbf{K} & \mathbf{G} \\ \mathbf{G}^{\mathrm{T}} & 0 \end{bmatrix} \begin{Bmatrix} \mathbf{U}^h \\ \boldsymbol{\Lambda} \end{Bmatrix} = \begin{Bmatrix} \mathbf{f} \\ \mathbf{q} \end{Bmatrix} \tag{3.6.33}$$

where $\mathbf{U}^h$ is the global vector of XIGA degrees of freedom that includes the displacement control variables $\boldsymbol{u}$ and additional Heaviside and crack tip enrichment degrees of freedom, $\boldsymbol{d}$ and $\mathbf{c}_\alpha$, respectively,

$$\mathbf{U}^h = \begin{Bmatrix} \boldsymbol{u} & \boldsymbol{d} & \mathbf{c}_1 & \mathbf{c}_2 & \mathbf{c}_3 & \mathbf{c}_4 \end{Bmatrix}^{\mathrm{T}} \tag{3.6.34}$$

and $\boldsymbol{\lambda}$ is the vector of Lagrange multipliers associated with n_λ control points on the essential boundary.

$\mathbf{K}$ is the stiffness matrix assembled from the contributions of the stiffnesses of elements,

$$\mathbf{K}_{ij}^{\mathbf{e}} = \begin{bmatrix} \mathbf{K}_{ij}^{uu} & \mathbf{K}_{ij}^{ud} & \mathbf{K}_{ij}^{uc} \\ \mathbf{K}_{ij}^{du} & \mathbf{K}_{ij}^{dd} & \mathbf{K}_{ij}^{dc} \\ \mathbf{K}_{ij}^{cu} & \mathbf{K}_{ij}^{cd} & \mathbf{K}_{ij}^{cc} \end{bmatrix} (i, j = 1, 2, \ldots, n_{\mathrm{en}}) \tag{3.6.35}$$

with

$$\mathbf{K}_{ij}^{rs} = \int_{\Omega_e} (\mathbf{B}_i^r)^{\mathrm{T}} \mathbf{D} \mathbf{B}_j^s \, \mathrm{d}\Omega \, (r, s = \boldsymbol{u}, \boldsymbol{d}, \mathbf{c}) \tag{3.6.36}$$

where $\mathbf{B}_i$ matrices are defined as

$$\mathbf{B}_i^{\boldsymbol{u}} = \begin{bmatrix} \left(R_i\right)_{,X_1} & 0 \\ 0 & \left(R_i\right)_{,X_2} \\ \left(R_i\right)_{,X_2} & \left(R_i\right)_{,X_1} \end{bmatrix} \tag{3.6.37}$$

$$\mathbf{B}_i^{\boldsymbol{d}} = \begin{bmatrix} \left(R_i\right)_{,X_1} H & 0 \\ 0 & \left(R_i\right)_{,X_2} H \\ \left(R_i\right)_{,X_2} H & \left(R_i\right)_{,X_1} H \end{bmatrix} \tag{3.6.38}$$

$$\mathbf{B}_i^{\boldsymbol{c}} = \begin{bmatrix} \mathbf{B}_i^{c_1} & \mathbf{B}_i^{c_2} & \mathbf{B}_i^{c_3} & \mathbf{B}_i^{c_4} \end{bmatrix} \tag{3.6.39}$$

$$\mathbf{B}_i^{c_l} = \begin{bmatrix} \left(R_i F_l\right)_{,X_1} & 0 \\ 0 & \left(R_i F_l\right)_{,X_2} \\ \left(R_i F_l\right)_{,X_2} & \left(R_i F_l\right)_{,X_1} \end{bmatrix} (l = 1,2,3,4) \tag{3.6.40}$$

and $\mathbf{f}$ is the force vector assembled from the contributions of elements

$$\mathbf{f}_i^{\mathbf{e}} = \left\{ \mathbf{f}_i^{\boldsymbol{u}} \quad \mathbf{f}_i^{\boldsymbol{d}} \quad \mathbf{f}_i^{c_1} \quad \mathbf{f}_i^{c_2} \quad \mathbf{f}_i^{c_3} \quad \mathbf{f}_i^{c_4} \right\}^{\mathrm{T}} \tag{3.6.41}$$

$$\mathbf{f}_i^{u} = \int_{\Omega_e} R_i^{\mathrm{T}} \boldsymbol{f}^b \, \mathrm{d}\Omega + \int_{\Gamma_{t_e}} R_i^{\mathrm{T}} \boldsymbol{f}^t \, \mathrm{d}\Gamma \tag{3.6.42}$$

$$\mathbf{f}_i^{d} = \int_{\Omega_e} R_i^{\mathrm{T}} H \boldsymbol{f}^b \, \mathrm{d}\Omega + \int_{\Gamma_{t_e}} R_i^{\mathrm{T}} H \boldsymbol{f}^t \, \mathrm{d}\Gamma \tag{3.6.43}$$

$$\mathbf{f}_i^{c_l} = \int_{\Omega_e} R_i^{\mathrm{T}} F_l \boldsymbol{f}^b \, \mathrm{d}\Omega + \int_{\Gamma_{t_e}} R_i^{\mathrm{T}} F_l \boldsymbol{f}^t \, \mathrm{d}\Gamma \quad (l = 1,2,3,4) \tag{3.6.44}$$

Finally, the terms associated with the Lagrange multiplier approach can be written as

$$\mathbf{G}_{ij}^{\mathrm{T}} = -\int_{\Gamma_u} (\mathbf{N}_\lambda)_i^{\mathrm{T}} R_j \mathrm{d}\Gamma \tag{3.6.45}$$

$$\mathbf{q}_i = -\int_{\Gamma_u} (\mathbf{N}_\lambda)_i^{\mathrm{T}} \bar{\boldsymbol{u}} \mathrm{d}\Gamma \tag{3.6.46}$$

3.6.5 Numerical Integration

The Gauss quadrature rule is usually adopted to compute IGA integrals (3.6.36) and (3.6.42) to (3.6.46). Similar to XFEM integrations, the accuracy of XIGA computations may

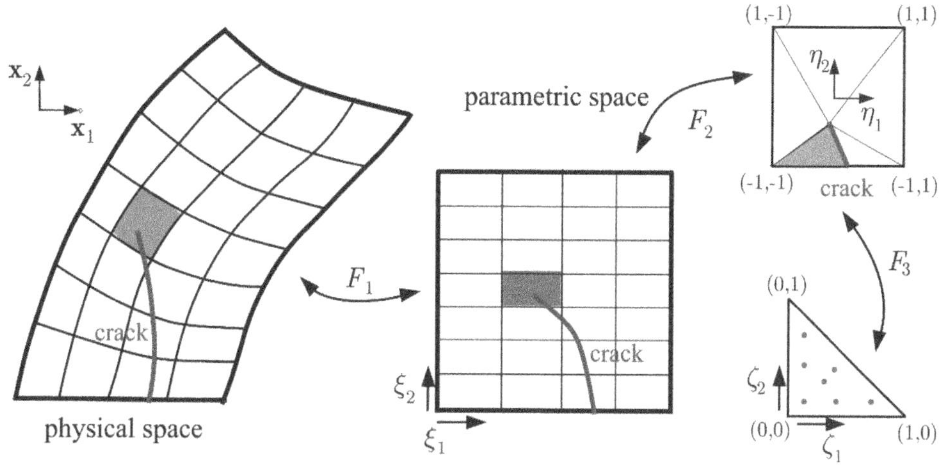

Figure 3.29 Transformations for sub-triangle numerical integration in XIGA. *Source:* Adapted from Ghorashi et al. 2012.

significantly decrease for crack problems, where the conventional Gauss quadrature rules largely deteriorate. An efficient and simple remedy is to use the sub-triangles technique, illustrated in Figure 3.29, which subdivides the elements intersected with a crack into sub-triangles whose edges conform with crack faces.

Sub-triangulation of the cracked element is performed on the parent element with $[-1,1] \times [-1,1]$ domain. Identification of the intersection points of the crack and the parent element edges is determined by the signed distances of the element vertices from the crack face in the physical space. Accordingly, a simple nonlinear equation has to be solved to determine the position of the crack tip in the parent element.

Necessary transformations between the physical and parametric coordinates must be considered in the numerical integration. According to Figure 3.29, while ordinary elements only require IGA conventional transformations F_2 and F_1 (Cottrell et al. 2009), elements with straight cracks require an additional transformation F_3.

$$F_3 : \begin{cases} \eta_1 = \eta_1^1 (1 - \zeta_1 - \zeta_2) + \eta_1^2 (\zeta_1) + \eta_1^3 (\zeta_2) \\ \eta_2 = \eta_2^1 (1 - \zeta_1 - \zeta_2) + \eta_2^2 (\zeta_1) + \eta_2^3 (\zeta_2) \end{cases} \tag{3.6.47}$$

$$\mathbf{J}_3 = \begin{bmatrix} \dfrac{\partial \eta_1}{\partial \zeta_1} & \dfrac{\partial \eta_2}{\partial \zeta_1} \\ \dfrac{\partial \eta_1}{\partial \zeta_2} & \dfrac{\partial \eta_2}{\partial \zeta_2} \end{bmatrix} = \begin{bmatrix} -\eta_1^1 + \eta_1^2 & -\eta_2^1 + \eta_2^2 \\ -\eta_1^1 + \eta_1^3 & -\eta_2^1 + \eta_2^3 \end{bmatrix} \tag{3.6.48}$$

Another approach for numerical integration is to apply the polar integration technique (Laborde et al. 2005), where an almost polar distribution is adopted for the locations of the

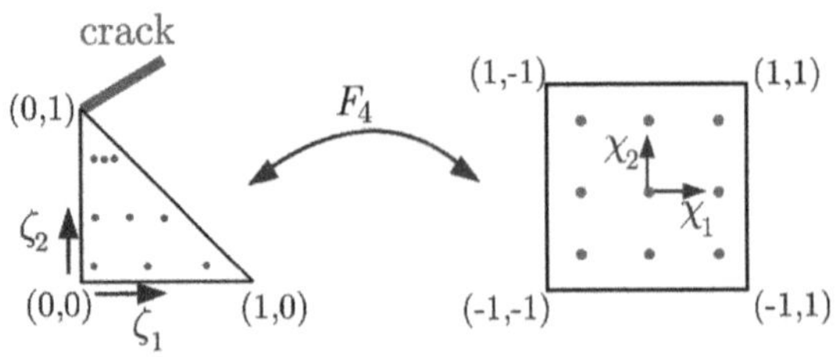

Figure 3.30 Transformation F_4 from a square into a triangle with the crack tip on its vertex. *Source:* Adapted from Laborde et al. 2005.

integration points in the crack tip element, as shown in Figure 3.30. The method further requires another transformation, F_4,

$$F_4 : \begin{cases} \zeta_1 = \dfrac{1}{4}(1 + \chi_1 - \chi_2 - \chi_1\chi_2) \\ \zeta_2 = \dfrac{1}{2}(1 + \chi_2) \end{cases} \tag{3.6.49}$$

$$\mathbf{J}_4 = \begin{bmatrix} \dfrac{\partial \zeta_1}{\partial \chi_1} & \dfrac{\partial \zeta_2}{\partial \chi_1} \\ \dfrac{\partial \zeta_1}{\partial \chi_2} & \dfrac{\partial \zeta_2}{\partial \chi_2} \end{bmatrix} = \begin{bmatrix} \dfrac{1}{4}(1 - \chi_2) & 0 \\ \dfrac{1}{4}(-1 - \chi_1) & \dfrac{1}{2} \end{bmatrix} \tag{3.6.50}$$

3.7 Meshless Methods

Meshless methods, also called meshfree methods, are among the advanced methods for numerical analysis of engineering and physical systems. Moreover, some of the excellent meshless concepts have influenced other computational methodologies. In this section, a number of main meshless methods are briefly reviewed.

3.7.1 Why Going Meshless

The finite element method has developed well over the decades and many mature powerful commercial FEM softwares are available for almost any engineering application. Moreover, several tailored softwares based on other numerical techniques are available for specific applications.

Meshless methods have been developed mainly in the 1990s, in a bid to overcome the shortcomings of the finite element method and, for some researchers, to replace it. Since then, however, the concepts of meshless methods have substantially transformed many existing numerical methods into more powerful approaches.

Extremely large number of meshless methods are available, which may share almost nothing except than the fact of not being a finite element method. As a result, the concepts of meshless methods vary significantly from one meshless method to another. Nevertheless,

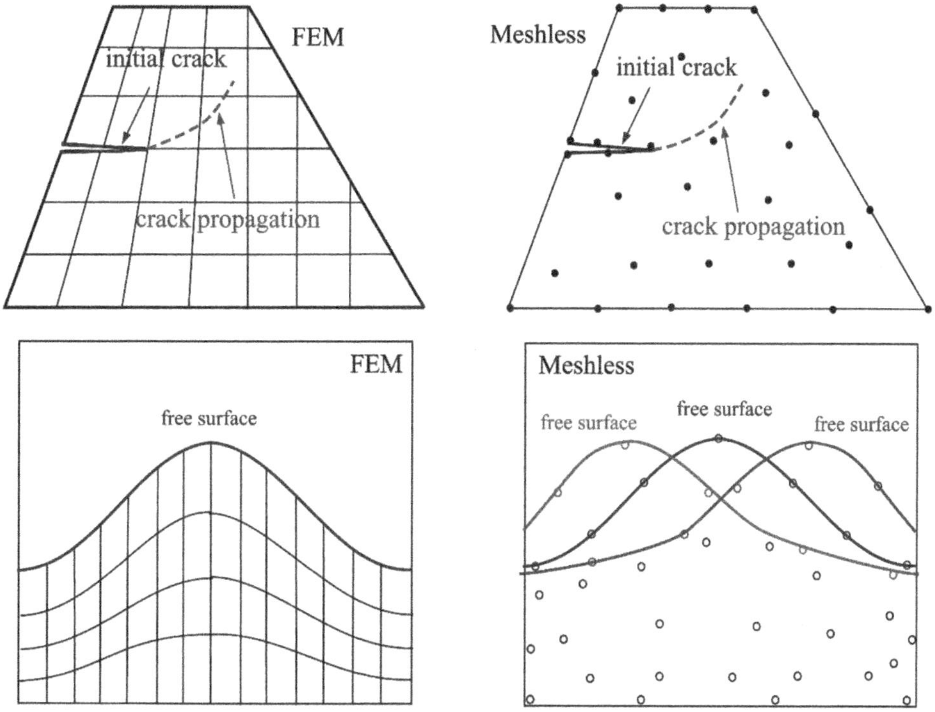

Figure 3.31 Comparison of the finite element and meshless methods for a crack problem and a free-surface fluid simulation.

most of the meshless methods allow for arbitrary/moving/propagating internal boundaries (Sadeghirad and Mohammadi 2007). In contrast to the finite element method, which may suffer from element distortions, meshless methods allow for large deformations in solids and free surface changes in fluid flows, as depicted in Figure 3.31.

This section includes two main parts. The first part is dedicated to discussing major meshless approximation techniques and the way meshless shape functions are constructed. The second part is devoted to solving boundary value problems using meshless techniques.

3.7.2 Meshless Approximations

In meshless methods, the domain of the problem Ω and its boundary Γ are represented by a number of predefined nodal points (nodes) positioned at $\mathbf{x}_j$, as depicted in Figure 3.32. It should be noted that while the number of nodes in a meshless modelling is assumed finite, the meshless approximation is expected to define the geometry or field variable in all infinite points $\mathbf{x}$ of the domain and its boundary.

3.7.2.1 Moving Least Square (MLS)

The moving least square (MLS) technique is an extension of the original weighted least square (WLS) approximation to ensure continuity of the approximation in the whole domain. Consider a typical meshless domain, as illustrated in Figure 3.32. In MLS, the

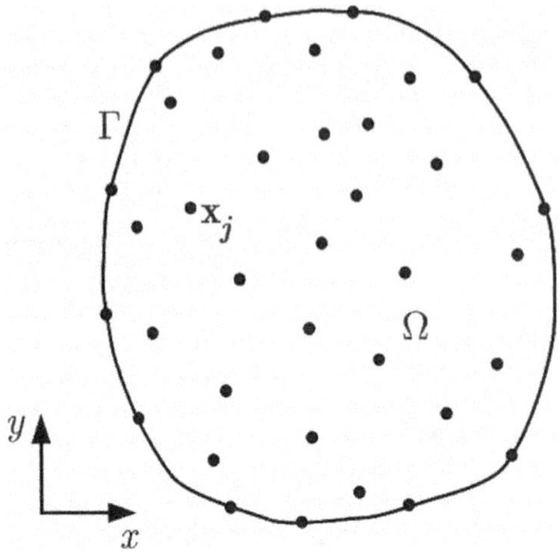

Figure 3.32 A typical meshless domain.

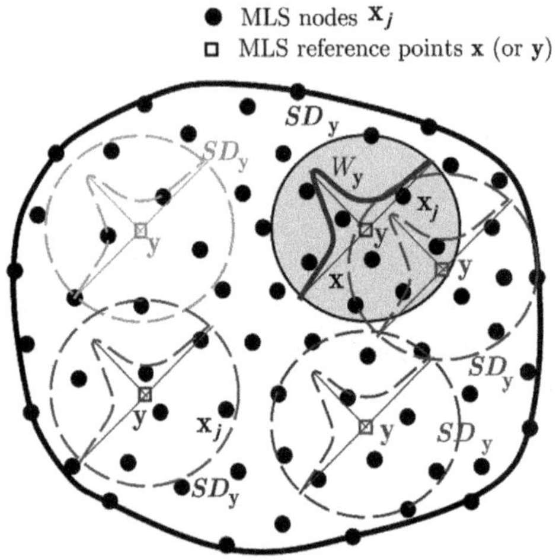

Figure 3.33 MLS subdomain and the corresponding weight function associated with the point x.

nodes that directly influence a specific point $\mathbf{x}$ of the domain are determined by defining a local support domain associated with that point, as shown by SD_x in Figure 3.33. Any point has its own independent support domain with the corresponding nodes $\mathbf{x}_j$. Any node $\mathbf{x}_j$ directly influences the field approximation in all points $\mathbf{x}$ whose support domains include that node.

The aim of this section is to define a numerical approximation for the field variable (displacement) based on the known nodal values of the variable. Later, the concept of MLS approximation is extended to the element free Galerkin (EFG) method to solve the partial differential equations by approximating the field variable in terms of the unknown nodal degrees of freedom (DOFs) (Liu 2003; Liu and Gu 2005).

Assume that $\boldsymbol{u}(\mathbf{x})$ is a continuous field variable (displacement) in the domain. The MLS method for defining $\boldsymbol{u}^h(\mathbf{x})$ as the approximation of $\boldsymbol{u}(\mathbf{x})$ at an arbitrary point $\mathbf{x}$ can be written as

$$\boldsymbol{u}^h(\mathbf{x}) = \mathbf{p}^{\mathrm{T}}(\mathbf{x})\boldsymbol{a}(\mathbf{x}) = \sum_{k=1}^{m} p_k(\mathbf{x}) a_k(\mathbf{x}) \tag{3.7.1}$$

where

$$\boldsymbol{a}^{\mathrm{T}}(\mathbf{x}) = \{a_0(\mathbf{x}), a_1(\mathbf{x}), \ldots, a_m(\mathbf{x})\} \tag{3.7.2}$$

$$\mathbf{p}^{\mathrm{T}}(\mathbf{x}) = \{p_0(\mathbf{x}), p_1(\mathbf{x}), \ldots, p_m(\mathbf{x})\} \tag{3.7.3}$$

$\mathbf{p}(\mathbf{x})$ is the basis function, which includes the polynomial functions

$$\mathbf{p}^{\mathrm{T}}(\mathbf{x}) = \{1, x, y, x^2, xy, y^2\} \tag{3.7.4}$$

or specific functions such as the crack tip enrichment functions for crack problems, as described in the XFEM formulation (Section 3.5):

$$\mathbf{p}^{\mathrm{T}}(\mathbf{x}, r, \theta) = \left\{1, x, y, \sqrt{r}\cos\frac{\theta}{2}, \sqrt{r}\sin\frac{\theta}{2}, \ldots\right\} \tag{3.7.5}$$

The vector of unknowns $\boldsymbol{a}(\mathbf{x})$ depends on the position $\mathbf{x}$. In fact, for each point $\mathbf{x}$, a new vector of constant unknowns $\boldsymbol{a}$ is computed. In other words, $\boldsymbol{a}$ is a constant vector for the support domain of any arbitrary point $\mathbf{x}$, but it changes for different points $\mathbf{x}$ (and so for different support domains). Approximation $\boldsymbol{u}^h(\mathbf{x}, \mathbf{x} = \mathbf{x}_j)$ of the field function at the nodal point $\mathbf{x}_j$ can then be defined as

$$\boldsymbol{u}^h(\mathbf{x}, \mathbf{x}_j) = \mathbf{p}^{\mathrm{T}}(\mathbf{x}_j)\boldsymbol{a}_{\mathbf{x}} = \mathbf{p}^{\mathrm{T}}(\mathbf{x}_j)\boldsymbol{a}(\mathbf{x}), \qquad j = 1, 2, \ldots, n \tag{3.7.6}$$

where n is the number of nodes inside the support domain of point $\mathbf{x}$.

A functional is defined in terms of the weighted square of residuals of the approximated values of the field function $\boldsymbol{u}^h(\mathbf{x}_j)$ at nodes of the support domain and the nodal parameters $\bar{\boldsymbol{u}}_j$,

$$\Pi = \sum_{j=1}^{n} W_{\mathbf{x}}(\mathbf{x} - \mathbf{x}_j)\left[\boldsymbol{u}^h(\mathbf{x} - \mathbf{x}_j) - \bar{\boldsymbol{u}}_j\right]^2 \tag{3.7.7}$$

$$\Pi = \sum_{j=1}^{n} W_{\mathbf{x}}(\mathbf{x} - \mathbf{x}_j)\left[\mathbf{p}^{\mathrm{T}}(\mathbf{x}_j)\boldsymbol{a}(\mathbf{x}) - \bar{\boldsymbol{u}}_j\right]^2 \tag{3.7.8}$$

The non-negative MLS weight function $W_{\mathbf{x}}$ is defined on a compact support domain around the point $\mathbf{x}$ and provides weightings at different nodes of the local support domain $SD_{\mathbf{x}}$ for evaluation of the residual functional (3.7.8). While the weight function should not be confused with the shape function, it does directly contribute in deriving the MLS shape functions. To obtain the MLS shape functions with a certain order of continuity, the weight functions should be continuous accordingly.

The role of the MLS weight function is vital in creating the moving nature of MLS. The MLS weight function $W_{\mathbf{x}}(\mathbf{x}-\mathbf{x}_j)$ vanishes (or approaches zero) on the boundary of the support domain, so it allows for nodes to enter or leave the support domain gradually as point $\mathbf{x}$ continuously moves, thus ensuring the continuity of the MLS approximation.

At an arbitrary point $\mathbf{x}$, $a(\mathbf{x})$ is chosen to minimize the weighted residual functional (3.7.8),

$$\frac{\partial \prod}{\partial \boldsymbol{a}_k} = 0, k = 1,\ldots,m \tag{3.7.9}$$

which results in

$$\mathbf{A}(\mathbf{x})\boldsymbol{a}(\mathbf{x}) = \mathbf{B}(\mathbf{x})\bar{\mathbf{U}} \tag{3.7.10}$$

where

$$\mathbf{A}(\mathbf{x}) = \sum_{j=1}^{n} W_{\mathbf{x}}(\mathbf{x}-\mathbf{x}_j)\mathbf{p}(\mathbf{x}_j)\mathbf{p}^{\mathrm{T}}(\mathbf{x}_j) \tag{3.7.11}$$

$$\mathbf{B}(\mathbf{x}) = \left[W_{\mathbf{x}}(\mathbf{x}-\mathbf{x}_1)\mathbf{p}(\mathbf{x}_1),\ldots, W_{\mathbf{x}}(\mathbf{x}-\mathbf{x}_n)\mathbf{p}(\mathbf{x}_n)\right] \tag{3.7.12}$$

$$\bar{\mathbf{U}} = \left[\bar{\boldsymbol{u}}_1, \bar{\boldsymbol{u}}_2,\ldots,\bar{\boldsymbol{u}}_n\right]^{\mathrm{T}} \tag{3.7.13}$$

Therefore,

$$\boldsymbol{u}^h(\mathbf{x}) = \mathbf{p}^T(\mathbf{x})\boldsymbol{a}(\mathbf{x}) = \mathbf{p}^{\mathrm{T}}(\mathbf{x})\mathbf{A}^{-1}(\mathbf{x})\mathbf{B}(\mathbf{x})\bar{\mathbf{U}} \tag{3.7.14}$$

Or, in terms of the MLS shape functions (see Figure 3.34),

$$\boldsymbol{u}^h(\mathbf{x}) = \varphi(\mathbf{x})\bar{\mathbf{U}} \tag{3.7.15}$$

$$\varphi(\mathbf{x}) = \mathbf{p}^{\mathrm{T}}(\mathbf{x})\mathbf{A}^{-1}(\mathbf{x})\mathbf{B}(\mathbf{x}) \tag{3.7.16}$$

and the vector of the MLS shape functions $\varphi(\mathbf{x})$ is composed of the shape functions of all nodes within the support domain of point $\mathbf{x}$ (and computed at $\mathbf{x}$):

$$\varphi(\mathbf{x}) = \left[\varphi_1(\mathbf{x}), \varphi_2(\mathbf{x}),\ldots,\varphi_n(\mathbf{x})\right] \tag{3.7.17}$$

The MLS shape functions do not satisfy the Kronecker delta property

$$\varphi_j(\mathbf{x}_l) \neq \delta_{jl} \tag{3.7.18}$$

Therefore, the nodal approximations $\boldsymbol{u}^h(\mathbf{x}_j)$ are not the same as the nodal parameters $\bar{\boldsymbol{u}}_j$:

$$\boldsymbol{u}^h(\mathbf{x}_j) \neq \left[u(\mathbf{x}_j) = \bar{\boldsymbol{u}}_j\right] \tag{3.7.19}$$

As a result, the method is not an interpolation.

Moreover, approximation of the displacement at a node, $\boldsymbol{u}^h(\mathbf{x}_j)$, depends on all nodes within the support domain of that node, which complicates enforcement of the boundary condition rather than using the simple procedure in the FEM (see Figure 3.35).

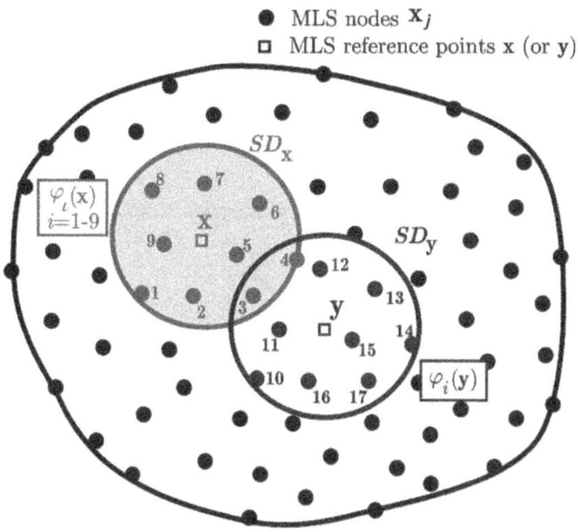

Figure 3.34 A simple procedure to determine the values of shape functions at nodes. For instance, $\varphi_{16}(\mathbf{x}_3) \neq 0$, $\varphi_{17}(\mathbf{x}_8) = 0$.

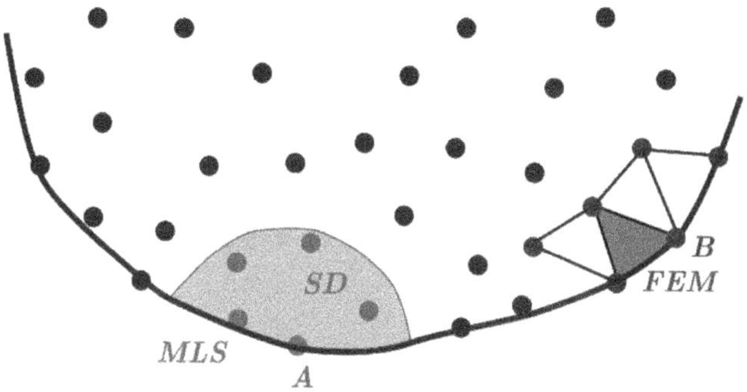

Figure 3.35 Enforcement of the boundary condition at node ***A*** by the MLS requires the internal nodes inside the support domain ($u^h(\mathbf{x}_A) \neq \bar{u}_A$), whereas it is only related to the boundary node ***B*** in the FEM ($u^h(\mathbf{x}_B) = \bar{u}_B$).

Definitions of the shape and size of the support domain and the corresponding weight function around concave corners or crack edges become more complicated because the computation of the straight distance between nodes $\mathbf{x}_j$ or the reference point $\mathbf{x}$ of the support domain may no longer remain valid. Figure 3.36 illustrates a number of concepts for redefining the distance that may be used in defining the MLS shape function. For further details, see Goudarzi and Mohammadi (2014).

The non-polynomial MLS shape functions $\varphi(\mathbf{x})$ can be constructed with any desired level of smoothness and continuity, depending on the order of the basis function, the support domain and the continuity of the weight function. They may even be constructed with different orders in the vicinity of each point $\mathbf{x}$ within the domain.

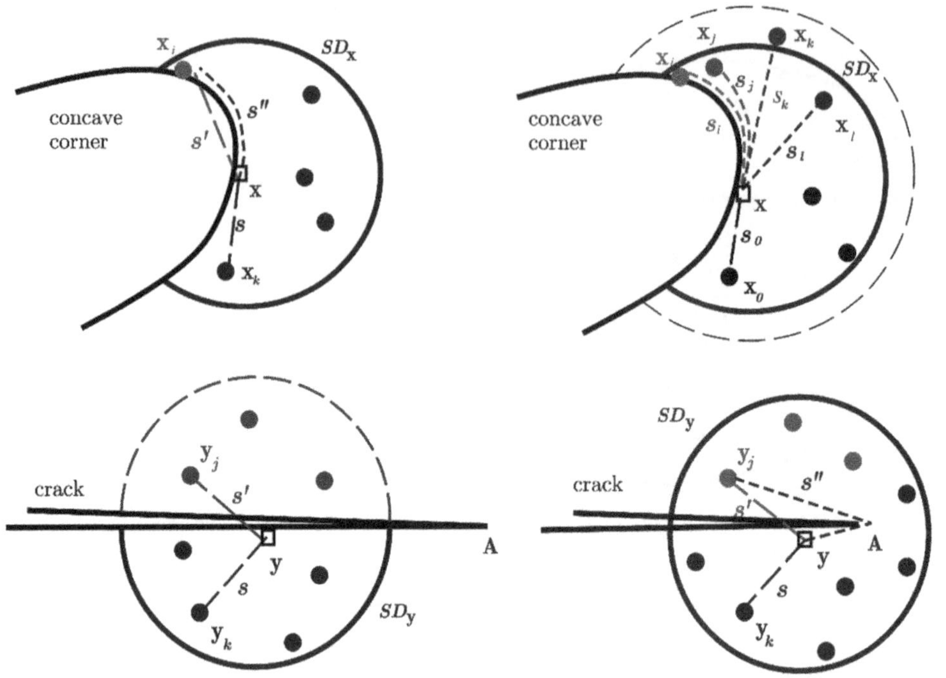

Figure 3.36 Definitions of the nodal distance of the weight function for concave corners and sharp cracks.

Computations of derivatives of MLS shape functions are more complicated and require derivatives of all parameters that affect the shape function:

$$\varphi_{,x}(\mathbf{x}) = \left(\mathbf{P}^{\mathrm{T}}\mathbf{A}^{-1}\mathbf{B}\right)_{,x} = \mathbf{p}_{,x}^{\mathrm{T}}\mathbf{A}^{-1}\mathbf{B} + \mathbf{p}^{\mathrm{T}}\left(\mathbf{A}^{-1}\right)_{,x}\mathbf{B} + \mathbf{p}^{\mathrm{T}}\mathbf{A}^{-1}\mathbf{B}_{,x} \tag{3.7.20}$$

More complicated and expensive computations are necessary for determining higher order derivatives of MLS shape functions, which constitute a major drawback of the MLS approximation and the MLS-based meshless methodologies for solving the governing partial differential equations of engineering problems.

3.7.2.2 Point Interpolation Method (PIM)

Point interpolation methods (PIMs) are alternative meshless methods for approximation on a set of discrete nodes within a domain.

Consider a typical point $\mathbf{x}_Q$ of the domain and its corresponding local support domain, as depicted in Figure 3.37. $\mathbf{x}_Q$ may represent one of the independent nodes or a quadrature point for integration of the governing equations of the problem.

There are three major categories of point interpolation methods. PPIM and RPIM use polynomial and radial basis functions, respectively, to derive the interpolations and RPPIM combines the concepts of both approaches.

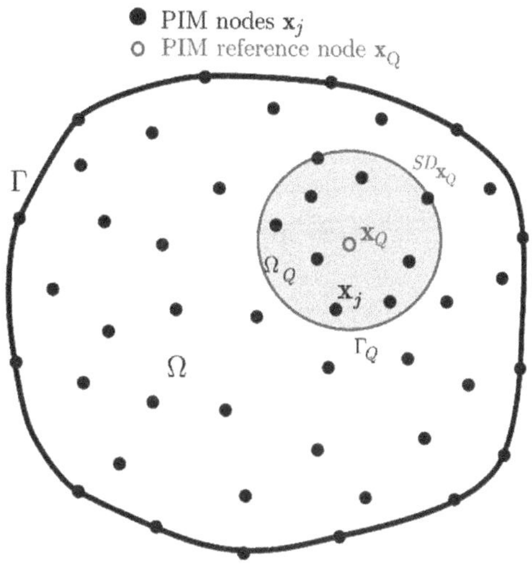

Figure 3.37 A local support domain associated with a typical point $\mathbf{x}_Q$.

3.7.2.2.1 Polynomial PIM (PPIM)

Beginning with the polynomial PIM (PPIM), the PIM approximation in the local support domain associated with the point $\mathbf{x}_Q$ can be defined as (Wang and Liu 2002)

$$\boldsymbol{u}^h(\mathbf{x},\mathbf{x}_Q) = \mathbf{p}^{\mathrm{T}}(\mathbf{x})\boldsymbol{a}(\mathbf{x}_Q) = \sum_{j=1}^{n} p_j(\mathbf{x})a_j(\mathbf{x}_Q) \tag{3.7.21}$$

or simply for a local support domain associated with $\mathbf{x}_Q$,

$$\boldsymbol{u}^h(\mathbf{x}) = \mathbf{p}^{\mathrm{T}}(\mathbf{x})\boldsymbol{a} = \sum_{j=1}^{n} p_j(\mathbf{x})a_j \tag{3.7.22}$$

where $\mathbf{p}(\mathbf{x})$ is the vector of polynomial basis functions, similar to Equation (3.7.3) for the MLS formulation. In PPIM, the number of polynomial terms is always equal to the number of nodes within the local support domain.

The main concept of PIM is to enforce interpolation at all nodes $\mathbf{x}_j$ of the local support domain of $\mathbf{x}_Q$:

$$\boldsymbol{u}^h(\mathbf{x} = \mathbf{x}_j) = \bar{\boldsymbol{u}}_j \tag{3.7.23}$$

which leads to a set of n simultaneous equations to solve for n unknown coefficients a_j. Equation (3.7.23) can be re-written as

$$\mathbf{p}^{\mathrm{T}}(\mathbf{x}_j)\boldsymbol{a} = \bar{\boldsymbol{u}}_j, \quad j = 1,\ldots,n \tag{3.7.24}$$

Assembling the equations for all nodes of the support domain leads to

$$\mathbf{P}_Q\boldsymbol{a} = \bar{\mathbf{U}} \tag{3.7.25}$$

where $\bar{\mathbf{U}}$ is the vector of nodal displacements,

$$\bar{\mathbf{U}}^{\mathrm{T}} = \{\bar{\boldsymbol{u}}_1, \bar{\boldsymbol{u}}_2, \ldots, \bar{\boldsymbol{u}}_n\} \tag{3.7.26}$$

and

$$\mathbf{P}_Q^{\mathrm{T}} = \{\mathbf{p}^{\mathrm{T}}(\mathbf{x}_1), \mathbf{p}^{\mathrm{T}}(\mathbf{x}_2), \ldots, \mathbf{p}^{\mathrm{T}}(\mathbf{x}_n)\} \tag{3.7.27}$$

$$\mathbf{P}_Q = \begin{bmatrix} p_1(\mathbf{x}_1) & p_2(\mathbf{x}_1) & \cdots & p_n(\mathbf{x}_1) \\ p_1(\mathbf{x}_2) & p_2(\mathbf{x}_2) & \cdots & p_n(\mathbf{x}_2) \\ \vdots & \vdots & \ddots & \vdots \\ p_1(\mathbf{x}_n) & p_2(\mathbf{x}_n) & \cdots & p_n(\mathbf{x}_n)) \end{bmatrix} \tag{3.7.28}$$

Solving Equation (3.7.25) for $\boldsymbol{a}$ results in

$$\boldsymbol{a} = \mathbf{P}_Q^{-1}\bar{\mathbf{U}} \tag{3.7.29}$$

and the final PIM interpolation $\boldsymbol{u}^h(\mathbf{x})$ can be written in terms of the PPIM shape functions $\varphi(\mathbf{x})$,

$$\boldsymbol{u}^h(\mathbf{x}) = \varphi(\mathbf{x})\bar{\mathbf{U}} \tag{3.7.30}$$

$$\varphi(\mathbf{x}) = \mathbf{p}^{\mathrm{T}}(\mathbf{x})\mathbf{P}_Q^{-1} \tag{3.7.31}$$

$$\varphi(\mathbf{x}) = [\varphi_1(\mathbf{x}), \varphi_2(\mathbf{x}), \ldots, \varphi_n(\mathbf{x})] \tag{3.7.32}$$

$\varphi(\mathbf{x})$ satisfies the partition of unity (PU) concept

$$\sum_{i=1}^{n} \varphi_i(\mathbf{x}) = 1 \tag{3.7.33}$$

and the Kronecker delta function property, which is important for imposing the essential boundary conditions,

$$\varphi_i(\mathbf{x}_j) = \begin{cases} 1, & i = j \\ 0, & i \neq j \end{cases} \tag{3.7.34}$$

Moreover, different orders of derivatives of PPIM shape functions can be readily computed:

$$\varphi_{,x}(\mathbf{x}) = \mathbf{p}_{,x}^{\mathrm{T}}(\mathbf{x})\mathbf{P}_Q^{-1} \tag{3.7.35}$$

There are, however, a number of major drawbacks of PPIM interpolation. For instance, for higher order of polynomials, oscillating solutions are unavoidable. More importantly, PPIM does not have the moving nature of MLS and lacks the smoothing effect of a weight function, which leads to discontinuous fields when multiple local support domains are adopted to approximate a global problem (see Figure 3.38). While it may somehow be

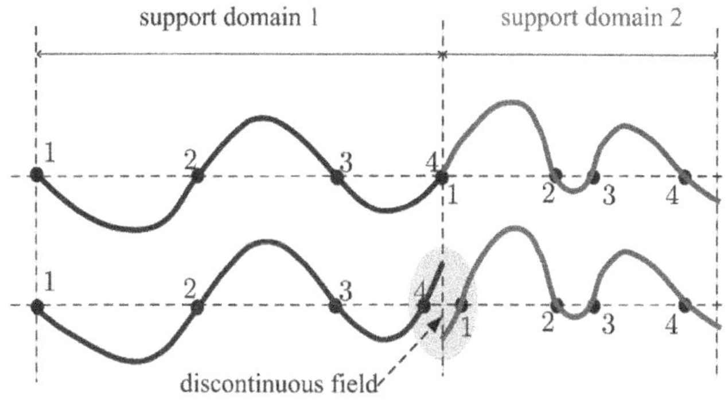

Figure 3.38 Potential discontinuity between neighbour local support domains.

avoided in one dimension by using common boundary nodes for neighbour support domains, discontinuous patches always occur in two-dimensional and three-dimensional overlapping support domains.

Moreover, $\mathbf{P}_Q$ may become singular or ill-conditioned in certain nodal distributions. A number of remedies have been proposed, such as the random nodal movement and coordinate transformation (Liu 2003), the coordinate normalization/mapping (Wang and Liu 2000), the singular value decomposition (SVD) technique (Emdadi et al. 2008) and the radial point interpolation method (RPIM) (Kansa 1990; Libre et al. 2009).

3.7.2.2.2 Radial PIM (RPIM)

The radial point interpolation method (RPIM) adopts the same procedure of PIM, but based on the radial basis function (RBF) $\mathbf{R}(\mathbf{x})$ (Kansa 1990):

$$\boldsymbol{u}^h(\mathbf{x},\mathbf{x}_Q) = \sum_{j=1}^{n} R_j(\mathbf{x}) a_j(\mathbf{x}_Q) = \mathbf{R}^{\mathrm{T}}(\mathbf{x}) \boldsymbol{a}(\mathbf{x}_Q) \tag{3.7.36}$$

or in terms of the radial distances from the nodes, $r(\mathbf{x})$:

$$\boldsymbol{u}^h(\mathbf{x},\mathbf{x}_Q) = \sum_{j=1}^{n} R_j(r_j(\mathbf{x})) a_j(\mathbf{x}_Q) = \mathbf{R}^{\mathrm{T}}(r(\mathbf{x})) \boldsymbol{a}(\mathbf{x}_Q) \tag{3.7.37}$$

$$\mathbf{R}^{\mathrm{T}}(\boldsymbol{r}) = \{R_1(r_1), R_2(r_2), \ldots, R_n(r_n)\} \tag{3.7.38}$$

$$r_j(\mathbf{x}) = r_j(x,y) = \left[(x-x_j)^2 + (y-y_j)^2\right]^{\frac{1}{2}} \tag{3.7.39}$$

Figure 3.39 illustrates the way the radial distances $r_j(\mathbf{x}_j)$ are defined for determination of the radial basis functions $\boldsymbol{R}_j(r_j)$.

A number of frequently used radial basis functions (RBFs) are defined as (Liu 2003):

$$\begin{cases} \textit{Multiquadrics} & R_j(x,y) = [r_j^2 + c^2]^q = [(x-x_j)^2 + (y-y_j)^2 + c^2]^q \\ \textit{Gaussian} & R_j(x,y) = \exp\left(-cr_j^2\right) = \exp[-c((x-x_j)^2 + (y-y_j)^2)] \\ \textit{TPS} & R_j(x,y) = r_j^\eta = ((x-x_j)^2 + (y-y_j)^2)^\eta \\ \textit{Logarithmic} & R_j(r_j) = r_j^\eta \ln r_j \end{cases} \tag{3.7.40}$$

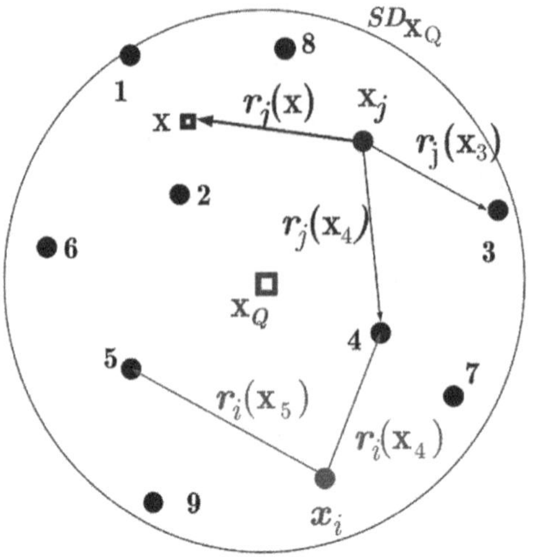

Figure 3.39 Definitions of radial distances.

Similar to PIM, RPIM interpolation is enforced by satisfying the nodal values at all nodes within the local support domain:

$$\mathbf{R}_Q \boldsymbol{a} = \bar{\mathbf{U}} \tag{3.7.41}$$

Solving Equation (3.7.41) for $\boldsymbol{a}$ results in

$$\boldsymbol{a} = \mathbf{R}_Q^{-1} \bar{\mathbf{U}} \tag{3.7.42}$$

where

$$\mathbf{R}_Q = \begin{bmatrix} R_1(r_1(\mathbf{x}_1)) & R_2(r_2(\mathbf{x}_1)) & \cdots & R_n(r_n(\mathbf{x}_1)) \\ R_1(r_1(\mathbf{x}_2)) & R_2(r_2(\mathbf{x}_2)) & \cdots & R_n(r_n(\mathbf{x}_2)) \\ \vdots & \vdots & \ddots & \vdots \\ R_1(r_1(\mathbf{x}_n)) & R_2(r_2(\mathbf{x}_n)) & \cdots & R_n(r_n(\mathbf{x}_n)) \end{bmatrix} \tag{3.7.43}$$

$$r_j(\mathbf{x}_k) = r_j(x_k, y_k) = \left[(x_k - x_j)^2 + (y_k - y_j)^2\right]^{\frac{1}{2}} \tag{3.7.44}$$

Numerical studies have shown that if certain rules for choosing the shape parameters of the radial basis functions are respected, RPIM guarantees the existence of the inverse of the moment matrix $\mathbf{R}_Q^{-1}$. The final solution is then obtained as

$$\boldsymbol{u}^h(\mathbf{x}) = \mathbf{R}^{\mathrm{T}}(\mathbf{x}) \mathbf{R}_Q^{-1} \bar{\mathbf{U}} \tag{3.7.45}$$

or in terms of the RPIM shape functions,

$$\boldsymbol{u}^h(\mathbf{x}) = \varphi(\mathbf{x}) \bar{\mathbf{U}} \tag{3.7.46}$$

$$\varphi(\mathbf{x}) = \mathbf{R}^{\mathrm{T}}(\mathbf{x}) \mathbf{R}_Q^{-1} \tag{3.7.47}$$

Derivatives of RPIM shape functions are readily obtained from the derivatives of RBF functions,

$$\varphi_{,x}(\mathbf{x}) = \mathbf{R}_{,x}^{\mathrm{T}}(\mathbf{x})\mathbf{R}_Q^{-1} \tag{3.7.48}$$

3.7.2.2.3 Radial-Polynomial PIM (RPPIM)

RPIM fails to reconstruct a linear field exactly. It is not consistent and cannot pass the standard patch test. Adding polynomials into the radial basis functions ensures the consistency of RPPIM shape functions.

Approximation $\boldsymbol{u}^h(\mathbf{x})$ of the RPPIM for a field function $\boldsymbol{u}(\mathbf{x})$ over the local support domain of a reference/integration point $\mathbf{x}_Q$ can be written as (Liu 2003)

$$\boldsymbol{u}^h(\mathbf{x}) = \sum_{\mathrm{i}=1}^{\mathrm{n}} R_{\mathrm{i}}(\mathbf{x})a_{\mathrm{i}} + \sum_{\mathrm{j}=1}^{\mathrm{m}} p_{\mathrm{j}}(\mathbf{x})b_{\mathrm{j}} \tag{3.7.49}$$

$$\boldsymbol{u}^h(\mathbf{x}) = \mathbf{R}^{\mathrm{T}}(\mathbf{x})\boldsymbol{a} + \mathbf{p}^T(\mathbf{x})\boldsymbol{b} \tag{3.7.50}$$

where n is the number of nodes of the support domain and m is the number of polynomial terms. Moreover,

$$\boldsymbol{a}^{\mathrm{T}} = \{a_1, a_2, \ldots, a_n\} \tag{3.7.51}$$

$$\boldsymbol{b}^T = \{b_1, b_2, \ldots, b_m\} \tag{3.7.52}$$

$$\mathbf{R}^{\mathrm{T}}(\mathbf{x}) = \{R_1(\mathbf{x}), R_2(\mathbf{x}), \ldots, R_n(\mathbf{x})\} \tag{3.7.53}$$

$$\mathbf{p}^T(\mathbf{x}) = \{\mathbf{p}_1(\mathbf{x}), \mathbf{p}_2(\mathbf{x}), \ldots, \mathbf{p}_m(\mathbf{x})\} \tag{3.7.54}$$

The coefficients $\boldsymbol{a}$ and $\boldsymbol{b}$ are determined by enforcing the interpolation (3.7.50) to reproduce all nodal values, which leads to

$$\mathbf{R}_Q\boldsymbol{a} + \mathbf{P}_m\boldsymbol{b} = \bar{\mathbf{U}} \tag{3.7.55}$$

An extra set of equations is required to guarantee a unique approximation (Liu 2003)

$$\mathbf{P}_m^{\mathrm{T}}\boldsymbol{a} = 0 \tag{3.7.56}$$

The final set of simultaneous equations can be written as

$$\begin{bmatrix} \mathbf{R}_Q & \mathbf{P}_m \\ \mathbf{P}_m^{\mathrm{T}} & 0 \end{bmatrix} \begin{bmatrix} \boldsymbol{a} \\ \boldsymbol{b} \end{bmatrix} = \begin{bmatrix} \bar{\mathbf{U}} \\ \mathbf{0} \end{bmatrix} \tag{3.7.57}$$

where

$$\mathbf{P}_m = \begin{bmatrix} p_1(\mathbf{x}_1) & p_2(\mathbf{x}_1) & \cdots & p_m(\mathbf{x}_1) \\ p_1(\mathbf{x}_2) & p_2(\mathbf{x}_2) & \cdots & p_m(\mathbf{x}_2) \\ \vdots & \vdots & \ddots & \vdots \\ p_1(\mathbf{x}_n) & p_2(\mathbf{x}_n) & \cdots & p_m(\mathbf{x}_n) \end{bmatrix}_{n\times m} \tag{3.7.58}$$

The solution can then be obtained as

$$\boldsymbol{u}^h(\mathbf{x}) = \left[\mathbf{R}^{\mathrm{T}}(\mathbf{x})\mathbf{S}^a + \mathbf{p}^T(\mathbf{x})\mathbf{S}^b\right]\bar{\mathbf{U}} \tag{3.7.59}$$

where

$$\mathbf{S}^a = \mathbf{R}_Q^{-1}\left[1 - \mathbf{P}_m S^b\right] \tag{3.7.60}$$

$$\mathbf{S}^b = \left[\mathbf{P}_m^{\mathrm{T}}\mathbf{R}_Q^{-1}\mathbf{P}_m\right]^{-1}\mathbf{P}_m^{\mathrm{T}}\mathbf{R}_Q^{-1} \tag{3.7.61}$$

or in terms of the RPPIM shape functions $\varphi(\mathbf{x})$,

$$\boldsymbol{u}^h(\mathbf{x}) = \varphi(\mathbf{x})\bar{\mathbf{U}} \tag{3.7.62}$$

where

$$\varphi(\mathbf{x}) = \mathbf{R}^{\mathrm{T}}(\mathbf{x})\mathbf{S}^a + \mathbf{p}^T(\mathbf{x})\mathbf{S}^b \tag{3.7.63}$$

and the shape function derivatives are

$$\varphi_{,x}(\mathbf{x}) = \mathbf{R}_{,x}^{\mathrm{T}}(\mathbf{x})\mathbf{S}^a + \mathbf{p}_{,x}^{\mathrm{T}}(\mathbf{x})\mathbf{S}^b \tag{3.7.64}$$

3.7.2.3 Smoothed Particle Hydrodynamics (SPH)

The smoothed particle hydrodynamics (SPH) is one of the first meshless methods, originally developed for astrophysical studies by Lucy (1974). It was further developed by Gingold and Monagan (1977, 1983) in more or less the same subject, and many of its initial instabilities and inaccuracies were removed over the 1970s and 1980s. They proposed/adopted the SPH methodology to transform the real discrete astrophysical system into a local continuous field. In contrast to the initial stage of development, a continuum physical system is now discretized into a discrete set of particles by a computational SPH approach. SPH has been successfully adopted for various computational mechanics problems, such as solution of general free surface fluid problems, gas dynamics and shock waves, fragmentation in solid dynamic problems, and even micro- and nanoscale simulations (Liu and Liu 2003).

The SPH approximation is based on the analytical approximation of the function $\boldsymbol{u}(\mathbf{x})$ by the Dirac delta function $\delta(\mathbf{x})$:

$$\boldsymbol{u}(\mathbf{x}) = \int_{-\infty}^{\infty} \delta(\mathbf{x} - \mathbf{y})\boldsymbol{u}(\mathbf{y})\mathrm{d}\mathbf{y} \tag{3.7.65}$$

While Equation (3.7.65) is exact, it is numerically difficult to compute in practical applications. SPH adopts a simplified approximation of Equation (3.7.65) in the form of

$$\boldsymbol{u}^h(\mathbf{x}) = \int_{\Omega} W_h(\mathbf{x} - \mathbf{y}, h)\boldsymbol{u}(\mathbf{y})\mathrm{d}\mathbf{y} \tag{3.7.66}$$

replacing the Dirac delta function $\delta(\mathbf{x})$ with a smoothing kernel or weight function $W_h(\mathbf{x})$ to facilitate the numerical computations. h is the smoothing length, which controls the size

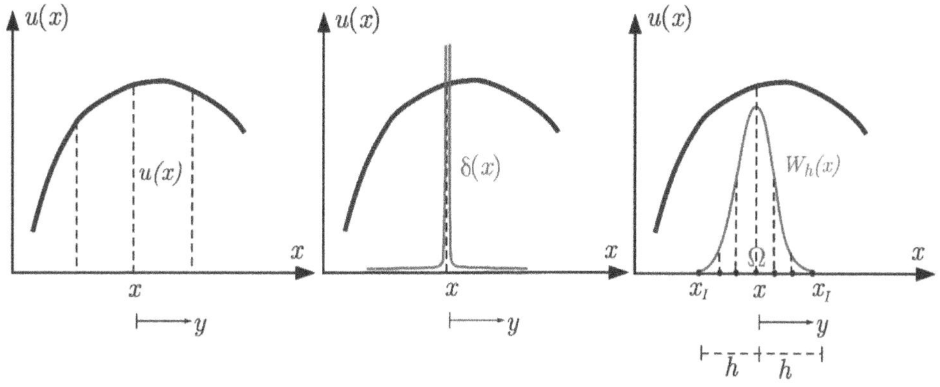

Figure 3.40 Integral approximation of a continuous function.

of the compact local support domain. Figure 3.40 illustrates the way a one-dimensional SPH approximation has evolved from a Dirac delta approximation.

The kernel function $W_h(\mathbf{x}-\mathbf{y},h)$ has a smoothing effect on the interpolation field and should be positive, compact, bell shaped, differentiable and satisfy the unity condition to resemble the Dirac delta property,

$$\int_{\Omega} W_h(\mathbf{x}-\mathbf{y},h)\mathbf{dy}=1 \tag{3.7.67}$$

In SPH, the kernel approximation around a point $\mathbf{x}$ is approximated by a discrete set of n material particles (nodes) $\mathbf{x}_I$ inside the local smoothing domain (support domain),

$$\boldsymbol{u}^h(\mathbf{x})=\sum_{I=1}^{n} W_h(\mathbf{x}-\mathbf{x}_I)\boldsymbol{u}(\mathbf{x}_I)\Delta V_I \tag{3.7.68}$$

or in terms of the SPH shape functions $\varphi(\mathbf{x})$,

$$\boldsymbol{u}^h(\mathbf{x})=\sum_{I=1}^{n} \varphi_I(\mathbf{x})\boldsymbol{u}(\mathbf{x}_I) \tag{3.7.69}$$

$$\varphi_I(\mathbf{x})=W_h(\mathbf{x}-\mathbf{x}_I)\Delta V_I \tag{3.7.70}$$

Computation of the derivative of $\boldsymbol{u}$ can be performed directly on Equation (3.7.68). Assuming a symmetric kernel W_h eventually leads to (Ostad and Mohammadi 2012)

$$\nabla.\boldsymbol{u}^h(\mathbf{x})=-\sum_{I=1}^{n} \nabla\cdot W_h(\mathbf{x}-\mathbf{x}_I)\boldsymbol{u}(\mathbf{x}_I)\Delta V_I \tag{3.7.71}$$

The SPH shape functions do not satisfy the Kronecker delta function,

$$\varphi_i(\mathbf{x}_j)\neq\begin{cases}1, & i=j\\ 0, & i\neq j\end{cases} \rightarrow \boldsymbol{u}_I\neq\boldsymbol{u}^h(\mathbf{x}_I) \tag{3.7.72}$$

Therefore, the method is an approximation and does not directly satisfy the essential boundary conditions.

Another major disadvantage of SPH is its relatively poor accuracy (particularly in boundaries), which requires a large number of particles to achieve reasonable accuracy in practical applications. Moreover, the motion of particles under certain tensile (hydrostatic) stress states may become unstable in the form of patches/clusters of particles instead of uniformly moving apart.

SPH approximations can be used to solve a boundary value problem by directly applying the partial differential equations at nodes $\mathbf{x}_I$ in a finite difference fashion.

3.7.2.3.1 Reproducing Kernel Particle Method (RKPM)

The reproducing kernel particle method (RKPM) is one of the remedies for removing the major drawbacks of the original SPH approximation by the idea of constructing a corrective kernel to ensure consistency and completeness of the SPH approximation. The SPH kernel is replaced by the product of the correction function C_h with the original kernel W_h (Liu et al. 1995):

$$\boldsymbol{u}^h(\mathbf{x}) = \int_\Omega \bar{W}_h(\mathbf{x}-\mathbf{y},h)\boldsymbol{u}(\mathbf{y})d\mathbf{y} = \sum_{I=1}^{n} \bar{W}_h(\mathbf{x}-\mathbf{x}_I)\boldsymbol{u}(\mathbf{x}_I)\Delta V_I \tag{3.7.73}$$

$$\bar{W}_h(\mathbf{x}-\mathbf{x}_I) = C_h(\mathbf{x}-\mathbf{x}_I,\mathbf{x})W_h(\mathbf{x}-\mathbf{x}_I) \tag{3.7.74}$$

$C_h(\mathbf{x}-\mathbf{x}_I,\mathbf{x})$ is the correction function for the original kernel function to satisfy a required order of consistency. For instance, the correction function C_h can be expressed as

$$C_h(\mathbf{x}-\mathbf{x}_I,\mathbf{x}) = b_0(\mathbf{x},h) + b_1(\mathbf{x},h)\frac{|\mathbf{x}-\mathbf{x}_I|}{h} + b_2(\mathbf{x},h)\left(\frac{|\mathbf{x}-\mathbf{x}_I|}{h}\right)^2 + \cdots \tag{3.7.75}$$

where $b_0(\mathbf{x})$, $b_1(\mathbf{x})$,..., $b_k(\mathbf{x})$ are unknown functions, which should be determined according to the required level of consistency k. The coefficients $b_k(\mathbf{x})$ can be computed from (Liu et al. 1995)

$$\begin{bmatrix} m_0(\mathbf{x}) & m_1(\mathbf{x}) & \dots & m_k(\mathbf{x}) \\ m_1(\mathbf{x}) & m_2(\mathbf{x}) & \dots & m_{k+1}(\mathbf{x}) \\ \vdots & \vdots & \vdots & \\ m_k(\mathbf{x}) & m_{k+1}(\mathbf{x}) & \dots & m_{2k}(\mathbf{x}) \end{bmatrix} \begin{bmatrix} b_0(\mathbf{x},h) \\ b_1(\mathbf{x},h) \\ \vdots \\ b_k(\mathbf{x},h) \end{bmatrix} = \begin{bmatrix} 1 \\ 0 \\ \vdots \\ 0 \end{bmatrix} \tag{3.7.76}$$

$$m_k(\mathbf{x}) = \sum_{I=1}^{n} \left(\frac{|\mathbf{x}-\mathbf{x}_I|}{h}\right)^k \bar{W}_h(\mathbf{x}-\mathbf{x}_I)\Delta V_I = 0 \tag{3.7.77}$$

The corrected kernel function $\bar{W}_h$ is conceptually different from the standard SPH kernel W_h and does not follow its specifications such as positivity, symmetry, decay and even unity.

3.7.2.3.2 Corrected SPH (CSPM)

Another remedy for deficiencies of SPH is the corrected SPH method (CSPM). In this method, the Taylor expansion is adopted to derive the kernel approximation (Chen and Beraun, 2000). The kernel approximation for function $f(\mathbf{x})$ can be expanded as

$$\begin{aligned}\int_\Omega f(\mathbf{x})W_h\,\mathrm{d}A &= f(\mathbf{x}_I)\int_\Omega W_h\,\mathrm{d}A + f_{,x}(\mathbf{x})\int_\Omega (x-x_I)W_h\,\mathrm{d}A + f_{,y}(\mathbf{x}_I)\int_\Omega (y-y_I)W_h\,\mathrm{d}A \\ &+\frac{1}{2}f_{,xx}(\mathbf{x}_I)\int_\Omega (x-x_I)^2W_h\,\mathrm{d}A + f_{,xy}(\mathbf{x}_I)\int_\Omega (x-x_I)(y-y_I)W_h\,\mathrm{d}A \\ &+\frac{1}{2}f_{,yy}(\mathbf{x}_I)\int_\Omega (y-y_I)^2W_h\,\mathrm{d}A + \cdots\end{aligned} \tag{3.7.78}$$

As a first approximation, all the derivative terms are neglected and a corrective version of the kernel is generated,

$$f_I = f(\mathbf{x}_I) = \frac{\int_\Omega f(\mathbf{x})W_I(\mathbf{x})dA}{\int_\Omega W_I(\mathbf{x})dA} \tag{3.7.79}$$

or in a discrete form,

$$f_I = \frac{\sum_{J=1}^{N} f_J W_{IJ}\Delta A_J}{\sum_{J=1}^{N} W_{IJ}\Delta A_J} = \frac{\sum_{J=1}^{N} f_J W_{IJ}\frac{m_J}{\rho_J}}{\sum_{J=1}^{N} W_{IJ}\frac{m_J}{\rho_J}} \tag{3.7.80}$$

Note that any symmetric function can be adopted for the kernel W_I in Equations (3.7.79) and (3.7.80). Clearly, CSPM allows for automatic and accurate enforcement of essential boundary conditions.

Approximations of the derivatives in one dimension can be similarly constructed by simply neglecting the higher order derivative terms of the Taylor series, leading to the following first- and second-order derivatives, f_x and f_{xx}, respectively,

$$f_x(x_I) = f_{xI} = \frac{\int_\Omega [f(x)-f_I]W_I^x(x)dx}{\int_\Omega (x-x_I)W_I^x(x)dx} \tag{3.7.81}$$

$$f_{xx}(x_I) = f_{xxI} = \frac{\int_\Omega [f(x)-f_I]W_I^{xx}(x)dx - f_{xI}\int_\Omega [x-x_I]W_I^{xx}(x)dx}{\frac{1}{2}\int_\Omega (x-x_I)^2W_I^{xx}(x)dx} \tag{3.7.82}$$

Kernels W_I^x and W_I^{xx} can be any anti-symmetric and symmetric functions for CSPM approximations of f_x and f_{xx}, respectively.

It is noted that CSPM formulations for derivatives do not even need the derivatives of the kernel. They only require the values of the function at particles (nodes).

3.7.3 Meshless Solutions for the Boundary Value Problems

The meshless methods described in previous sections have been widely employed for solving the governing partial differential equations of different physical problems. Among them, the element-free Galerkin (EFG), the meshless local Petrov–Galerkin (MLPG), the finite point method (FPM) and the smoothed particle hydrodynamics (SPH) methods are briefly reviewed.

3.7.3.1 Element-Free Galerkin Method (EFG)

The element-free Galerkin method (EFG) was proposed by Belytschko et al. (1994) as a meshless method. Many ideas of EFG have been extended to other meshless methods and even to other numerical techniques, such as the extended finite element method. EFG has been one of the first meshless methods being implemented in the commercial general purpose softwares.

EFG adopts the MLS approximation for the construction of the shape functions. The constrained Galerkin weak formulation is employed to derive the discretized system of equations.

Consider a domain and its boundary represented by a finite number of nodes, as depicted in Figure 3.41. The body is assumed to be subjected to the prescribed body force $\boldsymbol{f}^{\mathrm{b}}$ on Ω and traction $\boldsymbol{f}^{\mathrm{t}}$ on Γ_t. The essential boundary conditions $\boldsymbol{u} = \bar{\boldsymbol{u}}$ are prescribed on Γ_u.

The strong form of the equilibrium equation in terms of the stress tensor $\boldsymbol{\sigma}$ can be written as

$$\nabla.\sigma + \boldsymbol{f}^{\mathrm{b}} = \mathbf{0} \text{ on } \Omega \tag{3.7.83}$$

with the following displacement and traction boundary conditions:

$$\boldsymbol{u} = \bar{\boldsymbol{u}} \text{ on } \Gamma_u \tag{3.7.84}$$

$$\boldsymbol{\sigma} \cdot \mathbf{n} = \boldsymbol{f}^t \text{ on } \Gamma_t \tag{3.7.85}$$

The MLS approximation ensures a continuous field over the whole domain, but it cannot directly satisfy the displacement boundary conditions. Having adopted a Lagrange

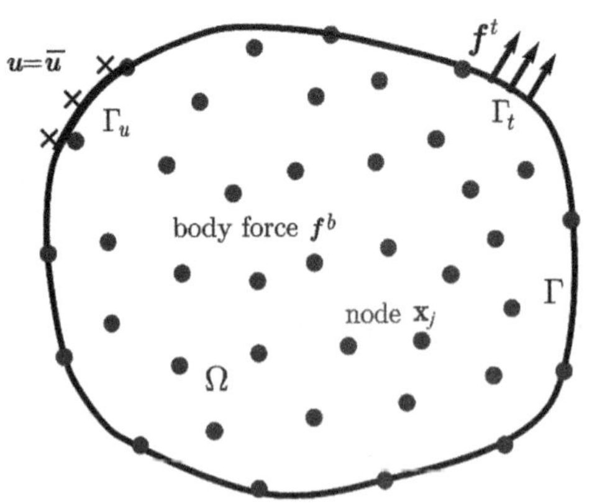

Figure 3.41 Definition of a meshless domain with boundary conditions.

multiplier approach, the constrained Galerkin weak form of the governing boundary value problem can then be defined as (Goudarzi and Mohammadi 2014, 2015; Khazal et al. 2016)

$$\underbrace{\int_{\Omega}\delta\varepsilon^{\mathrm{T}}\sigma d\Omega}_{\text{internal vitual work}} - \underbrace{\int_{\Omega}\delta\boldsymbol{u}^{\mathrm{T}}\boldsymbol{f}^{\mathrm{b}}d\Omega}_{\text{virtual work of body force}} - \underbrace{\int_{\Gamma_t}\delta\boldsymbol{u}^{\mathrm{T}}\boldsymbol{f}^{\mathrm{t}}d\Gamma}_{\text{virtual work of tractions}} \underbrace{-\int_{\Gamma_u}\delta\lambda^{\mathrm{T}}(\boldsymbol{u}-\bar{\boldsymbol{u}})d\Gamma - \int_{\Gamma_u}\delta\boldsymbol{u}^{\mathrm{T}}\lambda d\Gamma}_{\substack{\text{enforcing essential boundary conditions} \\ (u-\bar{u}=0)}} = 0 \tag{3.7.86}$$

The MLS approximation is now adopted to express the displacement field $\boldsymbol{u}^h$ at any point of interest $\mathbf{x}$ in terms of the displacements $\boldsymbol{u}_I$ of n nodes in the support domain of $\mathbf{x}$:

$$\boldsymbol{u} = \boldsymbol{u}^h(\mathbf{x}) = \boldsymbol{\Phi}(\mathbf{x})\mathbf{U} = \sum_{I=1}^{n}\varphi_I(\mathbf{x})\boldsymbol{u}_I \tag{3.7.87}$$

where $\boldsymbol{\Phi}$ is the MLS shape function. The strain tensor ε can then be evaluated from

$$\varepsilon = \mathbf{B}\bar{\mathbf{U}} = \sum_{j=1}^{n}\mathbf{B}_j\mathbf{u}_j \tag{3.7.88}$$

$$\mathbf{B}_j = \begin{bmatrix} \varphi_{j,x} & 0 \\ 0 & \varphi_{j,y} \\ \varphi_{j,y} & \varphi_{j,x} \end{bmatrix} \tag{3.7.89}$$

and

$$\sigma = \mathbf{D}\varepsilon = \mathbf{D}\mathbf{B}\bar{\boldsymbol{U}} \tag{3.7.90}$$

The Lagrange multiplier λ is an unknown function of the coordinates, which is interpolated using n_λ nodal values λ_k on the essential boundary Γ_u:

$$\lambda(\mathbf{x}) = \sum_{k=1}^{n_\lambda}N_k(s)\lambda_k, \quad \mathbf{x}\in\Gamma_u \tag{3.7.91}$$

The arc length along the essential boundary (s) is used to define the Lagrange interpolation $N_k(s)$:

$$N_k(s) = \frac{(s-s_0)(s-s_1)\cdots(s-s_{k-1})(s-s_{k+1})\cdots(s-s_{n_\lambda})}{(s_k-s_0)(s_k-s_1)\cdots(s_k-s_{k-1})(s_k-s_{k+1})\cdots(s_k-s_{n_\lambda})} \tag{3.7.92}$$

The final discretized set of EFG equations can be obtained as

$$\begin{bmatrix} \mathbf{K} & \mathbf{G} \\ \mathbf{G}^{\mathrm{T}} & 0 \end{bmatrix}\begin{bmatrix} \bar{\mathbf{U}} \\ \boldsymbol{\Lambda} \end{bmatrix} = \begin{bmatrix} \mathbf{f} \\ \mathbf{q} \end{bmatrix} \tag{3.7.93}$$

where $\bar{\mathbf{U}}$ is the vector of main nodal displacement DOFs, $\boldsymbol{\Lambda}$ represents the vector of Lagrange multipliers and

$$\mathbf{K}_{IJ} = \int_{\Omega} \mathbf{B}_I^{\mathrm{T}} \mathbf{D} \mathbf{B}_J \mathrm{d}\Omega \tag{3.7.94}$$

$$\mathbf{G}_{IJ}^{\mathrm{T}} = -\int_{\Gamma_u} \mathbf{N}_I^{\mathrm{T}} \boldsymbol{\Phi}_J \mathrm{d}\Gamma \tag{3.7.95}$$

$$\mathbf{f}_I = \int_{\Omega} \boldsymbol{\Phi}_I^{\mathrm{T}} \boldsymbol{f}^b \mathrm{d}\Omega + \int_{\Gamma_t} \boldsymbol{\Phi}_I^{\mathrm{T}} \boldsymbol{f}^t \mathrm{d}\Gamma \tag{3.7.96}$$

$$\mathbf{q}_I = -\int_{\Gamma_u} \mathbf{N}_I^{\mathrm{T}} \bar{\boldsymbol{u}} \mathrm{d}\Gamma \tag{3.7.97}$$

It should be noted that the components $\mathbf{G}_{ij}$ of matrix $\mathbf{G}$ depend on both the MLS and Lagrange shape functions based on nodes I and J inside the support domain of the integration points on the essential boundary Γ_u.

Numerical integrations are required to evaluate Equations (3.7.94) to (3.7.97). Unlike the finite element, the meshless model does not have any specific element with a known integration rule. Practically, integrations are performed on a predefined structured grid of quadrature cells, called the background cells, as illustrated in Figure 3.42, to compute the nodal matrices. Each cell has a well-established desired order of Gauss quadrature rule.

Figure 3.43 shows the way the MLS should be performed for each Gauss quadrature point $\mathbf{x}_Q$ to evaluate the MLS shape functions and their derivatives for every node inside the support domain of the Gauss point.

Alternative quadrature rules have also been proposed to reduce the huge numerical costs or to avoid definition of a background cell by the nodal integration rule, thus ensuring a truly meshless EFG procedure. Nevertheless, they usually encounter oscillations or may lead to a reduced rate of convergency and even a low level of accuracy and cannot be recommended for general EFG solutions (Forouzan-sepehr and Mohammadi 2010).

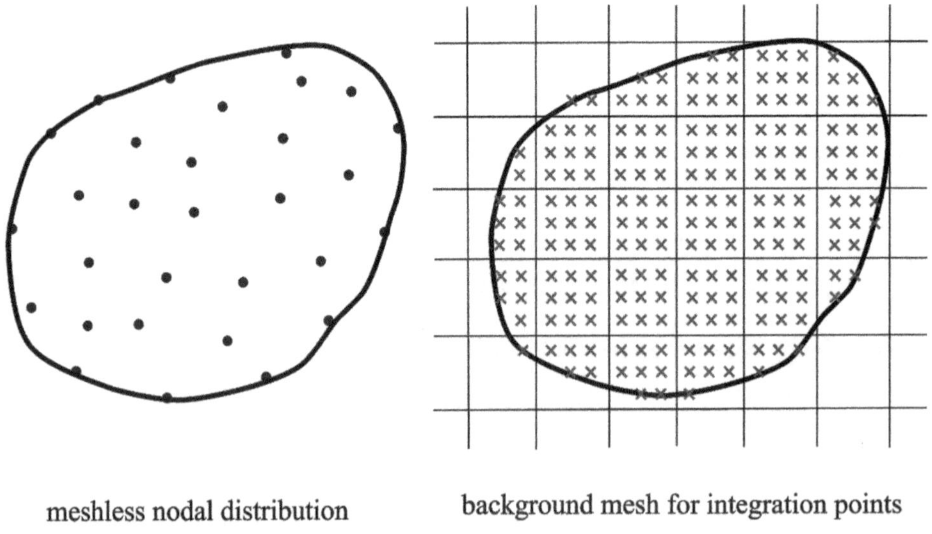

Figure 3.42 Background cell integration for EFG.

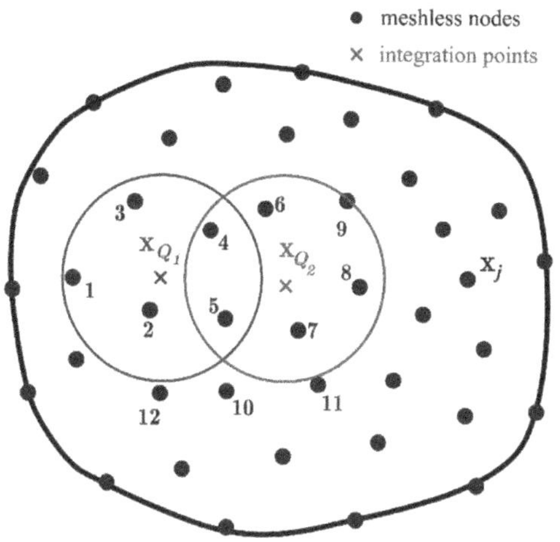

Figure 3.43 EFG integration and MLS support domain procedure.

Finally, the constrained weak form of EFG in Equation (3.7.86) may be reformulated by the penalty approach:

$$\underbrace{\int_{\Omega} \delta\varepsilon^{\mathrm{T}}\sigma d\Omega}_{\text{internal vitual work}} - \underbrace{\int_{\Omega} \delta \boldsymbol{u}^{\mathrm{T}} \boldsymbol{f}^{b} d\Omega}_{\text{virtual work of body force}} - \underbrace{\int_{\Gamma_t} \delta \boldsymbol{u}^{\mathrm{T}} \boldsymbol{f}^{t} d\Gamma}_{\text{virtual work of tractions}} \\ - \underbrace{\delta\left[\int_{\Gamma_u} \frac{1}{2}(\boldsymbol{u}-\bar{\boldsymbol{u}})^{T} \alpha (\boldsymbol{u}-\bar{\boldsymbol{u}}) \mathrm{d}\Gamma\right]}_{\text{imposition of essential boundary conditions by the penalty approach}} = 0 \tag{3.7.98}$$

where α is a diagonal matrix of penalty factors. Following the same procedure as EFG with Lagrange multipliers, the final discretized form of Equation (3.7.98) can be written as

$$\left[\mathbf{K}+\mathbf{K}^{\alpha}\right]\bar{\mathbf{U}} = \mathbf{f}+\mathbf{f}^{\alpha} \tag{3.7.99}$$

with

$$\mathbf{K}_{IJ}^{\alpha} = \int_{\Gamma_u} \mathbf{\Phi}_I^{\mathrm{T}} \alpha \mathbf{\Phi}_J \mathrm{d}\Gamma \tag{3.7.100}$$

$$\mathbf{f}_I^{\alpha} = \int_{\Gamma_u} \mathbf{\Phi}_I^{\mathrm{T}} \alpha \bar{\boldsymbol{u}} \mathrm{d}\Gamma \tag{3.7.101}$$

The procedure of solving a problem by EFG can be summarized as:

- Set the background integration cell and the corresponding quadrature points $\mathbf{x}_Q$.
- For each quadrature point $\mathbf{x}_Q$,
  - Set the support domain Ω_s of the quadrature point $\mathbf{x}_Q$.
  - Compute the MLS shape functions $\mathbf{\Phi}$ and their derivatives.

  - Numerically evaluate integrals in Equations (3.7.94) to (3.7.97) (or (3.7.100) and (3.7.101)).
  - Assemble in global $\mathbf{K}$, $\mathbf{f}$, $\mathbf{G}$ and $\mathbf{q}$ (or $\mathbf{K}^\alpha$ and $\mathbf{f}^\alpha$).
- End.
- Solve $\bar{\mathbf{U}}$ from the global Equation (3.7.93) or (3.7.99).

3.7.3.2 Meshless Local Petrov–Galerkin Method (MLPG)

Atluri and Zhu (1998) developed the concept of the meshless local Petrov–Galerkin method (MLPG) to satisfy equilibrium of each node based on a local weak form around the node. MLPG can somehow be regarded as a compromised methodology in between the accurate and complex EFG and the less accurate simple collocation techniques.

Consider a domain and its boundary represented by a finite number of nodes, as depicted in Figure 3.44. The body is assumed to be subjected to a prescribed body force $\boldsymbol{f}^{\mathrm{b}}$ on Ω and traction $\boldsymbol{f}^t$ on Γ_t. The essential boundary conditions $\boldsymbol{u} = \bar{\boldsymbol{u}}$ are prescribed on Γ_u.

The strong form of the equilibrium equation and the corresponding boundary conditions are assumed to be similar to Equations (3.7.83) to (3.7.85).

The local weighted residual Petrov–Galerkin formulation for a local equilibrium domain Ω_I around the node I with the penalty method to enforce the boundary condition (see Figure 3.44) is written as (Atluri and Shen 2002):

$$\underbrace{\int_{\Omega_I} (\boldsymbol{\nabla}.\boldsymbol{\sigma} + \boldsymbol{f}^b) W_I^e \mathrm{d}\Omega}_{\text{weighted residual of force balance}} - \underbrace{\alpha \int_{\Gamma_{Iu}} (\mathbf{u} - \bar{\mathbf{u}}) W_I^e \mathrm{d}\Gamma}_{\text{weighted residual of boundary-force balance}} = \mathbf{0} \tag{3.7.102}$$

where W_I^e is the test function of the local Petrov–Galerkin formulation, defined on the local equilibrium domain Ω_I. W_I^e must vanish on the boundary Γ_I of the local equilibrium domain. For local domains near a displacement boundary, Ω_I intersects with the essential boundary on Γ_{Iu} (see Figure 3.44).

Using the divergence theorem and after some manipulations, Equation (3.7.102) leads to

$$\int_{\Omega_I} \boldsymbol{\sigma} \boldsymbol{\nabla} W_I^e \mathrm{d}\Omega + \alpha \int_{\Gamma_{Iu}} \boldsymbol{u} W_I^e \mathrm{d}\Gamma - \int_{\Gamma_{Iu}} \boldsymbol{\sigma n} W_I^e \mathrm{d}\Gamma = \int_{\Gamma_{It}} \boldsymbol{f}^t W_I^e \mathrm{d}\Gamma + \alpha \int_{\Gamma_{Iu}} \bar{\boldsymbol{u}} W_I^e \mathrm{d}\Gamma + \int_{\Omega_I} \boldsymbol{f}^b W_I^e \mathrm{d}\Omega \tag{3.7.103}$$

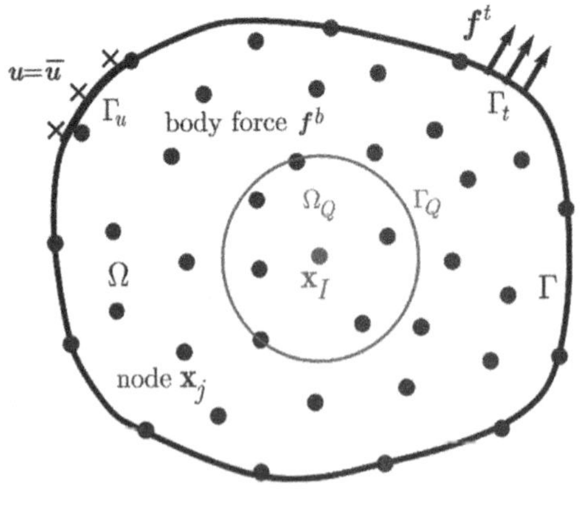

Figure 3.44 Definition of equilibrium domains in MLPG.

Equation (3.7.103) is further simplified if Ω_I is located entirely inside the domain:

$$\int_{\Omega_I} \sigma \nabla W_I^e \mathrm{d}\Omega = \int_{\Omega_I} \boldsymbol{f}^b W_I^e \mathrm{d}\Omega \tag{3.7.104}$$

Therefore, the strong form of Equation (3.7.83) on a typical node is replaced by a relaxed weak form over a local equilibrium domain around that node.

Discretization of the weak form Equation (3.7.103) is performed by the MLS approximation,

$$\boldsymbol{u}^h = \sum_{j=1}^{n} \varphi_j \boldsymbol{u}_j \tag{3.7.105}$$

and the final matrix form of the local equilibrium Equation (3.7.103) for node I can be written as (Atluri and Zhu 1998; Atluri and Shen 2002)

$$\sum_{j=1}^{n} \mathbf{K}_{Ij} \boldsymbol{u}_j = \mathbf{f}_I \tag{3.7.106}$$

with

$$\mathbf{K}_{Ij} = \int_{\Omega_I} (\nabla W_I^e)^{\mathrm{T}} \mathbf{D}\mathbf{B}_j \mathrm{d}\Omega + \alpha \int_{\Gamma_{Iu}} W_I^e \boldsymbol{\Phi}_j \mathrm{d}\Gamma - \int_{\Gamma_{Iu}} W_I^e \boldsymbol{n}\mathbf{D}\mathbf{B}_j \mathrm{d}\Gamma \tag{3.7.107}$$

$$\mathbf{f}_I = \int_{\Omega_I} W_I^e \boldsymbol{f}^b \mathrm{d}\Omega + \int_{\Gamma_{It}} W_I^e \boldsymbol{f}^t \mathrm{d}\Gamma + \alpha \int_{\Gamma_{Iu}} W_I^e \bar{\boldsymbol{u}} \mathrm{d}\Gamma \tag{3.7.108}$$

Computation of integrals in Equations (3.7.107) and (3.7.108) are performed on the local background cell, as depicted in Figure 3.45. The Gauss quadrature rule is then adopted for each background cell. Evaluation of integrands at each Gauss quadrature point $\mathbf{x}_Q$ requires the MLS support domain and the corresponding shape functions on the set of nodes $\mathbf{x}_j$ inside the support domain. Note that according to the MLS procedure, some nodes $\mathbf{x}_j$ may not be part of the local equilibrium domain around the node I (see Figure 3.45).

The procedure of solving a problem by MLPG can be summarized as:

- For any node I, determine the weight W_I^e on the local equilibrium domain Ω_I.
  - Set the background integration cell and the corresponding quadrature points $\mathbf{x}_Q$
  - For each quadrature point $\mathbf{x}_Q$:
    - o Set the support domain Ω_s of the quadrature point.
    - o Compute MLS shape functions $\boldsymbol{\Phi}$ and their derivatives.
    - o Numerically evaluate integrals in Equations (3.7.107) and (3.7.108).
    - o Assemble in global $\mathbf{K}$ and $\mathbf{f}$.
  - End.
- End.

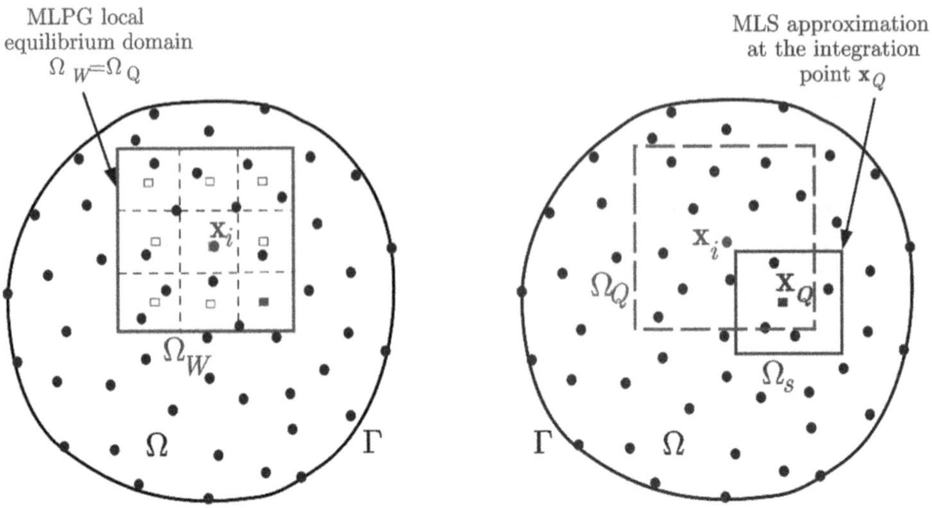

Figure 3.45 MLPG integration procedure.

- Solve $\bar{\mathbf{U}}$ from the global $\mathbf{K}\bar{\boldsymbol{u}} = \mathbf{f}$.

A major drawback of the standard MLPG is its non-symmetric stiffness matrix. For a review on available remedies, see Atluri and Zhu (1998), which can be used for alternative MLPG formulations based on different choices of the test function, types of the weak form and other meshless discretization techniques.

3.7.3.3 Finite Point Method (FPM)

The finite point method (FPM) is a strong form truly meshless solution technique based on the MLS approximation around each node (Onate et al. 1999; Boroomand et al. 2005, 2009). Figure 3.46 illustrates one of the typical MLS support domains for node $\mathbf{x}_i$ and the corresponding nodes $\mathbf{x}_j$ inside the support domain.

The strong form of a general equilibrium equation can be considered as

$$A(\boldsymbol{u}_j) = 0 \quad \text{in } \Omega \tag{3.7.109}$$

with the following essential and natural boundary conditions:

$$\boldsymbol{u}_j - \bar{\boldsymbol{u}}_j = 0 \text{ on } \Gamma_u \tag{3.7.110}$$

$$B(\boldsymbol{u}_j) = 0 \text{ on } \Gamma_t \tag{3.7.111}$$

Similar to other strong form solutions, FPM directly applies the equilibrium equation on all nodes j:

$$\begin{cases} A(\boldsymbol{u}_j) = 0 \text{ in } \Omega \\ \boldsymbol{u}_j - \bar{\boldsymbol{u}}_j = 0 \text{ on } \Gamma_u \\ B(\boldsymbol{u}_j) = 0 \text{ on } \Gamma_t \end{cases} \rightarrow \begin{cases} \left[A(\boldsymbol{u}_j)\right]_p = 0, p = 1,2,\ldots,N_r \\ \left[\boldsymbol{u}_j\right]_s - \bar{\boldsymbol{u}}_j = 0, s = 1,2,\ldots,N_u \\ \left[B(\boldsymbol{u}_j)\right]_r = 0, r = 1,2,\ldots,N_t \end{cases} \tag{3.7.112}$$

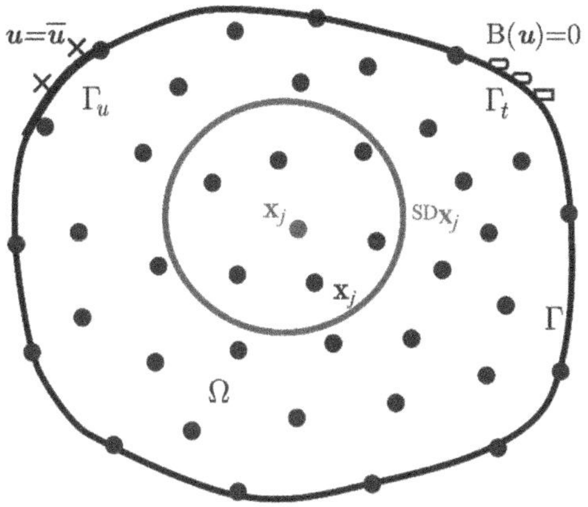

Figure 3.46 FPM support domain.

FPM can be assumed to be a generalized finite difference solution that uses MLS shape functions to approximate the field value around a node:

$$\boldsymbol{u}(\mathbf{x}) = \boldsymbol{u}^h(\mathbf{x}) = \mathbf{N}(\mathbf{x})\bar{\mathbf{U}} = \mathbf{p}^{\mathrm{T}}\mathbf{A}^{-1}\mathbf{B}\bar{\mathbf{U}} \tag{3.7.113}$$

Then, the discretized governing equations are modified to

$$\begin{cases} \left[A\left(\mathbf{N}\left(\mathbf{x}_j\right)\bar{\mathbf{U}}\right)\right]_p = 0,\ p = 1,2,\ldots,N_r \\ \left[\mathbf{N}\left(\mathbf{x}_j\right)\bar{\mathbf{U}}\right]_s - \bar{\boldsymbol{u}}_j = 0,\ s = 1,2,\ldots,N_u \\ \left[B(\mathbf{N}\left(\mathbf{x}_j\right)\bar{\mathbf{U}}\right]_r = 0,\ r = 1,2,\ldots,N_t \end{cases} \tag{3.7.114}$$

The final set of simultaneous equations can then be obtained as (Bitaraf and Mohammadi 2008, 2010)

$$\mathbf{K}\bar{\mathbf{U}} = \mathbf{f} \tag{3.7.115}$$

In order to avoid ill-conditioning of FPM (similar to other strong form solutions), the following stabilized forms in terms of the characteristic stabilization term h are added to the stress/pressure terms (Onate et al. 1999):

$$\begin{cases} A\left(\boldsymbol{u}_j\right) - \dfrac{1}{2}h_k\dfrac{\partial A\left(\boldsymbol{u}_j\right)}{\partial x_k} = 0 \text{ in } \Omega \\ \boldsymbol{u}_j - \bar{\boldsymbol{u}}_j = 0 \text{ on } \Gamma_u \\ B\left(\boldsymbol{u}_j\right) - \dfrac{1}{2}h_k n_k A\left(\boldsymbol{u}_j\right) = 0 \text{ on } \Gamma_t \end{cases} \tag{3.7.116}$$

and the discretized governing equations are obtained as

$$[\mathbf{K} + \mathbf{K}_S(h_k)]\bar{\mathbf{U}} = \mathbf{f} \tag{3.7.117}$$

3.7.3.4 Smoothed Particle Hydrodynamics (SPH)

In the SPH method, the kernel approximation is approximated by imagining that the domain is divided into small volume particles. For a sub-domain with n particles around the point $\mathbf{x}$,

$$\boldsymbol{u}^h(\mathbf{x}) = \sum_{j=1}^{n} W_h(\mathbf{x} - \mathbf{x}_j)\boldsymbol{u}(\mathbf{x}_j)\Delta V_j \tag{3.7.118}$$

or in terms of the SPH shape functions,

$$\boldsymbol{u}^h(\mathbf{x}) = \sum_{j=1}^{n} \varphi_j(\mathbf{x})\boldsymbol{u}(\mathbf{x}_j) \tag{3.7.119}$$

$$\varphi_j(\mathbf{x}) = W_h(\mathbf{x} - \mathbf{x}_j)\Delta V_j \tag{3.7.120}$$

and its derivative

$$\frac{\partial \boldsymbol{u}}{\partial x} = -\sum_{j=1}^{n} \boldsymbol{u}_j \frac{\partial W_{hij}}{\partial x} \Delta V_j \tag{3.7.121}$$

which is proved to be equivalent to (Ostad and Mohammadi 2009, 2010)

$$\frac{\partial \boldsymbol{u}}{\partial x} = \sum_{j=1}^{n} (\boldsymbol{u}_i - \boldsymbol{u}_j) \frac{\partial W_{hij}}{\partial x} \Delta V_j \tag{3.7.122}$$

In SPH, all spatial derivatives $\boldsymbol{u}_{j,x}$ are transformed into algebraic values $\boldsymbol{u}_j$ on particles inside the smoothing domain. As a result, the governing partial differential equations are transformed into a set of nonlinear algebraic simultaneous equations.

Similar to FPM, a strong form solution procedure can be followed in SPH and CSPM. A similar stabilization technique is usually required to ensure the stability of the numerical solution (Ostad and Mohammadi, 2012; Ostad-Hossein and Mohammadi, 2008).

3.8 Variable Node Element (VNE)

The variable node element (VNE) was developed based on a combination of finite element and meshless concepts (Alizadeh and Mohammadi 2019). In VNE, the domain is modelled by a mesh of finite elements with conventional finite element corner nodes that also include a number of internal nodes inside each element, as depicted in Figure 3.47. The number and positions of internal nodes may vary in different elements depending on the type of local approximation and the required level of accuracy (Figure 3.48).

VNE holds the consistency of the field solution at element boundaries and on the whole domain. It facilitates adaptivity by readily allowing nodes to be inserted, removed or re-positioned inside an element.

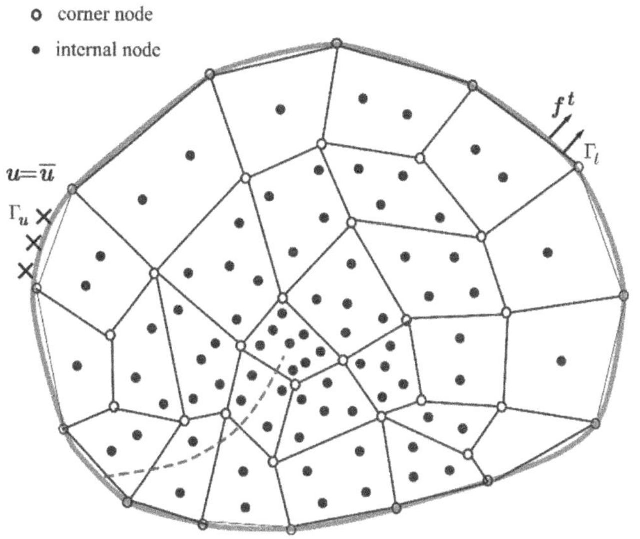

Figure 3.47 Meshing a model with variable node elements.

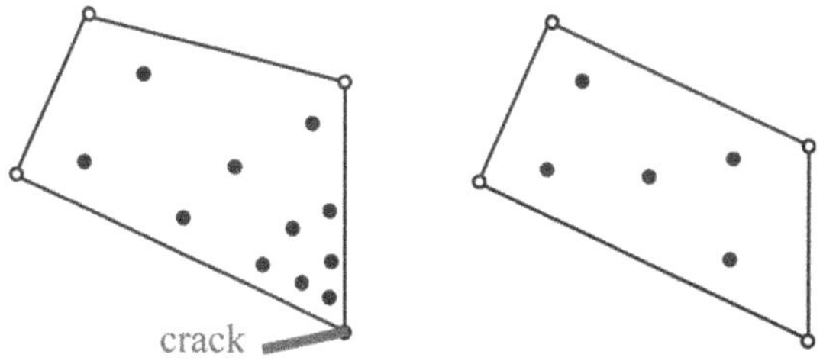

Figure 3.48 The number and positions of the internal nodes may vary in each variable node element.

The element was originally designed in the context of the variable node multiscale method (VNMM) and later adopted in the enriched multiscale method (EMM) to automatically handle concurrent atomistic and continuum formulations by setting the internal nodes as atoms. In this section, only the continuum description of the variable node element is discussed; the multiscale characteristics are described in Section 4.5 in Chapter 4.

Formulation of VNE is based on a PIM meshless approach with a support domain equivalent to the parent finite element. It is then modified to satisfy the necessary requirements of the finite element interpolation. To avoid potential singularities and sensitivity to nodal positions of the polynomial basis functions in PPIM, the radial point interpolation (RPIM) approach is adopted to generate the initial VNE shape functions.

In VNE, the shape functions of all internal nodes are computed by the well-developed RPIM technique of Equation (3.7.37), based on the radial basis functions $R_j(\mathbf{x})$:

$$\boldsymbol{u}^h(\mathbf{x},\mathbf{x}_Q)=\sum_{j=1}^{n}R_j(r_j(\mathbf{x}))\boldsymbol{a}_j(\mathbf{x}_Q)=\mathbf{R}^{\mathrm{T}}(r(\mathbf{x}))\boldsymbol{a}(\mathbf{x}_Q) \tag{3.8.1}$$

or in terms of the shape functions $\varphi(\mathbf{x})$:

$$\boldsymbol{u}^h(\mathbf{x})=\varphi(\mathbf{x})\bar{\mathbf{U}}_s=\mathbf{R}^{\mathrm{T}}(\mathbf{x})\mathbf{R}_Q^{-1} \tag{3.8.2}$$

where $\mathbf{R}_Q$ is defined in Equation (3.7.43).

The shape function $\varphi_i(\mathbf{x})$ of node i for the variable node element can be similarly written as

$$\varphi_i(\mathbf{x})=\sum_{j=1,j\neq i}^{N_b+N_i}R_j(\mathbf{x})\lambda_j(\mathbf{x}_i)=\mathbf{R}^{\mathrm{T}}(\mathbf{x})\ \boldsymbol{\Lambda}(\mathbf{x}_i) \tag{3.8.3}$$

where N_b and N_i are the number of nodes on the boundary and inside of the variable node element, respectively, and the vector of nodal parameters $\boldsymbol{\Lambda}$ is obtained from Equation (3.8.4):

$$\boldsymbol{\Lambda}=\mathbf{R}_Q^{-1}=\mathbf{R}^{-1}(\mathbf{x})\varphi(\mathbf{x}) \tag{3.8.4}$$

or for a node m:

$$\lambda(\mathbf{x}_m)=R^{-1}(\mathbf{x}_m)\ \varphi_i(\mathbf{x}_m)=\bar{R}^{-1}(\mathbf{x}_m)\ \delta_{mi} \tag{3.8.5}$$

where δ_{mi} is the Kronecker delta function.

It is known that the RPIM interpolation cannot satisfy continuity in between the local support domains. Here it means that the shape functions of Equation (3.8.3) and the corresponding VNE interpolation are not continuous across the edges of adjacent elements. As a necessary condition to ensure overall C^0 continuity, the formulation is modified by a ramp function $\chi(\mathbf{x})$ to make a smooth transition from RPIM functions $R_j(\mathbf{x})$ (or shape functions $\varphi_i(\mathbf{x})$) for the internal nodes to the conventional finite element shape functions $N_i(\mathbf{x})$ for the corner nodes,

$$\bar{R}_i(\mathbf{x})=\chi(\mathbf{x})R_i(\mathbf{x})+(1-\chi(\mathbf{x}))\ N_i(\mathbf{x}) \tag{3.8.6}$$

The ramp function $\chi(\mathbf{x})$ should at least be a C^0 continuous function, such as

$$\chi(\mathbf{x})=\chi(x.y)=(1-x^2)(1-y^2) \tag{3.8.7}$$

The modified VNE shape function $\bar{\varphi}_i(\mathbf{x})$, which is continuous all over the domain, can then be written as

$$\bar{\varphi}_i(\mathbf{x})=\sum_{j=1,j\neq i}^{N_b+N_i}\bar{R}_i(\mathbf{x})\ \lambda_j(\mathbf{x}_i)=\bar{\mathbf{R}}^T(\mathbf{x})\ \boldsymbol{\Lambda}(\mathbf{x}_i) \tag{3.8.8}$$

Equation (3.8.8) allows for a conventional finite element interpolation on the VNE edges, including the corner nodes, and performs as an RPIM mesh-free approximation for VNE internal nodes.

Furthermore, the VNE approximation is expected to satisfy the partition of unity (PU) condition. This is achieved by the modified $\bar{\bar{\varphi}}_i(\mathbf{x})$ (Alizadeh 2019):

$$\bar{\bar{\varphi}}_i(\mathbf{x}) = \bar{\varphi}_i(\mathbf{x}) + \frac{1}{N_b + N_i}\left[1 - \sum_{i=1}^{N_b+N_i} \bar{\varphi}_i(\mathbf{x})\right] \tag{3.8.9}$$

Figure 3.49 illustrates a sample VNE for simulation of a graphene structure (Zhu 2018) with 128 internal nodes, as well as four conventional finite element corner nodes. It also depicts the shape functions associated with a corner node and an internal one. Clearly, both shape functions provide linear variations along the edges, which satisfy the continuity across adjacent element edges.

The element formulation can be modified to an extended version of VNE to accommodate for the crack edge enrichment functions,

$$\boldsymbol{u}(\mathbf{x}) = \left[\sum_{i=1}^{N_b+N_i} \bar{\bar{\varphi}}_i(\mathbf{x})\, \boldsymbol{u}_i\right] + \left[\sum_{h=1}^{N_b+N_i} \bar{\bar{\varphi}}_h(\mathbf{x})\left[H(\mathbf{x}) - H(\mathbf{x}_h)\right]\boldsymbol{a}_h\right] \tag{3.8.10}$$

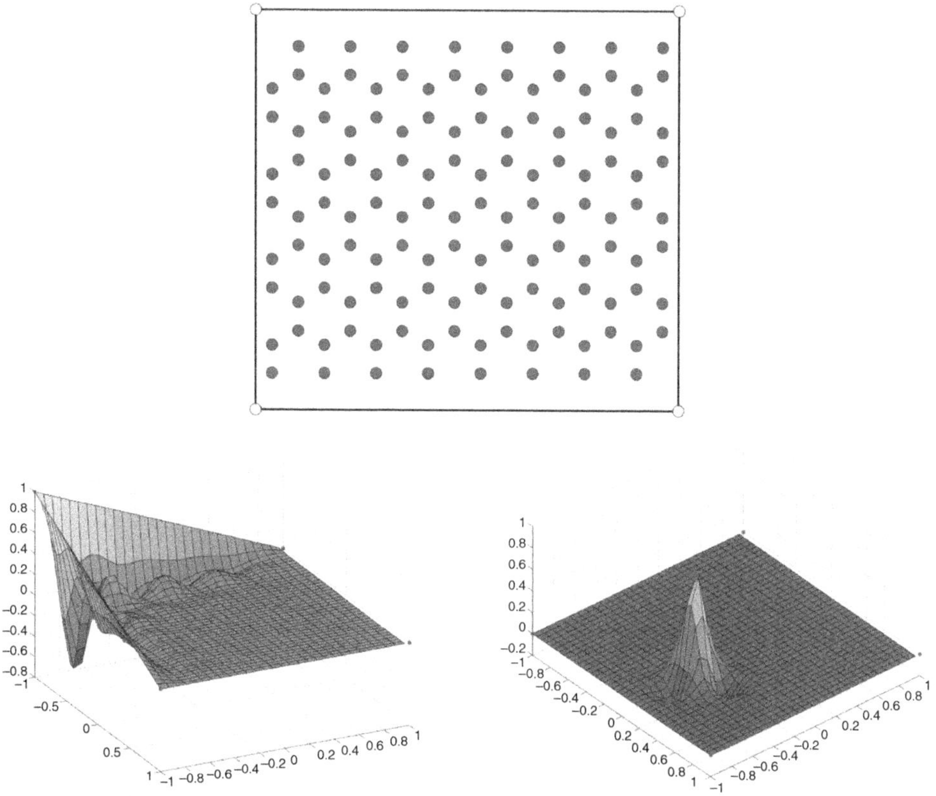

Figure 3.49 Shape functions of two nodes of a variable node element of graphene.

where

$$H(\mathbf{x})=\begin{cases}+1, & \mathbf{x}\geq 0\\ -1, & \mathbf{x}<0\end{cases} \tag{3.8.11}$$

Similarly, for the crack tip enrichment functions,

$$\boldsymbol{u}(\mathbf{x})=\left[\sum_{\mathrm{i}=1}^{N_b+N_i}\overline{\overline{\varphi}}_i(\mathbf{x})\,\boldsymbol{u}_{\mathrm{i}}\right]+\left[\sum_{k=1}^{N_b+N_i}\overline{\overline{\varphi}}_k(\mathbf{x})(\sum_{l=1}^{mf=4}\left[F_l(\mathbf{x})-F_l(\mathbf{x}_k)\right]\boldsymbol{b}_k^l)\right] \tag{3.8.12}$$

where

$$F_l(x)=\left\{\sqrt{r}\sin\left(\frac{\theta}{2}\right),\sqrt{r}\cos\left(\frac{\theta}{2}\right),\sqrt{r}\sin\left(\frac{\theta}{2}\right)\sin(\theta),\sqrt{r}\cos\left(\frac{\theta}{2}\right)\sin(\theta)\right\} \tag{3.8.13}$$

For details of the formulation of the variable node element and its performance in single scale applications, see Alizadeh and Mohammadi (2019).

4

Multiscale Methods

4.1 Introduction

This chapter is dedicated to a review of major categories of multiscale methods. Multiscale methods constitute a wide set of diverse analysis methods that bridge across various length scales (and sometimes time scales) to provide computationally affordable techniques for analysis of very complex problems with microscale characteristics or complicated physical phenomena that would require virtually unlimited computational resources for a direct numerical analysis.

After this brief introduction, the homogenization techniques are reviewed. These methods are mainly adopted to analyse heterogeneous materials with a repeating microstructure. While a direct numerical analysis of such microstructures requires an expensive numerical model with a very fine mesh of finite elements, the homogenization technique may efficiently analyse the same problem to obtain similar levels of accuracy by solving the microscale representative volume element (RVE) to compute a set of homogenized properties, which will then be used to solve the coarse macroscale model. The overall number of elements remains far lower than a direct analysis.

The molecular dynamic (MD) solutions are then comprehensively discussed. Molecular or atomistic modelling has been widely used in analysis of various physical and engineering applications. MD is a fundamental approach based on the simulation of the building structure of the material using the classical dynamics of particles/objects. The molecules/atoms interact with each other through the interatomic potentials and are subject to a number of specific macroscopic requirements based on the concepts of statistical mechanics.

MD models may require an extremely huge number of atoms or molecules to create a sufficiently accurate model for representing a complex nanoscale problem. Analysis of such complicated models requires massive high-performance hardware and software resources. While it is noted that an MD analysis cannot be considered as a multiscale method, the MD solutions are essential parts of major sequential and concurrent multiscale methods. In other words, many multiscale techniques cannot be discussed without an in-depth knowledge of MD solutions.

The chapter continues with a review of sequential multiscale methods. Sequential multiscale methods deal with multiple scale problems in a sequential uncoupled procedure.

Multiscale Biomechanics: Theory and Applications, First Edition. Soheil Mohammadi.

The main concept is to obtain the necessary response at one scale and to export it directly to another scale for a more accurate analysis.

The procedure may be designed from lower scales to higher scales, or vice versa. In a lower to higher scale simulation, the lower scale specimen is analysed with all details to determine its overall stress–strain response. It can then be used as a constitutive relation to analyse the higher scale model. The same procedure may be systematically followed to obtain the response of further higher scale problems.

In a higher to lower scale analysis, the sequential multiscale solution begins with a simulation at the higher scale (macro) to determine the critical points of the domain. Then, a lower scale model (micro) based on the microstructure of the critical point is constructed and analysed to study the effects on microingredients. The procedure may be repeated further down to lower scales.

The final part of the chapter deals with the set of concurrent multiscale methods. In current multiscale methods, both atomistic and continuum formulations constitute the coupled governing equations of the multiscale method. They may be employed simultaneously to describe the response of a specific region, or they may be used for different regions of the multiscale problem, with an interface or transition zone to connect the regions. A variety of concurrent methods, including the quasi-continuum approach (QC), the bridging domain method (BDM), the bridging scale method (BSM), the disordered concurrent multiscale method (DCMM), the variable node multiscale method (VNMM) and the enriched multiscale method (EMM) are examined in more detail.

4.2 Homogenization Methods

4.2.1 Introduction

The homogenization technique is a powerful approach for efficient analysis of heterogeneous media with a repeating microstructure.

Direct numerical modelling of microstructures requires a mesh of very fine finite elements, which leads to huge computational costs. The homogenization techniques are designed to solve such problems in a fast and efficient fashion by solving a representative part of the microscale to compute the homogenized properties of the coarse macroscale model.

For instance, consider a beam constructed from a porous shape memory alloy (SMA), as depicted in Figure 4.1. A direct numerical analysis requires a mesh of fine elements to achieve the necessary accuracy to capture the superelastic response of the SMA beam. In contrast, a homogenization technique may efficiently obtain the same level of accuracy by solving the microscale representative volume element (RVE) of the SMA microstructure, as presented in Figure 4.1.

The concept of homogenization is based on the modification of the material modules by considering the microeffects. It includes mathematical and computational homogenization techniques. Alternatively, it may be formulated by adding the microdisplacement terms into the macrodisplacement fields through the variational multiscale method. Recently,

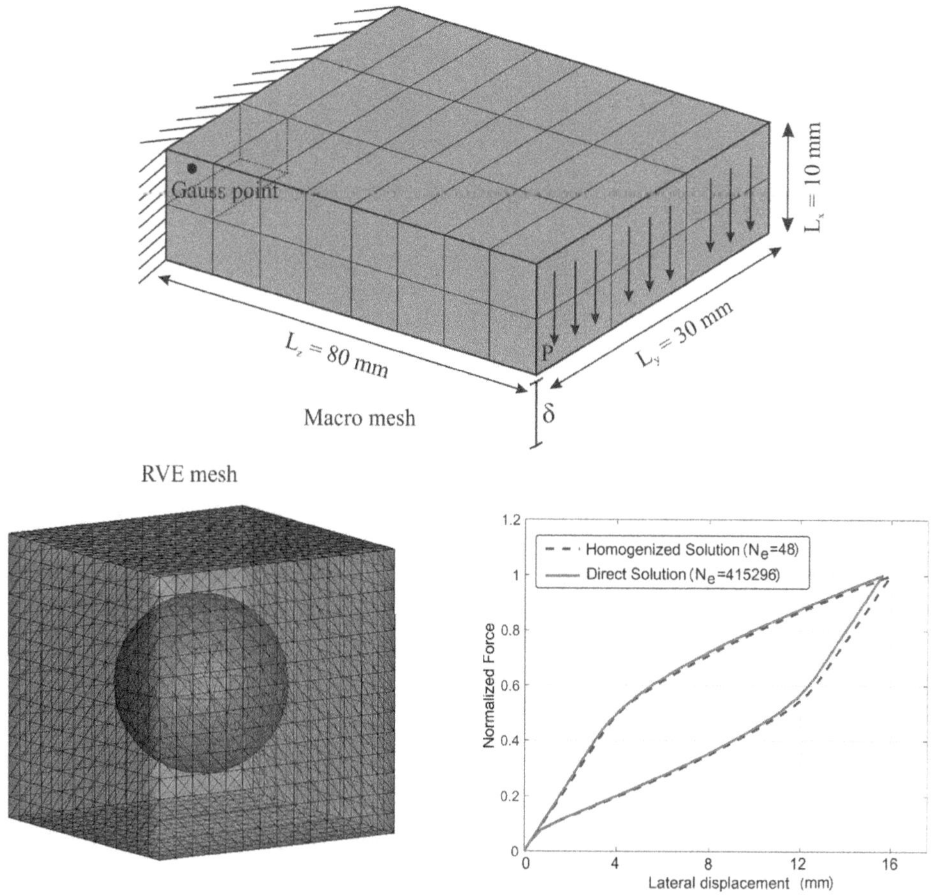

Figure 4.1 A porous SMA beam solved by direct and homogenized algorithms. *Source:* Adapted from Fatemi et al. 2017.

the concept has been extended to enrichment of conventional continuum formulations through the additional micro-based kinematic variables to account for the microeffects and the micro–macro coupling in softening problems.

4.2.2 Representative Volume Element (RVE)

Homogenization techniques adopt the concept of the representative volume element (RVE) to describe the microstructure of the model. It is in fact a statistical approach in description of the repeating microstructure in a global or local periodic fashion, as presented in Figure 4.2. In a global assumption, RVE is assumed to be repeated everywhere in the model, whereas in a local periodic description, the microstructure should remain the same in the vicinity of a point.

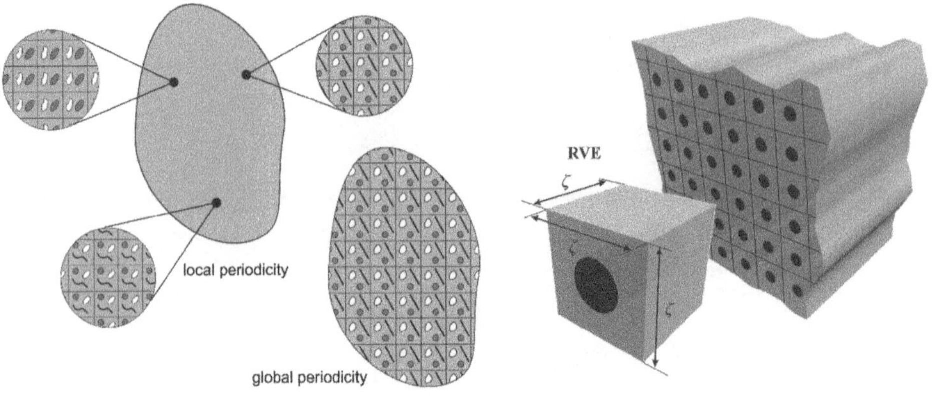

Figure 4.2 The concept of a representative volume element, and the global and local periodicity of a microstructure. *Source:* Adapted from Fatemi et al. 2017.

It is important to note that a periodic description is not equivalent to the assumption of a constant response everywhere. For any variable Φ, an approximation Φ^{ζ} in terms of the spatial macro- $\mathbf{x}^{\mathrm{M}}$ and micro- $\mathbf{x}^{\mathrm{m}}$ variables can be written as

$$\Phi^{\zeta}\left(\mathbf{x}^{\mathrm{M}},\mathbf{x}^{\mathrm{m}}\right)=\Phi\left(\mathbf{x}^{\mathrm{M}},\frac{\mathbf{x}^{\mathrm{M}}}{\zeta}\right) \tag{4.2.1}$$

$$\mathbf{x}^{\mathrm{M}}=\zeta\mathbf{x}^{\mathrm{m}} \tag{4.2.2}$$

where $\mathbf{x}^{\mathrm{M}}$ represents the position of a point in the macrodomain Ω^{M}, $\mathbf{x}^{\mathrm{m}}$ refers to the position of the point in the microscale model of RVE and ζ is an index relating the two scales and the level of oscillations of a field variable inside the RVE.

4.2.3 Mathematical Homogenization

The roots of mathematical homogenization can be related to the asymptotic solution and perturbation analysis for solving algebraic and differential equations. Here, the concept of the method is described for general boundary value problems. It is then applied to the solution of differential equations of general elasticity problems.

4.2.3.1 Mathematical Homogenization for General Boundary Value Problems

A general boundary value problem in terms of the field variable T can be written as

$$A^{\zeta}T^{\zeta}=f \text{ in } \Omega^{\mathrm{M}} \tag{4.2.3}$$

with the boundary conditions

$$T^{\zeta}=0 \text{ on } \Gamma^{\mathrm{M}} \tag{4.2.4}$$

where

$$A^{\zeta} = \frac{\partial}{\partial \mathbf{x}^{\mathrm{M}}}\left(a\left(\mathbf{x}^{\mathrm{m}}\right)\frac{\partial}{\partial \mathbf{x}^{\mathrm{M}}}\right) \tag{4.2.5}$$

A double-scale asymptotic expansion is defined as

$$T^{\zeta}\left(\mathbf{x}^{\mathrm{M}}\right) = \zeta^{0}T^{0}\left(\mathbf{x}^{\mathrm{M}},\mathbf{x}^{\mathrm{m}}\right) + \zeta^{1}T^{1}\left(\mathbf{x}^{\mathrm{M}},\mathbf{x}^{\mathrm{m}}\right) + \zeta^{2}T^{2}\left(\mathbf{x}^{\mathrm{M}},\mathbf{x}^{\mathrm{m}}\right) + \cdots \tag{4.2.6}$$

or simply

$$T^{\zeta}\left(\mathbf{x}^{\mathrm{M}}\right) = T^{0} + \zeta T^{1} + \zeta^{2}T^{2} + \cdots \tag{4.2.7}$$

where functions $T^{i}\left(\mathbf{x}^{\mathrm{M}},\mathbf{x}^{\mathrm{m}}\right)$ are periodic in $\mathbf{x}^{\mathrm{m}}$.

It is noted that the derivative of a periodic function remains periodic with the same period. Moreover, the integral of the derivative of a periodic function over its period becomes zero.

Differentiation of any $\phi = \phi\left(\mathbf{x}^{\mathrm{M}},\mathbf{x}^{\mathrm{m}}\right)$ can be written as

$$\frac{\mathrm{d}\phi}{\mathrm{d}\mathbf{x}^{\mathrm{M}}} = \frac{\partial \phi}{\partial \mathbf{x}^{\mathrm{M}}} + \frac{1}{\zeta}\frac{\partial \phi}{\partial \mathbf{x}^{\mathrm{m}}} \tag{4.2.8}$$

Therefore, the operator A^{ζ} can be expanded by the asymptotic solution to (Hasani and Hinton 1998a, 1998b)

$$A^{\zeta} = \frac{1}{\zeta^{2}}A^{1} + \frac{1}{\zeta^{2}}A^{2} + A^{3} \tag{4.2.9}$$

where

$$A^{1} = \frac{\partial}{\partial \mathbf{x}^{\mathrm{m}}}\left(a\left(\mathbf{x}^{\mathrm{m}}\right)\frac{\partial}{\partial \mathbf{x}^{\mathrm{m}}}\right) \tag{4.2.10}$$

$$A^{2} = \frac{\partial}{\partial \mathbf{x}^{\mathrm{m}}}\left(a\left(\mathbf{x}^{\mathrm{m}}\right)\frac{\partial}{\partial \mathbf{x}^{\mathrm{M}}}\right) + \frac{\partial}{\partial \mathbf{x}^{\mathrm{M}}}\left(a\left(\mathbf{x}^{\mathrm{m}}\right)\frac{\partial}{\partial \mathbf{x}^{\mathrm{m}}}\right) \tag{4.2.11}$$

$$A^{3} = \frac{\partial}{\partial \mathbf{x}^{\mathrm{M}}}\left(a\left(\mathbf{x}^{\mathrm{m}}\right)\frac{\partial}{\partial \mathbf{x}^{\mathrm{M}}}\right) \tag{4.2.12}$$

Equation (4.2.3) is then transformed to

$$\left(\frac{1}{\zeta^{2}}A^{1} + \frac{1}{\zeta^{2}}A^{2} + A^{3}\right)T^{0} + \zeta T^{1} + \zeta^{2}T^{2} + \cdots = f \tag{4.2.13}$$

Equating terms with the same power of ζ on both sides leads to

$$\zeta^{-2}:\ A^{1}T^{0} = 0 \tag{4.2.14}$$

$$\varsigma^{-1}: A^1T^1 + A^2T^0 = 0 \tag{4.2.15}$$

$$\varsigma^{0}: A^1T^2 + A^2T^1 + A^3T^0 = f \tag{4.2.16}$$

It can be shown that an equation in the form of $AT = F$ in a period Ω^{m} of the periodic function T has a unique solution if

$$\bar{F} = \frac{1}{\Omega^{\mathrm{m}}} \int_{\Omega^{\mathrm{m}}} F \, \mathrm{d}\Omega = 0 \tag{4.2.17}$$

Equation (4.2.17) along with Equations (4.2.14) to (4.2.16) result in

$$A^1T^0 = 0 \rightarrow T^0 = T^0\left(\mathbf{x}^{\mathrm{M}}\right) \tag{4.2.18}$$

which shows that $T^0 = T^0(\mathbf{x}^{\mathrm{M}})$ can be obtained independently of the microstructure $\mathbf{x}^{\mathrm{m}}$.

Equation (4.2.15) can now be used to determine T^1 from (Hassani and Hinton 1999)

$$A^1T^1 + A^2T^0 = 0 \rightarrow A^1T^1 = -\frac{\partial a\left(\mathbf{x}^{\mathrm{m}}\right)}{\partial \mathbf{x}^{\mathrm{m}}} \frac{\partial T\left(\mathbf{x}^{\mathrm{M}}\right)}{\partial \mathbf{x}^{\mathrm{M}}} \tag{4.2.19}$$

or

$$T^1\left(\mathbf{x}^{\mathrm{M}}, \mathbf{x}^{\mathrm{m}}\right) = \chi\left(\mathbf{x}^{\mathrm{m}}\right) \frac{\partial T\left(\mathbf{x}^{\mathrm{M}}\right)}{\partial \mathbf{x}^{\mathrm{M}}} + \xi\left(\mathbf{x}^{\mathrm{M}}\right) \tag{4.2.20}$$

where $\xi(\mathbf{x}^{\mathrm{M}})$ is a constant in terms of the macroscale $\mathbf{x}^{\mathrm{M}}$, and $\chi(\mathbf{x}^{\mathrm{m}})$ is a periodic solution of the governing equation of the microscale RVE,

$$A^1\chi\left(\mathbf{x}^{\mathrm{m}}\right) = \frac{\partial a\left(\mathbf{x}^{\mathrm{m}}\right)}{\partial \mathbf{x}^{\mathrm{m}}} \quad \text{in } \Omega^{\mathrm{m}} \tag{4.2.21}$$

or

$$\frac{\partial}{\partial \mathbf{x}^{m}}\left(a\left(\mathbf{x}^{\mathrm{m}}\right)\frac{\partial \chi}{\partial \mathbf{x}^{\mathrm{m}}} - a\left(\mathbf{x}^{\mathrm{m}}\right)\right) = 0 \tag{4.2.22}$$

The following homogenized (macroscopic) equation for $T^0(\mathbf{x}^{\mathrm{M}})$ is finally obtained:

$$A^hT^0\left(\mathbf{x}^{\mathrm{M}}\right) = f \quad \text{on } \Omega^{\mathrm{M}} \tag{4.2.23}$$

where the effective homogenized operator A^h,

$$A^h = \frac{\partial}{\partial \mathbf{x}^{\mathrm{M}}}\left(a^h \frac{\partial}{\partial \mathbf{x}^{\mathrm{M}}}\right) \tag{4.2.24}$$

is obtained from the homogenized modulus a^h:

$$a^h = \frac{1}{\Omega^{\mathrm{m}}} \int_{\Omega^{\mathrm{m}}} \left(a\left(\mathbf{x}^{\mathrm{m}}\right) + a\left(\mathbf{x}^{\mathrm{m}}\right)\frac{\partial \chi}{\partial \mathbf{x}^{\mathrm{m}}}\right) \mathrm{d}\Omega \tag{4.2.25}$$

The homogenization solution procedure can be summarized as:

1) Determine $\chi(\mathbf{x}^{\mathrm{m}})$ from Equation (4.2.21) on the RVE level Ω^{m}.

2) Solve Equation (4.2.23) on the macroscopic level Ω^{M} (with its boundary conditions). The homogenized coefficients a^h and the homogenized operator A^h are obtained from Equations (4.2.25) and (4.2.24), respectively.

In mathematical homogenization (the asymptotic technique), the domain is assumed to be composed of periodic RVEs (unit cells), but there is no explicit presumed assumption to relate the RVE solution with the macroresponse.

It is proved (and not assumed) that the first term of the macrodisplacement $T^0(\mathbf{x}^{\mathrm{M}},\mathbf{x}^{\mathrm{m}})$ is only a function of $\mathbf{x}^{\mathrm{M}}$, $T^0(\mathbf{x}^{\mathrm{M}})$. Similarly, it is mathematically illustrated that the coefficients of the homogenized governing equations can be determined from an averaging procedure on the microstructure level.

4.2.3.2 Homogenization for General Elasticity Problems

Consider a body governed by the general equations of elasticity. The body is assumed to be composed of a periodic microstructure, as depicted in Figure 4.3.

Having considered the effect of microscale fluctuations ζ, the strong form of the governing equation for a macroscale point $\mathbf{x}^{\mathrm{M}}$ can be written as

$$\nabla^{\mathrm{M}} \cdot \boldsymbol{\sigma}^{\zeta}\left(\mathbf{x}^{\mathrm{M}}\right) + \boldsymbol{f}^{\mathrm{b}\zeta}\left(\mathbf{x}^{\mathrm{M}}\right) = \mathbf{0},\ \mathbf{x}^{\mathrm{M}} \in \Omega^{\mathrm{M}} \tag{4.2.26}$$

with the conventional displacement and traction boundary conditions:

$$\boldsymbol{u}^{\zeta} = \bar{\boldsymbol{u}} \qquad \text{on } \Gamma_u^{\mathrm{M}} \tag{4.2.27}$$

$$\boldsymbol{\sigma}^{\zeta} \cdot \boldsymbol{n} = \boldsymbol{f}^t \qquad \text{on } \Gamma_t^{\mathrm{M}} \tag{4.2.28}$$

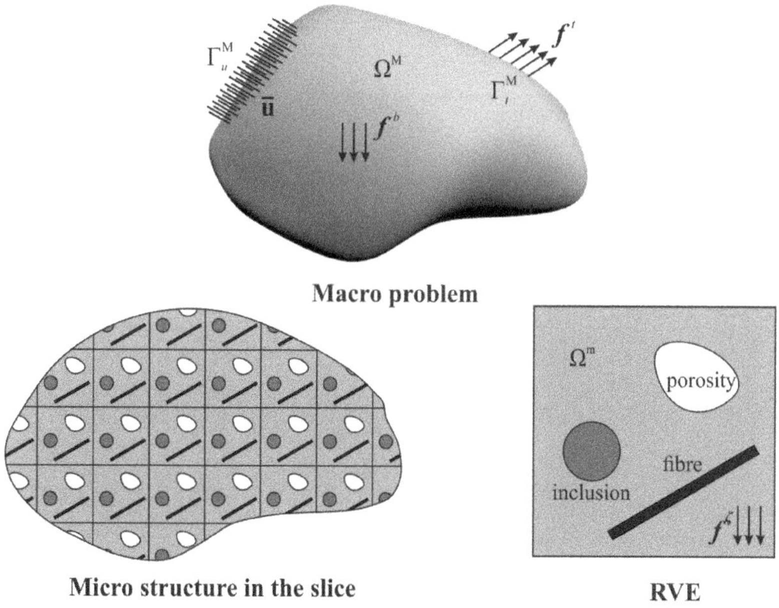

Figure 4.3 An elastic domain with a periodic microstructure.

and the corresponding weak form can be written as (Hassani and Hinton 1998b)

$$\int_{\Omega^{M}} \boldsymbol{\sigma}^{\varsigma}\left(\mathbf{x}^{M}\right) \cdot \delta\boldsymbol{\varepsilon}\, d\Omega = \int_{\Omega^{M}} \boldsymbol{f}^{b\varsigma}\left(\mathbf{x}^{M}\right) \cdot \delta\boldsymbol{u} d\Omega + \int_{\Gamma_{t}^{M}} \boldsymbol{f}^{t}\left(\mathbf{x}^{M}\right) \cdot \delta\boldsymbol{u}\, d\Gamma = 0 \tag{4.2.29}$$

where the stress–strain constitutive relation is defined in terms of the elasticity modulus $\mathbf{D}^{\zeta}$,

$$\boldsymbol{\sigma}^{\varsigma}\left(\mathbf{x}^{M}\right) = \mathbf{D}^{\varsigma}\left(\mathbf{x}^{M}\right)\boldsymbol{\varepsilon}^{\varsigma}\left(\mathbf{x}^{M}\right) \tag{4.2.30}$$

and the strain-displacement relation can be written as

$$\boldsymbol{\varepsilon}^{\zeta} = \frac{1}{2}\left[\nabla \boldsymbol{u}^{\zeta} + \nabla^{T}\boldsymbol{u}^{\zeta}\right] \tag{4.2.31}$$

A double-scale asymptotic expansion for $\boldsymbol{u}^{\zeta}$ can be written in terms of functions $\boldsymbol{u}^{i}\left(\mathbf{x}^{M},\mathbf{x}^{m}\right)$, which are assumed to be periodic in $\mathbf{x}^{m}$:

$$\boldsymbol{u}^{\varsigma}\left(\mathbf{x}^{M}\right) = \zeta^{0}\boldsymbol{u}^{0}\left(\mathbf{x}^{M},\mathbf{x}^{m}\right) + \zeta^{1}\boldsymbol{u}^{1}\left(\mathbf{x}^{M},\mathbf{x}^{m}\right) + \zeta^{2}\boldsymbol{u}^{2}\left(\mathbf{x}^{M},\mathbf{x}^{m}\right) + \cdots = u^{0} + \zeta\boldsymbol{u}^{1} + \zeta^{2}\boldsymbol{u}^{2} + \cdots \tag{4.2.32}$$

Here, only the first-order expansion is assumed, but the same procedure can be followed to derive the formulation for higher orders of the expansion. In general, while all functions $\boldsymbol{u}^{i}(\mathbf{x}^{M},\mathbf{x}^{m})$ may affect the macroscopic and microscopic solutions, it is shown that $\boldsymbol{u}^{0}$ corresponds to the macroscopic behaviour and $\boldsymbol{u}^{1}$ represents the microscopic response. The effects of microstructure fluctuations are considered in the evaluation of the homogenized material modulus and the homogenized body force vector.

The homogenization procedure satisfies the governing weak form of the equilibrium equations at both the microscale and macroscale levels with their relevant boundary conditions. They remain consistent with the material constitutive relation.

Without going into the details and proofs, the homogenization procedure for solving the general form of an elasticity problem can be summarized as:

1) Find $\boldsymbol{\varsigma}$ within the base cell by solving the following integrals on the base cell:

$$\int_{\Omega^{m}} \mathbf{D}\left(\mathbf{x}^{m}\right)\frac{\partial\boldsymbol{\varsigma}\left(\mathbf{x}^{m}\right)}{\partial\mathbf{x}^{m}} \cdot \delta\boldsymbol{\varepsilon}\left(\mathbf{x}^{m}\right) d\Omega = \int_{\Omega^{m}} \mathbf{D}\left(\mathbf{x}^{m}\right)\delta\boldsymbol{\varepsilon}\left(\mathbf{x}^{m}\right) d\Omega \tag{4.2.33}$$

2) Find $\mathbf{D}^{h}$ and $\boldsymbol{f}^{bh}$ for each macroscale point $\mathbf{x}^{M}$ associated with the microscale RVE:

$$\mathbf{D}^{h}\left(\mathbf{x}^{M}\right) = \frac{1}{\Omega^{m}}\int_{\Omega^{m}}\left(\mathbf{D}\left(\mathbf{x}^{m}\right) - \mathbf{D}\left(\mathbf{x}^{m}\right)\frac{\partial\boldsymbol{\varsigma}\left(\mathbf{x}^{m}\right)}{\partial\mathbf{x}^{m}}\right) d\Omega \tag{4.2.34}$$

$$\boldsymbol{f}^{bh}\left(\mathbf{x}^{M}\right) = \frac{1}{\Omega^{m}}\int_{\Omega^{m}} \boldsymbol{f}^{b}\, d\Omega \tag{4.2.35}$$

Note that for linear elastic problems, this step can be performed once for all macroscale points.

3) Solve the macroscopic weak form to obtain the macro displacement field $\boldsymbol{u}^0(\mathbf{x}^{\mathrm{M}})$

$$\int_{\Omega^{\mathrm{M}}} \mathbf{D}^h \varepsilon^0\left(\mathbf{x}^{\mathrm{M}}\right) \cdot \delta\varepsilon\left(\mathbf{x}^{\mathrm{M}}\right) d\Omega = \int_{\Omega^{\mathrm{M}}} \boldsymbol{f}^{bh}\left(\mathbf{x}^{\mathrm{M}}\right) \cdot \delta\boldsymbol{u}\left(\mathbf{x}^{\mathrm{M}}\right) d\Omega + \int_{\Gamma_t^{\mathrm{M}}} \boldsymbol{f}^t\left(\mathbf{x}^{\mathrm{M}}\right) \cdot \delta\boldsymbol{u}\left(\mathbf{x}^{\mathrm{M}}\right) d\Gamma = 0 \quad (4.2.36)$$

$$\varepsilon^0\left(\mathbf{x}^{\mathrm{M}}\right) = \frac{\partial \boldsymbol{u}^0\left(\mathbf{x}^{\mathrm{M}}\right)}{\partial \mathbf{x}^{\mathrm{M}}} \quad (4.2.37)$$

It is important to note that the expensive first step can be performed only once if the microstructure is similarly repeated everywhere in the macromodel. It should be re-computed for every point $\mathbf{x}^{\mathrm{M}}$ of the macromodel Ω^{M}, however, if the microstructure is changed or if a nonlinear behaviour is assumed.

Equation (4.2.34) is analytically sufficient for computing the homogenized material modulus $\mathbf{D}^h$. In practice, however, the procedure may numerically be simplified by a closer look at the concept of $\varsigma(\mathbf{x}^{\mathrm{m}})$. It is interesting to note that computation of $\varsigma(\mathbf{x}^{\mathrm{m}})$ in Equation (4.2.33) can be interpreted as a unit strain ε^{f_i} imposed on the microscale model (RVE). This concept numerically facilitates the way the components of the homogenized material modulus $\mathbf{D}^h$ are computed (in two dimensions) (Hassani and Hinton 1998a):

$$\mathbf{D}^h = \begin{bmatrix} D_{11}^h & D_{12}^h & 0 \\ D_{12}^h & D_{22}^h & 0 \\ 0 & 0 & D_{66}^h \end{bmatrix} \quad (4.2.38)$$

Application of a unit initial strain loading ε^{f_i} can be performed as a force vector on RVE,

$$f^{\mathrm{m}_i} = \int_{\Omega^{\mathrm{m}}} \mathbf{B}^{\mathrm{T}} \mathbf{D}\left(\mathbf{x}^{\mathrm{m}}\right) \varepsilon^{f_i} d\Omega \quad (4.2.39)$$

$$\mathbf{K}^{\mathrm{m}} \varsigma^{\mathrm{m}_i} = \boldsymbol{f}^{\mathrm{m}_i} \quad (4.2.40)$$

After solving Equation (4.2.40) for ς^{m_i} and computing RVE strains $\varepsilon^{\varsigma^{\mathrm{m}_i}}(\mathbf{x}^{\mathrm{m}})$, the components of $\mathbf{D}^h$ are determined.

For the first set of initial strain ε^{f_1} on RVE, ς^{m_1} and the microstructure strain field $\varepsilon^{\varsigma^{\mathrm{m}_1}}$ are computed:

$$\varepsilon^{f_1} = \begin{Bmatrix} \varepsilon_{11}^{f_1} \\ \varepsilon_{22}^{f_1} \\ 2\varepsilon_{12}^{f_1} \end{Bmatrix} = \begin{Bmatrix} 1 \\ 0 \\ 0 \end{Bmatrix} \xrightarrow{\mathbf{K}^{\mathrm{m}} \varsigma^{\mathrm{m}_1} = \boldsymbol{f}^{\mathrm{m}_1}} \varsigma^{\mathrm{m}_1} \rightarrow \varepsilon^{\varsigma^{\mathrm{m}_1}}\left(\mathbf{x}^{\mathrm{m}}\right) \quad (4.2.41)$$

and the D_{11}^h component can be determined from

$$\mathrm{D}_{11}^h = \frac{1}{\Omega^{\mathrm{m}}} \int_{\Omega^{\mathrm{m}}} \left(\mathrm{D}_{11}\left(\mathbf{x}^{\mathrm{m}}\right) - \left(\mathbf{D}\left(\mathbf{x}^{\mathrm{m}}\right) \varepsilon^{f_1}\right)^{\mathrm{T}} \varepsilon^{\varsigma^{\mathrm{m}_1}} \right) d\Omega \quad (4.2.42)$$

Similarly, for the second set of initial straining of RVE,

$$\boldsymbol{\varepsilon}^{f_2}=\begin{Bmatrix}\varepsilon_{11}^{f_2}\\ \varepsilon_{22}^{f_2}\\ 2\varepsilon_{12}^{f_2}\end{Bmatrix}=\begin{Bmatrix}0\\1\\0\end{Bmatrix}\xrightarrow{\mathbf{K}^{\mathrm{m}}\varsigma^{\mathrm{m}_2}=\boldsymbol{f}^{\mathrm{m}_2}}\varsigma^{\mathrm{m}_2}\rightarrow\boldsymbol{\varepsilon}^{\varsigma^{\mathrm{m}_2}}\left(\mathbf{x}^{\mathrm{m}}\right) \tag{4.2.43}$$

and the components D_{12}^h and D_{22}^h are obtained from

$$D_{12}^h=\frac{1}{\Omega^{\mathrm{m}}}\int_{\Omega^{\mathrm{m}}}\left[D_{12}\left(\mathbf{x}^{\mathrm{m}}\right)-\left(\mathbf{D}\left(\mathbf{x}^{\mathrm{m}}\right)\boldsymbol{\varepsilon}^{f_1}\right)^{\mathrm{T}}\boldsymbol{\varepsilon}^{\varsigma^{\mathrm{m}_2}}\right]\mathrm{d}\Omega \tag{4.2.44}$$

$$\mathrm{D}_{22}^h=\frac{1}{\Omega^{\mathrm{m}}}\int_{\Omega^{\mathrm{m}}}\left[\mathrm{D}_{22}\left(\mathbf{x}^{\mathrm{m}}\right)-\left(\mathbf{D}\left(\mathbf{x}^{\mathrm{m}}\right)\boldsymbol{\varepsilon}^{f_2}\right)^{\mathrm{T}}\boldsymbol{\varepsilon}^{\varsigma^{\mathrm{m}_2}}\right]\mathrm{d}\Omega \tag{4.2.45}$$

Finally, the remaining component is obtained from the third set of initial straining:

$$\boldsymbol{\varepsilon}^{f_{12}}=\begin{Bmatrix}\varepsilon_{11}^{f_{12}}\\ \varepsilon_{22}^{f_{12}}\\ 2\varepsilon_{12}^{f_{12}}\end{Bmatrix}=\begin{Bmatrix}0\\0\\1\end{Bmatrix}\xrightarrow{\mathbf{K}^{\mathrm{m}}\varsigma^{\mathrm{m}_{12}}=\boldsymbol{f}^{\mathrm{m}_{12}}}\varsigma^{\mathrm{m}_{12}}\rightarrow\boldsymbol{\varepsilon}^{\varsigma^{\mathrm{m}_{12}}}\left(\mathbf{x}^{\mathrm{m}}\right) \tag{4.2.46}$$

$$D_{66}^h=\frac{1}{\Omega^{\mathrm{m}}}\int_{\Omega^{\mathrm{m}}}\left[D_{66}\left(\mathbf{x}^{\mathrm{m}}\right)-\left(\mathbf{D}\left(\mathbf{x}^{\mathrm{m}}\right)\boldsymbol{\varepsilon}^{f_{12}}\right)^{\mathrm{T}}\boldsymbol{\varepsilon}^{\varsigma^{\mathrm{m}_{12}}}\right]\mathrm{d}\Omega \tag{4.2.47}$$

Figure 4.4 illustrates the typical solutions of RVE for different initial longitudinal strains along the *x*, *y* directions and the shear straining.

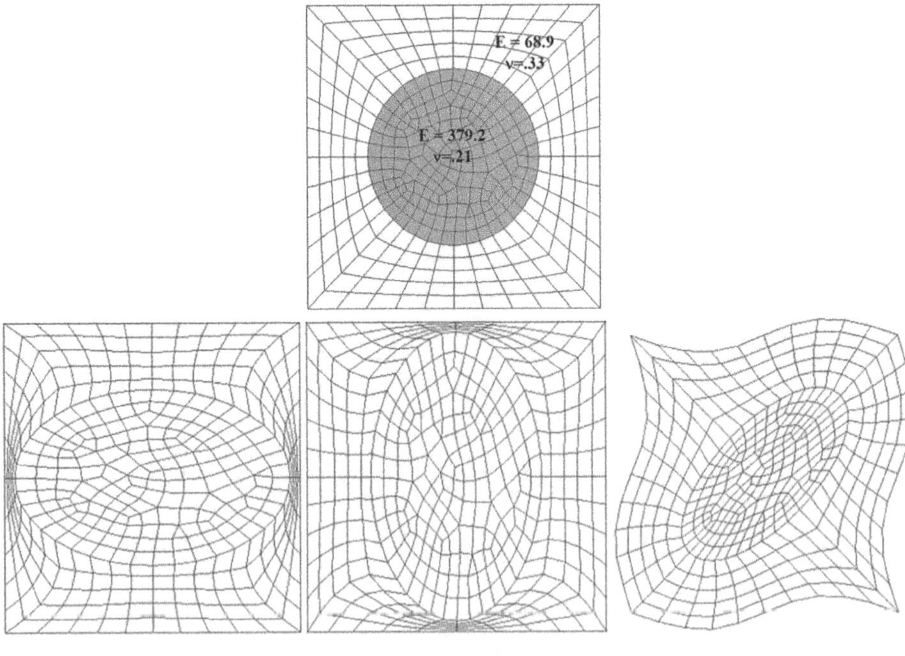

Figure 4.4 Typical solutions of RVE for initial strains along the *x*, *y* directions and the shear straining. *Source:* Reproduced from Bayesteh 2018/University of Tehran.

4.2.4 Computational Homogenization

4.2.4.1 Computational Homogenization in Small Deformations

Computational homogenization is based on the concept of averaging microscale parameters to obtain an estimation for the macroscale in an average sense. The averaging operator $\langle f \rangle$ for a microscopic parameter f is defined as (Miehe and Koch 2002)

$$\bar{f} = \langle f \rangle = \frac{1}{V} \int_V f \, dV \tag{4.2.48}$$

where $\bar{f}$ is the average value of f on the volume V.

In computational homogenization, it is assumed that the averaging operator can be employed to determine the stress $\boldsymbol{\sigma}^h(\mathbf{x}^{\mathrm{M}})$ and strain $\boldsymbol{\varepsilon}^h(\mathbf{x}^{\mathrm{M}})$ tensors at the macroscale point $\mathbf{x}^{\mathrm{M}}$ based on the RVE average of the stress $\boldsymbol{\sigma}^{\mathrm{m}}(\mathbf{x}^{\mathrm{m}})$ and the strain $\boldsymbol{\varepsilon}^{\mathrm{m}}(\mathbf{x}^{\mathrm{m}})$ distributions, respectively:

$$\boldsymbol{\sigma}^h\left(\mathbf{x}^{\mathrm{M}}\right) = \langle \boldsymbol{\sigma}^{\mathrm{m}} \rangle = \frac{1}{\Omega^{\mathrm{m}}} \int_{\Omega^{\mathrm{m}}} \boldsymbol{\sigma}^{\mathrm{m}}\left(\mathbf{x}^{\mathrm{m}}\right) d\Omega \tag{4.2.49}$$

$$\boldsymbol{\varepsilon}^h\left(\mathbf{x}^{\mathrm{M}}\right) = \langle \boldsymbol{\varepsilon}^{\mathrm{m}} \rangle = \frac{1}{\Omega^{\mathrm{m}}} \int_{\Omega^{\mathrm{m}}} \boldsymbol{\varepsilon}^{\mathrm{m}}\left(\mathbf{x}^{\mathrm{m}}\right) d\Omega \tag{4.2.50}$$

Moreover, the computational homogenization assumes the same averaging procedure for the internal power density,

$$\boldsymbol{\sigma}^h : \dot{\boldsymbol{\varepsilon}}^h = \langle \boldsymbol{\sigma}^{\mathrm{m}} : \dot{\boldsymbol{\varepsilon}}^{\mathrm{m}} \rangle \tag{4.2.51}$$

which remains valid only if certain boundary conditions are imposed on RVE (Miehe and Koch 2002):

1) The linear deformation prescribed on RVE boundaries ($\boldsymbol{\varepsilon}^h$ is constant on RVE):

$$\boldsymbol{u}^{\mathrm{m}}\left(\mathbf{x}^{\mathrm{m}}\right) = \boldsymbol{\varepsilon}^h \mathbf{x}^{\mathrm{m}}, \ \mathbf{x}^{\mathrm{m}} \in \Gamma^{\mathrm{m}} \tag{4.2.52}$$

2) The constant traction prescribed on RVE boundaries ($\boldsymbol{\sigma}^h$ is constant on RVE):

$$\boldsymbol{t}^{\mathrm{m}}\left(\mathbf{x}^{\mathrm{m}}\right) = \boldsymbol{\sigma}^h \boldsymbol{n}\left(\mathbf{x}^{\mathrm{m}}\right), \ \mathbf{x}^{\mathrm{m}} \in \Gamma^{\mathrm{m}} \tag{4.2.53}$$

3) Periodic boundary conditions ensure periodic deformations and anti-periodic surface tractions on opposite RVE boundaries:

$$\boldsymbol{u}^{\mathrm{m}}\left(\mathbf{x}^{\mathrm{m}+}\right) - \boldsymbol{u}^{\mathrm{m}}\left(\mathbf{x}^{\mathrm{m}-}\right) = \boldsymbol{\varepsilon}^h\left(\mathbf{x}^{\mathrm{m}+} - \mathbf{x}^{\mathrm{m}-}\right), \ \mathbf{x}^{\mathrm{m}+}, \mathbf{x}^{\mathrm{m}-} \in \Gamma^{\mathrm{m}} \tag{4.2.54}$$

$$\boldsymbol{t}^{\mathrm{m}}\left(\mathbf{x}^{\mathrm{m}+}\right) = -\boldsymbol{t}^{\mathrm{m}}\left(\mathbf{x}^{\mathrm{m}-}\right), \ \mathbf{x}^{\mathrm{m}+}, \mathbf{x}^{\mathrm{m}-} \in \Gamma^{\mathrm{m}} \tag{4.2.55}$$

In fact, the averaging theorem requires the average of the microscopic power to be equal to the macroscopic power.

Analysis of RVE subject to each boundary condition leads to the corresponding macroscale material modulus $\mathbf{D}^h$, which is computed over RVE.

The macroscale governing equations can be written as

$$\boldsymbol{\nabla}\cdot\boldsymbol{\sigma}^h\left(\mathbf{x}^{\mathrm{M}}\right)+\boldsymbol{f}^{\mathrm{b}}\left(\mathbf{x}^{\mathrm{M}}\right)=\mathbf{0},\ \mathbf{x}^{\mathrm{M}}\in\Gamma^{\mathrm{M}} \tag{4.2.56}$$

with the constitutive relation of

$$\boldsymbol{\sigma}^h\left(\mathbf{x}^{\mathrm{M}}\right)=\mathbf{D}^h\boldsymbol{\varepsilon}^h\left(\mathbf{x}^{\mathrm{M}}\right) \tag{4.2.57}$$

or in a rate form,

$$\dot{\boldsymbol{\sigma}}^h\left(\mathbf{x}^{\mathrm{M}}\right)=\mathbf{D}^h\dot{\boldsymbol{\varepsilon}}^h\left(\mathbf{x}^{\mathrm{M}}\right) \tag{4.2.58}$$

where $\mathbf{D}^h$ is the homogenized macroscale material modulus and

$$\boldsymbol{\varepsilon}^h\left(\mathbf{x}^{\mathrm{M}}\right)=\frac{1}{2}\left[\nabla\boldsymbol{u}^{\mathrm{M}}\left(\mathbf{x}^{\mathrm{M}}\right)+\nabla^{\mathrm{T}}\boldsymbol{u}^{\mathrm{M}}\left(\mathbf{x}^{\mathrm{M}}\right)\right] \tag{4.2.59}$$

In nonlinear problems, an iterative procedure must be performed to achieve a converged solution. First, the macro solution is obtained based on an assumed material modulus. Then, the microscale is solved by an appropriate RVE boundary condition and the average stress $\boldsymbol{\sigma}^h$ and the homogenized material modulus $\mathbf{D}^h$ are computed based on averaging the microscale variables. If the computed values of the macroscale variables are sufficiently close to the values at the beginning of each iteration, the convergence is achieved. Otherwise, the procedure is repeated with the updated values and continues until the convergence.

4.2.4.2 First Order Computational Homogenization

The same procedure can be performed in large deformation regimes. The strong form of the equilibrium equation in terms of the first Piola–Kirchhoff stress on the initial/reference configuration Ω_0^m of the microscale RVE can be written as

$$\boldsymbol{\nabla}_0^{\mathrm{m}}\cdot\boldsymbol{P}^{\mathrm{m}}\left(\mathbf{x}^{\mathrm{m}}\right)+\boldsymbol{f}^{\mathrm{b}}\left(\mathbf{x}^{\mathrm{m}}\right)=\mathbf{0},\ \mathbf{x}^{\mathrm{m}}\in\Omega_0^{\mathrm{m}} \tag{4.2.60}$$

In single-scale computations, the current $\mathbf{x}$ and initial $\mathbf{X}$ positions can be simply related by the deformation gradient $\boldsymbol{F}$,

$$\mathrm{d}\mathbf{x}=\boldsymbol{F}\mathrm{d}\mathbf{X} \tag{4.2.61}$$

Here, however, the microscale fluctuations should be considered. Therefore, the current position $\mathbf{x}^{\mathrm{m}}$ of any point in the microscale can be obtained from its position $\mathbf{X}^{\mathrm{m}}$ in the initial configuration of the microscale (computed by the macroscale deformation gradient $\boldsymbol{F}^{\mathrm{M}}$ at this point) and the microscale fluctuations $\mathrm{d}\boldsymbol{\omega}$ caused by inhomogeneity inside the RVE,

$$\mathrm{d}\mathbf{x}^{\mathrm{m}}=\boldsymbol{F}^{\mathrm{M}}\,\mathrm{d}\mathbf{X}^{\mathrm{m}}+\mathrm{d}\boldsymbol{\omega} \tag{4.2.62}$$

Moreover, the deformation gradient $\boldsymbol{F}^{\mathrm{m}}$ in the microscale can be defined as

$$\boldsymbol{F}^{\mathrm{m}}=\boldsymbol{\nabla}^{\mathrm{m}}\mathrm{d}\mathbf{x}^{\mathrm{m}} \tag{4.2.63}$$

$$\mathrm{d}\mathbf{x}^{\mathrm{m}}=\boldsymbol{F}^{\mathrm{m}}\mathrm{d}\mathbf{X}^{\mathrm{m}} \tag{4.2.64}$$

Computational homogenization in the large deformation regime is based on the assumption of averaging the microdeformation gradient $\boldsymbol{F}^{\mathrm{m}}$ to obtain the macrodeformation gradient $\boldsymbol{F}^{\mathrm{M}}$

$$\boldsymbol{F}^{\mathrm{M}}\left(\mathbf{X}^{\mathrm{M}}\right)=\frac{1}{\Omega_0^{\mathrm{m}}}\int_{\Omega_0^{\mathrm{m}}}\boldsymbol{F}^{\mathrm{m}}\left(\mathbf{X}^{\mathrm{m}}\right)\mathrm{d}\Omega \tag{4.2.65}$$

where Ω_0^{m} is the volume of RVE at the initial configuration. Equation (4.2.65) is the fundamental assumption of the computational homogenization technique, which relates the kinematic variables across the scales. This assumption is valid if one of the following boundary conditions is adopted for solving the microscale RVE model (see Equation (4.2.62) and Kouznetsova 2002):

1) Taylor boundary conditions: $\mathrm{d}\boldsymbol{\omega}=0$ on Ω_0^{m}
2) Linear boundary conditions: $\mathrm{d}\boldsymbol{\omega}=0$ on Γ_0^{m}
3) Periodic boundary conditions: $\mathrm{d}\boldsymbol{\omega}^{+}=\mathrm{d}\boldsymbol{\omega}^{-}$
4) Minimal boundary conditions: $\int_{\Gamma_0}\boldsymbol{n}\mathrm{d}\boldsymbol{\omega}\,d\Gamma=0$

Imposition of each set of RVE boundary conditions results in the corresponding macroscale material modulus $\mathbf{D}^h$ based on the computed variables of the microscale.

Another important assumption in computational homogenization is the Hill–Mandel principle, which relates the rates of virtual work of macro- and microscales (Kouznetsova et al. 2004):

$$\boldsymbol{P}^{\mathrm{M}}\left(\mathbf{X}^{\mathrm{M}}\right):\delta\boldsymbol{F}^{\mathrm{M}}\left(\mathbf{X}^{\mathrm{M}}\right)=\frac{1}{\Omega_0^{\mathrm{m}}}\int_{\Omega_0^{\mathrm{m}}}\boldsymbol{P}^{\mathrm{m}}\left(\mathbf{X}^{\mathrm{m}}\right):\delta\boldsymbol{F}^{\mathrm{m}}\left(\mathbf{X}^{\mathrm{m}}\right)\mathrm{d}\Omega \tag{4.2.66}$$

which leads to the principle of averaging of the second Piola–Kirchhoff stress state $\boldsymbol{P}^{\mathrm{m}}$ at the microscale to obtain the macroscale stress $\boldsymbol{P}^{\mathrm{M}}$,

$$\boldsymbol{P}^{\mathrm{M}}\left(\mathbf{X}^{\mathrm{M}}\right)=\frac{1}{\Omega_0^{\mathrm{m}}}\int_{\Omega_0^{\mathrm{m}}}\boldsymbol{P}^{\mathrm{m}}\left(\mathbf{X}^{\mathrm{m}}\right)\mathrm{d}\Omega \tag{4.2.67}$$

and the corresponding homogenized constitutive material modules at the macroscale, $\mathbf{D}^{hP}$ and $\mathbf{D}^{h\sigma}$, in terms of the first Piola–Kirchhoff and Cauchy stress tensors, respectively, can be written as

$$\delta\boldsymbol{P}^{\mathrm{M}}=\mathbf{D}^{hP}:\delta\boldsymbol{F}^{\mathrm{M}} \tag{4.2.68}$$

$$\delta\boldsymbol{\sigma}^{\mathrm{M}}=\mathbf{D}^{h\sigma}:\delta\boldsymbol{F}^{\mathrm{M}} \tag{4.2.69}$$

The solution procedure for the first-order computational homogenization can be briefly explained as (see Figure 4.5):

1) The macroscale model is initialized
   a) For all integration points
      i) Analyse RVE for $\boldsymbol{F}^{\mathrm{M}}=\mathbf{I}$
         1) Prescribe the boundary conditions
         2) Solve RVE and compute initial $\mathbf{D}^{hP}$

2) Load increment
   a) Apply the load increment
   b) Solution iterations
      i) Solve the macromodel
      ii) For all integration points
         1) Determine the deformation gradient $\boldsymbol{F}^{\mathrm{M}}$
         2) Solve RVE subject to prescribed boundary conditions
            a) Compute the micro states $\boldsymbol{P}^{\mathrm{m}}$
            b) Compute the average macro states, $\boldsymbol{P}^{\mathrm{M}}$ and $\mathbf{D}^{hP}$
      iii) Compute the macroscale internal forces
   c) If not converged, re-iterate
3) Next load increment

4.2.4.3 Second Order Computational Homogenization

In micro localization problems or in modelling problems with moderate gradients on large RVEs, the first order computational homogenization technique cannot be employed. In such cases, the current position $\mathbf{x}^{\mathrm{m}}$ of any point in the microscale can be obtained from its position $\mathbf{X}^{\mathrm{m}}$ in the initial configuration of the microscale (computed by the first and second order macroscale deformation gradients, $\boldsymbol{F}^{\mathrm{M}}$ and $\boldsymbol{G}^{\mathrm{M}}$, respectively, at this point), and the microscale fluctuations $\mathrm{d}\boldsymbol{\omega}$ caused by the inhomogeneity inside the RVE,

$$\mathrm{d}\mathbf{x}^{\mathrm{m}} = \boldsymbol{F}^{\mathrm{M}}\mathrm{d}\mathbf{X}^{\mathrm{m}} + \frac{1}{2}\mathrm{d}\mathbf{X}^{\mathrm{m}}\ \boldsymbol{G}^{\mathrm{M}}\mathrm{d}\mathbf{X}^{\mathrm{m}} + \mathrm{d}\boldsymbol{\omega}, \quad \mathbf{x}^{\mathrm{m}} \in \Omega_0^{\mathrm{m}} \tag{4.2.70}$$

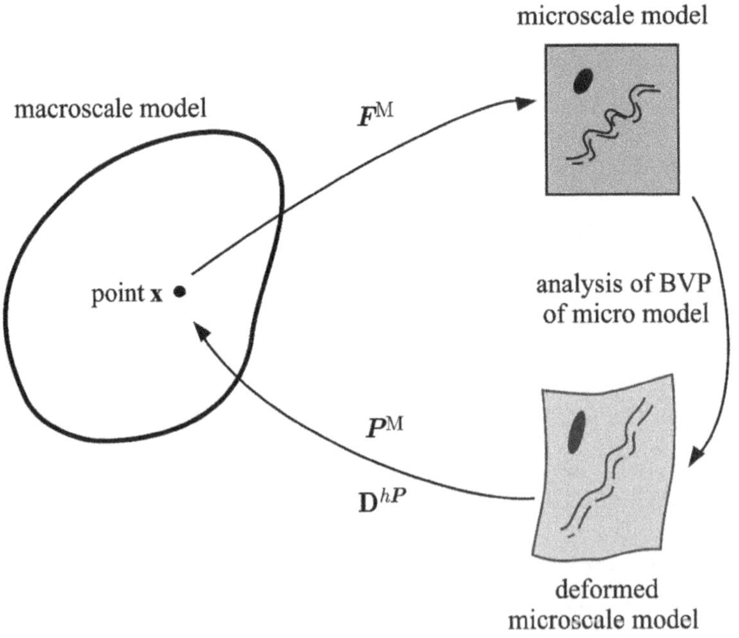

Figure 4.5 The first order computational homogenization.

where

$$\boldsymbol{G}^{\mathrm{M}} = \boldsymbol{\nabla}_0^{\mathrm{M}} \boldsymbol{F}^{\mathrm{M}} \tag{4.2.71}$$

The microscale deformation gradient $\boldsymbol{F}^{\mathrm{m}}$ can then be computed from

$$\boldsymbol{F}^{\mathrm{m}} = \boldsymbol{F}^{\mathrm{M}} + \mathrm{d}\mathbf{X}^{\mathrm{m}}\ \boldsymbol{G}^{\mathrm{M}} + \boldsymbol{\nabla}_0^{\mathrm{M}} \mathrm{d}\boldsymbol{\omega} \tag{4.2.72}$$

Adopting the averaging relation for the deformation gradient allows for determination of $\boldsymbol{G}^{\mathrm{M}}$ for any specified RVE.

The Hill–Mandel principle can then be written as (Kouznetsova et al. 2004)

$$\boldsymbol{P}^{\mathrm{M}}\left(\mathbf{X}^{\mathrm{M}}\right) : \delta\boldsymbol{F}^{\mathrm{M}}\left(\mathbf{X}^{\mathrm{M}}\right) + \boldsymbol{Q}^{\mathrm{M}}\left(\mathbf{X}^{\mathrm{M}}\right) \vdots \delta\boldsymbol{G}^{\mathrm{M}}\left(\mathbf{X}^{\mathrm{M}}\right) = \frac{1}{\Omega_0^{\mathrm{m}}} \int_{\Omega_0^{\mathrm{m}}} \boldsymbol{P}^{\mathrm{m}}\left(\mathbf{X}^{\mathrm{m}}\right) : \delta\boldsymbol{F}^{\mathrm{m}}\left(\mathbf{X}^{\mathrm{m}}\right) \mathrm{d}\Omega \tag{4.2.73}$$

where $\boldsymbol{Q}^{\mathrm{M}}$ is the work conjugate of $\boldsymbol{G}^{\mathrm{M}}$ and

$$\boldsymbol{P}^{\mathrm{M}}\left(\mathbf{X}^{\mathrm{M}}\right) = \frac{1}{\Omega_0^{\mathrm{m}}} \int_{\Omega_0^{\mathrm{m}}} \boldsymbol{P}^{\mathrm{m}}\left(\mathbf{X}^{\mathrm{m}}\right) \mathrm{d}\Omega \tag{4.2.74}$$

$$\boldsymbol{Q}^{\mathrm{M}}\left(\mathbf{X}^{\mathrm{M}}\right) = \frac{1}{2\Omega_0^{\mathrm{m}}} \int_{\Omega_0^{\mathrm{m}}} \left(\boldsymbol{P}^{\mathrm{m}}\left(\mathbf{X}^{\mathrm{m}}\right)\mathbf{X}^{\mathrm{m}} + \mathbf{X}^{\mathrm{m}}\boldsymbol{P}^{\mathrm{m}}\left(\mathbf{X}^{\mathrm{m}}\right)\right) \mathrm{d}\Omega \tag{4.2.75}$$

The four consistent material tensors $\mathbf{D}^{h_i\boldsymbol{P}}$ can then be computed from (Kouznetsova 2002)

$$\delta\boldsymbol{P}^{\mathrm{M}} = \mathbf{D}^{h_1\boldsymbol{P}} : \delta\boldsymbol{F}^{\mathrm{M}} + \mathbf{D}^{h_2\boldsymbol{P}} : \delta\boldsymbol{G}^{\mathrm{M}} \tag{4.2.76}$$

$$\delta\boldsymbol{Q}^{\mathrm{M}} = \mathbf{D}^{h_3\boldsymbol{P}} : \delta\boldsymbol{F}^{\mathrm{M}} + \mathbf{D}^{h_4\boldsymbol{P}} : \delta\boldsymbol{G}^{\mathrm{M}} \tag{4.2.77}$$

Note that $\mathbf{D}^{h_1\boldsymbol{P}}$ is a fourth order tensor, $\mathbf{D}^{h_2\boldsymbol{P}}$ and $\mathbf{D}^{h_3\boldsymbol{P}}$ are fifth order tensors and $\mathbf{D}^{h_4\boldsymbol{P}}$ is a sixth order tensor.

The solution procedure for the second order computational homogenization can be briefly explained as (see Figure 4.6):

1) The macroscale model is initialized
   a) For all integration points
      i) Analyse RVE for $\boldsymbol{F}^{\mathrm{M}} = \mathbf{I}$ and $\boldsymbol{G}^{\mathrm{M}} = \mathbf{0}$
         1) Prescribe the boundary conditions
         2) Solve RVE and compute initial $\mathbf{D}^{h\boldsymbol{P}}$
2) Load increment
   a) Apply the load increment
   b) Solution iterations
      i) Solve the macro model
      ii) For all integration points
         1) Determine $\boldsymbol{F}^{\mathrm{M}}$ and $\boldsymbol{G}^{\mathrm{M}}$
         2) Solve RVE subject to prescribed boundary conditions
            a) Compute the micro states $\boldsymbol{P}^{\mathrm{m}}$
            b) Compute the average macro states, $\boldsymbol{P}^{\mathrm{M}}$, $\boldsymbol{Q}^{\mathrm{M}}$ and $\mathbf{D}^{h_i\boldsymbol{P}}$
      iii) Compute the macroscale internal forces
   c) If not converged, re-iterate.
3) Next load increment

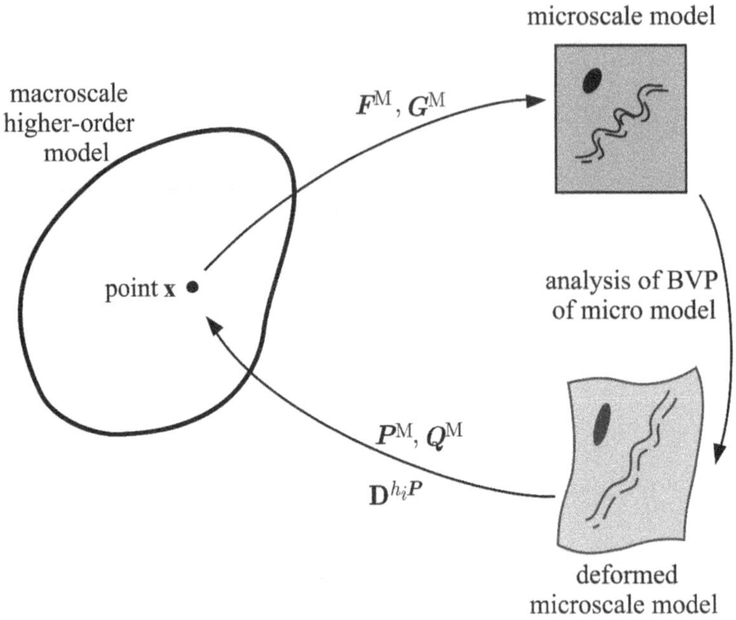

Figure 4.6 The second order computational homogenization.

The second order computational homogenization clearly requires a higher order of continuity of the finite element model. It involves complicated boundary conditions and is computationally expensive, as the whole procedure should be repeated for each integration point in each iteration. Its major advantage is that it does not require any constitutive relation at the macroscale and acts on the RVE level and may capture non-local phenomena.

4.2.4.4 Macro-based Framework for Computational Homogenization

As discussed in the first order computational homogenization technique, there are problems that the assumption of a uniform macro field on the microstructure may not remain valid. For instance, in points near a crack, the high stress gradient near the crack tip or the discontinuous field across a crack edge cannot be represented by a uniform field over the microstructure model (see Figure 4.7).

The conventional approach in the so-called macro-based approach is to express a highly oscillating field $f^{\zeta}(\mathbf{x}^{\mathrm{M}},\mathbf{x}^{\mathrm{m}})$ in a heterogeneous medium in terms of the macroscale field $f^{\mathrm{M}}(\mathbf{x}^{\mathrm{M}})$, microscale fluctuations $f^{\mathrm{m}}(\mathbf{x}^{\mathrm{m}})$ and the additional macroscale kinematic variable $\Theta(\mathbf{x}^{\mathrm{M}})$ (Kouznetsova 2002; Kouznetsova et al. 2004, Sanchez and Zhang 2010; Sánchez et al. 2013),

$$f^{\zeta}\left(\mathbf{x}^{\mathrm{M}},\mathbf{x}^{\mathrm{m}}\right)=f^{\mathrm{M}}\left(\mathbf{x}^{\mathrm{M}}\right)+L\left(\Theta\left(\mathbf{x}^{\mathrm{M}}\right),\mathbf{x}^{\mathrm{m}}\right)+f^{\mathrm{m}}\left(\mathbf{x}^{\mathrm{m}}\right) \tag{4.2.78}$$

where L is an operator in terms of Θ.

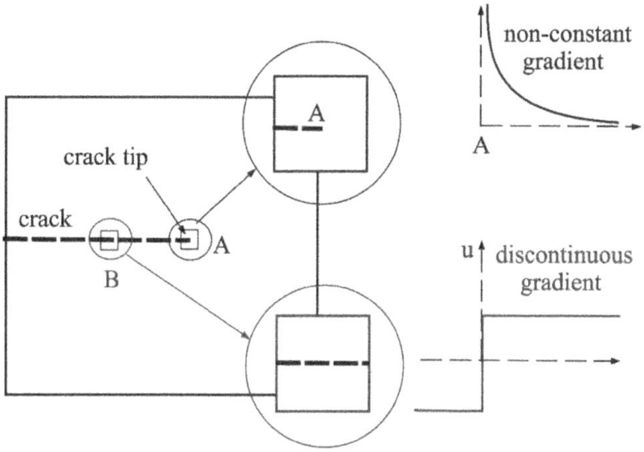

Figure 4.7 Problems with non-constant gradients.

The next step is to consider appropriate forms of the macroscopic fields f^{M} and Θ^{M}. For specific problems (and not generally), the following assumptions can be made (Bayesteh 2018):

$$f^{\mathrm{M}}\left(\mathbf{x}^{\mathrm{M}}\right)=\frac{1}{\Omega^{\mathrm{m}}}\int_{\Omega^{\mathrm{m}}} f^{\varsigma}\left(\mathbf{x}^{\mathrm{m}}\right)\mathrm{d}\Omega \tag{4.2.79}$$

$$\Theta^{\mathrm{M}}\left(\mathbf{x}^{\mathrm{M}}\right)=\frac{1}{\Omega^{\mathrm{m}}}\int_{\Omega^{\mathrm{m}}} \Theta\left(\mathbf{x}^{\mathrm{m}}\right)\mathrm{d}\Omega \tag{4.2.80}$$

The third step is to define the necessary boundary conditions on RVE, which should be consistent with the kinematic variables. This stage is more complicated than the first order computational homogenization approach.

Finally, in order to employ the Hill–Mandel principle, the work conjugates of the macroscopic kinematic variables should be defined.

In the second order computational homogenization, the field variable f^{M} and the kinematic variable Θ are assumed to be the deformation gradient $\boldsymbol{F}^{\mathrm{M}}$ and the gradient of deformation gradient $\boldsymbol{G}^{\mathrm{M}}$ (Kouznetsova et al. 2004),

$$\boldsymbol{F}^{\varsigma}\left(\mathbf{x}^{\mathrm{M}},\mathbf{x}^{\mathrm{m}}\right)=\boldsymbol{F}^{\mathrm{M}}\left(\mathbf{x}^{\mathrm{M}}\right)+\boldsymbol{G}^{\mathrm{M}}\left(\mathbf{x}^{\mathrm{M}}\right)\mathbf{x}^{\mathrm{m}}+\boldsymbol{F}^{\mathrm{m}}\left(\mathbf{x}^{\mathrm{m}}\right) \tag{4.2.81}$$

For a macro crack, the field variable f^{M} is assumed to be the linear strain ε, and the kinematic variable Θ can be written in terms of the crack opening $\left[\!\left[\boldsymbol{u}\left(\mathbf{x}^{\mathrm{M}}\right)\right]\!\right]$ (Sánchez et al. 2013, Bosco et al. 2014),

$$\varepsilon^{\varsigma}\left(\mathbf{x}^{\mathrm{M}},\mathbf{x}^{\mathrm{m}}\right)=\varepsilon^{\mathrm{M}}\left(\mathbf{x}^{\mathrm{M}}\right)+\frac{\left(\left[\!\left[\boldsymbol{u}\left(\mathbf{x}^{\mathrm{M}}\right)\right]\!\right]\boldsymbol{n}^{\mathrm{c}}\right)^{\mathrm{sym}}}{l_{\mu}}+\varepsilon^{\mathrm{m}}\left(\mathbf{x}^{\mathrm{m}}\right) \tag{4.2.82}$$

where l_{μ} is the size of the crack opening and $\boldsymbol{n}^{\mathrm{c}}$ is the normal to the crack edge.

4.2.4.5 Micro-based Framework for Softening Microstructures

In order to extend the application of the computational homogenization to analysis of microscale localization problems, the concept of a micro-based framework can be employed (Bayesteh and Mohammadi 2017). In this method, a procedure similar to Equation (4.2.78) is adopted for the strain field ε^{ζ}, but the additional kinematic variable $\Theta^{m}(\mathbf{x}^{M},\mathbf{x}^{m})$ is defined inside the RVE and is based on the microscale solution:

$$\varepsilon^{\zeta}\left(\mathbf{x}^{M},\mathbf{x}^{m}\right)=\varepsilon^{M}\left(\mathbf{x}^{M}\right)+\Theta^{m}\left(\mathbf{x}^{M},\mathbf{x}^{m}\right)+\varepsilon^{m}\left(\mathbf{x}^{m}\right) \tag{4.2.83}$$

or

$$\varepsilon^{\zeta}\left(\mathbf{x}^{M},\mathbf{x}^{m}\right)=\varepsilon^{M}\left(\mathbf{x}^{M}\right)+\Theta^{m}\left(\mathbf{x}^{M},\mathbf{x}^{m}\right)+\nabla\boldsymbol{u}^{1}\left(\mathbf{x}^{m}\right) \tag{4.2.84}$$

where $\varepsilon^{m}(\mathbf{x}^{m})$ represents the usual fluctuations in the microscale solution, whereas $\Theta^{m}(\mathbf{x}^{M},\mathbf{x}^{m})$ accounts for the severe strain gradients in RVE; $\nabla\boldsymbol{u}^{1}$ describes the conventional strain fluctuation inside the RVE.

Figure 4.8 compares the numerical procedure for macro- and micro-based frameworks of computational homogenization methods. In the macro-based framework, the macroscopic governing equations are described with the appropriate known macroscopic kinematic variables. Then the additional macroscopic field variable $\Theta(\mathbf{x}^{M})$ is adopted, and the micro strain field over RVE, $\varepsilon^{\zeta}(\mathbf{x}^{M},\mathbf{x}^{m})$, is defined as Equation (4.2.83). Next, the RVE is solved with appropriate governing equations and boundary conditions.

In the micro-based framework, the steps are somehow reversed. First, the RVE is considered and the micro strain field over RVE, $\varepsilon^{\zeta}(\mathbf{x}^{M},\mathbf{x}^{m})$, with intended enrichments and appropriate degrees of freedom, is defined. It leads to the adoption of the additional microscopic kinematic variable $\Theta^{m}(\mathbf{x}^{M},\mathbf{x}^{m})$ and its appropriate work conjugate. Finally, the macroscopic model is solved by derivation of the macroscopic governing equations.

In the first-order homogenization, the average rate of power δp^{ζ} can be written as

$$\langle\delta p^{\zeta}\left(\varepsilon^{M},\varepsilon^{m}\right)\rangle=\delta p^{M}\left(\varepsilon^{M}\right) \tag{4.2.85}$$

based on

$$\varepsilon^{\zeta}\left(\mathbf{x}^{M},\mathbf{x}^{m}\right)=\varepsilon^{M}\left(\mathbf{x}^{M}\right)+\varepsilon^{m}\left(\mathbf{x}^{m}\right) \tag{4.2.86}$$

The rate of power in the present formulation can now be expressed based on Equation (4.2.83) with the important note that only the independent micro-based kinematic variable contributes to the microscale power,

$$\langle\delta p^{\zeta}\left(\varepsilon^{M},\tilde{\varepsilon}\right)\rangle=\delta p^{M}\left(\varepsilon^{M}\right)+\delta p^{\chi}\left(\chi^{m}\right) \tag{4.2.87}$$

$$\langle\delta p^{\zeta}\left(\varepsilon^{M},\Theta^{m}\right)\rangle=\frac{1}{\Omega^{m}}\int_{\Omega^{m}}\sigma^{m}\left(\mathbf{x}^{M},\mathbf{x}^{m}\right)\delta\varepsilon^{\zeta}\left(\mathbf{x}^{M},\mathbf{x}^{m}\right)d\Omega \tag{4.2.88}$$

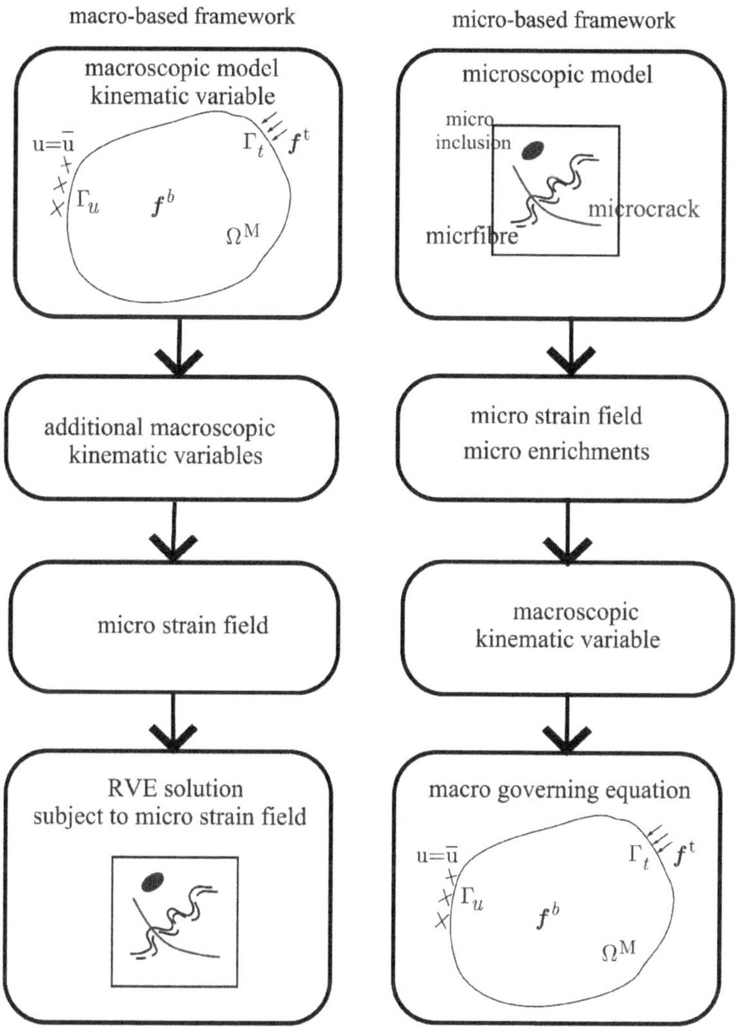

Figure 4.8 Comparison of micro- and macro-based frameworks. *Source:* Adapted from Bayesteh and Mohammadi 2017.

with

$$\delta p^{\mathrm{M}}\left(\boldsymbol{\varepsilon}^{\mathrm{M}}\right)=\frac{1}{\Omega^{\mathrm{m}}}\int_{\Omega^{\mathrm{m}}}\boldsymbol{\sigma}^{\mathrm{m}}\left(\mathbf{x}^{\mathrm{M}},\mathbf{x}^{\mathrm{m}}\right)\delta\boldsymbol{\varepsilon}^{\mathrm{M}}\left(\mathbf{x}^{\mathrm{M}}\right)\mathrm{d}\Omega=\boldsymbol{\sigma}^{\mathrm{M}}\left(\mathbf{x}^{\mathrm{M}}\right)\delta\boldsymbol{\varepsilon}^{\mathrm{M}}\left(\mathbf{x}^{\mathrm{M}}\right) \tag{4.2.89}$$

$$\delta p^{\chi}\left(\Theta^{\mathrm{m}}\right)=\frac{1}{\Omega^{\mathrm{m}}}\int_{\Omega^{\mathrm{m}}}\boldsymbol{\sigma}^{\mathrm{m}}\left(\mathbf{x}^{\mathrm{M}},\mathbf{x}^{\mathrm{m}}\right)\delta\Theta^{\mathrm{m}}\left(\mathbf{x}^{\mathrm{M}},\mathbf{x}^{\mathrm{m}}\right)\mathrm{d}\Omega \tag{4.2.90}$$

which define the work conjugates of the kinematic variables $\boldsymbol{\varepsilon}^{\mathrm{M}}$ and χ^{m}. Equation (4.2.89) leads to the definition of $\boldsymbol{\sigma}^{\mathrm{M}}$ as

$$\boldsymbol{\sigma}^{\mathrm{M}}\left(\mathbf{x}^{\mathrm{M}}\right)=\frac{1}{\Omega^{\mathrm{m}}}\int_{\Omega^{\mathrm{m}}}\boldsymbol{\sigma}^{\mathrm{m}}\left(\mathbf{x}^{\mathrm{M}},\mathbf{x}^{\mathrm{m}}\right)\mathrm{d}\Omega \tag{4.2.91}$$

Moreover, the weak form of the microscale equilibrium equation,

$$\frac{1}{\Omega^{\mathrm{m}}}\int_{\Omega^{\mathrm{m}}}\boldsymbol{\sigma}^{\mathrm{m}}\left(\mathbf{x}^{\mathrm{M}},\mathbf{x}^{\mathrm{m}}\right)\nabla\boldsymbol{u}^{1}\left(\mathbf{x}^{\mathrm{m}}\right)\mathrm{d}\Omega=0 \tag{4.2.92}$$

along with the appropriate boundary conditions on RVE,

$$\int_{\Gamma^{\mathrm{m}}}\boldsymbol{\sigma}^{\mathrm{m}}\left(\mathbf{x}^{\mathrm{M}},\mathbf{x}^{\mathrm{m}}\right)\boldsymbol{n}\delta\boldsymbol{u}^{1}\,\mathrm{d}\Gamma=0 \tag{4.2.93}$$

allow for evaluation of the microscale fluctuation field $\varepsilon^{\mathrm{m}}(\mathbf{x}^{\mathrm{m}})=\nabla\boldsymbol{u}^{1}(\mathbf{x}^{\mathrm{m}})$.

The micro-based kinematic framework allows the introduction of microstructure enrichment in the homogenization technique. Beginning with Equation (4.2.83), the micro-based kinematic variable can be written in the form of an enriched solution:

$$\Theta^{\mathrm{m}}\left(\mathbf{x}^{\mathrm{M}},\mathbf{x}^{\mathrm{m}}\right)=\mathbf{E}\left(\mathbf{x}^{\mathrm{m}}\right)\odot\psi\left(\mathbf{x}^{\mathrm{M}}\right) \tag{4.2.94}$$

where $\psi(\mathbf{x}^{\mathrm{M}})$ is the additional kinematic variable, $\mathbf{E}(\mathbf{x}^{\mathrm{m}})$ is a distribution function in RVE and $\odot$ is an appropriate operator. $\psi(\mathbf{x}^{\mathrm{M}})$ is independent of other macroscale variables and can be selected with great flexibility in different problems (Bayesteh and Mohammadi 2017).

Equation (4.2.90) can now be re-written as

$$\delta p^{\chi}\left(\chi^{\mathrm{m}}\right)=\frac{1}{\Omega^{\mathrm{m}}}\int_{\Omega^{\mathrm{m}}}\sigma^{\mathrm{m}}\left(\mathbf{x}^{\mathrm{M}},\mathbf{x}^{\mathrm{m}}\right):\mathbf{E}\left(\mathbf{x}^{\mathrm{m}}\right)\odot\delta\psi\left(\mathbf{x}^{\mathrm{M}}\right)\mathrm{d}\Omega \tag{4.2.95}$$

and with the assumption of the constant $\psi(\mathbf{x}^{\mathrm{M}})$ on RVE, finally yields to

$$\delta p^{\chi}\left(\chi^{\mathrm{m}}\right)=\boldsymbol{\Sigma}^{\mathrm{M}}\left(\mathbf{x}^{\mathrm{M}}\right)\odot\delta\psi\left(\mathbf{x}^{\mathrm{M}}\right) \tag{4.2.96}$$

with

$$\boldsymbol{\Sigma}^{\mathrm{M}}\left(\mathbf{x}^{\mathrm{M}}\right)=\frac{1}{\Omega^{\mathrm{m}}}\int_{\Omega^{\mathrm{m}}}\sigma^{\mathrm{m}}\left(\mathbf{x}^{\mathrm{M}},\mathbf{x}^{\mathrm{m}}\right):\mathbf{E}\left(\mathbf{x}^{\mathrm{m}}\right)\mathrm{d}\Omega \tag{4.2.97}$$

where the stress state $\boldsymbol{\Sigma}^{\mathrm{M}}(\mathbf{x}^{\mathrm{M}})$ is the work conjugate to $\delta\psi(\mathbf{x}^{\mathrm{M}})$ and accounts for the effect of enrichment on the homogenized stress. The enrichment function $\mathbf{E}(\mathbf{x}^{\mathrm{m}})$ is chosen based on the type of localization problem within the RVE.

The final weak form of the governing equation on the macroscale can be written as

$$\delta W^{\mathrm{int}}=\delta W^{\mathrm{ext}} \tag{4.2.98}$$

with

$$\begin{aligned}\delta W^{\mathrm{int}}&=\int_{\Omega^{\mathrm{M}}}\left[\delta p^{\varsigma}\left(\varepsilon^{\mathrm{M}},\Theta^{\mathrm{m}}\right)\right]\mathrm{d}\Omega=\int_{\Omega^{\mathrm{M}}}\left[\delta p^{\mathrm{M}}\left(\varepsilon^{\mathrm{M}}\right)+\delta p^{\Theta}\left(\Theta^{\mathrm{m}}\right)\right]\mathrm{d}\Omega\\&=\int_{\Omega^{\mathrm{M}}}\left[\boldsymbol{\sigma}^{\mathrm{M}}\left(\mathbf{x}^{\mathrm{M}}\right)\delta\varepsilon^{\mathrm{M}}\left(\mathbf{x}^{\mathrm{M}}\right)+\boldsymbol{\Sigma}^{\mathrm{M}}\left(\mathbf{x}^{\mathrm{M}}\right)\odot\delta\psi\left(\mathbf{x}^{\mathrm{M}}\right)\right]\mathrm{d}\Omega\end{aligned} \tag{4.2.99}$$

$$\delta W^{\mathrm{ext}}=\int_{\Omega^{\mathrm{M}}}\boldsymbol{f}^{\mathrm{b}}\left(\mathbf{x}^{\mathrm{M}}\right)\delta\boldsymbol{u}^{\mathrm{M}}\mathrm{d}\Omega+\int_{\Gamma_{t}}\boldsymbol{f}^{\mathrm{t}}\delta\boldsymbol{u}^{\mathrm{M}}\mathrm{d}\Gamma+\int_{\Omega^{\mathrm{M}}}\boldsymbol{f}_{\mathrm{c}}^{\mathrm{b}}\left(\mathbf{x}^{\mathrm{M}}\right)\delta\psi\mathrm{d}\Omega \tag{4.2.100}$$

In order to explicitly define an enrichment case, a cracked RVE is considered. Equations (4.2.99) and (4.2.100) are modified to

$$\delta W^{\text{int}} = \int_{\Omega^{\text{M}}} \left[\boldsymbol{\sigma}^{\text{M}}\left(\mathbf{x}^{\text{M}}\right) \delta\boldsymbol{\varepsilon}^{\text{M}}\left(\mathbf{x}^{\text{M}}\right) + \boldsymbol{\Sigma}_{\text{c}}^{\text{M}}\left(\mathbf{x}^{\text{M}}\right) \delta\boldsymbol{\beta}\left(\mathbf{x}^{\text{M}}\right) \right] \mathrm{d}\Omega \tag{4.2.101}$$

$$\delta W^{\text{ext}} = \int_{\Omega^{\text{M}}} \boldsymbol{f}^{\text{b}}\left(\mathbf{x}^{\text{M}}\right) \delta\boldsymbol{u}^{\text{M}} \mathrm{d}\Omega + \int_{\Gamma_t} \boldsymbol{f}^{\text{t}} \delta\boldsymbol{u}^{\text{M}}\, \mathrm{d}\Gamma + \int_{\Omega^{\text{M}}} \boldsymbol{f}_{\text{c}}^{\text{b}}\left(\mathbf{x}^{\text{M}}\right) \delta\boldsymbol{\beta} \mathrm{d}\Omega \tag{4.2.102}$$

where $\boldsymbol{\beta}$ is the crack edge displacement degrees of freedom at the microscale, which can be interpreted as the average of the crack opening inside RVE (and not the macroscale crack opening),

$$\boldsymbol{\beta} = \frac{1}{\Gamma_{\text{c}}^{\text{m}}} \int_{\Gamma_{\text{c}}^{\text{m}}} \left[\!\left[\boldsymbol{u}^{\text{m}}\left(\mathbf{x}^{\text{m}}\right) \right]\!\right] \mathrm{d}\Gamma \tag{4.2.103}$$

$\boldsymbol{f}_{\text{c}}^{\text{b}}(\mathbf{x}^{\text{M}})$ is the average body force generated by the crack surface tractions $\boldsymbol{f}^{\text{c}}(\mathbf{x}^{\text{m}})$ in the microscale,

$$\boldsymbol{f}_{\text{c}}^{\text{b}}\left(\mathbf{x}^{\text{M}}\right) = \frac{1}{\Omega^{\text{m}}} \int_{\Gamma_{\text{c}}^{\text{m}}} \boldsymbol{f}^{\text{c}}\left(\mathbf{x}^{\text{m}}\right) \mathrm{d}\Gamma \tag{4.2.104}$$

and $\boldsymbol{\Sigma}_{\text{c}}^{\text{M}}$ is the enriched stress tensor conjugate to $\delta\boldsymbol{\beta}$,

$$\boldsymbol{\Sigma}_{\text{c}}^{\text{M}}\left(\mathbf{x}^{\text{M}}\right) = \frac{1}{\Omega^{\text{m}}} \int_{\Omega^{\text{M}}} \left[-\boldsymbol{\nabla}\varphi\left(\mathbf{x}^{\text{m}}\right) \boldsymbol{\sigma}^{\text{m}}\left(\mathbf{x}^{\text{m}}\right) \right] \mathrm{d}\Omega \tag{4.2.105}$$

where φ is defined as

$$\varphi\left(\mathbf{x}^{\text{m}}\right) = \sum_{k=1}^{n} N_k\left(\mathbf{x}^{\text{m}}\right) H\left(\xi\left(\mathbf{x}_k^{\text{m}}\right)\right) \tag{4.2.106}$$

where H is the Heaviside function (3.5.9), ξ is the signed distance function (3.5.4) and N is the conventional finite element shape function.

The microscale displacement field can be decomposed into

$$\boldsymbol{u}^{\text{m}}\left(\mathbf{x}^{\text{m}}\right) = \varsigma^{\text{m}}\left(\mathbf{x}^{\text{m}}\right) + \boldsymbol{u}^{1}\left(\mathbf{x}^{\text{m}}\right) \tag{4.2.107}$$

where $\varsigma^{\text{m}}(\mathbf{x}^{\text{m}})$ represents the localization displacement field, which is associated with the localization strain field $\chi^{\text{m}}(\mathbf{x}^{\text{m}})$,

$$\varsigma^{\text{m}}\left(\mathbf{x}^{\text{m}}\right) = \left[H\left(\xi\left(\mathbf{x}^{\text{m}}\right)\right) - \varphi\left(\mathbf{x}^{\text{m}}\right) \right] \boldsymbol{\beta} \tag{4.2.108}$$

and

$$\boldsymbol{u}^{1}\left(\mathbf{x}^{\text{m}}\right) = \sum_{k=1}^{n} N_k \boldsymbol{u}_k^{\text{m}} + \sum_{k=1}^{ne} N_k \left[H\left(\xi\left(\mathbf{x}^{\text{m}}\right)\right) - H\left(\xi\left(\mathbf{x}_k^{\text{m}}\right)\right) \right] \boldsymbol{a}_k^{\text{m}} \tag{4.2.109}$$

where $\boldsymbol{u}_k^{\mathrm{m}}$ and $\boldsymbol{a}_k^{\mathrm{m}}$ represent the standard and enriched degrees of freedom in RVE.

It can be shown that $\boldsymbol{u}^1(\mathbf{x}^{\mathrm{m}})$ satisfies the constraint of a weak zero crack opening,

$$\int_{\Gamma_c^{\mathrm{m}}} \left[\!\left[\boldsymbol{u}^1\left(\mathbf{x}^{\mathrm{m}}\right) \right]\!\right] \mathrm{d}\Gamma = 0 \tag{4.2.110}$$

which should be considered as an additional condition in solving the governing equations of the microstructure.

Figure 4.9 Shows the procedure for solving a cracked problem using the enriched homogenization technique.

The solution procedure can be briefly presented as:

1) The kinematic variables ε^{M} and β are obtained from the initial solution of the macroscale model, $\boldsymbol{u}^1\left(\mathbf{x}^{\mathrm{m}}\right)$ from Equation (4.2.92), boundary conditions (4.2.93) and the constraint (4.2.110). An iterative Newton–Raphson solution procedure may be required.
2) The stress conjugates $\boldsymbol{\Sigma}^{\mathrm{M}}$ and $\boldsymbol{\Sigma}_{\mathrm{c}}^{\mathrm{M}}$ are determined from Equations (4.2.91) and (4.2.105), respectively.
3) The material modus can be computed by a linear asymptotic perturbation analysis.
4) The new macroscale solution is obtained.
5) If the convergence criterion is not satisfied, the procedure is repeated.

The same strategy can be extended to include crack tip enrichments within RVE. For a full description, refer to Bayesteh and Mohammadi (2017).

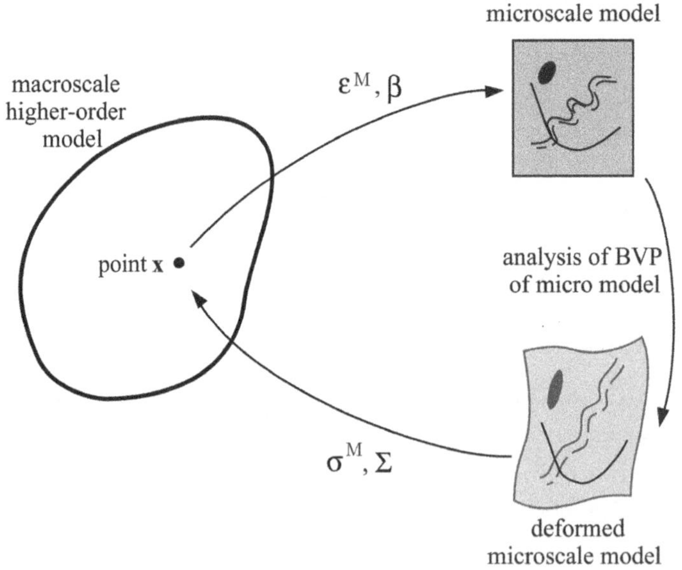

Figure 4.9 Homogenization with a micro-based enrichment strategy. *Source:* Adapted from Bayesteh and Mohammadi 2017.

4.2.4.5.1 Numerical Example

In this section, a numerical example according to Bayesteh and Mohammadi (2017) is revisited to examine the performance of the micro-based framework for the analysis of the complex problem of a cracked heterogeneous plate with a severe strain gradient, as depicted in Figure 4.10 (the inclusion volume fraction in the RVE is assumed to be 0.267).

A direct modelling is adopted to assess the obtained results. Figure 4.11 shows the finite element mesh for the homogenization and direct analyses. The direct model is constructed by over 140,160 elements, whereas the micro-based and the first order homogenization models use 30 elements on the macromodel and 1,168 elements for the RVE simulation.

The proposed micro-based model adopts the singular distribution of the strain field in a local region around the macroscopic crack in the form of the crack-tip enrichment (4.2.111):

$$f^{\text{enr}}\left(\mathbf{x}^{\text{M}}\right) = \sqrt{r}\sin\frac{\theta}{2} \tag{4.2.111}$$

The corresponding basic modes, typically shown in Figure 4.12, should be re-evaluated for each enriched RVE (depending on the macro location of the crack tip $\mathbf{x}^{\text{M}}$).

Figure 4.13 compares the contours of the von Mises stress on an RVE near the crack tip for three cases of direct modelling, micro-based homogenization and the first order homogenization. The maximum values are 510, 467 and 123 MPa, respectively. Clearly, the first order solution is not even close to the reference direct solution and does not predict any singular stress field. Moreover, the error of the first order homogenization reaches an unacceptable level of 75 percent, whereas the error of the micro-based technique is limited to less than 10 percent.

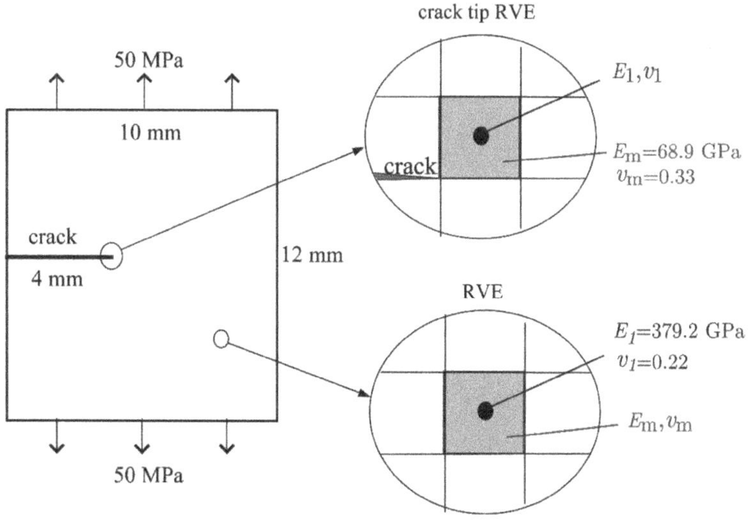

Figure 4.10 A cracked heterogeneous plate. Geometry and material properties. *Source:* Adapted from Bayesteh and Mohammadi 2017.

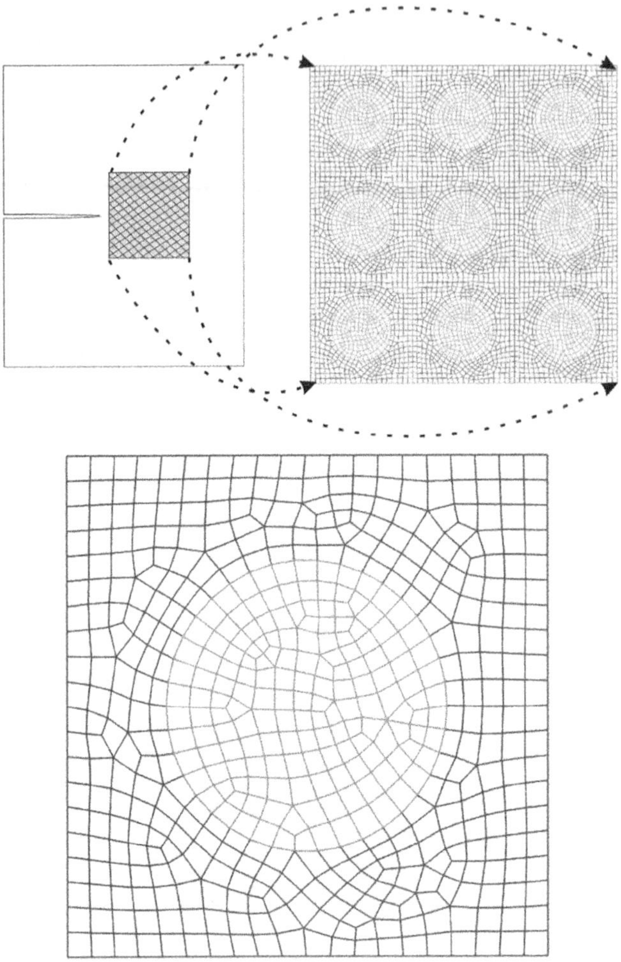

Figure 4.11 FE mesh for the homogenization and direct simulations. *Source:* Adapted from Bayesteh and Mohammadi 2017.

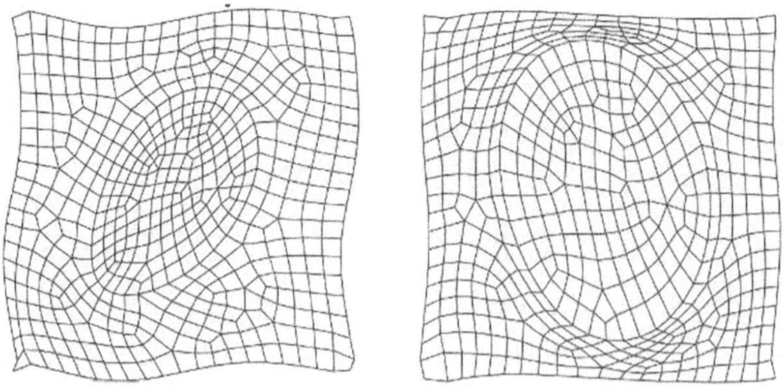

Figure 4.12 Mode shapes of RVE. *Source:* Reproduced from Bayesteh and Mohammadi 2017/with Elsevier.

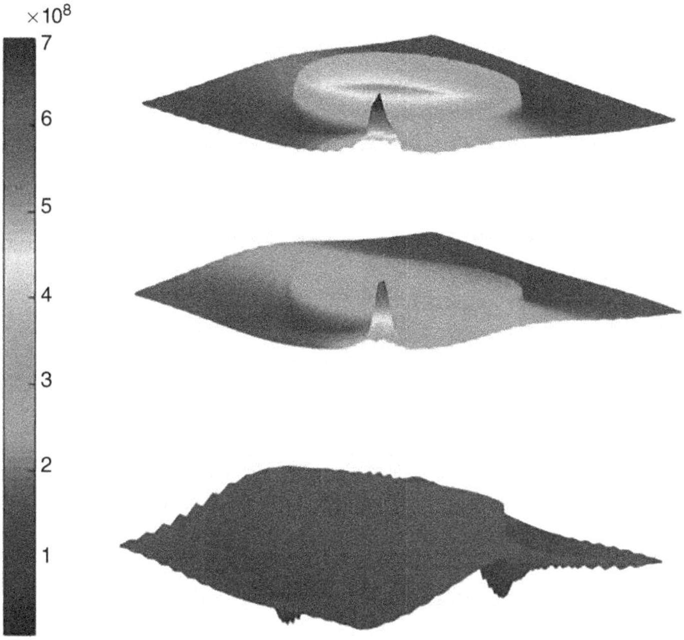

Figure 4.13 Contours of the von Mises stress for the direct model (top), the micro-based homogenization (middle) and the first order homogenization (bottom). *Source:* Adapted from Bayesteh and Mohammadi 2017.

4.3 Molecular Dynamics (MD)

4.3.1 Introduction

Molecular modelling has been extensively used in the analysis of various physical phenomena and high-tech engineering applications, including simulation of defects in crystalline structures to determine the damage and plastic behaviours, material phase transition, non-equilibrium heat processes, analysis of fracture and micro cracks, study of roughness and nano surface properties, lattice defects, chemical reactions (using quantum mechanics), micro electronic devices and MEMS, and, particularly, investigation of macro biomolecules such as proteins, DNAs, membranes and the neurological system.

Classical and quantum molecular dynamics are the two main categories of molecular simulations. In the classical MD simulation, molecules are viewed as conventional objects whose dynamics and interaction are governed by the classical laws of mechanics (Petrenko and Meller 2010). The quantum simulation considers the quantum characteristics of the chemical bonds (Car and Parrinello 1985, 1987) and provides significant improvement over the classical method, especially in simulation of many biological systems. However, it requires extremely greater computational resources, which limits it to simulation of only very small systems. At present, only the classical MD is practical for simulations of large systems of atoms over time scales of nanoseconds (Petrenko and Meller 2010).

Molecular or atomistic modelling is a fundamental approach based on simulation of the building structure of the matter, as typically presented for solid and liquid materials in

Figure 4.14. Molecular simulations are based on the classical Newtonian dynamics of particles/objects. The objects are assumed to be dimensionless spheres representing atoms, molecules, or a cluster of atoms or molecules. The mass of each object is lumped at its centre.

An MD system is in fact an idealized subdomain of a macro problem. The molecules/atoms interact with each other through interatomic potentials and with the surrounding environment through a number of specific thermodynamic constraints, which define the macroscopic requirements.

MD simulations may provide a general computational approach for resembling experimental tests to determine constitutive relations for new material structures, as depicted in Figure 4.15.

MD models may require well over hundreds of millions of atoms or molecules to create a sufficiently accurate model for representing a complex nanoscale problem. Analysis of such complex models requires high-performance hardware and consumes massive computational resources.

The governing equations are derived based on the concepts of statistical mechanics, which are reviewed in the next section.

4.3.2 Statistical Mechanics

4.3.2.1 Thermodynamics

The principles of thermodynamics govern physical and engineering phenomena (see Section 2.3.1). If a system is in the state of thermodynamic equilibrium, its state can be defined by thermodynamic macroscopic state parameters (V, T, P) or in terms of internal

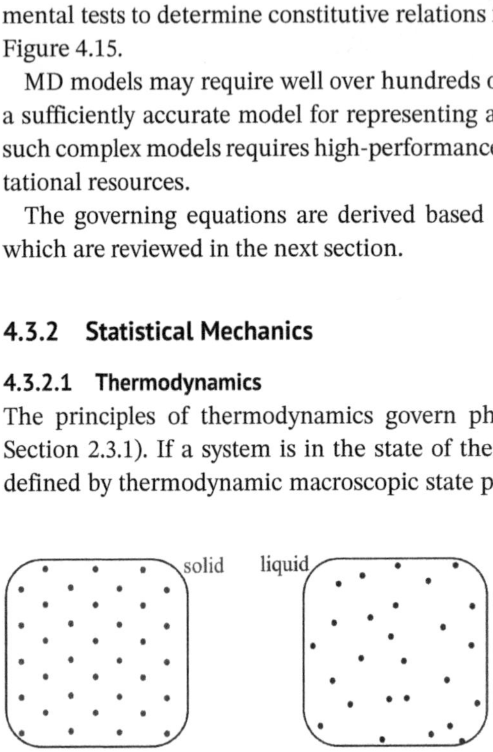

Figure 4.14 Illustrations of basic molecular models for solid/crystal and liquid/gas materials.

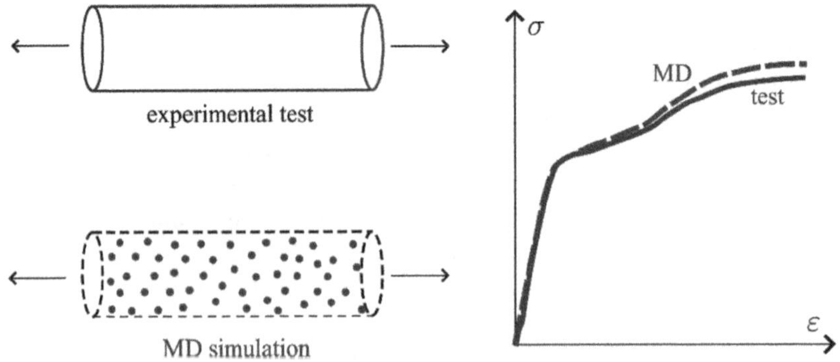

Figure 4.15 Experimental and MD tests to determine the constitutive relation.

energy or entropy. It should be noted that the thermodynamic parameters may vary with time, but their average values at the thermodynamic equilibrium state are assumed to be independent of time,

$$f\left(V,T,P,a^{\text{ext}}\right)=0 \tag{4.3.1}$$

where a^{ext} are the set of system parameters.

The equations of state may be derived from various forms. Beginning with the internal energy $U=U(S,V)$:

$$\mathrm{d}U=T\,\mathrm{d}S-\delta W=T\,\mathrm{d}S-P\,\mathrm{d}V \tag{4.3.2}$$

Or, alternatively, in terms of the Helmholtz free energy $\psi_{\mathrm{H}}(T,V)$, the Gibbs potential $\psi_{\mathrm{G}}(T,P)$ and the enthalpy $I(S,P)$:

$$\psi_{\mathrm{H}}(T,V)=U-ST\rightarrow\ \mathrm{d}\psi_{\mathrm{H}}=-S\ \mathrm{d}T-P\ \mathrm{d}V \tag{4.3.3}$$

$$\psi_{\mathrm{G}}(T,P)=\psi_{\mathrm{H}}+PV\rightarrow\ \mathrm{d}\psi_{\mathrm{G}}=-S\ \mathrm{d}T+V\ \mathrm{d}P \tag{4.3.4}$$

$$I(S,P)=U+PV\rightarrow\ \mathrm{d}\psi_{\mathrm{H}}=T\ \mathrm{d}S+V\mathrm{d}P \tag{4.3.5}$$

and the thermodynamic variables can be computed accordingly,

$$T=\left.\frac{\partial U}{\partial S}\right|_V=\left.\frac{\partial I}{\partial S}\right|_P \tag{4.3.6}$$

$$P=-\left.\frac{\partial U}{\partial V}\right|_S=-\left.\frac{\partial \psi_{\mathrm{H}}}{\partial V}\right|_T \tag{4.3.7}$$

$$V=\left.\frac{\partial \psi_{\mathrm{G}}}{\partial P}\right|_T=\left.\frac{\partial I}{\partial P}\right|_S \tag{4.3.8}$$

$$S=-\left.\frac{\partial \psi_{\mathrm{G}}}{\partial T}\right|_P=-\left.\frac{\partial \psi_{\mathrm{H}}}{\partial T}\right|_V \tag{4.3.9}$$

4.3.2.2 Mechanics of a System of Particles

The mechanics of a system of particles is described in terms of the generalized coordinates $\boldsymbol{r}$ of particles, which is defined based on the geometric specifications (such as position, etc.) of each particle,

$$\boldsymbol{r}=\left(\boldsymbol{r}_1,\boldsymbol{r}_2,\ldots,\boldsymbol{r}_{n_{\mathrm{p}}}\right) \tag{4.3.10}$$

where $\boldsymbol{r}_\alpha$ is the position of particle α,

$$\boldsymbol{r}_\alpha=\left(r_\alpha^1,r_\alpha^2,r_\alpha^3\right) \tag{4.3.11}$$

and n_p is the total number of particles of the system. The equation of motion of the system of particles may be derived based on the Lagrangian or Hamiltonian formulations.

4.3.2.2.1 Lagrangian Formulation

The generalized coordinates in the Lagrangian formulation include the position $\boldsymbol{r}$ and the velocity $\dot{\boldsymbol{r}}$ of each particle:

$$\left(\boldsymbol{r}_1,\boldsymbol{r}_2,\ldots,\boldsymbol{r}_{n_p},\dot{\boldsymbol{r}}_1,\dot{\boldsymbol{r}}_2,\ldots,\dot{\boldsymbol{r}}_{n_p}\right) \tag{4.3.12}$$

where $\dot{\boldsymbol{r}}_\alpha$ is the velocity of particle α. The time-dependent Lagrangian $L(\boldsymbol{r},\dot{\boldsymbol{r}};t)$ is defined as

$$L(\boldsymbol{r},\dot{\boldsymbol{r}};t)=L\left(\boldsymbol{r}_1,\boldsymbol{r}_2,\ldots,\boldsymbol{r}_{n_p},\dot{\boldsymbol{r}}_1,\dot{\boldsymbol{r}}_2,\ldots,\dot{\boldsymbol{r}}_{n_p};t\right) \tag{4.3.13}$$

The Lagrangian form for a material point can be written in terms of the kinetic energy U^{k} and the potential energy U^{p},

$$L=U^{\mathrm{k}}(\dot{\boldsymbol{r}})-U^{\mathrm{p}}(\boldsymbol{r}) \tag{4.3.14}$$

For conservative systems, the kinetic energy depends only on velocities $\dot{\boldsymbol{r}}$ and the potential energy depends only on coordinates $\boldsymbol{r}$.

The Lagrangian equation of motion for each degree of freedom $\alpha=1,2,\ldots,n_{\mathrm{p}}$ is written as

$$\frac{\mathrm{d}}{\mathrm{d}t}\frac{\partial L}{\partial \dot{\boldsymbol{r}}_\alpha}-\frac{\partial L}{\partial \boldsymbol{r}_\alpha}=0 \tag{4.3.15}$$

4.3.2.2.2 Hamiltonian Formulation

Beginning with the definition of the linear momentum $\mathbf{p}$,

$$\mathbf{p}=\left(\mathbf{p}_1,\mathbf{p}_2,\ldots,\mathbf{p}_{n_p}\right) \tag{4.3.16}$$

where $\mathbf{p}_\alpha=m_\alpha\dot{\boldsymbol{r}}_\alpha$ is the generalized momentum of particle α,

$$\mathbf{p}_\alpha=\left(p_\alpha^1,p_\alpha^2,p_\alpha^3\right) \tag{4.3.17}$$

the Hamiltonian formulation is based on the independent coordinates of position $\boldsymbol{r}$ and momentum $\mathbf{p}$ to define the generalized coordinates system $(\boldsymbol{r},\mathbf{p})$:

$$(\boldsymbol{r},\mathbf{p})=\left(\boldsymbol{r}_1,\boldsymbol{r}_2,\ldots,\boldsymbol{r}_{n_p},\mathbf{p}_1,\mathbf{p}_2,\ldots,\mathbf{p}_{n_p}\right) \tag{4.3.18}$$

The time-dependent energy function Hamiltonian $H(\boldsymbol{r},\mathbf{p};t)$ is defined as

$$H(\boldsymbol{r},\mathbf{p};t)=H\left(\boldsymbol{r}_1,\boldsymbol{r}_2,\ldots,\boldsymbol{r}_{n_p},\mathbf{p}_1,\mathbf{p}_2,\ldots,\mathbf{p}_{n_p};t\right) \tag{4.3.19}$$

The Hamiltonian can be related to the Lagrangian description by

$$H(\boldsymbol{r},\mathbf{p};t)=\sum_{\alpha=1}^{n_p}\mathbf{p}_\alpha\dot{\boldsymbol{r}}_\alpha-L \tag{4.3.20}$$

and

$$\dot{\boldsymbol{r}}_\alpha=\frac{\partial H}{\partial \mathbf{p}_\alpha} \tag{4.3.21}$$

$$\dot{\mathbf{p}}_\alpha=-\frac{\partial H}{\partial \boldsymbol{r}_\alpha} \tag{4.3.22}$$

The Hamiltonian function can be substituted into the Lagrangian equation of motion (4.3.15), leading to a system of $2n_p$ first-order differential equations.

4.3.2.3 Phase Space

The $6n_p$ phase space Λ is represented by the two generalized coordinates $(\boldsymbol{r}, \boldsymbol{p})$ (Figure 4.16).

Each phase point $(\boldsymbol{r}, \boldsymbol{p})$ on a phase trajectory corresponds to the state of the system at a given time t. In other words, a trajectory represents the evolution of the equilibrium state $(\boldsymbol{r}(t), \boldsymbol{p}(t))$.

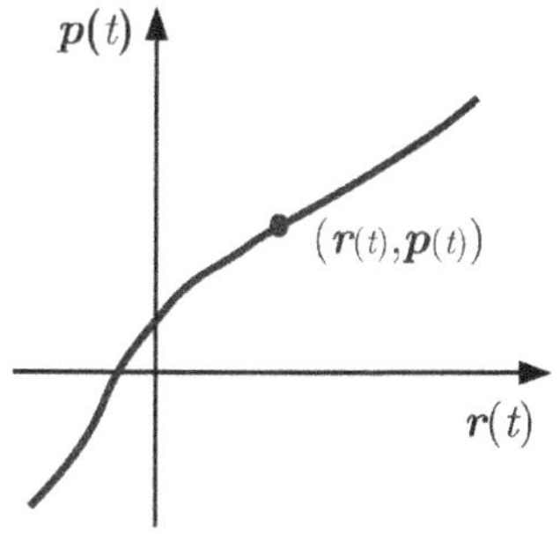

Figure 4.16 Definition of the phase space $\Lambda(r,p)$.

Evolution of a system of particles in a conservative force field can be expressed by the Hamilton's equations (for $\alpha = 1, \ldots, n_p$):

$$H = U^k(\boldsymbol{p}) - U^p(\boldsymbol{r}) \tag{4.3.23}$$

with

$$\dot{\boldsymbol{r}}_\alpha = \frac{\partial H}{\partial \boldsymbol{p}_\alpha} = \frac{\boldsymbol{p}_\alpha}{m_\alpha} \tag{4.3.24}$$

$$\dot{\boldsymbol{p}}_\alpha = -\frac{\partial H}{\partial \boldsymbol{r}_\alpha} = -\frac{\partial U^p}{\partial \boldsymbol{r}_\alpha} = \boldsymbol{f}_\alpha \tag{4.3.25}$$

where $\boldsymbol{f}_\alpha$ represents the force on atom α.

Theoretically, a unique solution $(\boldsymbol{r}(t), \boldsymbol{p}(t))$ is obtained for the Hamilton equations. It represents a conserved total energy with a constant Hamiltonian along a trajectory, which corresponds to the equilibrium state.

Another characteristic of the phase space is due to the Liouville theorem, which states that the material time derivative of the phase space volume Λ_R occupied by a set of systems is zero:

$$\frac{\mathrm{D}\Lambda_R}{\mathrm{D}t} = 0 \tag{4.3.26}$$

According to Heisenberg's principle, the smallest cell in phase space is

$$\Lambda_{\min} = \hbar^3 \tag{4.3.27}$$

where $\hbar = 1.054 \times 10^{-34}$ Js is the reduced Planck constant (the Planck constant $h = 6.626 \times 10^{-34}$ J Hz^{-1}).

Moreover, the Liouville operator $\mathcal{L}$ on any function $G(\boldsymbol{r}(t), \boldsymbol{p}(t))$ in the phase space can be defined from the Hamiltonian H as (Schofield and van Zon 2008; Wahnstrom 2018)

$$\mathcal{L}G = \{G, H\} = \frac{\partial G[\boldsymbol{r}(t), \boldsymbol{p}(t)]}{\partial t} = \sum_{\alpha=1}^{n_p} \left(\frac{\partial G}{\partial \boldsymbol{r}_\alpha} \frac{\partial H}{\partial \boldsymbol{p}_\alpha} - \frac{\partial G}{\partial \boldsymbol{p}_\alpha} \frac{\partial H}{\partial \boldsymbol{r}_\alpha} \right) \tag{4.3.28}$$

or, more explicitly,

$$\{G,H\}=\sum_{\alpha=1}^{n_p}\left(\dot{\boldsymbol{r}}_\alpha\frac{\partial G}{\partial \boldsymbol{r}_\alpha}-\dot{\boldsymbol{p}}_\alpha\frac{\partial G}{\partial \boldsymbol{p}_\alpha}\right) \tag{4.3.29}$$

where the Poisson bracket $\{A,B\}$ in the phase space $(\boldsymbol{r},\boldsymbol{p})$ is generally defined as,

$$\{A,B\}=\frac{\partial A}{\partial \boldsymbol{r}}\cdot\frac{\partial B}{\partial \boldsymbol{p}}-\frac{\partial A}{\partial \boldsymbol{p}}\cdot\frac{\partial B}{\partial \boldsymbol{r}} \tag{4.3.30}$$

Evolution of $G(\boldsymbol{r}(t),\boldsymbol{p}(t))$ in time can be stated as

$$\frac{\mathrm{d}G(\boldsymbol{r}(t),\boldsymbol{p}(t))}{\mathrm{d}t}=\frac{\partial G(\boldsymbol{r}(t),\boldsymbol{p}(t))}{\partial t}+\mathcal{L}G(\boldsymbol{r}(t),\boldsymbol{p}(t)) \tag{4.3.31}$$

with the solution of evolution equation in the following general form in terms of its initial state $G(\boldsymbol{r}(0), \boldsymbol{p}(0))$:

$$G(\boldsymbol{r}(t),\boldsymbol{p}(t))=e^{\mathcal{L}t}G(\boldsymbol{r}(0),\boldsymbol{p}(0)) \tag{4.3.32}$$

Therefore, the position $(\boldsymbol{r},\boldsymbol{p})$ in the phase space can be formally expressed as

$$(\boldsymbol{r}(t),\boldsymbol{p}(t))=e^{\mathcal{L}t}(\boldsymbol{r}(0),\boldsymbol{p}(0)) \tag{4.3.33}$$

The necessary condition for a conserved evolution of G in the phase space (steady state of G) can then be expressed as

$$\frac{\mathrm{d}G}{\mathrm{d}t}=\frac{\partial G}{\partial t}=\mathcal{L}G=\{G,H\}=0 \tag{4.3.34}$$

4.3.2.4 Probabilistic Determination of Micro States

Thermodynamics equations deal with a limited number of macroscopic parameters such as volume V, temperature T, pressure P, etc. In deterministic micromodelling, however, the micro state of the system is determined by a set of micro-level variables at each micro particle or object, which constitute the micro state. Clearly, while any microscopic state $(\boldsymbol{r},\boldsymbol{p})$ leads to a specific macro state $X^M(V,P,T,\ldots)$, no specific unique micro state can be computed from a predefined macro state,

$$X^M \nrightarrow (\boldsymbol{r},\boldsymbol{p}) \tag{4.3.35}$$

The reason can be attributed to the fact that Hamilton's equations for a macroscopic system cannot be directly integrated to obtain the micro state $(\boldsymbol{r}(t),\boldsymbol{p}(t))$. In other words, any macro state corresponds to infinite different micro states (Figure 4.17).

Consequently, no deterministic procedure can be performed to compute the microscale variables from the limited macroscale parameters and, therefore, a statistical approach should be adopted for computing the microscale variables. The statistical approach allows for the macroscale parameters to be reproduced from the micro variables by an average framework.

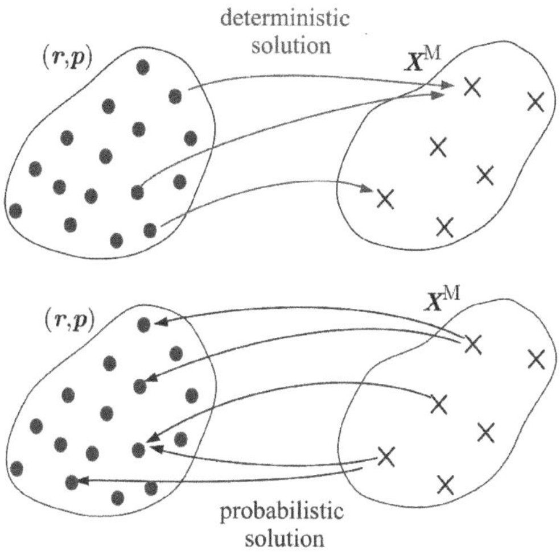

Figure 4.17 Deterministic and probabilistic macro–micro state relations.

In other words, statistical mechanics establishes a deterministic approach for computing the macroscale parameters from the microscale phase vector $(\boldsymbol{r}(t),\boldsymbol{p}(t))$ based on its initial conditions, whereas it provides a time-dependent probabilistic approach based on macroscale parameters to determine, in a phase average framework, the microscale phase vector $(\boldsymbol{r},\boldsymbol{p})$ based on the initial value of the probability density $w(\boldsymbol{r},\boldsymbol{p};t)$.

4.3.2.5 Phase Averages

A macroscale parameter $\overline{f}$ in terms of the time average of the corresponding phase function $f(\boldsymbol{r},\boldsymbol{p})$ is defined as

$$\overline{f} = \lim_{t\to\infty}\frac{1}{t}\int_0^t f\left(\boldsymbol{r}(\tau),\boldsymbol{p}(\tau)\right)\mathrm{d}\tau \tag{4.3.36}$$

Despite the fact that the macro state specifications $X^M(V,P,T,\ldots)$ cannot determine the micro state $(\boldsymbol{r},\boldsymbol{p};t)$, a probability density function (or distribution function) $w(\boldsymbol{r},\boldsymbol{p};t)$ can be computed:

$$w(\boldsymbol{r},\boldsymbol{p};t) = w\left(\boldsymbol{r}_1,\boldsymbol{r}_2,\ldots,\boldsymbol{r}_{n_p},\boldsymbol{p}_1,\boldsymbol{p}_2,\ldots,\boldsymbol{p}_{n_p};t\right) \tag{4.3.37}$$

where $w(\boldsymbol{r},\boldsymbol{p};t)$ is the probability of finding the micro state system in a given phase volume $G([\boldsymbol{r},\boldsymbol{r}+\mathrm{d}\boldsymbol{r}],[\boldsymbol{p},\boldsymbol{p}+\mathrm{d}\boldsymbol{p}])$, with the normalization condition over the whole phase space $\Lambda(\boldsymbol{r},\boldsymbol{p})$

$$\int_{\Lambda(\boldsymbol{r},\boldsymbol{p})} w(\boldsymbol{r},\boldsymbol{p};t)\mathrm{d}\boldsymbol{r}\ \mathrm{d}\boldsymbol{p} = 1 \tag{4.3.38}$$

A statistical phase average for the phase function $f(\boldsymbol{r},\boldsymbol{p})$ is then defined in terms of the probability density function $w(\boldsymbol{r},\boldsymbol{p})$:

$$\langle f;w\rangle = \int_{\Lambda(\boldsymbol{r},\boldsymbol{p})} f(\boldsymbol{r},\boldsymbol{p})w(\boldsymbol{r},\boldsymbol{p})\mathrm{d}\boldsymbol{r}\ \mathrm{d}\boldsymbol{p} \tag{4.3.39}$$

The average $\langle f;w\rangle$ provides a so-called statistical expectation of the phase function $f(\mathbf{r},\mathbf{p})$. This probabilistic view is established through the concept of the phase space ensemble, which is the set of all micro states $(\mathbf{r},\mathbf{p})$ that are consistent with the required macroscopic constraints V, P, T, E, etc. If the concept of the probability density function is adopted, a statistical ensemble is obtained.

4.3.2.5.1 Ergodic Hypothesis and the Time Average

The ergodic hypothesis assumes that the time average $\bar{f}$ of a phase function $f(\mathbf{r}(t),\mathbf{p}(t))$ over a theoretically infinite time interval $(t \to \infty)$ is equivalent to its statistical phase average $\langle f;w\rangle$:

$$\bar{f} = \langle f;w\rangle \tag{4.3.40}$$

In ergodic systems, trajectories pass all points in the phase space to reach equilibrium with constant energy and equal probability (Micoulaut 2013). On the contrary, non-ergodic systems, such as glasses, do not cover the whole constant energy surface in their corresponding timescales.

Evaluation of the ensemble average $\langle f;w\rangle$ in Equation (4.3.39) for any given system requires an explicit definition of the probability density function $w(\mathbf{r},\mathbf{p})$.

The probability of finding the evolution of a system of particles in time can be described by the Liouville operator $\mathcal{L}$:

$$\frac{\partial w}{\partial t} = -\mathcal{L}w = -\{w,H\} \tag{4.3.41}$$

with its formal solution as

$$\frac{\partial w(\mathbf{r},\mathbf{p},t)}{\partial t} = \mathrm{e}^{-\mathcal{L}t} w(\mathbf{r},\mathbf{p},0) \tag{4.3.42}$$

In a stationary equilibrium, $w(\mathbf{r},\mathbf{p},t) \to w(\mathbf{r},\mathbf{p})$ and

$$\frac{\partial w(\mathbf{r},\mathbf{p})}{\partial t} = -\mathcal{L}w(\mathbf{r},\mathbf{p}) = -\{w,H\} = 0 \tag{4.3.43}$$

Therefore, computation of $w(\mathbf{r},\mathbf{p})$ for a thermodynamic equilibrium state is performed by an equivalent form of the equation of motion:

$$\frac{\partial w}{\partial t} = \{w,H\} = \sum_{\alpha=1}^{n_p} \left(\frac{\partial H}{\partial \mathbf{r}_\alpha} \frac{\partial w}{\partial \mathbf{p}_\alpha} - \frac{\partial H}{\partial \mathbf{p}_\alpha} \frac{\partial w}{\partial \mathbf{r}_\alpha} \right) = 0 \tag{4.3.44}$$

This shows that the Hamiltonian H is a stationary solution of the Liouville evolution equation. It is also noted that both NVE and NVT distribution functions satisfy the Liouville equation.

As a consequence of the ergodic hypothesis, the probability density function w depends only on the total energy (Hamiltonian) H in the phase space of the system at equilibrium:

$$w(\mathbf{r},\mathbf{p},t) \xrightarrow{\text{ergodic hypothesis}} w^{\mathrm{e}}\left(H\left(\mathbf{r},\mathbf{p},a^{\mathrm{ext}}\right)\right) \tag{4.3.45}$$

where a^{ext} represents the additional external parameters. The distribution function then satisfies the equilibrium equation:

$$\left\{w^{\text{e}}(H),H\right\}=0 \tag{4.3.46}$$

4.3.2.5.2 Calculation of Phase Averages

Conventional numerical integration schemes cannot be used for numerical evaluation of $\langle f;w\rangle$ in Equation (4.3.39) due to complexity of the $6n_p$ dimensional phase space $(\boldsymbol{r},\boldsymbol{p})$. In practice, MD can be used for numerical integration of the equations of motion to generate n_c configurations $(\boldsymbol{r}^{(i)},\boldsymbol{p}^{(i)})$. Then Equation (4.3.39) can be computed from

$$\langle f;w\rangle \approx \frac{1}{n_c}\sum_{i=1}^{n_c} f\left(\boldsymbol{r}^{(i)},\boldsymbol{p}^{(i)}\right)w\left(\boldsymbol{r}^{(i)},\boldsymbol{p}^{(i)}\right) \tag{4.3.47}$$

Alternatively, a similar formulation may be obtained if the Monte Carlo (MC) approach is used based on selection of n_r random values $(\boldsymbol{r}^{(i)},\boldsymbol{p}^{(i)})$ from the distribution function w:

$$\langle f;w\rangle \approx \frac{1}{n_r}\sum_{i=1}^{n_r} f\left(\boldsymbol{r}^{(i)},\boldsymbol{p}^{(i)}\right)w\left(\boldsymbol{r}^{(i)},\boldsymbol{p}^{(i)}\right) \tag{4.3.48}$$

4.3.2.5.3 Restricted Phase Average

The restricted phase average $\langle f;w\rangle_\chi$ of the phase function $f(\boldsymbol{r},\boldsymbol{p})$ of a system can be written as

$$\langle f;w\rangle_\chi = \int_{\Lambda(\boldsymbol{r},\boldsymbol{p})} f(\boldsymbol{r},\boldsymbol{p})w(\boldsymbol{r},\boldsymbol{p})\chi(\boldsymbol{r})\mathrm{d}\boldsymbol{r}\ \mathrm{d}\boldsymbol{p} \tag{4.3.49}$$

where $\chi(\boldsymbol{r})$ is the characteristic function imposing a restriction.

4.3.2.5.4 Non-equilibrium Statistical Mechanics

In this case, the distribution function $w(\boldsymbol{r},\boldsymbol{p};t)$ depends on time,

$$\frac{\mathrm{d}w}{\mathrm{d}t}=\frac{\partial w}{\partial t}+\sum_{\alpha=1}^{n_p}\left(\frac{\partial w}{\partial \boldsymbol{r}_\alpha}\dot{\boldsymbol{r}}_\alpha+\frac{\partial w}{\partial \boldsymbol{p}_\alpha}\dot{\boldsymbol{p}}_\alpha\right) \tag{4.3.50}$$

and the non-equilibrium phase average is defined as

$$\langle f(t);w\rangle = \int_{\Lambda(\boldsymbol{r},\boldsymbol{p})} f(\boldsymbol{r},\boldsymbol{p})w(\boldsymbol{r},\boldsymbol{p};t)\mathrm{d}\boldsymbol{r}\ \mathrm{d}\boldsymbol{p} \tag{4.3.51}$$

with

$$\frac{\partial}{\partial t}\langle f(t);w\rangle=\sum_{\alpha=1}^{n_p}\langle\frac{\boldsymbol{p}_\alpha}{m_\alpha}\cdot\frac{\partial f}{\partial \boldsymbol{r}_\alpha}-\frac{\partial U^{\boldsymbol{p}}}{\partial \boldsymbol{r}_\alpha}\cdot\frac{\partial f}{\partial \boldsymbol{p}_\alpha};w\rangle \tag{4.3.52}$$

4.3.2.6 Ensembles

An ensemble is the set of all micro states $(\boldsymbol{r},\boldsymbol{p})$ that are consistent with the set of applied macroscopic constraints V, P, T, E, etc., through the concept of statistical averaging. It only requires an explicit definition for the probability density function $w^e(H)$ based on the type of thermodynamic constraint, which governs the way the particle system and its surrounding domain interact.

The thermodynamic constraint is enforced by fixing a specific thermodynamic variable through the statistical concepts described earlier. Before going into the details, a brief description of major ensembles is presented:

- Fixing energy (E) is the simplest ensemble, which can be achieved by ignoring the external forces in the equations of motion.
- Fixing the number of particles (N) requires a periodic boundary condition that allows for a conserved total number of particles. A system with fixed N does not exchange any particles with the environment.
- Fixing volume (V) in the physical space is accomplished indirectly by tuning the total pressure. The procedure begins with an instantaneous pressure-dependent coefficient to scale the dimensions of the simulation box. Then the average of all instantaneous volumes determines the fixed volume.
- Fixing temperature (T) is practically one of the most difficult ensembles. There are a number of methods for imposing the fixed temperature. This usually requires computation of energy of a combination of the system and a fictitious constant temperature heat-bath.

In the following, the two frequently used NVE and NVT ensembles are further reviewed.

4.3.2.6.1 NVE Ensemble (Microcanonical Ensemble)

In this ensemble, an isolated Hamiltonian system with a fixed number of particles (N), fixed volume (V) and fixed energy (E) is considered, as depicted in Figure 4.18.

In an NVE ensemble, the total energy does not vary over the time,

$$H\left(\mathbf{r},\mathbf{p},a^{\text{ext}}\right)=E=\text{constant} \tag{4.3.53}$$

where a^{ext} is the set of external parameters.

The statistical averaging of an NVE ensemble on the micro system is obtained from

$$\langle f;w\rangle_{\text{NVE}}=\frac{1}{\Omega(N,V,E)}\int_{\Lambda(\mathbf{r},\mathbf{p})} f(\mathbf{r},\mathbf{p})w(\mathbf{r},\mathbf{p},E)\,d\mathbf{r}\;d\mathbf{p} \tag{4.3.54}$$

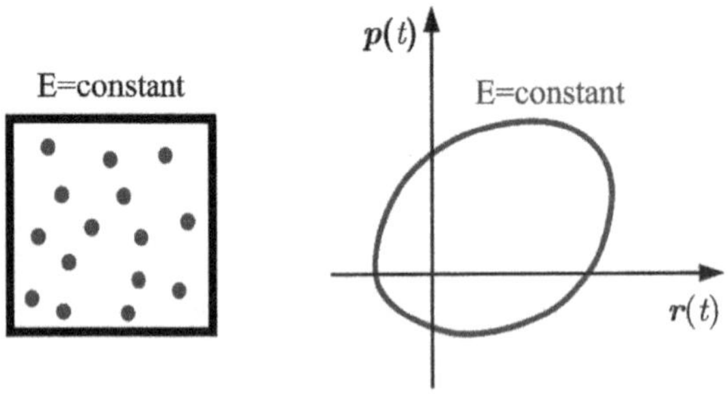

Figure 4.18 Definition of an NVE ensemble.

where the NVE microcanonical probability density function (or partition function) $w(\boldsymbol{r},\boldsymbol{p})$ can be expressed as

$$w(\boldsymbol{r},\boldsymbol{p},E)=\delta\left(E-H(\boldsymbol{r},\boldsymbol{p})\right) \tag{4.3.55}$$

and the microcanonical partition function Ω is the density of states of the system. It is defined as the number of different possible NVE microscopic states with the same probability in a small finite energy interval around E (Edholm 2014; Strachan 2017):

$$\Omega(N,V,E)=\frac{1}{\hbar^{3N}N!}\int_{\Lambda(\boldsymbol{r},\boldsymbol{p})}\delta\left(E-H(\boldsymbol{r},\boldsymbol{p})\right)\mathrm{d}\boldsymbol{r}\ \mathrm{d}\boldsymbol{p} \tag{4.3.56}$$

$\Omega(N,V,E)$ can be defined in terms of the normalized phase volume $\Upsilon\left(E,a^{\mathrm{ext}}\right)$

$$\Omega\left(N,V,E,a^{\mathrm{ext}}\right)=\left(\frac{\partial\Upsilon\left(E,a^{\mathrm{ext}}\right)}{\partial E}\right)_{a^{\mathrm{ext}}} \tag{4.3.57}$$

with

$$\Upsilon\left(E,a^{\mathrm{ext}}\right)=\frac{1}{(2\pi\hbar)^{3N}N!}\int_{H(\boldsymbol{r},\boldsymbol{p})<E}\mathrm{d}\boldsymbol{r}\ \mathrm{d}\boldsymbol{p} \tag{4.3.58}$$

The NVE statistical averaging Equation (4.3.54) can be rewritten as

$$\langle f;w\rangle_{\mathrm{NVE}}=\frac{\int_{\Lambda(\boldsymbol{r},\boldsymbol{p})}f(\boldsymbol{r},\boldsymbol{p})\delta\left(E-H(\boldsymbol{r},\boldsymbol{p})\right)\mathrm{d}\boldsymbol{r}\ \mathrm{d}\boldsymbol{p}}{\int_{\Lambda(\boldsymbol{r},\boldsymbol{p})}\delta\left(E-H(\boldsymbol{r},\boldsymbol{p})\right)\mathrm{d}\boldsymbol{r}\ \mathrm{d}\boldsymbol{p}} \tag{4.3.59}$$

The thermodynamic parameters entropy S, temperature T and pressure P of the system can then be computed from the normalized phase volume Υ (Edholm 2014):

$$S=k\ln\Upsilon=k\ln\Omega(E,V,N) \tag{4.3.60}$$

$$T=\left(\frac{\partial S}{\partial E}\right)^{-1} \tag{4.3.61}$$

$$P=\frac{1}{\Omega(N,V,E)}\left(\frac{\partial\Upsilon}{\partial V}\right)_E \tag{4.3.62}$$

The numerical procedure for solving an NVE ensemble can be summarized as:

1) The physical model is analysed and the NVE ensemble is justified.
2) The micromodel of particles with corresponding boundaries is constructed.
3) The micromodel is solved directly for the phase space trajectories $(\boldsymbol{r},\boldsymbol{p})$ based on the initial conditions.
4) A strategy of perturbed energy is followed. The total energy E and the normalized phase volume integral Υ are computed for the initial and perturbed models.

5) Numerically compute entropy from Equation (4.3.60), temperature from Equation (4.3.61) and other thermodynamic parameters (based on the perturbation strategy).
6) Check the computed temperature with benchmark values.

4.3.2.6.2 NVT Ensemble (Canonical Ensemble)

In this ensemble, a Hamiltonian system with a fixed number of particles (N), fixed volume (V) and fixed temperature (T) is considered. In contrast to the NVE ensemble, the total energy does not remain constant.

It is known that the following micro–macro relation defines the temperature of a system in terms of the kinetic energy of the system of particles:

$$\frac{1}{2}N_{\text{eff}}k_{\text{B}}T(t)=U^{k}(t)=\sum_{\alpha=1}^{n_p}\frac{1}{2}m_{\alpha}(v_{\alpha})^{2} \tag{4.3.63}$$

where $N_{\text{eff}}=3n_p-3$ and k_{B} is the Boltzman constant ($k_{\text{B}}=1.380649\times10^{-23}$ J K^{-1}).

The temperature T is then set with the average kinetic energy $\langle U^{k}\rangle$,

$$\langle U^{k}\rangle=\frac{3N}{2}k_{\text{B}}T \tag{4.3.64}$$

Equation (4.3.64) schematically shows one of the ways for fixing the temperature by tuning the velocity of particles appropriately.

Two classes of deterministic and stochastic approaches are available to enforce a fixed temperature in an NVT ensemble. The first set includes the velocity scaling technique, the Brendsen approach and the Nose–Hoover thermostat. The Langevin thermostat is examined as a stochastic approach.

Scaling Method In this method, the total kinetic energy of the system, as a measure of temperature, is conserved by scaling the velocity of particles. The velocity scaling algorithm is usually adopted to apply the temperature T_0,

$$\boldsymbol{v}_i\rightarrow s\boldsymbol{v}_i,\ s=\sqrt{\frac{T_0}{T(t)}} \tag{4.3.65}$$

The velocities are scaled accordingly at each time step to gradually reach the required temperature T_0. The instantaneous temperature $T(t)$ is obtained from

$$T(t)=\frac{2}{3Nk_B}\sum_{\alpha=1}^{n_p}\frac{1}{2}m_{\alpha}(v_{\alpha})^{2}=\frac{2}{3Nk_B}\sum_{\alpha=1}^{n_p}\boldsymbol{r}_{\alpha}\cdot\boldsymbol{f}_{\alpha} \tag{4.3.66}$$

which is expected to lead asymptotically in time to the required temperature T_0 in an average way:

$$\langle T(t)\rangle=\frac{2}{3Nk_B}\sum_{\alpha=1}^{n_p}\frac{1}{2}m_{\alpha}\langle(v_{\alpha})^{2}\rangle=T_0 \tag{4.3.67}$$

Clearly, this approach does not analytically guarantee the system to reach the temperature T_0.

Berendsen introduced the concept of weak coupling to an external bath in the form of exponential velocity scaling at each time step Δt (Grotendorst et al. 2009):

$$\lambda_T = \sqrt{1 - \frac{\Delta t}{\tau_T}\left(1 - \frac{T_0}{T(t)}\right)} \tag{4.3.68}$$

where the constant τ_T determines the influence of the external bath in the form of the time scale on which the temperature T_0 is expected to reach.

Brendsen Method The Brendsen thermostat is a simple approach frequently used in MD simulations. This approach is in fact a constraint method for imposition of a thermostat by adding an interatomic friction term to the equation of motion:

$$\boldsymbol{f}^{\alpha} - \xi \dot{\boldsymbol{r}}_{\alpha} = m_{\alpha} \ddot{\boldsymbol{r}}_{\alpha} \tag{4.3.69}$$

$$\xi = m_{\alpha} \mu_T \left(1 - \frac{T_0}{T(t)}\right) \tag{4.3.70}$$

where μ_T is the thermostat constant and $T(t)$ and T_0 are the system and reference temperatures, respectively. The thermostat constant may be tuned to conserve the kinetic energy of the system.

The interparticle force is reduced by a positive friction coefficient for system temperatures larger than the reference ensemble temperature T_0.

The Brendsen concept may also be used to simulate at constant pressure by exponential scaling of the volume of the simulation box at each time step by the factor λ_V (pressure bath):

$$\lambda_V = 1 - \alpha_P \left(P_A - P_{\text{bath}}\right) \tag{4.3.71}$$

Nose–Hoover Thermostat In the Nose–Hoover thermostat, the fixed temperature of the system A is achieved if the system is confined by a large heat bath region (thermostat system R) with the fixed temperature T_R, as depicted in Figure 4.19. The system A is assumed to be in thermal equilibrium with the system R. If the temperature of the system A decreases below or increases above the assigned temperature T_R, the thermostat provides or absorbs heat accordingly in order to maintain the temperature at the fixed value of T_R.

While the heat bath reference system R can be simulated as a many-particle system $(\boldsymbol{r}_R, \boldsymbol{p}_R)$, it can be simply considered as a single particle system of the heat bath, characterized by its position R_R, momentum P_R and mass M_R. Assuming a weak interaction $(H^{A-R} = 0)$, the Hamiltonian H^e of the combined system can then be written as

$$H^e = H^A + H^R \tag{4.3.72}$$

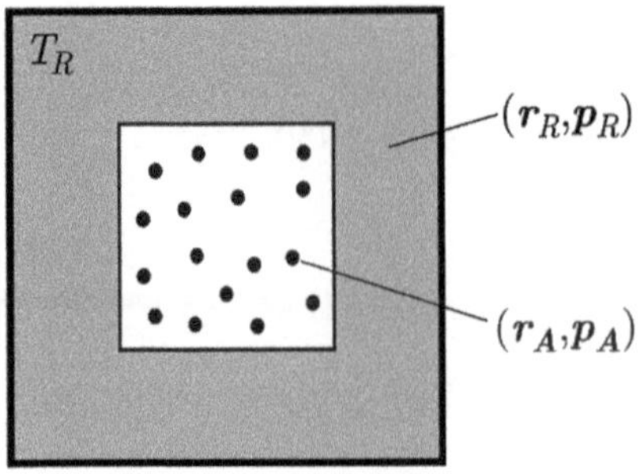

Figure 4.19 Nose–Hoover thermostat for the NVT ensemble.

$$H^e = \underbrace{\frac{2}{3Nk_B}\sum_{\alpha=1}^{n_p}\frac{\left\|\boldsymbol{p}_\alpha{}^2\right\|}{2m_\alpha} + U^p(\boldsymbol{r})}_{H^A} + \underbrace{\frac{P_R^2}{2M_R} + 3Nk_BT_R \ln R_R}_{H^R} \tag{4.3.73}$$

Similar to the Brendsen method, a friction term is added to the equation of motion,

$$\boldsymbol{f}_\alpha - \gamma(t)m_\alpha\dot{\boldsymbol{r}}_\alpha = m_\alpha\ddot{\boldsymbol{r}}_\alpha \tag{4.3.74}$$

where

$$\dot{\gamma}(t) = \frac{3Nk_B}{M_R}\left(T(t) - T_R\right) \tag{4.3.75}$$

Adding the friction coefficient usually leads to significant temperature oscillations, which need to be treated using numerical means, such as the modified velocity Verlet technique.

The micro states of both individual systems A and R can be described by their corresponding phase vectors $(\boldsymbol{r}_A,\boldsymbol{p}_A)$ and (R_R,P_R), respectively. Since the combined system $A + R$ is adiabatically isolated, an NVE ensemble can be considered with the probability density function $w(\boldsymbol{r}_A,\boldsymbol{p}_A,R_R,P_R)$:

$$w(\boldsymbol{r}_A,\boldsymbol{p}_A,R_R,P_R,E) = \frac{1}{\Omega\left(E,a^{\text{ext}}\right)}\delta\left(E - H^e(\boldsymbol{r}_A,\boldsymbol{p}_A)\right) \tag{4.3.76}$$

For the combined set of A and R, the equivalent statistical average can be written as

$$\langle f;w\rangle_{\text{NVT}} = \frac{1}{Q(N,V,T)}\int_{\Lambda(\boldsymbol{r}_A,\boldsymbol{p}_A)} f(\boldsymbol{r}_A,\boldsymbol{p}_A)w(\boldsymbol{r}_A,\boldsymbol{p}_A,R_R,P_R;T)\mathrm{d}\boldsymbol{r}\,\mathrm{d}\boldsymbol{p} \tag{4.3.77}$$

and the NVT canonical probability density function $w(\boldsymbol{r}_A,\boldsymbol{p}_A,R_R,P_R;T)$ allows variation of the combined systems around a fixed average value. It is defined by the Gibbs distribution function with the assumption that the system R is much larger than the system A $(w(\boldsymbol{r}_A,\boldsymbol{p}_A,R_R,P_R;T) \rightarrow w(\boldsymbol{r}_A,\boldsymbol{p}_A;T))$,

$$w(\boldsymbol{r}_A,\boldsymbol{p}_A;T) = \frac{1}{N!\hbar^{3N}Z}\mathrm{e}^{-\frac{H(\boldsymbol{r}_A,\boldsymbol{p}_A)}{k_BT}} = \frac{1}{N!\hbar^{3N}}\mathrm{e}^{-\frac{H(\boldsymbol{r}_A,\boldsymbol{p}_A)-\psi_H}{k_BT}} \tag{4.3.78}$$

where N is the number of particles comprising the system and the term $N!$ comes from the enumeration of all degenerated micro states.

The function $Q(N,V,T)$ is the canonical partition function (the Boltzman distribution function) defined as

$$Q(N,V,T)=\frac{1}{(2\pi\hbar)^{3N}N!}\int_{\Lambda(\boldsymbol{r}_A,\boldsymbol{p}_A)}\mathrm{e}^{-\frac{H(\boldsymbol{r}_A,\boldsymbol{p}_A)}{k_BT}}\,\mathbf{d}\boldsymbol{r}\,d\boldsymbol{p} \tag{4.3.79}$$

and Z is the partition function defined in terms of the Helmholtz energy ψ_H:

$$Z=\mathrm{e}^{-\frac{\psi_H}{k_BT}}=\mathrm{e}^{-\beta\psi_H} \tag{4.3.80}$$

or

$$\psi_H=-k_BT\ln Z=-k_BT\ln Q \tag{4.3.81}$$

Definitions of the partition functions $Z(N,V,T)$ and $Q(N,V,T)$ can be related through the following relations:

$$Q(N,V,T)=\int\mathrm{e}^{-\frac{\Phi(\boldsymbol{r})}{k_BT}}\,\mathbf{d}\boldsymbol{r} \tag{4.3.82}$$

$$Z(N,V,T)=\frac{1}{\hbar^{3N}N!}\int\mathrm{e}^{-\frac{H(\boldsymbol{r}_A,\boldsymbol{p}_A)}{k_BT}}\,\mathbf{d}\boldsymbol{r}\,\mathbf{d}\boldsymbol{p} \tag{4.3.83}$$

which leads to

$$Z(N,V,T)=\left(\frac{2\pi k_BT}{\hbar^2}\right)^{\frac{3N}{2}}\frac{1}{N!}Q(N,V,T) \tag{4.3.84}$$

For a complete discussion on the necessity for the coefficients in front of $Z(N,V,T)$ or $Q(N,V,T)$ for computing the phase averages, see Edholm (2014).

From Equation (4.3.78), the NVT statistical average (4.3.77) can be expressed as

$$\langle f;w\rangle_{\mathrm{NVT}}=\frac{1}{N!\hbar^{3N}Q(N,V,T)}\int_{\Lambda(\boldsymbol{r}_A,\boldsymbol{p}_A)}f(\boldsymbol{r}_A,\boldsymbol{p}_A)\mathrm{e}^{-\frac{H(\boldsymbol{r}_A,\boldsymbol{p}_A)-\psi_H}{k_BT}}\,\mathbf{d}\boldsymbol{r}\,d\boldsymbol{p} \tag{4.3.85}$$

Moreover, the canonical partition function $Q(N,V,T)$ in Equation (4.3.79) can be related to the microcanonical density/partition function $\Omega(N,V,E,a^{\mathrm{ext}})$ (Strachan 2017),

$$Q(N,V,T)=\int_{\Lambda(\boldsymbol{r}_A,\boldsymbol{p}_A)}\Omega(N,V,E)\,\mathrm{e}^{-\frac{E}{k_BT}}\mathbf{d}\boldsymbol{r}\,\mathbf{d}\boldsymbol{p} \tag{4.3.86}$$

with

$$Q(N,V,T)=\frac{1}{\hbar^{3N}N!}\int_{\Lambda(\boldsymbol{r}_A,\boldsymbol{p}_A)}\mathrm{e}^{-\frac{H(\boldsymbol{r}_A,\boldsymbol{p}_A)}{k_BT}}\,\mathbf{d}\boldsymbol{r}\,\mathbf{d}\boldsymbol{p} \tag{4.3.87}$$

$$\Omega(N,V,E)=\frac{1}{\hbar^{3N}N!}\int_{\Lambda(\mathbf{r}_A,\mathbf{p}_A)}\delta\left[E-H(\mathbf{r}_A,\mathbf{p}_A)\right]\mathrm{d}\mathbf{r}\,\mathrm{d}\mathbf{p} \tag{4.3.88}$$

leading to (Strachan 2007)

$$Q(N,V,T)=\Omega(N,V,E)\ \mathrm{e}^{-\frac{E}{k_B T}} \tag{4.3.89}$$

$$\ln Q(N,V,T)=\ln\Omega(N,V,E)-\frac{E}{k_B T} \tag{4.3.90}$$

and the well-known thermodynamics relation (Section 4.3.2.1)

$$\underbrace{k_B T\ln Q(N,V,T)}_{\psi_H(N,V,T)}=\underbrace{Tk_B\ln\Omega(N,V,E)}_{S(N,V,E)}-E \tag{4.3.91}$$

Similarly, for an NPT ensemble,

$$Q(N,P,T)=\int_{\Lambda(\mathbf{r}_A,\mathbf{p}_A)}\mathrm{e}^{-\frac{E-PV}{k_B T}}\,\mathrm{d}\mathbf{r}\,\mathrm{d}\mathbf{p} \tag{4.3.92}$$

leads to the Gibbs energy $\psi_G(N,P,T)$ (Edholm 2014):

$$\psi_G(N,P,T)=-k_B T\ln Q(N,P,T) \tag{4.3.93}$$

The numerical procedure for solving an NVT ensemble by the Nose–Hoover thermostat can be summarized as:

1) The physical model is analysed and the NVT ensemble and the Gibbs distribution function are justified.
2) The micromodel of particles with corresponding boundaries is constructed.
3) The micromodel is solved directly for the phase space trajectories $(\boldsymbol{r},\boldsymbol{p})$ based on the initial conditions.
4) The average kinetic energy and average temperature are computed.
5) A perturbed temperature strategy is followed. The partition function is computed for the initial and perturbed models.
6) Numerically compute the Helmholtz free energy ψ_H from Equation (4.3.81) and other thermodynamic parameters (based on the perturbation strategy).
7) Check the accuracy of the computed internal energy.

Langevin Thermostat (Stochastic) The Langevin thermostat is based on the combination of molecular dynamics (MD) and the Monte Carlo stochastic approach to enforce a fixed temperature constraint. The Langevin thermostat is based on the main physical variables $(\boldsymbol{r},\boldsymbol{p})$ with the constraint of the invariant Gibbs distribution function $w(\boldsymbol{r}_A,\boldsymbol{p}_A;T)$.

The equation of motion contains the governing equation of the Nose–Hoover combined system of particles and the heat bath, and considers a set of random impacts of particles,

$$m_\alpha \ddot{\boldsymbol{r}}_\alpha = \boldsymbol{f}_\alpha - \gamma(t) m_\alpha \dot{\boldsymbol{r}}_\alpha + G_\alpha(t) \tag{4.3.94}$$

where γ is the damping coefficient (4.3.75) and $G_\alpha(t)$ is the force associated with the random impact of particles:

$$\langle G_\alpha \rangle = 0 \tag{4.3.95}$$

$$\langle G_{\alpha,i}(t) G_{\beta,i}(t') \rangle = 2\gamma m_\alpha k_B T \delta_{ij} \delta^{\alpha\beta} \delta(t - t') \tag{4.3.96}$$

For further information, see Schlick (2002).

4.3.3 MD Equations of Motion

The direct solution of a particle system of atoms, molecules, etc., as typically shown in Figure 4.20, is briefly reviewed in this section. It is based on the Newtonian multibody equations of motion based on the Hamiltonian formulation, and is subject to statistical ensemble constraints, as described in previous sections.

A numerical model for a molecular or atomistic system is usually defined based on a periodic boundary condition, as illustrated in Figure 4.21, so the finite model may represent the whole MD model.

A predetermined number of atoms move and interact within the representative cell with periodic boundary conditions. The general concepts of periodic boundary conditions for MD simulations are similar to periodic RVEs in the multiscale homogenization technique (see Section 4.2.2). The representative cell is surrounded by an environment that periodically repeats its own configurations. Consequently, atoms within the representative cell interact, not only with atoms within the same cell, but also with image atoms in nearby

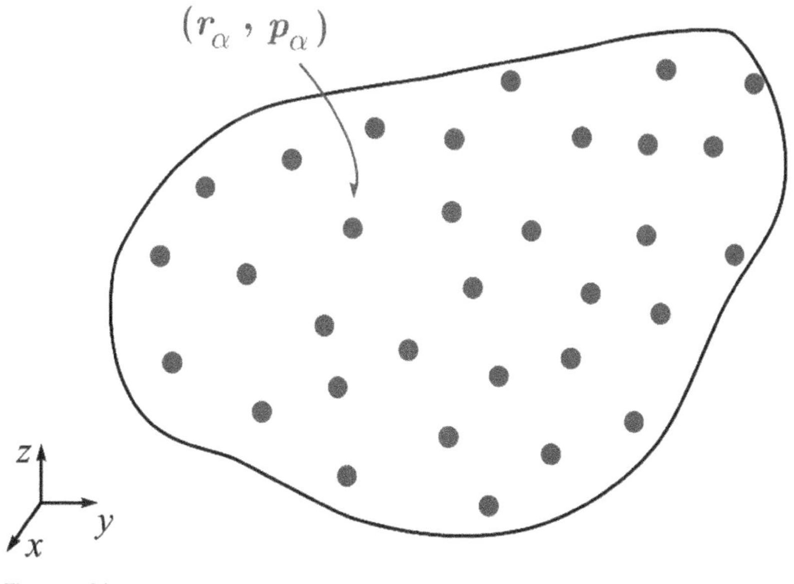

Figure 4.20 A typical MD model.

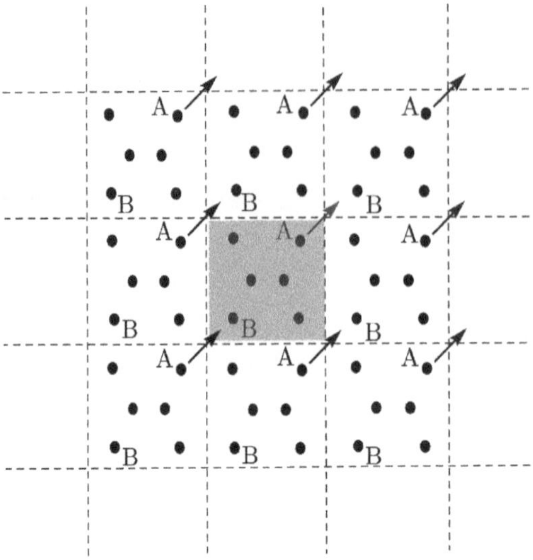

Figure 4.21 Periodic boundary conditions of a two dimensional MD model (RVE).

mapped cells. During the simulation, when an atom leaves the unit cell, its image will enter the cell from the other side (Figure 4.21). As a result, the number of atoms inside the unit cell is maintained. It is important to note that atoms that lie within a distance r_c of a particular boundary interact with atoms in the neighbouring periodic image of the system or, equivalently, with atoms near its opposite boundary (the wraparound effect).

Moreover, the MD model should be set to the initial equilibrium before it is subjected to the external effects and constraints. This is usually achieved by a statistical ensemble to ensure that the kinetic energy remains at a predefined value. At the same time, the total momentum should be kept at zero to avoid rigid motion of the whole MD system.

The main remaining parts include the definitions of interatomic potentials and the procedure for postprocessing computations, such as an evaluation of stress.

The total energy $E = U^k + U^p$ is the sum of kinetic energy U^k and the interatomic potential U^p:

$$U^k = \frac{1}{2}\sum_{\alpha=1}^{n_p} m_\alpha v_\alpha^2 \tag{4.3.97}$$

$$U^p = U^p\left(\boldsymbol{r}\right) \tag{4.3.98}$$

where $\boldsymbol{r}$ is the vector of positions of all particles,

$$\boldsymbol{r} = \left(\boldsymbol{r}_1, \boldsymbol{r}_2, \ldots, \boldsymbol{r}_{n_p}\right) \tag{4.3.99}$$

Utilizing the Lagrangian function $L = U^k + U^p$, the Newton governing equation for any particle α of a system of N particles can be written as

$$\boldsymbol{F}_\alpha = m_\alpha \ddot{\boldsymbol{r}}_\alpha = -\frac{\partial U^p\left(\boldsymbol{r}\right)}{\partial \boldsymbol{r}_\alpha}, \ \alpha = 1, \ldots, n_p \tag{4.3.100}$$

where $\boldsymbol{F}_\alpha = (F_{\alpha x}, F_{\alpha y}, F_{\alpha z})$

The set of Equations (4.3.100) for each particle are coupled through the interatomic potentials, which depend on different orders of interatomic distance functions. This highly coupled and nonlinear set of simultaneous differential equations can only be solved by numerical techniques.

Solution of Equation (4.3.100) by a time integration scheme (such as the velocity Verlet central difference method) leads to general forms similar to

$$\boldsymbol{r}_\alpha(t_0+\Delta t) = -\boldsymbol{r}_\alpha(t_0-\Delta t) + 2\boldsymbol{r}_\alpha(t_0)\Delta t + \boldsymbol{a}_\alpha(t_0)(\Delta t)^2 + \cdots \tag{4.3.101}$$

which requires acceleration $\boldsymbol{a}_\alpha(t_0)$ and, consequently, the forces acting on atoms. They, in turn, need interatomic or MD potentials $U^p(\boldsymbol{r})$, which define the potential energy of a system of atoms in terms of their positions. Note that according to Equation (4.3.101), the update procedure for positions $\boldsymbol{r}_\alpha$ does not require the velocities $\dot{\boldsymbol{r}}_\alpha$.

In a conservative system, the force on an atom α is computed by differentiation of the interatomic potential $U^p(\boldsymbol{r})$ with respect to the atom position $\boldsymbol{r}_\alpha$

$$\boldsymbol{f}_\alpha = -\frac{\partial U^p(\boldsymbol{r})}{\partial \boldsymbol{r}_\alpha} \tag{4.3.102}$$

Moreover, for a system of atoms with positions $\boldsymbol{r}$, which is subjected to external forces $\boldsymbol{f}^{\text{ext}}$, the potential energy of the system can be written as

$$\Pi = U^p(\boldsymbol{r}) - \mathrm{W}(\boldsymbol{f}^{\text{ext}}, \boldsymbol{r}) = U^p(\boldsymbol{r}) - \boldsymbol{f}^{\text{ext}} \cdot \boldsymbol{r} \tag{4.3.103}$$

and the force $\boldsymbol{f}^\alpha$ on an atom α is obtained from

$$\boldsymbol{f}_\alpha = -\frac{\partial \Pi}{\partial \boldsymbol{r}_\alpha} = -\frac{\partial U^p(\boldsymbol{r})}{\partial \boldsymbol{r}_\alpha} + \boldsymbol{f}_\alpha^{\text{ext}} \tag{4.3.104}$$

The force acting on an atom can always be expressed as the vector sum of all forces between the atom and its surrounding atoms. Note that, in general, the force $\boldsymbol{f}_{\alpha\beta}$ between two atoms α and β depends on the whole set of interatomic distances:

$$\boldsymbol{f}_{\alpha\beta} = \boldsymbol{f}_{\alpha\beta}\left(r_{12}, r_{13}, \ldots, r_{1n_p}, r_{23}, \ldots, r_{(n_p-1)n_p}\right) \tag{4.3.105}$$

4.3.3.1 Time Integration of MD Equations of Motion

Several time integration schemes may be used to solve the MD equation of motion from the Taylor expansion series,

$$\boldsymbol{r}(t+\Delta t) \approx \boldsymbol{r}(t) + \underbrace{\frac{\mathrm{d}\boldsymbol{r}(t)}{\mathrm{d}t}}_{\boldsymbol{v}(t)}\Delta t + \frac{1}{2}\underbrace{\frac{\mathrm{d}^2\boldsymbol{r}(t)}{\mathrm{d}t^2}}_{\boldsymbol{a}(t)=\boldsymbol{f}(t)/m}(\Delta t)^2 \tag{4.3.106}$$

The position-based Verlet algorithm is written in terms of the previous steps (third order accuracy in position $\boldsymbol{r}$),

$$\boldsymbol{a}(t) \approx \frac{\boldsymbol{f}(\boldsymbol{r}(t))}{m} \tag{4.3.107}$$

$$\boldsymbol{r}(t+\Delta t) \approx -\boldsymbol{r}(t-\Delta t) + 2\boldsymbol{r}(t) + \boldsymbol{a}(t)(\Delta t)^2 \tag{4.3.108}$$

$$\boldsymbol{v}(t) \approx \frac{\boldsymbol{r}(t+\Delta t) - \boldsymbol{r}(t-\Delta t)}{2\Delta t} \tag{4.3.109}$$

Alternatively, in the Leap-Frog algorithm, the procedure is based on the velocities at half timesteps, leading to (third-order accuracy for $\boldsymbol{r}$ and second-order accuracy for velocity $\boldsymbol{v}$ or momentum $\boldsymbol{p}$):

$$\boldsymbol{a}(t) \approx \frac{\boldsymbol{f}(\boldsymbol{r}(t))}{m} \tag{4.3.110}$$

$$\boldsymbol{v}\left(t+\frac{\Delta t}{2}\right) \approx \boldsymbol{v}\left(t-\frac{\Delta t}{2}\right) + \boldsymbol{a}(t)\Delta t \tag{4.3.111}$$

$$\boldsymbol{r}(t+\Delta t) \approx \boldsymbol{r}(t) + \boldsymbol{v}\left(t+\frac{\Delta t}{2}\right)\Delta t \tag{4.3.112}$$

Note that computation of the kinetic energy should be adapted in the Leap-Frog technique due to the fact that the velocities are determined at the half timesteps.

Alternatively, the efficient velocity–Verlet (VV) algorithm (Swope and Anderson 1984) is usually adopted:

$$\boldsymbol{r}_\alpha(t+\Delta t) = -\boldsymbol{r}_\alpha(t-\Delta t) + 2\boldsymbol{r}_\alpha(t)\Delta t + \boldsymbol{a}_\alpha(t)(\Delta t)^2 \tag{4.3.113}$$

In the VV algorithm, for given initial conditions $\boldsymbol{r}(0)$ and $\boldsymbol{v}(0) = \dot{\boldsymbol{r}}(0)$, the update procedure for the time $t+\Delta t$ in terms of $\boldsymbol{r}(t)$, $\boldsymbol{v}(t)$ and $\boldsymbol{a}(t)$ can be followed by (Momentum/Velocity Verlet)

$$\boldsymbol{a}(t) = \frac{\boldsymbol{f}(t)}{m} \tag{4.3.114}$$

$$\boldsymbol{r}(t+\Delta t) \approx \boldsymbol{r}(t) + \boldsymbol{v}(t)\Delta t + \frac{1}{2}\boldsymbol{a}(t)(\Delta t)^2 \tag{4.3.115}$$

$$\boldsymbol{f}(t+\Delta t) = \boldsymbol{f}(\boldsymbol{r}(t+\Delta t)) \tag{4.3.116}$$

$$\boldsymbol{a}(t+\Delta t) \approx \frac{\boldsymbol{f}(t+\Delta t)}{m} \tag{4.3.117}$$

$$\boldsymbol{v}(t+\Delta t) \approx \boldsymbol{v}(t) + \frac{\boldsymbol{a}(t) + \boldsymbol{a}(t+\Delta t)}{2}\Delta t \tag{4.3.118}$$

In all integration algorithms, the force field $\boldsymbol{f}(\boldsymbol{r}(t))$ is computed from the interatomic potential $U^P(\boldsymbol{r})$.

The timestep Δt must be sufficiently small to ensure the stability of the time integration scheme and the balance of energy of the system must be continuously controlled to guarantee the stability of the overall numerical solution.

4.3.4 Models for Atomic Interactions – MD Potentials

Classical atomistic simulations regulate the interatomic interactions of a system of n_p particles by means of the potential functions $U^p(\boldsymbol{r})$.

There are a wide range of atomic potentials (also called the force fields) based on the chemical bonds that exist between atoms or molecules. They constitute strong primary bonds, such as the covalent and ionic bonds, and weak or secondary bonds, such as the Van der Waals potential.

Proper interatomic potentials should be adopted to allow for accurate computation of forces on atoms and their accelerations within a time integration MD analysis. Based on a set of available fitting data for a specific material structure, the main requirements for interatomic or intermolecular potentials can be summarized as (Rafii-Tabar and Mansoori 2004):

- Precision to reproduce the data.
- Flexibility to sufficiently accommodate a broad range of other experimental results.
- Characterizing the structures that are not included in the fitting data.
- Computational efficiency in various size and time characteristics of the system.

It has been proved, in fundamental physics, that the energy of an atomic system can be expressed as a function of positions of the atoms, called the interatomic potential. Other necessary characteristics, such as symmetry and objectivity, further characterize the interatomic potential to be a function of distances of atoms.

The interatomic potential $U^p(\boldsymbol{r})$ defines the total potential energy of a system of atoms in terms of their positions $\boldsymbol{r} = (\boldsymbol{r}_1, \boldsymbol{r}_2, \ldots, \boldsymbol{r}_{n_p})$ (Figure 4.22):

$$U^p(\boldsymbol{r}) = U^p\left(\boldsymbol{r}_1, \boldsymbol{r}_2, \ldots, \boldsymbol{r}_{n_p}\right) \tag{4.3.119}$$

Obviously, it is impossible to derive analytical expressions for general interatomic potentials. As a result, semi-empirical potentials, with strong conceptual justifications, have been developed for various atomistic material structures.

The general form of the interatomic potential $U^p(\mathbf{r})$ can be written as

$$\begin{aligned} U^p = & \sum_{\alpha} \phi_1\left(\boldsymbol{r}_\alpha\right) + \frac{1}{2} \sum_{\substack{\alpha,\beta \\ \beta \neq \alpha}} \phi_2\left(\boldsymbol{r}_{\alpha\beta}\right) + \frac{1}{6} \sum_{\substack{\alpha,\beta,\gamma \\ \alpha \neq \beta \neq \gamma}} \phi_3\left(\boldsymbol{r}_{\alpha\beta}, \boldsymbol{r}_{\beta\gamma}, \boldsymbol{r}_{\gamma\alpha}\right) \\ & + \frac{1}{24} \sum_{\substack{\alpha,\beta,\gamma,\delta \\ \alpha \neq \beta \neq \gamma \neq \delta}} \phi_4\left(\boldsymbol{r}_{\alpha\beta}, \boldsymbol{r}_{\alpha\gamma}, \boldsymbol{r}_{\alpha\delta}, \boldsymbol{r}_{\beta\gamma}, \boldsymbol{r}_{\beta\delta}, \boldsymbol{r}_{\gamma\delta}\right) + \cdots \end{aligned} \tag{4.3.120}$$

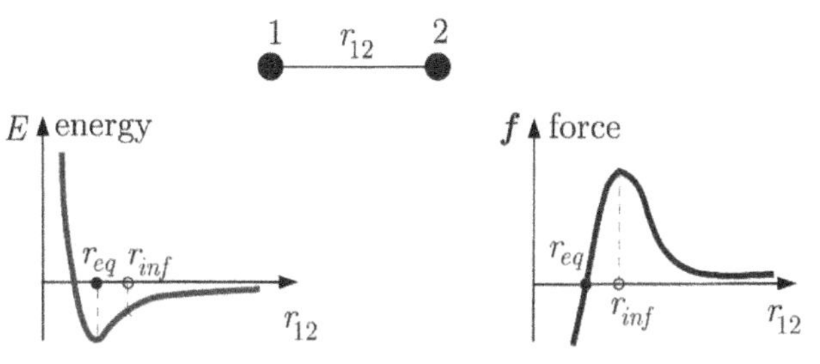

Figure 4.22 Variations of interatomic energy and force in terms of the distance between two atoms.

where ϕ_n represents the n-body potential.

The term $\phi_1(\boldsymbol{r}_\alpha)$, which represents externally applied potential (for instance, gravity, electric fields, etc.), usually from the container of MD system, is not required for bulk modelling with periodic boundary conditions.

Most of practical interatomic potentials are based on the second and third terms:

$$U^p = \frac{1}{2}\sum_{\substack{\alpha,\beta \\ \beta\neq\alpha}} \phi_2\left(\boldsymbol{r}_{\alpha\beta}\right) + \frac{1}{6}\sum_{\substack{\alpha,\beta,\gamma \\ \alpha\neq\beta\neq\gamma}} \phi_3\left(\boldsymbol{r}_{\alpha\beta},\boldsymbol{r}_{\beta\gamma},\boldsymbol{r}_{\gamma\alpha}\right) \tag{4.3.121}$$

with ϕ_2 and ϕ_3 as the two-body and three-body potentials, respectively (Figure 4.23):

$$\phi_2\left(\boldsymbol{r}_{\alpha\beta}\right) = \phi_2\left(\boldsymbol{r}_\beta - \boldsymbol{r}_\alpha\right) = \phi_2\left(r_{\alpha\beta}\right) \tag{4.3.122}$$

$$\phi_3\left(\boldsymbol{r}_{\alpha\beta},\boldsymbol{r}_{\beta\gamma},\boldsymbol{r}_{\gamma\alpha}\right) = \phi_3\left(\cos\theta_{\alpha\beta\gamma}\right) = \phi_3\left(\frac{\boldsymbol{r}_{\beta\alpha}\cdot\boldsymbol{r}_{\beta\gamma}}{\left|\boldsymbol{r}_{\beta\alpha}\right|\left|\boldsymbol{r}_{\beta\gamma}\right|}\right) \tag{4.3.123}$$

where

$$r_{\alpha\beta} = \left|\boldsymbol{r}_{\alpha\beta}\right| = \left|\boldsymbol{r}_\beta - \boldsymbol{r}_\alpha\right| \tag{4.3.124}$$

The interactions are assumed to be zero beyond a specific cut-off radius r^c (see Figure 4.24)

$$\phi(\boldsymbol{r}) = 0 \text{ for all } |\boldsymbol{r}| > r^c \tag{4.3.125}$$

In practice, such a cut-off distance may lead to a discontinuous variation of energy and force fields and possible convergence problems. As a result, shifted and smoothed cut-off solutions are available to keep the continuous nature of the energy/potential function and the corresponding interatomic force (see Figure 4.24).

4.3.4.1 Pair Potentials

The two-body or pair potentials are the simplest models for interatomic potentials. These potentials only account for atomic bond distances.

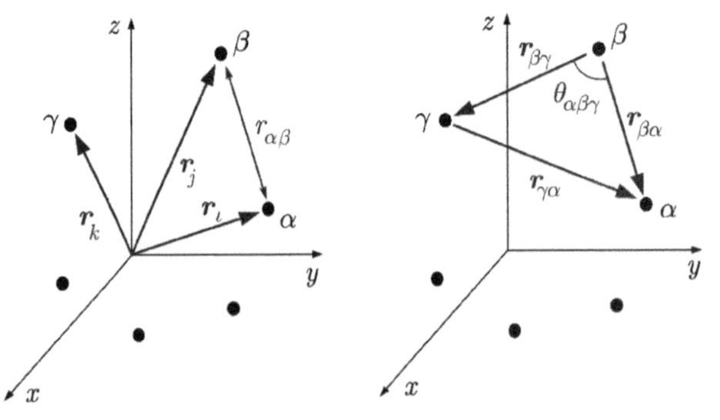

Figure 4.23 Definitions of variables for two and three body atomistic potentials.

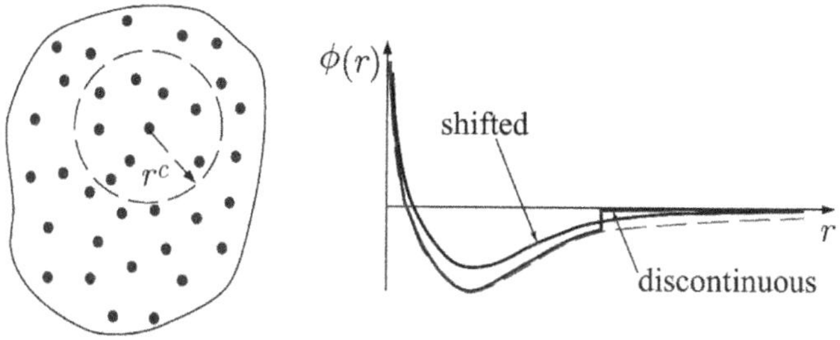

Figure 4.24 Definition of the cut off distance r^c.

The two-body potentials are defined in terms of the distance between the atoms $r^{\alpha\beta}$ and not their positions $\boldsymbol{r}_\alpha$:

$$U^p(\boldsymbol{r}) = \frac{1}{2}\sum_{\substack{\alpha,\beta \\ \beta\neq\alpha}} \phi_2(r_{\alpha\beta}) \tag{4.3.126}$$

The force $\boldsymbol{f}^\alpha$ on an atom α is obtained by differentiation of $U^p(\boldsymbol{r})$ with respect to its position $\boldsymbol{r}_\alpha$:

$$\boldsymbol{f}_\alpha = -\frac{\partial U^p(\boldsymbol{r})}{\partial \boldsymbol{r}_\alpha} \tag{4.3.127}$$

or

$$\begin{aligned} U^p(\boldsymbol{r}) &= \sum_{\substack{\alpha,\beta \\ \beta\neq\alpha}} \phi_2(r_{\alpha\beta}) \rightarrow \boldsymbol{f}_\alpha = -\sum_{\substack{\alpha,\beta \\ \beta\neq\alpha}} \frac{\partial \phi_2(r_{\alpha\beta})}{\partial \boldsymbol{r}_\alpha} \\ &= -\sum_{\substack{\alpha,\beta \\ \beta\neq\alpha}} \underbrace{\frac{\partial \phi_2(r_{\alpha\beta})}{\partial r_{\beta\gamma}}}_{\phi_2'(r_{\alpha\beta})} \frac{\partial r_{\beta\gamma}}{\partial \boldsymbol{r}_\alpha} = -\sum_{\substack{\alpha,\beta \\ \beta\neq\alpha}} \underbrace{\phi_2'(r_{\alpha\beta}) \frac{(\boldsymbol{r}_\beta - \boldsymbol{r}_\alpha)}{r_{\alpha\beta}}}_{\boldsymbol{f}_{\alpha\beta}} \end{aligned} \tag{4.3.128}$$

or, simply,

$$\boldsymbol{f}_\alpha = -\sum_{\substack{\alpha,\beta \\ \beta\neq\alpha}} \boldsymbol{f}_{\alpha\beta} \tag{4.3.129}$$

Equation (4.3.129) shows that, for a pair potential, the interatomic force $\boldsymbol{f}_{\alpha\beta}$ between two atoms α and β depends only on the distance between the two atoms,

$$\boldsymbol{f}_{\alpha\beta} = \boldsymbol{f}_{\alpha\beta}(r_{\alpha\beta}) \tag{4.3.130}$$

In the following, a number of major two-body potentials are briefly presented.

4.3.4.1.1 Lennard–Jones (LJ) Potential

The Lennard–Jones potential is one of the most basic pair potentials based on two distinct terms for the repulsion and attraction effects in between the atoms:

$$\phi_2\left(r_{\alpha\beta}\right) = 4\varepsilon\left[\underbrace{\left(\frac{\sigma}{r_{\alpha\beta}}\right)^{12}}_{\text{repulsion}} - \underbrace{\left(\frac{\sigma}{r_{\alpha\beta}}\right)^{6}}_{\text{attraction}}\right] \tag{4.3.131}$$

where ε and σ are the LJ energy and length scale parameters, respectively.

The interatomic force $f^{\alpha\beta}$ is obtained by differentiation of $\phi_2(r_{\alpha\beta})$

$$f_{\alpha\beta}\left(r_{\alpha\beta}\right) = -\frac{\partial\phi_2\left(r_{\alpha\beta}\right)}{\partial r_{\alpha\beta}} = -24\frac{\varepsilon}{\sigma}\left[2\left(\frac{\sigma}{r_{\alpha\beta}}\right)^{13} - \left(\frac{\sigma}{r_{\alpha\beta}}\right)^{7}\right] \tag{4.3.132}$$

Equations (4.3.131) and (4.3.132) allow for computing the distance r_0 between atoms and the maximum force $f_{\max}$ associated with the equilibrium state

$$\text{at equilibrium}: \begin{cases} r_0 = \sigma\sqrt[6]{2} \\ f_{\max} = 2.394\dfrac{\varepsilon}{\sigma} \end{cases} \tag{4.3.133}$$

4.3.4.1.2 Coulomb Potential

The Coulomb potential is used in MD problems with electrostatic effects. The electric potential of a single charge Q at a distance r from the charge can be defined as

$$U^p(r) = \frac{1}{4\pi\varepsilon_0}\frac{Q}{r} \tag{4.3.134}$$

where ε_0 is the electric permittivity of the space.

The Coulomb potential for two particles with electrostatic charges Q_α and Q_β with a distance r can then be obtained as

$$U^p(r) = \frac{1}{4\pi\varepsilon_0}\frac{Q_\alpha\, Q_\beta}{r_{\alpha\beta}} \tag{4.3.135}$$

and the corresponding force field:

$$f\left(r_{\alpha\beta}\right) = \frac{1}{4\pi\varepsilon_0}\frac{Q_\alpha\, Q_\beta}{r_{\alpha\beta}^2} \tag{4.3.136}$$

4.3.4.1.3 Morse Potential

The Morse potential is a simple pair potential in exponential form, which can be used to model atomic interactions, such as the interaction of an atom and a surface. It includes both the repulsion and attraction terms,

$$\phi_2\left(r_{\alpha\beta}\right) = d_\varepsilon\left[e^{2a_\varepsilon\left(r_0 - r_{\alpha\beta}\right)} - 2e^{a_\varepsilon\left(r_0 - r_{\alpha\beta}\right)}\right] \tag{4.3.137}$$

where r_0 represents the location associated with the equilibrium spacing, d_ε is the well depth and a_ε is defined in terms of the force constant k_ε:

$$a_\varepsilon = \sqrt{\frac{k_\varepsilon}{2d_\varepsilon}} \tag{4.3.138}$$

The Morse interatomic force becomes

$$f_{\alpha\beta}(r_{\alpha\beta}) = 2d_\varepsilon a_\varepsilon \left[e^{2a_\varepsilon (r_0 - r_{\alpha\beta})} - e^{a_\varepsilon (r_0 - r_{\alpha\beta})} \right] \tag{4.3.139}$$

4.3.4.1.4 Born–Mayer Potential

The Born–Mayer potential is a pair potential usually employed for modelling ionic crystals with atomic number Z and z electrons in the outer shell. Setting e_c as the electron charge, the Born–Mayer potential can be written as

$$\phi_2(r_{\alpha\beta}) = \frac{1}{2} \sum_{\substack{\alpha,\beta \\ \beta\neq\alpha}}^{n_p} \left(\frac{Z_\alpha Z_\beta}{r_\alpha} \right) e_c^{\,2} + \frac{1}{2} \sum_{\substack{\alpha,\beta \\ \beta\neq\alpha}}^{n_{near}} A_{\alpha\beta} \left(1 + \frac{Z_\alpha}{z_\alpha} \frac{Z_\beta}{z_\beta} \right) e^{\left[\frac{\sigma_\alpha + \sigma_\beta - r_{\alpha\beta}}{\rho_{\alpha\beta}} \right]} \tag{4.3.140}$$

where A, σ and ρ are the constant parameters of the Born–Mayer model.

The interatomic force $f^{\alpha\beta}$ is obtained by differentiation of (4.3.140).

4.3.4.2 Three-Body Potentials

Two-body potentials do not account for the bond angles and cannot properly represent the potentials of materials with directionality properties such as BCC crystals, polymers and organic molecular structures. Three-body potentials are designed accordingly to better represent the physics of more complicated crystalline structures.

In addition to the pair potential ϕ_2, the general form of atomic potential includes the three-body potential ϕ_3,

$$U^p = \frac{1}{2} \sum_{\substack{\alpha,\beta \\ \beta\neq\alpha}} \phi_2(r_{\alpha\beta}) + \frac{1}{6} \sum_{\substack{\alpha,\beta,\gamma\neq\alpha \\ \alpha\neq\beta\neq\gamma}} \phi_3(r_{\alpha\beta}, r_{\beta\gamma}, r_{\gamma\alpha}) \tag{4.3.141}$$

4.3.4.2.1 Stillinger–Weber (SW) Potential

The Stillinger–Weber three-body potential is described in terms of the angular dependence by

$$U^p = \frac{1}{2} \sum_{\substack{\alpha,\beta \\ \beta\neq\alpha}} \phi_2(r_{\alpha\beta}) + \frac{1}{6} \sum_{\substack{\alpha,\beta,\gamma\neq\alpha \\ \alpha\neq\beta\neq\gamma}} \phi_3(r_{\alpha\beta}, r_{\alpha\gamma}, \theta_{\beta\alpha\gamma}) \tag{4.3.142}$$

with

$$\phi_2\left(r_{\alpha\beta}\right)=f_c\left(r_{\alpha\beta}\right)\left[A_1 e^{-\lambda_1}-A_2 e^{-\lambda_2}\right] \tag{4.3.143}$$

$$\phi_3\left(r_{\alpha\beta},r_{\alpha\gamma},\theta_{\beta\alpha\gamma}\right)=Z\left[f_c\left(r_{\alpha\beta}\right)f_c\left(r_{\alpha\gamma}\right)\right]\left[\cos\theta_{\beta\alpha\gamma}+\frac{1}{3}\right]^2 \tag{4.3.144}$$

where Z, A_1, A_2, λ_1 and λ_2 are the Stillinger–Weber constants. The decaying function f_c is defined as

$$f_c\left(r_{\alpha\beta}\right)=\begin{cases} e^{\left[\frac{\mu}{r_{\alpha\beta}-r^c}\right]}, & r_{\alpha\beta}<r^c \\ 0, & r_{\alpha\beta}\geq r^c \end{cases} \tag{4.3.145}$$

where r^c is the cut-off radius and μ is a constant.

4.3.4.3 Bond-Order Tersoff Potential

The Tersoff potential is somehow a compromise between two and three body potentials in the form of

$$U^p=\frac{1}{2}\sum_{\substack{\alpha,\beta\\ \beta\neq\alpha}}\left[\phi_2^r\left(r_{\alpha\beta}\right)+b\left(z_{\alpha\beta}\right)\phi_2^a\left(r_{\alpha\beta}\right)\right] \tag{4.3.146}$$

where ϕ_2^r and ϕ_2^a are repulsive and attractive pair potentials, respectively. In the Tersoff potential, the attractive term ϕ_2^a is modified by the bond-order function $b(z_{\alpha\beta})$.

The function $z_{\alpha\beta}$ is the coordination function for a bond

$$z_{\alpha\beta}=\frac{1}{2}\sum_{\substack{\gamma\\ \gamma\neq\alpha,\beta}}\left[f_c\left(r_{\alpha\gamma}\right)g\left(\theta_{\alpha\beta\gamma}\right)e^{\left[\lambda_3\left(r_{\alpha\beta}-r_{\alpha\gamma}\right)\right]^3}\right] \tag{4.3.147}$$

where f_c represents the cut-off and g accounts for the bond-angle dependence (Tersoff 1988).

4.3.4.4 The Embedded Atom Potential (EAM)

The embedded atom model (EAM) can provide an accurate description based on the environment dependence of bondings and phase transformations. The concept has been used to develop a number of more sophisticated potentials such as the effective medium theory (EMT) and the Finnis–Sinclair (FS) potential. These potentials are generally suitable for most metals, but they are inconvenient for materials with covalent bonding.

In the EAM class of potentials, a pair potential may be generalized in the form of pair functionals or a glue model,

$$U^p=\frac{1}{2}\sum_{\substack{\alpha,\beta\\ \beta\neq\alpha}}\phi_2\left(r_{\alpha\beta}\right)+\sum_{\alpha}\phi_e\left(\rho_\alpha\right) \tag{4.3.148}$$

where ϕ_e represents the energy to embed atom α in the local electronic density ρ_α around atom α, which is contributed by the nearby atoms by $g(r_{\alpha\beta})$:

$$\rho_\alpha = \sum_{\substack{\beta \\ \beta\neq\alpha}} g\left(r_{\alpha\beta}\right) \tag{4.3.149}$$

4.3.4.5 Interatomic Potentials for Polymers

Polymers typically consist of repeated units of long macromolecules, linked to each other with mainly covalent forces. Various types of natural or synthetic polymers and organic or inorganic polymers are available.

Simulation of large polymeric materials may be conducted using the coarse grain multi-scale method, where each group of atoms is represented by an equivalent hypothetical pseudo-atom. On the other hand, atomistic modelling of polymers can be accomplished by general reactive interatomic potentials, such as MEAM, or by pre-set bonded potentials, such as the CHARRM models.

4.3.4.5.1 CHARMM Potential for Polymers

The CHARMM potential is defined as a combination of bonded and non-bonded potentials,

$$U^p = U^{\text{bonded}} + U^{\text{non-bonded}} \tag{4.3.150}$$

where the bonded part is a full four-body potential for the n_b set of bonded atoms in the form of

$$\begin{aligned} U^{\text{bonded}} &= \frac{1}{2}\sum_{\substack{\alpha,\beta \\ \alpha\neq\beta}}^{n_\text{b}} k_{\alpha\beta}\left(r_{\alpha\beta} - r^0_{\alpha\beta}\right)^2 + \frac{1}{6}\sum_{\substack{\alpha,\beta,\gamma \\ \alpha\neq\beta\neq\gamma}}^{n_\text{b}} k^\theta_{\alpha\beta\gamma} k^{\alpha\beta}\left(\theta_{\alpha\beta\gamma} - \theta^0_{\alpha\beta\gamma}\right)^2 \\ &+ \frac{1}{24}\sum_{\substack{\alpha,\beta,\gamma,\delta \\ \alpha\neq\beta\neq\gamma}}^{n_\text{b}} k^\phi_{\alpha\beta\gamma\delta} k_{\alpha\beta}\left(\phi_{\alpha\beta\gamma\delta} - \phi^0_{\alpha\beta\gamma\delta}\right)^2 \end{aligned} \tag{4.3.151}$$

where (Allen 2004)

$$\cos\theta_{\alpha\beta\gamma} = \left(\boldsymbol{r}_{\alpha\beta}\cdot\boldsymbol{r}_{\alpha\beta}\right)^{-\frac{1}{2}}\left(\boldsymbol{r}_{\beta\gamma}\cdot\boldsymbol{r}_{\beta\gamma}\right)^{-\frac{1}{2}}\left(\boldsymbol{r}_{\alpha\beta}\cdot\boldsymbol{r}_{\beta\gamma}\right) \tag{4.3.152}$$

$$\cos\phi_{\alpha\beta\gamma\delta} = -\left(\frac{\boldsymbol{r}_{\alpha\beta}\times\boldsymbol{r}_{\beta\gamma}}{\left|\boldsymbol{r}_{\alpha\beta}\times\boldsymbol{r}_{\beta\gamma}\right|}\right)\cdot\left(\frac{\boldsymbol{r}_{\beta\gamma}\times\boldsymbol{r}_{\gamma\delta}}{\left|\boldsymbol{r}_{\beta\gamma}\times\boldsymbol{r}_{\gamma\delta}\right|}\right) \tag{4.3.153}$$

The non-bonded part can be based on the Lennard-Jones pair potential (and the Coulomb interaction),

$$U^{\text{non-bonded}} = \sum_{\substack{\alpha,\beta \\ \beta\neq\alpha}}^{n_\text{nb}} 2\varepsilon\left[\left(\frac{\sigma}{r_{\alpha\beta}}\right)^{12} - \left(\frac{\sigma}{r_{\alpha\beta}}\right)^{6}\right] \tag{4.3.154}$$

where n_nb is the set of non-bonded atoms. For further details, see (Allen 2004).

4.3.5 Measures for Determining the State of MD Systems

4.3.5.1 Velocity Autocorrelation Function

Any oscillating MD system must be allowed to reach equilibrium. Therefore, a measure or criterion is required to quantify the time an equilibrium state can be considered for any particle system within a general dynamic time-stepping process.

The velocity autocorrelation function (VAF) is a simple measure to assess the state of equilibrium as it approaches zero at equilibrium:

$$\mathrm{VAF}(t) = \frac{1}{n_\mathrm{p}} \sum_{\alpha=1}^{n_\mathrm{p}} \boldsymbol{v}_\alpha(0) \cdot \boldsymbol{v}_\alpha(t) \tag{4.3.155}$$

For a system in equilibrium, the initial value of VAF is

$$\mathrm{VAF}(0) = \frac{3k_B T}{m} \tag{4.3.156}$$

where m is the mass of the particles or atoms.

4.3.5.2 Radial Distribution Function

The radial distribution function $g(r)$ assesses how the density of particles in a system varies as a function of distance with respect to a reference particle. It indirectly indicates the local structure of an atomistic system by determining the probability of finding a pair of atoms with a distance r from each other in a system of n_p particles with density ρ,

$$g(r) = \frac{1}{\rho n_\mathrm{p}} \sum_{\substack{\alpha,\beta \\ \beta \neq \alpha}}^{n_\mathrm{p}} \left\langle \delta\left(r - r_{\alpha\beta}\right) \right\rangle \tag{4.3.157}$$

where $\langle\rangle$ represents the averaging operator where

$$\langle A \rangle_{\mathrm{NVE}} = \frac{\int A(\boldsymbol{r},\boldsymbol{p})\delta\left(E - H(\boldsymbol{r},\boldsymbol{p})\right)\mathrm{d}\boldsymbol{r}\,\mathrm{d}\boldsymbol{p}}{\int \delta\left(E - H(\boldsymbol{r},\boldsymbol{p})\right)\mathrm{d}\boldsymbol{r}\,\mathrm{d}\boldsymbol{p}} \tag{4.3.158}$$

$$\langle A \rangle_{\mathrm{NVT}} = \frac{\int A(\boldsymbol{r},\boldsymbol{p})\mathrm{e}^{-\frac{H(\boldsymbol{r},\boldsymbol{p})}{k_B T}}\,\mathrm{d}\boldsymbol{r}\,\mathrm{d}\boldsymbol{p}}{\int \mathrm{e}^{-\frac{H(\boldsymbol{r},\boldsymbol{p})}{k_B T}}\,\mathrm{d}\boldsymbol{r}\,\mathrm{d}\boldsymbol{p}} \tag{4.3.159}$$

A major advantage of the radial distribution function (RDF) is that in addition to numerical evaluation, RDF can be measured experimentally by using the neutron scattering techniques. Any peak in the RDF–r curve indicates a preferred separation distance of the neighbour atoms to a given atom.

RDF depends on the temperature of the system and may indirectly represent the phase of a system on the edge of solid–liquid state. The radial distribution function for the crystalline structures at a temperature below the melting point is similar to a series of Dirac-delta functions, whereas it is transformed to Gaussian functions after melting, as typically depicted in Figure 4.25.

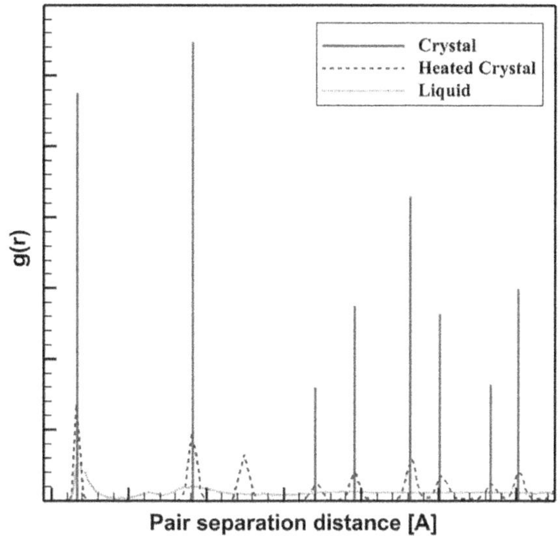

Figure 4.25 Typical RDF results for an amorphous polymer in a crystal condition, its heated state and liquid phase. *Source:* Adapted from Moslemzadeh and Mohammadi 2022.

4.3.5.3 Mean Square Displacement Function (MSD)

Another measure that may be used to distinguish the solid and fluid phases is the mean square displacement function (MSD), defined as a measure of the transport property:

$$\Delta r^2 = \frac{1}{n_p} \sum_{\alpha=1}^{n_p} \left[r_\alpha(t) - r_\alpha(t=0) \right]^2 \tag{4.3.160}$$

In solids, MSD oscillates around a mean value, whereas it varies linearly in time in liquids.

4.3.6 Stress Computation in MD

If a microscale MD solution is performed to determine the constitutive relation of a specific material, the stress tensor, which is a fundamental part of the macroscale constitutive relation, should be computed. However, the discrete nature of atomistic and molecular analysis does not allow for direct computation of stress, which is a concept at the continuum level. Analogous to other macroscale parameters, stress can be computed similar to a macro constraint using the averaging technique on microscale solutions.

Having known the position $\boldsymbol{r}_\alpha(t)$ from the MD solution, the force $\boldsymbol{f}_\alpha(t)$ of each atom α at time t can be computed from

$$\boldsymbol{f}_\alpha = -\frac{\partial \Pi}{\partial \boldsymbol{r}_\alpha} = -\frac{\partial U^p\left(\boldsymbol{r}_1, \boldsymbol{r}_2, \ldots, \boldsymbol{r}_{n_p}\right)}{\partial \boldsymbol{r}_\alpha} + \boldsymbol{f}_\alpha^{exp} \tag{4.3.161}$$

Computation of the stress measures in a large deformation problem is based on the known deformation gradient $\boldsymbol{F}$. Assuming $\boldsymbol{R}_\alpha$ and $\boldsymbol{r}_\alpha$ as the positions of atoms in the reference and deformed configurations, the first Piola–Kirchhof stress $\boldsymbol{P}_{iJ}$ can be computed from

$$\boldsymbol{P}_{iJ} = \frac{1}{V_0}\frac{\partial U}{\partial \boldsymbol{F}} = \frac{1}{V_0}\sum_{\alpha}^{n_p} \boldsymbol{f}_i^\alpha \boldsymbol{R}_J^\alpha \tag{4.3.162}$$

and the Cauchy stress is obtained from

$$\boldsymbol{\sigma}_{ij} = \frac{1}{\det \boldsymbol{F}} \boldsymbol{P}_{iJ}\boldsymbol{P}\boldsymbol{F}_{jJ} = -\frac{1}{V}\sum_{\alpha}^{n_p} \boldsymbol{f}_i^\alpha \boldsymbol{r}_j^\alpha \tag{4.3.163}$$

4.3.6.1 IKN Stress

For general non-equilibrium MD problems based on the statistical mechanics, the Irving–Kirkwood–Noll (IKN) stress derivation can be followed. Alternatively, the Hardy numerical sampling technique, the virial uniform weighting approach and the Tsai traction can be adopted (Wahnstrom 2018).

In the IKN derivation for all interatomic potentials φ, the symmetric stress tensor $\boldsymbol{\sigma}(\boldsymbol{x})$ is decomposed to

$$\boldsymbol{\sigma}(\boldsymbol{x}) = \boldsymbol{\sigma}_k(\boldsymbol{x}) + \boldsymbol{\sigma}_v(\boldsymbol{x}) \tag{4.3.164}$$

where $\boldsymbol{\sigma}_k$ and $\boldsymbol{\sigma}_v$ are, respectively, the average momentum transfer and force in bonds passing through point $\boldsymbol{x}$,

$$\boldsymbol{\sigma}_k(\boldsymbol{x};t) = -\sum_{\alpha} \langle m_\alpha (\Delta \boldsymbol{v}_\alpha \otimes \Delta \boldsymbol{v}_\alpha)\delta(\Delta \boldsymbol{r}_\alpha - \boldsymbol{x}); f\rangle \tag{4.3.165}$$

$$\boldsymbol{\sigma}_v(\boldsymbol{x}) = \frac{1}{2}\sum_{\substack{\alpha,\beta \\ \alpha\neq\beta}} \int_{\boldsymbol{z}} \left[\frac{\boldsymbol{z}\otimes\boldsymbol{z}}{\|\boldsymbol{z}\|} \int_{s=0}^{1} \left\langle \varphi_{\alpha\beta}\delta(\Delta \boldsymbol{r}_\alpha - \boldsymbol{x}^+)\delta(\Delta \boldsymbol{r}_\beta - \boldsymbol{x}^-); f \right\rangle \mathrm{d}s \right] \mathrm{d}\boldsymbol{z} \tag{4.3.166}$$

and

$$\begin{cases} \boldsymbol{x}^+(\boldsymbol{z},s) = \boldsymbol{x} + s\boldsymbol{z} \\ \boldsymbol{x}^-(\boldsymbol{z},s) = \boldsymbol{x} - (1-s)\boldsymbol{z} \end{cases} \tag{4.3.167}$$

where $\boldsymbol{z} = \boldsymbol{r}^{\alpha\beta}$ is the bond vector.

To obtain the macroscopic stress tensor, the spatial averaging technique is employed:

$$\bar{\boldsymbol{\sigma}}(\boldsymbol{x};t) = \bar{\boldsymbol{\sigma}}_k(\boldsymbol{x};t) + \bar{\boldsymbol{\sigma}}_v(\boldsymbol{x};t) \tag{4.3.168}$$

with

$$\bar{\boldsymbol{\sigma}}_k(\boldsymbol{x};t) = \int_{\boldsymbol{y}} w(\boldsymbol{y}-\boldsymbol{x})\boldsymbol{\sigma}_k(\boldsymbol{y},t)d\boldsymbol{y} \tag{4.3.169}$$

$$\bar{\boldsymbol{\sigma}}_v(\boldsymbol{x};t) = \int_{\boldsymbol{y}} w(\boldsymbol{y}-\boldsymbol{x})\boldsymbol{\sigma}_v(\boldsymbol{y},t)d\boldsymbol{y} \tag{4.3.170}$$

where $w(\boldsymbol{y}-\boldsymbol{x})$ is a pre-defined weight function based on the macroscale measurement characteristics, schematically shown in Figure 4.26.

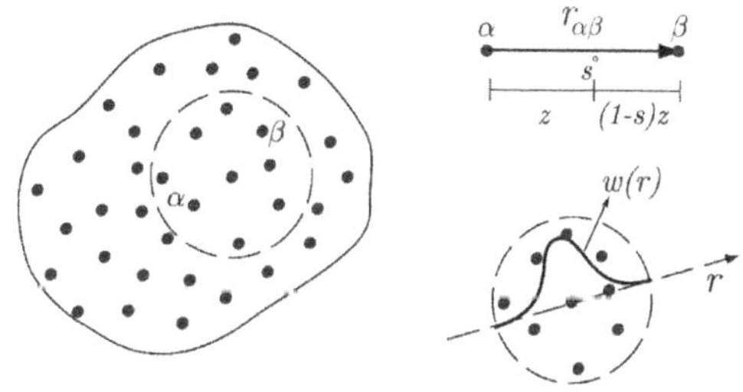

Figure 4.26 Schematic representation of the weighting $w(r)$ in the IKN stress measure.

4.3.6.2 Hardy Stress

In the Hardy stress, the full phase averaging process is replaced by a numerical sampling approach,

$$\langle A;f\rangle=\frac{1}{n_s}\sum_{i=1}^{n_s}A(\boldsymbol{r}_i,\boldsymbol{p}_i) \qquad (4.3.171)$$

where n_s represents the number of samples. The Hardy stress for a system of atoms is then obtained from the IKN concept as

$$\bar{\boldsymbol{\sigma}}_k(\boldsymbol{x})=-\frac{1}{n_s}\sum_{i=1}^{n_s}\sum_{\alpha}m_\alpha w\left(\boldsymbol{r}_\alpha^i-\boldsymbol{x}\right)\left(\Delta\boldsymbol{v}_\alpha^i\otimes\Delta\boldsymbol{v}_\alpha^i\right) \qquad (4.3.172)$$

$$\bar{\boldsymbol{\sigma}}_v(\boldsymbol{x})=\frac{1}{2n_s}\sum_{i=1}^{n_s}\sum_{\substack{\alpha,\beta\\ \alpha\neq\beta}}\frac{\boldsymbol{r}_{\beta\alpha}^i\otimes\boldsymbol{r}_{\beta\alpha}^i}{r_{\beta\alpha}^i}\varphi_{\alpha\beta}^{i}B\left(\boldsymbol{x};\boldsymbol{r}_\alpha^i,\boldsymbol{r}_\beta^i\right) \qquad (4.3.173)$$

where the bond function B is defined as

$$B\left(\boldsymbol{x};\boldsymbol{y}^+,\boldsymbol{y}^-\right)=\int_0^1 w\left[\boldsymbol{y}^+(1-s)+s\boldsymbol{y}^--\boldsymbol{x}\right]ds \qquad (4.3.174)$$

4.3.6.3 The Virial (Clausius) Stress

The virial stress is obtained from the Hardy stress by a further simplification of adopting a uniform weighting function $w(\boldsymbol{y}-\boldsymbol{x})=1/V$,

$$\bar{\boldsymbol{\sigma}}_k(\boldsymbol{x})=-\frac{1}{Vn_s}\sum_{i=1}^{n_s}\sum_{\alpha}m_\alpha\left(\Delta\boldsymbol{v}_\alpha^i\otimes\Delta\boldsymbol{v}_\alpha^i\right) \qquad (4.3.175)$$

$$\bar{\boldsymbol{\sigma}}_v(\boldsymbol{x})=\frac{1}{2Vn_s}\sum_{i=1}^{n_s}\sum_{\substack{\alpha,\beta\\ \alpha\neq\beta}}\frac{\boldsymbol{r}_{\beta\alpha}^i\otimes\boldsymbol{r}_{\beta\alpha}^i}{r_{\beta\alpha}^i}\varphi_{\alpha\beta}^{i} \qquad (4.3.176)$$

The virial stress is one of the most widely used descriptions of stress in atomistic simulations.

4.3.6.4 The Tsai Traction

The Tsai formulation for computing traction $\boldsymbol{t}(\boldsymbol{n}) = \bar{\sigma}\boldsymbol{n}$ on plane $\boldsymbol{n}$ can be expressed as

$$\boldsymbol{t}(\boldsymbol{n}) = \frac{1}{An_s}\sum_{i=1}^{n_s}\left\{-\frac{1}{\Delta\tau}\sum_{\alpha} m_\alpha \Delta\boldsymbol{v}_\alpha^i \frac{\Delta\boldsymbol{v}_\alpha^i \cdot \boldsymbol{n}}{\left\|\Delta\boldsymbol{v}_\alpha^i \cdot \boldsymbol{n}\right\|} + \sum_{\alpha\beta} \boldsymbol{f}_{\beta\alpha}^i \frac{\boldsymbol{r}_{\beta\alpha}^i \cdot \boldsymbol{n}}{\left\|\boldsymbol{r}_{\beta\alpha}^i \cdot \boldsymbol{n}\right\|}\right\} \tag{4.3.177}$$

where $\Delta\tau$ is the time increment.

4.3.7 Molecular Statics

The molecular statics (MS) is an alternative to the computationally expensive step-by-step molecular dynamics solution. It avoids solving the dynamic equations of motion by direct energy minimization in a quasi-static framework.

The potential energy surface is highly non-convex with numerous minima corresponding to different microscale characteristics of the bulk material (Stillinger and Weber 1983). Instead of a cumbersome computation of the global minimum, which is associated with the lowest minimum on the energy surface, a local minimum may be reached from any given initial guess $\boldsymbol{r}^{(0)}$:

$$\boldsymbol{r}^{(0)} \rightarrow \boldsymbol{r}_{\min} = \min_{\boldsymbol{r}\in C^0} U^p(\boldsymbol{r}) \tag{4.3.178}$$

and an iterative conjugate gradient approach may be adopted by computing a search direction $\boldsymbol{d}^{(i)}$ at iteration i:

$$\boldsymbol{d}^{(i)} = \boldsymbol{f}^{(i)} + \beta^{(i)}\boldsymbol{d}^{(i-1)} \tag{4.3.179}$$

$$\beta^{(i)} = \frac{\boldsymbol{f}^{(i)}\cdot\boldsymbol{f}^{(i)} - \boldsymbol{f}^{(i)}\cdot\boldsymbol{f}^{(i-1)}}{\left\|\boldsymbol{f}^{(i-1)}\right\|^2} \tag{4.3.180}$$

with

$$\boldsymbol{f}^{(i)} = -\frac{\partial U^p(\boldsymbol{r})}{\partial \boldsymbol{r}^{(i)}} \tag{4.3.181}$$

to improve the solution to $\boldsymbol{r}^{(i+1)}$ by a line minimization using the step length $\alpha^{(i)}$:

$$\boldsymbol{r}^{(i+1)} = \alpha^{(i)}\boldsymbol{d}^{(i)} \tag{4.3.182}$$

The iterative procedure terminates if a pre-defined convergence criterion is met:

$$\left\|\frac{\partial U^p}{\partial \boldsymbol{r}}\boldsymbol{r}^{(i+1)}\right\| < \text{tolerance} \tag{4.3.183}$$

4.3.8 Sample MD Simulation of a Polymer

As an example of molecular dynamics simulation of polymers, a system containing five chains of polyethylene, consisting of 100 monomers each (100 united atoms) is simulated. One methyl group (i.e., the CH_2 monomer) is considered as a pseudo-atom and the force field establishes the interactions between the pseudo-atoms.

The orthogonal simulation cell has an initial length of 5 nm in each direction. The initial chain structure is created by the Avogadro open-source molecular builder and visualization tool (Hanwell et al. 2012), and the five chains are packed inside the simulation box by the Packmol package (Martínez et al. 2009), as illustrated in Figure 4.27.

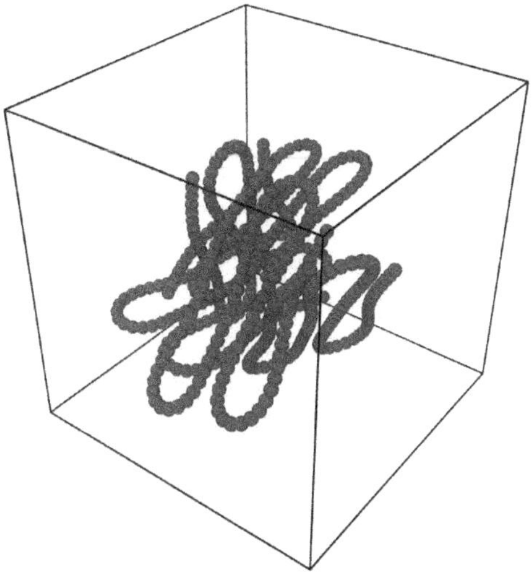

Figure 4.27 Initial configuration, pre-equilibrium (visualized by OVITO. *Source:* Reproduced from Vokhshoori and Mohammadi 2022/University of Tehran).

The inter-pseudoatomic interactions are described by the Dreiding potential (Mayo et al. 1990), which consists of four terms associated with the bond stretching, changes in the bond angle, changes in the dihedral rotation, and the van der Waals non-bonded interactions (Hossain et al. 2010):

$$\text{Bond length}: E_{\text{bond}}(r) = \frac{1}{2} K_b (r - r_0)^2 \tag{4.3.184}$$

$$\text{Bond angle}: E_{\text{angle}}(\theta) = \frac{1}{2} K_\theta (\theta - \theta_0)^2 \tag{4.3.185}$$

$$\text{Dihedral angle}: E_{\text{dihedral}}(\phi) = \sum_{i=0}^{3} K_i (\cos\phi)^i \tag{4.3.186}$$

$$\text{Non-bonded}: E_{\text{non-bonded}}(r) = 4\varepsilon\left[\left(\frac{\sigma}{r}\right)^{12} - \left(\frac{\sigma}{r}\right)^{6}\right], r \le r \tag{4.3.187}$$

The constants of the Dreiding potential are defined in Table 4.1.

Table 4.1 Constants of the Dreiding potential.

$K_b = 350$ kcal/mol	$r_0 = 1.53$ Å		
$K_\theta = 60$ kcal/mol/rad^2	$\theta_0 = 1.911$ rad		
$K_0 = 1.736$ kcal/mol	$K_1 = -4.490$ kcal/mol	$K_2 = 0.776$ kcal/mol	$K_3 = 6.990$ kcal/mol
$\sigma = 4.01$ Å	$\varepsilon = 0.112$ kcal/mol		

Source: Adapted from Hossain et al. 2010/John Wiley & Sons Ltd.

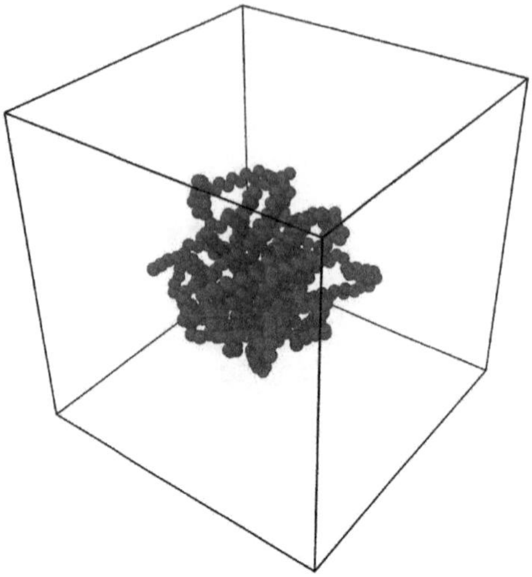

Figure 4.28 Post-relaxation configuration (equilibrium state at 100 K). *Source:* Reproduced from Vokhshoori and Mohammadi 2022/ University of Tehran.

The LAMMPS molecular dynamics (Plimpton 1995) is then adopted for MD analyses of relaxation and stretch simulations. After an initial minimization step of the conjugate gradient style, the relaxation phase involves the following five different stages from a uniform distribution of temperature at 500 K ($\Delta t = 0.001\text{fs}$):

- Relaxation at initial temperature of 500 K by NVT for 250,000 timesteps.
- Relaxation at 500 K by NPT for 550,000 timesteps.
- Cooling down the model to the desired temperature of 100 K by NPT over 700,000 timesteps.
- Relaxation at 100 K by the same NPT ensemble for 250,000 timesteps.
- Relaxation at 100 K temperature by the NVT ensemble.

The final state of post-relaxation configuration is depicted in Figure 4.28.

After the equilibrium is achieved, the polyethylene system is subjected to a uniaxial tensile deformation on the simulation box in the *x*-direction at a strain rate of $1\times10^{13}\text{s}^{-1}$. The deformed configuration of the chains of polyethylene is illustrated in Figure 4.29.

To examine the equilibrium process of the polyethylene structure, the glass transition temperature T_g, which indicates the onset of transition from the glassy structure to the rubbery state (Gee and Boyd 1998; Capaldi et al. 2004; Hossain et al. 2010), is determined from the

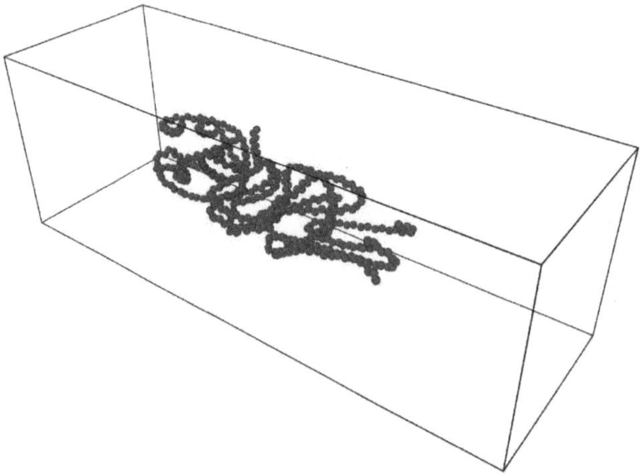

Figure 4.29 Post-deformation state, 100% uniaxial tension along the *x*-direction. *Source:* Reproduced from Vokhshoori and Mohammadi 2022/University of Tehran.

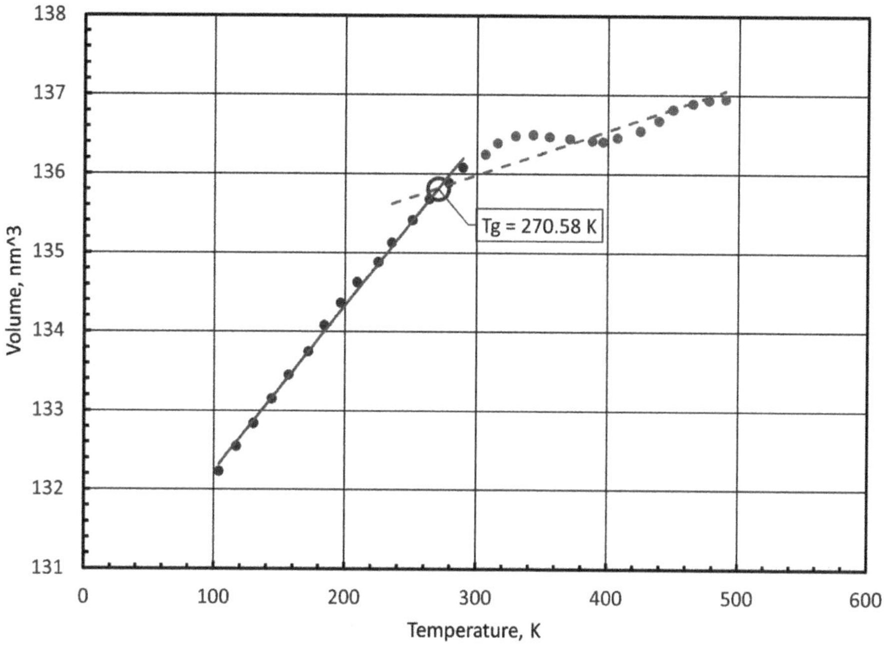

Figure 4.30 Variations of the volume vs. temperature during the cooling process. *Source:* Reproduced from Vokhshoori and Mohammadi 2022/University of Tehran.

variation of the specific volume with respect to the temperature, as presented in Figure 4.30. The intersection of the two linear lines of best fit of the data gives an estimate of 270 K for the glass transition temperature, which is in agreement with available data in the literature (Hossain et al. 2010).

4.4 Sequential Multiscale Method

4.4.1 Introduction

Sequential multiscale methods deal with multiple scale problems in a sequential uncoupled procedure. The main concept is to obtain the material response at a scale and export it to another scale for more accurate analysis.

The procedure may be designed from lower scales to higher ones, or vice versa. For instance, a nanoscale specimen with all details of the nano structure, and possibly with chemical reactions, is analysed to determine the overall stress–strain response at the nanoscale specimen. It can then be used as a constitutive relation to analyse a micro- or mesoscale model. The same procedure may be followed to obtain the response of a macro problem that consists of micro- or mesoscale components. Solution at each scale is based on the results of lower scales but is not related or coupled to higher scales. It can be regarded as a simple uncoupled homogenization procedure at each scale to obtain its equivalent overall response. Alternatively, a sequential multiscale solution may be designed from an upper scale to lower ones. For instance, a heart valve subject to blood pressure and other potential effects may be analysed in the macroscale to determine the critical points of the bio-organ or bio-tissue. Then, a microscale model based on the microbio-structure of the critical point may be constructed and analysed to study the effects on bio-constituents, such as collagen fibres, etc. The procedure may be repeated further down to the lower scales, i.e., the cell scale, to determine the corresponding deformation and other measures that are necessary, in biological concepts, to describe the state of the specific cell. Again, the solution for each scale is based on the conditions applied from higher scales, but is not related or coupled to lower scales. A full description of the sequential multiscale solution of a heart valve is presented in Section 5.5.

A sample analysis of a carbon nanotube (CNT)-reinforced concrete specimen is presented in this section to show the way a low-to-high sequential multiscale solution can be performed. No specific multiscale formulation is necessary, and only brief descriptions of the modelling for each scale are provided. The materials provided here follow the published works of Eftekhari et al. (2013, 2014, 2018), Eftekhari and Mohammadi (2016a, 2016b) and Eftekhari (2015).

4.4.2 Multiscale Modelling of CNT Reinforced Concrete

Figure 4.31 illustrates the general procedure of sequential multiscale modelling of a concrete specimen with CNT additives subject to extreme loading conditions at the macroscale. The procedure begins by molecular dynamic modelling of intact and defected CNTs. Then, at a nanoscale simulation, the combined behaviour of calcium silicate hydrate (CSH) and CNT is analysed to determine the CSH response and the bridging effects of CNT fibres. The microscale analysis is performed to predict the potential damages and stress–strain response of CNT-reinforced cement at the microscale. At the mesoscale, the composite

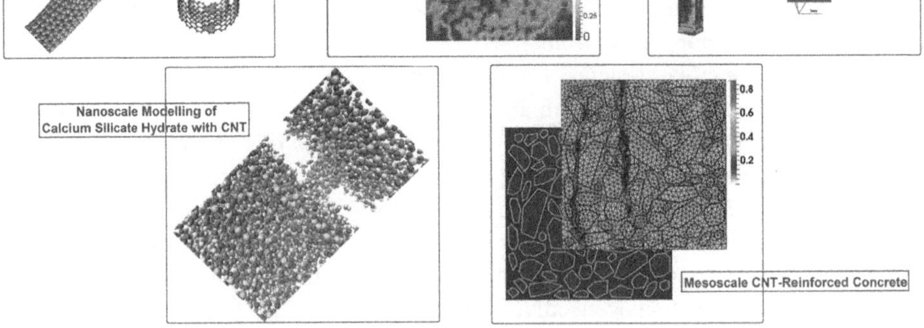

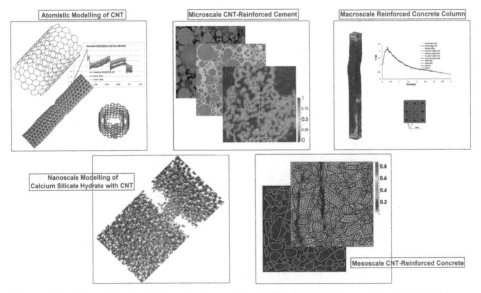

Figure 4.31 Schematic presentation for the multiscale simulation of the cyclic behaviour of a CNT-reinforced concrete column.

concrete paste and aggregates are modelled to determine the equivalent constitutive relation for modelling the macro concrete specimen to investigate its ductility and energy dissipative mechanisms in lateral cyclic loadings or its strength against extreme mechanical conditions.

CNTs are made from rolled graphene sheets in the forms of a zigzag, armchair, etc., as typically presented in Figure 4.32.

CNTs may suffer from a variety of defects in their nano structure, mostly in the forms of vacancy and SW defects, as illustrated in Figure 4.33. The tensile and compressive nano response of CNTs largely depend on the type of defects and may show forms of fracture or rupture and modes of buckling instability.

Moreover, CNTs may be constructed from a single tube (called SWCNT) or they may be multiwalled (MWCNT), as depicted in Figure 4.34, with different properties and more complex forms of defects.

4.4.3 Molecular Dynamics Simulation of CNTs

4.4.3.1 Basic Properties of CNTs

The first stage of a multiscale analysis of a CNT reinforced concrete specimen is to determine the nanoscale stress–strain response of CNT. The open-source code LAMMPS

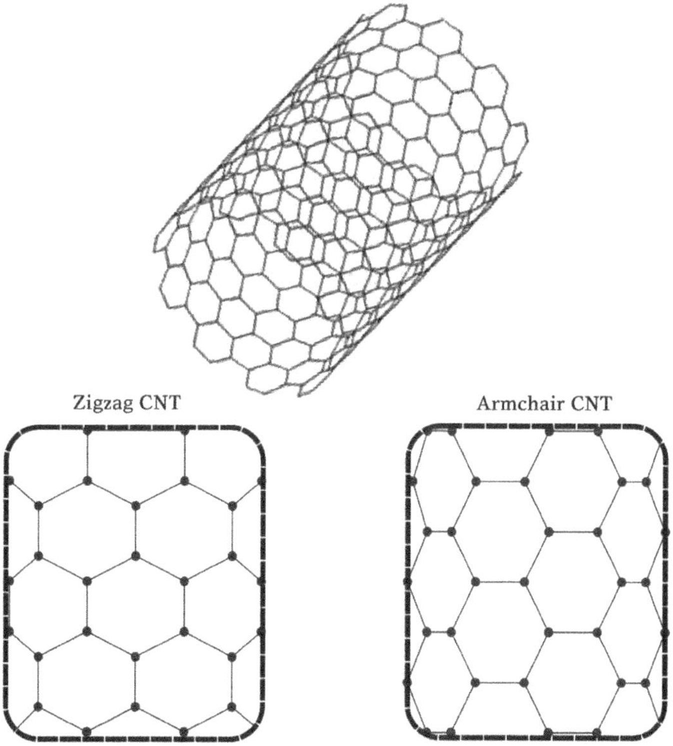

Figure 4.32 A CNT with zigzag and armchair rollings.

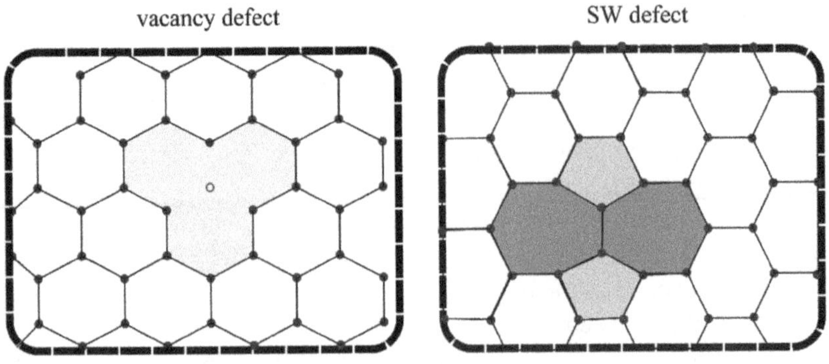

Figure 4.33 Topological vacancy and SW defects in CNTs.

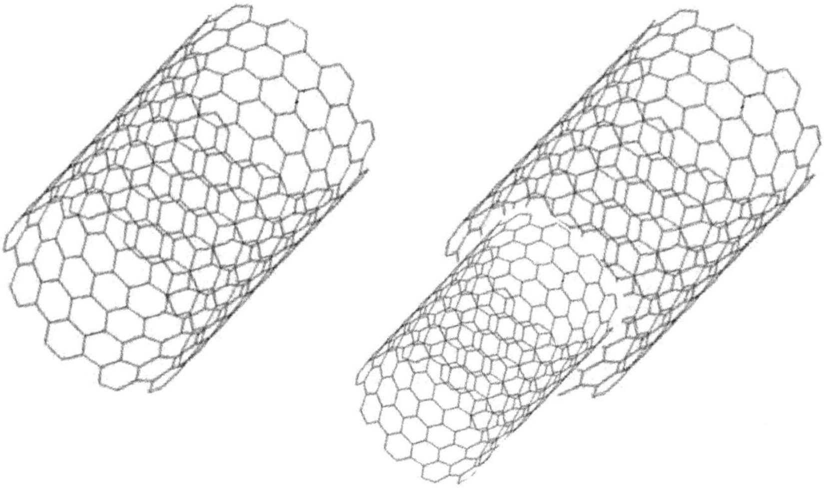

Figure 4.34 Single and multi (double)-walled CNTs.

(Plimpton 1995), among many others, is usually adopted to analyse the molecular structure of CNTs. Analyses are performed on several types of CNTs with lengths of 100 Å at room temperature (300 K). The Tersoff potential (Section 4.3.4.3) is adopted to define the interaction between the carbon atoms, while the Lennard–Jones (LJ) potential governs the non-bonding interactions between the different tubes of MWCNTs. The constant parameters of the Tersoff and LJ potentials are defined in Table 4.2. A rigid cut-off is considered for the Tersoff potential and the cut-off range for the LJ potential is assumed to be 8.5 Å.

MD simulations begin by relaxing the CNT model at 300 K temperature by the NVT ensemble to reach an initial equilibrium state. Having fixed the atoms at one side of CNT, independent tensile and compressive simulations are performed by applying the axial strains on the other end of the CNT with a constant velocity of 0.01 Å/fs. The time step is set to 1 fs.

Table 4.2 Constant parameters of the Tersoff and LJ potentials.

$A = 1.3936 \times 10^3$ eV	$\lambda = 3.4879$ Å^{-1}	$R = 1.8$ Å	$\varepsilon_{C-C} = 0.00239$ eV
$B = 3.467 \times 10^2$ eV	$\mu = 2.2119$ Å^{-1}	$S = 2.1$Å	$\sigma_{C-C} = 0.34$ nm

Source: Adapted from Xiao et al., 2007, Chandra and Namilae, 2006, Eftekhari et al., 2013.

In order to obtain the stress–strain constitutive responses of CNTs, an equivalent cross-section of the double wall carbon nanotube (DWCNT) is adopted according to

$$A = \pi\left[\left(r_{\text{out}} + \frac{t_c}{2}\right)^2 - \left(r_{\text{in}} - \frac{t_c}{2}\right)^2\right] \tag{4.4.1}$$

where r_{in} and r_{out} are the radius of the inner and outer tubes of DWCNT, respectively, and t_c is the assumed tube wall thickness. In this study, this thickness is set to 0.34 nm (Hao et al. 2008).

4.4.3.2 Tensile Response of CNTs with Defects

The tensile deformation of CNT may be localized in some parts of the CNT, especially where an initial defect exists. Such a localization may eventually lead to tensile fracture and rupture, as presented in Figure 4.35.

The typical tensile stress–strain responses of armchair and zigzag CNTs are presented in Figure 4.36. In practical simulations, the mean value of such results can be employed to define the tensile constitutive relation for CNT fibres at higher scale simulations.

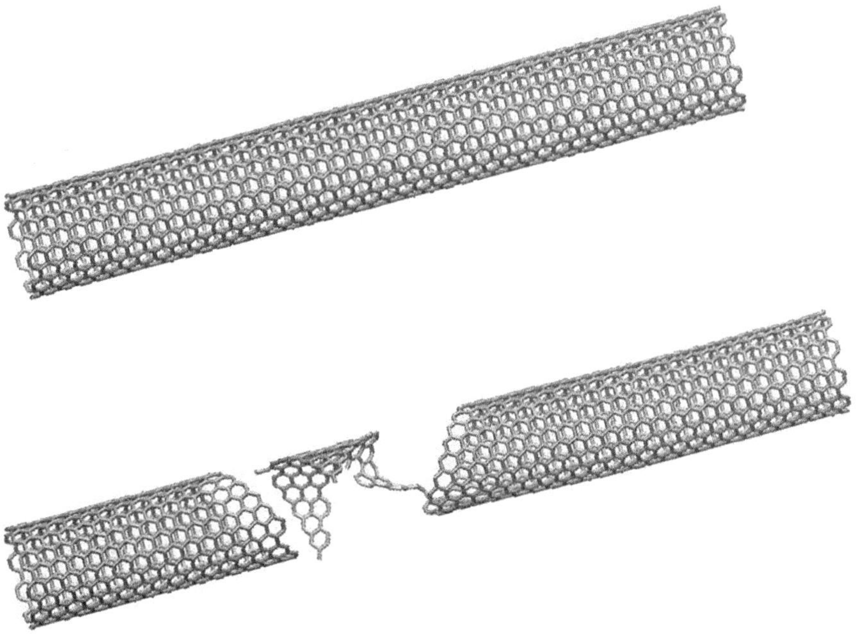

Figure 4.35 Typical tensile deformation and tearing of a CNT.

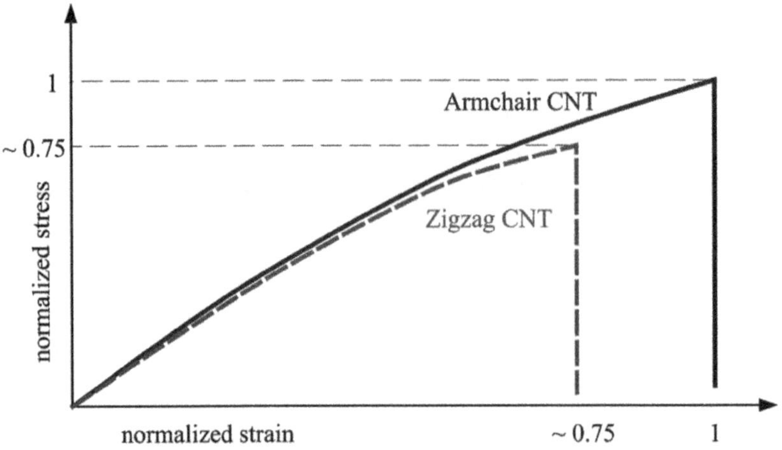

Figure 4.36 Tensile stress–strain curves for armchair and zigzag CNTs (normalized by the maximum values of the armchair CNT).

4.4.3.3 Compressive Response of DWCNTs with a Single Defect on Each Wall

Armchair [10,10]/[15,15] and zigzag [17,0]/[26,0] double-walled carbon nanotubes with the length of 100 Å are considered. A defect on each wall of the DWCNT is assumed, as illustrated in Figure 4.37.

Compressive CNTs are vulnerable to buckling in a wide variety of modes. Local or global buckling modes may dominate the response of CNTs. They could be initiated by the inner or outer wall deformation/buckling or may occur simultaneously. For embedded CNTs, however, numerical studies have shown that various local modes of buckling are likely to determine the overall compressive response of the material. Moreover, a second state of

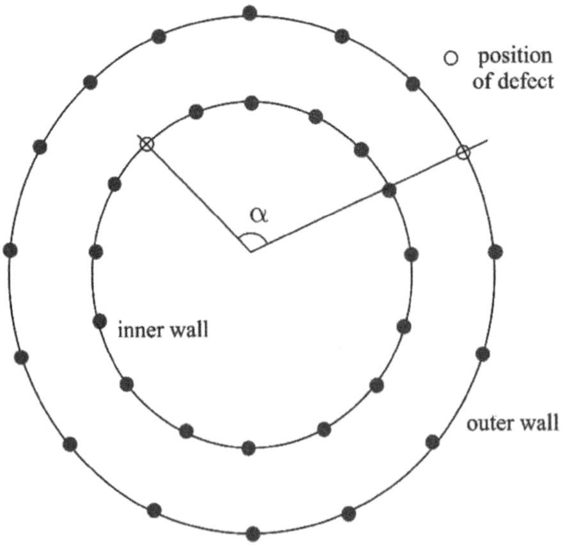

Figure 4.37 Definition of relative positions of two defects on a typical DWCNT.

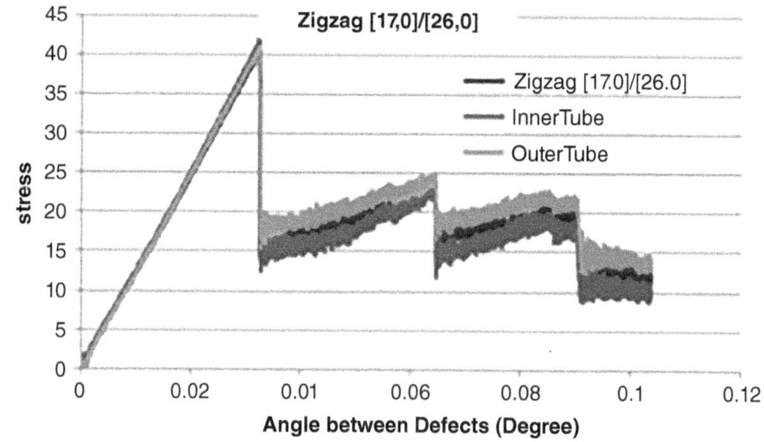

Figure 4.38 A typical compressive stress–strain response of CNT, showing multiple consecutive states of buckling. *Source:* Reproduced from Eftekhari 2015/Islamic Azad University.

buckling (and usually more) is observed by further compressive straining of the CNT (Figure 4.38). These additional states of buckling are related to the local modes of buckling that are triggered by the existence of defects.

The buckling response of MWCNTs becomes more complicated as the occurrence of each state of buckling virtually forms a new setup of connected CNTs, which may result in new local and even global modes of buckling for each part of the CNT (Figure 4.39). In fact, the basic configuration of the CNT is lost in buckling due to very large out-of-plane deformations.

The results of buckling stress–strain responses of armchair [10,10]/[15,15] and zigzag [17, 0]/[26,0] DWCNTs (length of 100 Å) for various angles of defects (either vacancy or SW) are presented in Figures 4.40 and 4.41.

Moreover, Figure 4.42 provides the ratio of buckling stress of the defective CNT to the intact one for the same armchair [10,10]/[15,15] and zigzag [17,0]/[26,0] DWCNTs for various angles of defects (either vacancy or SW). Major reductions are observed in most cases.

Similar comparisons can be made for the ratio of the secondary buckling stress of the defected CNT to the intact one for the same CNTs in terms of the angles of defects, as presented in Figure 4.43. Relatively lower levels of reduction than the main buckling stress are observed in most cases.

Finally, variations of the elastic modulus of the defected CNTS (in compression) are examined in Figure 4.44. Clearly, the elastic modulus, as a general mechanical characteristic, is only lightly affected by the defects.

4.4.3.4 Analysis of DWCNT with Multiple Defects

In this section, DWCNTs are assumed to contain multiple vacancy and SW defects, as illustrated in Figure 4.45. Other specifications are similar to the previous section.

The effects of a number of defects on buckling and post-buckling stress–strain responses of armchair [10,10]/[15,15] and zigzag [17, 0]/[26,0] DWCNTs are examined in Figures 4.46 and 4.47.

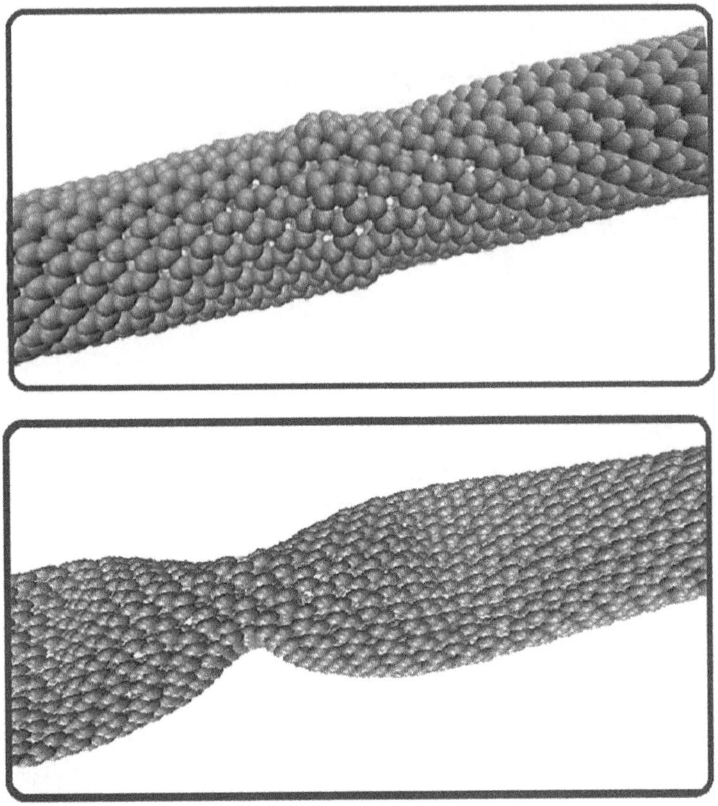

Figure 4.39 Typical deformation and buckling modes in compressive CNTs. *Source:* Reproduced from Eftekhari 2015/Islamic Azad University.

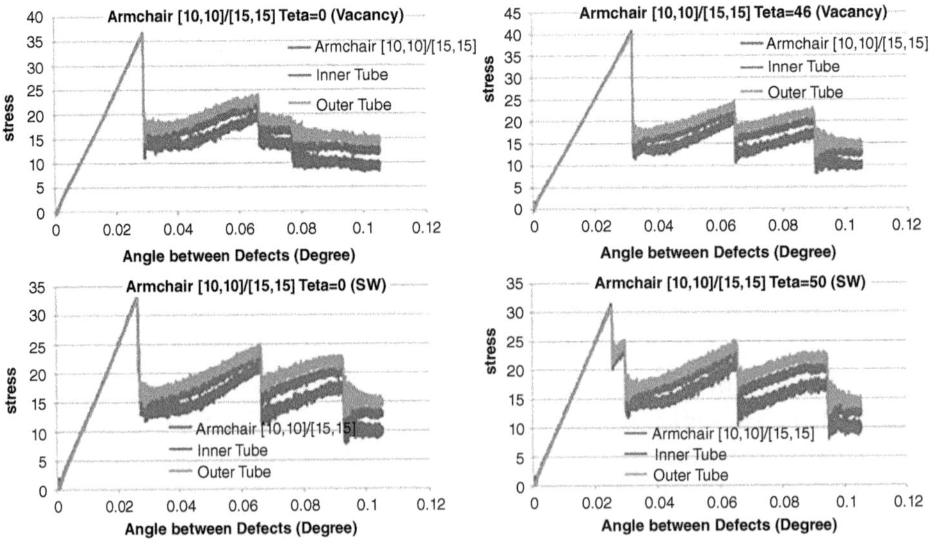

Figure 4.40 Compressive stress–strain response of an armchair [10,10]/[15,15] DWCNT for various angles of vacancy and SW defects. *Source:* Reproduced from Eftekhari 2015/Islamic Azad University.

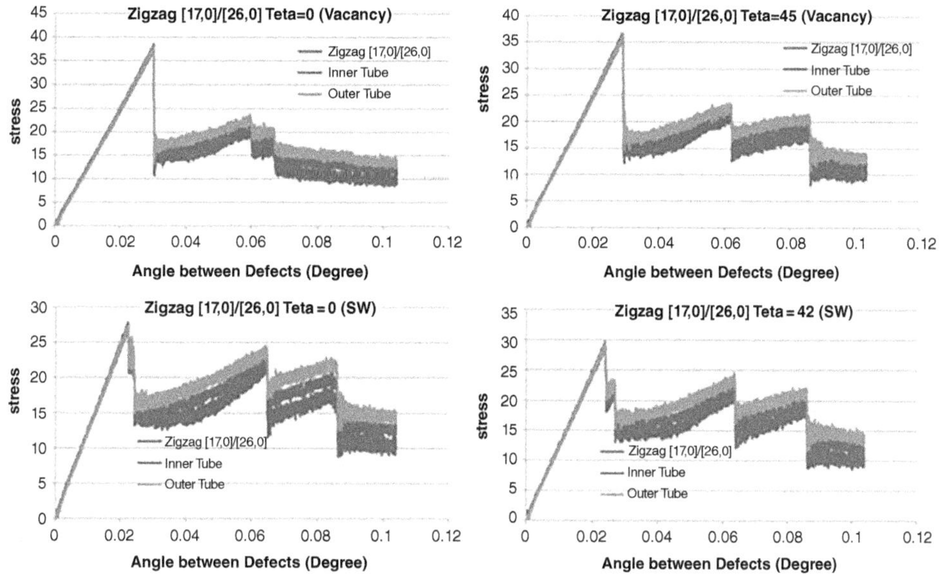

Figure 4.41 Compressive stress–strain response of a zigzag [17, 0]/[26,0] DWCNT for various angles of vacancy and SW defects. *Source:* Reproduced from Eftekhari 2015/Islamic Azad University.

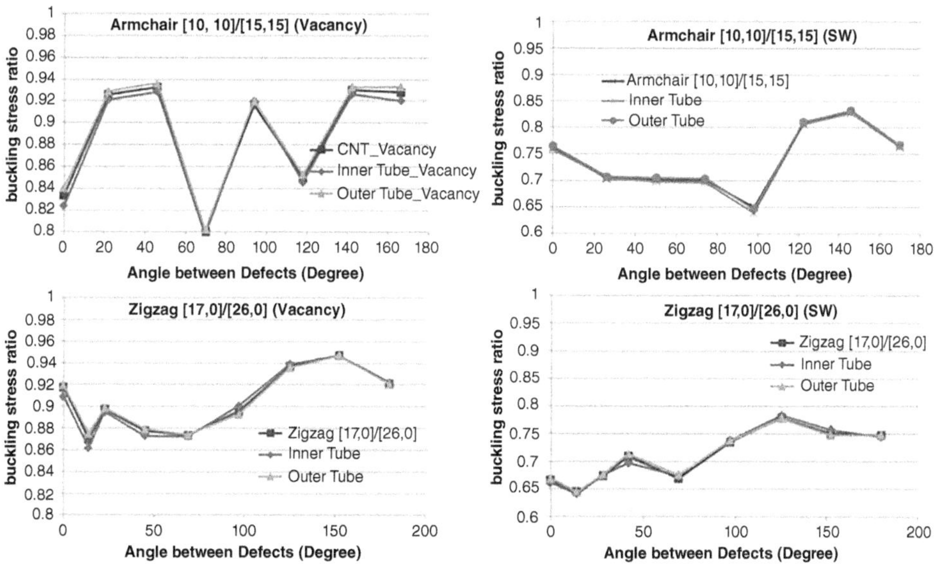

Figure 4.42 The ratio of buckling stress of defected armchair [10,10]/[15,15] and zigzag [17,0]/[26,0] DWCNTs to the intact one for various angles of defects. *Source:* Reproduced from Eftekhari 2015/Islamic Azad University.

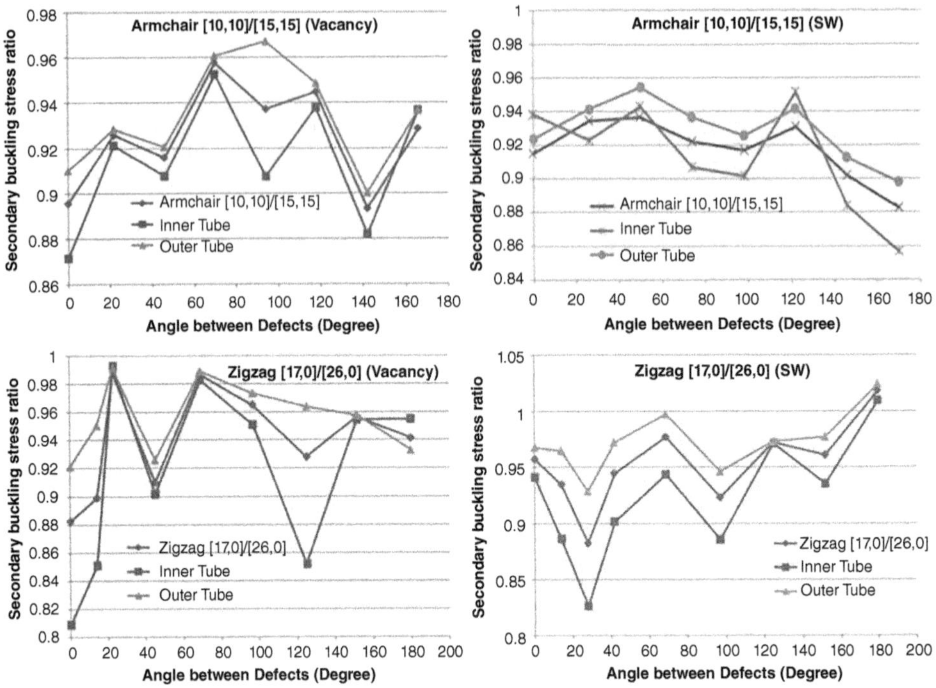

Figure 4.43 The ratio of secondary buckling stress of defected armchair [10,10]/[15,15] and zigzag [17, 0]/[26,0] DWCNTs to the intact one for various angles of defects. *Source:* Reproduced from Eftekhari 2015/Islamic Azad University.

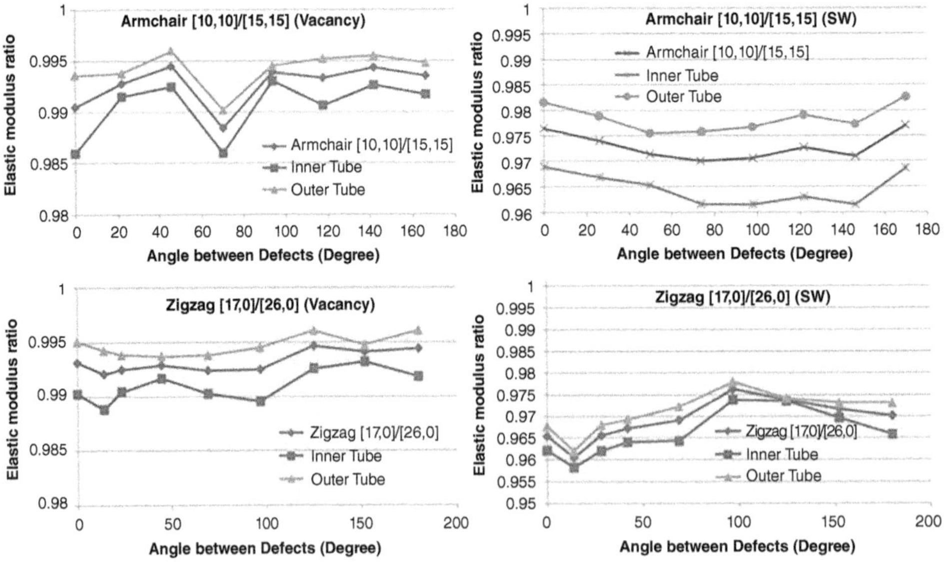

Figure 4.44 The ratio of the elastic modulus of defected armchair [10,10]/[15,15] and zigzag [17, 0]/[26,0] DWCNTs to the intact one for various angles of defects. *Source:* Reproduced from Eftekhari 2015/Islamic Azad University.

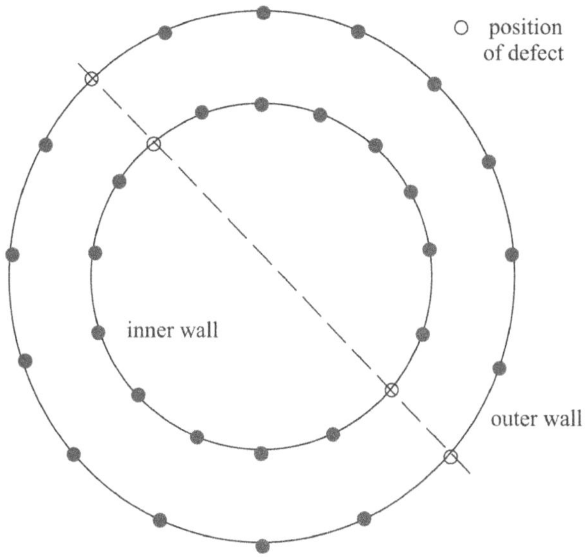

Figure 4.45 DWCNTs with multiple defects on outer and inner walls.

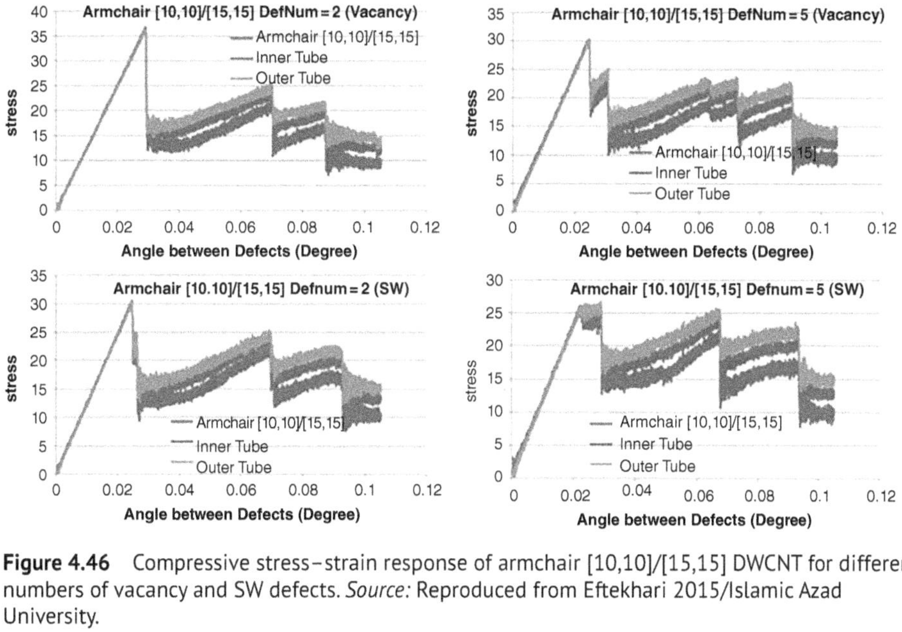

Figure 4.46 Compressive stress–strain response of armchair [10,10]/[15,15] DWCNT for different numbers of vacancy and SW defects. *Source:* Reproduced from Eftekhari 2015/Islamic Azad University.

Similar comparisons can be made for the ratio of the buckling stress of the defected CNT to the intact one, as presented in Figure 4.48 for the armchair [10,10]/[15,15] and zigzag [17, 0]/[26,0] DWCNTs.

Again, comparisons for the ratio of the secondary buckling stress of the defective CNT to the intact one presented in Figure 4.49 show that a relatively lower level of reduction is observed in most cases.

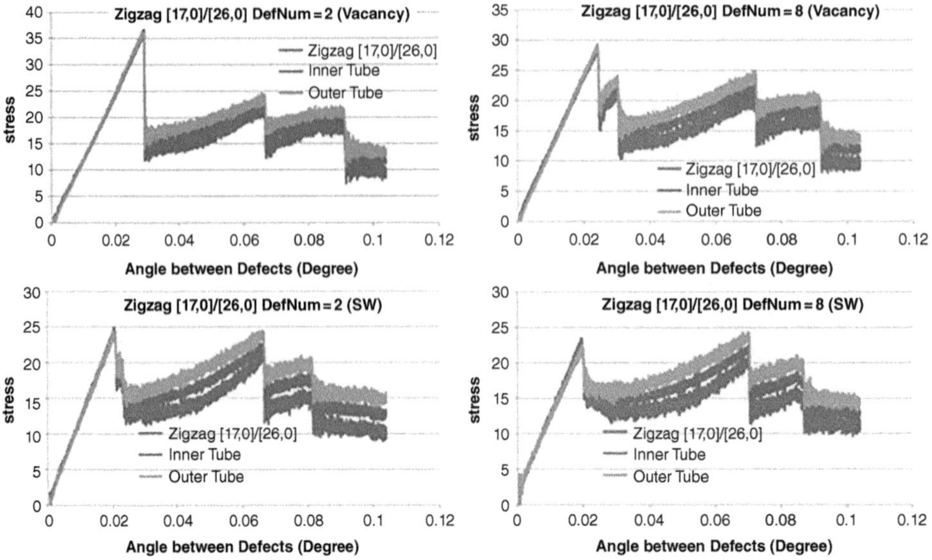

Figure 4.47 Compressive stress–strain response of zigzag [17,0]/[26,0] DWCNT for different numbers of vacancy and SW defects. *Source:* Reproduced from Eftekhari 2015/Islamic Azad University.

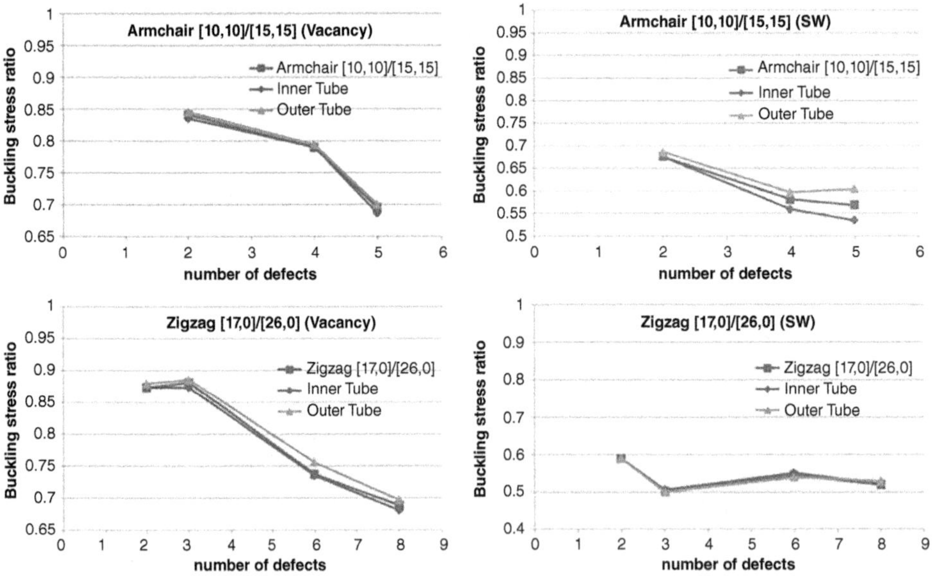

Figure 4.48 The ratio of buckling stress of defected armchair [10,10]/[15,15] and zigzag [17, 0]/[26,0] DWCNTs to the intact one for different numbers of defects. *Source:* Reproduced from Eftekhari 2015/Islamic Azad University.

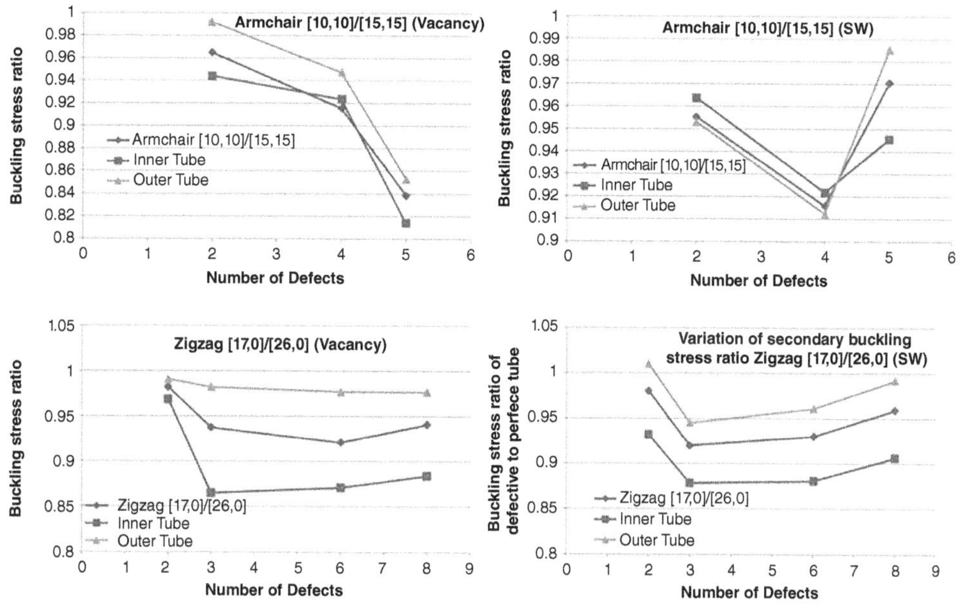

Figure 4.49 The ratio of the secondary buckling stress of defected armchair [10,10]/[15,15] and zigzag [17, 0]/[26,0] DWCNTs to the intact one for different numbers of defects. *Source:* Reproduced from Eftekhari 2015/Islamic Azad University.

4.4.3.5 Final Results of the CNT Response

The average mechanical properties of CNTs can be computed from the large set of nano analysis of CNTs, as summarized in Table 4.3. These CNT properties can be used in higher scale simulations.

Table 4.3 Atomistic scale tensile and compressive properties of CNTs.

		Tension			Compression		
Type	ν	E (GPa)	f_t (GPa)	G_{ft} (MN/m)	E (GPa)	f_{cb} (GPa)	G_{fc} (MN/m)
Armchair [10,10]	0.14	1,130	128.7	13.2	1,280	43.6	1.9
Armchair [6,6]	0.14	1,100	126.7		1,265	96	
Armchair[10,10]/ [15,15]	0.14	1,090	125.8	12.3	1,250	43.4	2.0
Zigzag [17,0]	0.13	1,120	102.3	8.0	1,250	45.5	2.0
Zigzag [10,0]	0.13	1,090	104.7		1,235	105.1	
Zigzag [17,0]/ [26,0]	0.13	1,070	99.9	7.7	1,220	40.4	1.9

Source: Adpated from Eftekhari et al., 2014, 2018, Eftekhari and Mohammadi, 2016b.

4.4.4 Simulation of CNT-Reinforced Calcium Silicate Hydrate

4.4.4.1 Description of Calcium Silicate Hydrate (CSH)

Cement is a porous medium composed of a variety of hydrated and unhydrated constituents. The nanostructure of the cement gel, or calcium silicate hydrate (CSH), which significantly affects the mechanical properties of cement, is assumed to be the building structure of the cement paste. CSH globules are composed of nano silicate chains, calcium oxides and water, as typically depicted in Figure 4.50.

4.4.4.2 CSH Modelling

To study CSH behaviour, a periodic unit cell of a CSH globule is constructed on the atomistic scale, as presented in Figure 4.51. The model contains a CNT perpendicular to the silicate layer to obtain the most reinforcement efficiency. It is also possible to consider random orientation of CNT reinforcement. MD simulations are performed to determine the tensile and compressive stress–strain mechanical properties of CNT-reinforced CHS.

The initial model for the CSH phase is constructed from a triclinic crystal box with periodic boundary conditions. The lattice parameters are defined as (Eftekhari and Mohammadi 2016b)

$$a=39.25\text{Å},\ b=36.26\text{Å},\ c=47.68\text{Å},\ \alpha=88.97^{\circ},\ \beta=92.81^{\circ},\ \gamma=88.57^{\circ}.$$

The complex MD modelling of CSH consists of harmonic bond stretching, harmonic bond angle bending, Coulomb interaction and LJ potential. The Tersoff potential is used for

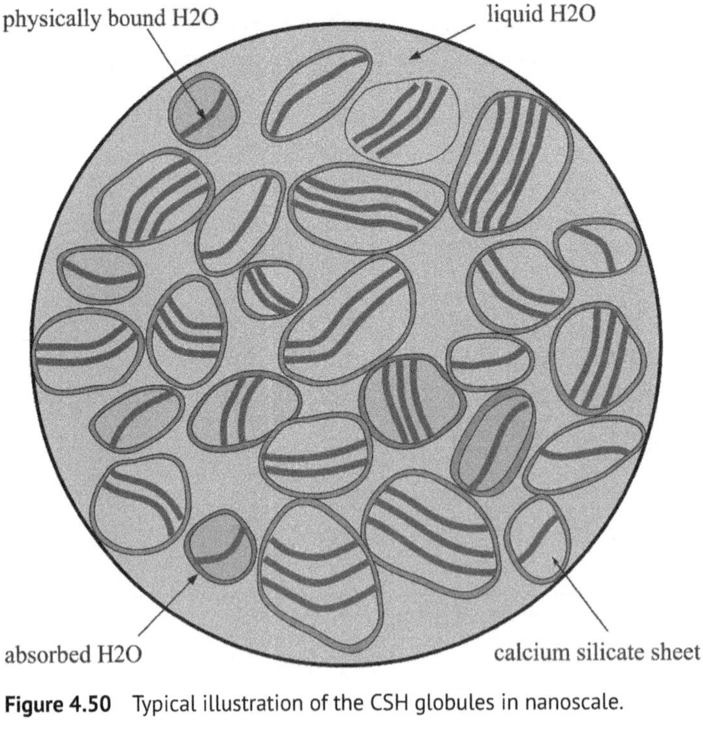

Figure 4.50 Typical illustration of the CSH globules in nanoscale.

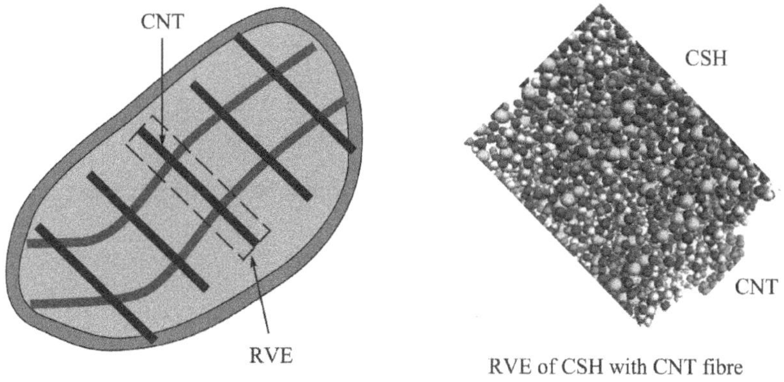

Figure 4.51 Typical nanoscale model of the CSH globule and the molecular structure of the CNT-reinforced CSH phase.

the CNT modelling. Interaction of CNT and CSH atoms are assumed to follow the LJ potential with the Lorentz–Bethelot mixing rules for LJ parameters (Sanchez and Zhang 2010). Finally, the water molecules are governed by the SPC model for hydrogen and oxygen atoms. For details of parameters for potentials of CNT, CSH and water, see Eftekhari and Mohammadi (2016b).

The model is constructed based on a 40 Å CNT and the molar Ca/Si ratio of 1.6. The simulation procedure begins with an NPT ensemble to enforce the MD model to reach an equilibrium state at 300 K temperature and the atmospheric pressure. Then, the cylindrical hole for positioning the CNT is gradually indented within the model. The hole is initially generated with a zero diameter and is progressively enlarged to the required size to reach equilibrium with the surrounding CSH atoms in a consistent manner without formation or breakage of any existing bonds. Subsequently, the CNT model is inserted inside the hole.

4.4.4.3 Computation of the Compliance Matrix

In order to obtain the mechanical properties of CNT reinforced CSH, the components of the compliance tensor are required. Therefore, a series of independent uniaxial 1, 2, 3 and shear 12, 23 and 31 deformations are imposed on the model with the constant velocity of $1\,\nu/\text{ps}$ and the NVT ensemble constraint.

The results for the elasticity components are presented in Table 4.4. Clearly, while the elasticity component along the CNT length is significantly increased, other components are reduced, leading to a strong transversely isotropic response of CNT reinforced CSH (compared to the isotropic response of CSH).

The typical resultant stress–strain curves are observed in Figure 4.52. Inclusion of CNT has changed the brittle tensile behaviour of CSH to a ductile mode. The reason can be partially attributed to the crack-bridging effect of CNT in the tensile response of CSH, as clearly depicted in Figure 4.53.

In contrast, the overall compressive response of CSH is less affected by CNT. It shows an increasing trend in the reinforcement direction, but with more or less the same stress–strain results, and with limited fluctuations to decrease the strengths in the unreinforced orientations. Similar results are observed in the shear conditions.

Table 4.4 Elasticity components C_{ij}(GPa) for the CSH models.

Component	CSH	CSH + CNT(armchair)	CSH + CNT(zigzag)
11	76.2	36.8	24.9
12	36.2	36.1	25.8
13	19.9	14.5	17.7
22	77.0	43.6	30.4
23	24.0	17.3	17.9
33	36.8	148.3	153.7
23–23	10.0	10.1	8.8
13–13	11.3	5.9	6.7
12–12	27.4	24.9	15.4

Source: Adapted from Eftekhari and Mohammadi, 2016b.

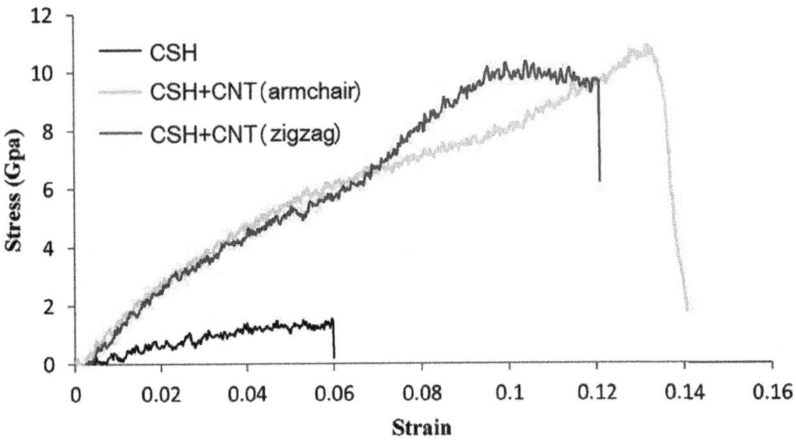

Figure 4.52 Typical longitudinal tensile stress–strain response (along the CNT reinforcement). *Source:* Reproduced from Eftekhari 2015/Islamic Azad University.

Finally, the elastic constants of the CSH and reinforced CSH models are summarized in Table 4.5.

Table 4.6 provides the percentage of reduction of mechanical properties of the cement paste RVE with defected CNTs in comparison to the cement paste with intact CNT.

4.4.4.4 CNT Pull-Out Test

A pull-out test can be performed to determine the interfacial strength in between the CNT and CSH matrix. It fundamentally affects the crack bridging response, which largely determines the cracking extent of CSH and its post-cracking softening behaviour.

Eftekhari and Mohammadi (2015) constructed an MD setup of a CNT pull-out test from a CSH matrix and examined two cases of constant velocities of $10\ \text{Å}/\text{ps}$ and $15\text{Å}/\text{ps}$. They concluded that the higher pull-out velocity leads to higher interfacial strength, which implies that such a bridging effect of CNT performs more efficiently in extreme dynamic conditions.

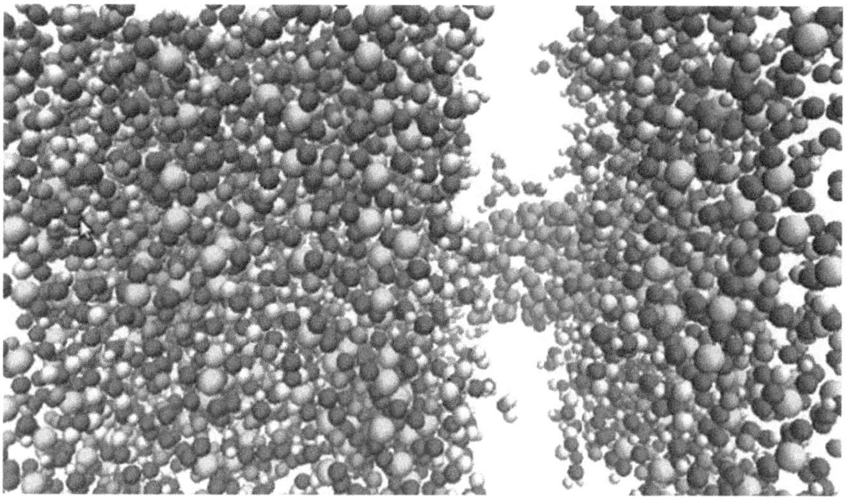

Figure 4.53 Typical CNT crack bridging in the CNT-reinforced CSH. *Source:* Reproduced from Eftekhari 2015/Islamic Azad University.

Table 4.5 Nanoscale mechanical properties of CSH and the CNT-reinforced CSH.

	Component	Elasticity modulus (GPa)	Poisson's ratio	Tensile strength (GPa)	Compressive strength (GPa)	Shear strength (GPa)
CSH	1	55.7	0.3	3.5	31.3	1.1 (23)
	2	52.7		3.3	40.7	1.2 (13)
	3	28.2		1.6	60.0	2.3 (12)
CSH + CNT (armchair)	1	6.8	0.3–0.4	2.2	19.5	0.9 (23)
	2	8.1		2.4	31.3	1.0 (13)
	3	142.3		6.7	40.8	1.6 (12)
CSH + CNT (zigzag)	1	1.9	0.3–0.5	2.1	22.3	1.0 (23)
	2	2.8		2.4	25.6	0.9 (13)
	3	138.4		5.5	36.2	1.7 (12)

Source: Adapted from Eftekhari and Mohammadi, 2016b.

Table 4.6 Reduction of mechanical properties of the cement paste RVE with defected CNTs in comparison to the cement paste with intact CNTs.

	Reduction in elastic modulus	Reduction in fracture strain	Reduction in fracture stress
CNT with vacancy defects	5.7%	40%	25%
CNT with SW defects	3%	53%	35%

Source: Reproduced from Eftekhari, M. and Mohammadi, S. (2015)/Iran Nano Technology Initiative Council.

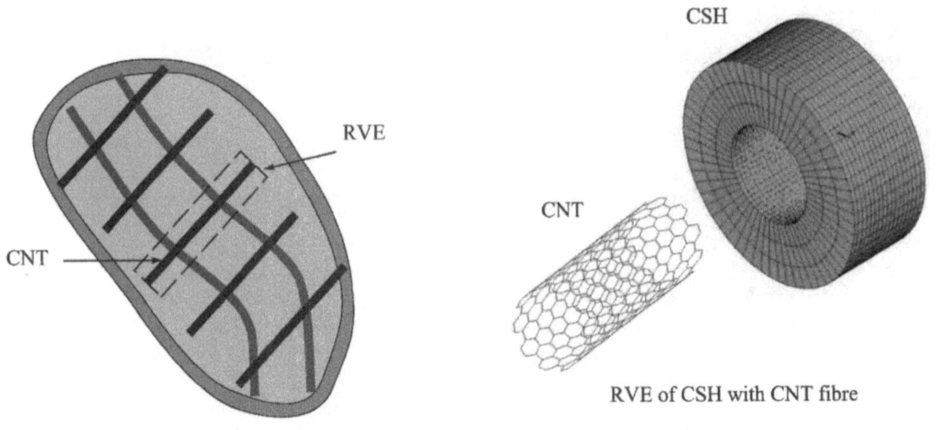

Figure 4.54 Continuum model of the pull-out test.

Alternatively, a continuum approach based on the concepts of the molecular characteristics of CNT can be adopted, as typically shown in Figure 4.54. Analysis of RVE is based on the specific assumptions for the CNT, the cement paste and the interface.

The CNT structure is modelled by structural molecular statics elements, where the carbons are nodes of the frames and their covalent bonds are represented by three-dimensional frame elements, whose cross-sections are determined by the concept of equivalence of modified Morse potential energy and beam strain energy (Cornell et al. 1995). The equivalent stress–strain response is presented in Figure 4.55.

The CSH is modelled by a continuum model with the elastic modulus of 25 GPa and Poisson's ratio of 0.3. Various plasticity, creep, hardening and large strains regimes may be considered.

The interface of CNT and cement is simulated by springs, whose response is defined by the LJ potential for the van der Waals bonding.

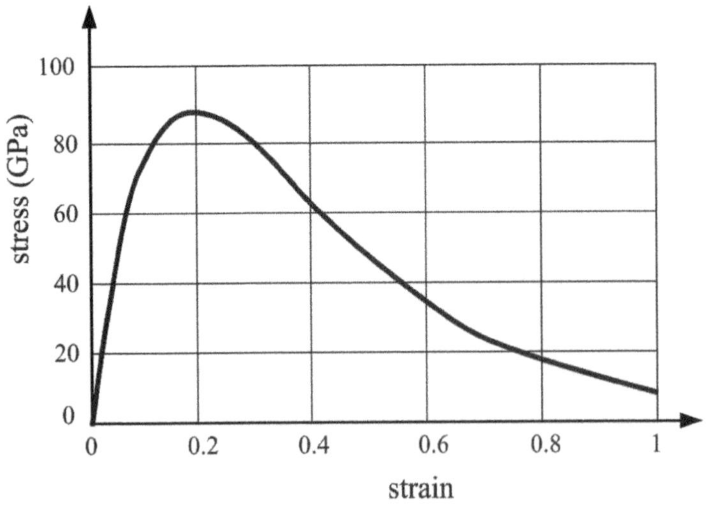

Figure 4.55 Morse-based stress–strain response for the CNT beam elements. *Source:* Adapted from Wenik and Meguid 2011; Eftekhari and Mohammadi 2015.

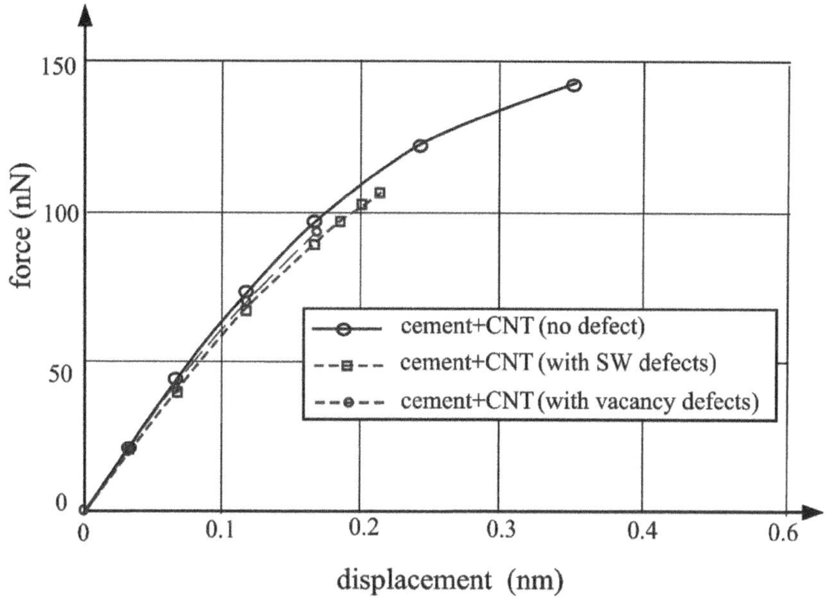

Figure 4.56 Comparison of the force-displacement responses of intact and defected CNTs in the cement paste. *Source:* Reproduced from Eftekhari and Mohammadi 2015/Iran Nano Technology Initiative Council.

An incremental tensile loading is imposed on the model and the final force-displacement response is obtained, similar to the typical results of Figure 4.56.

4.4.5 Micromechanical Simulation of CNT-Reinforced Cement

4.4.5.1 Microscale Description

The next step of the multiscale simulation is to study the response of CNT-reinforced cement at the microscale. The micro model is a porous cement composed of unhydrated and hydrated parts and randomly distributed CNTs, as typically depicted in Figure 4.57.

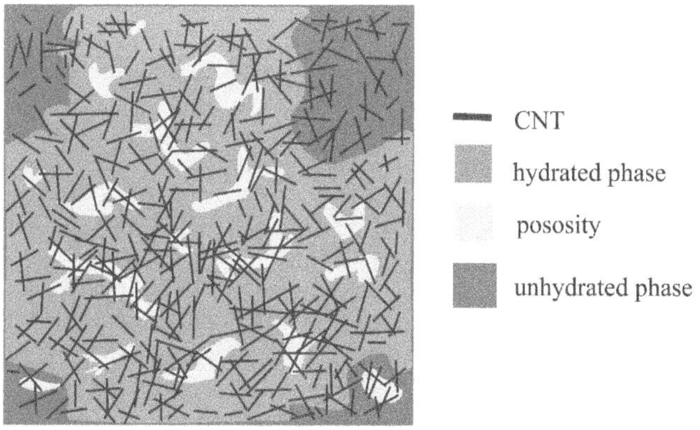

Figure 4.57 Schematic computational model of the cement paste reinforced by CNTs, including the unhydrated and hydrated phases and porosities.

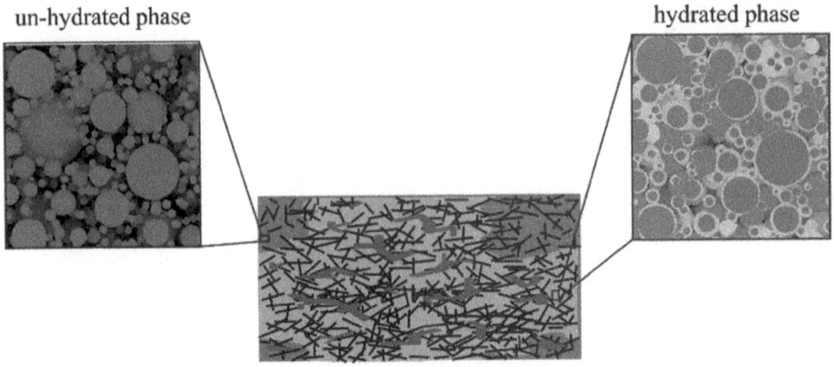

Figure 4.58 Sample illustration of the computational micro model from the micro scans of cement samples before and after hydration.

The necessary mechanical properties of the hydrated material, including the elastic modulus, the Poisson's ratio and the fracture energy are obtained by upscaling the values from the lower scale computations.

There are a number of open-source codes available for generation of a computational model from microscans (for instance, the particle kinetics chemical hydration model μic by Bishnoi and Scrivener (2009) and CEMHYD3D developed by the National Institute of Standards and Technology (NIST) (Bentz 2005)) (Figure 4.58). The size and distribution of the grains of the cement within the model can be obtained accordingly.

4.4.5.2 Modelling of Microscale Problem

The cement hydration model is considered as a $50 \times 50\ \mu m$ specimen with the thickness of $1\ \mu m$. CNTs are assumed to be randomly dispersed within the cement with different lengths and orientations.

Table 4.7 defines the mechanical properties of different phases of the cement paste (Eftekhari et al. 2014). Very small quantities are assigned to the elements of porosity to avoid numerical instability. The average mechanical properties of CNT in tension and compression are obtained from Table 4.3.

The two-dimensional plane stress finite element model is generated from a slice of the three-dimensional model of microscale hydrated cement. CNTs are simulated by bar/truss elements randomly embedded within the cement. Different CNT volume fractions are studied and the models are subjected to an applied strain rate to determine the

Table 4.7 Mechanical properties of the cement paste phases.

Phase	E(GPa)	ν	f_t (MPa)	G_f (N/m)
Hydrated materials	22	0.24	6	12
Unhydrated materials	135	0.30	1,800	120

Source: Adapted from Eftekhari et al., 2014, Eftekhari and Mohammadi, 2016a.

corresponding stress based on a damage analysis. An isotropic linear damage model is assumed for all constituents:

$$\sigma = (1-d)E\varepsilon = f_t\left(1 - \frac{hd\varepsilon f_t}{2G_f}\right) \tag{4.4.2}$$

where f_t is the uniaxial tensile strength, h is the effective width of the finite element and d is the damage parameter.

Figure 4.59 illustrates the typical distribution of damage of the model, with clear effects of CNTs on damage distribution. Larger values of dissipated energy are expected in longer CNTs.

The resultant mechanical properties of CNT-reinforced cement are summarized in Table 4.8.

The results of Table 4.8 can be used for the CNT-reinforced cement in simulations at higher mesoscale problems.

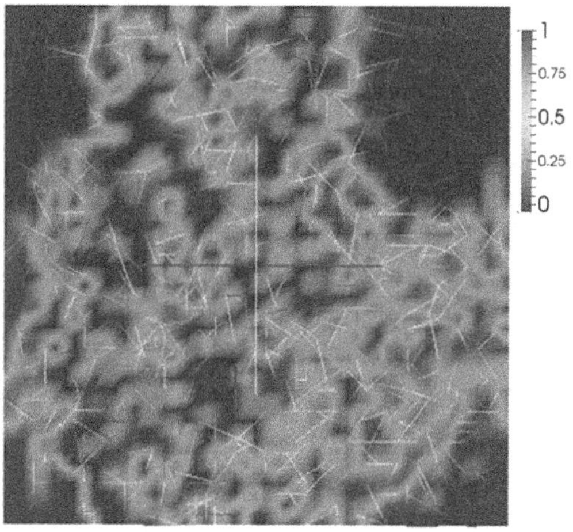

Figure 4.59 Damage distribution in the CNT-reinforced cement with 3% volume fraction and 5 μm CNT.

Table 4.8 Microscale mechanical properties of plain and CNT-reinforced cements in tension and compression.

	Tension			Compression		
Type	***E* (GPa)**	**f_t (MPa)**	**G_{ft} (N/m)**	***E* (MPa)**	**f_c (MPa)**	**G_{fc} (N/m)**
Cement	18.5–18.6	2.25–2.35	5–6	18.5–18.6	28.0–29.0	820–830
CNT-1%	18.5–18.6	2.45–2.55	9–10	18.5–18.6	28.5–29.5	840–860
CNT-2%	18.6–18.7	2.45–2.55	19–21	18.6–18.7	30.0–32.0	1,020–1,040
CNT- 3%	18.8–18.9	2.75–3.25	45–47	18.8–18.9	33.0–35.0	1,030–1,050

Source: Adapted from Eftekhari et al., 2014, 2018.

4.4.6 Mesoscale Simulation of CNT-Reinforced Concrete

4.4.6.1 Description of the Mesoscale Problem

The previous section provided the results for mechanical properties of CNT-reinforced cement in modelling microscale problems to determine the damage response of concrete samples made of a CNT-reinforced cement.

Samples with the size of 100×100 mm with a mean finite element size of 2mm are analysed subject to tensile and compressive loadings. All phases of CNT-reinforced cement, including the large aggregates (in the order of above 1 mm) and the interfacial zones are simulated. The aggregates are considered to be linear elastic with $E = 30$ GPa and $\nu = 0.2$, with different volume fractions. The effects of various shapes (irregularly angular, hexagonal and round, as depicted in Figure 4.60) and sizes of aggregates are considered. The 200Åm thickness ITZ layer is assumed to have 70% of the mechanical properties of the cement. No porosity is modelled as it has been modelled in the lower microscale to determine the microscale stress–strain behaviour of the CNT-reinforced cement.

The cement and ITZ follow the isotropic damage model for both tensile and compressive conditions. The mechanical properties of the cement are adopted from the microscale simulations (Table 4.8).

4.4.6.2 The Effects of Aggregate Shape

First, the effects of various shapes of aggregates for the case of the 50% volume fraction are studied. The samples include round, hexagonal and irregularly angular shaped aggregates, as presented in Figure 4.61.

Analysis of the axial tensile strain distribution contours (Figure 4.62) shows that while the aggregates remain elastic, the damage zones are initiated in the ITZ regions and propagated into the cement. The crack bridging phenomena of CNT reinforcement change the

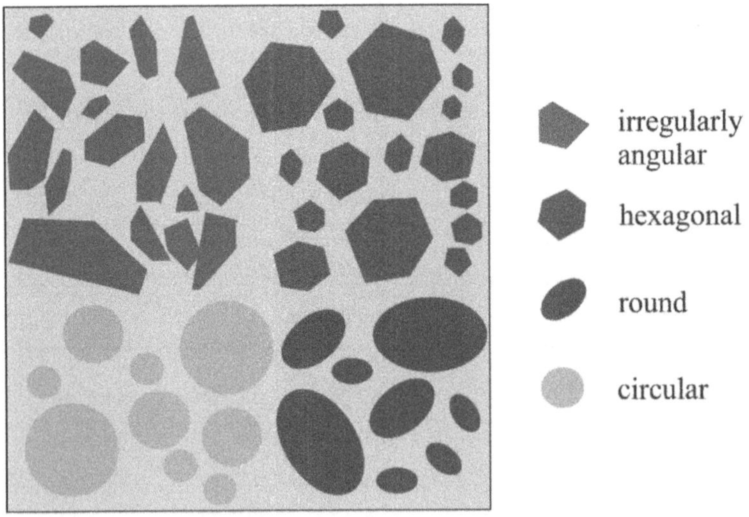

Figure 4.60 An illustration of the mesoscale CNT-reinforced concrete sample with different aggregate shapes and distributions.

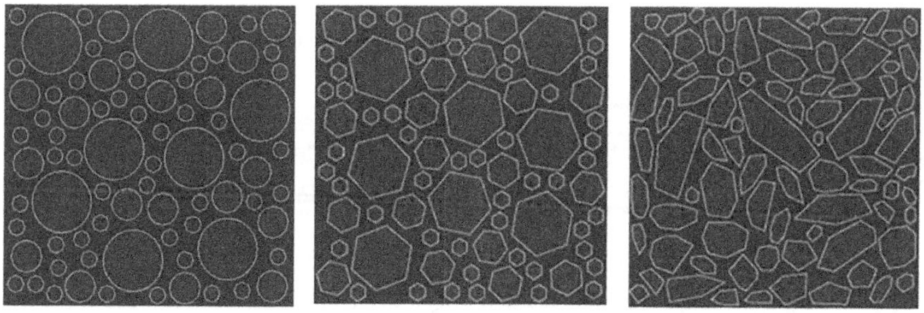

Figure 4.61 Mesoscale CNT-reinforced concrete samples with different aggregate shapes. *Source:* Adapted from Eftekhari et al. 2016a.

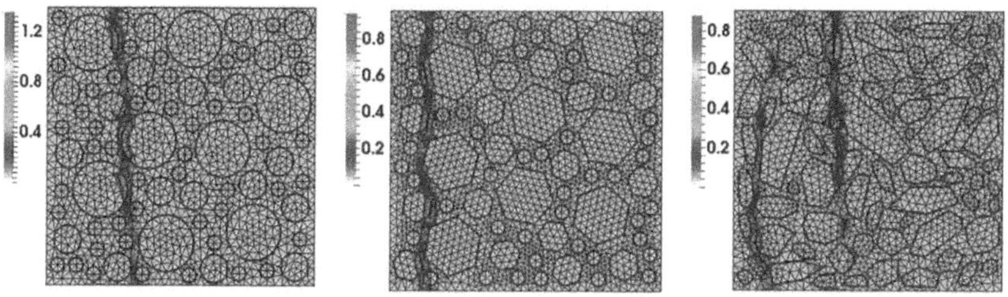

Figure 4.62 Axial strain contours in the tensile CNT-reinforced cement specimen. *Source:* Adapted from Eftekhari et al. 2016a.

response of concrete to a relatively ductile behaviour by increased ductility and energy dissipation and decreased crack propagations. In compression, the compressive strength is considerably increased, with a considerable effect of the shape of aggregates on the localization patterns (Figure 4.63) without affecting the overall mechanical properties considerably.

4.4.6.3 The Effects of Aggregate Distribution

It is possible to examine the effects of spatial distribution of aggregates. For instance, samples with two different round aggregate patterns, depicted in Figure 4.64, provide the same 50% volume content of aggregates.

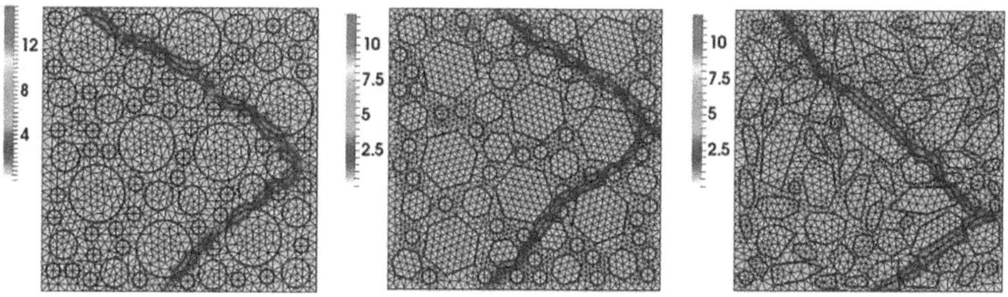

Figure 4.63 Axial strain contours in the compressive CNT-reinforced cement specimen. *Source:* Adapted from Eftekhari and Mohammadi 2016a.

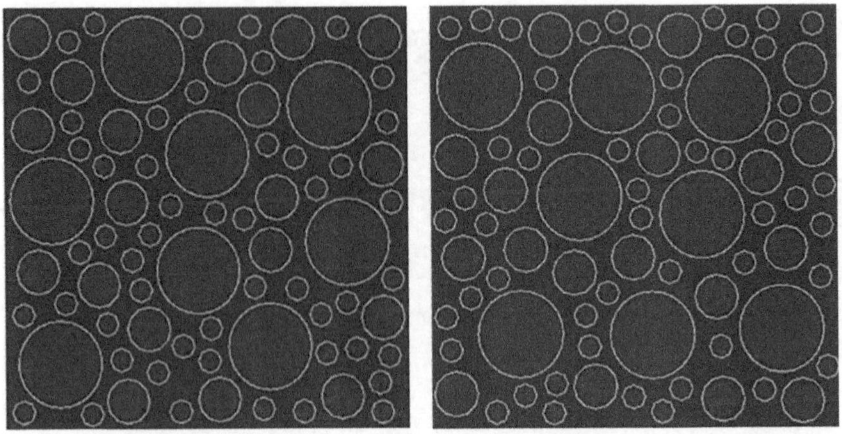

Figure 4.64 Spatial distributions of aggregates for the concrete model A (left) and model B (right) (with 50% volume content). *Source:* Reproduced from Eftekhari 2015/Islamic Azad University.

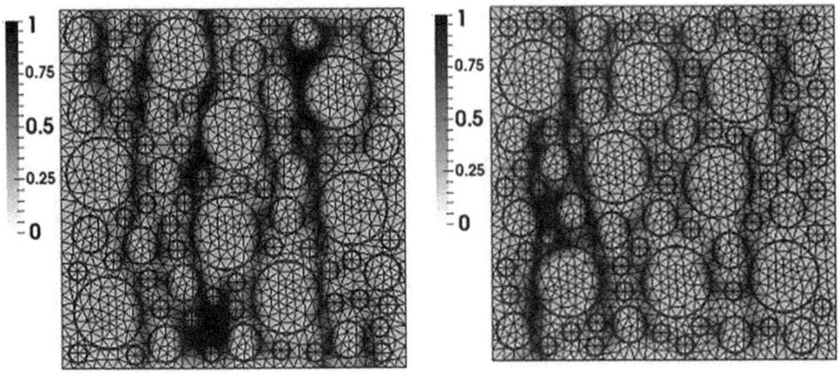

Figure 4.65 Spatial distribution of damage in the tensile specimen (with 50% volume content) with the CNT reinforcement.

Their tensile damage patterns (Figure 4.65) are locally different, but the overall strength remains more or less the same. The same conclusions can be made from the distribution of tensile strains (Figure 4.66), where different localization zones are observed. Moreover, the maximum levels of generated strains differ to some extent. Despite the local differences, the global responses of all specimen remain almost similar, as presented in Figure 4.67 for both the plain concrete and the CNT-reinforced concrete samples.

In compression, totally different patterns of damage distribution are observed in Figure 4.68. However, the general patterns remain similar for all cases. Distribution and the maximum values for the axial strain (Figure 4.69) are clearly different from the tensile cases. Finally, the global stress–strain responses for both aggregate distributions show similar trends for the plain concrete and the CNT-reinforced concrete samples (Figure 4.70).

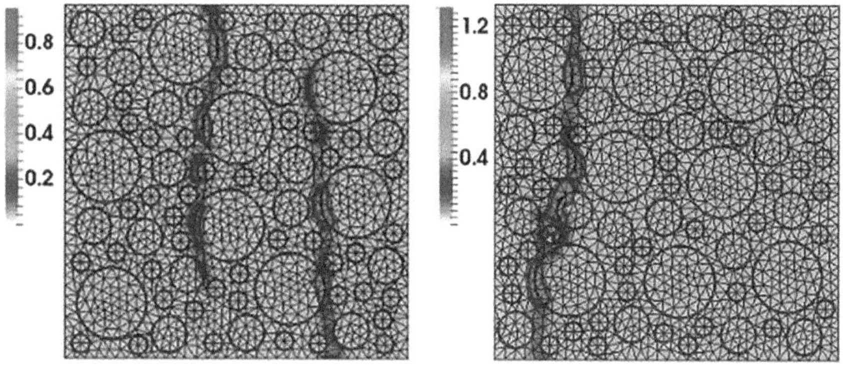

Figure 4.66 Spatial distribution of axial strain in the tensile specimen (with 50% volume content) with the CNT reinforcement.

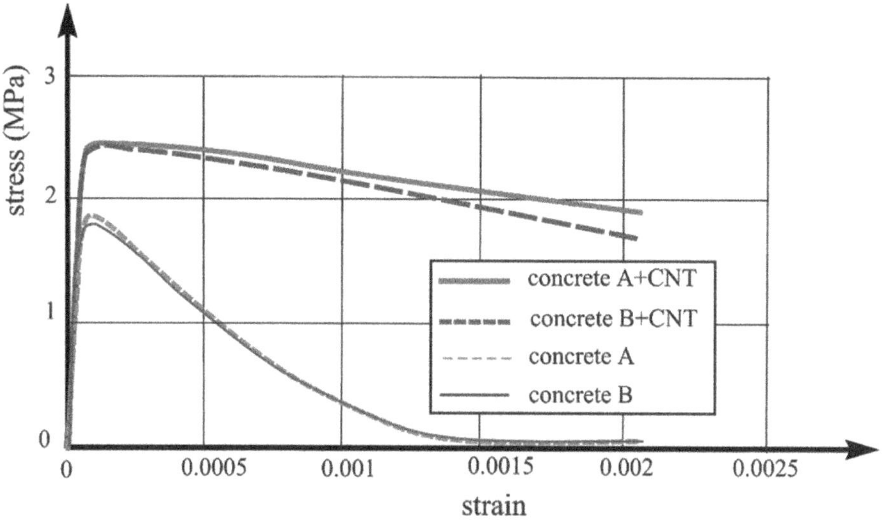

Figure 4.67 Stress–strain responses of the tensile specimen (with 50% volume content).

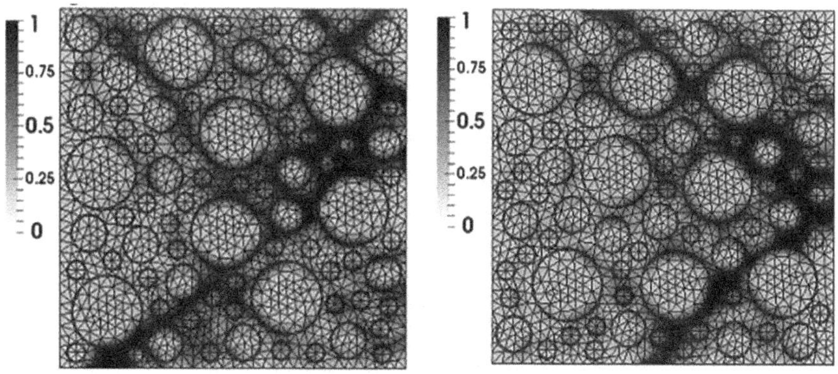

Figure 4.68 Spatial distribution of damage in the compressive specimen (with 50% volume content) with the CNT reinforcement.

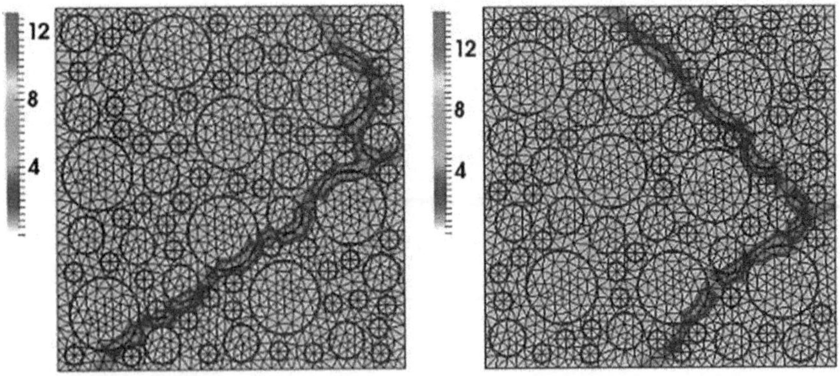

Figure 4.69 Spatial distribution of axial strain in the compressive specimen (with 50% volume content) with the CNT reinforcement.

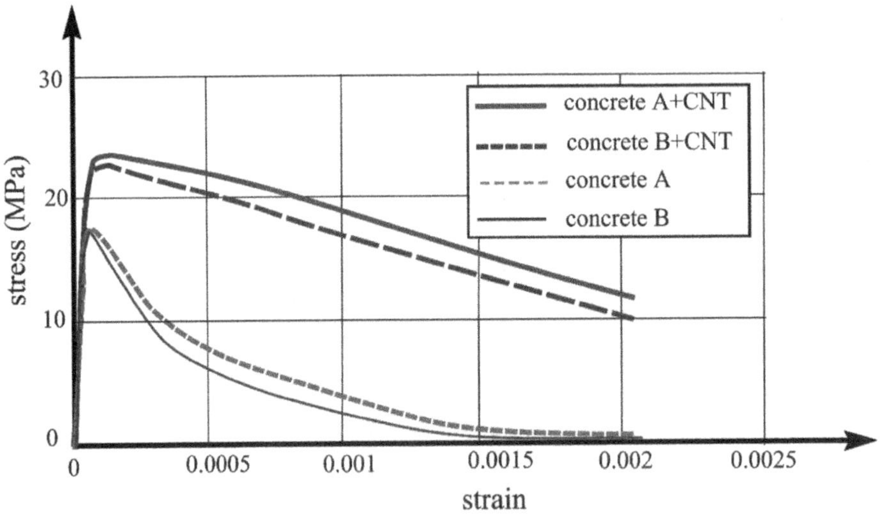

Figure 4.70 Stress–strain responses of the compressive specimen (with 50% volume content).

4.4.6.4 The Effects of the Aggregate Volume Fraction

The effect of volume content of aggregates is examined for the round shape aggregates of Figure 4.71, with 10, 30 and 50% volume contents.

Again, both tensile and compressive uniaxial loadings are considered. In the tensile condition, the strain contours show very similar patterns (Figure 4.72) and the overall stress–strain responses in Figure 4.73 confirms the small effect of the volume content of aggregates in the tensile responses for both the plain concrete and the CNT-reinforced concrete specimens.

In compressive loading, however, in addition to the different patterns in damage and strain distributions, as presented in Figure 4.74, the overall response of Figure 4.75 shows that the volume contents of aggregates have a considerable effect on the increased strength of the specimen.

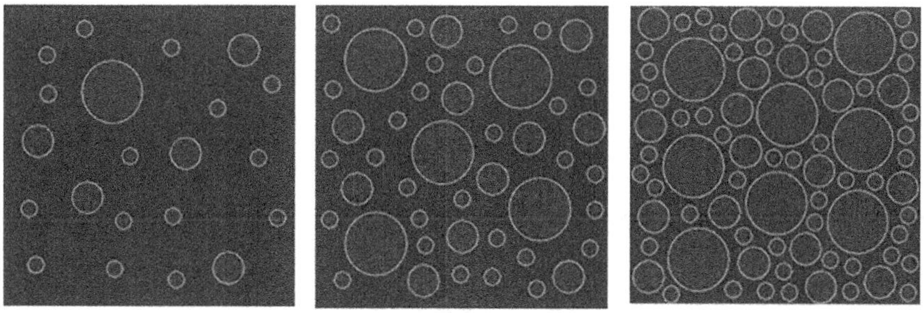

Figure 4.71 CNT-reinforced concrete samples with round aggregates with 10, 30 and 50% volume contents.

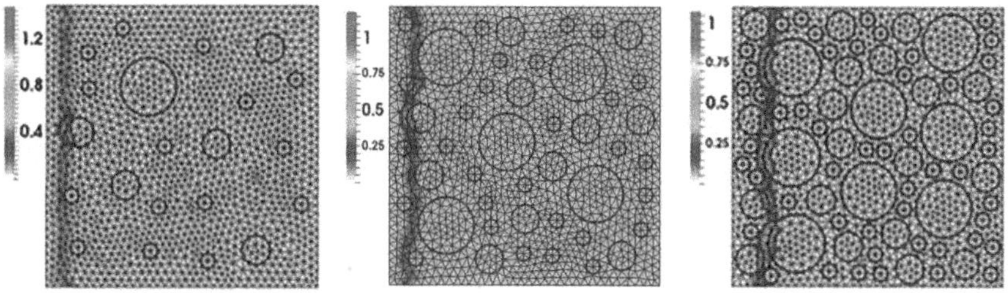

Figure 4.72 Spatial distribution of the axial strain in the tensile specimen with various volume contents of aggregates with the CNT reinforcement.

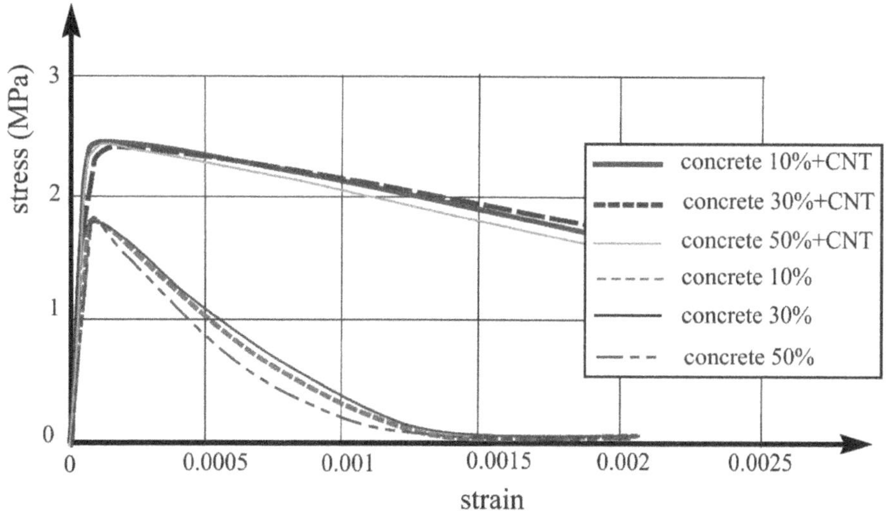

Figure 4.73 Stress–strain responses of the tensile specimen with various volume contents of aggregates.

4.4.6.5 Final Results of Mesoscale Simulations

The final obtained results of tensile and compressive strengths of the plain concrete and the CNT-reinforced concrete at the mesoscale are presented in Table 4.9.

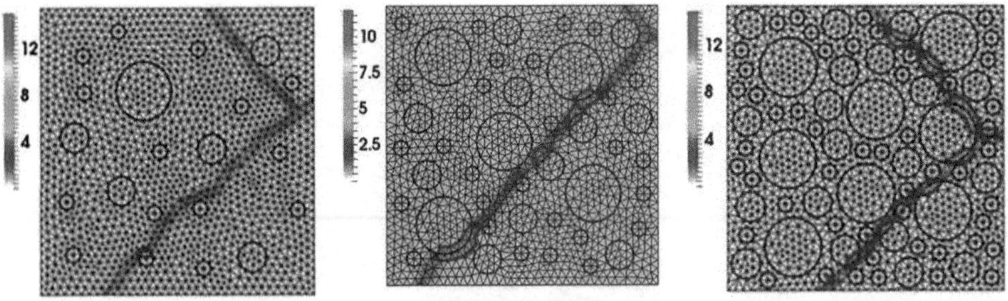

Figure 4.74 Spatial distribution of the axial strain in the compressive specimen with various volume contents of aggregates with the CNT reinforcement.

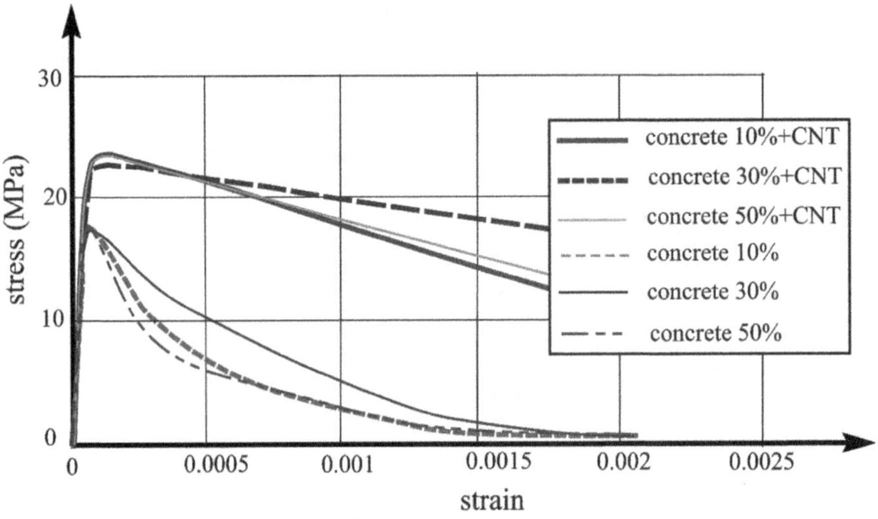

Figure 4.75 Stress–strain responses of the compressive specimen with various volume contents of aggregates.

Table 4.9 Mechanical properties of the concrete at mesoscale RVE with defected CNTs in comparison to the concrete paste reinforced with intact CNT.

	Tensile strength (MPa)	Compressive strength (MPa)
Concrete	1.82	18.0
CNT-reinforced concrete	2.43	24.5
Strength increase (%)	33	36

Source: Reproduced from Eftekhari, M. and Mohammadi, S. (2015)/Iran Nano Technology Initiative Council.

4.4.7 Macroscale Simulation of CNT-Reinforced Concrete

The previous section provided the results for mechanical properties of the CNT-reinforced cement in modelling mesoscale cement/aggregate problems. At this stage of the multiscale simulation, conventional numerical approaches such as the crack analysis by the extended finite element method (XFEM), the continuum damage mechanics for elastoplastic damage analysis and the quasi-static nonlinear cyclic analysis for seismic resistance simulations can be adopted for various macroscale engineering problems.

Sample material properties of the confined and unconfined plain and CNT-reinforced concrete specimens are compared in Table 4.10 (from the mesoscale properties of Table 4.9).

Eftekhari et al. (2018) performed a number of macroscale analyses of concrete structures with CNT-reinforced cements. The problems included plain concrete specimens (without steel bar reinforcements) and steel-reinforced concrete columns (with CNT-reinforced cement) subjected to various seismic and dynamic effects.

For instance, in analysis of a projectile impact on a plain concrete specimen, the ultimate penetration depth in the CNT-reinforced concrete may be limited to even half of the plain concrete without the CNT fibres, as typically presented in Figure 4.76.

Furthermore, Eftekhari et al. (2018) studied the effects of CNT fibre inclusion on the cyclic behaviour of steel-reinforced concrete columns, as typically illustrated in Figure 4.77. The model included longitudinal and transverse steel rebars and the effects of potential debonding and sliding of longitudinal rebars from concrete were included. Quasi-static cyclic lateral displacement loading and several seismic records were used to examine the

Table 4.10 Average macroscale material properties of the concrete and the CNT-reinforced concrete (3% CNT).

Material properties	Concrete	Concrete (confined)	CNT-concrete	CNT-concrete (confined)
E (GPa)	26.2	26.2	25.2	25.2
(MPa)	792.1	792.1	829.9	829.9
K (MPa)	1,716.2	1,716.2	1,797.9	1,797.9
f_c (MPa)	18.4	27.9	24.8	34.8
$f_{c,el}$ (MPa)	9.8	14.8	13.1	18.4
f_t (MPa)	1.9	1.9	2.4	2.4
$f_{t,el}$ (MPa)	1.3	1.3	1.7	1.7
ε_{c0}	0.0008	0.0029	0.0208	0.0063
ε_u	0.0070	0.0140	0.0400	0.0500
$h_{\text{hardening-slope}}$	2.0	2.0	2.0	2.0

Source: Adapted from Eftekari et al., 2016b, 2018.

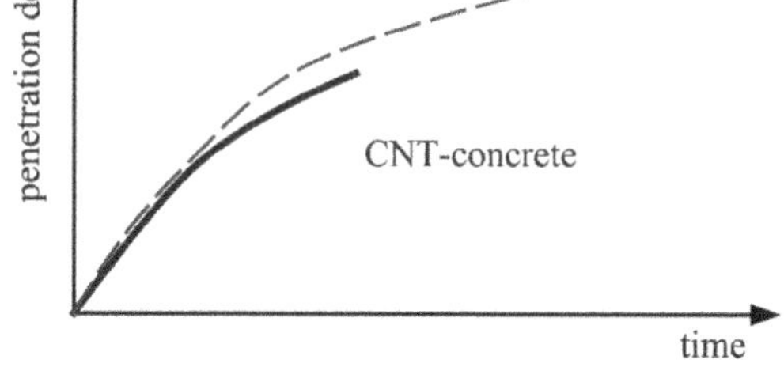

Figure 4.76 Typical reduction of the penetration depth in the CNT-concrete specimen. *Source:* Adapted from Eftekhari et al. 2018.

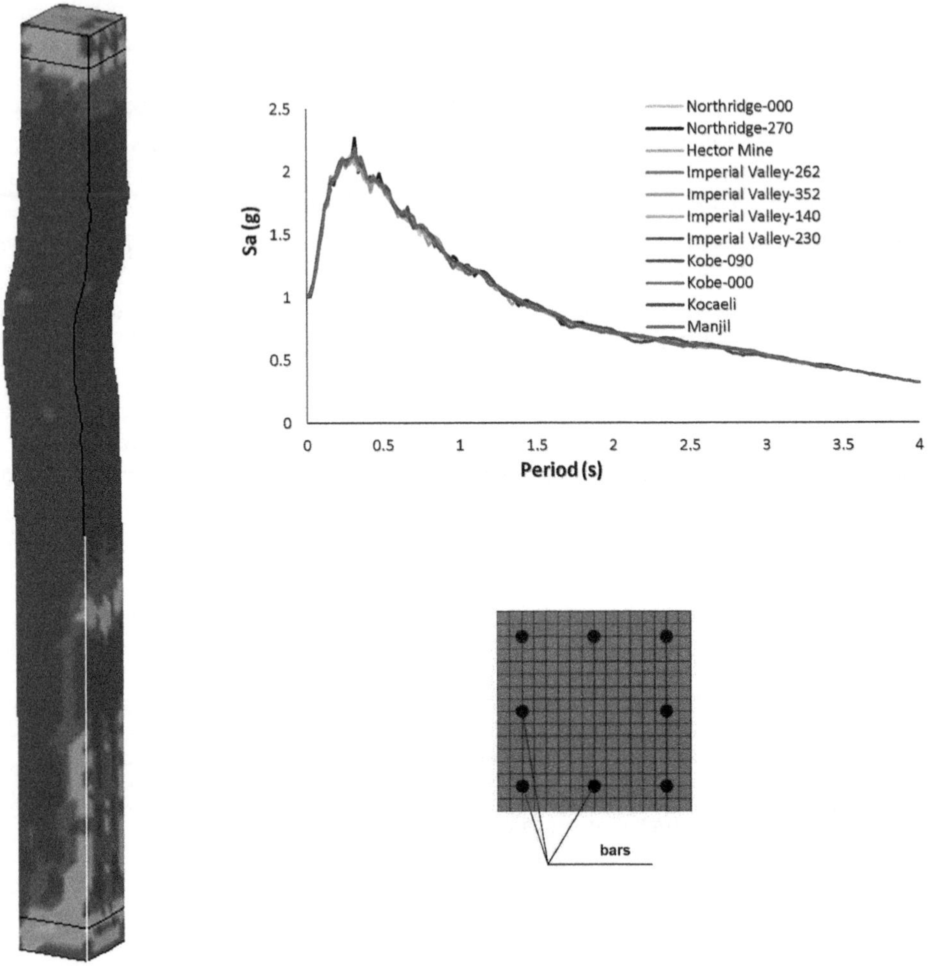

Figure 4.77 A typical illustration for the finite element macro model of a steel-reinforced CNT-concrete column subjected to cyclic lateral displacement loading.

hysteresis seismic performance of the CNT and plane steel-reinforced columns and determined their energy absorption capacities. As a sample conclusion, they reported a nearly 50% increase in the lateral load bearing capacity of the CNT concrete (Eftekhari et al. 2018). Analysis can be performed by commercial general-purpose finite element packages or by specialized open-source software for quasi-static cyclic analysis of structures.

For further details on hysteresis responses and ductility studies refer to (Eftekhari et al. 2018).

4.5 Concurrent Multiscale Methods

4.5.1 Introduction

In a fully atomistic modelling, typically shown in Figure 4.78, analysis of the model is involved with massive and extremely expensive molecular dynamics/statics solutions based on each individual atom of the model and at each time step of the simulation period.

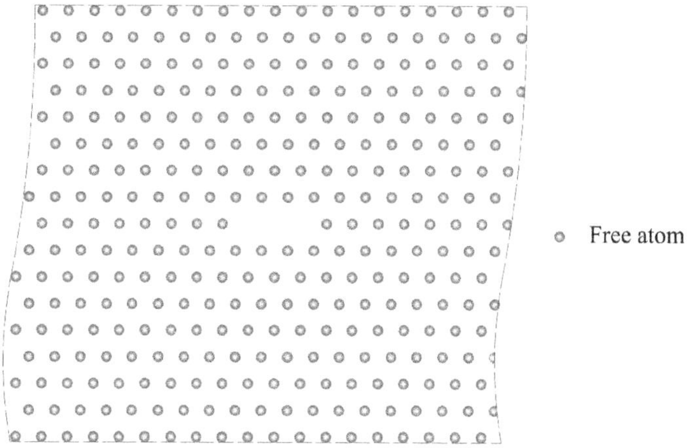

Figure 4.78 A typical fully atomistic modelling of a domain with a crack/hole. *Source:* Reproduced from Alizadeh 2019/University of Tehran.

The concurrent multiscale methods have been developed to reduce the computational time and to increase the efficiency of the numerical solution by using both atomistic and continuum (finite element) methods (although they are not limited only to these approaches). In these methods, both atomistic and continuum formulations constitute the coupled governing equations of the multiscale method. They may be employed concurrently to describe the response of a specific region, or they may be used for different regions of the multiscale problem, with an interface or transition zone to connect the regions (Figure 4.79).

Depending on the existence of the transition zone, the positions of atoms and nodes of elements significantly influence the way the concurrent multiscale models are coupled. Methods designed on coincident atoms and nodes provide a strong coupling between the two setups, but they require complex mesh generation algorithms with high computational costs. In contrast, weakly coupled methods do not need such a complex procedure, but they may become less accurate in imposing a consistent field in or around the transition zone.

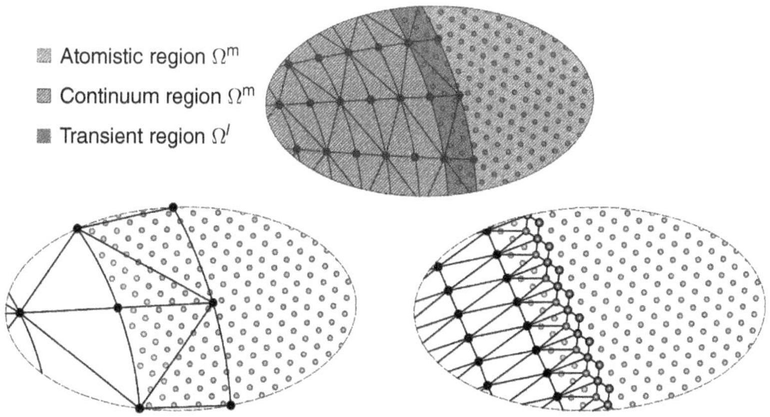

Figure 4.79 Multiscale domain partitioning into continuum and atomistic regions with different transition concepts. *Source:* Reproduced from Alizadeh 2019/University of Tehran.

Several multiscale methods have been developed since the pioneering work of Kohlhoff et al. (1991) but with different approaches in terms of energy or force-based formulations and the way the interface region is defined and formulated. Later, the macroscopic atomistic ab initio dynamics (MAAD) was developed by Abraham et al. (1998) as a coupled atomistic and continuum model with an overlap over a handshake region (Fish 2006, 2009). MAAD uses a finite element for the continuum region with its size being refined down to the lattice size at its interface with the atomic region.

In addition to the conventional concurrent multiscale methods with coinciding atoms and nodes (of elements), such as QC, new generations of strong coupling methods have been developed by adopting the concepts from the meshless techniques (Alizadeh and Mohammadi 2019). The variable node multiscale method (VNMM) and the enriched multiscale method (EMM) employ a novel variable node finite element, which is designed to perform as atomistic and continuum elements, simultaneously. The basic idea is to assign the nodes of the elements both the degrees of freedom of the element and the atoms concurrently. Moreover, the disordered concurrent multiscale method (DCMM) adopts the meshless approach to generate a continuous deformation gradient over the discrete set of atoms for accurate simulation of polymers and amorphous materials.

The energy-based methods, such as QC, CQC, BDM and BSM, are built on the summation of the total energy (E^{tot}) over the fine (E^{m}) and coarse (E^{M}) scales, by the atomistic potentials U_α^{p} and the finite element strain energy density w^{s}, respectively,

$$E^{\mathrm{tot}} = E^{\mathrm{m}} + E^{\mathrm{M}} = \sum_{\alpha=1}^{N_{\mathrm{a}}} U_\alpha^{\mathrm{p}}(\boldsymbol{u}) + \sum_{e=1}^{N_{\mathrm{a}}} \Omega_e w^{\mathrm{s}}\left(\boldsymbol{F}^e\right) \tag{4.5.1}$$

$$\boldsymbol{f} = -\frac{\partial E^{\mathrm{tot}}}{\partial \boldsymbol{u}} \tag{4.5.2}$$

where Ω_e is the volume of each finite element e and $\boldsymbol{F}^e$ is the corresponding deformation gradient. N_{a} and N_{e} are the total number of atoms and elements, respectively.

The force methods, such as Feat, CADD and the forced-based QC, directly describe the force vector in terms of the displacement field in different regions,

$$\boldsymbol{f} = \boldsymbol{f}(\boldsymbol{u}) \tag{4.5.3}$$

An advantage of the forced-based methods is to eliminate/reduce the unwanted ghost forces that are usually encountered in the energy-based methods.

Table 4.11 provides a brief comparison between a number of main concurrent multiscale methods (Alizadeh 2019).

4.5.2 Quasi-Continuum Method (QC)

The quasi-continuum (QC) method is an efficient multiscale technique with an integrated approach for transition from the atomistic scale to higher scales. QC was introduced by Tadmor, Philips and Ortiz in 1996 for simple crystalline structures to analyse dislocation core problems and nano indentations (Tadmor 1996; Tadmorc ct al. 1996, 1999). It was later

Table 4.11 Comparison of the main specifications of a number of concurrent multiscale methods.

Multiscale method	Coincidence of atoms and nodes	Transition zone	Governing equations	Continuum constitutive model
Quasi continuum (QC)	✔	–	Energy	Cauchy–Born
Bridging domain (BDM)	–	✔	Energy	Cauchy–Born
Atomistic to continuum (AtC)	✔	✔	Force	Linear elastic
Bridging scale (BSM)	–	–	Energy	Cauchy–Born
DCMM	✔	–	Energy	MLS-Smoothed Cauchy–Born
VNMM	✔	–	Energy	Cauchy–Born
EMM	✔	–	Energy	Cauchy–Born

extended to polycrystalline materials (Shenoy et al. 1998) and enhanced by adaptivity to solve nano fracture problems (Miller et al. 1998). Extension to three-dimensional (Shenoy et al. 1999), multilattice (Tadmor et al. 1999), ferroelectric crystal applications (Tadmor et al. 2002) and finite temperature solutions (Dupuy et al. 2005) were among the major developments of QC. The coupled atomistic and discrete dislocation (CADD) was developed as a combination of the forced-based QC with the discrete dislocation dynamics formulation (Shilkort et al. 2002).

QC is a concurrent multiscale method with a mixed atomistic and continuum formulation. It is among the most widely employed concurrent methods due to its efficiency for materials with simple atomistic/molecular structures, simplicity of formulation and availability of open-source codes.

The QC model consists of both atoms and finite elements over the whole model, but with different characteristics based on the level of deformation or strain gradient in each region. The first set includes atoms with independent degrees of freedom for regions with high gradients, such as domains that include a dislocation core. In these regions, the size of finite elements is reduced down to the lattice size of the crystal. The rest of the model with relatively small gradients includes the constrained atoms that are somehow related to the degrees of freedom of corresponding coarse finite elements (Figure 4.80). It is also possible to adopt the so-called cluster QC, where a set of atoms in the vicinity of an independent atom are selected for sampling and computation of the energy (Figure 4.81).

According to Figure 4.79, two main types of atoms can be considered in the QC modelling. Atoms in the high gradient region are assumed to be independent atoms that are subject to a highly variable atomic response. In fact, each atom in this region has its own degrees of freedom to accurately capture any existing atomic phenomena. In contrast, atoms far from the high gradient region are expected to follow a similar pattern of continuous response, and their degrees of freedom are constrained and governed by the corresponding finite element field in terms of the independent atoms positioned on the nodes of

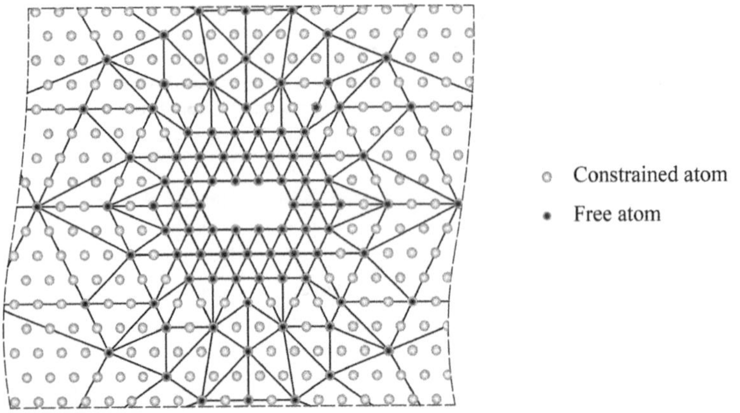

Figure 4.80 QC atomistic/FEM modelling strategy with constrained (white-filled circles) and independent/free atoms (black-filled circles). *Source:* Reproduced from Alizadeh 2019/University of Tehran.

finite elements. Such nodal atoms are called the representative atoms. A QC model is computationally more efficient if the number of independent atoms N_{ind} can be kept as low as possible ($N_{\text{ind}} \gg N_{\text{t}}$).

To explain the QC formulation, the total energy of the system can be computed by summation over the energy U_{α}^{p} of all N_{t} atoms of the domain (both constrained and independent atoms):

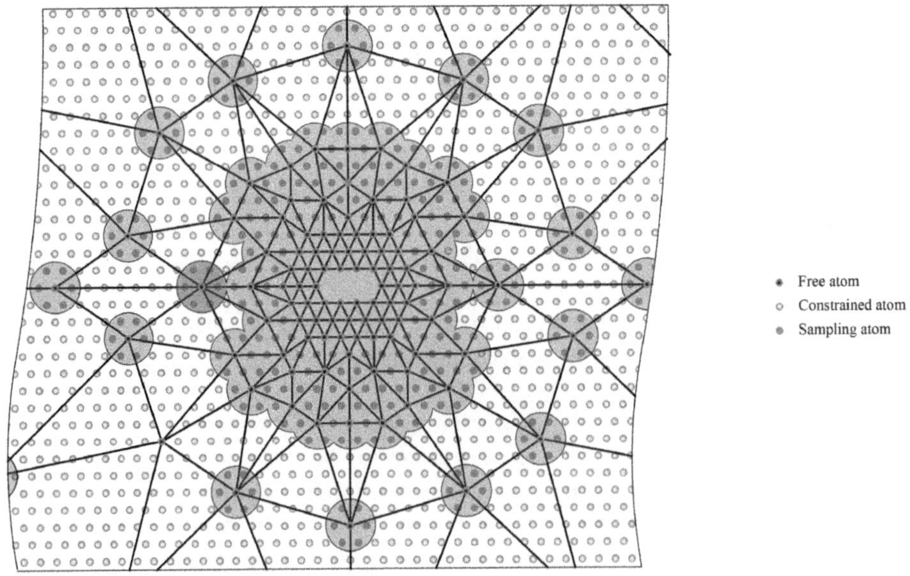

Figure 4.81 Cluster QC with constrained atoms (white-filled circles), independent (free) atoms (black-filled circles), and cluster/sampling atoms (grey-filled circles). *Source:* Reproduced from Alizadeh 2019/University of Tehran.

$$E^{\text{tot}} = \sum_{\alpha=1}^{N_t} U_\alpha^{\text{p}} \tag{4.5.4}$$

Equation (4.5.4) can be written explicitly in terms of the energies of independent and constrained atoms,

$$E^{\text{tot}} = \sum_{\alpha=1}^{N_{\text{ind}}} U_\alpha^{\text{p,ind}} + \sum_{\alpha=1}^{N_{\text{ind}}} U_\alpha^{\text{p,con}} \left(\boldsymbol{u}_1,\ \boldsymbol{u}_2, \ldots, \boldsymbol{u}_{N_{\text{ind}}} \right) \tag{4.5.5}$$

In general, the energy of a constraint atom $U_\alpha^{\text{p,con}}$ depends to all N_{ind} independent atoms.

In QC, computation of energy of independent atoms is performed by the molecular statics (MS) formulation. For each constrained atom, its displacement can be obtained from the finite element interpolation based on the independent atoms at corner nodes of the finite element that contains the atom.

The position $\mathbf{x}$ of each constrained atom can be computed by the finite element interpolation of the displacement field $\boldsymbol{u}_\alpha = \boldsymbol{u}_\alpha^{\text{ind}}$ of independent atoms:

$$\boldsymbol{u}^{\text{h}}(\mathbf{x}) = \sum_{\alpha=1}^{N_{\text{ind}}} N_\alpha(\mathbf{x}) \boldsymbol{u}_\alpha \tag{4.5.6}$$

where N_α represents the classical linear finite element shape function at node (independent atom) α.

The energy of constrained atoms can now be related to the finite element displacement field $\boldsymbol{u}^{\text{h}}$

$$E^{\text{tot}} = \sum_{\alpha=1}^{N_{\text{ind}}} U_\alpha^{\text{p,ind}} + \sum_{\alpha=1}^{N_{\text{con}}} U_\alpha^{\text{p,con}} \left(\boldsymbol{u}^{\text{h}} \right) \tag{4.5.7}$$

According to the Cauchy–Born rule, deformations of the constrained atoms are governed by the continuum (FEM) deformation gradient. In conventional QC modelling, linear triangular elements are employed, which only provide a constant deformation gradient $\boldsymbol{F}_e$ over the element e,

$$\mathbf{x} = \boldsymbol{F}_e(\mathbf{X}) \tag{4.5.8}$$

As a result, the energy of all constrained atoms within a finite element remains the same. Therefore, the total energy of the system (4.5.7) can now be computed by

$$E^{\text{tot}} = \sum_{\alpha=1}^{N_{\text{ind}}} U_\alpha^{\text{p,ind}} + \sum_{e=1}^{N_e} \Omega_e w^{\text{s}} \left(\boldsymbol{F}_e \right) \tag{4.5.9}$$

where N_e is the number of elements, Ω_e is the volume of element e and w^{s} is the strain energy density function.

Alternatively, the total energy can be simply written in terms of the energy of independent atoms $U_\alpha^{\mathrm{p,ind}}$ and the number of constrained atoms to each independent atom (n_α^{con}), determined by a Voronoi approach (Tadmor and Miller 2011; Tadmore et al. 2012):

$$E^{\mathrm{tot}} = \sum_{\alpha=1}^{N_{\mathrm{ind}}} n_\alpha^{\mathrm{con}} U_\alpha^{\mathrm{p,ind}} \tag{4.5.10}$$

The force $\boldsymbol{f}_{\alpha-\beta}$ on an independent atom α by its neighbour independent atom β can be computed from the derivative of the interatomic potential ϕ with respect to the distance $r_{\alpha\beta}$ of the two atoms (see Equation (4.3.129)):

$$\boldsymbol{f}_{\alpha-\beta} = -\frac{\partial \phi(r_{\alpha\beta})}{\partial r_{\alpha\beta}} \mathbf{e}_{\alpha\beta} \tag{4.5.11}$$

where $\mathbf{e}_{\alpha\beta}$ is the unit vector along $\boldsymbol{r}_{\alpha\beta}$. The corresponding so-called atomistic stiffness $\boldsymbol{K}_{\alpha-\beta\gamma}$,

$$\boldsymbol{K}_{\alpha-\beta\gamma} = \frac{\partial \boldsymbol{f}_{\alpha-\beta}}{\partial r_{\alpha\gamma}} \tag{4.5.12}$$

A drawback of the initial QC formulation is the generation of ghost forces. According to Figure 4.82, in computations related to the independent atom A, the constrained atom B is involved (because B is inside the influence domains of A), whereas computations of the constrained atom B is not related to atom A and is only related to the independent atoms at corner nodes of its corresponding element. In a simple description,

$$\frac{\partial U_{\mathrm{A}}^{p}}{\partial \boldsymbol{u}_{\mathrm{B}}} \neq 0, \quad \frac{\partial U_{\mathrm{B}}^{p}}{\partial \boldsymbol{u}_{\mathrm{A}}} = 0 \tag{4.5.13}$$

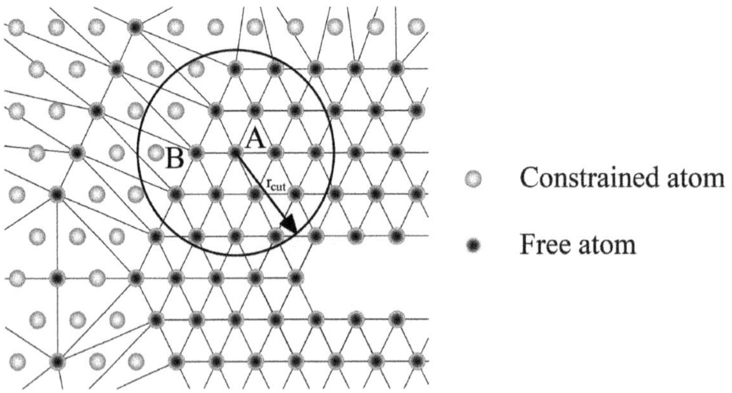

Figure 4.82 Generation of ghost forces due to different effects of independent (black-filled) and dependent (white-filled) atoms. *Source:* Reproduced from Alizadeh 2019/University of Tehran.

In order to take into account the effect of ghost forces, the potential energy of the system Π

$$\Pi(\boldsymbol{u}) = \sum_{\alpha=1}^{N_{\text{ind}}} U_{\alpha}^{\text{p,ind}} + \sum_{e=1}^{N_e} \Omega_e w^s(\boldsymbol{F}_e) - \sum_{\alpha=1}^{N_{\text{ind}}} \boldsymbol{f}_{\alpha}^{\text{ext}}.\boldsymbol{u}_{\alpha} \tag{4.5.14}$$

is modified to

$$\Pi^{\text{mod}}(\boldsymbol{u}) = \sum_{\alpha=1}^{N_{\text{ind}}} U_{\alpha}^{\text{p,ind}} + \sum_{e=1}^{N_e} \Omega_e w^s(\boldsymbol{F}_e) - \sum_{\alpha=1}^{N_{\text{ind}}} \left(f_{\alpha}^{\text{ext}} + f_{\alpha}^{\text{ghost}}\right).\boldsymbol{u}_{\alpha} \tag{4.5.15}$$

It should be noted that QC may lead to a conventional large deformation finite element model if independent atoms are ignored. Moreover, it may lead to a full MS solution if all atoms are considered to be independent.

In order to provide a simple example of QC modelling, the problem solved by Moslemzadeh et al. (2019) is re-examined. The model is related to the study of the roughness effect on nanoindentation of copper thin plates (films), as depicted in Figure 4.83. The atoms are oriented along the $X[1,1,1]$, $Y[\bar{1},1,0]$ and $Z[\bar{1},\bar{1},2]$ crystallographic directions with the lattice spacing parameter $a_0 = 3.615\,\text{Å}$. The model is assumed to be periodic in the Z direction. The EAM potential is adopted for the MS computations within the QC multiscale procedure.

Figure 4.84 Illustrates the QC and full atomic models around the indentation region.

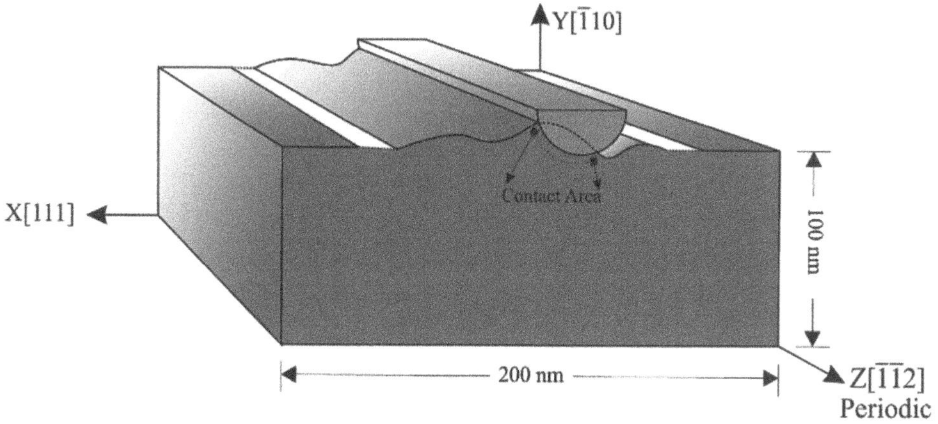

Figure 4.83 Definition of nanoindentation on a rough surface.

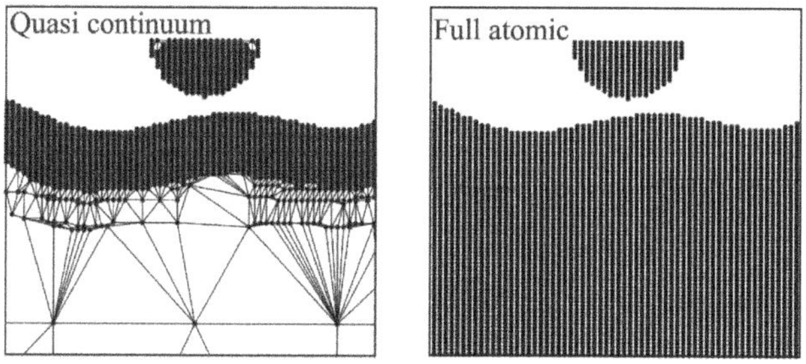

Figure 4.84 QC and full atomic models of the nanoindentation problem.

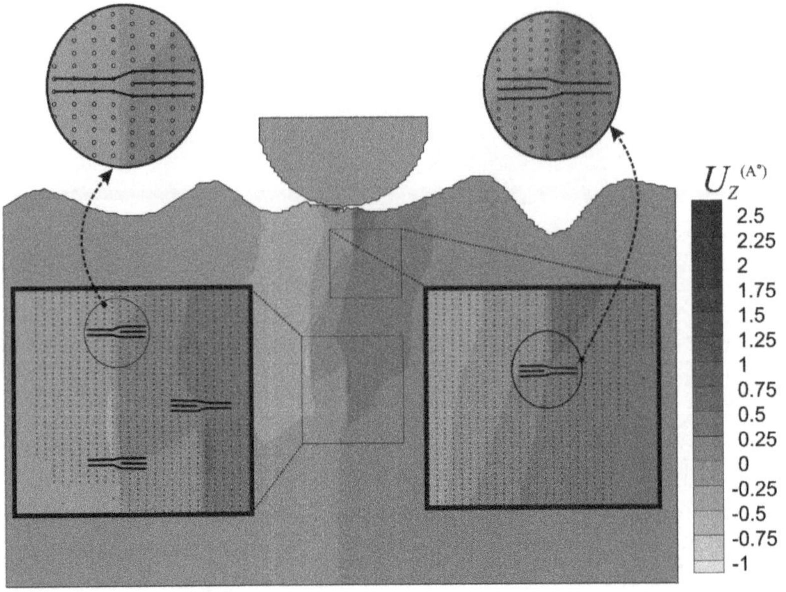

Figure 4.85 Displacement in the Z direction for the circular indenter. *Source:* Adapted from Moslemzadeh et al. 2019.

The atomic structure of the model and contours of the *Z*-displacements at the final step of the adaptive QC multiscale analysis is presented in Figure 4.85. Nucleation of dislocations are clearly observed in regions where stacking fault planes are created, as reported by Moslemzadeh et al. (2019). They are closely related to the jumps in the force-penetration curves of Figure 4.86.

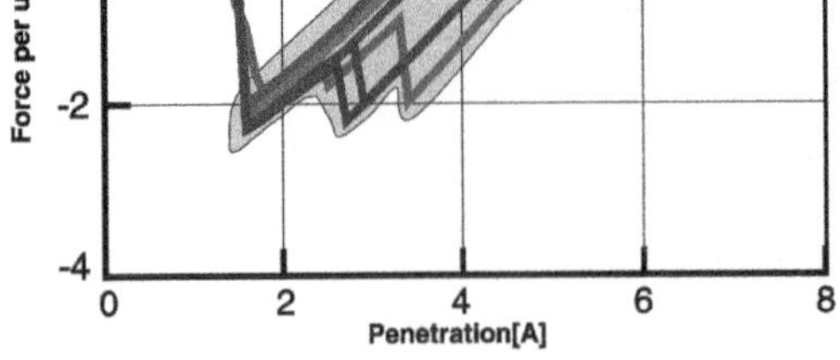

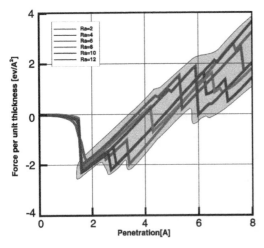

Figure 4.86 Load-penetration response for the circular indenter. *Source:* Adapted from Moslemzadeh et al. 2019.

4.5.3 Bridging Domain Method (BDM)

The bridging domain method (BDM) was developed by Xiao and Belytschko (2004) by partitioning the domain into atomistic and continuum regions, which share an overlap or bridging region.

Figure 4.87 Shows a typical description for partitioning a domain into coarse (continuum) and fine (atomistic) scale regions, Ω^{M} and Ω^{m}, respectively. The coarse region is usually handled by the finite element method, while the fine region is basically governed by the atomic formulation.

Moreover, the interface or bridging region Ω^{i} may be considered to facilitate imposition of consistency in between the two regions. The interface region is composed of both set of atoms and finite elements, which are related in different fashions depending on the assumptions of each specific multiscale method.

Figure 4.88 shows two different approaches for discretizing the interface region. In the first approach, the nodes of finite elements are refined towards the interatomic distance and match the atoms of the fine region. In the second approach, an additional set of atoms in the so-called handshaking/transient region Ω^{h} exists inside the interface region. The handshake region is a mix of fine and coarse descriptions. Independent atoms near the interface depend on the solution of atoms within the interface zone. The handshaking atoms are defined in terms of the independent and dependent atoms.

For computing the energy of atoms in both cases of Figure 4.88, all atoms within the range of the cut-off zone of each atom should be considered. For a typical atom A, the sum of energy on all free/independent neighbouring atoms can simply be calculated, but for a

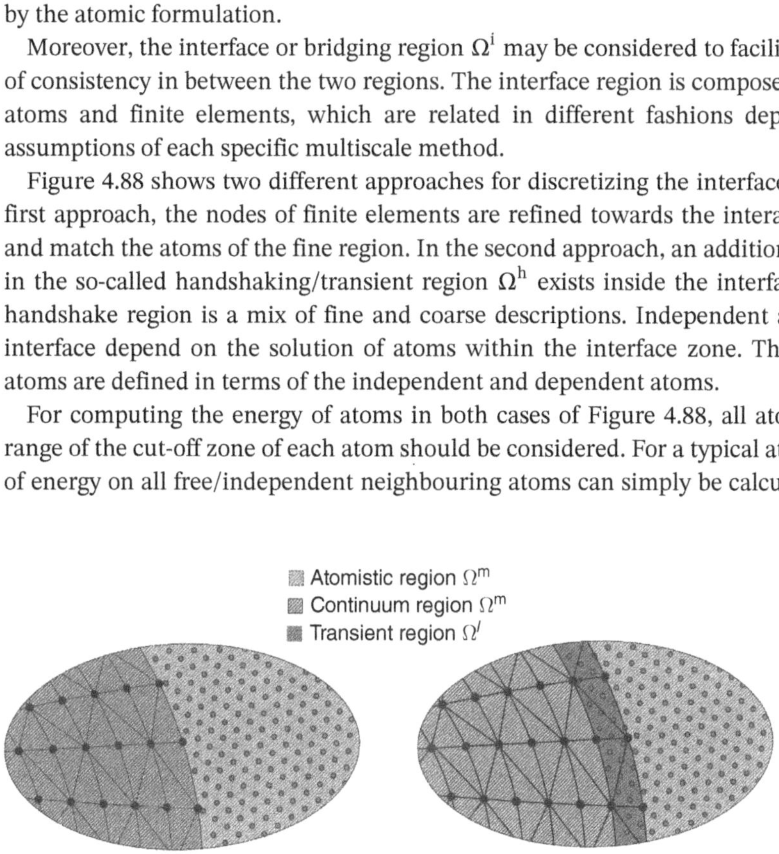

Figure 4.87 Domain partitioning into continuum and atomistic regions. *Source:* Reproduced from Alizadeh 2019/University of Tehran.

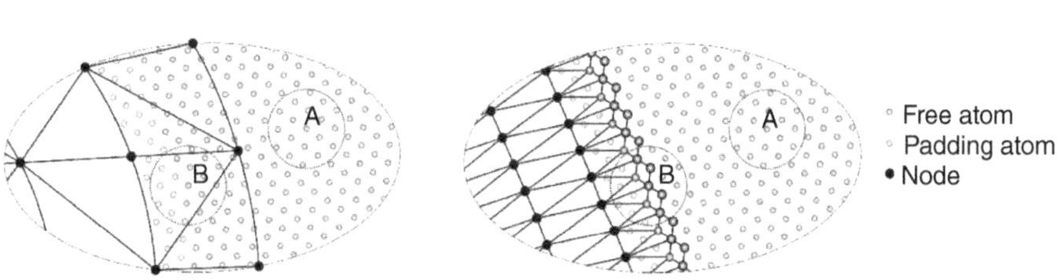

Figure 4.88 The interface zone composed of handshake and padding regions. *Source:* Reproduced from Alizadeh 2019/University of Tehran.

typical atom B near or on the transition zone, only some of the atoms inside the cut-off zone are free/independent, and a number of dependent atoms (pad atoms) should contribute in enforcing the necessary boundary conditions of nearby independent atoms. Pad atoms only follow the finite element displacement field. The size of padding zone Ω^{p}, however, depends on the size of the cut-off of the adopted atomic potential.

The bridging domain method may adopt two approaches in dealing with compatibilities of different regions. The first assumption is a strong compatibility by imposing the FEM displacements $\boldsymbol{u}_I$ to determine the displacements of the handshake and pad dependent atoms $\boldsymbol{u}_\alpha$ in the interface region,

$$\boldsymbol{u}_\alpha = \sum_{I=1}^{N_n} N_I(\mathbf{x}_\alpha)\boldsymbol{u}_I \tag{4.5.16}$$

where N_I is the finite element shape function.

The second type is a weak compatibility, where the displacements $\boldsymbol{u}_I$ of finite element nodes inside the fine/atomic region are determined from the displacements of independent atoms $\boldsymbol{u}_\alpha$. This can be achieved by a simple weighted averaging technique (Figure 4.89),

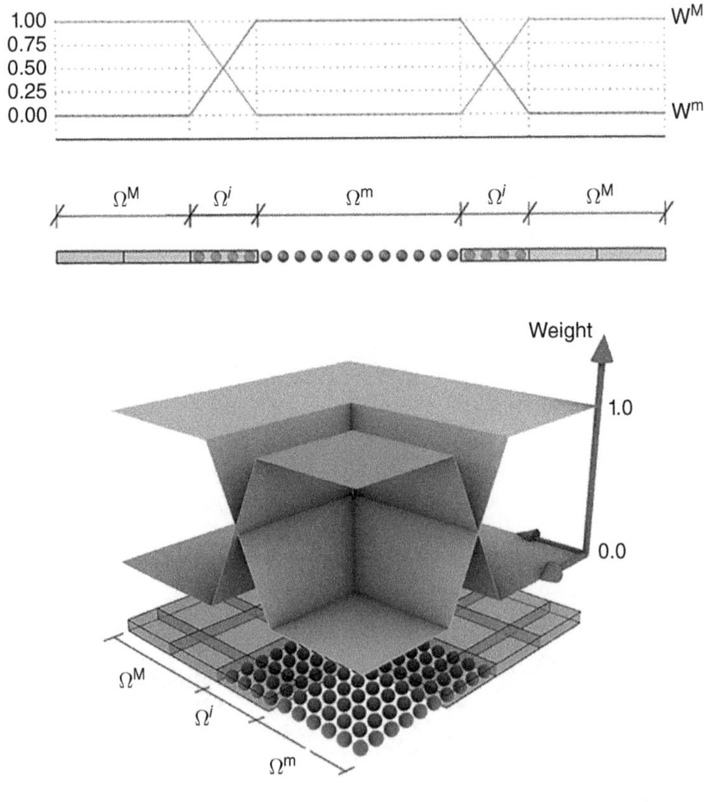

Figure 4.89 Illustration of the weight function for atomic and continuum regions. *Source:* Reproduced from Alizadeh 2019/University of Tehran.

$$\boldsymbol{u}_l = \sum_{\alpha=1}^{N_{\text{ind}}} W_l(\boldsymbol{R}_\alpha)\boldsymbol{u}_\alpha \tag{4.5.17}$$

$$W_l(\boldsymbol{R}_\alpha) = 1 - 3\left(\frac{r}{r_{\text{cut}}}\right)^2 + 2\left(\frac{r}{r_{\text{cut}}}\right)^3 \tag{4.5.18}$$

$$r = \left\|\mathbf{x}_l - \boldsymbol{R}_\alpha\right\| \tag{4.5.19}$$

The total potential energy is computed as the sum of fine and coarse scales, E^{m} and E^{M}, respectively, and the energy of handshake/interface region E^{i},

$$E^{\text{tot}} = E^{\text{m}} + E^{\text{M}} + E^{\text{i}} \tag{4.5.20}$$

where the energy of the fine scale E^{m} is determined from the sum of energy U_α^{p} of atoms α and the corresponding external work by $f^{\text{ext},\alpha}$ on the atomistic region,

$$E^{\text{m}} = \sum_{\alpha=1}^{N_a} U_\alpha^{\text{p}} - \sum_{e=1}^{N_e} \boldsymbol{f}^{\text{ext},\alpha} \cdot \boldsymbol{u}_\alpha \tag{4.5.21}$$

For the coarse scale, the energy E^{M} is determined from the strain energy density function $w^{\text{s}}(\boldsymbol{F})$ and the work of prescribed external tractions $\boldsymbol{f}^t$ with the continuum displacement field $\boldsymbol{u}(\mathbf{x})$,

$$E^{\text{M}} = \int_{\Omega^{\text{M}}} w^{\text{s}}(\boldsymbol{F})\,\mathrm{d}\Omega - \int_{\Gamma_{\text{t}}^{\text{M}}} \boldsymbol{f}^{\text{t}} \cdot \boldsymbol{u}\,\mathrm{d}\Gamma \tag{4.5.22}$$

or in a discrete form over the finite elements for the first term, and over the traction boundary $\Gamma_{\text{t}}^{\text{M}}$ for the second one,

$$E^{\text{M}} = \sum_{e=1}^{N_e} \Omega_e^{\text{M}} w_e^{\text{s}}(\boldsymbol{F}) - \sum_{t=1}^{N_{\text{tr}}} \left[\boldsymbol{f}^t \cdot \boldsymbol{u}\right]_t A_t^{\text{M}} \tag{4.5.23}$$

The numerical quadrature form of E^{M} on N_q quadrature points $\mathbf{x}_{eq}$ with weights W_q for the continuum finite elements and on N_{qt} quadrature points $\mathbf{x}_{tq}$ with weights W_{qt} on the traction boundary Γ_t^{M} can be written as

$$E^{\text{M}} = \sum_{e=1}^{N_e}\sum_{q=1}^{N_q}\left[W_q \Omega_e^{\text{M}} w_e^{\text{s}}\left(\boldsymbol{F}(\mathbf{x}_{eq})\right)\right] - \sum_{q=1}^{N_{qt}}\left[W_{qt} \boldsymbol{f}^t(\mathbf{x}_{tq}) \cdot \boldsymbol{u}(\mathbf{x}_{tq}) A_t^{\text{M}}\right] \tag{4.5.24}$$

where Ω_e^{M} is the volume of element e and A_t^{M} represents the area/length associated with each quadrature point on the traction boundary.

The energy of the handshake/interface region E^i is defined by a blending of atomic and continuum regions. The continuous blending function $\psi(\mathbf{x})$ is assumed to vary from 0 at the atomic edge to 1 at the continuum edge of the handshake region (Figure 4.89). Moreover, the handshake region is chosen in such a way to avoid any external loadings (which is a disadvantage of the approach). Therefore, the energy of the system in the handshake region can be written as (Liu et al. 2006a)

$$E^{\mathrm{i}} = [1-\psi(\boldsymbol{R}_\alpha)]E^{\mathrm{m}} + \psi(\mathbf{x})E^{\mathrm{M}} \tag{4.5.25}$$

or, more explicitly,

$$E^{\mathrm{i}} = \left[\sum_{\alpha=1}^{N_{\mathrm{ah}}}[1-\psi(\boldsymbol{R}_\alpha)]U_\alpha^{\mathrm{p}}(\boldsymbol{u})\right] + \int_{\Omega^{\mathrm{h}}}\psi(\mathbf{x})w^{\mathrm{s}}(F(\mathbf{x}))\mathrm{d}\Omega \tag{4.5.26}$$

In this region, the displacement $\boldsymbol{u}_\alpha$ of each atom is constrained to the displacement $\boldsymbol{u}(\boldsymbol{R}_\alpha)$, which is the finite element interpolated displacement at the reference position of atom α. The constrained potential energy of the handshake region can be written using a Lagrangian approach,

$$E^{\mathrm{i}} = \left[\sum_{\alpha=1}^{N_{\mathrm{aH}}}[1-\psi(\boldsymbol{R}_\alpha)]U_\alpha^{\mathrm{p}}(\boldsymbol{u})\right] + \left[\int_{\Omega^{\mathrm{h}}}\psi(\mathbf{x})w^{\mathrm{s}}(\boldsymbol{F}(\mathbf{x}))\mathrm{d}\Omega\right] + \left[\sum_{\alpha=1}^{N_{\mathrm{ah}}}[\gamma_\alpha.\boldsymbol{h}_\alpha]\right] \tag{4.5.27}$$

where γ_α is the vector of Lagrange additional degrees of freedom of atom α and the constraint $\boldsymbol{h} \equiv \boldsymbol{0}$ is defined as

$$\boldsymbol{h}_\alpha = \boldsymbol{u}(\boldsymbol{R}_\alpha) - \boldsymbol{u}_\alpha \tag{4.5.28}$$

with

$$\boldsymbol{u}(\boldsymbol{R}_\alpha) = \sum_{I=1}^{N_{\mathrm{n}}} N_I(\boldsymbol{R}_\alpha)\boldsymbol{u}_I \tag{4.5.29}$$

The coupled BDM formulation can finally be written in the general forms of

$$\mathcal{M}^{\mathrm{m}}\ddot{\boldsymbol{r}} = \boldsymbol{f}^{\mathrm{ext}} - \boldsymbol{f}^{\mathrm{int}} - \boldsymbol{f}^{\mathrm{h}} \tag{4.5.30}$$

$$\mathcal{M}^{\mathrm{M}}\ddot{\boldsymbol{u}} = \boldsymbol{F}^{\mathrm{ext}} - \boldsymbol{F}^{\mathrm{int}} - \boldsymbol{F}^{\mathrm{h}} \tag{4.5.31}$$

where $\boldsymbol{f}^{\mathrm{h}}$ and $\boldsymbol{F}^{\mathrm{h}}$ are the additional forces associated with the Lagrange enforcement of the constraint (4.5.28).

It is recommended to employ a relaxed compatibility constraint, so there will be no need for the atoms and finite element nodes to match along the interface. Computationally, however, it is difficult to enforce in dynamic solutions.

In a similar approach, the atomistic to continuum (AtC) method was developed as the force-based version of the bridging domain approach. AtC assumes that the energies of atomistic and continuum regions linearly vary inside the transition zone (Figure 4.90).

In AtC, the force $\boldsymbol{f}_i$ on free atoms and the residual force vector $\boldsymbol{f}_I$ on finite element nodes are computed from (Alizadeh 2019)

$$\boldsymbol{f}_i = \sum_{\alpha,\beta\neq\alpha} \frac{\partial\phi(r_{\alpha\beta})}{\partial r_{\alpha\beta}}\mathbf{e}_{\alpha\beta} \tag{4.5.32}$$

$$\boldsymbol{f}_I - \int_{\Omega_t}\mathbf{B}^{\mathrm{T}}\boldsymbol{P}\mathrm{d}\Omega \tag{4.5.33}$$

where $\mathbf{B}$ is the matrix of derivatives of the finite element shape functions.

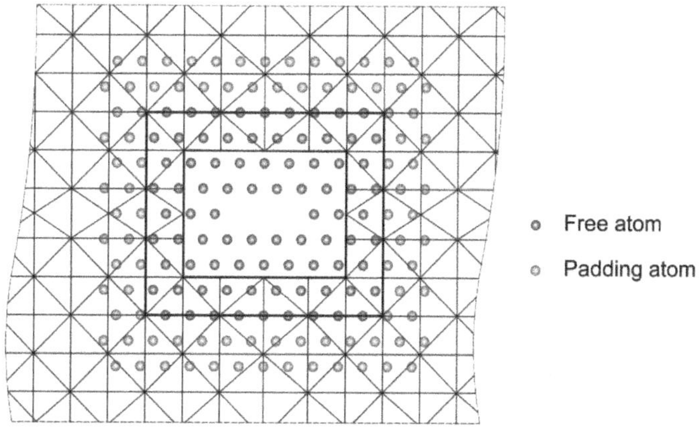

Figure 4.90 A schematic model of the atomistic to continuum method. *Source:* Reproduced from Alizadeh 2019/University of Tehran.

4.5.4 Bridging Scale Method (BSM)

In the bridging scale method, while MD is employed in high gradient regions, which require atomistic accuracy, FEM is used in the whole domain (Wagner and Liu 2003), as depicted in Figure 4.91. The fine scale region is naturally defined by the atomistic formulation. The continuum region is handled by the displacements of finite elements. The finite elements, however, do exist, even in the fine scale region. Compatibility of finite elements with the atomic solution inside the fine scale region is accomplished by constraining the finite element displacements with the atomic displacements by the least square technique.

BSM is an energy-based multiscale approach based on decomposing the displacement field $\boldsymbol{u}^{\mathrm{t}}$ into the coarse and fine scale fields, $\boldsymbol{u}^{\mathrm{M}}$ and $\boldsymbol{u}^{\mathrm{m}}$, respectively,

$$\boldsymbol{u}^{\mathrm{t}}(\mathbf{x}) = \boldsymbol{u}^{\mathrm{M}}(\mathbf{x}) + \boldsymbol{u}^{\mathrm{m}}(\mathbf{x}) \tag{4.5.34}$$

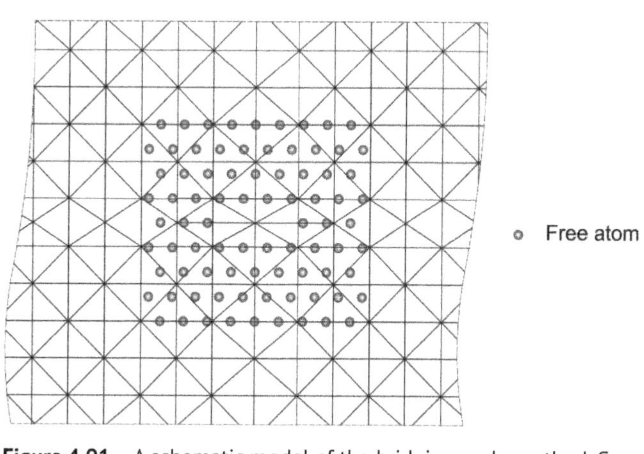

Figure 4.91 A schematic model of the bridging scale method. *Source:* Reproduced from Alizadeh 2019/University of Tehran.

$\boldsymbol{u}^{\mathrm{m}}(\mathbf{x})$ represents the fine scale fluctuation, which is assumed to be orthogonal to the coarse scale solution $\boldsymbol{u}^{\mathrm{M}}(\mathbf{x})$ (Hughes et al. 1998).

In the coarse region, $\boldsymbol{u}^{\mathrm{M}}$ is defined by the finite element approximation using the matrix of FE shape functions, $\mathbf{N}^{\mathrm{M}}$, as

$$\boldsymbol{u}^{\mathrm{M}}(\mathbf{x}) = \mathbf{N}^{\mathrm{M}}(\mathbf{x})\bar{\mathbf{d}} = \sum_{I} N_I^{\mathrm{M}}(\mathbf{x})\boldsymbol{d}_I \tag{4.5.35}$$

$\boldsymbol{u}^{\mathrm{M}}$ for the position $\mathbf{x}$ on the fine region can be related to the MD displacements $\boldsymbol{r}(\mathbf{x})$ by the projection operator $\mathbf{P}$,

$$\mathbf{P}\boldsymbol{r}(\mathbf{x}) = \sum_{I} N_I^{\mathrm{M}}(\mathbf{x})d_I \tag{4.5.36}$$

The projection operator $\mathbf{P}$ is computed by minimizing the weighted least square difference J between the atomic displacement $\boldsymbol{u}_\alpha\left(\boldsymbol{r}(\mathbf{x}_\alpha)\right)$ and the finite element displacement $\boldsymbol{d}_I$ at the position of atom α:

$$\boldsymbol{J} = \sum_{\alpha=1}^{n_a} m_\alpha \left[\boldsymbol{u}_\alpha - \sum_{I} N_I^{\mathrm{M}}(\mathbf{x}_\alpha)\boldsymbol{d}_I\right]^2 \tag{4.5.37}$$

with (Liu et al. 2006a)

$$\mathbf{P} = \mathbf{N}^{\mathrm{M}}\left(\mathcal{M}^{\mathrm{M}}\right)^{-1}\left(\mathbf{N}^{\mathrm{M}}\right)^{\mathrm{T}}\mathcal{M}^{\mathrm{m}} \tag{4.5.38}$$

where $\mathcal{M}^{\mathrm{m}}$ is the diagonal atomistic mass matrix and $\mathcal{M}^{\mathrm{M}}$ is defined as

$$\mathcal{M}^{\mathrm{M}} = \left(\mathbf{N}^{\mathrm{M}}\right)^{\mathrm{T}}\mathcal{M}^{\mathrm{m}}\mathbf{N}^{\mathrm{M}} \tag{4.5.39}$$

The solution procedure requires an iterative solution for the two independent sets of fine and coarse scale displacements. It should be noted that numerically induced wave reflections may be generated at the interface of atomic and continuum regions. Consequently, some un-physical phenomena may be generated from the MD oscillations due to spurious wave reflections. A remedy is to adopt a non-reflecting interface to allow for the waves to propagate freely into the coarse-scale region.

For further details of the BSM solution, refer to Liu et al. (2006a).

4.5.5 Disordered Concurrent Multiscale Method (DCMM)

Most conventional multiscale methods are incapable or ineffective in solving disordered or irregular atomistic structures. The so-called disordered concurrent multiscale method (DCMM) is designed to simulate any irregular structural forms encountered in polymers, glasses, amorphous materials, proteins, etc.

In contrast to most metals with a repetitive atomistic crystalline structure, polymers and amorphous metals are not usually formed from any repetitive basic structural block (see Figure 4.92). A wide range of materials with irregular structures are of interest in various

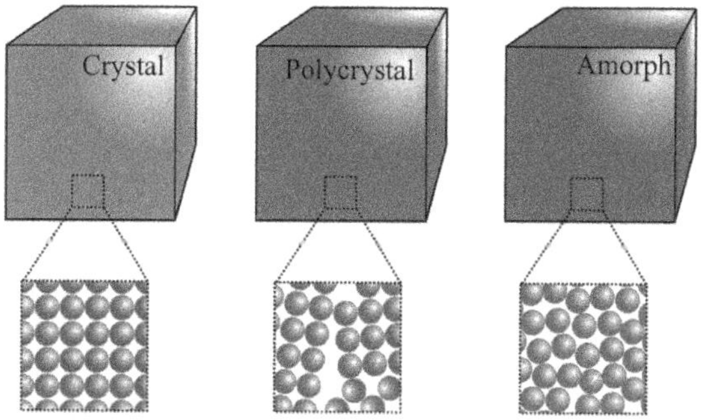

Figure 4.92 Different atomic structures.

engineering disciplines and in biomechanical applications. Even a structured crystalline metal can be converted into an amorphous structure if it is heated to melt into a liquid and then cooled by a high temperature rate to create a glassy/amorphous state.

4.5.5.1 Multiscale Model

The procedure for the disordered concurrent multiscale method (DCMM) simulation is schematically shown in Figure 4.93. The amorphous domain is made of complicated atomic amorphous atomic chains. A full atomic model requires a very large number of atoms. In DCMM, the model comprises both the macro and the atomic scales concurrently. The whole model is discretized by sufficiently fine irregular finite elements. In regions with low gradients (Ω^M), the continuum formulation is adopted to define the governing equations of the finite elements. The shape functions of irregular elements are constructed

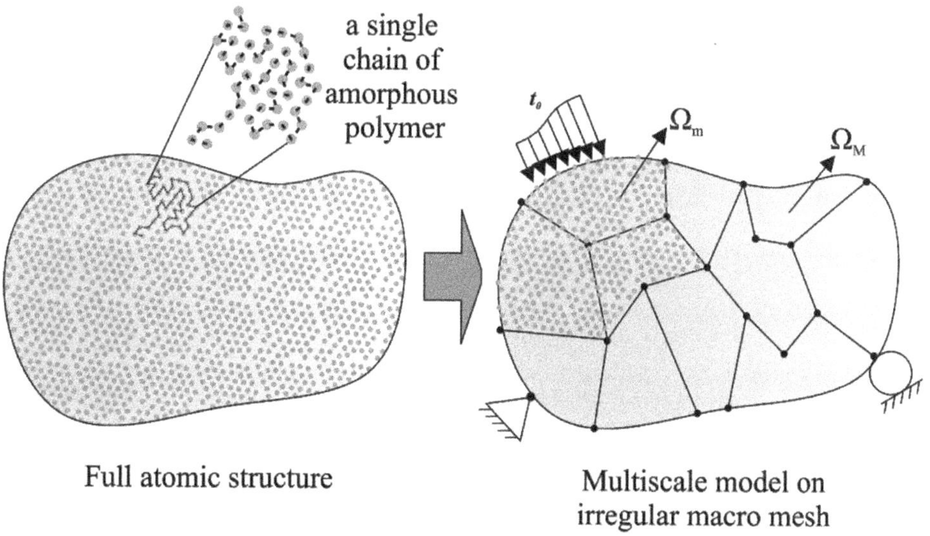

Figure 4.93 Schematic model of the micro Ω^m and macro Ω^M regions. *Source:* Adapted from Moslemzadeh and Mohammadi 2022.

using the concept of maximum entropy. In regions with high gradients or other severe conditions (Ω^{m}), the elements contain the explicit atomic model of the amorphous structure.

4.5.5.2 Governing Equations of the Whole Model

The governing formulation of the present concurrent problem is expressed in the form of minimization of the energy functional Π of the atomic and continuum regions,

$$\Pi = \Pi^{\mathrm{M}} + \Pi^{\mathrm{m}} \tag{4.5.40}$$

Π^{M} defines the total energy of the macro part Ω^{M} in terms of the macro strain energy density function w^{s}, the deformation gradient $\boldsymbol{F}$, the body force $\rho_0 \boldsymbol{f}^{\mathrm{b}}$ and the imposed traction $\boldsymbol{f}^{\mathrm{t}}$,

$$\Pi^{\mathrm{M}} = \int_{\Omega^{\mathrm{M}}} w^{\mathrm{s}}\left(\boldsymbol{F}(\boldsymbol{u})\right)\mathrm{d}\Omega - \int_{\Omega^{\mathrm{M}}} \rho_0 \boldsymbol{f}^{\mathrm{b}}.\boldsymbol{u}\mathrm{d}\Omega - \int_{\Gamma_{\mathrm{t}}^{\mathrm{M}}} \boldsymbol{f}^{\mathrm{t}}.\boldsymbol{u}\mathrm{d}\Gamma \tag{4.5.41}$$

In general, Π^{m} represents the energy of the atomistic region Ω^{m}, including both the potential and kinetic energies, U^{p} and U^{k}, respectively. For instance, Π^{m} for the three-body potential ϕ_{n} for a system of n_{t} atoms can be described by

$$\Pi^{\mathrm{m}} = \underbrace{\left[\frac{1}{2}\sum_{\substack{i,j\\ j\neq i}}^{n_{\mathrm{t}}} \phi_2\left(r_{ij}\right) + \frac{1}{6}\sum_{\substack{i,j,k\neq i\\ i\neq j\neq k}}^{n_{\mathrm{t}}} \phi_3\left(r_{ij}, r_{ik}, \theta_{jik}\right)\right]}_{U^{\mathrm{p}}} + \underbrace{\left[\sum_{i}^{n_{\mathrm{t}}} \frac{1}{2} m_i \left(v_i\right)^2\right]}_{U^{\mathrm{k}}} \tag{4.5.42}$$

Within a typical macro element α, the interatomic forces/stresses in the form of the first Piola–Kirchhoff stress tensor $\boldsymbol{P}^{\mathrm{m}}$ can be derived by differentiation of the potential energy U^{p} (in terms of the interatomic potential ϕ) of the microscale model with respect to the deformation gradient $\boldsymbol{F}_{\alpha}$ (see Section 4.3.6),

$$\boldsymbol{P}^{\mathrm{m}} = \frac{1}{\Omega_0}\frac{\partial U^{\mathrm{p}}}{\partial \boldsymbol{F}_{\alpha}} = \frac{1}{\Omega_0}\sum_{i=1}^{n_t}\left(\frac{\partial \phi}{\partial \boldsymbol{F}_{\alpha}}\right) \tag{4.5.43}$$

Since the atomic model is discrete, a continuous deformation gradient cannot be directly imposed. Therefore, DCMM adopts a meshless-based approach to redistribute atom positions based on the decomposition of the deformation gradient $\boldsymbol{F}_{\alpha}$ into an equivalent continuous deformation gradient $\boldsymbol{F}_{\alpha}^{\mathrm{m}}$ on the atomistic model and the macro deformation gradient $\boldsymbol{F}_{\alpha}^{\mathrm{M}}$, as will be discussed in Section 4.5.5.4.

An iterative solution procedure is followed to obtain the equilibrium state, based on the fact that the two scales share the same Piola–Kirchhoff stress $\boldsymbol{P}^{\mathrm{m}} = \boldsymbol{P}^{\mathrm{M}}$,

$$\boldsymbol{P}^{\mathrm{M}} = \frac{\partial w^{\mathrm{s}}}{\partial \boldsymbol{F}} \tag{4.5.44}$$

with the material modulus $\mathbf{D}^{\mathrm{M}}$

$$\mathbf{D}^{\mathrm{M}} = \frac{\partial^2 w^{\mathrm{s}}}{\partial \boldsymbol{F}^2} \tag{4.5.45}$$

The global force vector $\boldsymbol{f}$ is computed from

$$\boldsymbol{f} = \frac{\partial \Pi}{\partial \boldsymbol{u}} = \int_{\Omega^{\mathrm{m}}} \left(\boldsymbol{P}^{\mathrm{M}}.\nabla\mathbf{N}\right) \mathrm{d}\Omega + \int_{\Omega^{\mathrm{M}}} \left(\boldsymbol{P}^{\mathrm{M}}.\nabla\mathbf{N}\right) \mathrm{d}\Omega - \int_{\Omega^{\mathrm{M}}} \rho_0 \boldsymbol{f}^{\mathrm{b}}.\mathbf{N}\mathrm{d}\Omega - \int_{\Gamma_{\mathrm{t}}^{\mathrm{M}}} \boldsymbol{f}^{\mathrm{t}}.\mathbf{N}\mathrm{d}\Gamma \tag{4.5.46}$$

4.5.5.3 Entropy-Based Macro Scale Model for Irregular Discretization

Discretization of the governing Equation (4.5.46) for the macro displacements $\boldsymbol{u}$ is performed by the conventional finite element shape functions N_i,

$$\boldsymbol{u} = \sum_{i=1}^{n_n} N_i \boldsymbol{u}_i \tag{4.5.47}$$

Nevertheless, the highly irregular and potentially non-convex shapes of finite elements to match the irregular atomistic structure of amorphous materials does not allow for conventional FEM shape functions. The remedy is to use the concept of maximum entropy to derive the shape functions N_i on an irregular non-convex element with variable shapes and number of nodes and edges, as described by Norouzi et al. (2019).

The basic concept is to adopt the definition of the entropy H of a discrete probability function p_i for n_e events in a system:

$$H = -k_{\mathrm{b}} \sum_{i=1}^{n_{\mathrm{e}}} p_i \log p_i \tag{4.5.48}$$

where k_{b} is the Boltzmann coefficient.

In order to extend the concept to compute the shape functions of irregular macro elements, the probability functions p_i are considered to be the shape functions N_i (Norouzi et al. 2019). Therefore, minimization of the constrained functional L_{M} with respect to N_i leads to computation of n_{n} (number of nodes in an element) shape functions

$$L_{\mathrm{M}} = H_{\mathrm{M}}\left(N_i\right) + \sum_{k=1}^{4} \lambda_{\mathrm{M}}^{k} t_{\mathrm{M}}^{k}\left(N_i\right) \tag{4.5.49}$$

with

$$H_{\mathrm{M}}\left(N_i\right) = -k_b \sum_{i=1}^{n_n} N_i \log N_i \tag{4.5.50}$$

$$t_{\mathrm{M}}^{k}\left(N_i\right) = \chi^{\mathrm{k}} - \sum_{i=1}^{n_n} N_i \chi_i^{\mathrm{k}}, \quad \chi^{\mathrm{k}} = 1, x, y, z \tag{4.5.51}$$

For further details, see Norouzi et al. (2019).

4.5.5.4 Continuous Deformation Gradient on an Atomistic Model

Due to the lack of the Cauchy–Born rule for elements with atomic refinement, the positions of discrete atoms in any element α cannot be accurately estimated by the macro deformation gradient $\boldsymbol{F}_{\alpha}^{\mathrm{M}}$ and they should be updated according to an unknown consistent continuous deformation gradient $\boldsymbol{F}_{\alpha}$. Construction of a continuous field over a set of discrete points (atoms) is well accomplished by the meshless concepts. Moreover, the procedure is shown to avoid generation of excessive fictitious (ghost) forces (Moselmzadeh and Mohammadi 2022).

A multiplicative decomposition of the deformation gradient $\boldsymbol{F}_{\alpha}$ for an element α may be adopted to redefine the positions of atoms to ensure a consistent continuous deformation gradient is defined on the set of discrete atoms within a macro element (see Figure 4.94),

$$\boldsymbol{F}_{\alpha} = \boldsymbol{F}_{\alpha}^{\mathrm{M}}\, \boldsymbol{F}_{\alpha}^{\mathrm{m}} \tag{4.5.52}$$

The intermediate configuration is obtained by applying the macro deformation gradient $\boldsymbol{F}_{\alpha}^{\mathrm{M}}$ to the set of atoms within the element α. It is clearly an inconsistent procedure due to the discrete nature of the set of atoms. A meshless-modified continuous and consistent deformation gradient $\boldsymbol{F}_{\alpha}^{\mathrm{m}}$ can then be obtained to transform the inaccurate positions of atoms at the intermediate configuration into the final accurate and consistent positions.

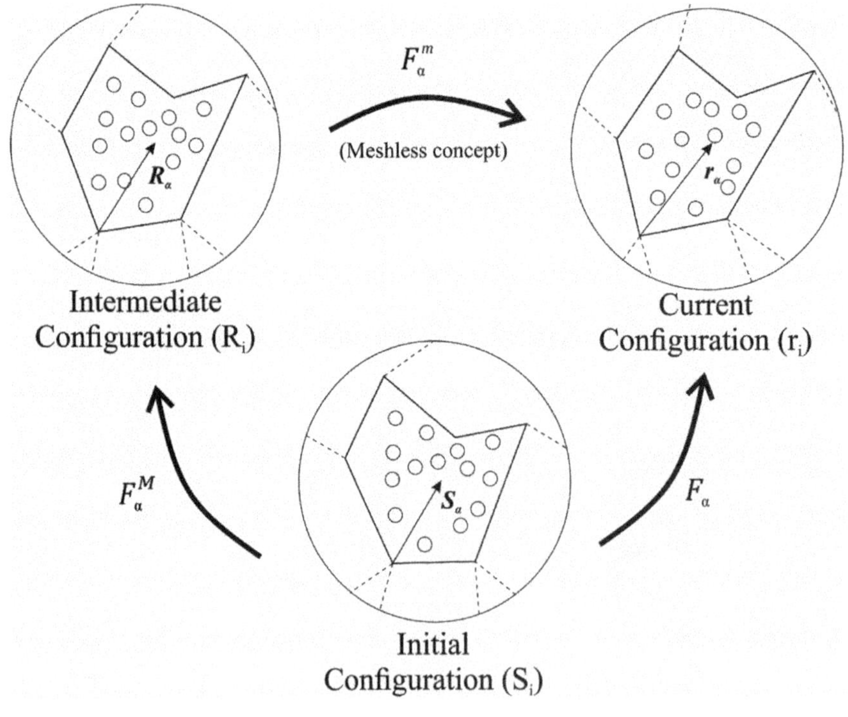

Figure 4.94 Initial and deformed positions of atoms in the element α.

The relative position $\boldsymbol{r}_{\alpha i}$ of atom i with respect to the position $\boldsymbol{r}_{\alpha}$ of the centre of mass of the element is determined from

$$\boldsymbol{r}_{\alpha i} = \boldsymbol{r}_i - \boldsymbol{r}_{\alpha} \tag{4.5.53}$$

where the position $\boldsymbol{r}_{\alpha}$ of the mass centre of n_{α} atoms in element α is obtained from

$$\boldsymbol{r}_{\alpha}(t) = \frac{1}{\sum_{i=1}^{n_{\alpha}} m_i} \sum_{i=1}^{n_{\alpha}} m_i \boldsymbol{r}_i(t) \tag{4.5.54}$$

At each increment of the solution, the position $\boldsymbol{r}_{\alpha i}$ at the current configuration is related to its position at the initial configuration $\boldsymbol{S}_{\alpha i}$ by the deformation gradient $\boldsymbol{F}_{\alpha}$ of the element α:

$$\boldsymbol{r}_{\alpha i} = \boldsymbol{F}_{\alpha}\, \boldsymbol{S}_{\alpha i} \tag{4.5.55}$$

Moreover, the current positions of atoms can be updated from the positions at the intermediate configuration by the deformation gradient $\boldsymbol{F}_{\alpha}$:

$$\boldsymbol{r}_i = \boldsymbol{F}_{\alpha}^{\mathrm{m}} \boldsymbol{R}_i \tag{4.5.56}$$

The same microscale Cauchy–Born rule with the continuous deformation gradient $\boldsymbol{F}_{\alpha}^{\mathrm{m}}$ can be applied to the mass centre of the element,

$$\boldsymbol{r}_{\alpha}^{\mathrm{h}} = \boldsymbol{F}_{\alpha}^{\mathrm{m}} . \boldsymbol{R}_{\alpha} \tag{4.5.57}$$

where the updated positions of atoms $\boldsymbol{r}_{\alpha}^{\mathrm{h}}$ are computed based on the adopted meshless technique.

The meshless procedure for updating the positions of atoms within an element α may be explained according to Figure 4.95. Deriving a continuous deformation gradient is performed on a support domain with radius R for the centre of element α. Accordingly, a

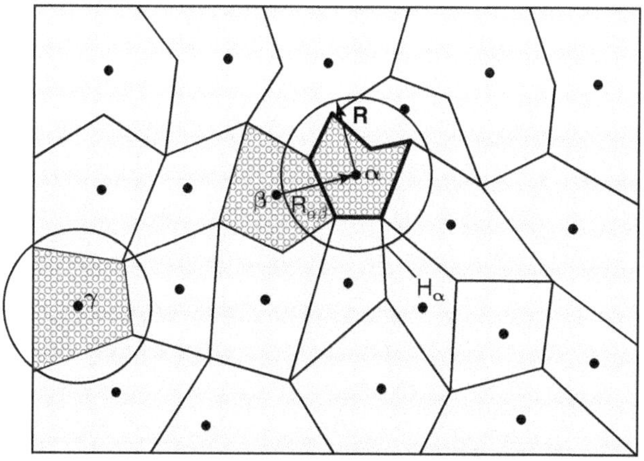

Figure 4.95 Definition of neighbour elements β.

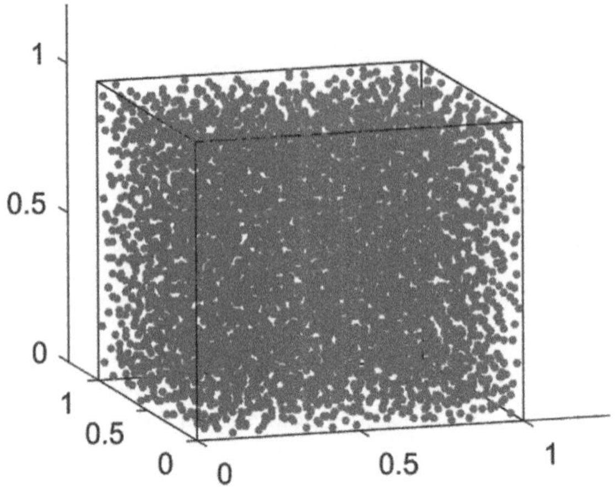

Figure 4.96 The MD model of amorphous material. *Source:* Adapted from Moslemzadeh and Mohammadi 2022.

number of atoms in neighbouring elements β influence the construction of $\boldsymbol{F}_{\alpha} = \boldsymbol{F}_{\alpha}^{\mathrm{M}}\,\boldsymbol{F}_{\alpha}^{\mathrm{m}}$ and the updated positions of atoms.

For any current time t, the relative position of the mass centre of the element with respect to the mass centre of a neighbour element β can be defined by (see Figure 4.95):

$$\boldsymbol{r}_{\alpha\beta}(t) = \boldsymbol{r}_{\beta}(t) - \boldsymbol{r}_{\alpha}(t) \tag{4.5.58}$$

The idea of the moving least square (MLS) technique is adopted to generate a continuous approximation of the deformation gradient field $\boldsymbol{F}_{\alpha}^{\mathrm{m}}$ on the set of atoms within the element α at the current configuration,

$$\boldsymbol{F}_{\alpha}^{\mathrm{m}} = \left(\sum_{\beta=1}^{n_{\mathrm{h}}} W\left(\left|\boldsymbol{R}_{\alpha\beta}\right|\right) \boldsymbol{r}_{\alpha\beta}\,\Omega\,\boldsymbol{R}_{\alpha\beta}\Omega_{\beta} \right) \boldsymbol{M}_{\alpha}^{-1} \tag{4.5.59}$$

with the moment matrix $\boldsymbol{M}_{\alpha}$,

$$\boldsymbol{M}_{\alpha} = \sum_{\beta=1}^{n_{\mathrm{h}}} W\left(\left|\boldsymbol{R}_{\alpha\beta}\right|\right) \boldsymbol{R}_{\alpha\beta}\,\Omega\,\boldsymbol{R}_{\alpha\beta}\Omega_{\beta} \tag{4.5.60}$$

where

$$\boldsymbol{R}_{\alpha\beta} = \boldsymbol{R}_{\beta} - \boldsymbol{R}_{\alpha} \tag{4.5.61}$$

and $\boldsymbol{R}_{\alpha}$ and $\boldsymbol{R}_{\beta}$ are the positions of mass centres of elements α and β in the intermediate configuration ($\boldsymbol{R}_{\alpha} = \boldsymbol{r}_{\alpha}(0)$). $W\left(\left|\boldsymbol{R}_{\alpha\beta}\right|\right)$ is the MLS weight function, n_{h} is the number of elements β surrounding the element α and Ω_{β} is the volume of element β.

It is important to note that construction of $\boldsymbol{F}_{\alpha}^{\mathrm{m}}$ is a key feature of DCMM that corelates the continuous and atomic scale configurations. Ignoring $\boldsymbol{F}_{\alpha}^{\mathrm{m}}$ would lead to a conventional Cauchy–Born rule, which is clearly inaccurate for modelling disordered amorphous structures.

Finally, Equation (4.5.43) can now be completed as

$$\boldsymbol{P}^{\mathrm{m}} = \frac{1}{\Omega_0}\frac{\partial U^{\mathrm{p}}}{\partial \boldsymbol{F}_\alpha} = \frac{1}{\Omega_0}\sum_{i=1}^{n_t}\left(\frac{\partial \phi}{\partial \boldsymbol{F}_\alpha}\right) = \frac{1}{\Omega_0}\sum_{i=1}^{n_t}\left(\frac{\partial \phi}{\partial \boldsymbol{F}_\alpha^{\mathrm{m}}}\right)\left(\boldsymbol{F}_\alpha^{\mathrm{M}}\right)^{-1} \tag{4.5.62}$$

4.5.5.5 Sample Simulation

In order to illustrate the performance of the DCMM approach, simulation of a $54\times54\times54$ Å^3 sample of amorphous silicon, analysed by Moslemzadeh and Mohammadi (2022), is followed. Simulation begins by relaxing the atomic model of crystalline silicon to reach equilibrium at the initial temperature of 300 K and at the atmospheric pressure. It is then heated to 3,000 K (the melting point) and rapidly cooled down to 300 K by the rate of 1 K/ps to create an amorphous structure of silicon.

MD simulations are performed by 8,000 atoms with an NPT ensemble (to allow for volume changes during the heating and cooling) and is based on the three-body Stillinger–Weber potential:

$$A = 7.049556,\ B = 0.6022245584,\ \lambda = 21.0,\ \gamma = 1.2,\ p = 6,\ q = 0,\ a = 1.8 \tag{4.5.63}$$

The obtained MD configuration is presented in Figure 4.96. Variations of energy with temperature in Figure 4.97 clearly show specific jumps associated with the solid–liquid phase changes. A similar conclusion can be made from Figure 4.98, which shows the variation of the volume of the sample in time.

In the next stage of simulation, the amorphous atomistic model is subjected to the tensile displacement loading. The meshless technique is adopted for correcting the deformation gradient and updating the positions of atoms accordingly, as presented in Figure 4.99.

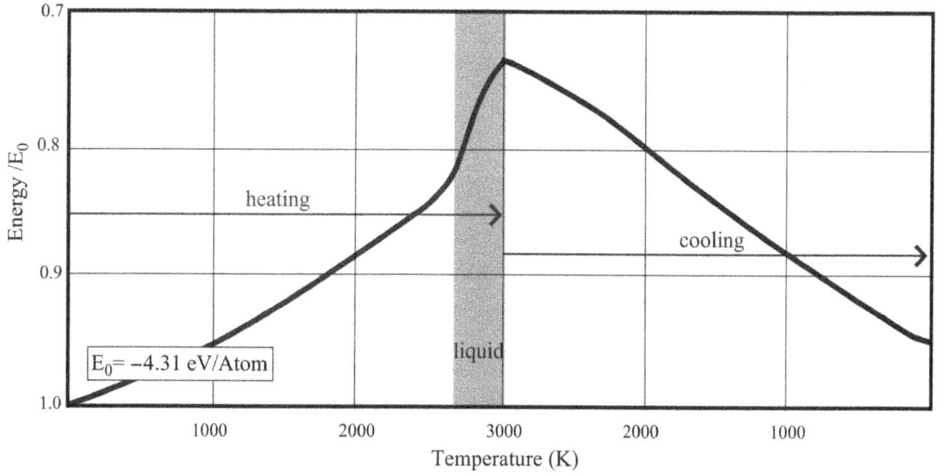

Figure 4.97 Energy variations during the heating and cooling processes. *Source:* Adapted from Moslemzadeh and Mohammadi 2022.

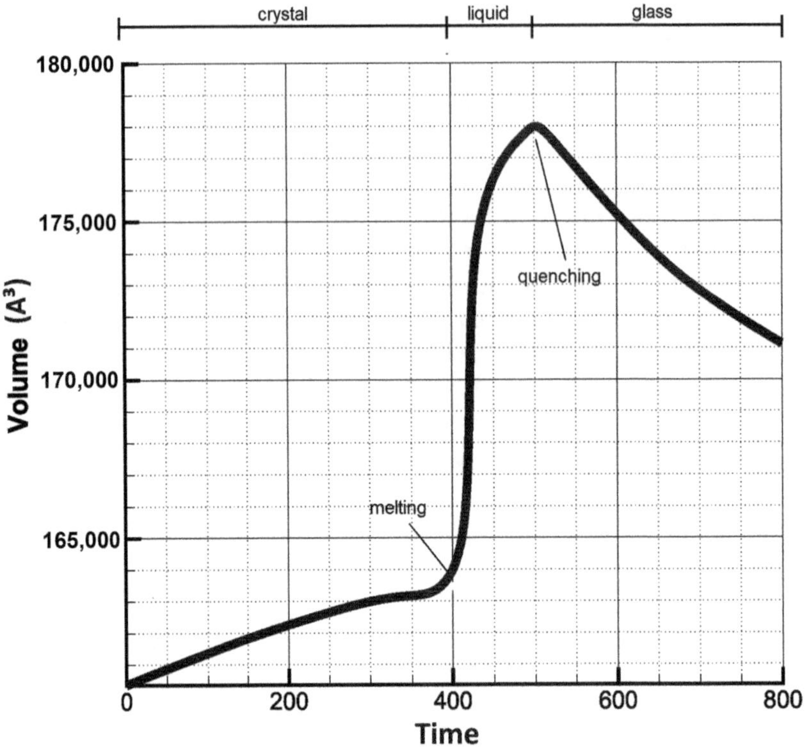

Figure 4.98 Variations of the volume in time during the heating and cooling processes. *Source:* Adapted from Moslemzadeh and Mohammadi 2022.

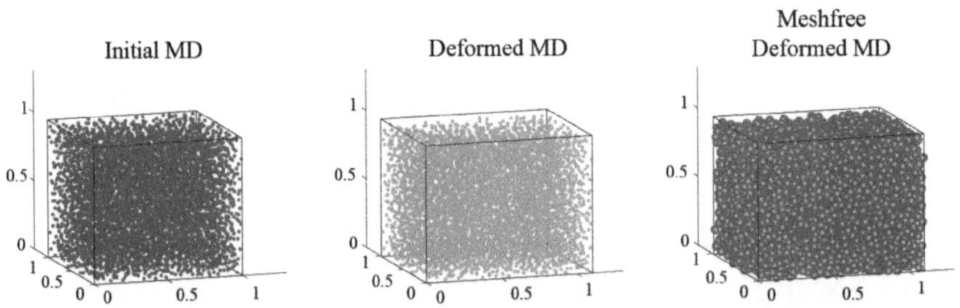

Figure 4.99 MD results of the tensile specimen. *Source:* Adapted from Moslemzadeh and Mohammadi 2022.

Finally, the multiscale procedure continues at the macro level by simulating the specimen shown in Figure 4.100 by two different regular and irregular initial finite element meshes. Nodes of finite elements match the positions of independent atoms. The model consists of about 250 elements and 16,000 atoms. Various elements of the irregular mesh have different number of atoms (up to 70 atoms) and consist of 5–12 nodes. All shape functions are generated by the maximum entropy approach.

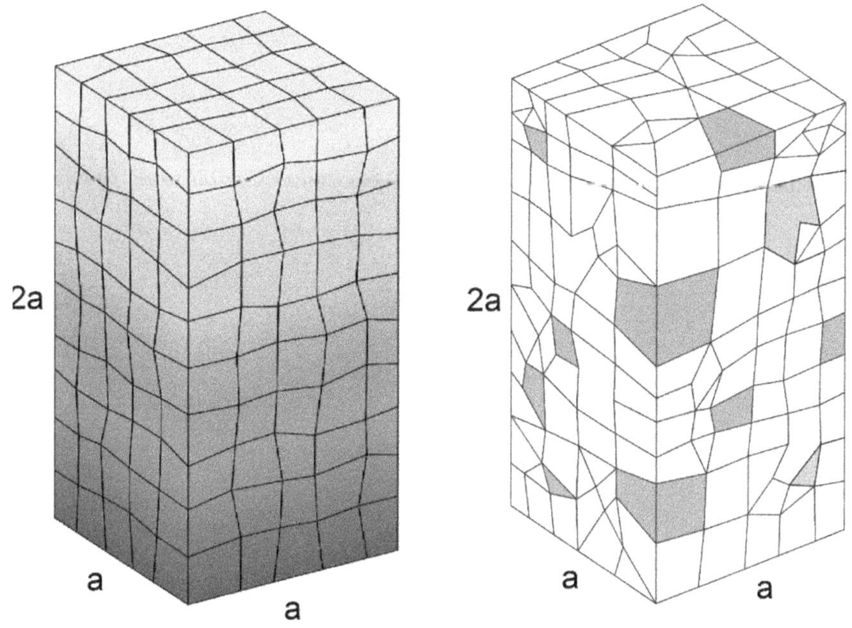

Figure 4.100 Different initial finite element meshes with different number of nodes/atoms. *Source:* Adapted from Moslemzadeh and Mohammadi 2022.

The results of displacements for the two meshes remain quite similar at all stages of the analysis. Moreover, Figure 4.101 compares the predicted stress–strain response with the full MD simulation. It shows that the results of the multiscale simulation without the meshless-based modification of the deformation gradient quickly deviates from the correct path and terminates abruptly.

Finally, the direct effect of the strain rate on the stress–strain response is presented in Figure 4.101. Clearly, a sample subjected to a high strain rate is expected to break at higher stresses (and strains).

For a complete discussion on the approach and the simulations results, see Moslemzadeh and Mohammadi (2022).

4.5.6 Variable Node Multiscale Method (VNMM)

The variable node multiscale method (VNMM) is a new concurrent multiscale approach that combines some existing concepts of earlier multiscale methods with the meshless techniques to develop a method with strong compatibility between atoms and finite elements. VNMM does not require any hand-shaking region and the adopted meshless concepts within the variable node element avoid the expensive adaptive remeshing by automatic re-positioning of atoms inside the elements (Alizadeh 2019).

Figure 4.102 illustrates a typical problem with a small localized zone. VNMM constructs a mesh of finite elements on a molecular/atomic setup, as typically illustrated in Figure 4.102a.

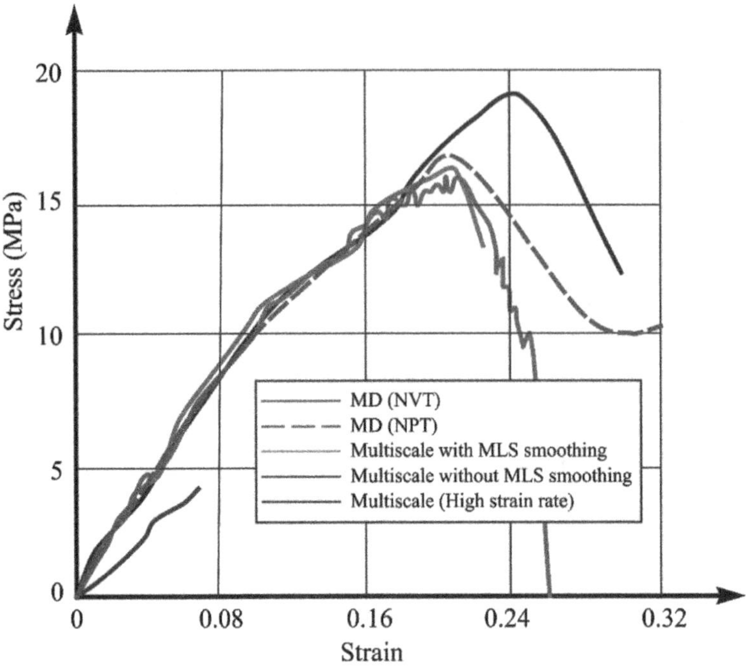

Figure 4.101 Different initial finite element meshes with different numbers of nodes/atoms. *Source:* Adapted from Moslemzadeh and Mohammadi 2022.

VNMM finite element discretization may be designed similar to various QC, BDM or BSM multiscale models. In a QC fashion, VNMM uses finite elements on all atoms (Figure 4.102c). VNEs in the localized zone include the independent atoms, whereas the constrained atoms are located beneath the finite elements elsewhere. If a BDM partition approach is preferred, only the localized region is modelled by the variable node element with independent atoms (Figure 4.102d). Due to cut-off zone effects, a set of variable node elements with dependent atoms should also be considered around the localized zone. Elsewhere, the conventional finite elements (without any type of atom) are adopted. All variable node finite elements, either with independent or dependent atoms, and even the conventional fixed node finite elements of the model, remain compatible along their edges.

A variable node element (VNE) consists of four corner nodes and a number of inside nodes, which may vary in number and distribution in different elements, as comprehensively discussed in Section 3.8. Each VNE in the localized region is composed of a number of independent atoms (either on edges or inside the elements) with full degrees of freedom (the variable node multiscale element, VNME). Distribution of the independent atoms inside any VNE complies with the atomic structure of the material in the localized zone and may well consider any defects or inclusion, as schematically illustrated in Figure 4.103 for a number of material structures. In fact, the inside nodes perform both as atoms and finite element nodes. The molecular statics formulation is adopted to compute and minimize the energy of the independent atoms.

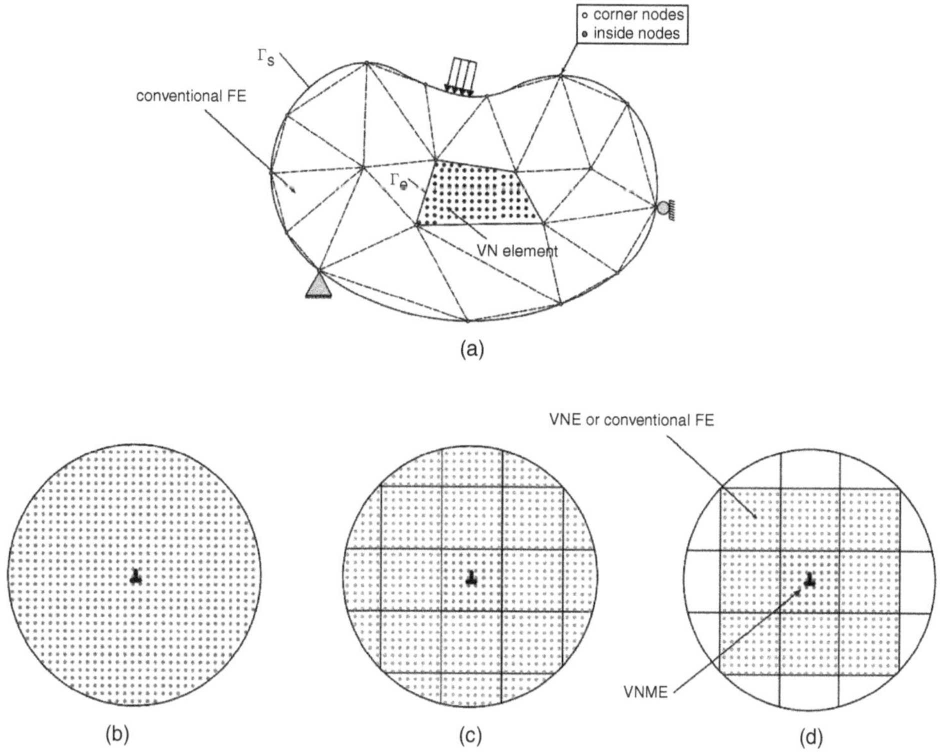

Figure 4.102 Definition of a multiscale problem (a), the MD setup (b), the VNMM with modelling of all atoms (c) and the VNMM with partial modelling of atoms (d). *Source:* Adapted from Alizadeh and Mohammadi 2019.

4.5.6.1 VNMM Governing Formulation

The VNMM formulation is presented for the case of Figure 4.102c. A similar methodology with some modifications can be followed for the case of Figure 4.102d.

The total energy $\prod$ of the system can be computed from the balance of energy of conventional finite elements $\prod^{\text{conv}}$, the energy of variable node elements $\prod^{\text{vnme}}$ and the work of external forces,

$$\Pi(\boldsymbol{u}) = \Pi^{\text{conv}}(\boldsymbol{u}) + \Pi^{\text{vnme}}(\boldsymbol{u}) - W^{\text{ext}}(\boldsymbol{u}) \tag{4.5.64}$$

where the external work W^{ext} for the total N_{n} nodes is defined as

$$W^{\text{ext}}(\boldsymbol{u}) = \sum_{k=1}^{N_{\text{n}}} \boldsymbol{f}_k^{\text{ext}} . \boldsymbol{u}_k \tag{4.5.65}$$

4.5.6.1.1 Energy of Conventional Elements

$\prod^{\text{conv}}(\boldsymbol{u})$ is defined in terms of the strain energy density w^{s} of the finite elements,

$$\prod^{\text{conv}}(\boldsymbol{u}) = \sum_{i=1}^{N_{\text{e}}^{\text{conv}}} w^{\text{s}}\left(\boldsymbol{F}_i(\boldsymbol{u})\right) \Omega_i \tag{4.5.66}$$

where $\boldsymbol{F}$ is the deformation gradient, Ω is the volume of the element and $N_{\mathrm{e}}^{\mathrm{conv}}$ is the number of conventional finite elements. Π^{conv} can be computed from the energy U_{α}^{p} of its N_{a}^{conv} inside constrained atoms (grey atoms in Figure 4.102),

$$\Pi^{\mathrm{conv}}(\boldsymbol{u}) \approx \sum_{\alpha=1}^{N_{\mathrm{a}}^{\mathrm{conv}}} U_{\alpha}^{\mathrm{p}} \tag{4.5.67}$$

with the assumption that the Cauchy–Born rule governs the deformation of the constrained atoms inside an element i to determine the final positions of atoms $\mathbf{x}_i^{\mathrm{atom}}$ from the initial position $\mathbf{X}_i^{\mathrm{atom}}$ according to the deformation gradient $\boldsymbol{F}$ of the element,

$$\mathbf{x}_i^{\mathrm{atom}} = \boldsymbol{F}\mathbf{X}_i^{\mathrm{atom}} \tag{4.5.68}$$

The corresponding deformation of atoms can be regarded as the finite element or coarse-based deformation, $\boldsymbol{u}^{\mathrm{coarse}}$.

4.5.6.1.2 Energy of VNMEs

The energy Π^{vnme} of VNMEs can be determined from the energy U_{α}^{p} of all N_{a}^{VNME} independent atoms in all VNMEs (based on the interatomic potentials):

$$\Pi^{\mathrm{vnme}}(\boldsymbol{u}) = \sum_{\alpha=1}^{N_{\mathrm{a}}^{\mathrm{VNME}}} U_{\alpha}^{\mathrm{p}} \tag{4.5.69}$$

Clearly, the continuum-based formulation cannot be directly adopted to determine the energy Π^{vnme} of the variable node multiscale elements due to the fact that VNMEs are fundamentally non-uniform and cannot follow the standard Cauchy–Born rule:

$$\Pi^{\mathrm{vnme}}(\boldsymbol{u}) \neq \sum_{j=1}^{\mathrm{N}_{\mathrm{e}}^{\mathrm{VNME}}} w^{\mathrm{s}}\left(\boldsymbol{F}_j(\boldsymbol{u})\right) \Omega_j \tag{4.5.70}$$

where $N_{\mathrm{e}}^{\mathrm{VNME}}$ is the number of variable node multiscale elements. Nevertheless, it is noted that the continuum energy represented by the coarse term $w^{\mathrm{s}}\left(\boldsymbol{F}(\boldsymbol{u})\right)$ constitutes the major part of the energy of the system. A modified formulation can be written be adding a correction term $\widetilde{\Pi}$,

$$\Pi^{\mathrm{vnme}}(\boldsymbol{u}) = \sum_{\alpha=1}^{N_{\mathrm{a}}^{\mathrm{VNME}}} U_{\alpha}^{\mathrm{p}} \approx \sum_{j=1}^{\mathrm{N}_{\mathrm{e}}^{\mathrm{VNME}}} w^{\mathrm{s}}\left(\boldsymbol{F}_j(\boldsymbol{u})\right) \Omega_j + \widetilde{\Pi} \tag{4.5.71}$$

Derivation of the correction energy term $\widetilde{\Pi}$ is based on defining the total displacement $\boldsymbol{u}$ of an independent inside node/atom of VNME in terms of the coarse (FE) and fine (MS) solutions, $\boldsymbol{u}^{\mathrm{M}}$ and $\boldsymbol{u}^{\mathrm{m}}$, respectively:

$$\boldsymbol{u} = \boldsymbol{u}^{\mathrm{M}} + \boldsymbol{u}^{\mathrm{m}} \tag{4.5.72}$$

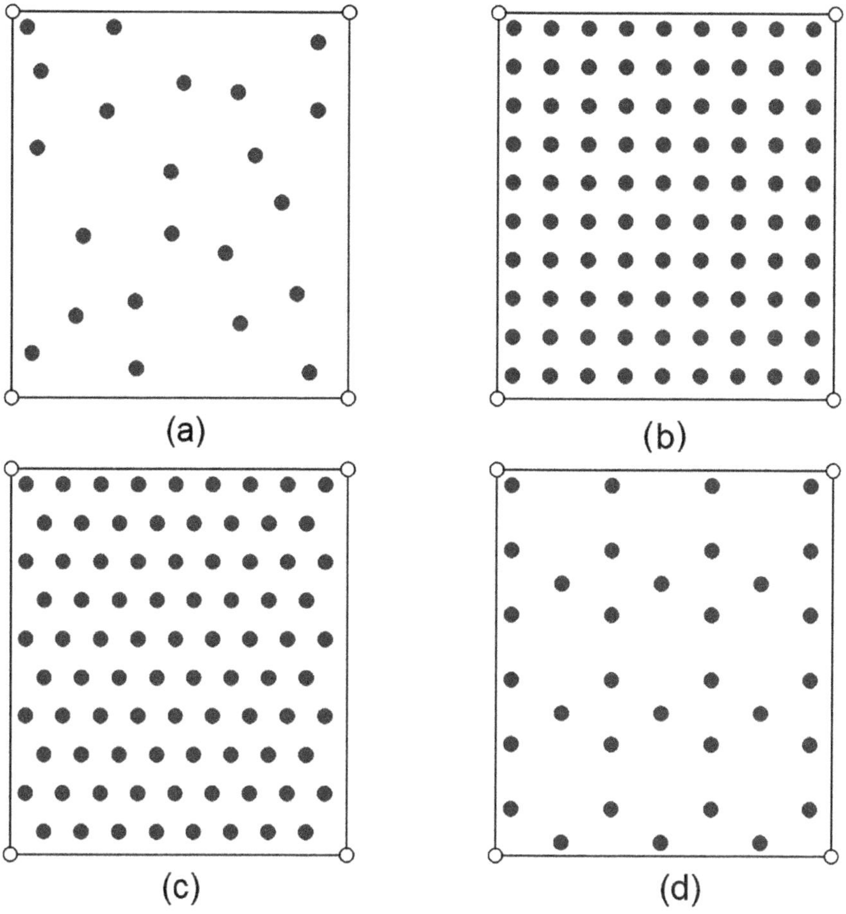

Figure 4.103 Variable node multiscale element for different distributions of the inside nodes: (a) random, (b) BCC structure, (c) FCC structure and (d) graphene.

where $\boldsymbol{u}^{\mathrm{M}}$ is the minimized solution $\boldsymbol{u}$ of the system with a fully continuum assumption for the energy of the system (ignoring $\widetilde{\Pi}$):

$$\Pi(\boldsymbol{u}) = \sum_{i=1}^{N_{\mathrm{e}}^{\mathrm{conv}}} w^{\mathrm{s}}\left(\boldsymbol{F}_i(\boldsymbol{u})\right) \Omega_i + \sum_{j=1}^{N_{\mathrm{e}}^{\mathrm{VNME}}} w^{\mathrm{s}}\left(\boldsymbol{F}_j(\boldsymbol{u})\right) \Omega_j - \sum_{k=1}^{N_{\mathrm{n}}} \boldsymbol{f}_k^{\mathrm{ext}} . \boldsymbol{u}_k \tag{4.5.73}$$

and $\boldsymbol{u}^{\mathrm{m}}$ represents the fine-tuned deformation of the independent inside atoms/nodes of VNME, which is the minimization solution of $\widetilde{\Pi}\left(\boldsymbol{u}^{\mathrm{m}}\right)$ based on the atomic model,

$$\widetilde{\Pi}\left(\boldsymbol{u}^{\mathrm{m}}\right) = \sum_{\alpha=1}^{N_{a}^{\mathrm{VNME}}} U_{\alpha}^{\mathrm{p}}\left(\mathbf{X}_{\alpha} + \boldsymbol{u}^{\mathrm{M}} + \boldsymbol{u}^{\mathrm{m}}\right) \tag{4.5.74}$$

Accordingly, the independent atoms are first deformed by $\boldsymbol{u}^{\mathrm{M}}$ based on the continuum deformation gradient and are then moved by $\boldsymbol{u}^{\mathrm{m}}$ to their correct positions. Figure 4.104

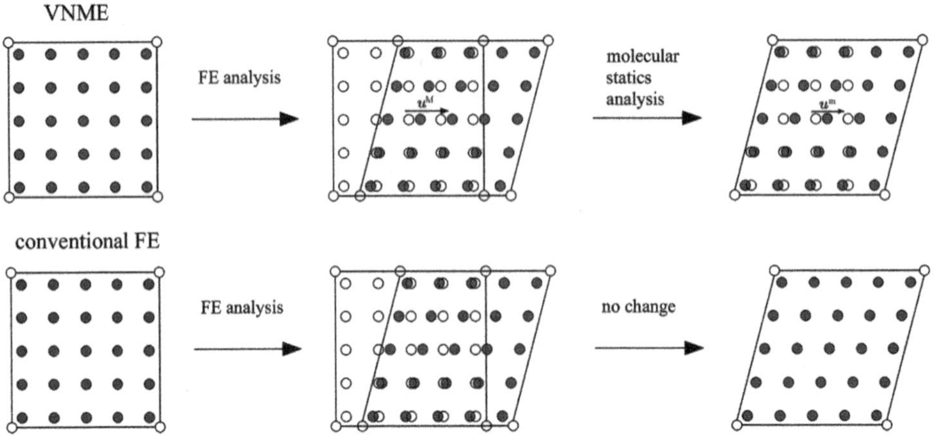

Figure 4.104 VNME procedure: initial setup, FEM analysis, and MS minimization of the energy of inside nodes.

illustrates the way conventional and variable node multiscale elements can be treated in the present variable node multiscale method.

The final form of the energy of the system $\Pi(\boldsymbol{u})$ can be written as

$$\Pi(\boldsymbol{u}) = \sum_{i=1}^{N_e^{\text{conv}}} w^{\text{s}}\left(\boldsymbol{F}_i\left(\boldsymbol{u}^{\text{M}}\right)\right)\Omega_i + \begin{bmatrix} \displaystyle\sum_{j=1}^{N_e^{\text{VNME}}} w^{\text{s}}\left(\boldsymbol{F}_j\left(\boldsymbol{u}^{\text{M}}\right)\right)\Omega_j + \\ \displaystyle\sum_{\alpha=1}^{N_a^{\text{VNME}}} U_\alpha^{\text{p}}\left(\mathbf{X}_\alpha + \boldsymbol{u}^{\text{M}} + \boldsymbol{u}^{\text{m}}\right) \end{bmatrix} - \sum_{k=1}^{N_n} \boldsymbol{f}_k^{\text{ext}}.\boldsymbol{u}_k \tag{4.5.75}$$

which should be simultaneously minimized with respect to $\boldsymbol{u}^{\text{M}}$ and $\boldsymbol{u}^{\text{m}}$,

$$\begin{aligned} \frac{\partial \Pi(\boldsymbol{u})}{\partial \boldsymbol{u}^{\text{M}}} = {} & \frac{\partial\left(\displaystyle\sum_{j=1}^{N_e^{\text{VNME}}} w^{\text{s}}\left(\boldsymbol{F}_j\left(\boldsymbol{u}^{\text{M}}\right)\right)\Omega_j\right)}{\partial \boldsymbol{u}^{\text{M}}} + \frac{\partial\left(\displaystyle\sum_{\alpha=1}^{N_a^{\text{VNME}}} U_\alpha^{\text{p}}\left(\mathbf{X}_\alpha + \boldsymbol{u}^{\text{M}} + \boldsymbol{u}^{\text{m}}\right)\right)}{\partial \boldsymbol{u}^{\text{M}}} + \\ & \frac{\partial\left(\displaystyle\sum_{i=1}^{N_e^{\text{conv}}} w^{\text{s}}\left(\boldsymbol{F}_i\left(\boldsymbol{u}^{\text{M}}\right)\right)\Omega_j\right)}{\partial \boldsymbol{u}^{\text{M}}} - \frac{\partial\left(\displaystyle\sum_{k=1}^{N_n} \boldsymbol{f}_k^{\text{ext}}.\boldsymbol{u}_k\right)}{\partial \boldsymbol{u}^{\text{M}}} = 0 \end{aligned} \tag{4.5.76}$$

$$\frac{\partial \Pi(\boldsymbol{u})}{\partial \boldsymbol{u}^{\text{m}}} = \frac{\partial\left(\displaystyle\sum_{\alpha=1}^{N_a^{\text{VNME}}} U_\alpha^{\text{p}}\left(\mathbf{X}_\alpha + \boldsymbol{u}^{\text{M}} + \boldsymbol{u}^{\text{m}}\right)\right)}{\partial \boldsymbol{u}^{\text{m}}} = 0 \tag{4.5.77}$$

It should be noted that the energy minimization procedure is limited to a relatively small number of inside nodes of VNMEs.

The positions of independent inside nodes should be updated, leading to the update of VNE shape functions and the corresponding stiffness matrix. For further details, see Alizadeh and Mohammadi (2019).

4.5.6.2 Enriched VNMM for Simulation of a Cracked Nano Specimen

VNMM is now adopted to simulate a 100 Å square plate with a 7 Å long edge crack. The bottom edge is fixed and the top edge is subjected to tensile displacement. The modelling process of Alizadeh and Mohammadi (2019) is followed.

The model consists of 10,201 atoms with the initial lattice dimension of 1 Å and an unstructured mesh of extended finite elements with the Heaviside enrichment is adopted to take into account the displacement discontinuity across the crack edge, as depicted in Figure 4.105. No crack tip enrichment is used due to the fact that the crack tip is located inside the VNME, which is expected to accurately simulate its complex response by molecular statics (MS). The localized cracked region is modelled by two VNMEs and linear triangular finite elements discretize the rest of the specimen. Its equivalent QC model is presented in Figure 4.106.

The results of VNMM are compared with the full MS solution and the results of QC simulation. Due to the lack of enrichment in QC, modelling of the crack in QC simulation is performed by simply creating a gap between the elements in the position of the crack. Basic specifications of MS, QC and VNMM models are presented in Table 4.12.

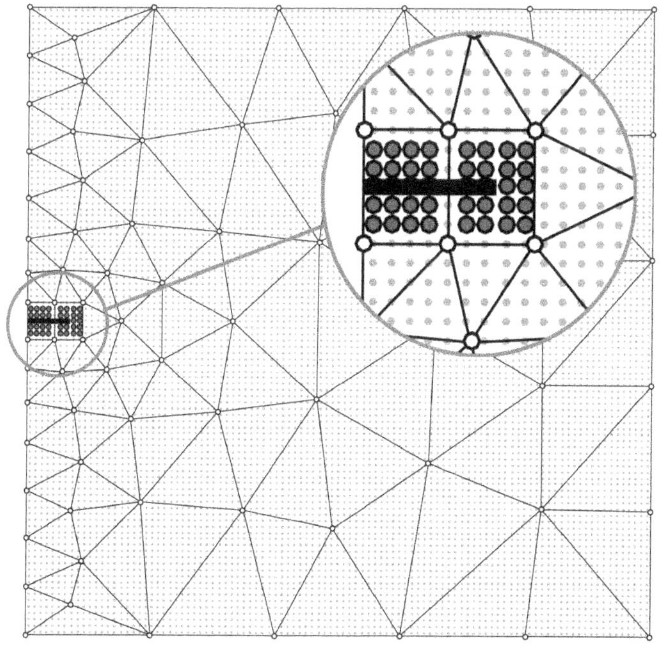

Figure 4.105 Initial configuration of the VNMM model. *Source:* Adapted from Alizadeh and Mohammadi 2019.

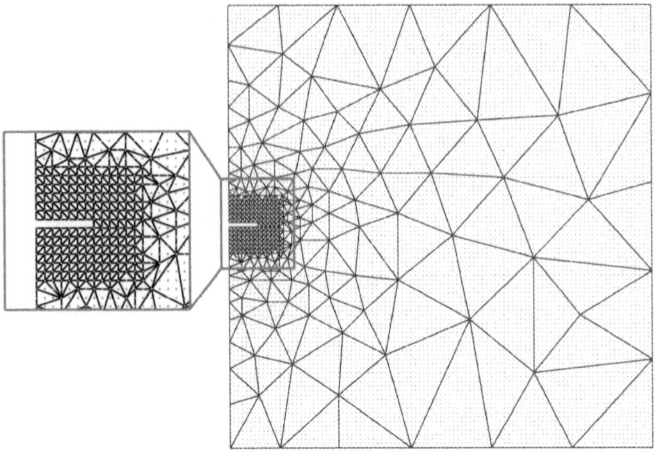

Figure 4.106 The QC model. *Source:* Adapted from Alizadeh and Mohammadi 2019.

Table 4.12 Specifications of MS, QC and VNMM models.

Multiscale method	Number of atoms	Number of nodes	Number of DOFs	Number of elements
MS	10,201	-	20,402	-
QC	10,201	300	600	404
VNMM	10,201	111	234	108

Source: Adapted from Alizadeh and Mohammadi, 2019.

Figure 4.107 compares the variations of the virial stress component S_{yy} near the crack tip for MS, QC and VNMM models. While close agreements are generally observed for this complex high gradient response, the relative error with respect to MS shows that the VNMM prediction for the S_{yy} component near the crack tip is far more accurate than QC (even with a third degrees of freedom), as briefly presented in Table 4.13.

4.5.7 Enriched Multiscale Method (EMM)

The enriched multiscale method (EMM) is a novel multiscale idea that adopts the concept of atomistic displacement enrichment to combine the formulations of energy minimization of atomic and continuum zones in the framework of the nonlinear finite element method (Alizadeh 2019).

4.5.7.1 EMM Concept

Consider a domain Ω subjected to the body force vector $\boldsymbol{f}_0^{\text{b}}$ and the traction vector $\boldsymbol{f}_0^{\text{t}}$ on its boundary Γ_{t}, which includes a typical atomistic defect, as depicted in Figure 4.108. In EMM,

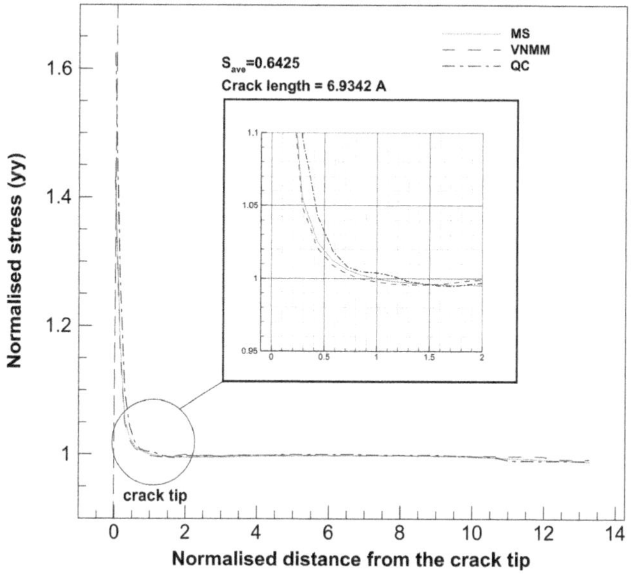

Figure 4.107 Variations of the virial S_{yy} along the crack. *Source:* Adapted from Alizadeh and Mohammadi 2019.

Table 4.13 Error of QC and VNMM (with respect to MS).

Multiscale method	Number of DOFs	Maximum error of vertical stress (%)
QC	600	11.52
VNMM	234	1.75

Source: Adapted from Alizadeh and Mohammadi, 2019.

while the entire domain is discretized by FEM and is governed by the continuum constitutive laws, the effects of atomistic solutions are included in the form of enrichments.

In the localized zone (for instance with a defect), the variable node element is adopted to increase the accuracy of the solution (see Figure 4.108). All inside and corner nodes of VNE and all corner nodes of the conventional finite elements are given full degrees of freedom.

The main concept of EMM is to increase the accuracy of the total displacement field $\boldsymbol{u}$ at the initial position $\mathbf{X}^{\text{node}}$ of any node in the material coordinate system by an enriched solution:

$$\boldsymbol{u}\left(\mathbf{X}^{\text{node}}\right) = \boldsymbol{u}^{\text{M}}\left(\mathbf{X}^{\text{node}}\right) + \boldsymbol{u}^{\text{m,enr}}\left(\mathbf{X}^{\text{node}}\right) \tag{4.5.78}$$

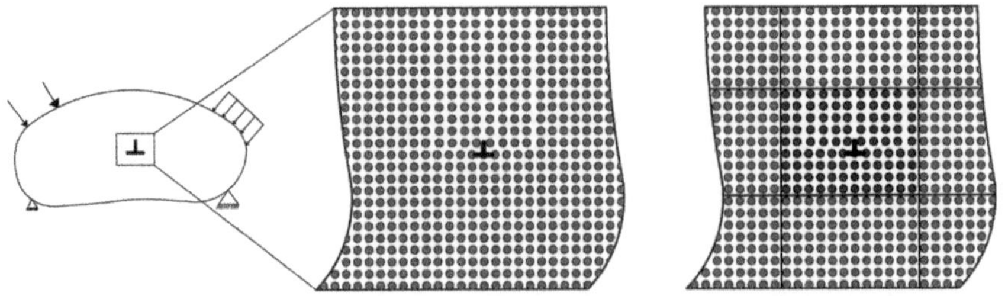

Figure 4.108 A typical domain with a defect modelled by MD and VNE. *Source:* Adapted from Alizadeh 2019.

where $\boldsymbol{u}^{\mathrm{M}}\left(\mathbf{X}^{\text{node}}\right)$ is the continuum-based displacement and $\boldsymbol{u}^{\mathrm{m,enr}}\left(\mathbf{X}^{\text{node}}\right)$ is the enrichment term computed from the atomistic-based displacements.

The deformed position $\mathbf{x}_{\mathrm{t}}^{\text{node}}$ in the spatial coordinate system can be derived from the mapping function Φ (Figure 4.109),

$$\mathbf{x}^{\text{node}} = \Phi\left(\mathbf{X}^{\text{node}}\right) \tag{4.5.79}$$

The continuum part of the displacement field, $\boldsymbol{u}^{\mathrm{M}}$, is derived from the VNME approximation:

$$\boldsymbol{u}^{\mathrm{M}}\left(\mathbf{X}^{\text{node}}\right) = \mathbf{N}^{\mathrm{M}}\left(\mathbf{X}^{\text{node}}\right)\bar{\mathbf{U}}^{\mathrm{M}} = \sum_{i\in S} N_i\left(\mathbf{X}^{\text{node}}\right)\boldsymbol{u}_i \tag{4.5.80}$$

where N_i is the VNME shape function and S is the set of all nodes in the element.

The enrichment displacement is defined by the radial basis function (R_i) approximation on atoms of VNME,

$$\boldsymbol{u}^{\mathrm{m,enr}}\left(\mathbf{X}^{\text{node}}\right) = \mathbf{R}^{m}\left(\mathbf{X}^{\text{node}}\right)\overline{\boldsymbol{\Lambda}}^{m} = \sum_{i\in A} \mathcal{R}_i\left(\mathbf{X}^{\text{node}}\right)\boldsymbol{\lambda}_i \tag{4.5.81}$$

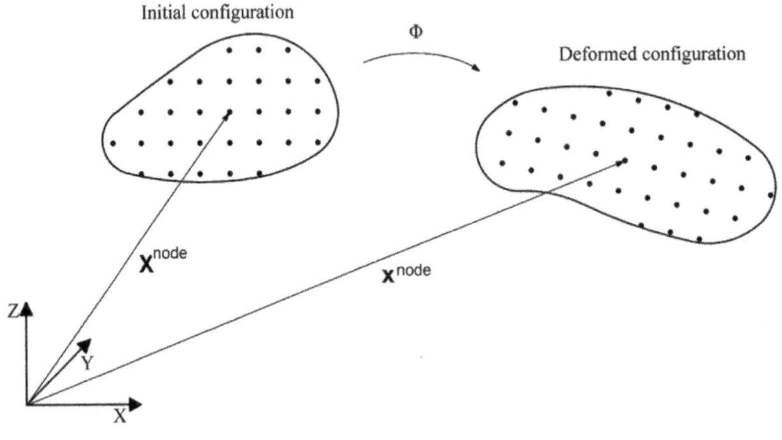

Figure 4.109 Mapping Φ from the initial to the deformed configuration.

where $\boldsymbol{\lambda}_i$ is the set of unknown coefficients and $\mathcal{A}$ is the set of all atoms inside the variable node element.

4.5.7.2 Governing Equation

The total energy of the domain can be written as

$$\Pi(\boldsymbol{u}(\mathbf{x})) = \int_{\Omega_0} w^{\mathrm{s}}(\boldsymbol{F}(\boldsymbol{u}(\mathbf{x})))\mathrm{d}\Omega - \int_{\Omega_0} \boldsymbol{u}(\mathbf{x}).\boldsymbol{f}_0^{\mathrm{b}}\mathrm{d}\Omega - \int_{\Gamma_{t0}} \boldsymbol{u}(\mathbf{x}) . \boldsymbol{f}_0^{\mathrm{t}} d\Gamma \tag{4.5.82}$$

where w^{s} is the strain energy function and $\boldsymbol{F}$ is the deformation gradient,

$$\boldsymbol{F} = \frac{\partial \Phi(\mathbf{X})}{\partial \mathbf{X}} = \frac{\partial \mathbf{x}}{\partial \mathbf{X}} \tag{4.5.83}$$

The minimum state of total energy is obtained from the derivative of the total energy with respect to the total displacement field $\boldsymbol{u}$,

$$\frac{\partial \Pi(\boldsymbol{u}(\mathbf{x}))}{\partial \boldsymbol{u}} = \boldsymbol{f}^{\mathrm{int}} - \boldsymbol{f}^{\mathrm{ext}} = \mathbf{0} \tag{4.5.84}$$

where $\boldsymbol{f}^{\mathrm{ext}}$ is the external force vector,

$$\boldsymbol{f}^{\mathrm{ext}} = \int_{\Omega_0} \mathbf{N}^{\mathrm{M}} . \boldsymbol{f}_0^{\mathrm{b}} \, \mathrm{d}\Omega + \int_{\Gamma_{t0}} \mathbf{N}^{\mathrm{M}} . \boldsymbol{f}_0^{\mathrm{t}} \, d\Gamma \tag{4.5.85}$$

and $\boldsymbol{f}^{\mathrm{int}}$ is the internal force vector, which can be expressed in terms of the first Piola–Kirchhoff stress tensor $\boldsymbol{P}$ (Belytschko et al. 2000),

$$\boldsymbol{f}^{\mathrm{int}} = \int_{\Omega_0} \frac{\partial w^{\mathrm{s}}}{\partial \boldsymbol{F}} \frac{\partial \boldsymbol{F}}{\partial \boldsymbol{u}} \mathrm{d}\Omega = \int_{\Omega_0} \boldsymbol{P} \frac{\partial \boldsymbol{F}}{\partial \boldsymbol{u}} \mathrm{d}\Omega \tag{4.5.86}$$

or in terms of the second Piola–Kirchhoff stress tensor $\boldsymbol{S}$ (with the help of the Green strain tensor $\boldsymbol{E}$) (Bonet and Wood 1997):

$$\boldsymbol{f}^{\mathrm{int}} = \int_{\Omega_0} \boldsymbol{F}\boldsymbol{S} \, \delta \boldsymbol{F} \, \mathrm{d}\Omega = \int_{\Omega_0} \boldsymbol{S} : \delta \boldsymbol{E} \, \mathrm{d}\Omega = \int_{\Omega_0} \boldsymbol{S} : \frac{1}{2}\left(\delta \boldsymbol{F}^{\mathrm{T}} \boldsymbol{F} + \boldsymbol{F}^{\mathrm{T}} \delta \boldsymbol{F}\right) \mathrm{d}\Omega \tag{4.5.87}$$

Linearization of (4.5.87) is required to extract the displacement field $\boldsymbol{u}$, as derived by Alizadeh and Mohammadi (2022). Assuming

$$\hat{\boldsymbol{u}} = \boldsymbol{u} + \epsilon \, \Delta \boldsymbol{u} = \boldsymbol{u} + \frac{\partial \boldsymbol{u}}{\partial \mathbf{x}} \, \Delta \boldsymbol{u} \tag{4.5.88}$$

and

$$\hat{\boldsymbol{F}} = \frac{\partial \boldsymbol{u}}{\partial \mathbf{x}} = \mathbf{I} - \boldsymbol{F} \tag{4.5.89}$$

$$\delta \hat{F} = \frac{\partial \hat{u}}{\partial \mathbf{x}} \tag{4.5.90}$$

$$\Delta \hat{F} = \frac{\partial \Delta u}{\partial \mathbf{x}} \tag{4.5.91}$$

the final form of f_i^{int} is obtained:

$$f^{\text{int}} = \int_{\Omega_0} (S : \delta E) \mathrm{d}\Omega + \int_{\Omega_0} (F^{\mathrm{T}} \delta \hat{F} : \mathbf{D} : F^{\mathrm{T}} \Delta \hat{F} + \delta \hat{F}^{\mathrm{T}} : \Delta \hat{F}^{\mathrm{T}} \; S) \mathrm{d}\Omega \tag{4.5.92}$$

Therefore, Equation (4.5.82) is re-written as

$$\begin{aligned} &\int_{\Omega_0} (\delta \hat{F} F^{\mathrm{T}} \mathbf{D} F^{\mathrm{T}} \Delta \hat{F} + \delta \hat{F}^{\;\mathrm{T}} \mathbf{S} \Delta \hat{F}) \, \mathrm{d}\Omega + \int_{\Omega_0} \delta(\hat{F} \; F^{\mathrm{T}} \mathbf{S}) d\Omega \\ &- \int_{\Omega_0} \mathbf{N}^{\mathrm{M}} . f_0^b \, \mathrm{d}\Omega - \int_{\Gamma_{t0}} \mathbf{N}^{\mathrm{M}} . f_0^t \, \mathrm{d}\Gamma = 0 \end{aligned} \tag{4.5.93}$$

After some manipulations of Equation (4.5.93), the final form of the governing equation can be written as (Alizadeh 2019; Alizadeh and Mohammadi 2022)

$$\left(\mathbf{K}_{\text{mat}} + \mathbf{K}_{\mathrm{G}}\right) \bar{\mathbf{U}}^M = -f_a - f_i + f_b + f_t \tag{4.5.97}$$

where $\mathbf{K}_{\text{mat}}$ is the material stiffness matrix, $\mathbf{K}_{\mathrm{G}}$ is the geometric stiffness matrix, f_a is the equivalent atomistic force vector, f_i is the nodal equivalent force vector, f_b is the equivalent body force vector and f_t is the traction vector applied on the nodes.

4.5.7.3 Derivation of Atomistic-Based Enrichment

To determine the atomistic-based enriched displacement $u^{\text{m,enr}}(\mathbf{x})$, a set of unlimited number of atoms is considered. The energy of the interaction between each pair of atoms i and j (with the distance r_{ij}) can be calculated by the interatomic bond potential ϕ_{ij}. Summation of the interatomic bond force F_{ij} and stiffness K_{ij} over the neighbours of atom i can be used to compute its force vector $\mathbf{F}_i$ and stiffness matrix $\mathbf{K}_i$,

$$\mathbf{F}_i = \sum_{j \in \bar{\mathcal{A}}_i} F_{ij} = \sum_{j \in \bar{\mathcal{A}}_i} \frac{\phi_{ij}}{\partial r_{ij}} \tag{4.5.98}$$

$$\mathbf{K}_i = \sum_{j \in \bar{\mathcal{A}}_i} K_{ij} = \sum_{j \in \bar{\mathcal{A}}_i} \frac{\partial^2 \phi_{ij}}{\partial r_{ij}^2} \tag{4.5.99}$$

where $\bar{\mathcal{A}}_i$ is the set of atoms in the cut-off zone of atom i.

Assuming a fixed position for the neighbours of an atom i while computing its displacement, u_i^{atom} along the vector e_k can be computed from

$$u_i^{\text{atom}} . e_k = \sum_{j \in \bar{\mathcal{A}}_i} \left(\frac{\partial^2 \phi}{\partial r_{ij} \partial r_{ij}} \right)^{-1} \times \left(\frac{\partial \phi}{\partial r_{ij}} \right) \frac{r . e_k}{r_{ij}} \tag{4.5.100}$$

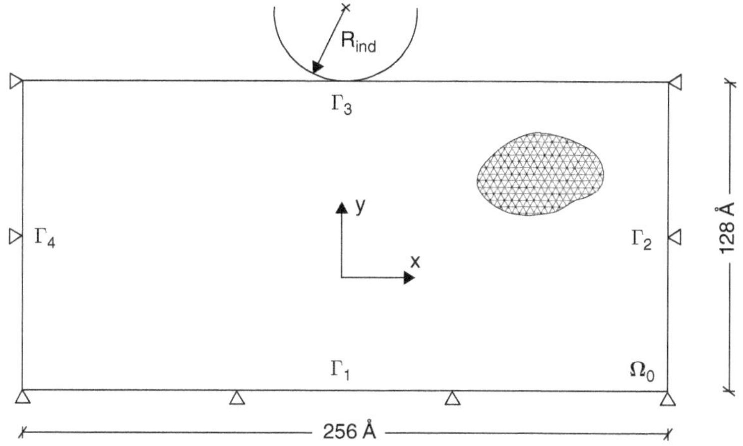

Figure 4.110 Nano-indentation modelling with EMM. *Source:* Reproduced from Alizadeh 2019/ University of Tehran.

4.5.7.4 Numerical Simulation of Indentation

A 256 Å × 128 Å plate subjected to indentation of a rigid circular indenter with the radius of 32 Å is considered, as depicted in Figure 4.110. The indenter penetrates 0.8 Å in the specimen in each step. The modelling procedure of Alizadeh and Mohammadi (2022) is followed.

The following simple interatomic potential is adopted:

$$\phi_{ij} = \frac{1}{2}\frac{EA}{r_{0ij}}\left(r_{ij} - r_{0ij}\right)^2 \tag{4.5.101}$$

where r_{ij} and r_{0ij} are the current and initial distances between atoms i and j, respectively, and EA is assumed to be 1 for the whole domain.

In order to assess the accuracy and performance of EMM, the problem is simulated by four different models: molecular statics (MS), quasi-continuum (QC), variable node multiscale method (VNMM) and enriched multiscale method (EMM). The models for QC and EMM (also VNMM) are depicted in Figure 4.111. Details of the models are presented in Table 4.14.

Figure 4.112 Illustrates the snapshots at different deformation configurations of the EMM model, which shows the extent of penetration of the indenter.

Table 4.14 Specifications of models of the indentation problem.

Model	No. of atoms	No. of DOFs	No. of elements
MS	33,153	66,306	-
QC	33,153	10,112	2,937
VNMM	33,153	1,542	458
EMM	33,153	1,542	458

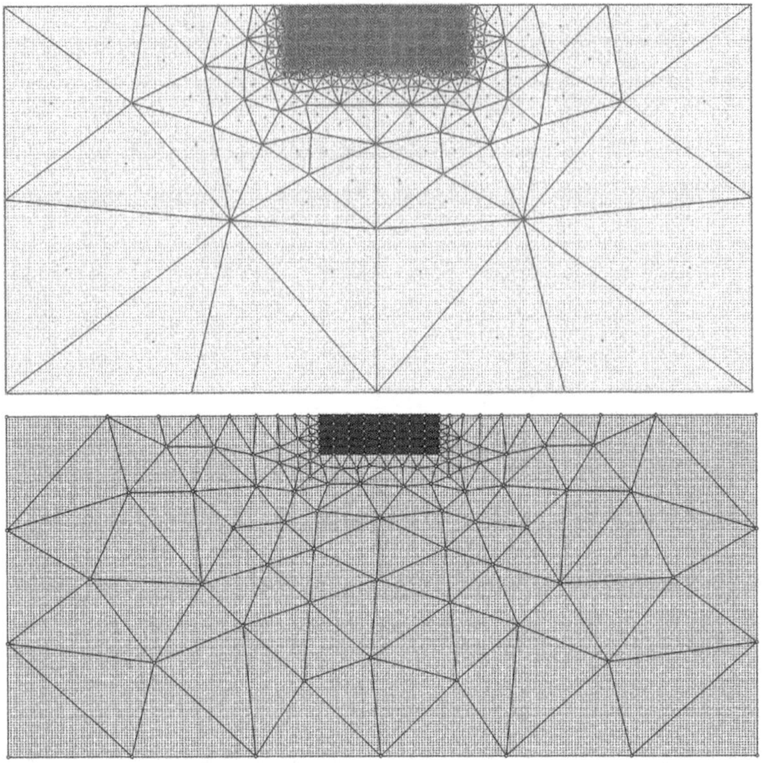

Figure 4.111 QC (top) and VNMM and EMM (bottom) models of the indentation problem. *Source:* Reproduced from Alizadeh 2019/University of Tehran.

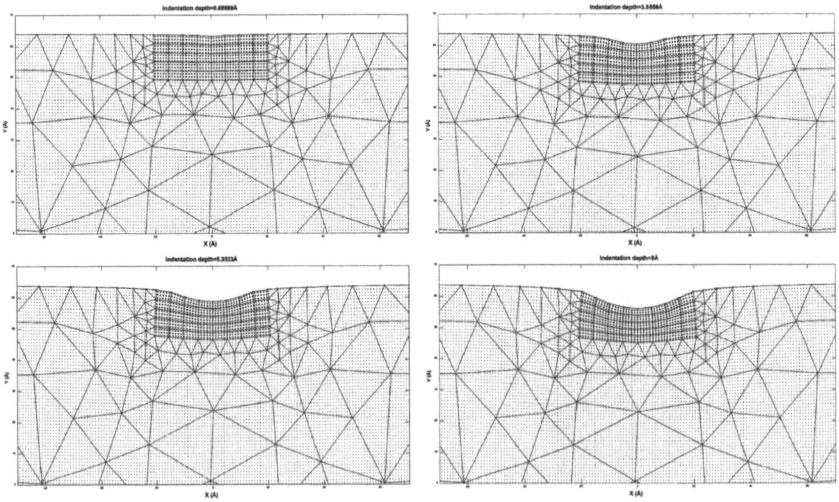

Figure 4.112 Deformed configurations of specimen in the indentation phenomena. *Source:* Reproduced from Alizadeh 2019/University of Tehran.

The vertical and horizontal displacement contours for all four models are compared in Figures 4.113 and 4.114, respectively, which show a very close agreement.

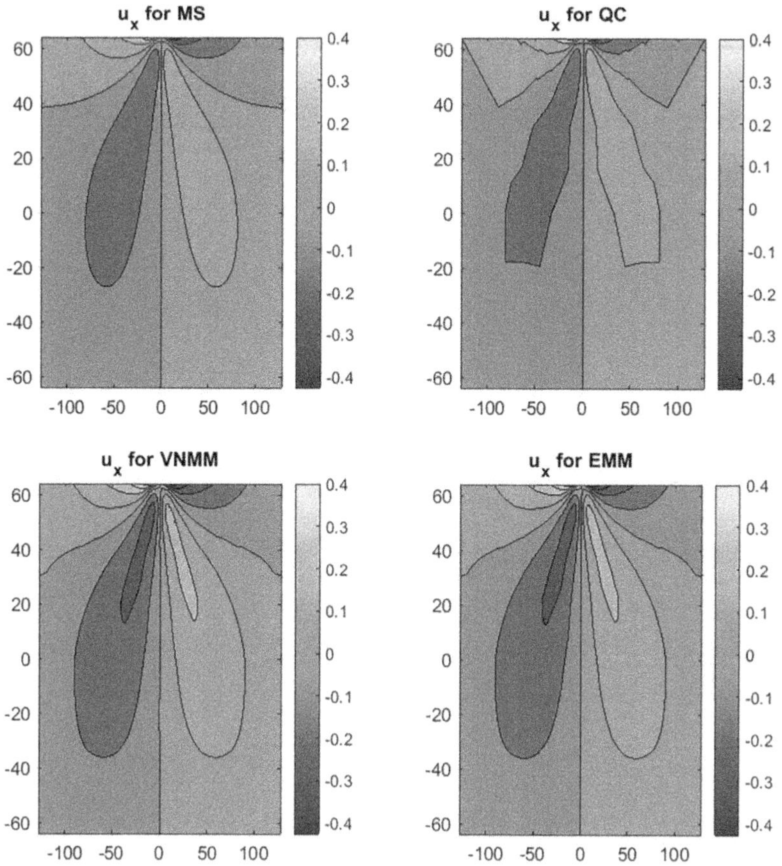

Figure 4.113 Comparison of horizontal displacement contours for all four MS, QC, VNMM and EMM models. *Source:* Reproduced from Alizadeh 2019/University of Tehran.

For further details on EMM, refer to Alizadeh and Mohammadi (2022).

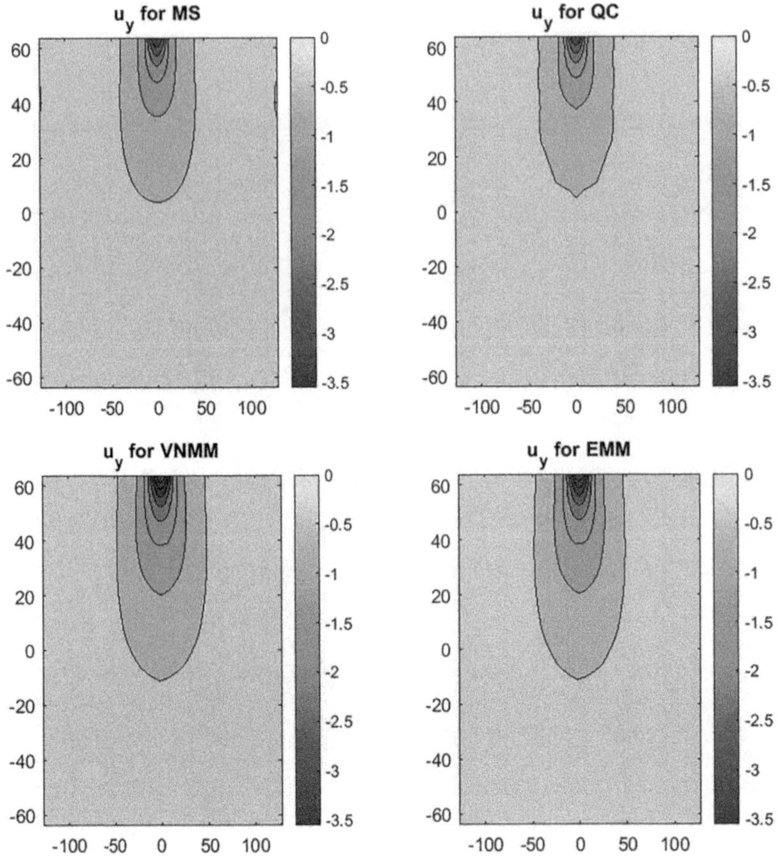

Figure 4.114 Comparison of vertical displacement contours for all four MS, QC, VNMM and EMM models. *Source:* Reproduced from Alizadeh 2019/University of Tehran.

Part III

Biomechanical Simulations

5

Biomechanics of Soft Tissues

5.1 Introduction

This comprehensive chapter is dedicated to the computational modelling of a number of major soft tissue problems. Both single and multiscale solutions are covered, depending on the necessities of the soft tissue simulation.

First, a brief review of the physiological aspects of the soft tissues is presented. It is only intended to provide very basic definitions and characteristics without going into the physiological details, which is beyond the scope of the present book. The provided material is required, to some extent, for defining the mechanical properties and simulation procedures in subsequent sections.

Section 5.3 comprehensively discusses the available hyperelastic constitutive formulations for various soft tissues. It provides basic large deformation definitions for the fibrous materials and extends the formulations from the simple isotropic neo-Hookean model to the advanced Holzapfel model for fibrous tissues, combined with an integral-type damage model. The provided formulations are used in the subsequent sections for simulation of each specific soft tissue problem.

Then, a multiscale analysis is performed on a tendon tissue. It covers the finer scale of fibrils to determine the constitutive relation at the upper fibre scale. Analysis of the fibre scale includes the fibrils and matrix. The macroscale model is constructed at the tissue level.

The heart valve simulation comprehensively examines the effects of internal blood pressure on a heart valve leaflet and studies the critical region of the leaflet by modelling, at a lower scale, the tissue with its constituent layers. The procedure continues by simulating the most critical points of the tissue scale in a cell-based microscale modelling.

The next section adopts the full coupled hyperelastic and damage models to analyse the ligament tissue in the form of a simple tensile specimen and a perforated tissue test. It examines the extents of damage in matrix and fibre at different stages of the loading simulation.

The peeling test is then covered in order to resemble the important test of dissection for skin or medial tissues. The Holzapfel hyperelastic constitutive law of fibrous tissues is adopted along with the integral-based non-local damage model in order to simulate the propagating damage in large deformation regimes.

Multiscale Biomechanics: Theory and Applications, First Edition. Soheil Mohammadi.

The comprehensive Section 5.8 is dedicated to influential factors, physiological mechanisms, corresponding formulations and a number of numerical simulations of the complicated coupled process of self-healing in soft tissues. The self-healing characteristics of living organs and tissues have been the source of inspiration for the development of new material technologies, not only for artificial organs/tissues but also for very wide conventional engineering applications: from pedestrian pavements to high-tech aerospace industries.

Another important topic in soft biomechanical problems is somehow related to the aneurysm disease mechanism. A region with reduced collagen fibre content is assumed in a model of the arterial wall. It is a presumed position of the damaged tissue of the artery, which deforms largely and experiences large strains in high blood pressure, resulting in the artery wall becoming thinner, with a high risk of rupture. A computational homogenization multiscale method is adopted with different types of representative volume elements to enable better resemblance of the microstructure of the fibrous tissue. The genetic algorithm is adopted to determine the necessary material parameters of the hyperelastic models of the soft tissues at different scales.

The final section is dedicated to comprehensive modelling of the brain tissue in macro- and microscales. Simulations are performed based on the isotropic neo-Hookean and Mooney–Rivlin hyperelastic constitutive models. Then, a viscoelastic model is developed to examine the short- and long-term effects on the brain (macromodel) and on the critical RVE (micromodel).

5.2 Physiology of Soft Tissues

A brief review of some of the physiological characteristics of soft tissues are presented here. The focus is on the characteristics that are required in the biomechanical studies covered in this chapter. For comprehensive physiological and biological reviews of the soft tissues refer to the major textbooks on the subject (Griffiths et al. 2016; McGrath and Uitto 2010; Torto and Derrickson 2017).

5.2.1 Soft Tissues, Skin

Skin is regarded as the first protection shield for the body and its internal organs. It may be subjected to various types of damage, which need continuous healing and replacement. It also helps in controlling various physiological characteristics that regulate the functioning of the human body.

The living composite structure of the skin is composed of two main layers: the top layer (epidermis) and the bottom layer (dermis), as depicted in Figure 5.1. An impact-absorbing layer of lipid connects the skin to the muscles. Epidermis has a thickness of 50–100 μm and constitutes a defence-alarm mechanism against external effects, such as microorganisms, UV rays, etc., for the nerve system. The dermis (or the so-called real skin) has a thickness of about 0.5–5 mm and includes nerve fibres, the arteriovenous system, the sweat network and the set of hair growth systems.

In most numerical modellings of the skin, the lipid and muscles under the dermis are ignored or their effects are indirectly included.

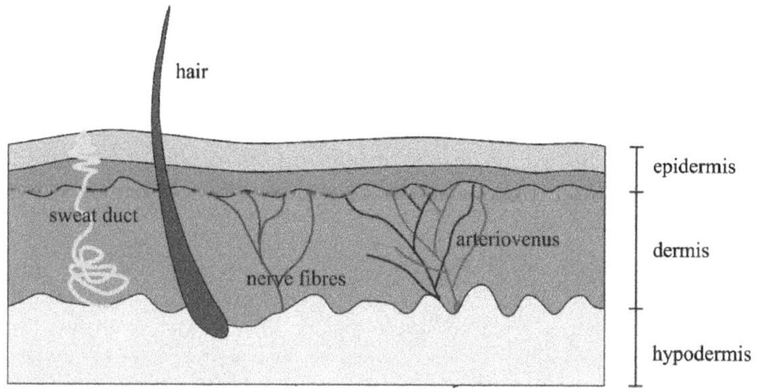

Figure 5.1 Typical composition of the human skin. *Source:* Adapted from Komorniczak 2012; Griffiths et al. 2016.

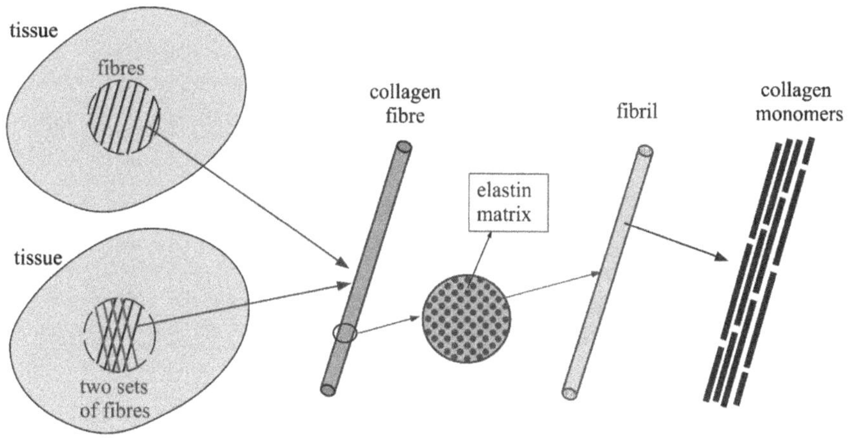

Figure 5.2 Hierarchical structure of soft tissues, collagen fibres and fibrils.

Dermis is a ground matrix for the principal constituents of the skin, such as fibroblasts, proteins and macromolecules. It contains the main proteins of the collagen, which constitutes the extracellular matrix (ECM). Soft tissues, in general, may include one or more sets of fibres. Soft tissues such as tendons include only one set of parallel fibres with major uniaxial tensile strength, whereas muscles usually include two sets of inclined fibres to provide sufficient tensile strengths in different orientations.

Tropocollagen monomers are the structural components that make the fibrils, as presented in Figure 5.2. Bundles of high-strength fibrils within an elastin matrix create the collagen fibres, which are responsible for the tensile strength and flexibility of the skin tissue.

The helical pattern of collagen fibres provides both sufficient flexibility and high tensile strength of the soft tissues. According to Figure 5.3, a sample soft tissue may contain a natural helical-shape collagen fibre. If the tissue is exposed to a tensile condition, the fibre shows a very low strength and begins to be reoriented and stretched to a smoother shape. On further extension, the fibre is fully stretched (similar to a straight rod) along the loading direction, gaining its highest tensile strength.

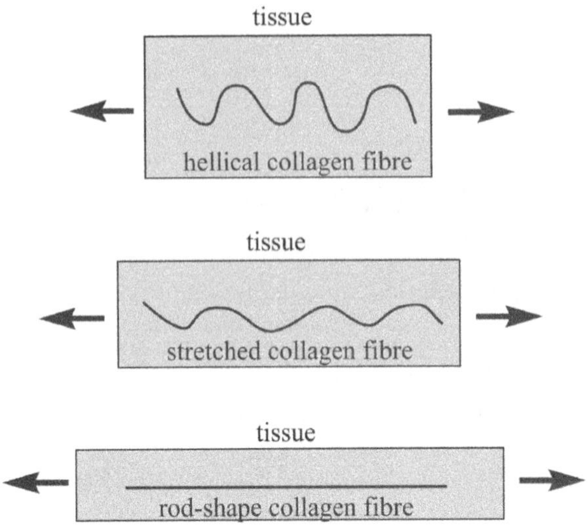

Figure 5.3 Deformation of a collagen fibre in different stretches.

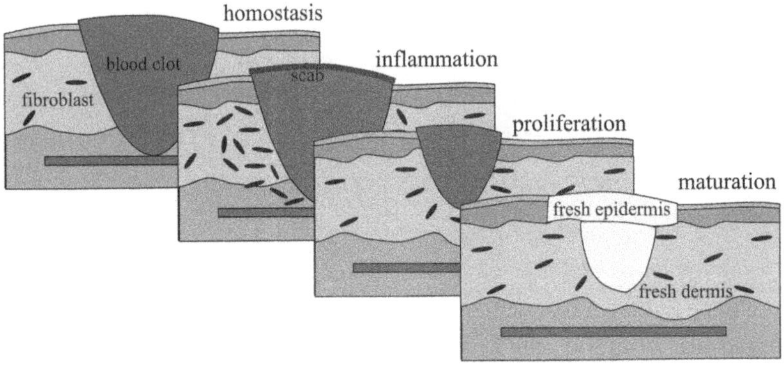

Figure 5.4 Various stages of healing of damaged skin.

A major source of degradation in soft tissues is the reduction in volume content of the collagen fibres, which may be due to the effect of some enzymes or other physiological effects (Sakakura et al. 2013). As a result, the tensile stiffness and strength of the tissue may be substantially reduced.

Another important aspect of the skin is its characteristic to begin to repair itself naturally and immediately after the occurrence of damage (such as scratches and tears, etc.). The process of healing of the human skin tissue consists of various stages over time, which includes homeostasis (bleeding stoppage), inflammation, proliferation and maturation, as presented in Figure 5.4.

Wound contraction occurs after the inflammation stage in approximately one week after the injury (Stadelmann et al. 1998; Kordestani 2019). At this stage, some of the fibroblasts moving towards the damaged zone are physiologically transformed into the myofibroblast cells due to the effects of growth factors. Myofibroblasts contain muscle proteins (Clark 1989; Stadelmann et al. 1998; Kordestani 2019) and can establish intracellular and intercellular bindings with the extracellular matrix (Clark 1989; Kordestani 2019). Simultaneously, collagen is produced and distributed by cells due to growth factors. It forms the extracellular matrix (ECM), which retains the tissue in place (Kordestani 2019).

5.2.2 Artery

An artery has a composite biological structure with a tubular form. It consists of three main layers of intima, media and adventitia, as illustrated in Figure 5.5. Each layer has its own microstructural components, mainly from one or two sets of fibres within a ground matrix.

Similar to other soft biological tissues, collagen fibres play a fundamental role in physiological functioning and mechanical characteristics of an artery. The matrix metalloproteinase (MMP) enzyme may have a significant degradation effect on collagen fibres and the strength and deformations of arteries subject to internal blood pressures (Sakakura et al. 2013). Other effects, such as abnormal thickness or accumulated plaque, may reduce the collagen fibre content and reduce the strength of an arterial wall (Sakakura et al. 2013), ultimately leading to a number of life-threatening diseases such as the aortic aneurysm, etc. (Zohdi et al. 1996; Milton 2002; Lukkassen and Milton 2008).

5.2.3 Heart Leaflet

Heart valve diseases require careful and fast treatment, either by medical prescription procedures to properly treat the main cause of damage or malfunctioning or by replacement with prosthetic valves, in one of mechanical, bioprosthetic or tissue engineered forms.

The aortic heart valve system is composed of three leaflets. Each leaflet is formed from a double-curved shell with a microstructure of ventricularis, spongiosa, and fibrosa layers, as presented in Figure 5.6. The leaflet is subject to blood flow pressure from the ventricularis side.

Variably oriented sets of collagen fibres are dispersed in various parts of the leaflet, which should be determined from clinical scans and accurately assigned to a computational model.

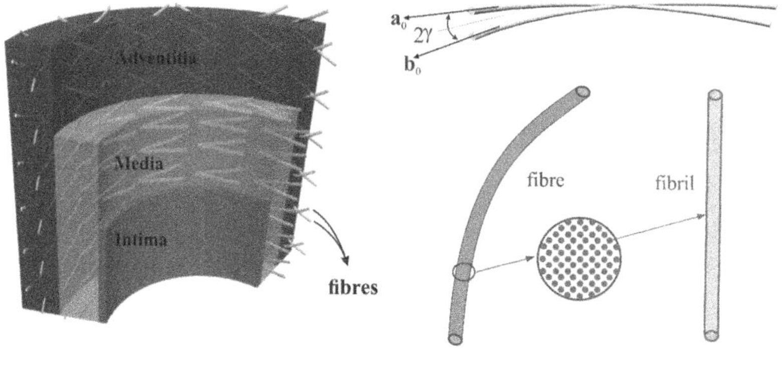

Figure 5.5 Schematic illustration of the composite microstructure of an artery.

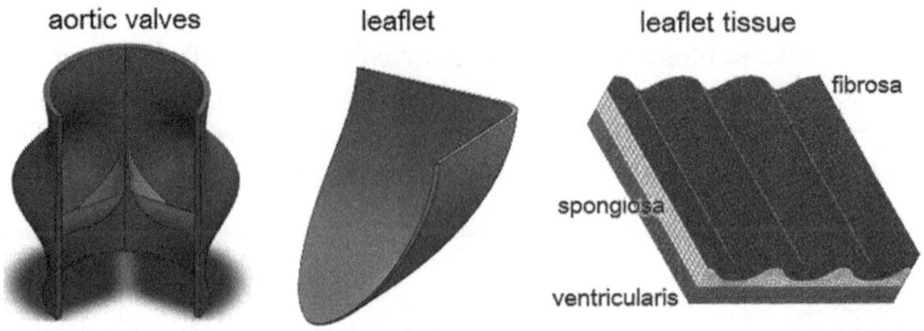

Figure 5.6 Aortic valve leaflets. *Source:* Adapted from Shahi 2013; Shahi and Mohammadi 2013.

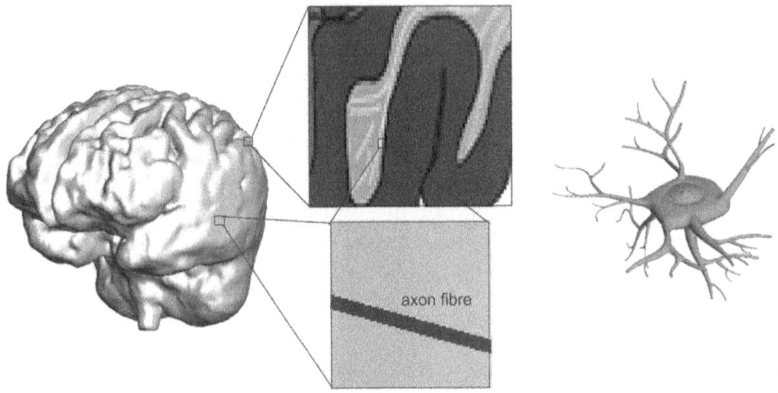

Figure 5.7 A typical brain and its microstructure down to the axons.

5.2.4 Brain Tissue

Inside the head skull, the meninges membrane separates the brain from the skull and/or other bones. Moreover, the cerebrospinal fluid inside the meninges membrane fully covers the brain and further protects it against moderate mechanical impacts (Franceschini 2006).

Figure 5.7 shows a typical human brain and its microstructures down to the axon level. It shows three successive scales of organ, tissue and cell levels. The brain tissue is similar to a composite material composed of a matrix and axon fibres.

Alternatively, the soft tissue of the brain can be classified into five parts of cerebrum, cerebellum, midbrain (mesencephalon), poles and medulla oblongata.

Grey and white matters constitute the most part of the brain. The grey matter is mainly located in the outer part of the brain and is controlled by the nerve cell system. The white matter makes up most of the internal part of the brain and constitutes the material structure of the thalamus and hypothalamus. While most of the human senses, including speaking, hearing, feeling, sight and memory, and the system of control of muscles, are conducted in the grey matter, many unintentional performances of the body, such as temperature, blood pressure, release of hormones, and expression of feelings, are moderated by the white matter (Minamel 2018).

The brain has billions of nerve cells (neurons) and a massive grid of nerve chains, so is probably one of the most complex structures of the natural world (Malekmohammadi 2012).

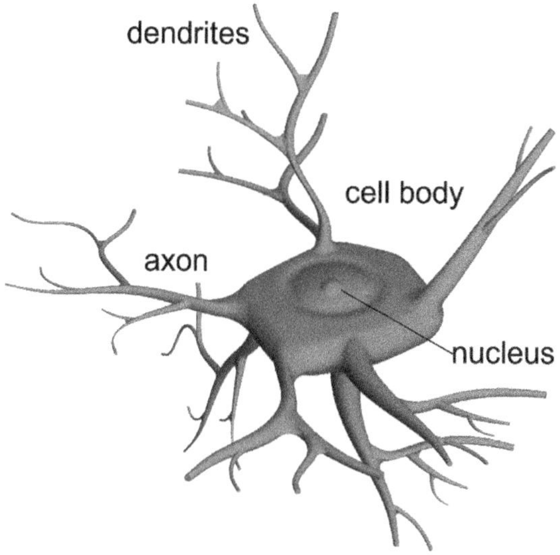

Figure 5.8 Microstructure of a typical neuron.

Neurons are the cells responsible for signal transmissions (send and receive). Each neuron is composed of axons for signal transmission to other neurons and dendrites for collecting the synopsis signals from other neurons (Figure 5.8). Axons are in long cylindrical shapes with circular or ellipsoidal cross-sections. Axons with a myelin cover form the nerve fibre. The myelin sheath is a lipoprotein layer, which covers long dendrites and axons.

Apart from biomedical causes, the brain may be damaged or malfunctioning due to daily practices such as running or falling objects or may be exposed to extreme conditions, such as the effects of car accidents on the driver and passengers and the blast waves on occupants of a building or an industrial complex. Fracture of the skull, excessive acceleration, extreme levels of tensile and shear stress and strains and internal pressures may lead to disproportionate deformations between the brain and skull and between different parts of the brain tissue, which damages the nerve cells.

5.3 Hyperelastic Models of Soft Tissues

5.3.1 Introduction

Soft biological tissues can be considered as composites made of an isotropic matrix and anisotropic dispersed reinforced fibres. One, two or even three independent sets or families of fibres may be assumed in an analysis of various soft tissues. Aligned or randomly distributed fibres can be considered depending on the physiological composition of the soft tissue (Balzani et al. 2006).

Figure 5.9 shows the typical response of a soft tissue, composed of a fibrous microstructure, in various levels of tensile loading. At the early stages, the folded or helical collagen fibres do not contribute to the load-bearing capacity of the tissue. Then, the fibres gradually unfold and contribute to the stiffness and hardening of the tissue. At the final stage of the

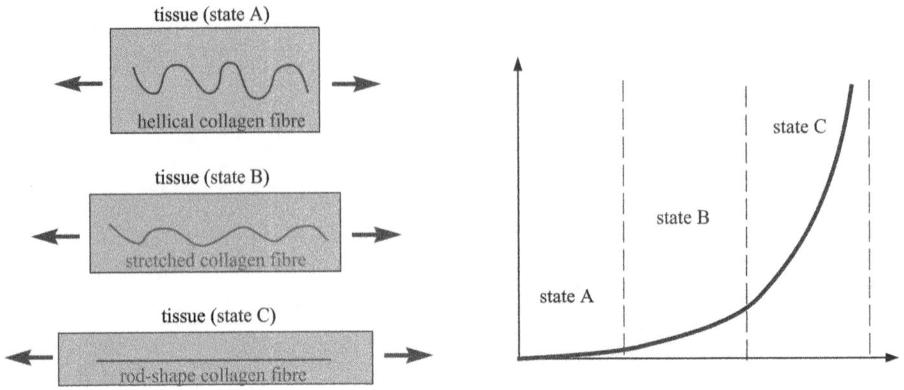

Figure 5.9 Schematic illustration of folded (helical), unfolded and stretched fibres in load bearing and stiffness of the fibrous tissue.

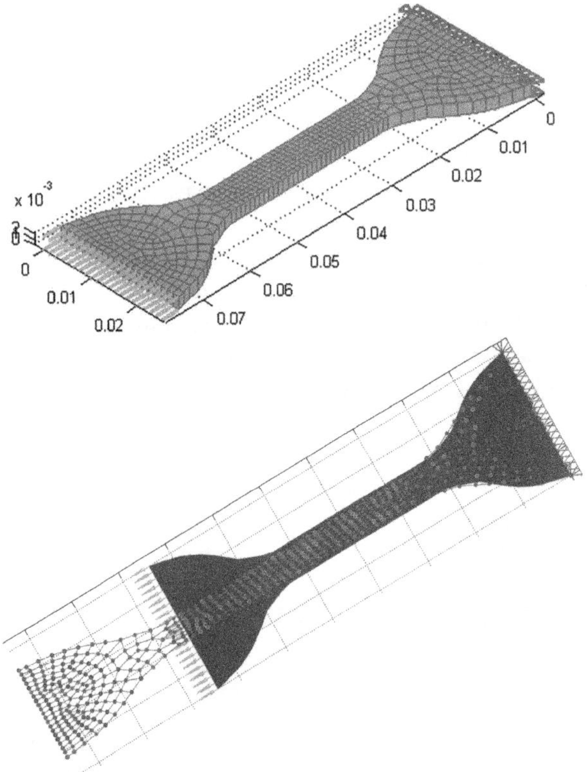

Figure 5.10 A typical tissue sample in the uniaxial tensile test.

deformation, the fully stretched fibres constitute the main resisting part of the tissue, with highly increased overall stiffness and strength of the soft tissue. Such a trend can be formulated by the hyperelastic constitutive formulation.

In order to examine typical tensile responses of a soft tissue, a dog-bone tissue sample, depicted in Figure 5.10, is considered. Typical stress–strain responses of the tissue for different directions of the Langer lines (Langer 1861) are illustrated in Figure 5.11. Clearly, the

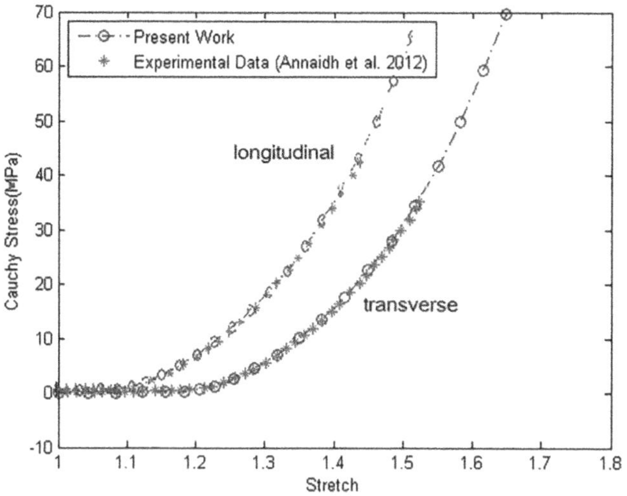

Figure 5.11 Typical tensile stress–strain responses for different directions of the Langer lines.

direction of fibres largely affects the overall response of the tissue and a harder response is observed for the fibres along the loading direction. Moreover, the hyperelastic-type constitutive laws can generally be adopted to best fit such responses for various directions of the fibrous tissue.

5.3.2 Description of Deformation and Definition of Invariants

Consider a soft biological organ/body defined by position $\mathbf{X} \in \Omega_0$ in the initial configuration Ω_0 and its counterpart position $\mathbf{x} \in \Omega$ in the current configuration Ω. The deformation gradient $\boldsymbol{F}$ defines the relation or mapping in between these two configurations (see Figure 5.12),

$$\mathrm{d}\mathbf{x} = \boldsymbol{F}\mathrm{d}\boldsymbol{X} \tag{5.3.1}$$

$$\boldsymbol{F}(\boldsymbol{X}) = \boldsymbol{\nabla}\varphi_t(\boldsymbol{X}) \tag{5.3.2}$$

Another important term, which is used in the definition of material models, is the right Cauchy–Green deformation tensor $\boldsymbol{C}$, defined as (Bathe 2006)

$$\boldsymbol{C} = \boldsymbol{F}^{\mathrm{T}}\boldsymbol{F} \tag{5.3.3}$$

In soft biological tissues, the volume change directly affects the hyperelastic constitutive relation. The Jacobian J, which is a measure of the volume change, can be defined as (Bathe 2006)

$$J = \frac{\mathrm{d}v}{\mathrm{d}V} = \det \boldsymbol{F} > 0 \tag{5.3.4}$$

5.3.2.1 Invariants for a Fibrous Tissue

The three principal invariants I_1, I_2 and I_3 of the second-order tensor $\boldsymbol{C}$ are defined as (Bonet and Wood 1997; Belytschko et al. 2000)

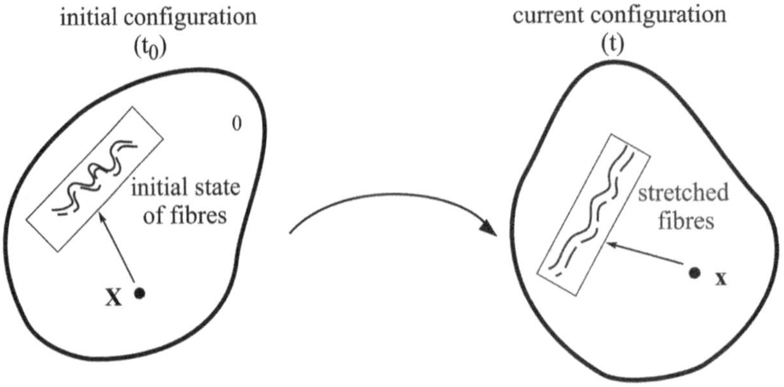

Figure 5.12 A biomechanical model in the initial (undeformed) and current (deformed) configurations.

$$I_1 = \text{tr}(\boldsymbol{C}) = \boldsymbol{C} : \mathbf{I} = \lambda_1^2 + \lambda_2^2 + \lambda_3^2 \tag{5.3.5}$$

$$I_2 = \text{tr}(\boldsymbol{CC}) = \boldsymbol{C} : \boldsymbol{C} = \lambda_1^2\lambda_2^2 + \lambda_2^2\lambda_3^2 + \lambda_3^2\lambda_1^2 \tag{5.3.6}$$

$$I_3 = \det(\boldsymbol{C}) = J^2 = \lambda_1^2\lambda_2^2\lambda_3^2 \tag{5.3.7}$$

where λ_i are the eigenvalues of $\boldsymbol{C}$ and $\mathbf{I}$ is the second-order tensor,

$$\mathbf{I}_{ij} = \begin{cases} 1, & \text{if } i = j \\ 0, & \text{otherwise} \end{cases} \tag{5.3.8}$$

The pseudo or generalized invariant I_{f_1} (also termed I_4) for a single set of fibres with the unit vector $\boldsymbol{a}_{f_1}$ in the tissue (as depicted in Figure 5.13) is defined as (Holzapfel et al. 2000)

$$I_{f_1} = \text{tr}(\boldsymbol{C}\boldsymbol{N}_{f_1}) \tag{5.3.9}$$

where $\boldsymbol{N}_{f_1}$ is the structural tensor associated with the direction of the set of fibres, as presented in Figure 5.13:

$$\boldsymbol{N}_{f_1} = \boldsymbol{a}_{f_1} \otimes \boldsymbol{a}_{f_1} \tag{5.3.10}$$

The fibre stretch λ_{f_1} represents the amount of elongation in the preferred direction $\boldsymbol{a}_{f_1}$:

$$\lambda_{f_1}\boldsymbol{N}_{f_1} = \boldsymbol{F}.\boldsymbol{a}_{f_1} \tag{5.3.11}$$

$$\lambda_{f_1}^2 = \boldsymbol{a}_{f_1}.\boldsymbol{C}.\boldsymbol{a}_{f_1} \tag{5.3.12}$$

If the tissue contains a second set of fibres with the unit vector $\boldsymbol{a}_{f_2}$, as shown in Figure 5.13, the pseudo invariant I_{f_2} (or I_6) is defined as

$$I_{f_2} = \text{tr}(\boldsymbol{C}\boldsymbol{N}_{f_2}) \tag{5.3.13}$$

where

$$\boldsymbol{N}_{f_2} = \boldsymbol{a}_{f_2} \otimes \boldsymbol{a}_{f_2} \tag{5.3.14}$$

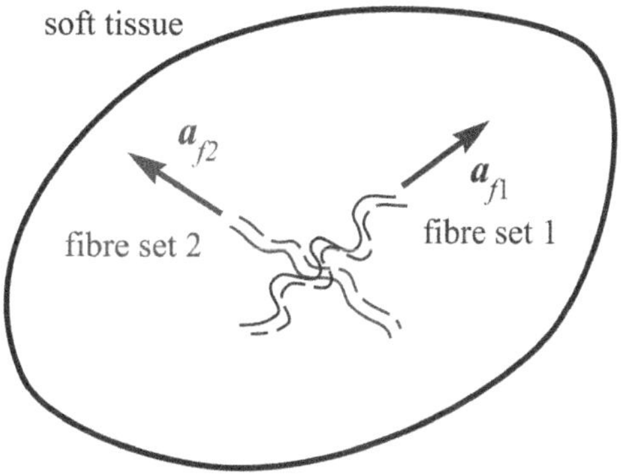

Figure 5.13 Definition of two sets of fibres in a tissue, represented by $\boldsymbol{a}_{f_1}$ and $\boldsymbol{a}_{f_2}$ unit vectors.

The structural tensors $\boldsymbol{N}_{f_1}$ and $\boldsymbol{N}_{f_2}$ account for the significant anisotropy generated by one or two sets of fibres.

Adopting the multiplicative decomposition assumption for the incompressible materials, the deformation gradient $\boldsymbol{F}$ is decomposed into volumetric $\boldsymbol{F}_v$ and deviatoric $\boldsymbol{F}_d$ parts (Bathe 2006):

$$\boldsymbol{F} = \boldsymbol{F}_v \boldsymbol{F}_d \tag{5.3.15}$$

The deviatoric part of the deformation gradient is defined by (Bathe 2006)

$$\boldsymbol{F}_d = J^{-\frac{1}{3}} \boldsymbol{F} \tag{5.3.16}$$

and the corresponding right Cauchy tensor ($\boldsymbol{C}_d$) by

$$\boldsymbol{C}_d = J^{-\frac{2}{3}} \boldsymbol{C} \tag{5.3.17}$$

Accordingly, the reduced (or deviatoric) invariants can be defined as

$$J_1 = \mathrm{tr}(\boldsymbol{C}_d) = I_1 (I_3)^{-\frac{1}{3}} \tag{5.3.18}$$

$$J_2 = \mathrm{tr}(\boldsymbol{C}_d \boldsymbol{C}_d) = I_2 (I_3)^{-\frac{2}{3}} \tag{5.3.19}$$

$$J_3 = \det(\boldsymbol{C}_d) = (I_3)^{\frac{1}{2}} \tag{5.3.20}$$

$$J_{f_1} = I_{f_1} (I_3)^{-\frac{1}{3}} \tag{5.3.21}$$

$$J_{f_2} = I_{f_2} (I_3)^{-\frac{1}{3}} \tag{5.3.22}$$

5.3.3 Isotropic neo-Hookean Hyperelastic Model

The neo-Hookean hyperelastic model has been extensively used for analysis of a number of soft tissues such as the brain. The brain tissue is similar to a composite material composed of a matrix and axon fibres (Janfada 2018; Zolghadr 2022). While some reports believe that the white matter is highly anisotropic (Pfefferbaum et al. 2000), some others have adopted isotropic hyperelastic models, at least in low strains (Arbogast et al. 1997; Donnelly and Medige 1997).

In isotropic hyperelastic problems, the strain energy function w^s is written in terms of $\boldsymbol{C}$,

$$w^s(\boldsymbol{C}) = w^s(I_1, I_2, I_3) \tag{5.3.23}$$

and the second Piola–Kirchhoff stress is computed from

$$\begin{aligned}
\boldsymbol{S} &= 2\frac{\partial w^s}{\partial \boldsymbol{C}} = 2\frac{\partial w^s}{\partial I_1}\frac{\partial I_1}{\partial \boldsymbol{C}} + 2\frac{\partial w^s}{\partial I_2}\frac{\partial I_2}{\partial \boldsymbol{C}} + 2\frac{\partial w^s}{\partial I_3}\frac{\partial I_3}{\partial \boldsymbol{C}} \\
&= 2\left[I_3 \frac{\partial w^s}{\partial I_3}\boldsymbol{C}^{-1} + \left(\frac{\partial w^s}{\partial I_1} + I_1 \frac{\partial w^s}{\partial I_2} \right)\mathbf{I} - \frac{\partial w^s}{\partial I_2}\boldsymbol{C} \right] \\
&= 2\frac{\partial w^s}{\partial I_1}\mathbf{I} + 4\frac{\partial w^s}{\partial I_2}\boldsymbol{C} + 2J^2\frac{\partial w^s}{\partial I_3}\boldsymbol{C}^{-1}
\end{aligned} \tag{5.3.24}$$

Beginning from the general form of the neo-Hookean model,

$$w^s(\boldsymbol{C}) = \frac{1}{2}\mu(I_1 - 3) - \mu \ln J + \frac{1}{2}\lambda(\ln J)^2 \tag{5.3.25}$$

where λ and μ are the elastic material properties. The second Piola–Kirchhoff stress $\boldsymbol{S}$ is obtained from Equation (2.2.238):

$$\boldsymbol{S} = \frac{\partial w^s}{\partial \boldsymbol{E}} = 2\frac{\partial w^s}{\partial \boldsymbol{C}} = \mu(\mathbf{I} - \boldsymbol{C}^{-1}) + \lambda(\ln J)\boldsymbol{C}^{-1} \tag{5.3.26}$$

and the corresponding elasticity tensor $\mathbf{D}^{SE}$ (associated with the second Piola–Kirchhoff stress $\boldsymbol{S}$ and the Green–Lagrange strain tensor $\boldsymbol{E}$),

$$\mathbf{D}^{SE} = \frac{\partial \boldsymbol{S}}{\partial \boldsymbol{E}} = \frac{\partial^2 w^s}{\partial \boldsymbol{E} \partial \boldsymbol{E}} = 2\frac{\partial \boldsymbol{S}}{\partial \boldsymbol{C}} = 4\frac{\partial^2 w^s}{\partial \boldsymbol{C} \partial \boldsymbol{C}} = \lambda \boldsymbol{C}^{-1} \otimes \boldsymbol{C}^{-1} - 2(\mu - \lambda \ln J)\frac{\partial \boldsymbol{C}^{-1}}{\partial \boldsymbol{C}} \tag{5.3.27}$$

The spatial description of Equations (5.3.25) and (5.3.26) are required if an updated Lagrangian formulation is adopted (Balzani et al. 2012). The Cauchy stress $\boldsymbol{\sigma}$ can be computed from

$$\boldsymbol{\sigma} = J^{-1}\boldsymbol{F}\boldsymbol{S}\boldsymbol{F}^{\mathrm{T}} \tag{5.3.28}$$

and the spatial elasticity tensor $\mathbf{D}^{PF}$ (corresponding to the first Piola–Kirchhoff stress $\boldsymbol{P}$ and the deformation gradient $\boldsymbol{F}$) can be derived as (Simo and Pister 1984; Belytschko et al. 2001; Bathe 2006)

$$\mathbf{D}^{PF} = \frac{\partial^2 w^s}{\partial \boldsymbol{F} \partial \boldsymbol{F}} = \boldsymbol{F}\mathbf{D}^{SE}\boldsymbol{F} + \boldsymbol{S} \otimes \mathbf{g}^{-1} \tag{5.3.29}$$

where $\boldsymbol{g}$ is the metric tensor in the current configuration Ω, defined by Simo and Pister (1984) and Clayton (2010). For further details refer to Bonet and Wood (1997), Belytschko et al. (2001) and Bathe (2006).

In nearly incompressible soft tissues ($J \approx 1$), the incompressibility is enforced numerically by a penalty method to impose an external volumetric pressure to preserve the volume (Balzani et al. 2006; Alastrué et al. 2007; Calvo et al. 2007). Therefore, an isotropic incompressible neo-Hookean model for modelling of the incompressible tissue can be written in terms of the deviatoric (isochoric) $w_{\mathrm{d}}^{\mathrm{s}}$ and volumetric $w_{\mathrm{v}}^{\mathrm{s}}$ strain energies,

$$w^{\mathrm{s}}(\boldsymbol{C}) = \underbrace{w_{\mathrm{d}}^{\mathrm{s}}(\boldsymbol{C})}_{\text{deviatoric}} + \underbrace{w_{\mathrm{v}}^{\mathrm{s}}(J)}_{\text{volumetric}} \tag{5.3.30}$$

where

$$w_{\mathrm{d}}^{\mathrm{s}}(\boldsymbol{C}) = \frac{1}{2}\mu(J_1 - 3) \tag{5.3.31}$$

$$w_{\mathrm{v}}^{\mathrm{s}}(J) = \frac{1}{2}\kappa_0(J-1)^2 \tag{5.3.32}$$

and κ_0 is the penalty coefficient to impose the pressure p,

$$p = \kappa_0(J-1) \tag{5.3.33}$$

The second Piola–Kirchhoff stress $\boldsymbol{S}$ can be computed from

$$\boldsymbol{S} = 2\frac{\partial w^{\mathrm{s}}}{\partial \boldsymbol{C}} = 2\frac{\partial w_{\mathrm{d}}^{\mathrm{s}}}{\partial \boldsymbol{C}} + 2\frac{\partial w_{\mathrm{v}}^{\mathrm{s}}}{\partial \boldsymbol{C}} \tag{5.3.34}$$

leading to

$$\boldsymbol{S} = \mu I_3^{-1/3}(\mathbf{I} - \frac{1}{3}I_1\boldsymbol{C}^{-1}) + pJ\boldsymbol{C}^{-1} \tag{5.3.35}$$

and the elasticity tensor $\mathbf{D}^{SE}$ becomes (Janfada 2018; Zolghadr 2022)

$$\begin{aligned}
\mathbf{D}^{SE} &= 2\frac{\partial \boldsymbol{S}}{\partial \boldsymbol{C}} = \left\{2\frac{\partial}{\partial \boldsymbol{C}}\left[\mu I_3^{-\frac{1}{3}}\left(\mathbf{I} - \frac{1}{3}I_1\boldsymbol{C}^{-1}\right)\right]\right\} + \left\{2\frac{\partial}{\partial \boldsymbol{C}}\left(pJ\boldsymbol{C}^{-1}\right)\right\} \\
&= \left\{2\mu\frac{\partial\left(I_3^{-1/3}\right)}{\partial \boldsymbol{C}}\left(\mathbf{I} - \frac{1}{3}I_1\boldsymbol{C}^{-1}\right) + 2\mu I_3^{-1/3}\frac{\partial}{\partial \boldsymbol{C}}\left(\mathbf{I} - \frac{1}{3}I_1\boldsymbol{C}^{-1}\right)\right\} \\
&\quad + \left\{2\frac{\partial p}{\partial J}\frac{\partial J}{\partial \boldsymbol{C}}\left(J\boldsymbol{C}^{-1}\right) + 2\frac{\partial J}{\partial \boldsymbol{C}}\left(p\boldsymbol{C}^{-1}\right) + 2\frac{\partial \boldsymbol{C}^{-1}}{\partial \boldsymbol{C}}(pJ)\right\} \\
&= \left\{2\mu\left(\mathbf{I} - \frac{1}{3}I_1\boldsymbol{C}^{-1}\right)\left(-\frac{1}{3}I_3^{-1/3}\boldsymbol{C}^{-1}\right) + 2\mu I_3^{-1/3}\left(-\frac{1}{3}\boldsymbol{C}^{-1}\frac{\partial I_1}{\partial \boldsymbol{C}} - \frac{1}{3}I_1\frac{\partial \boldsymbol{C}^{-1}}{\partial \boldsymbol{C}}\right)\right\} \\
&\quad + \left\{2(\kappa)\left(\frac{1}{2}J\boldsymbol{C}^{-1}\right)\left(J\boldsymbol{C}^{-1}\right) + 2\left(\frac{1}{2}J\boldsymbol{C}^{-1}\right)\left(p\boldsymbol{C}^{-1}\right) - 2(\mathbf{Y})(pJ)\right\} \\
&= \left\{\frac{2}{3}\mu I_3^{-1/3}\left(I_1\mathbf{Y} - \mathbf{I}\otimes\boldsymbol{C}^{-1} - \boldsymbol{C}^{-1}\otimes\mathbf{I} + \frac{1}{3}I_1\boldsymbol{C}^{-1}\otimes\boldsymbol{C}^{-1}\right)\right\} \\
&\quad + \left\{pJ\left(\boldsymbol{C}^{-1}\otimes\boldsymbol{C}^{-1} - 2\mathbf{Y}\right) + \kappa J^2\boldsymbol{C}^{-1}\otimes\boldsymbol{C}^{-1}\right\}
\end{aligned} \tag{5.3.36}$$

where the components of **Y** are defined as

$$\mathbf{¥}_{ijkl} = -\frac{\partial\left(\boldsymbol{C}^{-1}\right)_{ij}}{\partial\left(\boldsymbol{C}\right)_{kl}} = \frac{1}{2}\left[(\boldsymbol{C}^{-1})_{ik}(\boldsymbol{C}^{-1})_{jl} + (\boldsymbol{C}^{-1})_{il}(\boldsymbol{C}^{-1})_{jk}\right] \tag{5.3.37}$$

Equation (5.3.30) can be adopted to govern the constitutive response of an interstitial cell (IC) in a cell-scale modelling of a biomechanical multiscale simulation.

5.3.4 Isotropic Mooney–Rivlin Hyperelastic Model

Alternatively, the Mooney–Rivlin hyperelastic constitutive relation for nearly incompressible materials can be adopted,

$$w^{s} = A_1(J_1 - 3) + A_2(J_2 - 3) + \frac{1}{2}\kappa(J_3 - 1)^2 \tag{5.3.38}$$

where A_1 and A_2 are the material constants and κ is the penalty coefficient to impose the pressure of the nearly incompressible constraint.

The second Piola–Kirchhoff stress can be obtained by differentiation of Equation (5.3.38),

$$\boldsymbol{S} = \frac{\partial w^{s}}{\partial \boldsymbol{E}} = \frac{\partial w^{s}}{\partial J_1}\frac{\partial J_1}{\partial \boldsymbol{E}} + \frac{\partial w^{s}}{\partial J_2}\frac{\partial J_2}{\partial \boldsymbol{E}} + \frac{\partial w^{s}}{\partial J_3}\frac{\partial J_3}{\partial \boldsymbol{E}} = A_1 J_{1,\boldsymbol{E}} + A_2 J_{2,\boldsymbol{E}} + \kappa J_{3,\boldsymbol{E}}(J_3 - 1) \tag{5.3.39}$$

where (Zolghadr and Mohammadi 2022)

$$J_{1,\boldsymbol{E}} = \underbrace{I_{1,\boldsymbol{E}}}_{2\mathbf{I}}(I_3)^{-\frac{1}{3}} - \frac{1}{3}I_1(I_3)^{-\frac{4}{3}}\underbrace{I_{3,\boldsymbol{E}}}_{2I_3\boldsymbol{C}^{-1}} = 2\mathbf{I}(I_3)^{-\frac{1}{3}} - \frac{2}{3}I_1(I_3)^{-\frac{1}{3}}\boldsymbol{C}^{-1} \tag{5.3.40}$$

$$J_{2,\boldsymbol{E}} = \underbrace{I_{2,\boldsymbol{E}}}_{2(I_1\mathbf{I}-\boldsymbol{C})}(I_3)^{-\frac{2}{3}} - \frac{2}{3}I_2(I_3)^{-\frac{5}{3}}\underbrace{I_{3,\boldsymbol{E}}}_{2I_3\boldsymbol{C}^{-1}} = 2(I_1\mathbf{I}-\boldsymbol{C})(I_3)^{-\frac{2}{3}} - \frac{4}{3}I_2(I_3)^{-\frac{2}{3}}\boldsymbol{C}^{-1} \tag{5.3.41}$$

$$J_{3,\boldsymbol{E}} = \frac{1}{2}(I_3)^{-\frac{1}{2}}\underbrace{I_{3,\boldsymbol{E}}}_{2I_3\boldsymbol{C}^{-1}} = (I_3)^{\frac{1}{2}}\boldsymbol{C}^{-1} \tag{5.3.42}$$

The corresponding elasticity tensor $\mathbf{D}^{SE}$ is then obtained as

$$\mathbf{D}^{SE} = \frac{\partial \boldsymbol{S}}{\partial \boldsymbol{E}} = A_1 J_{1,\boldsymbol{EE}} + A_2 J_{2,\boldsymbol{EE}} + \kappa J_{3,\boldsymbol{EE}}(J_3 - 1) + 2J_{3,\boldsymbol{E}} \otimes J_{3,\boldsymbol{E}} \tag{5.3.43}$$

where a further set of differentiation is required:

$$\begin{aligned} J_{1,\boldsymbol{EE}} = \frac{\partial J_{1,\boldsymbol{E}}}{\partial \boldsymbol{E}} &= I_{1,\boldsymbol{EE}}(I_3)^{-\frac{1}{3}} - \frac{1}{3}(I_3)^{-\frac{4}{3}}(I_{1,\boldsymbol{E}} \otimes I_{3,\boldsymbol{E}} + I_{3,\boldsymbol{E}} \otimes I_{1,\boldsymbol{E}}) \\ &+ \frac{4}{9}I_1 I_3^{-\frac{7}{3}} I_{3,\boldsymbol{E}} \otimes I_{3,\boldsymbol{E}} - \frac{1}{3}I_1 I_3^{-\frac{4}{3}} I_{3,\boldsymbol{EE}} \end{aligned} \tag{5.3.44}$$

$$J_{2,EE} = \frac{\partial J_{2,E}}{\partial \mathbf{E}} = I_{1,EE}(I_3)^{-\frac{2}{3}} - \frac{2}{3}(I_3)^{-\frac{5}{3}}(I_{2,E} \otimes I_{3,E} + I_{3,E} \otimes I_{2,E})$$
$$+ \frac{10}{9} I_2 I_3^{-\frac{8}{3}} I_{3,E} \otimes I_{3.E} - \frac{1}{3} I_2 I_3^{-\frac{5}{3}} I_{3,EE} \qquad (5.3.45)$$

$$J_{3,EE} = \frac{\partial J_{3,E}}{\partial \mathbf{E}} = -\frac{1}{4}(I_3)^{-\frac{3}{2}} I_{3,E} \otimes I_{3,E} + \frac{1}{2}(I_3)^{-\frac{1}{2}}(I_{3,EE}) \qquad (5.3.46)$$

with

$$I_{1,EE} = 0 \qquad (5.3.47)$$

$$I_{2,EE} = 4(\mathbf{I} \otimes \mathbf{I} - \mathcal{I}) \qquad (5.3.48)$$

$$I_{3,EE} = 4I_3\left\{\boldsymbol{C}^{-1} \otimes \boldsymbol{C}^{-1} - \mathbf{Y}\right\} \qquad (5.3.49)$$

where $\mathbf{I}$ is defined in Equation (5.3.8) and $\mathcal{I}$ is a fourth-order symmetric tensor,

$$\mathcal{I}_{ijkl} = \frac{1}{2}(\delta_{ik}\delta_{jl} + \delta_{il}\delta_{jk}) \qquad (5.3.50)$$

and $\mathbf{Y}$ is defined in Equation (5.3.37).

5.3.5 Hyperelastic Models for Multiscale Simulation of Tendon

Collagen is the main protein that constitutes the extracellular matrix. Fibrils and fibres are polymers constructed from collagen monomers. Collagen fibres are made of very strong fibrils and elastin, which create the high tensile strength of the skin.

The multiscale study of tendons and ligaments requires employment of appropriate hyperelastic constitutive models for the soft tissue at different scales, which include the fibril, the fibre and the tissue, as presented in Figure 5.14 (Fathi and Mohammadi 2015).

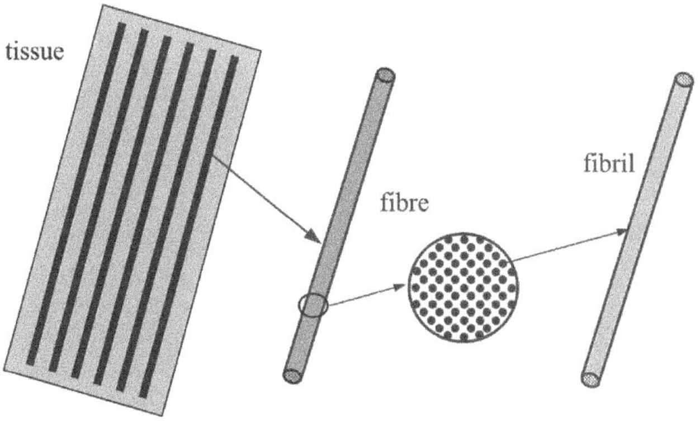

Figure 5.14 The hierarchical multiscale structure of a fibrous soft tissue (tendon).

5.3.5.1 Elastic Strain Energy Function for Fibril

The hyperelastic constitutive equation for fibril is assumed to be a combination of the simple neo-Hookean model and a slightly simplified exponential function (Guo et al. 2006; Tang et al. 2009):

$$w_{fl}^{s} = \frac{1}{2}\mu_{fl}(J_{f_1})(I_1(\boldsymbol{F}_{fl}) - 3) \tag{5.3.51}$$

where $I_1 = I_1(\boldsymbol{F}_{fl})$ is the first invariant of the deformation gradient of the fibril, $\boldsymbol{F}_{fl}$,

$$I_1(\boldsymbol{F}_{fl}) = \mathrm{tr}(\boldsymbol{F}_{fl}^{T}\boldsymbol{F}_{fl}) \tag{5.3.52}$$

and

$$\mu_{fl}(I_4) = \mu_{fl}^{0} f(I_4) \tag{5.3.53}$$

$$f(I_4) = a_1 + a_2 \exp\left[a_3(I_{f_1} - I_0)\right] \tag{5.3.54}$$

where a_1, a_2 and a_3 are dimensionless material parameters and I_0 characterizes the secondary stiffening of the fibres (Guo et al. 2006).

Decomposition of w_{fl}^{s} into the volumetric $w_{fl,v}^{s}$ and deviatoric $w_{fl,d}^{s}$ terms (Bonet and Wood 1997; Holzapfel 2000) can now be written as

$$w_{fl}^{s} = w_{fl,v}^{s} + w_{fl,d}^{s} \tag{5.3.55}$$

The second Piola–Kirchhoff stress is computed by differentiating the strain energy function w_{fl}^{s},

$$\boldsymbol{S}_{fl,d} = 2\frac{\partial w_{fl,d}^{s}}{\partial \boldsymbol{C}} = (J_1 - 3)\frac{\partial \mu_{fl}\left(J_{f_1}\right)}{\partial \boldsymbol{C}} + \mu_{fl}(J_{f_1})\frac{\partial J_1}{\partial \boldsymbol{C}} \tag{5.3.56}$$

The corresponding elasticity tensor $\mathbf{D}_{fl,d}^{SE}$ is then computed by further differentiation,

$$\mathbf{D}_{fl,d}^{SE} = 4\frac{\partial^2 w_{fl,d}^{s}}{\partial \boldsymbol{C}\partial \boldsymbol{C}} = 2\left\{(J_1 - 3)\frac{\partial^2 \mu_{fl}\left(J_{f_1}\right)}{\partial \boldsymbol{C}\partial \boldsymbol{C}} + \frac{\partial J_1}{\partial \boldsymbol{C}} \otimes \frac{\partial \mu_{fl}\left(J_{f_1}\right)}{\partial \boldsymbol{C}} + \frac{\partial \mu_{fl}\left(J_{f_1}\right)}{\partial \boldsymbol{C}} \otimes \frac{\partial J_1}{\partial \boldsymbol{C}} + \mu_{fl}(J_{f_1})\frac{\partial^2 J_1}{\partial \boldsymbol{C}\partial \boldsymbol{C}}\right\} \tag{5.3.57}$$

5.3.5.2 Elastic Strain Energy Function for Fibre

Fibres in soft tissues contain two main materials: fibrils and the surrounding matrix. The strain energy function of fibre w_{fl}^{s} consists of a combination of the fibrils, w_{fl}^{s}, the matrix, w_{m}^{s}, and their shear interactions w_{s}^{s} subject to a fibre-scale deformation gradient $\boldsymbol{F}_{fr}$ (Tang et al. 2009),

$$w_{fr}^{s} = \upsilon_{fl} w_{fl}^{s}(I_1(\boldsymbol{F}_{fr}), I_{f_1}) + \upsilon_m w_m^{s}(I_1(\boldsymbol{F}_{fr})) + w_s^{s} \tag{5.3.58}$$

where v_{fl} and v_m are the volume ratios of fibrils and matrix in a single fibre, respectively,

$$v_m + v_{fl} = 1 \tag{5.3.59}$$

Furthermore (Fathi and Mohammadi 2015),

$$I_1(\boldsymbol{F}_{fr}) = I_{f_1} + 2I_{f_1}^{-\frac{1}{2}} \tag{5.3.60}$$

The strain energy function of fibrils, w^s_{fl}, is defined earlier in Equation (5.3.51). The strain energy function of matrix, w^s_m, is defined by a simple incompressible neo-Hookean constitutive law,

$$w^s_m = \frac{1}{2}\mu_m(I_1 - 3) \tag{5.3.61}$$

where μ_m is the shear modulus of the matrix.

The shear interaction strain energy function w^s_s of the fibril–matrix interface is given by

$$w^s_s = \frac{1}{2}\mu^{\text{eff}}_{\text{fr}}\left[I_1 - I_1(\boldsymbol{F}_{\text{fr}})\right] \tag{5.3.62}$$

where (Guo. et al. 2006; Tang et al. 2009)

$$\mu^{\text{eff}}_{\text{fr}} = \mu_m \frac{\left(1 + v_{fl}\right) f\left(I_{f_1}\right) + \left(1 - v_{fl}\right)}{\left(1 - v_{fl}\right) f\left(I_{f_1}\right) + \left(1 + v_{fl}\right)} \tag{5.3.63}$$

5.3.5.3 Elastic Strain Energy Function for Tissue

Tissue is a fibre-reinforced composite in which the fibres are embedded randomly (skin) or aligned (tendon) in the ground matrix, depending on the type of tissue.

Tang et al. (2009) proposed the strain energy function w^s_t for tissue in terms of its fibre and matrix parts, w^s_{tf} and w^s_{tm}, respectively, and the shear interaction term, w^s_{ts},

$$w^s_t = v_{tf} w^s_{tf} + v_{tm} w^s_{tm} + w^s_{ts} \tag{5.3.64}$$

where w_{tf} is the strain energy of the fibre, defined as w^s_{fr} in the previous section. v_{tf} and v_{tm} are the volume contents of the fibre and matrix in the tissue.

Similar to the fibre, the strain energy function for the ground matrix of tissue follows the neo-Hookean type as

$$w^s_{tm} = \frac{1}{2}\mu_{tm}(I_1 - 3) \tag{5.3.65}$$

and the shear term w^s_{ts} (Tang et al. 2009),

$$w^s_{ts} = \frac{1}{2}\mu^{\text{eff}}_{\text{t}}\left[I_1 - I_1(\boldsymbol{F}_{\text{fr}})\right] \tag{5.3.66}$$

where

$$\mu_{t}^{\text{eff}} = \mu_{tm} \frac{\left(1+\upsilon_{tf}\right)\mu_{\text{fr}}^{\text{eff}} + \left(1-\upsilon_{tf}\right)\mu_{tm}}{\left(1-\upsilon_{tf}\right)\mu_{\text{fr}}^{\text{eff}} + \left(1+\upsilon_{tf}\right)\mu_{tm}} \tag{5.3.67}$$

At each scale, the second Piola–Kirchhoff stress and the associated elasticity tensor are computed from the first and second derivatives of the strain energy function.

5.3.6 Anisotropic Hyperelastic Models for Fibrous Tissues

Multiscale simulation of fibrous tissues, such as the heart valve, requires appropriate hyperelastic models at various scales, including the organ, the tissue and the cell level (Sacks and Yoganathan 2007; Weinberg and Kazempur Mofrad 2007a, 2007b).

5.3.6.1 Organ Scale

It is known that most tissues exhibit incompressible characteristics along with a strong anisotropic behaviour due to the existence and distribution of collagen fibres (Billiar and Sacks 2000; Koch et al. 2010).

Fung (1993) developed one of the fundamental frameworks on constitutive models of soft tissues by proposing a hyperelastic model with an exponential strain energy function, paving the way for many other more accurate and more complicated models for representing the behaviour of tissues with collagen fibres with damage, viscosity, etc. (Balzani et al. 2006; Martins et al. 2012; Holzapfel and Gasser 2001).

According to the material model presented by Ferrara and Pandolfi (2008) and Holzapfel et al. (2000), the mechanical response of collagen fibres is assumed to be governed by a combination of the anisotropic response of fibrils embedded in an isotropic gel-like matrix. As a result, the strain energy function w^s has volumetric w_v^s and deviatoric w_d^s parts, where the deviatoric part is defined in terms of the isotropic and anisotropic terms (Weinberg and Kazempur Mofrad 2007a, 2007b; Holzapfel 2000; Huang 2005; Ferrara and Pandolfi 2008),

$$w^s(\boldsymbol{C}) = w^s(J_1, J_{f_i}, J) = \underbrace{w_v^s(J)}_{\text{volumetric}} + \underbrace{w_d^s(J_1, J_{f_1}, J_{f_2})}_{\text{deviatoric}} \tag{5.3.68}$$

The volumetric part $w_v^s(J)$, which takes into account the incompressibility, is based on the positive penalty parameter κ_o (for the organ) (Holzapfel 2000; Bonet and Wood 2008):

$$w_v^s(J) = \frac{1}{2}\kappa_o (J-1)^2 \tag{5.3.69}$$

Alternatively, it can be defined by (Ferrara and Pandolfi 2008)

$$w_v^s(J) = \frac{1}{2}\lambda(\ln J)^2 \tag{5.3.70}$$

The isochoric/deviatoric part of the strain energy function w_d^s is written as the sum of two exponential terms,

$$w_d^s(J_1, J_{f_i}) = w_{di}^s(J_1) + w_{da}^s(J_{f_1}, J_{f_2}) \tag{5.3.71}$$

where

$$w_{di}^{s}(J_1)=\frac{c_2}{2c_1}\left(e^{c_1(J_1-3)}-1\right) \tag{5.3.72}$$

or alternatively (Ferrara and Pandolfi 2008),

$$w_{di}^{s}(J_1)=\frac{1}{2}\mu(J_1-3) \tag{5.3.73}$$

and the second term accounts for the transverse isotropy of two sets of fibres (Koch et al. 2010):

$$w_{da}^{s}(J_{f_1},J_{f_2})=w_{da}^{s}(J_{f_1})+w_{da}^{s}(J_{f_2}) \tag{5.3.74}$$

with

$$w_{da}^{s}(J_{f_1})=\begin{cases}\dfrac{c_{2f_1}}{2c_{1f_1}}\left(e^{c_{1f_1}(J_{f_1}-1)^2}-1\right), & J_{f_1}>1\\ 0, & J_{f_1}\le 1\end{cases} \tag{5.3.75}$$

$$w_{da}^{s}(J_{f_2})=\begin{cases}\dfrac{c_{2f_2}}{2c_{1f_2}}\left(e^{c_{1f_2}(J_{f_2}-1)^2}-1\right), & J_{f_2}>1\\ 0, & J_{f_2}\le 1\end{cases} \tag{5.3.76}$$

where λ and μ are the bulk and shear modulus of the ground matrix, c_1 and c_2 are the experimental-based isotropic material parameters, c_{2f_1} and c_{2f_2} denote the experimental constants associated with the stiffness of the first and second set of fibres, respectively and c_{1f_1} and c_{1f_2} are the additional constant parameters of the first and second set of fibres, respectively.

The second Piola–Kirchhoff stress $\boldsymbol{S}_d$ can be obtained by differentiation of w_d^{s}

$$\boldsymbol{S}_d=2\frac{\partial w_d^{s}}{\partial \boldsymbol{C}}=2\left(\frac{\partial w_{di}^{s}}{\partial J_1}\frac{\partial J_1}{\partial \boldsymbol{C}}+\frac{\partial w_{da}^{s}}{\partial J_{f_1}}\frac{\partial J_{f_1}}{\partial \boldsymbol{C}}+\frac{\partial w_{da}^{s}}{\partial J_{f_2}}\frac{\partial J_{f_2}}{\partial \boldsymbol{C}}\right) \tag{5.3.77}$$

with

$$\begin{aligned}\frac{\partial J_1}{\partial \boldsymbol{C}}&\xrightarrow{J_1=J^{-\frac{2}{3}}\times I_1}\frac{\partial J_1}{\partial \boldsymbol{C}}=\frac{\partial\left(J^{-\frac{2}{3}}I_1\right)}{\partial \boldsymbol{C}}=\frac{\partial\left(I_3^{-\frac{1}{3}}I_1\right)}{\partial \boldsymbol{C}}=\frac{\partial\left(I_3^{-\frac{1}{3}}\right)}{\partial \boldsymbol{C}}I_1+I_3^{-\frac{1}{3}}\frac{\partial(I_1)}{\partial \boldsymbol{C}}\\ &=\left(-\frac{1}{3}I_3^{-\frac{4}{3}}J^2\boldsymbol{C}^{-1}\right)I_1+I_3^{-\frac{1}{3}}\mathbf{I}=\left(-\frac{1}{3}J^{-\frac{2}{3}}\boldsymbol{C}^{-1}\right)I_1+I_3^{-\frac{1}{3}}\mathbf{I}=I_3^{-\frac{1}{3}}\left(\mathbf{I}-\frac{1}{3}I_1\boldsymbol{C}^{-1}\right)\end{aligned} \tag{5.3.78}$$

and for $i=1,2$,

$$\begin{aligned}\frac{\partial J_{f_i}}{\partial \boldsymbol{C}}&\xrightarrow{J_{f_i}=J^{-\frac{2}{3}}\times I_{f_i}}\frac{\partial J_{f_i}}{\partial \boldsymbol{C}}=\frac{\partial\left(J^{-\frac{2}{3}}I_{f_i}\right)}{\partial \boldsymbol{C}}=\frac{\partial\left(I_3^{-\frac{1}{3}}I_{f_i}\right)}{\partial \boldsymbol{C}}=\frac{\partial\left(I_3^{-\frac{1}{3}}\right)}{\partial \boldsymbol{C}}I_{f_i}+I_3^{-\frac{1}{3}}\frac{\partial(I_{f_i})}{\partial \boldsymbol{C}}\\ &=\left(-\frac{1}{3}J^{-\frac{2}{3}}\boldsymbol{C}^{-1}\right)I_{f_i}+I_3^{-\frac{1}{3}}\left(\boldsymbol{a}_{f_i}\otimes\boldsymbol{a}_{f_i}\right)=I_3^{-\frac{1}{3}}\left(\boldsymbol{a}_{f_i}\otimes\boldsymbol{a}_{f_i}-\frac{1}{3}I_{f_i}\boldsymbol{C}^{-1}\right)\end{aligned} \tag{5.3.79}$$

and the material tensors $\mathbf{D}_v^{SE}$ and $\mathbf{D}_d^{SE}$

$$\mathbf{D}_v^{SE} = 4\frac{\partial^2 w_v^s}{\partial \boldsymbol{C}\partial \boldsymbol{C}} = 2\kappa_o\left(J^2 - J\right)\frac{\partial \boldsymbol{C}^{-1}}{\partial \boldsymbol{C}} + 2\kappa_o\left(J^2 - \frac{J}{2}\right)\boldsymbol{C}^{-1}\otimes\boldsymbol{C}^{-1} \tag{5.3.80}$$

$$\begin{aligned}\mathbf{D}_d^{SE} &= 4\frac{\partial^2 w_d^s}{\partial \boldsymbol{C}\partial \boldsymbol{C}} = 4\frac{\partial w_d^s}{\partial J_1\partial J_1}\left(\frac{\partial J_1}{\partial \boldsymbol{C}}\otimes\frac{\partial J_1}{\partial \boldsymbol{C}}\right) + 4\frac{\partial w_d^s}{\partial J_1\partial J_{f_1}}\left(\frac{\partial J_1}{\partial \boldsymbol{C}}\otimes\frac{\partial J_{f_1}}{\partial \boldsymbol{C}} + \frac{\partial J_{f_1}}{\partial \boldsymbol{C}}\otimes\frac{\partial J_1}{\partial \boldsymbol{C}}\right)\\ &+4\frac{\partial w_d^s}{\partial J_{f_1}\partial J_{f_1}}\left(\frac{\partial J_4}{\partial \boldsymbol{C}}\otimes\frac{\partial J_{f_1}}{\partial \boldsymbol{C}}\right) + 4\frac{\partial w_d^s}{\partial J_1}\frac{\partial^2 J_1}{\partial \boldsymbol{C}\partial \boldsymbol{C}} + 4\frac{\partial w_d^s}{\partial J_{f_1}}\frac{\partial^2 J_{f_1}}{\partial \boldsymbol{C}\partial \boldsymbol{C}} + 4\frac{\partial w_d^s}{\partial J_1\partial J_{f_2}}\\ &\left(\frac{\partial J_1}{\partial \boldsymbol{C}}\otimes\frac{\partial J_{f_2}}{\partial \boldsymbol{C}} + \frac{\partial J_{f_2}}{\partial \boldsymbol{C}}\otimes\frac{\partial J_1}{\partial \boldsymbol{C}}\right) + 4\frac{\partial w_d^s}{\partial J_{f_2}\partial J_{f_2}}\left(\frac{\partial J_4}{\partial \boldsymbol{C}}\otimes\frac{\partial J_{f_2}}{\partial \boldsymbol{C}}\right) + 4\frac{\partial w_d^s}{\partial J_{f_2}}\frac{\partial^2 J_{f_2}}{\partial \boldsymbol{C}\partial \boldsymbol{C}}\end{aligned} \tag{5.3.81}$$

with

$$\frac{\partial^2 J_1}{\partial \boldsymbol{C}\partial \boldsymbol{C}} = -\frac{1}{3}J^{-\frac{2}{3}}\left(\boldsymbol{C}^{-1}\otimes\mathbf{I} + \mathbf{I}\otimes\boldsymbol{C}^{-1}\right) + \frac{1}{9}\left(\overbrace{I_3^{-\frac{1}{3}}I_1}^{J_1}\right)\boldsymbol{C}^{-1}\otimes\boldsymbol{C}^{-1} - \left(\overbrace{I_3^{-\frac{1}{3}}I_1}^{J_1}\right)\frac{\partial \boldsymbol{C}^{-1}}{\partial \boldsymbol{C}} \tag{5.3.82}$$

$$\begin{aligned}\frac{\partial^2 J_{f_i}}{\partial \boldsymbol{C}\partial \boldsymbol{C}} &= -\frac{1}{3}J^{-\frac{2}{3}}\left(\boldsymbol{C}^{-1}\otimes\boldsymbol{a}_{f_i}\otimes\boldsymbol{a}_{f_i} + \boldsymbol{a}_{f_i}\otimes\boldsymbol{a}_{f_i}\otimes\boldsymbol{C}^{-1}\right) + \frac{1}{9}\left(\overbrace{I_3^{-\frac{1}{3}}I_{f_i}}^{J_{f_i}}\right)\\ &\boldsymbol{C}^{-1}\otimes\boldsymbol{C}^{-1} - \frac{1}{3}\left(\overbrace{I_3^{-\frac{1}{3}}I_{f_i}}^{J_{f_i}}\right)\frac{\partial \boldsymbol{C}^{-1}}{\partial \boldsymbol{C}}\end{aligned} \tag{5.3.83}$$

and

$$\frac{\partial w_d^s}{\partial J_1} = \frac{c_2}{2}\left(e^{c_1(J_1-3)}\right) \tag{5.3.84}$$

$$\frac{\partial w_d^s}{\partial J_{f_i}} = c_{2f_i}(J_{f_i}-1)e^{c_{1f_i}(J_{f_i}-1)^2} \tag{5.3.85}$$

$$\frac{\partial w_d^e}{\partial J_1\partial J_1} = \frac{c_1 c_2}{2}\left(e^{c_1(J_1-3)}\right) \tag{5.3.86}$$

$$\frac{\partial w_d^s}{\partial J_1\partial J_{f_i}} = 0 \tag{5.3.87}$$

$$\frac{\partial w_d^s}{\partial J_{f_i}\partial J_{f_i}} = c_{2f_i}(1 + 2c_{1f_i}(J_{f_i}-1)^2)e^{c_{1f_i}(J_{f_i}-1)^2} \tag{5.3.88}$$

For the ground matrix, an incompressible isotropic neo-Hookean model can be adopted. This can also be achieved by a model similar to the fibres if the stiffness terms are omitted ($c_{2f_1} = c_{2f_2} = 0$).

5.3.6.2 Tissue Scale

The fibrous tissues/layers are considered to be composed of an isotropic matrix reinforced with one or two fibre families in different directions. The complete strain energy function (5.3.68) can be used to govern the fibrous layer (Weinberg and Kazempur Mofrad 2007a, 2007b; Holzapfel 2000):

$$w^s(J_1, J_{f_i}, J) = \underbrace{w_v^s(J)}_{\text{volumetric}} + \underbrace{w_d^s(J_1, J_{f_i})}_{\text{deviatoric}} \tag{5.3.89}$$

where the volumetric part $w_v^s(J)$ is defined based on the positive penalty parameter κ_t for the tissue,

$$w_v^s(J) = \frac{1}{2}\kappa_t (J-1)^2 \tag{5.3.90}$$

and the deviatoric part $w_d^s(J_1, J_{f_i})$ contains three independent terms,

$$w_d^s(J_1, J_{f_i}) = \underbrace{\frac{c_{2m}}{2c_{1m}}\left[e^{c_{1m}(J_1-3)} - 1\right]}_{w_{dm}^s} + \underbrace{\frac{c_{2f_i}}{2c_{1f_i}}\left[e^{c_{1f_i}(J_{f_i}-1)^2} - 1\right]}_{w_{df}^s} + \underbrace{\frac{1}{2}\mu_i (J_1 - 3)}_{w_{di}^s} \tag{5.3.91}$$

where c_{1m} and c_{2m} are the constant parameters of the matrix, c_{1f_i} and c_{2f_i} are the material constants of the fibre and μ_i represents the shear modulus of the fibre–matrix interface.

5.3.6.3 Cell Scale

The cell-level RVE consists of a single interstitial cell (IC) surrounded by the extracellular matrix (ECM), either fibrosa or ventricularis. The cell follows Equation (5.3.30) with the shear modulus μ_c of the cell,

$$w^s(\boldsymbol{C}) = \frac{1}{2}\mu_c (J_1 - 3) \tag{5.3.92}$$

and the constitutive model for the ECM (either fibrosa or ventricularis) is assumed similar to the related tissue-level model (Section 5.3.6.2).

5.3.7 Polyconvex Undamaged Functions for Fibrous Tissues

The model proposed by Balzani et al. (2006), for abdominal aorta, decomposes the strain energy into the isotropic and anisotropic terms,

$$w^s = w_i^s + w_a^s \tag{5.3.93}$$

where the generalized neo-Hookean term represents the isotropic response and the second term enforces the nearly incompressible condition:

$$w_i^s = c_1\left(\frac{I_1}{I_3^{1/3}} - 3\right) + \varepsilon\left(I_3^\gamma + \frac{1}{I_3^\gamma} - 2\right), c_1 > 0, \varepsilon > 0, \gamma > 1 \tag{5.3.94}$$

where c_1, ε and γ are material constants.

The anisotropic part w_a^s is characterized by the existence of two sets of fibres, as proposed by a convex stored-energy function (Balzani et al. 2006),

$$w_a^s = \begin{cases} \sum_{k=1}^{2} \alpha_1 \left(J_{f_k} - 1\right)^{\alpha_2}, & J_{f_1} \geq 1 \cap J_{f_2} \geq 1 \\ \alpha_1 \left(J_{f_1} - 1\right)^{\alpha_2}, & J_{f_1} \geq 1 \cap J_{f_2} < 1 \\ \alpha_1 \left(J_{f_2} - 1\right)^{\alpha_2}, & J_{f_1} < 1 \cap J_{f_2} \geq 1 \\ 0, & J_{f_1} < 1 \cap J_{f_2} < 1 \end{cases} \tag{5.3.95}$$

where $\alpha_1 \geq 0$ and $\alpha_2 > 1$ are the material constants. A number of alternative definitions for w_a^s can be found in Balzani et al. (2006).

The second Piola–Kirchhoff stress $\boldsymbol{S}$ is derived as (Fathi 2015)

$$\boldsymbol{S}_i = 2\frac{\partial w_i^s}{\partial \boldsymbol{C}} = 2c_1\left(\frac{\mathbf{I}}{I_3^{1/3}} - \frac{1}{3}\frac{I_1 \boldsymbol{C}^{-1}}{I_3^{1/3}}\right) + 2\varepsilon\gamma(I_3^{\gamma} - I_3^{-\gamma})\boldsymbol{C}^{-1} \tag{5.3.96}$$

$$\boldsymbol{S}_a = \begin{cases} \sum_{k=1}^{2} 2\alpha_1\alpha_2 \left(J_{f_k} - 1\right)^{\alpha_2 - 1} J^{-2/3} \bar{\boldsymbol{N}}_{f_k}, & J_{f_1} \geq 1 \cap J_{f_2} \geq 1 \\ 2\alpha_1\alpha_2 \left(J_{f_1} - 1\right)^{\alpha_2 - 1} J^{-2/3} \bar{\boldsymbol{N}}_{f_1}, & J_{f_1} \geq 1 \cap J_{f_2} < 1 \\ 2\alpha_1\alpha_2 \left(J_{f_2} - 1\right)^{\alpha_2 - 1} J^{-2/3} \bar{\boldsymbol{N}}_{f_2}, & J_{f_1} < 1 \cap J_{f_2} \geq 1 \\ 0, & J_{f_1} < 1 \cap J_{f_2} < 1 \end{cases} \tag{5.3.97}$$

where

$$\bar{\boldsymbol{N}}_{f_k} = \boldsymbol{N}_{f_k} - \frac{1}{3} I_{f_k} \boldsymbol{C}^{-1} \tag{5.3.98}$$

By further derivation with respect to $\boldsymbol{C}$, the two terms of elasticity tensor $\mathbf{D}^S$ can be derived:

$$\begin{aligned} \mathbf{D}_i^S = {} & 4c_1\left(-\frac{1}{3}\frac{\boldsymbol{C}^{-1} \otimes \mathbf{I}}{I_3^{1/3}} - \frac{1}{3}\frac{\mathbf{I} \otimes \boldsymbol{C}^{-1}}{I_3^{1/3}} + \frac{1}{9}\frac{I_1}{I_3^{1/3}} \boldsymbol{C}^{-1} \otimes \boldsymbol{C}^{-1} + \frac{1}{3}\frac{I_1}{I_3^{1/3}} \mathbf{Y}\right) \\ & + 4\varepsilon\gamma\left(\gamma I_3^{\gamma} \boldsymbol{C}^{-1} \otimes \boldsymbol{C}^{-1} - I_3^{\gamma}\mathbf{Y} + \gamma I_3^{-\gamma} \boldsymbol{C}^{-1} \otimes \boldsymbol{C}^{-1} + I_3^{-\gamma}\mathbf{Y}\right) \end{aligned} \tag{5.3.99}$$

$$\mathbf{D}_a = \begin{cases} \sum_{k=1}^{2} 4\alpha_1\alpha_2 \left[\frac{1}{3}\left(J_{f_k} - 1\right)^{\alpha_2 - 1} J^{-2/3} \chi^k + \left(\alpha_2 - 1\right)\left(J_{f_k} - 1\right)^{\alpha_2 - 2} \frac{1}{3} J^{-4/3} \xi^k\right], & J_{f_1} \geq 1 \cap J_{f_2} \geq 1 \\ 4\alpha_1\alpha_2 \left[\frac{1}{3}\left(J_{f_1} - 1\right)^{\alpha_2 - 1} J^{-2/3} \chi^1 + \left(\alpha_2 - 1\right)\left(J_{f_1} - 1\right)^{\alpha_2 - 2} J^{-4/3} \xi^1\right], & J_{f_1} \geq 1 \cap J_{f_2} < 1 \\ 4\alpha_1\alpha_2 \left[\frac{1}{3}\left(J_{f_2} - 1\right)^{\alpha_2 - 1} J^{-2/3} \chi^2 + \left(\alpha_2 - 1\right)\left(J_{f_2} - 1\right)^{\alpha_2 - 2} J^{-4/3} \xi^2\right], & J_{f_1} < 1 \cap J_{f_2} \geq 1 \\ 0, & J_{f_1} < 1 \cap J_{f_2} < 1 \end{cases} \tag{5.3.100}$$

where

$$\chi^k = \frac{1}{3} I_{f_k} \boldsymbol{C}^{-1} \otimes \boldsymbol{C}^{-1} - \boldsymbol{C}^{-1} \otimes \boldsymbol{N}_{f_k} - \boldsymbol{N}_{f_k} \otimes \boldsymbol{C}^{-1} + I_{f_k} \mathcal{I} \tag{5.3.101}$$

$$\xi^k = \frac{1}{9} (I_{f_k})^2 \boldsymbol{C}^{-1} \otimes \boldsymbol{C}^{-1} - \frac{1}{3} I_{f_k} \boldsymbol{C}^{-1} \otimes \boldsymbol{N}_{f_k} - \frac{1}{3} I_{f_k} \boldsymbol{N}_{f_k} \otimes \boldsymbol{C}^{-1} + \boldsymbol{N}_{f_k} \otimes \boldsymbol{N}_{f_k} \tag{5.3.102}$$

where $\mathcal{I}$ is defined in Equation (5.3.50).

5.3.8 Damaged Soft Tissue

Collagen fibres are the main mechanical load-bearing constituent of the soft tissues. It is known that in addition to the external effects, the collagen fibres may become exposed to degradation to some extent due to the effect of the matrix metalloproteinase (MMP) enzyme, or accumulated plagues may reduce the content of collagen fibre and decrease the overall mechanical strength of the tissue (Sakakura et al. 2013; Kugo et al. 2017).

Despite the importance of degradation of soft tissues in the study of various biomechanical problems, a limited work is available on the subject (Holzapfel et al. 2002, 2005; Alastrué et al. 2007; Pericevic et al. 2009; Fathi et al. 2017; Hatefi et al. 2022).

Incorporation of damage mechanics into biomechanical problems can be performed by a wide range of available and well-developed techniques of damage mechanics in general engineering problems (Ju 1989; Jirásek 2004; Comi et al. 2007; Simo and Ju 1987; Miehe et al. 2010; Balzani et al. 2006; Calvo et al. 2007; Peña 2011a, 2011b; Alastrué et al. 2007; Waffenschmidt et al. 2014; Arruda and Boyce 1993; Ehret and Itskov 2009; Fathi et al. 2017).

Based on the definition of Calvo et al. (2007), the hyperelastic constitutive model for a soft tissue with two sets of fibres can be defined as

$$w^{\mathrm{s}}(\boldsymbol{C}, \boldsymbol{N}_{f_1}, \boldsymbol{N}_{f_2}) = w_v^{\mathrm{s}}(J) + w_{di}^{\mathrm{s}}(\boldsymbol{C}_d) + w_{da}^{\mathrm{s}}(\boldsymbol{C}_d, \boldsymbol{N}_{f_1}, \boldsymbol{N}_{f_2}) \tag{5.3.103}$$

and the second Piola–Kirchhoff stress $\boldsymbol{S}$ and the corresponding elasticity tensor $\mathbf{D}^{SE}$ are obtained by differentiation of w^{s} (Calvo et al. 2007):

$$\boldsymbol{S} = \frac{\partial w^{\mathrm{s}}}{\partial \boldsymbol{E}} = 2\frac{\partial w^{\mathrm{s}}}{\partial \boldsymbol{C}} = \boldsymbol{S}_v + \boldsymbol{S}_{di} + \boldsymbol{S}_{da} \tag{5.3.104}$$

$$\mathbf{D}^{SE} = \frac{\partial^2 w^{\mathrm{s}}}{\partial \boldsymbol{E} \partial \boldsymbol{E}} = 4\frac{\partial^2 w^{\mathrm{s}}}{\partial \boldsymbol{C} \partial \boldsymbol{C}} = \boldsymbol{S}_v + \boldsymbol{S}_{di} + \boldsymbol{S}_{da} \tag{5.3.105}$$

5.3.8.1 Anisotropic Damage Model

A scalar damage model with uncoupled formulation can be used to account for degradation of soft tissues. The non-local method requires the damage parameter D_k to be applied to the initial (intact) deviatoric part of the energy density ($w_{dk}^{\mathrm{s}0}$) of each part/constituent k of the soft tissues,

$$w^{\mathrm{s}}(\boldsymbol{C}, \boldsymbol{N}_{f_1}, \boldsymbol{N}_{f_2}, D_k) = w_v^{\mathrm{s}}(J) + (1 - D_k) w_{dk}^{\mathrm{s}0}(\boldsymbol{C}_d, \boldsymbol{N}_{f_1}, \boldsymbol{N}_{f_2}) \tag{5.3.106}$$

where $k = m, f_1, f_2$ represents the matrix and the two sets of fibres, respectively. As usual, the matrix and fibres are assumed to be isotropic and anisotropic, respectively.

Equation (5.3.106) can be expanded more explicitly as (Calvo et al. 2207; Fathi et al. 2017)

$$w^s = \frac{1}{2}\lambda(\ln J)^2 + (1 - D_m)\left\{\frac{1}{2}\mu_{1m}(J_1 - 3) + \frac{1}{2}\mu_{2m}(J_2 - 3)\right\} + (1 - D_{f_1}) \frac{c_{2f_1}}{2c_{1f_1}}\left\{e^{c_{1f_1}(\eta_{f_1})^2} - 1\right\} + (1 - D_{f_2})\frac{c_{2f_2}}{2c_{1f_2}}\left\{e^{c_{1f_2}(\eta_{f_2})^2} - 1\right\} \tag{5.3.107}$$

with

$$\eta_{f_i} = \begin{cases} 1, & \kappa_{f_i} J_1 + [1 - 3\kappa_{f_i}] J_{f_i} \geq I_{0_i} \quad i = 1,2 \\ 0, & \text{otherwise} \end{cases} \tag{5.3.108}$$

where μ_{1m} and μ_{2m} are the elastic material parameters of the matrix, c_{1f_i} and c_{2f_i} are the material parameters of the fibre set i, and I_{0_i} determines the state of folding and mobilization of the fibre set i. The intensity of anisotropy of the collagen fibre set i is expressed by the dispersion factor κ_{f_i} in the range of $0 \leq \kappa_{f_i} \leq 1/3$ (Gasser et al. 2006; Fathi et al. 2017).

The Clausius–Duham inequality (in an isothermal process) can be expressed in terms of the derivative of w^s from Equation (5.3.106),

$$\boldsymbol{S} : \dot{\boldsymbol{E}} - \dot{w}^s \geq 0 \rightarrow \boldsymbol{S} : \dot{\boldsymbol{E}} - \left\{\frac{dw_v^s(J)}{dJ}\dot{J} - w_{dk}^{s0}\dot{D}_k + (1 - D_k)\frac{\partial w_{dk}^{s0}}{\partial \boldsymbol{C}_d} : \dot{\boldsymbol{C}}_d\right\} \geq 0 \tag{5.3.109}$$

which eventually yields to (Fathi et al. 2017):

$$\frac{1}{2}\left\{\boldsymbol{S} - J\frac{dw_v^s(J)}{dJ}\boldsymbol{C}^{-1} - J^{-\frac{2}{3}}\boldsymbol{Q} : (1 - D_k)2\frac{\partial w_{dk}^{s0}}{\partial \boldsymbol{C}_d}\right\} : \dot{\boldsymbol{C}}_d + w_{dk}^{s0}\dot{D}_k \geq 0 \tag{5.3.110}$$

where

$$\boldsymbol{Q} = \mathcal{I} - \frac{1}{3}\boldsymbol{C}_d^{-1} \otimes \boldsymbol{C}_d^{-1} \tag{5.3.111}$$

Additionally, the nearly incompressibility constraint requires

$$J \approx 1 \rightarrow \dot{J} = \frac{1}{2}J\boldsymbol{C}^{-1} : \dot{\boldsymbol{C}} = 0 \tag{5.3.112}$$

Satisfying the two sets of constraints (5.3.110) and (5.3.112) leads to the computation of the damage-induced second Piola–Kirchhoff stress $\boldsymbol{S}$:

$$\boldsymbol{S} = \boldsymbol{S}_v + (1 - D_k)\boldsymbol{S}_{dk}^0; \quad k = m, f_1, f_2 \tag{5.3.113}$$

$$\boldsymbol{S}_{dk}^0 = \frac{\partial w_{dk}^{s0}}{\partial \boldsymbol{E}} = 2\frac{\partial w_{dk}^{s0}}{\partial \boldsymbol{C}} \tag{5.3.114}$$

and the thermodynamic driving force f_k,

$$f_k \dot{D}_k = w_{dk}^{s0}(\boldsymbol{C}_d)\dot{D}_k = -\frac{\partial w^s}{\partial D_k}\dot{D}_k \tag{5.3.115}$$

The thermodynamic work conjugate form of Equation (5.3.115) along with the necessary condition of

$$f_k \dot{D}_k \geq 0 \tag{5.3.116}$$

are used to formulate the damage evolution, as discussed in Section 5.3.8.2.

Derivation of the strain energy function w^s in Equation (5.3.107) is required for evaluation of $\boldsymbol{S}$ in Equation (5.3.114),

$$\begin{aligned}\boldsymbol{S} = {} & \frac{1}{2}\lambda I_3^{-1}(\ln J)I_{3,\mathbf{E}} + (1-D_m)\left\{\frac{1}{2}\mu_{1m}J_{1,\mathbf{E}} + \frac{1}{2}\mu_{2m}J_{2,\mathbf{E}}\right\} + \eta_{f_1}(1-D_{f_1})c_{2f_1} \\ & (J_{f_1,\mathbf{E}} - I_{0_1})e^{c_{1f_1}\eta_{f_1}}\left[\kappa_{f_1}J_{1,\mathbf{E}} + (1-3\kappa_{f_1})J_{f_1,\mathbf{E}}\right] + \eta_{f_2}(1-D_{f_2})c_{2f_2}(J_{f_2,\mathbf{E}} - I_{0_2})e^{c_{1f_2}\eta_{f_2}} \\ & \left[\kappa_{f_2}J_{2,\mathbf{E}} + (1-3\kappa_{f_2})J_{f_2,\mathbf{E}}\right]\end{aligned} \tag{5.3.117}$$

where $J_{1,\mathbf{E}}$ and $J_{2,\mathbf{E}}$ are defined in Equations (5.3.40) and (5.3.41), respectively. Moreover,

$$J_{f_1,\mathbf{E}} = I_3^{-1/3} I_{f_1,\mathbf{E}} - \frac{1}{3} I_4 I_3^{-4/3} I_{3,\mathbf{E}} \tag{5.3.118}$$

$$J_{f_2,\mathbf{E}} = I_3^{\frac{-1}{3}} I_{f_2,\mathbf{E}} - \frac{1}{3} I_{f_2} I_3^{\frac{-4}{3}} I_{3,\mathbf{E}} \tilde{I}_{f_2,\mathbf{E}} = I_3^{\frac{-1}{3}} I_{f_2,\mathbf{E}} - \frac{1}{3} I_{f_2} I_3^{\frac{-4}{3}} I_{3,\mathbf{E}} \tag{5.3.119}$$

with

$$I_{3,\mathbf{E}} = 2I_3\boldsymbol{C}^{-1} \tag{5.3.120}$$

$$I_{f_1,\mathbf{E}} = 2\boldsymbol{N}_{f_1} \tag{5.3.121}$$

$$I_{f_2,\mathbf{E}} = 2\boldsymbol{N}_{f_2} \tag{5.3.122}$$

5.3.8.2 Damage Surface/Criterion

The damage surface is defined by the damage criterion Φ_k,

$$\Phi_k(\boldsymbol{C}(t), \epsilon_{dk}^t) = \sqrt{2w_{dk}^{s0}(\boldsymbol{C}_d(t))} - \epsilon_{dk}^t \leq 0; \;\; k = m, f_1, f_2 \tag{5.3.123}$$

where $\epsilon_{\boldsymbol{dk}}^{\boldsymbol{t}}$ is the maximum experienced value of the damage evolution parameter $\epsilon_{\boldsymbol{dk}}^{\tau}$ (Simo 1987),

$$\epsilon_{dk}^t = \max_{\tau\in(-\infty,t)} \epsilon_{dk}^{\tau} = \max_{\tau\in(-\infty,t)} \sqrt{2w_{dk}^{s0}(\boldsymbol{C}_d(\tau))}; \; k = m, f_1, f_2 \tag{5.3.124}$$

The damage surface and the corresponding loading or unloading conditions are then defined by (Fathi et al. 2017):

$$\begin{cases} \Phi_k < 0 & \text{elastic} \\ \Phi_k = 0 & \begin{cases} \boldsymbol{N}_k^{\Phi} : \dot{\boldsymbol{E}} < 0, & \text{elastic unloading} \\ \boldsymbol{N}_k^{\Phi} : \dot{\boldsymbol{E}} = 0, & \text{neutral loading} \\ \boldsymbol{N}_k^{\Phi} : \dot{\boldsymbol{E}} > 0, & \text{damage loading} \end{cases} \end{cases} ; k = m, f_1, f_2 \tag{5.3.125}$$

with the normal $\boldsymbol{N}_k^{\Phi}$ defined as (Simo 1987)

$$\boldsymbol{N}_k^{\Phi} = \frac{\partial \Phi_k}{\partial \boldsymbol{E}} = \frac{1}{\varepsilon_{dk}^{\tau}} \frac{\partial w_{dk}^{s0}}{\partial \boldsymbol{E}} \tag{5.3.126}$$

5.3.8.3 Damage Evolution Law

The damage evolution law describes the damage variable D_k by an irreversible rate form of the evolution function $H_{dk}(\varepsilon_{dk}, D_k)$ (Simo 1987; Bathe 2006; Calvo et al. 2007; Comi et al. 2007; Peña 2011a),

$$\frac{dD_k}{dt} = \begin{cases} H_{dk}\left(\varepsilon_{dk}, D_k\right)\dot{\varepsilon}_{dk}, & \text{if } \Phi_k = 0 \text{ and } \boldsymbol{N}_k^{\Phi} : \dot{\boldsymbol{E}} > 0 \\ 0, & \text{otherwise} \end{cases} \tag{5.3.127}$$

A more appropriate form can be obtained if H_{dk} is assumed to be independent (uncoupled) of the damage variable D_k (Simo 1987; Calvo et al. 2007; Peña 2011a, 2011b; Waffenschmidt et al. 2014),

$$H_{dk}(\varepsilon_{dk}) = -\frac{dG_{dk}(\varepsilon_{dk})}{d\varepsilon_{dk}} \tag{5.3.128}$$

where $G_{dk}(\varepsilon_{dk})$ is the damage function. An exponential damage function can be adopted (Simo 1987; Bathe 2006; Calvo et al. 2007; Comi et al. 2007; Peña 2011a; Gültekin et al. 2016),

$$G_{dk}(\varepsilon_{dk}) = 1 - D_k = e^{\beta_k(\varepsilon_{dk}^0 - \varepsilon_{dk})} \tag{5.3.129}$$

where β_k represents the so-called damage saturation. Equation (5.3.129) can be written in an explicit form in terms of the total and initial equivalent damage evolution parameters, $\varepsilon_{dk}^{\max}$ and $\varepsilon_{dk}^{\min}$, respectively (Fathi et al. 2017)

$$G_{dk}(\varepsilon_{dk}^{t}) = 1 - \mathrm{D}_k = \begin{cases} 1, & \varepsilon_{dk}^{t} < \varepsilon_{dk}^{\min} \\ \dfrac{\varepsilon_{dk}^{\max} - \varepsilon_{dk}^{t}}{\varepsilon_{dk}^{\max} - \varepsilon_{dk}^{\min}} e^{\left[\beta_k \frac{\varepsilon_{dk}^{t} - \varepsilon_{dk}^{\min}}{\varepsilon_{dk}^{\min} - \varepsilon_{dk}^{\max}}\right]}, & \varepsilon_{dk}^{\min} \le \varepsilon_{dk}^{t} \le \varepsilon_{dk}^{\max} \\ 0, & \varepsilon_{dk}^{t} > \varepsilon_{dk}^{\max} \end{cases} \tag{5.3.130}$$

5.3.8.4 Damage Induced Elasticity Tensor

Following Equation (5.3.130), the second Piola–Kirchhoff stress $\boldsymbol{S}$ in Equation (5.3.113) can be rewritten in terms of $G_{dk}(\varepsilon_{dk}^{t})$,

$$\boldsymbol{S} = \boldsymbol{S}_v + G_{dk}(\varepsilon_{dk}^{t})\boldsymbol{S}_{dk}^{0}; \quad k = m, f_1, f_2 \tag{5.3.131}$$

The rate form of $\boldsymbol{S}_{dk}$ can now be written as (Simo 1987)

$$\dot{\boldsymbol{S}}_{dk} = \mathbf{D}_{dk}^{SE} : \dot{\boldsymbol{E}} = \begin{cases} \left[G_{dk}\mathbf{D}_{dk}^{0} - H_{dk}\boldsymbol{S}_{dk}^{0} \otimes \boldsymbol{S}_{dk}^{0}\right] : \dot{\boldsymbol{E}}, & \text{if } \Phi_k = 0 \text{ and } \boldsymbol{N}_k^{\Phi} : \dot{\boldsymbol{E}} > 0 \\ G_{dk}\mathbf{D}_{dk}^{0} : \dot{\boldsymbol{E}}, & \text{otherwise} \end{cases} \tag{5.3.132}$$

The damage-induced deviatoric part of the elasticity tensor is

$$\mathbf{D}_{dk}^{SE} = \begin{cases} G_{dk}\mathbf{D}_{dk}^{0} - H_{dk}\boldsymbol{S}_{dk}^{0} \otimes \boldsymbol{S}_{dk}^{0}, & \text{if } \Phi_k = 0 \text{ and } \boldsymbol{N}_k : \dot{\boldsymbol{E}} > 0 \\ G_{dk}\mathbf{D}_{dk}^{0}, & \text{otherwise} \end{cases} \tag{5.3.133}$$

For the explicit form of the strain energy function (5.3.107) and the second Piola–Kirchhoff stress (5.3.117), the material description of the elasticity tensor $\mathbf{D}^{SE}$ can be written as

$$\begin{aligned} \mathbf{D}^{SE} = {} & \frac{1}{2}\lambda\left\{I_3^{-2}\left(\frac{1}{2} - \ln J\right)I_{3,\boldsymbol{E}} \otimes I_{3,\boldsymbol{E}} + I_3^{-1}\ln(J)I_{3,\boldsymbol{EE}}\right\} \\ & + (1 - D_m)\left\{\frac{1}{2}\mu_{1m}J_{1,\boldsymbol{EE}} + \frac{1}{2}\mu_{2m}J_{2,\boldsymbol{EE}}\right\} \\ & + \eta_{f_1}c_{2f_1}(1 - D_{f_1})\mathrm{e}^{c_{1f_1}\eta_{f_1}}\left\{k_{f_1}J_{f_1,\boldsymbol{E}} \otimes J_{1,\boldsymbol{E}} + (1 - 3k_{f_1})J_{f_1,\boldsymbol{E}} \otimes J_{f_1,\boldsymbol{E}}\right. \\ & + c_{1f_1}(J_{f_1,\boldsymbol{E}} - I_{0_1})\left[k_{f_1}{}^2 J_{1,\boldsymbol{E}} \otimes J_{1,\boldsymbol{E}}\right. \\ & + k_{f_1}(1 - 3k_{f_1})\left[J_{1,\boldsymbol{E}} \otimes J_{f_i,\boldsymbol{E}} + J_{f_i,\boldsymbol{E}} \otimes J_{1,\boldsymbol{E}}\right] \\ & \left.\left. + (1 - 3k_{f_1})^2 J_{f_1,\boldsymbol{E}} \otimes J_{f_1,\boldsymbol{E}}\right] + (J_{f_1,\boldsymbol{E}} - I_{0_1})\left[k_{f_1}J_{1,\boldsymbol{EE}} + (1 - 3k_{f_1})J_{f_1,\boldsymbol{EE}}\right]\right\} \\ & + \eta_{f_2}c_{2f_2}(1 - D_{f_2})\mathrm{e}^{c_{1f_2}\eta_{f_2}}\left\{k_{f_2}J_{f_2,\boldsymbol{E}} \otimes J_{1,\boldsymbol{E}} + (1 - 3k_{f_2})J_{f_2,\boldsymbol{E}} \otimes J_{f_2,\boldsymbol{E}}\right. \\ & + c_{1f_2}(J_{f_2,\boldsymbol{E}} - I_{0_2})\left[k_{f_2}{}^2 J_{1,\boldsymbol{E}} \otimes J_{1,\boldsymbol{E}}\right. \\ & + k_{f_2}(1 - 3k_{f_2})\left[J_{1,\boldsymbol{E}} \otimes J_{f_2,\boldsymbol{E}} + J_{f_2,\boldsymbol{E}} \otimes J_{1,\boldsymbol{E}}\right] \\ & \left.\left. + (1 - 3k_{f_2})^2 J_{f_2,\boldsymbol{E}} \otimes J_{f_2,\boldsymbol{E}}\right] + (J_{f_2,\boldsymbol{E}} - I_{0_2})\left[k_{f_2}J_{1,\boldsymbol{EE}} + (1 - 3k_{f_2})J_{f_2,\boldsymbol{EE}}\right]\right\} \end{aligned} \tag{5.3.134}$$

with

$$J_{1,EE} = -\frac{2}{3} I_3^{\frac{-1}{2}} (J_{1,E} \otimes J_{3,E} + J_{3,E} \otimes J_{1,E}) + I_1 I_3^{\frac{-4}{3}} \left\{ \frac{8}{9} J_{3,E} \otimes J_{3,E} - \frac{1}{3} I_{3,EE} \right\} \quad (5.3.135)$$

$$J_{2,EE} = \frac{4}{3} I_3^{-1/2} (J_{2,E} \otimes J_{3,E} + J_{3,E} \otimes J_{2,E}) + \frac{8}{9} I_2 I_3^{-5/3} (J_{3,E} \otimes J_{3,E})$$
$$+ I_3^{-2/3} I_{2,EE} - \frac{2}{3} I_2 I_3^{-5/3} I_{3,EE} \quad (5.3.136)$$

$$J_{3,EE} = -I_3^{-1/2} \left\{ J_{3,E} \otimes J_{3,E} - \frac{1}{2} I_{3,EE} \right\} \quad (5.3.137)$$

$$J_{f_1,EE} = -\frac{2}{3} I_3^{-1/2} (J_{f_1,E} \otimes J_{3,E} + J_{3,E} \otimes J_{f_1,E}) + I_{f_1} I_3^{-4/3} \left\{ \frac{8}{9} J_{3,E} \otimes J_{3,E} - \frac{1}{3} I_{3,EE} \right\} \quad (5.3.138)$$

$$J_{f_2,EE} = -\frac{2}{3} I_3^{-1/2} (J_{f_2,E} \otimes J_{3,E} + J_{3,E} \otimes J_{f_2,E}) + I_{f_2} I_3^{-4/3} \left\{ \frac{8}{9} J_{3,E} \otimes J_{3,E} - \frac{1}{3} I_{3,EE} \right\} \quad (5.3.139)$$

where $I_{2,EE}$ and $I_{3,EE}$ are defined in Equations (5.3.48) and (5.3.50), respectively.

5.3.8.5 Integral-Type Non-local Damage Continuum

A number of techniques have been employed to avoid mesh dependency of damage solutions in biological studies, including the viscoelastic model (Peña 2011b), the non-local gradient (Gültekin et al. 2016) and the non-local integral-type framework (Peña 2011b). The present uncoupled damage formulation is better fitted with the non-local integral framework (Fathi et al. 2017).

The non-local equivalent damage evolution parameter $\varepsilon_{dk}^{\text{non}}$ around a reference point ζ is expressed as a weighted ($W(\zeta,\xi)$) average of the damage evolution parameter ε_{dk}^{τ} in its vicinity (defined by ξ) (Bazant and Jirasek 2002; Jirasek 2007),

$$\varepsilon_{dk}^{\text{non}}(\zeta) = \int_v W(\zeta,\xi) \varepsilon_{dk}^{\tau}(\xi) d\xi \quad (5.3.140)$$

$$W(\zeta,\xi) = \frac{a_0(\|\zeta - \xi\|)}{\int_v a_0(\|\zeta - \vartheta\|) d\vartheta} \quad (5.3.141)$$

where a_0 is a radial polynomial function in terms of the distance $r = \|\zeta - \xi\|$ and the intrinsic length scale R (the largest distance r) (Jirasek 2007):

$$a_0(r) = \left\langle 1 - \frac{r^2}{R^2}^2 \right\rangle \quad (5.3.142)$$

where $\langle \bullet \rangle$ denotes the Macaulay bracket.

5.3.8.6 Uniaxial Test

This simple simulation resembles the experimental test on a sample of the muscular rectus sheath in an abdomen (Figure 5.15) by Martins et al. (2012), based on the damaged hyperelastic constitutive model. Due to the microstructure of the muscular tissue, only one set of parallel fibres can be assumed to accurately represent the mechanical response of the tissue. The material properties are defined in Table 5.1.

Figure 5.16 compares the uniaxial stress–stretch response of the model with the experimental data (Martins et al. 2012). Very good agreement has been observed and the present model may follow the experimental trend well to the very last stage prior to rupture of the tissue.

5.3.8.7 Dog Bone Sample

The damage hyperelastic model is adopted to analyse a dog bone sample of a soft tissue with a finite element model of 4,500 nodes and 2,136 elements, as presented in Figure 5.17.

The material constants of the damage hyperelastic model are defined in Table 5.2.

Figure 5.18 illustrates the distribution of damage in the matrix and fibres at the final stages of the analysis. Different patterns and magnitudes of damage in the matrix and fibres indicate that a non-uniform damage response governs the microstructure of the soft tissue.

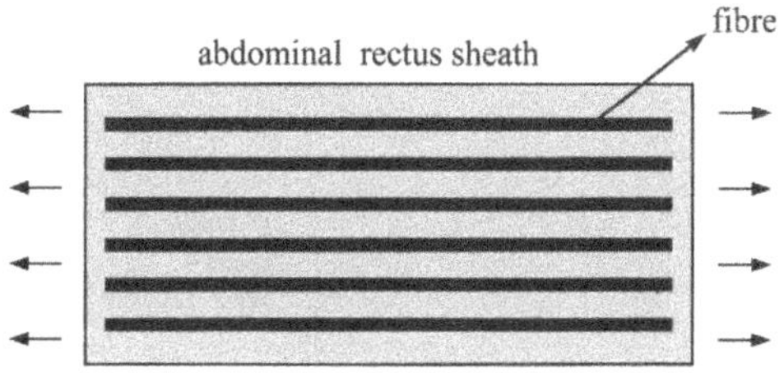

Figure 5.15 The numerical model of the sample of the abdominal rectus sheath.

Table 5.1 Material properties of the abdominal rectus sheath sample.

Tissue			
λ(MPa)	5,016.4		
Matrix		**Set of fibres $\theta = 0$**	
μ_{1m}(MPa)	1.8	c_{1f_1}	0.001
μ_{2m}(MPa)	0	c_{2f_1} (MPa)	1.45
		κ_{f_1}	0
		I_{0_1}	1
β_m	0.1	β_{f_1}	0.8
ε_{dm}^{min} ($\sqrt{\text{MPa}}$)	0.84	$\varepsilon_{df_1}^{min}$ ($\sqrt{\text{MPa}}$)	2.38
ε_{dm}^{max} ($\sqrt{\text{MPa}}$)	1.15	$\varepsilon_{df_1}^{max}$ ($\sqrt{\text{MPa}}$)	3.9

Source: Adapted from Fathi et al. 2017.

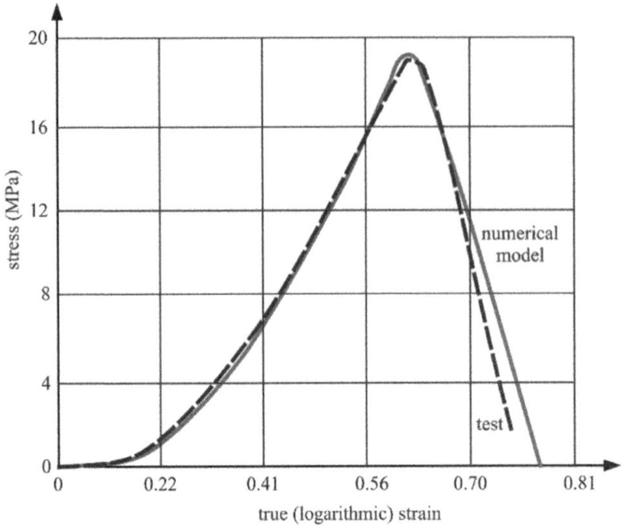

Figure 5.16 Stress–stretch response of the abdominal rectus sheath. *Source:* Adapted from Waffenschmidt et al. 2014; Martins et al. 2012; Fathi et al. 2017.

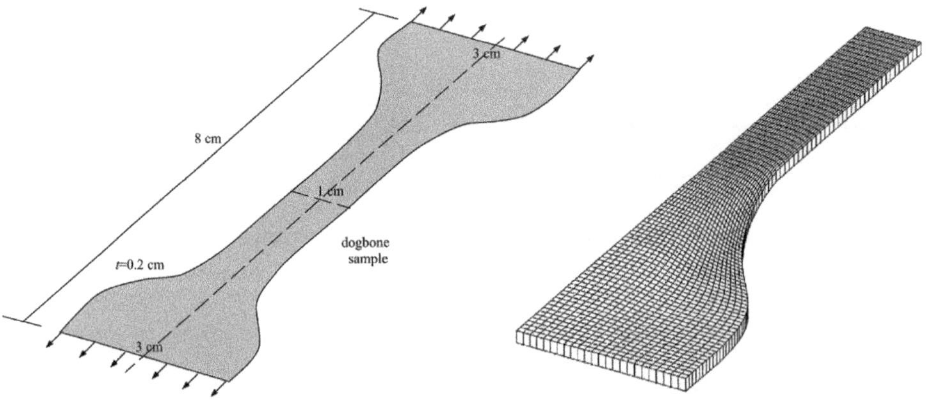

Figure 5.17 Geometry and finite element model of the dog bone soft tissue sample. *Source:* Adapted from Fathi et al. 2017.

5.4 Multiscale Modelling of Undamaged Tendon

The hierarchical structure of the tendon is presented in Figure 5.19. It is composed of a parallel set of fibres. Each fibre is formed by a set of parallel fibrils within a matrix. The set of hyperelastic material models of Section 5.3.5 are adopted to govern the behaviour of each constituent in different scales (Fathi and Mohammadi 2015). They include models for the three scales of the tissue (matrix and fibres), the fibre (matrix and parallel fibrils) and the fibril. Clearly, fibrils are the main microstructural constituent that provides the most significant contribution in the load-bearing capacity of the tissue.

Table 5.2 Material properties of the dog bone sample.

Tissue			
λ(MPa)	5,016.42		
Matrix		**Set of fibres $\theta = 0$**	
μ_{1m}(MPa)	2	c_{1f_1}	3.54
μ_{2m}(MPa)	0	c_{2f_1} (MPa)	20
		κ_{f_1}	0
		I_{0_i}	1
β_m	1	β_{f_1}	1
$\varepsilon_{dm}^{min}(\sqrt{\text{MPa}})$	0.336	$\varepsilon_{df_1}^{min}(\sqrt{\text{MPa}})$	0.7
$\varepsilon_{dm}^{max}(\sqrt{\text{MPa}})$	1.5	$\varepsilon_{df_1}^{max}(\sqrt{\text{MPa}})$	5

Source: Adapted from Fathi et al. 2017.

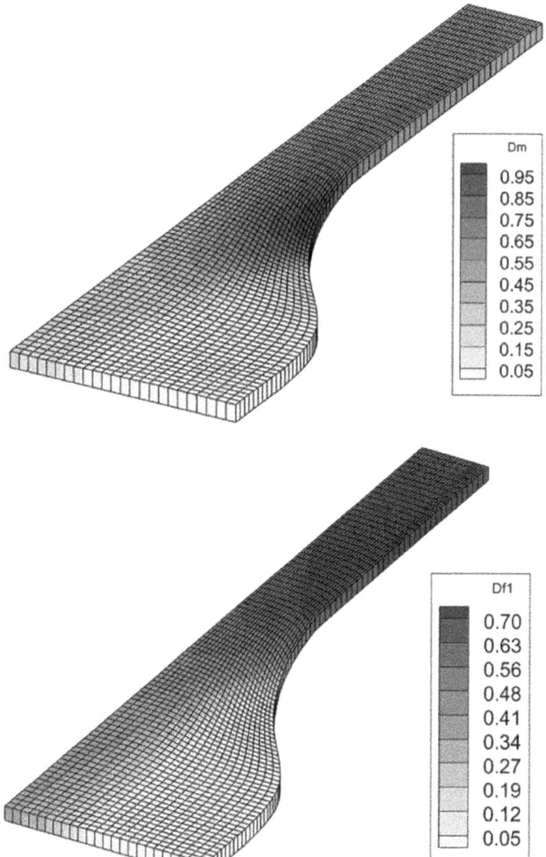

Figure 5.18 Distribution of damage in the matrix and fibre. *Source:* Adapted from Fathi et al. 2017.

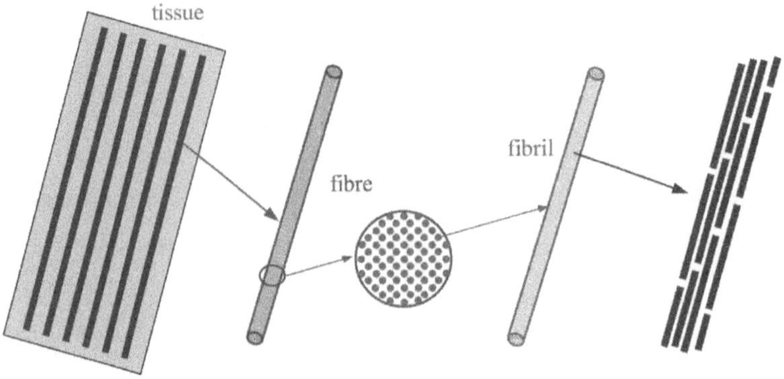

Figure 5.19 The hierarchical structure of a tendon.

In general, soft tissues include fibres that are not necessarily parallel. In a tendon, however, the fibres may be assumed to be parallel.

After choosing the proper functions for each scale, they are implemented into a large deformation finite element analysis code with the aim of simulating the soft materials precisely, as presented by Fathi and Mohammadi (2015). The material constant for the constituents at three scales are obtained from Tang et al. (2009) and Gou et al. (2006).

5.4.1 Fibril Scale

Fibrils are assumed to be homogeneous long rods with a diameter of 100 nm and a length of 1,000 nm (Fung 1993), as depicted in Figure 5.20. The material constants for the fibrils are presented in Table 5.3. The fibril is simulated subject to a uniaxial stretch and deforms axially, as presented in Figure 5.20.

Figure 5.21 shows the axial stress in the fibril. Moreover, the predicted stress–strain response conforms quite well with the reference experimental results (Tang et al. 2009). The contours of axial and shear stress distributions are presented in Figure 5.22.

5.4.2 Fibre Scale

Modelling at the fibre scale requires proper modelling of the behaviour of its constituents: fibrils and matrix. A fibre is simulated as a long rod with a diameter of $8\,\mu\text{m}$ and a length of $240\,\mu\text{m}$, as depicted in Figure 5.23 (Fung 1993; Maceri et al. 2010).

Material parameters at the fibre scale are a combination of the hyperelastic models of fibrils and matrix (see Section 5.3.5.2). The additional constants are defined in Table 5.4.

An RVE of a single fibril and its surrounding matrix is considered to allow for determination of the response of the fibre. Figure 5.24 compares the predicted stress–strain response of the fibre with the reference experimental results (Tang et al. 2009), which shows a very good agreement.

Moreover, comparison of the stress–strain response of the fibre and its constituent fibril in Figure 5.25 shows that almost half of the load-bearing capacity of the fibre is provided by the fibrils (despite its low volume content of about 20%).

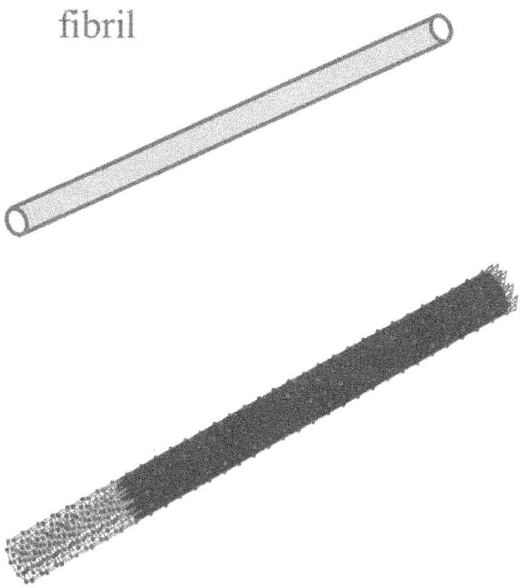

Figure 5.20 Modelling of the fibril scale and the deformed finite element model. *Source:* Adapted from Fathi and Mohammadi 2015.

Table 5.3 Material constants of the fibril scale.

μ_{fl}^{0} (MPa)	l_0	a_1	a_2	a_3
2000	1	1	1	1.57

Source: Adapted from Fathi and Mohammadi 2015.

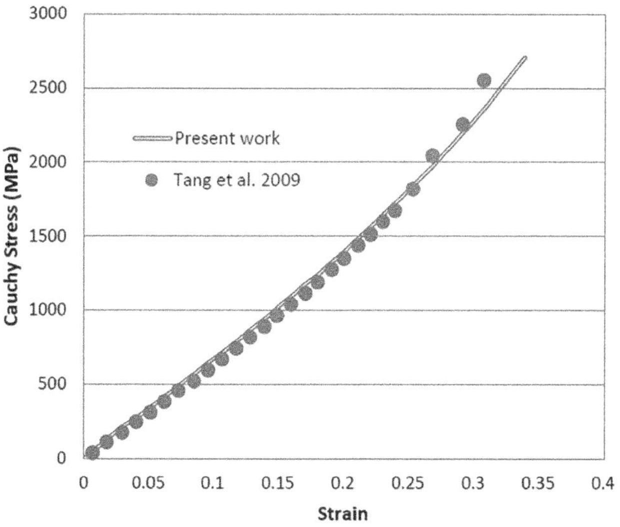

Figure 5.21 Stress–strain response of the fibril. *Source:* Adapted from Fathi and Mohammadi 2015.

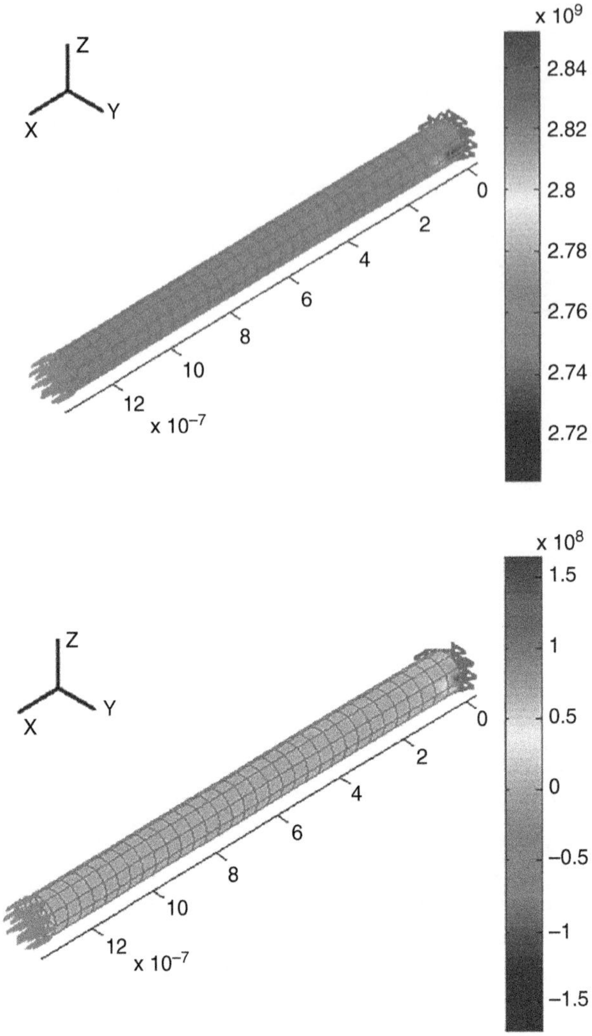

Figure 5.22 Axial (top) and shear (bottom) stress contours of the fibril. *Source:* Adapted from Fathi and Mohammadi 2015.

The axial and shear stress distributions of a representative model of collagen and fibre is presented in Figure 5.26.

5.4.3 Tissue Scale

The tissue is simulated by the hyperelastic model of Tang et al. (2009) (see Section 5.3.5.3), as presented in Figure 5.27.

The material response of the tissue is assumed to be similar to the hyperelastic constitutive model at the fibre scale. Only the material constants associated with the matrix and fibre at the tissue scale are adopted (Table 5.5).

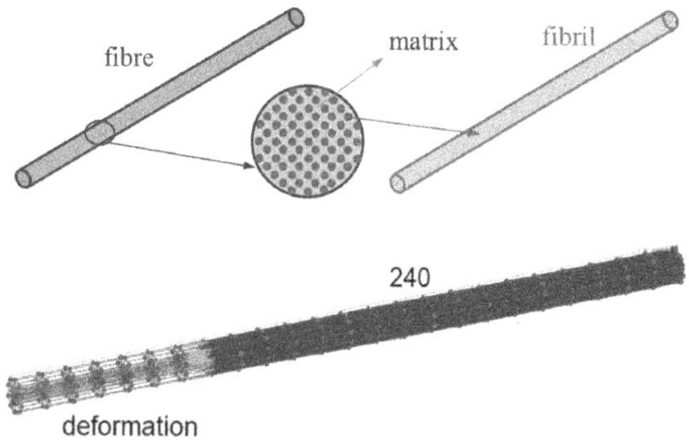

Figure 5.23 Modelling at the fibre scale and its deformed finite element mesh. *Source:* Adapted from Fathi and Mohammadi 2015.

Table 5.4 Material constants for the fibre scale.

Matrix		Fibril				
v_{fl}	μ_m (MPa)	μ_{fl}^0 (MPa)	I_0	a_1	a_2	a_3
0.15	1.0	2000	1	1	1	1.57

Source: Adapted from Fathi and Mohammadi 2015.

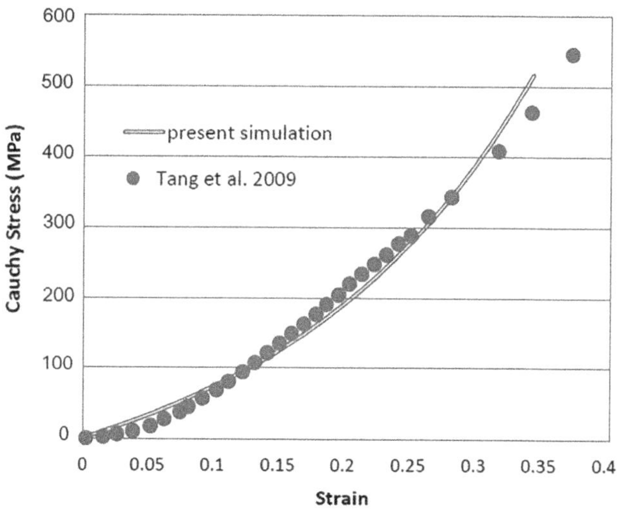

Figure 5.24 Stress–strain response of the fibre. *Source:* Adapted from Fathi and Mohammadi 2015.

Comparison of the predicted stress–strain response with the reference experimental results by Tang et al. (2009) in Figure 5.28 shows a very good agreement. Moreover, variations of the stress–strain response for different fibre orientations indicates the major effects of fibre directions. Clearly, the tissue with the highest strength is obtained for the case of fibres parallel to the loading direction, whereas for fibres normal to the loading direction, a significant reduction of strength (up to 60%) is anticipated.

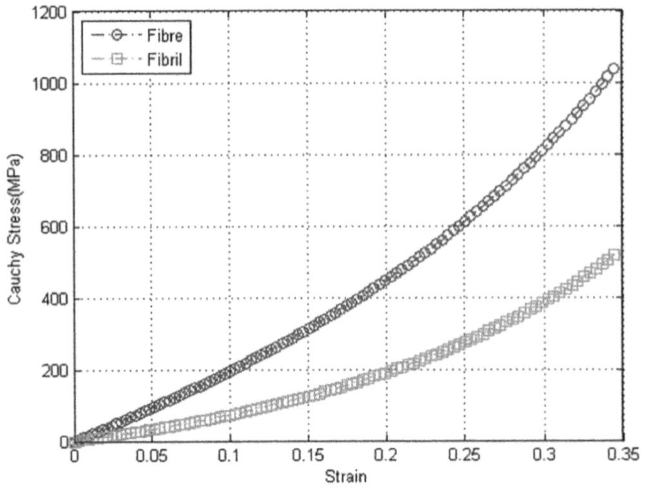

Figure 5.25 Stress–strain response of the fibre and its constituent fibril. *Source:* Adapted from Fathi and Mohammadi 2015.

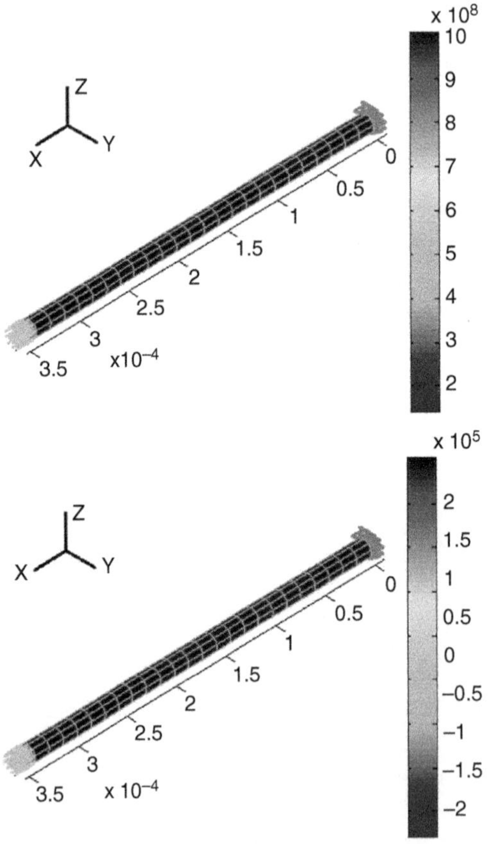

Figure 5.26 Axial (left) and shear (right) stress contours of a representative model of a collagen/matrix and fibre. *Source:* Reproduced from Fathi 2015/University of Tehran.

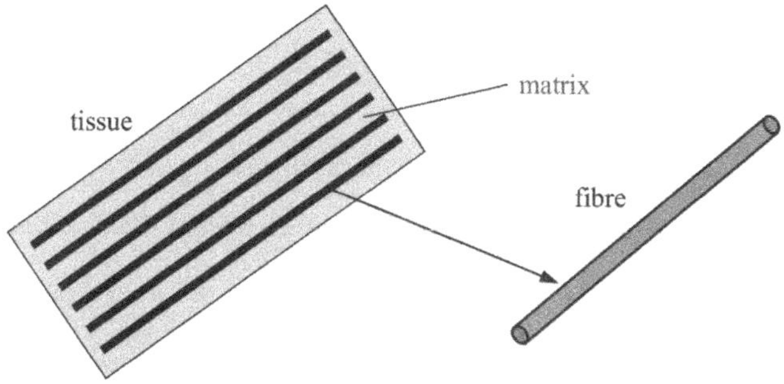

Figure 5.27 Modelling at the tissue scale. *Source:* Adapted from Fathi and Mohammadi 2015.

Table 5.5 Material constants for the tissue scale.

υ_{tf}	μ_{tm} (MPa)
0.15	1.0

Source: Adapted from Fathi and Mohammadi 2015.

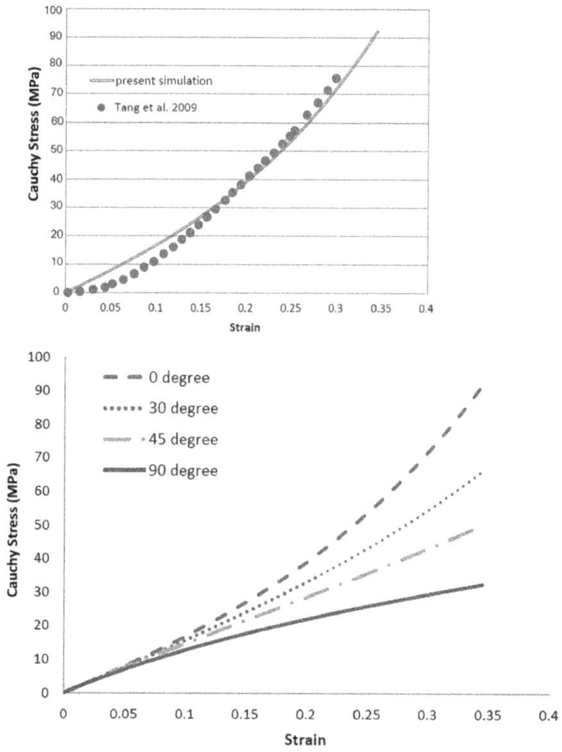

Figure 5.28 Stress–strain response of the tissue in comparison with the test results and for various fibre orientations. *Source:* Adapted from Fathi and Mohammadi 2015.

5.5 Multiscale Analysis of a Human Aortic Heart Valve

5.5.1 Introduction

Heart valve diseases require careful and fast treatment, either by surgical procedures to properly treat the main medical cause of damage or malfunctioning, or by replacement with prosthetic valves, in one of mechanical, bioprosthetic or tissue engineered forms. In any case, a detailed understanding of the undergoing phenomena in the valve, as an organ, or its tissue layers or even down to the cell level, is of paramount importance.

The multiscale technique is an efficient systematic numerical procedure that allows for efficient and accurate analysis of the very wide length scale of the problem, from the organ level (macromodel) to the very fine cellular scale.

A number of earlier attempts for multiscale simulations include embedding the valve motion to the motion of the left ventricle (Carmody et al. 2006), analytically linking the cell-scale to the pressure applied at the organ level (Huang 2005) and a comprehensive multi-scale simulation of the heart valve system of the organ, tissue and cell scales by applying the critical element deformations at upper scales to the lower scale models as the boundary conditions (Weinberg and Kaazempur Mofrad 2007b).

Here, a hierarchical simulation of the aortic valve is adopted to investigate the way a mechanical stimulus (blood pressure) in the organ scale can ultimately affect the interstitial cell in the cell scale. The tissue scale is analysed by considering a rather accurate composition of ventricularis, spongiosa and fibrosa layers and the existence of variably oriented collagen fibres. The adopted multiscale model, from the organ level to the cell scale, is schematically presented in Figure 5.29.

In the adopted one-way multiscale procedure, an analysis is performed at a macro level and the results at the critical points are passed, as boundary conditions, to a representative volume element (RVE) of the smaller scale with finer details and more complex constitutive relations for the constituents. The procedure continues until the required fine organ/tissue/cell data are determined (Weinberg and Kaazempur Mofrad 2007b, 2008; Huang 2005; Billiar and Sacks 2000; Koch et al. 2010; Cataloglu et al. 1977; Grande et al. 1998).

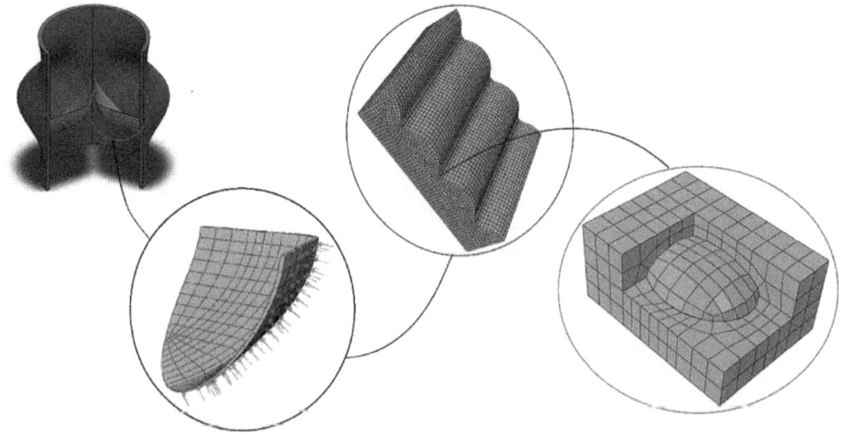

Figure 5.29 The multiscale model, from the organ level to the cell scale. *Source:* Reproduced from Shahi 2013/the University of Tehran.

At each scale and stage of simulation, the models are examined to verify the numerical procedure with available experimental data.

5.5.2 Organ Scale Simulation

The geometry of the aortic valve (AV) is obtained from Koch et al. (2010), as presented in Figure 5.30. One separate leaflet in the initial unloaded configuration is considered as the organ level.

Typical mechanical properties of soft tissues such as incompressibility and strong anisotropy are expected in aortic valve leaflets. Aa a result, the tissue of leaflets exhibits a much stiffer behaviour in the circumferential direction (Billiar and Sacks 2000; Koch et al. 2010).

An anisotropic hyperelastic constitutive model is expected to accurately describe the main mechanical characteristics of the leaflet soft tissue (Huang 2005; Weinberg and Kaazempur Mofrad 2007b; Koch et al. 2010). Accordingly, the anisotropic hyperelastic strain energy function w^s of Holzapfel (2000) is adopted (see Section 5.3.6):

$$w^s(\boldsymbol{C}) = w^s(J_1, J_{f_1}, J) = \underbrace{w_v^s(J)}_{\text{volumetric}} + \underbrace{w_{di}^s(J_1) + w_{da}^s(J_{f_1})}_{\text{deviatoric}} \tag{5.5.1}$$

with

$$w_v^s(J) = \frac{1}{2}\kappa_o (J-1)^2 \tag{5.5.2}$$

$$w_{di}^s(J_1) = \frac{c_2}{2c_1}\left(e^{c_1(J_1-3)} - 1\right) \tag{5.5.3}$$

$$w_{da}^e(J_{f_1}) = \begin{cases} \dfrac{c_{2f_1}}{2c_{1f_1}}\left(e^{c_{1f_1}(J_{f_1}-1)^2} - 1\right), & J_{f_1} > 1 \\ 0, & J_{f_1} \leq 1 \end{cases} \tag{5.5.4}$$

where κ_o is a positive penalty parameter for imposing the volumetric incompressibility (Bonet and Wood 1997; Holzapfel 2000), c_1 and c_2 are the experimental-based isotropic

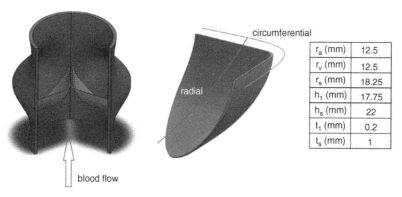

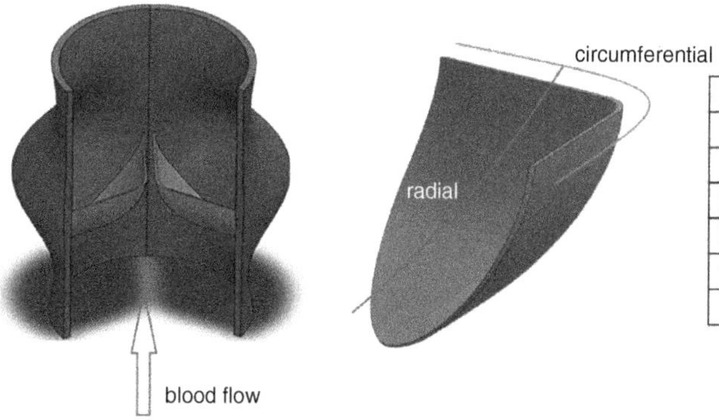

r_a (mm)	12.5
r_v (mm)	12.5
r_s (mm)	18.25
h_1 (mm)	17.75
h_s (mm)	22
t_1 (mm)	0.2
t_s (mm)	1

Figure 5.30 Geometry of the aortic valve leaflet subjected to the uniform pressure loading on the ventricular side. *Source:* Adapted from Shahi 2013; Shahi and Mohammadi 2013.

material parameters and c_{1f_1} and c_{2f_1} denote the experimental constants associated with the set of fibrils/fibres. J_1 is the first invariant of $\boldsymbol{C}_d$ and $J_{f_1} = a_0 \boldsymbol{C}_d a_0$ is known as the pseudo-invariant in terms of the unit vector of the collagen fibre orientation θ. Note that w_{da}^{s} accounts for the transverse isotropic response (Koch et al. 2010). For details, see Section 5.3.6.

According to the uniaxial experimental curve fitting on the aortic left leaflet by Koch et al. (2010), the hyperelastic material constants are defined in Table 5.6.

Uniaxial tests similar to the work of Koch et al. (2010) are performed with one single element, as depicted in Figure 5.31, and provide similar responses to the reference results (Figure 5.32).

Table 5.6 Hyperelastic material parameters for the organ scale.

c_1	c_2 (KPa)	c_{1f_1}	c_{2f_1} (KPa)
4.7483	27.3732	80.4291	11.9752

Source: Adapted from Koch et al. 2010; Shahi 2013.

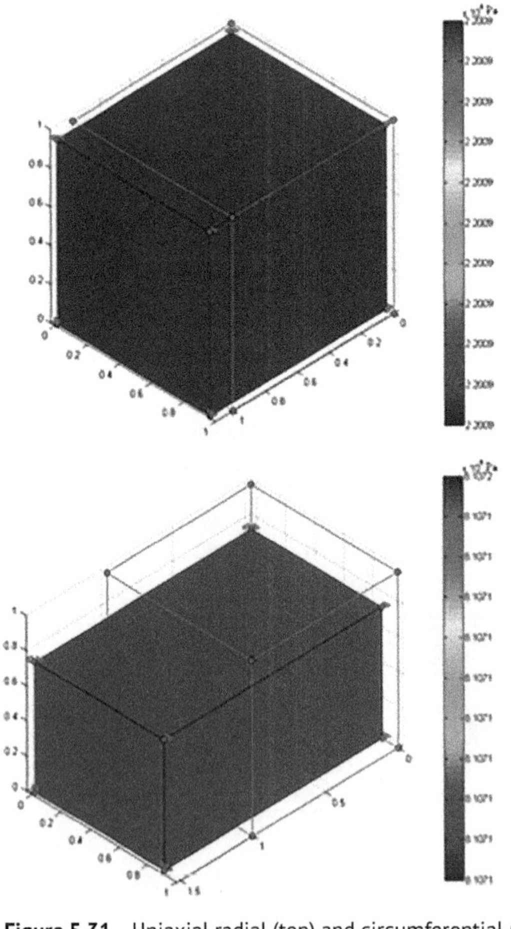

Figure 5.31 Uniaxial radial (top) and circumferential (bottom) tests. *Source:* Reproduced from Shahi 2013/the University of Tehran.

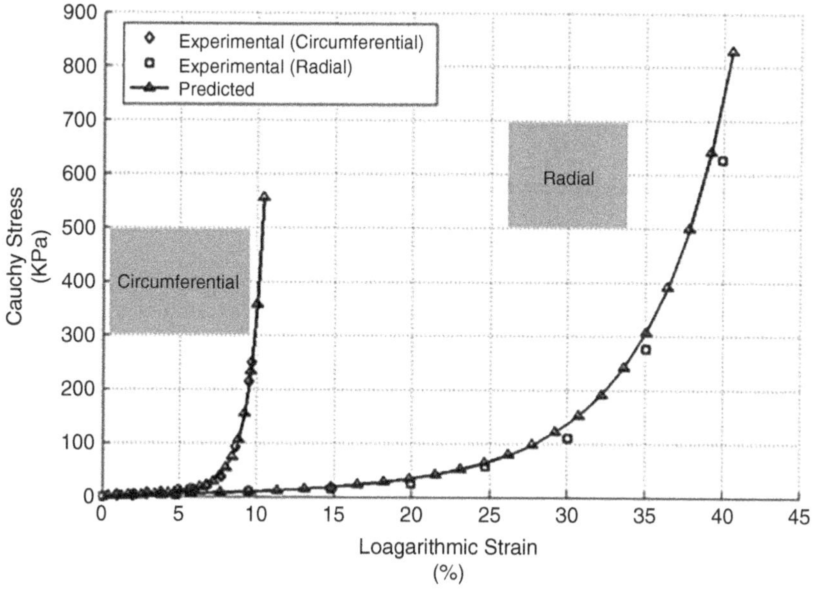

Figure 5.32 Comparison of the results of the present study and the references. *Source:* Adapted from Koch et al. 2010; Shahi 2013.

5.5.2.1 Collagen Orientations

The constitutive relation requires the unit vector a_0 of the orientation of collagen fibres in different parts of the aortic leaflet. According to Kim (2009), image processing techniques can be efficiently used to determine the collagen orientation.

The orientation θ of the collagen fibre is defined by the unit vector a_0,

$$a_0 = (0, \cos\theta, \sin\theta)^T \tag{5.5.5}$$

which is defined for each partition of the leaflet, as illustrated in Figure 5.33 and Table 5.7. A mesh of 280 eight-node brick elements and 622 nodes are adopted for analysis of the macro model, as presented in Figure 5.34.

Constant uniform pressures in the range of 1 to 90 mmHg (peak mean diastole blood pressure) are applied on the ventricular side of the leaflet, and the deformation, strain and stress states are computed. Figure 5.35 illustrates the initial and final deformed configurations of the leaflet at 90 mmHg pressure.

Figure 5.36 presents the deformations of the leaflet at different pressures of 1–90 mmHg.

Figure 5.37 illustrates a better view of the deformed shapes of the leaflets. Shahi (2013) has demonstrated that the numerical results are in good agreement with the reference results from scans (Thomas 2010) and numerical results by Kim et al. (2008) at different blood pressures.

Figure 5.38 shows the contours of principal stresses at 90 mmHG, which agree with the available reports that the maximum principal stress is between 320 and 550 kPa (Cataloglu et al. 1977; Grande et al. 1998; Koch et al. 2010).

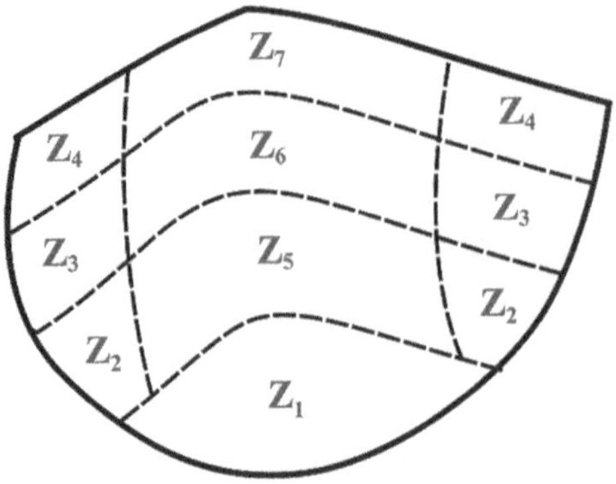

Figure 5.33 Partitioning of the leaflet for collagen fibre orientation. *Source:* Reproduced from Shahi 2013/the University of Tehran.

Table 5.7 Orientation of collagen fibres in different parts of the aortic valve leaflet.

Part	Z_1	Z_2	Z_3	Z_4	Z_5	Z_6	Z_7
Collagen orientation θ	5°	5°	10°	10°	0°	0°	0°

Source: Adapted from Huang 2005; Koch et al. 2010; Shahi 2013.

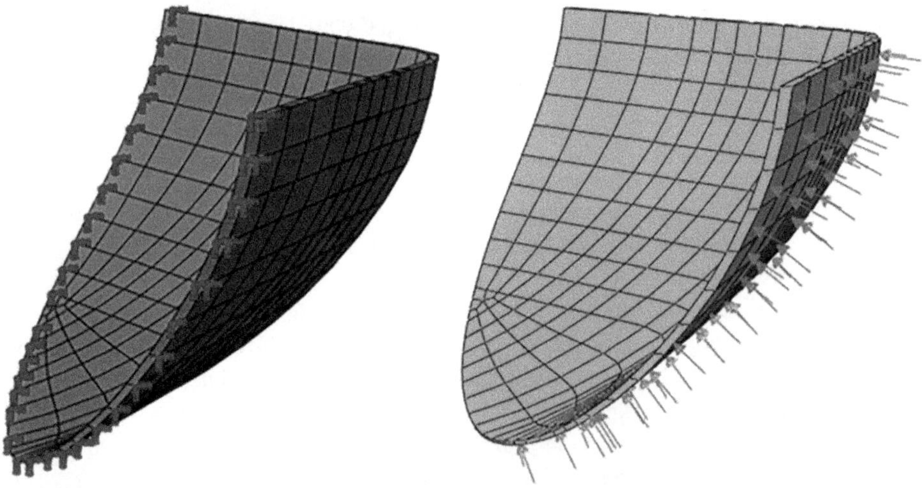

Figure 5.34 Finite element model of the aortic valve leaflet subjected to the uniform pressure loading. *Source:* Adapted from Shahi 2013; Shahi and Mohammadi 2013.

Deformation of middle parts of the leaflet may affect the rate of calcification. It is an important factor in designing the tissue engineered heart valves. Such deformations ultimately influence the aspect ratio of the cells in this region. Therefore, the critical points in this region are chosen for further multiscale analysis (Figure 5.39).

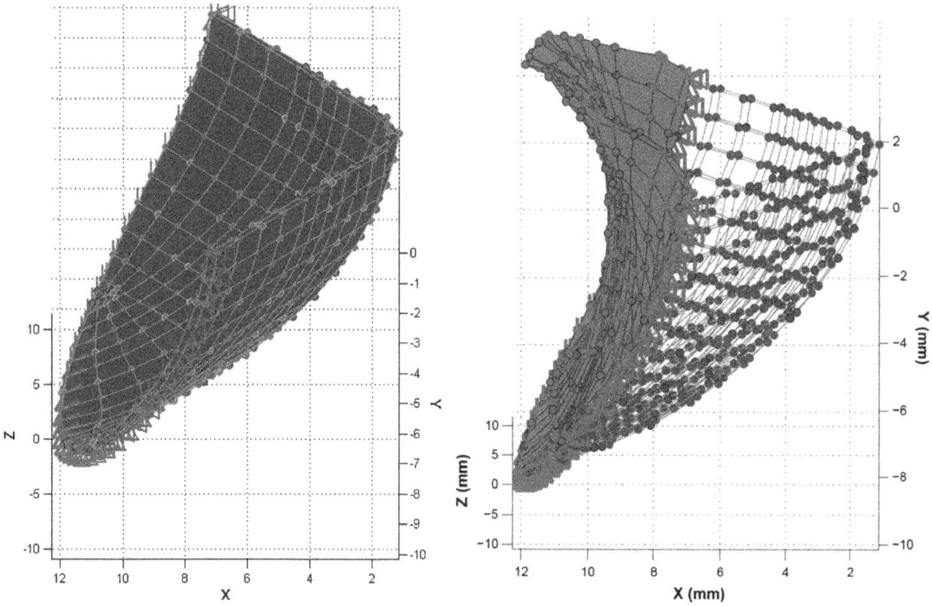

Figure 5.35 Finite element model and the deformed shape of the organ (leaflet) at 90 mmHG. *Source:* Reproduced from Shahi 2013/the University of Tehran.

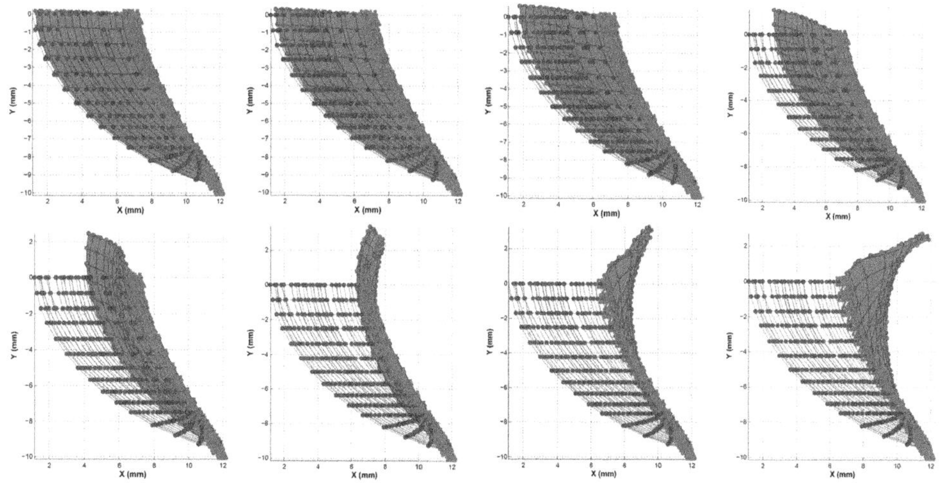

Figure 5.36 Deformations of the leaflet at different pressures. *Source:* Reproduced from Shahi 2013/the University of Tehran.

Table 5.8 compares the stretch of different parts of the leaflet with the reference results (Weinberg and Kazempur Mofrad 2007a, 2007b), which used a different hierarchical multiscale model. Regions A, B and C are illustrated in Figure 5.39. Very good agreements are observed.

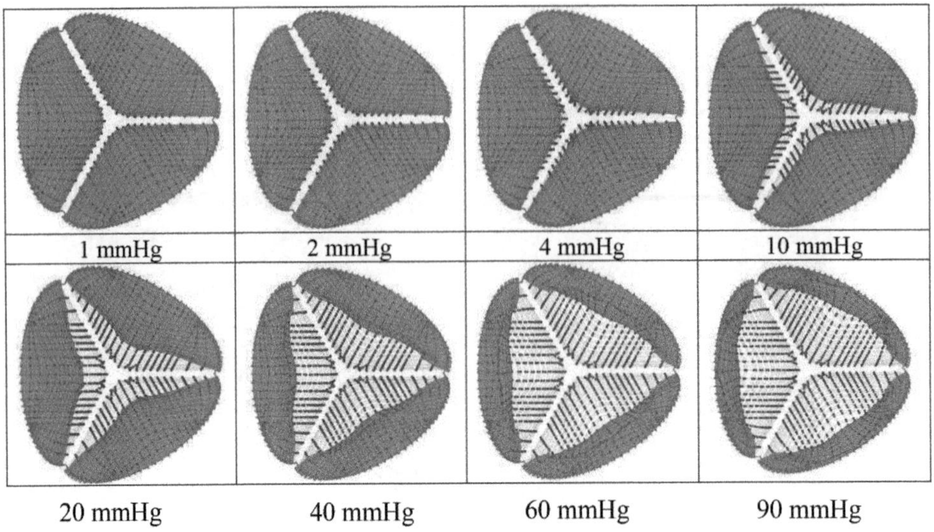

Figure 5.37 Deformed shapes of the leaflet at different pressures. *Source:* Reproduced from Shahi 2013/the University of Tehran.

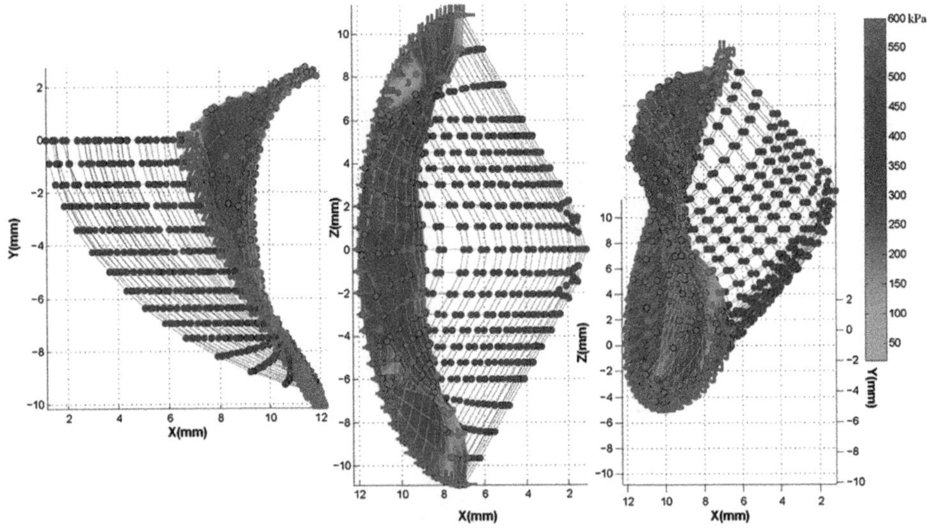

Figure 5.38 Contours of principal stress at 90 mmHG. *Source:* Reproduced from Shahi 2013/the University of Tehran.

5.5.3 Simulation in the Tissue Scale

Analysis of organ-scale does not provide sufficient detailed information on the local deformation of the tissue at various parts of the leaflet, especially across the thickness. Therefore, a lower-scale simulation (tissue scale) is performed based on the multilayered model of the tissue. The tissue-scale model is adopted to further refine the numerical model of the

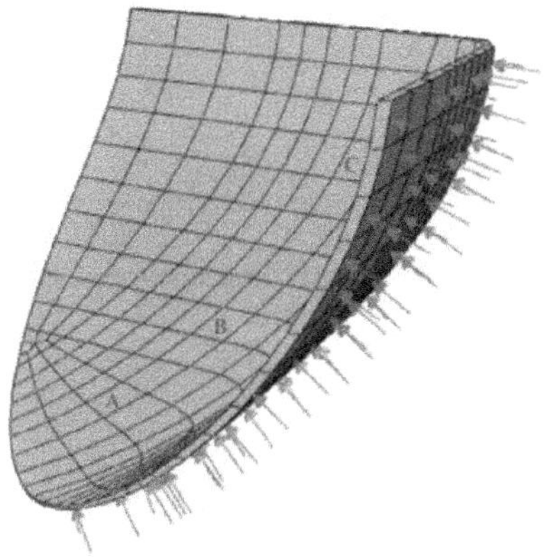

Figure 5.39 Critical points of the leaflet for micro analysis. *Source:* Adapted from Weinberg and Kazempur Mofrad 2007 and Shahi 2013.

Table 5.8 Comparison of the present and reference maximum principal stresses at different regions (C: circumferential, R: radial).

Pressure	Weinberg and Kaazempur Mofrad (2007a, 2007b)		Present study								Difference (%)	
	Average		Region A		Region B		Region C		Average			
(mmHg)	C	R	C	R	C	R	C	R	C	R	C	R
0	1	1	1	1	1	1	1	1	1	1	0.00	0.00
1	1.05	1.01	1.048	1.000	1.046	1.000	1.048	1.001	1.0470	1.0003	0.29	0.96
2	1.06	1.04	1.059	1.061	1.084	1.079	1.043	1.006	1.0622	1.0487	0.21	0.84
4	1.07	1.1	1.086	1.094	1.079	1.063	1.068	1.088	1.0778	1.0819	0.73	1.64
10	*NA*	*NA*	1.088	1.117	1.092	1.088	1.099	1.106	1.0929	1.1036		
20	*NA*	*NA*	1.099	1.184	1.114	1.091	1.119	1.118	1.1110	1.1307		
40	*NA*	*NA*	1.095	1.325	1.144	1.105	1.123	1.132	1.1207	1.1874		
60	1.11	1.22	1.117	1.439	1.199	1.138	1.135	1.172	1.1503	1.2495	3.63	2.42
90	1.12	1.23	1.148	1.526	1.263	1.144	1.139	1.191	1.1834	1.2871	5.66	4.64

Source: Reproduced from Shahi 2013/University of Tehran.

macromodel to obtain the necessary results for simulating the cell-scale model. The tissue-scale model with details of the tissue is illustrated in Figure 5.40.

The tissue of the aortic valve leaflet is comprised of three distinct layers: ventricularis, spongiosa and fibrosa. Fibrosa and ventricularis layers have embedded families of dense aligned collagen fibres and elastin, and are the main contributors in the load-bearing

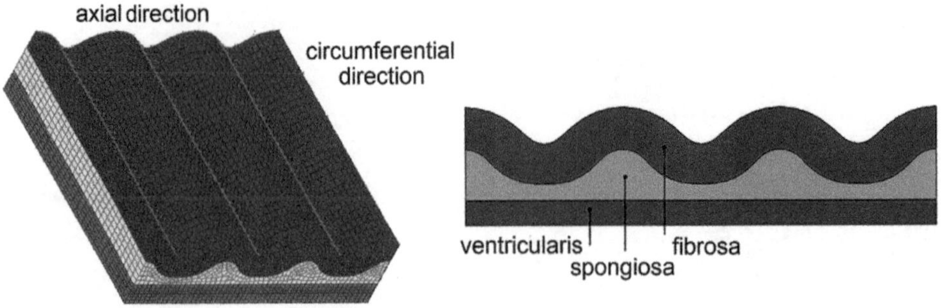

Figure 5.40 Geometry of the tissue-scale RVE model. *Source:* Reproduced from Shahi 2013/the University of Tehran.

capacity of the tissue. Spongiosa is a gel-like material that only contributes as a partitioner and adhesive layer without any major structural role (Weinberg and Kaazempur Mofrad 2007b, 2008). The overall thickness of the tissue varies between 0.2 and 2.0 mm.

In the tissue scale, the fibrous layers are assumed to be composed of an isotropic matrix, governed by the Fung-type hyperelastic model (w^s_{dm}), reinforced with one fibre family in the circumferential tissue direction, represented by the Holzapfel model (w^s_{df}) (Holzapfel et al. 2000). Moreover, the initial response of these layers can be described by a single-term Mooney–Rivlin formulation. The complete strain energy function w^s_d for a fibrous layer can be expressed as (Weinberg and Kaazempur Mofrad 2007b)

$$w^s_d(J_1, J_{f_1}) = \underbrace{\frac{c_{2m}}{2c_{1m}}\left[e^{c_{1m}(J_1-3)} - 1\right]}_{w^s_{dm}} + \underbrace{\frac{c_{2f_1}}{2c_{1f_1}}\left[e^{c_{1f_1}(J_{f_1}-1)^2} - 1\right]}_{w^s_{df}} + \underbrace{\frac{1}{2}\mu_i(J_1 - 3)}_{w^s_{di}} \qquad (5.5.6)$$

where c_{1m}, c_{2m}, c_{1f}, c_{2f} and μ_1 are the experimental constants of the hyperelastic model.

The first and second terms of the strain energy function (5.5.6) are similar to the strain energy function of the organ scale and so a similar methodology for computing the second Piola–Kirchhoff stress tensor and the material modulus is followed. The third term is a neo-Hookean hyperelastic material model (Holzapfel 2000; Bonet and Wood 2008) for representing the initial modulus.

The material constants are computed by fitting the biaxial experimental results of Sacks and Yoganathan (2007) for fibrosa and ventricularis in the radial and circumferential directions, as summarized in Table 5.9. For the weak spongiosa layer, a low coefficient is adopted.

Similar to the organ scale, a single element model (Figure 5.41) for the tissue scale is examined in the biaxial state to assess the adopted constitutive models (Figure 5.42).

Again, a number of critical points or elements should be selected for further analysis at the cell scale. Due to load-bearing characteristic of fibrosa and ventricularis layers, the critical elements are selected from these layers, as presented in Figure 5.43.

The three-dimensional finite element model of the tissue scale RVE is constructed of 26,832 nodes and 10,261 elements. Proper periodic boundary conditions are adopted for the tissue RVE model (Figure 5.44). The deformation field from the macroscale analysis is transformed and applied on the tissue scale model.

Table 5.9 Material properties of tissue constituents.

Layer	c_{1m}	c_{2m} (kPa)	c_{1f_1}	c_{2f_1} (kPa)	μ_i (kPa)
fibrosa	16.502	2.878	297.5	7.790	4.200
spongiosa	–	–	–	–	0.04
ventricularis	11.030	0.257	5.401	0.208	4.200

Source: Reproduced from Shahi 2013/University of Tehran.

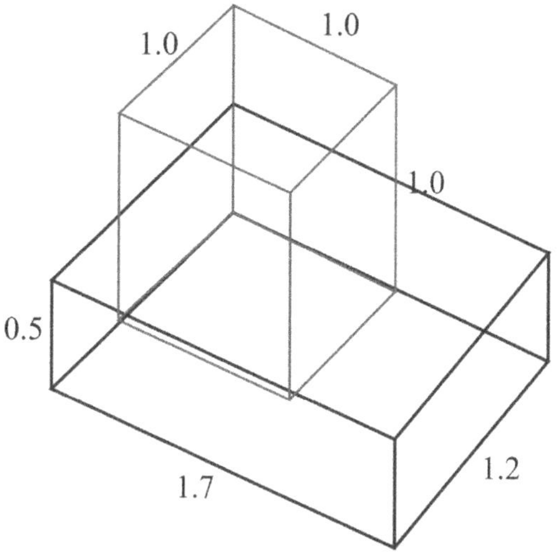

Figure 5.41 Single element model of biaxial tissue. Deformation and stress. *Source:* Reproduced from Shahi 2013/the University of Tehran.

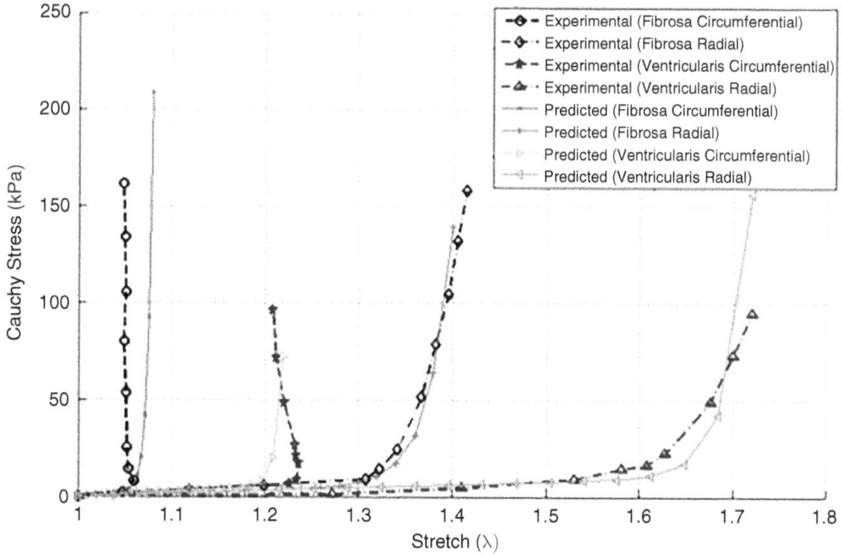

Figure 5.42 Comparison of the biaxial model of tissue with the reference results. *Source:* Adapted from Sacks and Yoganathan 2007; Shahi 2013.

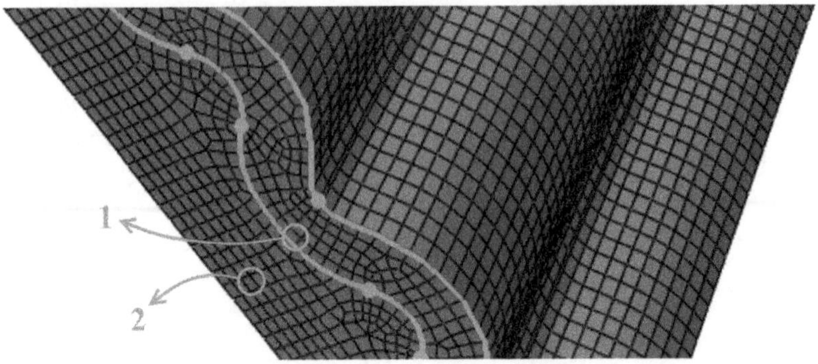

Figure 5.43 Critical elements in fibrosa and ventricularis layers for analysis at the cell level. *Source:* Reproduced from Shahi 2013/the University of Tehran.

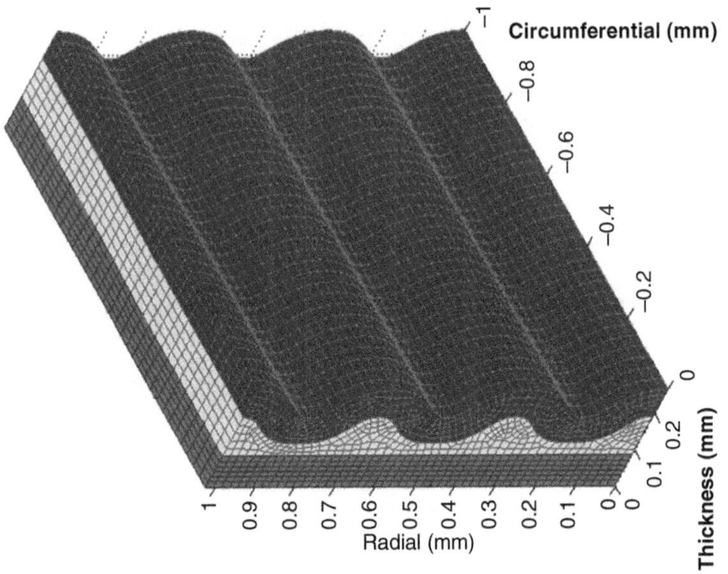

Figure 5.44 Finite element model for RVE of the tissue scale. *Source:* Reproduced from Shahi 2013/the University of Tehran.

At the tissue level, the contours of deformation and maximum principal stress for the region B (Figure 5.39) at 90 mmHG are presented in Figures 5.45 and 5.46, respectively.

In order to examine the critical states of various parts of the tissue model, points 1, 2 and 3 are selected (Figure 5.47), which is in accordance with the reference results (Weinberg and Kaazempur Mofrad 2007).

The results of predicted and reference elongations are compared in Table 5.10. The comparisons are made in circumferential and radial directions and for thc fibrosa and ventriculris layers. Very good agreements are observed.

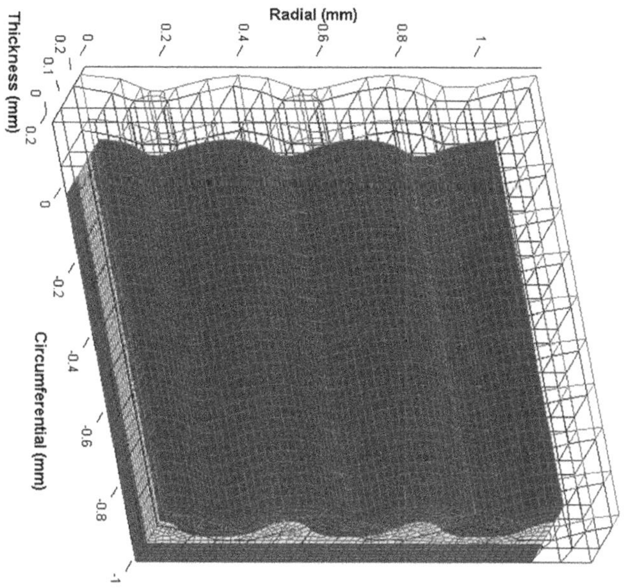

Figure 5.45 Contour of the deformation for the region B at 90 mmHG. *Source:* Reproduced from Shahi 2013/the University of Tehran.

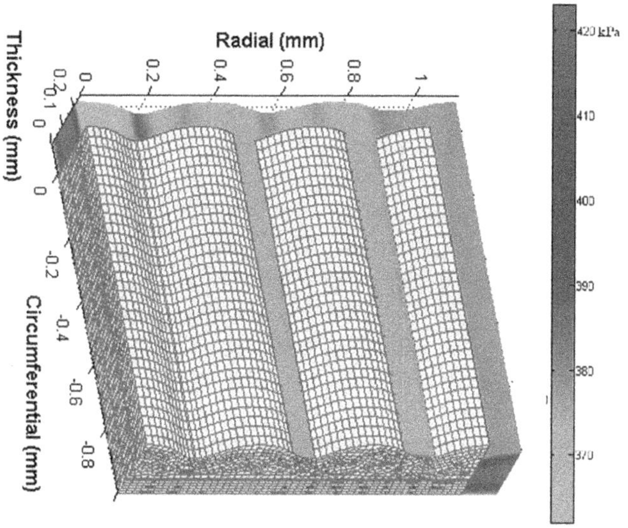

Figure 5.46 Contour of the maximum principal stress for the region B at 90 mmHG. *Source:* Reproduced from Shahi 2013/the University of Tehran.

5.5.4 Cell Scale Analysis

The final stage of the present multiscale simulation is the cell-scale analysis based on the results of the critical points of the tissue scale. Again, a periodic RVE at the cell level is adopted for each critical point. The RVE consists of a single interstitial cell (IC) in the form of an ellipsoidal aggregate surrounded by the extracellular matrix (ECM), either in fibrosa or ventricularis layers. The geometric dimensions $9.75 \times 7.5 \times 5.4\,\mu m$ of the hypercube of

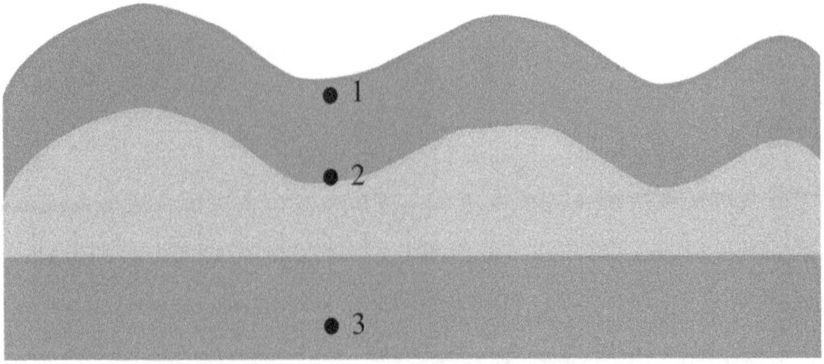

Figure 5.47 Points 1, 2 and 3 on the tissue model for examining the cell-scale analysis. *Source:* Adapted from Weinberg and Kaazempur Mofrad 2007; Shahi 2013.

Table 5.10 Comparison of the predicted elongations of the tissue model by an average of points 1, 2 and 3.

Pressure (mmHg)	Circumferential			Radial					
				fibrosa			ventricularis		
	Ref	present	diff %	Ref	present	diff %	Ref	present	diff %
0	1	1.000	0.000	1	1.000	0.000	1	1.000	0.000
1	1.05	1.056	0.605	1.005	1.006	0.115	1.01	1.009	3
2	1.06	1.067	0.613	1.03	1.031	0.097	1.05	1.067	1.638
4	1.07	1.079	0.841	1.1	1.123	2.091	1.08	1.066	1.306
60	1.1	1.089	0.987	1.29	1.310	1.550	1.22	1.242	1.803
90	1.12	1.161	3.661	1.32	1.360	3.030	1.24	1.290	4.032

Source: Adapted from Weinberg and Kaazempur Mofrad 2007; Shahi 2013.

the ellipsoidal cell are obtained from Huang (2005) and the ECM hypercube is assumed to bound the ellipsoidal cell, as presented in Figure 5.48.

The finite element model, composed of 336 eight-node solid elements and 586 nodes is typically depicted in Figure 5.49.

The ECM material of the cell scale is assumed to be similar to the ECM of the tissue model. Clearly, different values of ECM material properties are adopted for critical elements in the fibrosa and ventricularis layers. The cell is assumed to follow the single-term Mooney–Rivlin material with the value $\mu_1 = 800$ Pa (Shahi and Mohammadi 2013).

The deformation field of the tissue scale model is applied to the RVE model of the cell scale. The final results at the cell level at 90 mmHG for the fibrosa and ventricularis layers are presented in Figure 5.50.

At the final stage of analysis and discussion, the cell aspect ratio (CAR) is obtained by the ratio of the circumferential diameter to the thickness diameter, as proposed by Huang (2005). The predicted results are compared with the reference results of Weinberg and Kaazempur Mofrad (2007) and Huang (2005).

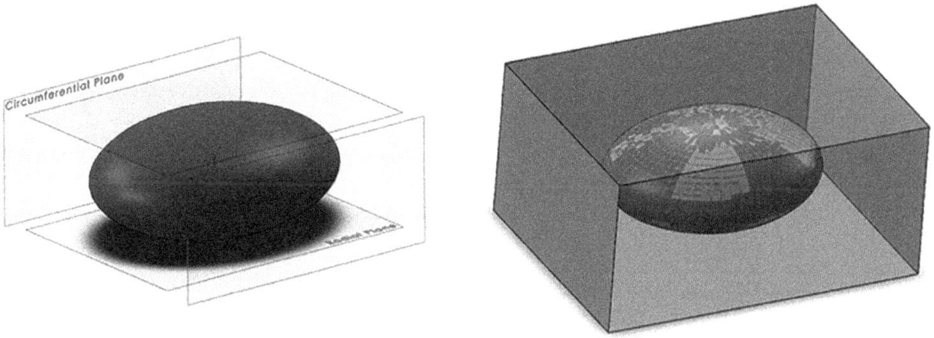

Figure 5.48 Cell-scale RVE consisting of the cell and ECM. *Source:* Reproduced from Shahi 2013/the University of Tehran.

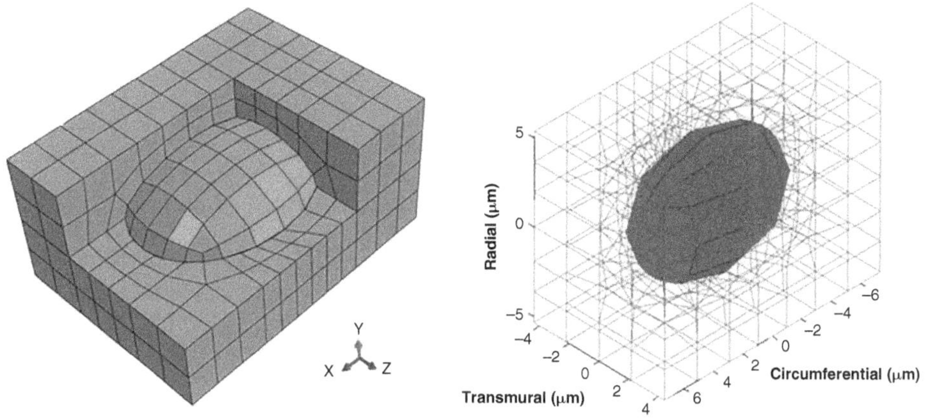

Figure 5.49 Finite element model of cell-scale RVE. *Source:* Reproduced from Shahi 2013/the University of Tehran.

The results of Table 5.11 can be better illustrated in Figure 5.51. Clearly, the present numerical results provide more accurate solutions.

5.6 Modelling of Ligament Damage

This simulation is aimed to model the response of the muscular ligament, which is mainly composed of one set of parallel fibres. The soft tissue can accurately be represented by a transversely isotropic behaviour ($\kappa = 0$) (Weise 1995; Calvo et al. 2007) and with the material constants defined in Table 5.12. The non-local integral type of the damage model is accompanied with the anisotropic nearly incompressible hyperelastic constitutive model, as explained in Section 5.3.8. The section follows the original work of Fathi et al. (2017).

First, the uniaxial test on a ligament specimen, conducted by Weiss (1995) is examined, as presented in Figure 5.52.

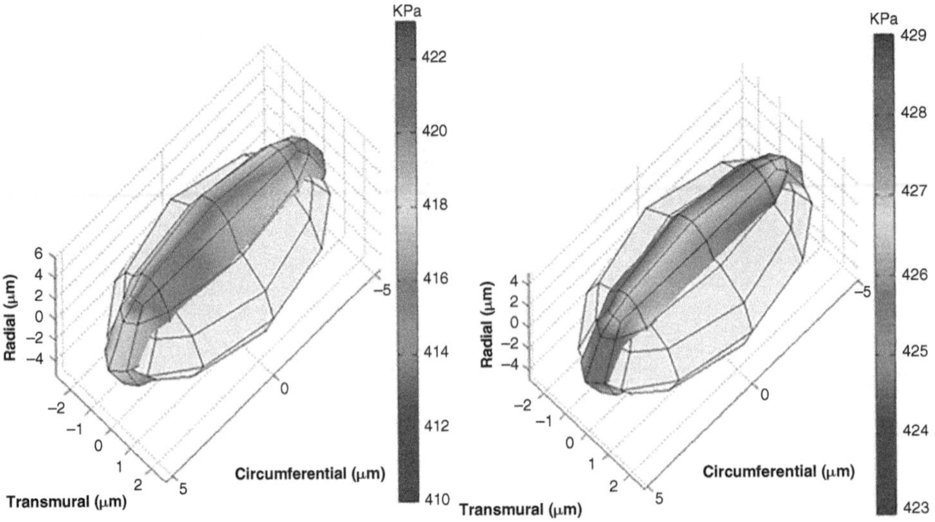

Figure 5.50 Maximum principal stress of the fibrosa and ventricularis layers at 90 mmHG. *Source:* Reproduced from Shahi 2013/the University of Tehran.

Table 5.11 Comparison of the predicted cell aspect ratio with the reference results.

	Huang (2005)		Weinberg and Kaazempur Mofrad (2007)		Present numerical model			Diff (%)
Pressure (mmHg)	test	numerical	fibrosa	ventricularis	fibrosa	ventricularis	average	
0	1.800	1.800	1.800	1.800	1.800	1.800	1.800	0.00
1	2.020	1.955	2.011	1.913	1.970	1.840	1.905	5.69
2	2.676	2.080	2.160	2.044	2.346	1.960	2.153	19.53
4	2.673	2.268	2.553	2.396	2.337	2.276	2.306	13.72
60	4.442	3.349	4.121	3.418	3.828	3.708	3.768	15.17
90	4.200	3.510	4.381	3.542	4.283	3.799	4.041	3.79

Source: Reproduced from Shahi 2013/University of Tehran.

The predicted stress–stretch responses of the model for four consecutive cycles of loading and unloading are presented in Figure 5.53. The loading paths of cycles 2, 3 and 4 cover the unloading paths of cycles 1, 2 and 3, respectively. Fathi et al. (2017) have shown that these results are generally comparable with the available experimental (Weiss 1995) and numerical (Calvo et al. 2007) results. The resultant damages at the end of loading of each cycle are reported in Table 5.13.

In the second stage, the damage analysis is performed on a perforated sample of the ligament. Geometry, boundary conditions, loading and the adopted finite element mesh are presented in Figure 5.54. The material properties are assumed to be similar to Table 5.12.

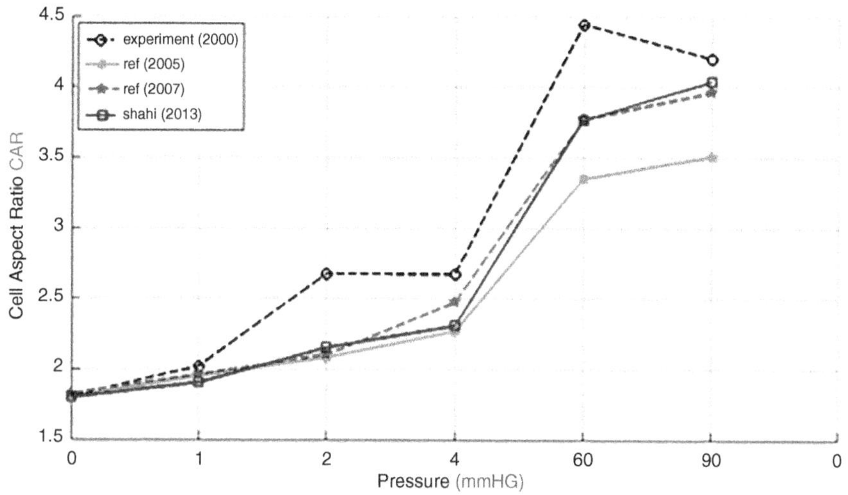

Figure 5.51 Comparison of the predicted aspect ratio of the cell with the reference results. *Source:* Adapted from Shahi 2013; Weinberg and Kaazempur Mofrad 2007; Huang 2005.

Table 5.12 Material properties of the ligament sample.

Tissue			
λ(MPa)	5016.4		
Matrix		**Set of fibres $\theta = 0$**	
μ_{1m}(MPa)	10.1	c_{1f_1}	150.2
μ_{2m}(MPa)	0	c_{2f_1}(MPa)	46.0
		κ_{f_1}	0
		I_{0_i}	1
β_m	0.001	β_{f_1}	0.0001
$\varepsilon_{dm}^{min}(\sqrt{\text{MPa}})$	0.16353	$\varepsilon_{df_1}^{min}(\sqrt{\text{MPa}})$	0.5
$\varepsilon_{dm}^{max}(\sqrt{\text{MPa}})$	0.2974	$\varepsilon_{df_1}^{max}(\sqrt{\text{MPa}})$	1.3342

Source: Adapted from Fathi et al. 2017.

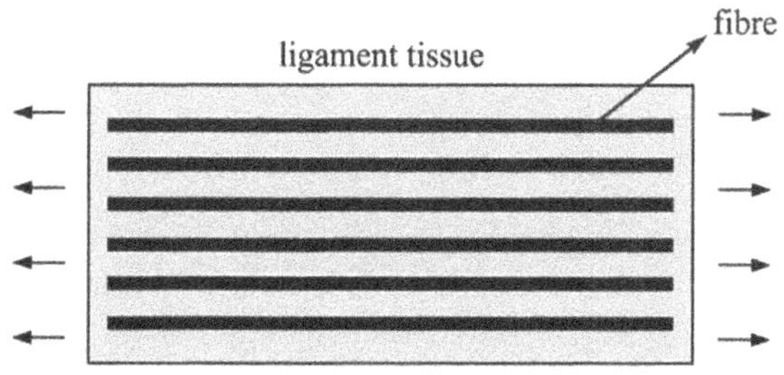

Figure 5.52 The sample of the ligament and the numerical model.

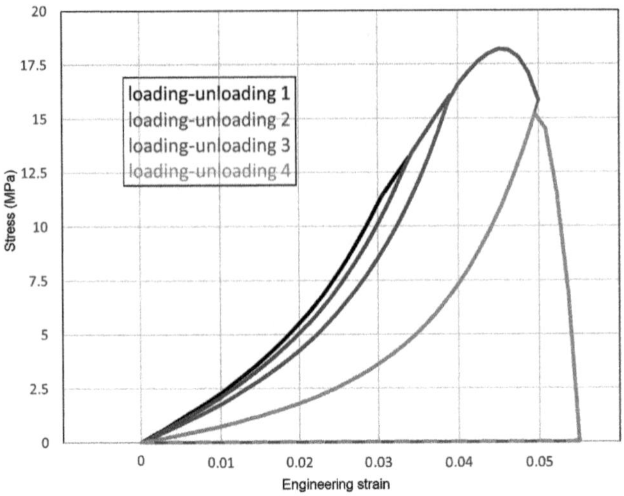

Figure 5.53 Stress–stretch response of the ligament. *Source:* Adapted from Weiss 1995; Calvo et al. 2007; Fathi et al. 2017.

Table 5.13 Damages in matrix and fibre at the end of the loading path of each cycle.

cycle	D_m	D_f
1	0.03	0.09
2	0.11	0.24
3	0.53	0.69
4	1	0.99

Source: Adapted from Fathi et al. 2017.

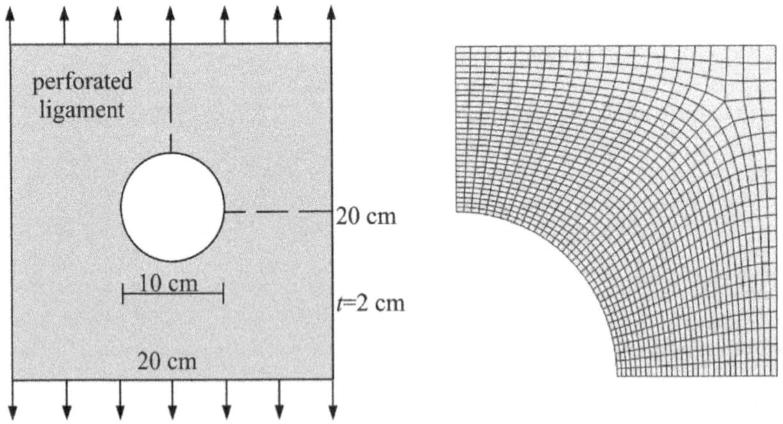

Figure 5.54 Geometry, boundary conditions and finite element mesh of the perforated ligament.

Figures 5.55 and 5.56 illustrate the contours of damage distribution in the matrix and fibre, respectively, at the final stages of the analysis. The level of damage index is clearly higher in the matrix.

Due to the fact that many damage simulations are vulnerable to the finite element mesh size, four different unstructured finite element meshes with 1,059, 1,576, 2,225 and 2,759 elements are examined. The global force-displacement response of all four meshes remain similar, as depicted in Figure 5.57, showing the mesh independency of the overall results.

Another important factor that may substantially affect the results of a damage analysis is the selection of the damage characteristic length (integration radius). Figures 5.58 and 5.59 compare the contours of the damage parameter in the matrix and fibre, D_m and D_f, respectively, for two different radii of integration: $R = 0.01, 0.02$ mm. While more or less similar results are observed for the damage in the matrix, the damage in fibre is changed by the change of the characteristic length, both in magnitude and in pattern.

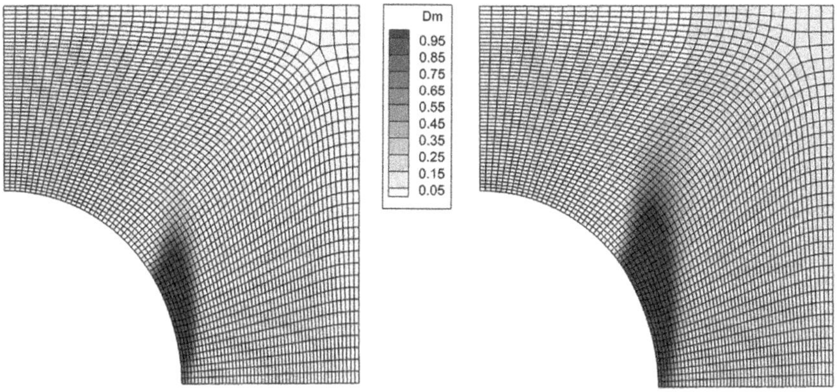

Figure 5.55 Distribution of damage in the matrix in different steps of the analysis. *Source:* Adapted from Fathi et al. 2017.

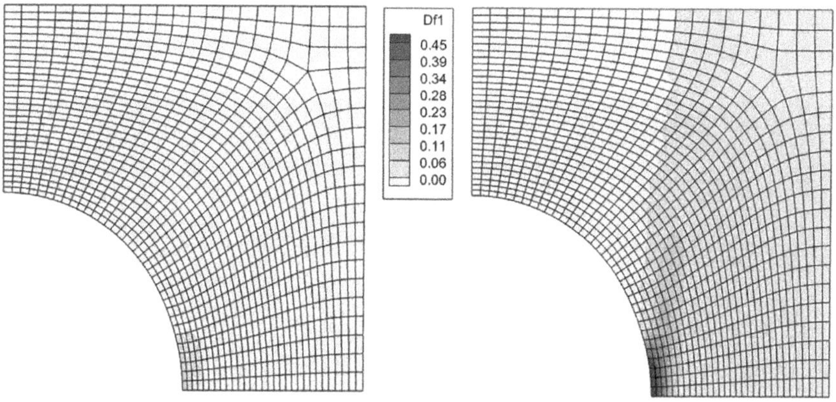

Figure 5.56 Distribution of damage in the fibres in different steps of the analysis. *Source:* Adapted from Fathi et al. 2017.

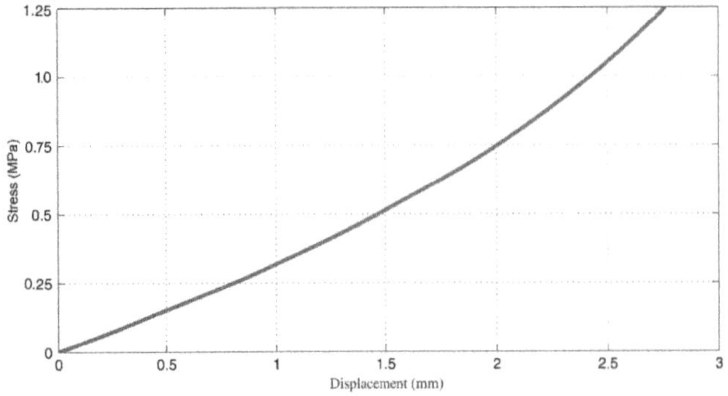

Figure 5.57 Force-displacement response of the perforated ligament for different finite element meshes. *Source:* Adapted from Fathi et al. 2017.

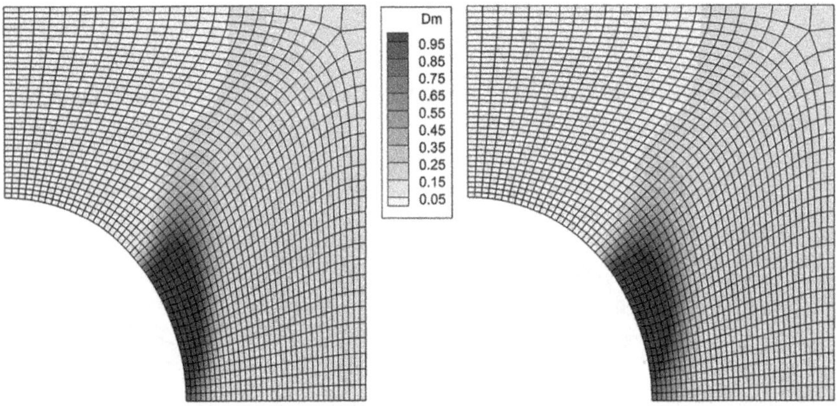

Figure 5.58 The effect of the characteristic lengths $R = 0.01, 0.02$mm (left and right, respectively) on the damage of the matrix. *Source:* Adapted from Fathi et al. 2017.

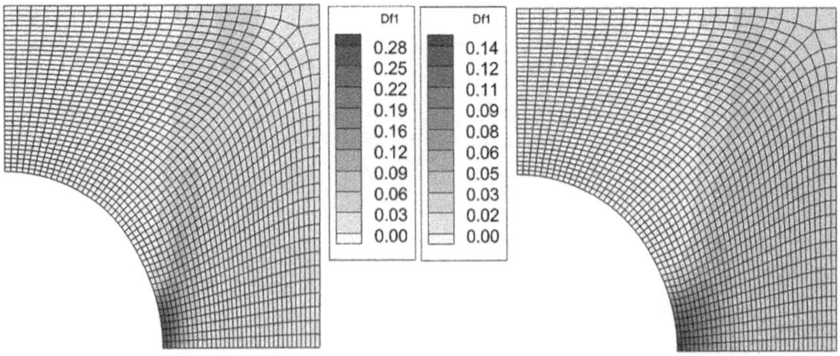

Figure 5.59 The effect of the characteristic lengths R = 0.01,0.02mm (left and right, respectively) on the damage of the fibre. *Source:* Adapted from Fathi et al. 2017.

5.7 Modelling of the Peeling Test: Dissection of the Medial Tissue

Aortic dissection is an important clinical topic due to its high risk of fatality (Gasser and Holzapfel 2006; Ferrara and Pandolfi 2008, 2010). In the event of aortic dissection, the blood may enter and flow in between the microstructural layers of the artery – intima, media and adventitia (Gasser and Holzapfel 2006; Criado 2011) – with an eventual rupture of the soft tissue and immediate death.

The hyperelastic material model accounts for the anisotropic nature of fibre reinforcement and the matrix of the tissue by a combination of the neo-Hookean and an exponential type of strain energy functions. Similar to most biomechanical problems, the soft tissue is assumed to follow a nearly incompressible response. The hyperelastic constitutive model is accompanied by a non-local integral type of damage model, as discussed in Section 5.3.8.

The large deformation numerical simulation is performed in the form of the peeling test, which is regarded as a relatively accurate introductory model for the full model of aortic dissection (Sommer et al. 2008). It allows for investigation of damage, fracture and rupture phenomena of the soft fibrous tissue. The section follows the original work of Fathi et al. (2017).

Here, only the media layer, which is the main load-bearing part of an artery, is considered to resemble the reference peeling experiments (Sommer et al. 2008). The geometry of the model and the finite element mesh are presented in Figure 5.60. the finite element model is composed of 3306 nodes and 2048 eight-noded solid elements.

The damage hyperelastic constants are based on the combination of intact model by Gasser and Holzapfel (2006) and the integral-based damage approach (Section 5.3.8), as defined in Table 5.14. Two sets of fibres with similar material properties but with different orientations are considered (see Figure 5.60).

The analysis is performed well into the large deformation regime, where the local orientations of microstructural components (fibres) gradually change with the extent of large deformation and strain. This is accompanied by potentially large softening and damage growth in the soft tissue, which further complicates the overall response of the media layer.

Figure 5.61 illustrates the mesh independency of the global force-displacement results for two different finite element meshes.

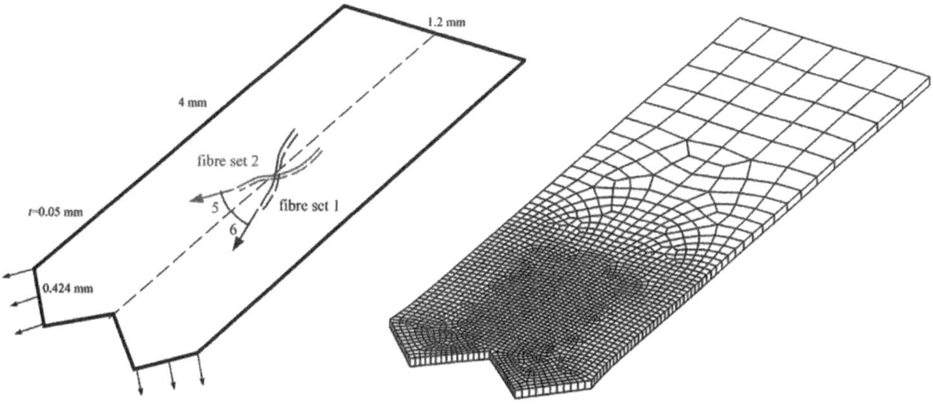

Figure 5.60 Geometry, boundary conditions and the finite element mesh of the medial dissection. *Source:* Adapted from Fathi et al. 2017.

Table 5.14 Material properties of the media sample.

Tissue					
λ(MPa)	5016.42				
Matrix		**First set of fibres $\theta = 5$**		**Second set of fibres $\theta = -5$**	
μ_{m1}(MPa)	32.4	c_{1f_1}	10	c_{1f_2}	0.001
μ_{m2}(MPa)	0	c_{2f_1}(MPa)	98.1	c_{2f_2}(MPa)	1.45
		κ_{f_1}	0	κ_{f_2}	0
		I_{0_1}	1	I_{0_2}	1
β_m	0.1	β_{f_1}	0.1	β_{f_2}	0.1
$\varepsilon_{dm}^{min}(\sqrt{\text{MPa}})$	1.2	$\varepsilon_{df_1}^{min}(\sqrt{\text{MPa}})$	1.5	$\varepsilon_{df_2}^{min}(\sqrt{\text{MPa}})$	1.5
$\varepsilon_{dm}^{max}(\sqrt{\text{MPa}})$	9	$\varepsilon_{df_1}^{max}(\sqrt{\text{MPa}})$	25	$\varepsilon_{df_2}^{max}(\sqrt{\text{MPa}})$	25

Source: Adapted from Fathi et al. 2017; Gasser and Holzapfel 2006.

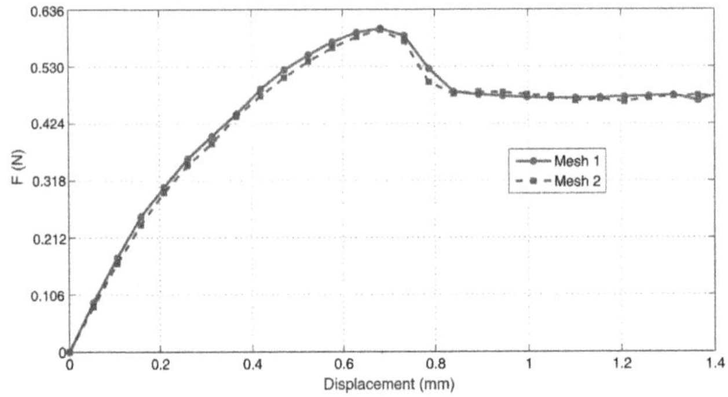

Figure 5.61 Global response of the medial dissection for different meshes (Mesh 1: 1,228 elements, Mesh 2: 2,048 elements). *Source:* Adapted from Fathi et al. 2017.

Distribution of the damage parameter in the matrix and the first set of fibres are illustrated in Figure 5.62.

Quite different patterns are observed and the distribution of critical damage regions differ for the microstructural components of matrix and fibre. The damage in the matrix is expected to initiate from the maximum shear stress regions. In contrast, such an explicit conclusion cannot be drawn for the fibre component due to their anisotropic dispersal, the fibre rotating mechanisms by the level of experienced deformation, which highly complicates the microstructural response.

These complex effects can be used to categorize the fibres into three types of inactive, active and damaged fibres, based on their level of mobilization and the damage threshold:

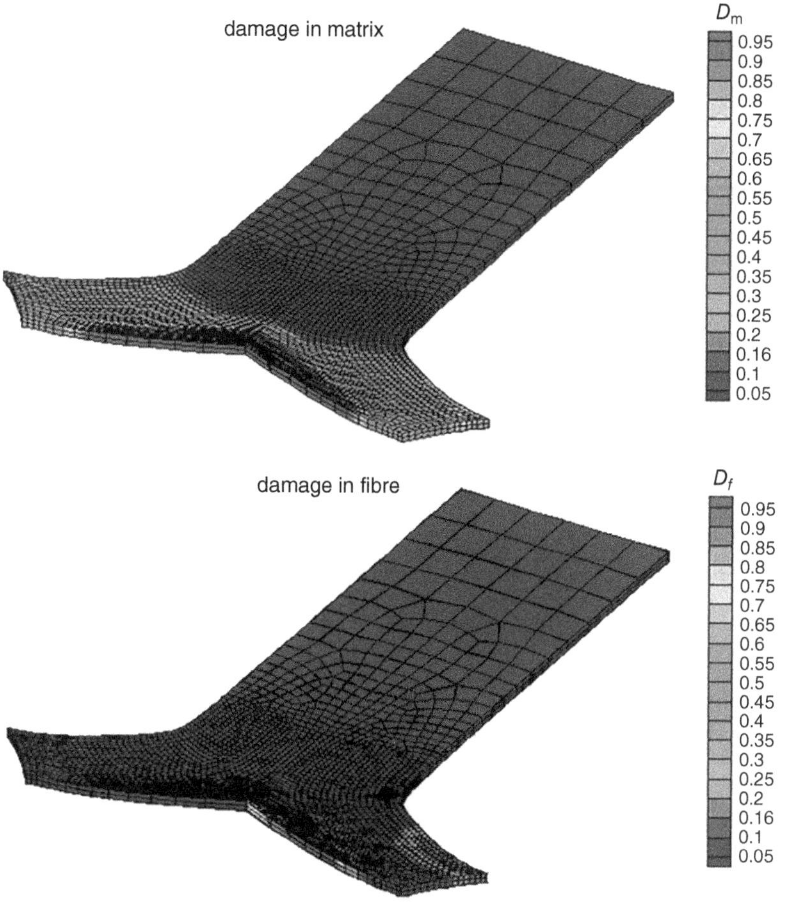

Figure 5.62 Distribution of damage in the matrix and the first family of fibres. *Source:* Adapted from Fathi et al. 2017.

$$\begin{cases} \text{Inactive}: & \text{shortened and do not contribute in load bearing} \\ \text{Active}: & \text{elongated but have not reached the initial damage threshold} \\ \text{Damaged}: & \text{elongated, but have passed the damage threshold} \end{cases} \tag{5.7.1}$$

Figure 5.63 presents the state of inactive, active and damaged fibres for the first set of fibres for various stages of the loading (defined in the global response of Figure 5.61). Clearly, major rotations are observed for the fibres, making them more aligned with the loading direction at each stage of the solution.

Finally, distributions of the von Mises stress at different stages of the large deformation of the tissue are presented in Figure 5.64. The bubble of stress is propagated from the edges of dissection towards the interior of the damaged soft tissue.

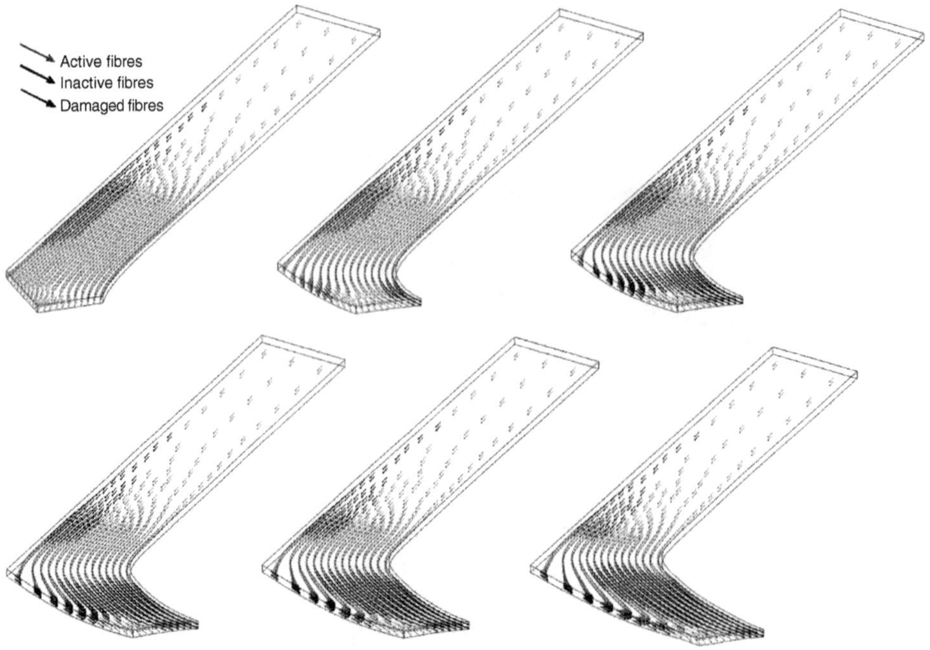

Figure 5.63 Orientations of inactive, active and damaged first set of fibres at different stages of the loading. *Source:* Adapted from Fathi et al. 2017.

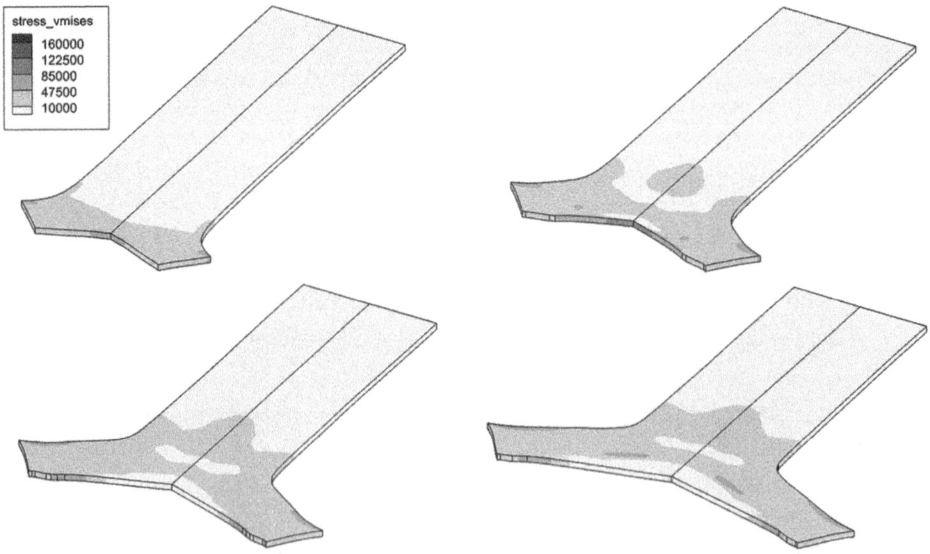

Figure 5.64 Distribution of the von Mises stress at different stages of the loading. *Source:* Adapted from Fathi et al. 2017.

5.8 Healing in Damaged Soft Tissue

5.8.1 Introduction

The analysis of healing processes has been an active research topic in various fields such as structures, pavements, aerospace and biomechanics. Moreover, the healing mechanism in living organisms can be adopted to design new prosthetic tissues and even engineering materials and applications with external healing or self-healing properties (Griffiths et al. 2016).

Human skin begins to repair itself naturally after the occurrence of a damage in the form of scratch, scar and tear. The process of skin healing consists of various interconnected stages over various periods of time. Generally, the healing process in a soft tissue can be categorized into four steps of homeostasis (or bleeding stoppage), inflammation, proliferation and maturation, as depicted in Figure 5.65 (Olsen et al. 1995; Kordestani 2019).

The homeostasis stage is the first phase of healing in the form of bleeding, immediately after occurrence of a wound. The region is filled with the blood, leading to the fibrin clot, which is the appropriate ground for movement of cells and other physiological mechanisms inside the wound. Later, the so-called fibrinolysis mechanism gradually dissolves the clot to create space for subsequent evolving healing steps (Kordestani 2019).

The second stage is inflammation, which is one of the most critical stages of wound healing. This stage is substantially contributed by the important role of secretion of various growth factors in cell reproduction, cell movement, protein production in cells, and enzyme

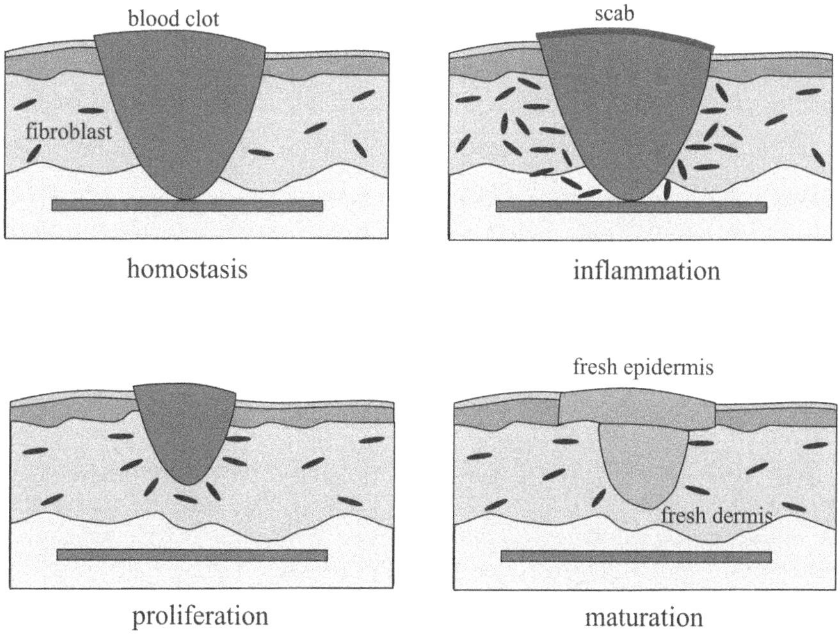

Figure 5.65 Stages of healing a wound in a soft tissue. *Source:* Adapted from Olsen et al. 1995; Kordestani 2019.

production by cells. They affect the regulation of collagen and myofibroblast production (Clark 1989; Stadelmann et al. 1998). Macrophages also affect the secretion of growth factors and necessary enzymes for a healthy repair process (Singer and Clark 1999). Growth factors are the primary driver of various repair processes. For example, cells are transported to the injury site because of the accumulation of growth factors at the injury site. Due to the presence of these factors, cell reproduction is higher than the healthy tissue state during the healing process.

The proliferation stage occurs in the period of 4 to 21 days from the beginning of healing. Wound contraction, which is an essential process for successful healing, occurs in this phase and after the completion of inflammation and the plateau form of the concentration of growth factors in the wounded area (Kordestani 2019). Contraction begins approximately one week after injury (Clark 1989; Stadelmann et al. 1998). Then, some of the fibroblasts chemotactically move in the granular tissue and turn into another cell type while moving to the injury site. This new cell type is called myofibroblast and contains muscle proteins (Clark 1989; Stadelmann et al. 1998; Singer and Clark 1999). The conversion of fibroblasts into myofibroblasts is mediated by the growth factor stimulation of cells, such as TGF-β1 and mechanical stimulation caused by the resistance to contraction of wound edges (Stadelmann et al. 1998). Due to the presence of muscle proteins, myofibroblasts can establish an intracellular and intercellular binding with the extracellular matrix (Stadelmann et al. 1998; Singer and Clark 1999). Unlike fibroblasts, these cells are not mobile and are spatially fixed (Olsen et al. 1995).

Simultaneously, collagen is produced and distributed by cells due to the growth factors, which leads to the formation of a skeleton, which retains the tissue in place (Stadelmann et al. 1998). Producing new fibres in the cell is due to the fibroblast chemotaxis, influenced by the growth factors (Singer and Clark 1999). Chemotaxis occurs when cells react to a chemical gradient; for instance, movement or reproduction of cells in a particular direction due to the growth factor gradient in that particular direction (Zigmond 1981; Zigmond et al. 1982; Oster et al. 1983).

Because myofibroblasts are attached to the wound borders, contraction of these cells leads to the contraction of the entire granular tissue and the closure of the wound borders (Stadelmann et al. 1998; Singer and Clark 1999). Fibroblasts also exert a force on the matrix due to its creep movement in the matrix (Tranquillo and Murray 1992). Wound contraction occurs due to cell-surface receptor stimulation by various growth factors and attachment of fibroblasts to the collagen matrix (Singer and Clark 1999). In some sources, the leading cause of contraction is believed to be the myofibroblasts (Stadelmann et al. 1998; Singer and Clark 1999).

The final stage is the remodelling, which is the slowest process of the healing and may take several days/weeks and even up to a year. In this stage, collagens are re-formed and obtain most of their strengths. Moreover, the contraction is finalized and the remaining extra cells of fibroblast and myofibroblast are gradually removed.

5.8.2 Physical Foundation of Tissue Healing

Modelling each stage of the tissue healing process requires the investigation of differential equations governing the phenomenon. Almost all mesenchymal cells (cells that react to chemicals) exist in the embryonic processes (processes related to the growth and healing of living organisms) and can move spontaneously. Each cell moves to its surroundings, which is usually a fibrous extracellular matrix, by applying force to keep the cell connected to the matrix.

The equations in this section are based on two main assumptions. First, the cells move and disperse in a substrate, which contains the fibrous extracellular matrix. Second, the mobile cells can exert a significant force on their surroundings through their movement (Harris et al. 1981). The main variables of the governing equations are:

ρ_{fb}: fibroblast density
ρ_{mfb}: myofibroblast density
ρ_{ECM}: ECM/collagen density
ρ_{gf}: growth factor concentration
$\boldsymbol{u}$: tissue displacement

5.8.2.1 Viscoelastic Deformation/Contraction of Tissue

The first necessary equation to investigate the tissue contraction phenomenon during healing is the equation that governs the tissue displacement as the first instance of contraction. The main reason for the tissue deformation is the movement of cells during the healing, which can be expressed in terms of the general equation of the motion (Lai et al. 2010):

$$\nabla\cdot(\boldsymbol{\sigma}-\boldsymbol{\sigma}_0)+\boldsymbol{f}^{\mathrm{b}}=\rho\ddot{\boldsymbol{u}} \tag{5.8.1}$$

where $\boldsymbol{\sigma}$ is the internal stress tensor, $\boldsymbol{f}^{\mathrm{b}}$ is the vector of volumetric forces acting on the element, $\ddot{\boldsymbol{u}}$ represents the acceleration and $\boldsymbol{\sigma}_0$ is the initial stress state of the pre-stretched tissue.

Equation (5.8.1) can be reduced to a quasi-static form by neglecting the inertial (or acceleration) term for the very slow physiological process of healing in soft tissues (Odell et al. 1981; Oster et al. 1983),

$$\nabla\cdot(\boldsymbol{\sigma}-\boldsymbol{\sigma}_0)+\boldsymbol{f}^{b}=\mathbf{0} \tag{5.8.2}$$

It should be noted that significantly different scales, both spatially and temporally, are present in a complete healing process. The spatial lengths of a wound are usually very small (on the orders of millimetres). On the other hand, a full healing process is very slow, with the orders of days, weeks and sometimes months.

The tissue can well be represented by a viscous constitutive model (Oster et al. 1983). This is consistent with the quasi-static assumption of Equation (5.8.2), as the viscous forces govern the motion by being in equilibrium with the elastic forces.

A soft tissue typically consists of two main parts of the extracellular matrix (ECM) and cells. The extracellular matrix contains collagen fibres, which determine its main mechanical response. The overall response can be described by the Kelvin–Voight viscoelastic model (see Figure 5.66) (Wijn et al. 1981).

In this model, the resultant force is obtained from the sum of forces of components of the model. Therefore, the governing equation can be written as (Oster et al. 1983)

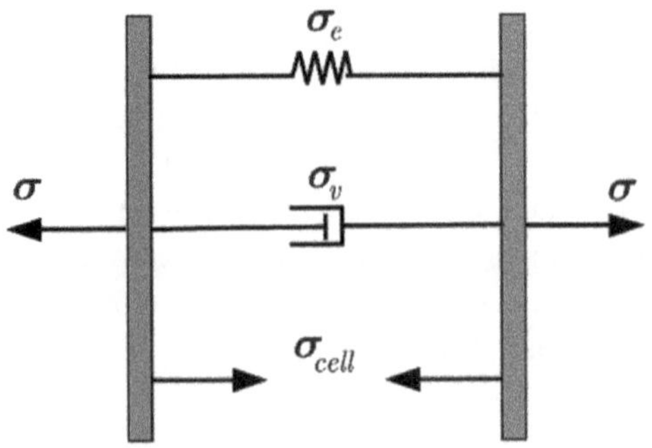

Figure 5.66 Force interactions in an element with cells and ECM. *Source:* Reproduced from Khaksar 2020/the University of Tehran.

$$\nabla \cdot \left[\boldsymbol{\sigma}_{\text{tissue}} + \boldsymbol{\sigma}_{\text{cell}} - \boldsymbol{\sigma}_0 \right] - \boldsymbol{f}^{\text{b}} = \mathbf{0} \tag{5.8.3}$$

or

$$\nabla \cdot \left[\boldsymbol{\sigma}_{\text{e}} + \boldsymbol{\sigma}_{\text{v}} + \boldsymbol{\sigma}_{\text{cell}} - \boldsymbol{\sigma}_0 \right] - \boldsymbol{f}^{\text{b}} = \mathbf{0} \tag{5.8.4}$$

5.8.2.1.1 Viscoelastic Stress (σ_{tissue})

The stress $\boldsymbol{\sigma}_{\text{tissue}}$ is caused by the reaction of the extracellular matrix to deformation and is composed of elastic and viscous terms (Oster et al. 1983):

$$\boldsymbol{\sigma}_{\text{e}} = \frac{E}{1+\nu}\left(\varepsilon + \frac{\nu}{1-2\nu}\varepsilon_v \mathbf{I}\right) \tag{5.8.5}$$

$$\boldsymbol{\sigma}_v = \mu_1 \frac{\partial \varepsilon}{\partial t} + \mu_2 \frac{\partial \varepsilon_v}{\partial t}\mathbf{I} \tag{5.8.6}$$

where ε_v is the volumetric strain,

$$\varepsilon_v = \nabla \cdot \boldsymbol{u} \tag{5.8.7}$$

and E, ν, μ_1 and μ_2 are the elasticity modulus, Poisson's ratio, the shear viscosity and the bulk viscosity of the intact (healthy) tissue.

5.8.2.1.2 Stress of the Cell Movements (σ_{cell})

The effect of cells in contraction during the healing is through the cell creep movement or, equivalently, through the corresponding cell tractions $\boldsymbol{\sigma}_{\text{cell}}$ on their surroundings. The cells attach themselves to the surrounding ECM and contract the healthy tissue towards the inside of the damaged area (Figure 5.67).

The stress/traction associated with the cells, $\boldsymbol{\sigma}_{\text{cell}}$, which should be applied on the ECM environment, can be determined from the experimentally derived relation by Bell et al.

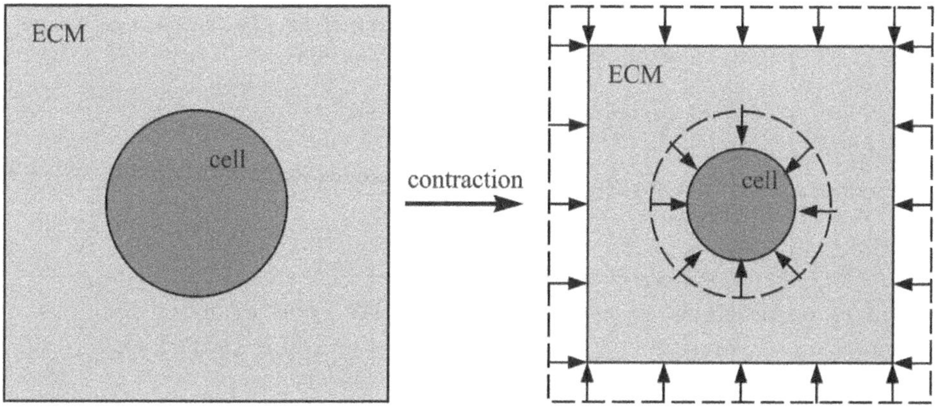

Figure 5.67 The state of internal cells in ECM before and after the contraction of ECM. *Source:* Reproduced from Khaksar 2020/the University of Tehran.

(1979) in terms of the cell and ECM densities (ρ_{cell} and ρ_{ECM}, respectively), the effect of myofibroblasts (ρ_{mfb}) (Olsen et al. 1995) and a stress/pressure function of volumetric strain ε_v (Oster et al. 1983; Tranquillo and Murray 1992),

$$\sigma_{\text{cell}} = p_{\text{cell}}(\varepsilon_v)\frac{\left(1+\zeta\rho_{\text{mfb}}\right)\rho_{\text{fb}}\rho_{\text{ECM}}}{R_\tau^2 + \rho_{\text{ECM}}^2}\mathbf{I} \tag{5.8.8}$$

with (Moreo et al. 2008)

$$p_{\text{cell}}(\varepsilon_v) = \begin{cases} K_{pas}\varepsilon_v, & \varepsilon_v < \varepsilon_v^1 \\ \dfrac{K_{act}\ p_{\max}}{K_{act}\ \varepsilon_v^1 - p_{\max}}\left(\varepsilon_v^1 - \varepsilon_v\right), & \varepsilon_v^1 \le \varepsilon_v \le \varepsilon_v^* \\ \dfrac{K_{act}\ p_{max}}{K_{act}\ \varepsilon_v^2 - p_{\max}}\left(\varepsilon_v^2 - \varepsilon_v\right), & \varepsilon_v^* < \varepsilon_v \le \varepsilon_v^2 \\ K_{pas}\varepsilon_v, & \varepsilon_v > \varepsilon_v^2 \end{cases} \tag{5.8.9}$$

$$\varepsilon_v^* = \frac{p_{\max}}{K_{act}} \tag{5.8.10}$$

where ζ is the myofibroblast enhancement of traction per fibroblasts density, R_τ is the traction inhibition collagen density, $p_{\max}$ is the maximum exerted stress of cells per unit of ECM, K_{pas} and K_{act} are the passive and active components of the volumetric stiffness of cell, respectively, and ε_v^1 and ε_v^2 are the shortening and stretching volumetric strains of filaments, respectively.

5.8.2.1.3 Volumetric Force ($\boldsymbol{f}^{\text{b}}$)

Volumetric forces $\boldsymbol{f}^{\text{b}}$ are created by the connection of the tissue from its surroundings tissues, which are assumed to be intact and healthy. The volumetric force acts opposite to the contraction. It can be computed from the density of the extracellular matrix ρ_{ECM} in a linear elastic form in terms of the deformation $\boldsymbol{u}$ (Oster et al. 1983; Olsen et al. 1995; Javierre et al. 2009):

$$\boldsymbol{f}^{b} = \rho_{ECM} s \boldsymbol{u} \tag{5.8.11}$$

where s is the so-called dermis attachment factor.

5.8.2.2 Diffusion of Densities of Cells and Growth Factor

Various factors such as fibroblast, myofibroblast and ECM densities affect the balances of concentrations and densities in the healing process. The governing equations for the distribution of variables in time are coupled and cannot be obtained independently, as they are all related to the deformation of the tissue.

For a given control volume dA (Figure 5.68), the governing diffusion equation for time variations of a specific variable Q can be expressed in terms of the flux $\boldsymbol{J}_Q$ and the rate of change of the variable in the control volume (f_Q),

$$\frac{\partial Q}{\partial t} = -\nabla \cdot \boldsymbol{J}_Q + f_Q \tag{5.8.12}$$

The governing equations used for the study of densities and concentrations are the corrected set of equations by Javierre et al. (2009).

5.8.2.2.1 Growth Factor Density ρ_{gf}

The diffusion equation for the growth factor density ρ_{gf} can be written as (Javierre et al. 2009)

$$\frac{\partial \rho_{gf}}{\partial t} = -\nabla \cdot \left(\boldsymbol{J}_{rdf} + \boldsymbol{J}_{pcf}\right) + \left(f_{pc} + f_{nd}\right) \tag{5.8.13}$$

where the four contributing terms are (Olsen et al. 1995; Sherratt et al. 1993)

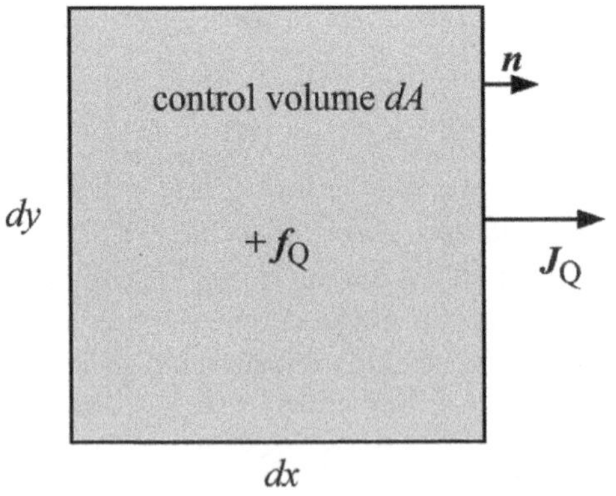

Figure 5.68 The control volume used in the diffusion equation. *Source:* Reproduced from Khaksar 2020/the University of Tehran.

$$\text{Random dispersal flux} = \boldsymbol{J}_{\mathrm{rdf}} = -D_{\mathrm{gf}} \boldsymbol{\nabla} \rho_{\mathrm{gf}} \tag{5.8.14}$$

$$\text{Passive convection flux} = \boldsymbol{J}_{\mathrm{pcf}} = \rho_{\mathrm{gf}} \frac{\partial \boldsymbol{u}}{\partial t} \tag{5.8.15}$$

$$\text{Production/consumption rate of chemicals} = f_{\mathrm{pc}} = k_{\mathrm{gf}} \frac{\left(\rho_{\mathrm{fb}} + \xi \rho_{\mathrm{mfb}}\right) \Gamma \rho_{\mathrm{gf}}}{C_{\mathrm{gf}} + \rho_{\mathrm{gf}}} \tag{5.8.16}$$

$$\text{Natural decay rate} = f_{\mathrm{nd}} = -d_{\mathrm{gf}} \rho_{\mathrm{gf}} \tag{5.8.17}$$

where D_{gf} is the growth factor diffusion rate, k_{gf} is the growth factor production rate per unit of cell density, ξ is the ratio of growth factor production rates of fibroblast to myofibroblast, C_{gf} is the Michaelis–Menten constant for the collagen production and d_{gf} is the growth factor deterioration rate.

5.8.2.2.2 *Collagen/ECM Density* ρ_{ECM}

The diffusion equation governing the collagen/ECM density ρ_{ECM} can be written as

$$\frac{\partial \rho_{\mathrm{ECM}}}{\partial t} = -\boldsymbol{\nabla} \cdot (\boldsymbol{J}_{\mathrm{pcf}}) + (f_{\mathrm{es}} + f_{\mathrm{ed}}) \tag{5.8.18}$$

where the three contributing terms are (Clark 1989; Olsen et al. 1995; Pierce et al. 1991)

$$\text{Passive convection flux} = \boldsymbol{J}_{\mathrm{pcf}} = \rho_{\mathrm{ECM}} \frac{\partial \boldsymbol{u}}{\partial t} \tag{5.8.19}$$

Growth factor enhanced synthesis by both cell types =

$$f_{\mathrm{es}} = \left(r_{\mathrm{ECM}} + \frac{r_{\mathrm{ECM}}^{\max} \rho_{\mathrm{gf}}}{C_{\mathrm{ECM}} + \rho_{\mathrm{gf}}} \right) \frac{\rho_{\mathrm{fb}} + \eta \rho_{\mathrm{mfb}}}{R_{\mathrm{ECM}}^2 + \rho_{\mathrm{ECM}}^2} \tag{5.8.20}$$

$$\text{Enzymatic degradation by both cell types} = f_{\mathrm{ed}} = -d_{\mathrm{ECM}} (\rho_{\mathrm{fb}} + \eta \rho_{\mathrm{mfb}}) \rho_{\mathrm{ECM}} \tag{5.8.21}$$

where r_{ECM} is the collagen production rate, $r_{\mathrm{ECM}}^{\max}$ is the maximal collagen production rate due to chemotaxis, C_{ECM} is the Michaelis–Menten constant for chemotaxis-induced collagen production, η is the collagen synthesis of fibroblasts to the myofibroblasts ratio, R_{ECM}^2 is the half-maximal collagen enhancement of the ECM, and d_{ECM} is the collagen deterioration rate per cell.

5.8.2.2.3 *Fibroblast Density* ρ_{fb}

The diffusion equation for the fibroblast density ρ_{fb} can be written as

$$\frac{\partial \rho_{\mathrm{fb}}}{\partial t} = -\boldsymbol{\nabla} \cdot (\boldsymbol{J}_{\mathrm{rdf}} + \boldsymbol{J}_{\mathrm{ctf}} + \boldsymbol{J}_{\mathrm{pcf}}) + (f_{\mathrm{m}} + f_{\mathrm{t}} + f_{\mathrm{r}} + f_{\mathrm{d}}) \tag{5.8.22}$$

where the seven contributing terms are (Olsen et al. 1995; Moreo et al. 2008; Javierre et al. 2009)

$$\text{Random dispersal flux} = \boldsymbol{J}_{\mathrm{rdf}} = -D_{\mathrm{fb}}\boldsymbol{\nabla}\rho_{\mathrm{fb}} \tag{5.8.23}$$

$$\text{Chemotaxis flux} = \boldsymbol{J}_{\mathrm{ctf}} = \frac{a_{\mathrm{fb}}}{\left(b_{\mathrm{fb}}+\rho_{\mathrm{gf}}\right)^2}\rho_{\mathrm{fb}}\boldsymbol{\nabla}\rho_{\mathrm{gf}} \tag{5.8.24}$$

$$\text{Passive convection flux} = \boldsymbol{J}_{\mathrm{pcf}} = \rho_{\mathrm{fb}}\frac{\partial \boldsymbol{u}}{\partial t} \tag{5.8.25}$$

$$\text{Mitosis} = f_{\mathrm{m}} = \left(r_{\mathrm{fb}} + \frac{r_{\mathrm{fb}}^{\max}\rho_{\mathrm{gf}}}{C_{1/2}+\rho_{\mathrm{gf}}}\right)\rho_{\mathrm{fb}}\left(1-\frac{\rho_{\mathrm{fb}}}{K}\right) \tag{5.8.26}$$

$$\text{Differentiation into myofibroblast} = f_{\mathrm{t}} = -\frac{k_{1_{\max}}\rho_{\mathrm{gf}}}{C_k+\rho_{\mathrm{gf}}}\frac{p_{\mathrm{cell}}\left(\varepsilon_v\right)}{\tau_d+p_{\mathrm{cell}}\left(\varepsilon_v\right)}\rho_{\mathrm{fb}} \tag{5.8.27}$$

$$\text{Reverse differentiation} = f_{\mathrm{r}} = k_2\rho_{\mathrm{mfb}} \tag{5.8.28}$$

$$\text{Death} = f_{\mathrm{d}} = -d_{\mathrm{fb}}\rho_{\mathrm{fb}} \tag{5.8.29}$$

where D_{fb} is the fibroblast diffusion rate, $a_{\mathrm{fb}}, b_{\mathrm{fb}}$ are the constants governing the chemotaxis rate of fibroblasts, r_{fb} is the fibroblasts proliferation rate, $r_{\mathrm{fb}}^{\max}$ is the maximum fibroblasts proliferation rate due to chemotaxis, $C_{1/2}$ is the Michaelis–Menten constant, K is the maximum fibroblast capacity of the dermis, $k_{1_{\max}}$ is the maximal fibroblast conversion rate, C_k is the Michaelis–Menten constant for the conversion of fibroblasts into myofibroblasts, τ_d is the half-maximum mechanical improvement of fibroblast conversion, k_2 is the inverse conversion rate of fibroblasts, d_{fb} is the fibroblasts death rate and $p_{\mathrm{cell}}(\varepsilon_v)$ is defined in Equation (5.8.9).

5.8.2.2.4 Myofibroblast Density ρ_{mfb}

The diffusion equation for the myofibroblast density ρ_{mfb} can be written as

$$\frac{\partial \rho_{\mathrm{mfb}}}{\partial t} = -\boldsymbol{\nabla}\cdot(\boldsymbol{J}_{\mathrm{pcf}}) + (f_{\mathrm{m}}+f_{\mathrm{t}}+f_{\mathrm{r}}+f_{\mathrm{d}}) \tag{5.8.30}$$

where the five contributing terms are (Javierre et al. 2009)

$$\text{Passive convection flux} = \boldsymbol{J}_{\mathrm{pcf}} = \rho_{\mathrm{mfb}}\frac{\partial \boldsymbol{u}}{\partial t} \tag{5.8.31}$$

$$\text{Mitosis} = f_{\mathrm{m}} = \epsilon_{\mathrm{r}}\left(r_{fb} + \frac{r_{\mathrm{fb}}^{\max}\rho_{\mathrm{gf}}}{C_{1/2}+\rho_{\mathrm{gf}}}\right)\rho_{\mathrm{mfb}}\left(1-\frac{\rho_{\mathrm{mfb}}}{K}\right) \tag{5.8.32}$$

$$\text{Differentiation into fibroblast} = f_{\mathrm{t}} = \frac{k_{1_{\max}}\rho_{\mathrm{gf}}}{C_k+\rho_{\mathrm{gf}}}\frac{p_{\mathrm{cell}}\left(\varepsilon_v\right)}{\tau_d+p_{\mathrm{cell}}\left(\varepsilon_v\right)}\rho_{\mathrm{fb}} \tag{5.8.33}$$

$$\text{Reverse differentiation} = f_{\mathrm{r}} = -k_2 \rho_{\mathrm{mfb}} \tag{5.8.34}$$

$$\text{Death} = f_{\mathrm{d}} = -d_{\mathrm{mfb}} \rho_{\mathrm{mfb}} \tag{5.8.35}$$

and ϵ_r is the ratio of the logistic growth factors of fibroblasts to myofibroblasts, d_{m} is the myofibroblasts death rate and $p_{\mathrm{cell}}(\varepsilon_v)$ is defined in Equation (5.8.9).

5.8.2.3 Set of Coupled Equations

The full set of five coupled differential equations in terms of ρ_{fb}, ρ_{mfb}, ρ_{ECM}, ρ_{gf} and $\boldsymbol{u}$ are summarized as follows.

5.8.2.3.1 Fibroblast and Myofibroblast Density ρ_{fb} and ρ_{mfb}

$$\begin{aligned}\frac{\partial \rho_{\mathrm{fb}}}{\partial t} &= \left\{ \nabla \cdot \left(D_{\mathrm{fb}} \nabla \rho_{\mathrm{fb}} - \frac{a_{\mathrm{fb}}}{\left(b_{\mathrm{fb}} + \rho_{\mathrm{gf}}\right)^2} \rho_{\mathrm{fb}} \nabla \rho_{\mathrm{gf}} - \rho_{\mathrm{fb}} \frac{\partial \boldsymbol{u}}{\partial t} \right) \right\} \\ &+ \left\{ \left(r_{\mathrm{fb}} + \frac{r_{\mathrm{fb}}^{\max} \rho_{\mathrm{gf}}}{C_{1/2} + \rho_{\mathrm{gf}}} \right) \rho_{\mathrm{fb}} \left(1 - \frac{\rho_{\mathrm{fb}}}{K} \right) - \frac{k_{1_{\max}} \rho_{\mathrm{gf}}}{C_k + \rho_{\mathrm{gf}}} \frac{p_{\mathrm{cell}}(\varepsilon_v)}{\tau_d + p_{\mathrm{cell}}(\varepsilon_v)} \rho_{\mathrm{fb}} + k_2 \rho_{\mathrm{mfb}} - d_{\mathrm{fb}} \rho_{\mathrm{fb}} \right\}\end{aligned} \tag{5.8.36}$$

$$\frac{\partial \rho_{\mathrm{mfb}}}{\partial t} = \left\{ \nabla \cdot \left(-\rho_{\mathrm{mfb}} \frac{\partial \boldsymbol{u}}{\partial t} \right) \right\} + \left\{ \begin{aligned} &\epsilon_r \left(r_{fb} + \frac{r_{\mathrm{fb}}^{\max} \rho_{\mathrm{gf}}}{C_{1/2} + \rho_{\mathrm{gf}}} \right) \rho_{\mathrm{mfb}} \left(1 - \frac{\rho_{\mathrm{mfb}}}{K} \right) \\ &+ \frac{k_{1_{\max}} \rho_{\mathrm{gf}}}{C_k + \rho_{\mathrm{gf}}} \frac{p_{\mathrm{cell}}(\varepsilon_v)}{\tau_d + p_{\mathrm{cell}}(\varepsilon_v)} \rho_{\mathrm{fb}} - k_2 \rho_{\mathrm{mfb}} - d_{\mathrm{mfb}} \rho_{\mathrm{mfb}} \end{aligned} \right\} \tag{5.8.37}$$

where the constants are defined in Table 5.15.

5.8.2.3.2 Collagen/ECM Density ρ_{ECM} and Growth Factor Concentration ρ_{gf}

$$\begin{aligned}\frac{\partial \rho_{\mathrm{ECM}}}{\partial t} &= -\nabla \cdot \left(\rho_{\mathrm{ECM}} \frac{\partial \boldsymbol{u}}{\partial t} \right) + \left(r_{\mathrm{ECM}} + \frac{r_{\mathrm{ECM}}^{\max} \rho_{\mathrm{gf}}}{C_{\mathrm{ECM}} + \rho_{\mathrm{gf}}} \right) \frac{\rho_{\mathrm{fb}} + \eta \rho_{\mathrm{mfb}}}{R_{\mathrm{ECM}}^2 + \rho_{\mathrm{ECM}}^2} \\ &\quad - d_{\mathrm{ECM}} (\rho_{\mathrm{fb}} + \eta \rho_{\mathrm{mfb}}) \rho_{\mathrm{ECM}}\end{aligned} \tag{5.8.38}$$

$$\frac{\partial \rho_{\mathrm{gf}}}{\partial t} = \left\{ \nabla \cdot \left(D_{\mathrm{gf}} \nabla \rho_{\mathrm{gf}} - \rho_{\mathrm{gf}} \frac{\partial \boldsymbol{u}}{\partial t} \right) \right\} + \left\{ k_g \frac{(\rho_{\mathrm{fb}} + \xi \rho_{\mathrm{mfb}}) \Gamma \rho_{\mathrm{gf}}}{C_{\mathrm{gf}} + \rho_{\mathrm{gf}}} - d_{\mathrm{gf}} \rho_{\mathrm{gf}} \right\} \tag{5.8.39}$$

where the constants are defined in Table 5.16.

Table 5.15 Constants of fibroblast and myofibroblast density diffusion equations.

Parameter	Unit	Description
D_{fb}	cm^2/day	Fibroblast diffusion rate
a_{fb}	μg/cm day	Constants governing the chemotaxis rate of fibroblasts
b_{fb}	μg/cm^3	
r_{fb}	day^{-1}	Fibroblasts proliferation rate
r_{fb}^{max}	day^{-1}	Maximum fibroblasts proliferation rate due to chemotaxis
$C_{1/2}$	μg/cm^3	Michaelis–Menten constant
K	cells/cm^3	Maximal fibroblast capacity of the dermis
$k_{1_{max}}$	day^{-1}	Maximal fibroblast conversion rate
C_k	μg/cm^3	Michaelis–Menten constant for the conversion of fibroblasts into myofibroblasts
k_2	day^{-1}	Inverse conversion rate of fibroblasts
d_{fb}	day^{-1}	Fibroblasts death rate
ϵ_r	-	The ratio of the logistic growth factor of fibroblasts to myofibroblasts
τ_d	N g/cm^2 cell	Half-maximal mechanical improvement of fibroblast conversion
p_{max}	N μg/cm^2 cell	Maximal exerted stress of cells per unit of ECM
K_{pas}	N μg/cm^2 cell	Cell's passive components volumetric stiffness
K_{act}	N g/cm^2 cell	Cell's active components volumetric stiffness
ε_v^1	–	Shortening volumetric strain of filaments
ε_v^2	–	Stretching volumetric strain of filaments
d_{mfb}	day^{-1}	Myofibroblasts death rate

Table 5.16 Constants of collagen/ECM density ρ_{ECM} and growth factor concentration ρ_{gf} diffusion equations.

Parameter	Unit	Description
r_{ECM}	$μg^3/cm^6$ cell day	Collagen production rate
r_{ECM}^{max}	$μg^3/cm^6$ cell day	Maximal collagen production rate due to chemotaxis
C_{ECM}	μg/cm^3	Michaelis–Menten constant for chemotaxis induced collagen production
η	-	Collagen synthesis of fibroblasts to myofibroblasts ratio
R_{ECM}	μg/cm^3	Half-maximal collagen enhancement of the ECM
d_{ECM}	cm^3/cell day	Collagen deterioration rate per cell
D_{gf}	cm^2/day	Growth factor diffusion rate
k_{gf}	cm^3/cell day	Growth factor production rate per unit of cell density
ξ	-	Ratio of growth factor production rates of fibroblast to myofibroblast
C_{gf}	μg/cm^3	Michaelis–Menten constant for collagen production
d_{gf}	day^{-1}	Growth factor deterioration rate

5.8.2.3.3 Equilibrium Equation u

$$\nabla\cdot\left[\begin{array}{l}\left[\mu_1\frac{\partial\varepsilon}{\partial t}+\mu_2\frac{\partial\varepsilon_v}{\partial t}\mathbf{I}\right]+\left[\frac{E}{1+\nu}(\varepsilon+\frac{\nu}{1-2\nu}\varepsilon_v\mathbf{I})\right]\\+\left[p_{\text{cell}}(\varepsilon_v)\frac{\left(1+\zeta\rho_{\text{mfb}}\right)\rho_{\text{fb}}\rho_{\text{ECM}}}{R_\tau^2+\rho_{\text{ECM}}^2}\mathbf{I}-\boldsymbol{\sigma}_0\right]\end{array}\right]-\rho_{\text{ECM}}s\boldsymbol{u}=0 \tag{5.8.40}$$

where the constants are defined in Table 5.17.

5.8.3 Solution Procedure

5.8.3.1 Solution in Time

The generalized trapezoidal rule can be adopted to formulate the time integration procedure of the coupled set of equations (Chawla et al. 1996). Denoting each variable H at time i by H^i, the recursive relation to compute the rate ΔH^i of variations of H^i in time can be defined by

$$\Delta H^i=\left(\frac{\partial H}{\partial t}\right)^i=\frac{H^i-\left(H^{i-1}+\Delta t\left(1-\alpha\right)V_H{}^{i-1}\right)}{\alpha\Delta t} \tag{5.8.41}$$

where $a=0.5$ for the trapezoidal rule:

$$\Delta H^i=(\frac{\partial H}{\partial t})^i=2\times\frac{H^i-H^{i-1}}{\Delta t}-\ \Delta H^{i-1} \tag{5.8.42}$$

The time marching procedure is simple and requires definitions of the initial distribution of variables ρ_{fb}, ρ_{mfb}, ρ_{ECM}, ρ_{gf} and $\boldsymbol{u}$, as presented in Table 5.18.

Table 5.17 Constants of the equilibrium equation.

Parameter	Unit	Description
μ_1	N day/cm^2	Shear viscosity of healthy tissue
μ_2	N day/cm^2	Bulk viscosity of healthy tissue
E	N/cm^2	Healthy tissue's Young's modulus
ν	-	Poison's ratio of healthy tissue
ζ	cm^3/cell	Ratio of myofibroblast enhancement of traction per fibroblasts density
R_τ	μg/cm^3	Traction inhibition collagen density
s	N/cm μg	Dermis attachment factor

Table 5.18 Initial values of the main variables.

Parameter	Unit	Description
$\boldsymbol{u}^0$	cm	Initial deformation field
ρ_{gf}^0	μg/cm^3	Initial growth factor concentration of the healthy tissue
ρ_{ECM}^0	μg/cm^3	Initial collagen (ECM) density in healthy skin
ρ_{fb}^0	cells/cm^3	Initial fibroblasts density in healthy tissue
ρ_{mfb}^0	cells /cm^3	Initial myofibroblasts density in healthy skin
ρ_{ECM}^{0d}	μg/cm^3	Initial collagen (ECM) density within the damaged region

5.8.3.2 Finite Element Solution (in Each Timestep)

The governing equations are in the two main general strong forms of the equilibrium and diffusion equations:

$$\nabla \cdot \boldsymbol{\sigma} + \boldsymbol{f}^{\text{b}} = 0 \tag{5.8.43}$$

$$\frac{\partial Q}{\partial t} = -\nabla \cdot \boldsymbol{J}_Q + f_Q \tag{5.8.44}$$

which can be expressed in the following weak forms:

$$\int_\Omega \boldsymbol{\sigma} : \frac{1}{2}(\nabla \boldsymbol{v} + \nabla \boldsymbol{v}^T) d\Omega = \int_\Omega (\boldsymbol{f}^{\text{b}} \cdot \boldsymbol{v}) d\Omega + \int_{\Gamma_t} (\boldsymbol{f}^{\text{t}} \cdot \boldsymbol{v}) d\Gamma \tag{5.8.45}$$

$$\int_\Omega \frac{\partial Q}{\partial t} v_Q \, d\Omega - \int_\Omega \boldsymbol{J}_Q \cdot \nabla v_Q d\Omega = \int_\Omega (f_Q v_Q) d\Omega - \int_{\Gamma_Q} (\boldsymbol{J}_Q \cdot \boldsymbol{n} v_Q) d\Gamma \tag{5.8.46}$$

where $\boldsymbol{v}$ is the weight function (virtual displacement) and v_Q is an arbitrary function (virtual density).

The finite element shape functions $\mathbf{N}_{\boldsymbol{u}}(\mathbf{x})$ and $\mathbf{N}_Q(\mathbf{x})$ can be used to discretize the field variables $\boldsymbol{u}(\mathbf{x}.t)$ and $Q(\mathbf{x}.t)$ to the nodal values $\boldsymbol{U}(t)$ and $\boldsymbol{Q}(t)$, respectively:

$$\boldsymbol{u}(\mathbf{x}.t) = \mathbf{N}_{\boldsymbol{u}}(\mathbf{x}) U(t) \tag{5.8.47}$$

$$Q(\mathbf{x}.t) = \mathbf{N}_Q(\mathbf{x}) \boldsymbol{Q}(t) \tag{5.8.48}$$

The final discretized form of the coupled set of equations at a time t can be written in a form suitable for nonlinear solution procedures:

$$F_Q = F_Q^{\text{int}} - F_Q^{\text{ext}} = 0 \tag{5.8.49}$$

Using the final set of strong form Equations (5.8.36) to (5.8.40) and assuming the domain boundaries are far enough to ignore the boundary terms, the internal force vectors can be expressed as

$$\boldsymbol{F}_{\rho_{\text{fb}}}^{\text{int}} = \int_{\Omega} \boldsymbol{N}_{\text{fb}}^{\text{T}} \frac{\partial \rho_{\text{fb}}}{\partial t} d\Omega + \int_{\Omega} \nabla \boldsymbol{N}_{\text{fb}}^{\text{T}} \left[D_{\text{fb}} \boldsymbol{\nabla} \rho_{\text{fb}} - \frac{a_{\text{fb}}}{\left(b_{\text{fb}} + \rho_{\text{gf}}\right)^2} \rho_{\text{fb}} \boldsymbol{\nabla} \rho_{\text{gf}} - \rho_{\text{fb}} \frac{\partial \boldsymbol{u}}{\partial t} \right] d\Omega \quad (5.8.50)$$

$$\boldsymbol{F}_{\rho_{\text{mfb}}}^{\text{int}} = \int_{\Omega} \boldsymbol{N}_{\text{mfb}}^{\text{T}} \frac{\partial \rho_{\text{mfb}}}{\partial t} d\Omega - \int_{\Omega} \nabla \boldsymbol{N}_{\text{mfb}}^{\text{T}} \rho_{\text{mfb}} \frac{\partial \boldsymbol{u}}{\partial t} d\Omega \quad (5.8.51)$$

$$\boldsymbol{F}_{\rho_{\text{ECM}}}^{\text{int}} = \int_{\Omega} \boldsymbol{N}_{\text{ECM}}^{\text{T}} \frac{\partial \rho_{\text{ECM}}}{\partial t} d\Omega - \int_{\Omega} \nabla \boldsymbol{N}_{\text{ECM}}^{\text{T}} \rho_{\text{ECM}} \frac{\partial \boldsymbol{u}}{\partial t} d\Omega \quad (5.8.52)$$

$$\boldsymbol{F}_{\rho_{\text{gf}}}^{\text{int}} = \int_{\Omega} \boldsymbol{N}_{\text{gf}}^{\text{T}} \frac{\partial \rho_{\text{gf}}}{\partial t} d\Omega + \int_{\Omega} \nabla \boldsymbol{N}_{\text{gf}}^{\text{T}} \left[D_{\text{gf}} \boldsymbol{\nabla} \rho_{\text{gf}} - \rho_{\text{gf}} \frac{\partial \boldsymbol{u}}{\partial t} \right] d\Omega \quad (5.8.53)$$

$$\boldsymbol{F}_{u}^{\text{int}} = \int_{\Omega} \boldsymbol{B}_{u}^{\text{T}} \left[D_e \frac{\rho}{\rho_0} B_u U + D_v B_u \dot{U} + p_{\text{cell}}(\varepsilon_v) \frac{\left(1 + \zeta \rho_{\text{mfb}}\right) \rho_{\text{fb}} \rho_{\text{ECM}}}{R_\tau^2 + \rho_{\text{ECM}}^2} \mathbf{I} \right] d\Omega \quad (5.8.54)$$

and the external force vectors:

$$\boldsymbol{F}_{\rho_{\text{fb}}}^{\text{ext}} = \int_{\Omega} \boldsymbol{N}_{\text{fb}}^{\text{T}} \left[\begin{array}{l} \left(r_{\text{fb}} + \dfrac{r_{\text{fb}}^{\max} \rho_{\text{gf}}}{C_{1/2} + \rho_{\text{gf}}} \right) \rho_{\text{fb}} \left(1 - \dfrac{\rho_{\text{fb}}}{K} \right) - \\ \dfrac{k_{1.\max} \rho_{\text{gf}}}{C_k + \rho_{\text{gf}}} \dfrac{p_{\text{cell}}\left(\varepsilon_v\right)}{\tau_d + p_{\text{cell}}\left(\varepsilon_v\right)} \rho_{\text{fb}} + k_2 \rho_{\text{mfb}} \end{array} \right] d\Omega - \int_{\Omega} \boldsymbol{N}_{\text{fb}}^{\text{T}} d_{\text{fb}} \rho_{\text{fb}} d\Omega \quad (5.8.55)$$

$$\boldsymbol{F}_{\rho_{\text{mfb}}}^{\text{ext}} = \int_{\Omega} \boldsymbol{N}_{\text{mfb}}^{\text{T}} \left[\begin{array}{l} \varepsilon_r \left(r_{\text{fb}} + \dfrac{r_{\text{fb}}^{\max} \rho_{\text{gf}}}{C_{1/2} + \rho_{\text{gf}}} \right) \rho_{\text{mfb}} \left(1 - \dfrac{\rho_{\text{mfb}}}{K} \right) \\ + \dfrac{k_{1.\max} \rho_{\text{gf}}}{C_k + \rho_{\text{gf}}} \dfrac{p_{\text{cell}}\left(\varepsilon_v\right)}{\tau_d + p_{\text{cell}}\left(\varepsilon_v\right)} \rho_{\text{fb}} \end{array} \right] d\Omega - \int_{\Omega} \boldsymbol{N}_{\text{mfb}}^{\text{T}} \left[\begin{array}{l} k_2 \rho_{\text{mfb}} \\ + d_{\text{mfb}} \rho_{\text{mfb}} \end{array} \right] d\Omega \quad (5.8.56)$$

$$\boldsymbol{F}_{\rho_{\text{ECM}}}^{\text{ext}} = \int_{\Omega} \boldsymbol{N}_{\text{ECM}}^{\text{T}} \left[\left(r_{\text{ECM}} + \frac{r_{\text{ECM}}^{\max} \rho_{\text{gf}}}{C_\rho + \rho_{\text{gf}}} \right) \frac{\rho_{\text{fb}} + \eta \rho_{\text{mfb}}}{R_{\text{ECM}}^2 + \rho_{\text{ECM}}^2} - d_{\text{ECM}} (\rho_{\text{fb}} + \eta \rho_{\text{mfb}}) \rho_{\text{ECM}} \right] d\Omega \quad (5.8.57)$$

$$\boldsymbol{F}_{\rho_{\text{gf}}}^{\text{ext}} = \int_{\Omega} \boldsymbol{N}_{\text{gf}}^{\text{T}} \left[k_{gf} \frac{\left(\rho_{\text{fb}} + \xi \rho_{\text{mfb}}\right) \Gamma \rho_{\text{gf}}}{C_{\text{gf}} + \rho_{\text{gf}}} - d_{\text{gf}} \rho_{\text{gf}} \right] d\Omega \quad (5.8.58)$$

$$\boldsymbol{F}_{u}^{\text{ext}} = -\int_{\Omega} \boldsymbol{N}_{u}^{\text{T}} \rho_{\text{ECM}} \boldsymbol{s} \boldsymbol{u} d\Omega \quad (5.8.59)$$

The final discretized nonlinear set of equations becomes

$$\begin{Bmatrix} -\boldsymbol{F}_{\rho_{\mathrm{fb}}} \\ -\boldsymbol{F}_{\rho_{\mathrm{mfb}}} \\ -\boldsymbol{F}_{\rho_{\mathrm{ECM}}} \\ -\boldsymbol{F}_{\rho_{\mathrm{gf}}} \\ -\boldsymbol{F}_{u} \end{Bmatrix} = \begin{Bmatrix} \boldsymbol{F}^{\mathrm{int}}_{\rho_{\mathrm{fb}}} \\ \boldsymbol{F}^{\mathrm{int}}_{\rho_{\mathrm{mfb}}} \\ \boldsymbol{F}^{\mathrm{int}}_{\rho_{\mathrm{ECM}}} \\ \boldsymbol{F}^{\mathrm{int}}_{\rho_{\mathrm{gf}}} \\ \boldsymbol{F}^{\mathrm{int}}_{u} \end{Bmatrix} - \begin{Bmatrix} \boldsymbol{F}^{\mathrm{ext}}_{\rho_{\mathrm{fb}}} \\ \boldsymbol{F}^{\mathrm{ext}}_{\rho_{\mathrm{mfb}}} \\ \boldsymbol{F}^{\mathrm{ext}}_{\rho_{\mathrm{ECM}}} \\ \boldsymbol{F}^{\mathrm{ext}}_{\rho_{\mathrm{gf}}} \\ \boldsymbol{F}^{\mathrm{ext}}_{u} \end{Bmatrix} = \begin{Bmatrix} \mathbf{0} \\ \mathbf{0} \\ \mathbf{0} \\ \mathbf{0} \\ \mathbf{0} \end{Bmatrix} \tag{5.8.60}$$

which should be solved by a proper nonlinear solver.

5.8.3.3 Nonlinear Solution

The iterative Newton–Raphson nonlinear solution for a general form of nonlinear equation $F(x) = 0$ can be expressed as (at the iteration number n)

$$\frac{dF(x_n)}{dx}\{x_{n+1} - x_n\} = -F(x_n) \tag{5.8.61}$$

The iterative solution continues until a specific convergence criterion is met.

Adopting the same procedure for the coupled set of Equation (5.8.60) leads to

$$\begin{bmatrix} \frac{\partial \boldsymbol{F}_{\rho_{\mathrm{fb}}}}{\partial \rho_{\mathrm{fb}}} & \frac{\partial \boldsymbol{F}_{\rho_{\mathrm{fb}}}}{\partial \rho_{\mathrm{mfb}}} & \frac{\partial \boldsymbol{F}_{\rho_{\mathrm{fb}}}}{\partial \rho_{\mathrm{ECM}}} & \frac{\partial \boldsymbol{F}_{\rho_{\mathrm{fb}}}}{\partial \rho_{\mathrm{gf}}} & \frac{\partial \boldsymbol{F}_{\rho_{\mathrm{fb}}}}{\partial \boldsymbol{u}} \\ \frac{\partial \boldsymbol{F}_{\rho_{\mathrm{mfb}}}}{\partial \rho_{\mathrm{fb}}} & \frac{\partial \boldsymbol{F}_{\rho_{\mathrm{mfb}}}}{\partial \rho_{\mathrm{mfb}}} & \frac{\partial \boldsymbol{F}_{\rho_{\mathrm{mfb}}}}{\partial \rho_{\mathrm{ECM}}} & \frac{\partial \boldsymbol{F}_{\rho_{\mathrm{mfb}}}}{\partial \rho_{\mathrm{gf}}} & \frac{\partial \boldsymbol{F}_{\rho_{\mathrm{mfb}}}}{\partial \boldsymbol{u}} \\ \frac{\partial \boldsymbol{F}_{\rho_{\mathrm{ECM}}}}{\partial \rho_{\mathrm{fb}}} & \frac{\partial \boldsymbol{F}_{\rho_{\mathrm{ECM}}}}{\partial \rho_{\mathrm{mfb}}} & \frac{\partial \boldsymbol{F}_{\rho_{\mathrm{ECM}}}}{\partial \rho_{\mathrm{ECM}}} & \frac{\partial \boldsymbol{F}_{\rho_{\mathrm{ECM}}}}{\partial \rho_{\mathrm{gf}}} & \frac{\partial \boldsymbol{F}_{\rho_{\mathrm{ECM}}}}{\partial \boldsymbol{u}} \\ \frac{\partial \boldsymbol{F}_{\rho_{\mathrm{gf}}}}{\partial \rho_{\mathrm{fb}}} & \frac{\partial \boldsymbol{F}_{\rho_{\mathrm{gf}}}}{\partial \rho_{\mathrm{mfb}}} & \frac{\partial \boldsymbol{F}_{\rho_{\mathrm{gf}}}}{\partial \rho_{\mathrm{ECM}}} & \frac{\partial \boldsymbol{F}_{\rho_{\mathrm{gf}}}}{\partial \rho_{\mathrm{gf}}} & \frac{\partial \boldsymbol{F}_{\rho_{\mathrm{gf}}}}{\partial \boldsymbol{u}} \\ \frac{\partial \boldsymbol{F}_{u}}{\partial \rho_{\mathrm{fb}}} & \frac{\partial \boldsymbol{F}_{u}}{\partial \rho_{\mathrm{mfb}}} & \frac{\partial \boldsymbol{F}_{u}}{\partial \rho_{\mathrm{ECM}}} & \frac{\partial \boldsymbol{F}_{u}}{\partial \rho_{\mathrm{gf}}} & \frac{\partial \boldsymbol{F}_{u}}{\partial \boldsymbol{u}} \end{bmatrix} \begin{Bmatrix} \delta\rho_{\mathrm{fb}} \\ \delta\rho_{\mathrm{mfb}} \\ \delta\rho_{\mathrm{ECM}} \\ \delta\rho_{\mathrm{gf}} \\ \delta\boldsymbol{u} \end{Bmatrix} = \begin{Bmatrix} -\boldsymbol{F}_{\rho_{\mathrm{fb}}} \\ -\boldsymbol{F}_{\rho_{\mathrm{mfb}}} \\ -\boldsymbol{F}_{\rho_{\mathrm{ECM}}} \\ -\boldsymbol{F}_{\rho_{\mathrm{gf}}} \\ -\boldsymbol{F}_{u} \end{Bmatrix} \tag{5.8.62}$$

5.8.4 Numerical Analysis

Real wounds in skin usually occur in various shapes. First, a simple case of a strip wound is considered. Then, a circular damaged skin is simulated. The material constants for the governing healing equations are defined in Table 5.19.

The initial distributions of densities of fibroblast, collagen (ECM) and growth factors are presented in Figure 5.69.

5.8.4.1 Investigating the Planar Wound

In the first simulation, a simple rectangular damaged skin is considered, as depicted in Figure 5.70. It is sufficient to model only a strip from the specimen, as illustrated in Figure 5.71. The size of the intact tissue is assumed to be four times the size of the wound, long enough to neglect the effects of far boundary conditions.

Table 5.19 Material constants for numerical simulations.

Parameter	Unit	Value	References
D_{fb}	cm^2/day	2e-2	Ghosh et al. (2007)
a_{fb}	μg/cm day	4e-4	Javierre et al. (2009)
b_{fb}	μg/cm^3	2e-3	Javierre et al. (2009)
r_{fb}	day^{-1}	0.832	Ghosh et al. (2007)
r_{fb}^{max}	day^{-1}	0.3	Javierre et al. (2009)
$C_{1/2}$	μg/cm^3	1e-2	Olsen et al. (1995)
K	cells/cm^3	1e7	Olsen et al. (1995)
$k_{1_{max}}$	day^{-1}	0.8	Javierre et al. (2009)
C_k	μg/cm^3	1e-2	Javierre et al. (2009)
k_2	day^{-1}	0.693	Javierre et al. (2009)
d_{mfb}	day^{-1}	0.831	Javierre et al. (2009)
ε_r	-	0.5	Olsen et al. (1995)
τ_d	N μg/cm^2 cell	1e1	Javierre et al. (2009)
p_{max}	N μg/cm^2 cell	1e2	Javierre et al. (2009)
K_{pas}	N μg/cm^2 cell	2e1	Moreo et al. (2008)
K_{act}	N g/cm^2 cell	1e-4	Moreo et al. (2008)
ε_v^1	-	−0.6	Javierre et al. (2009)
ε_v^2	-	0.5	Moreo et al. (2008)
d_{mfb}	day^{-1}	2.1e-2	Javierre et al. (2009)
r_{ECM}	μg^3/cm^6 cell day	7.59e8	Javierre et al. (2009)
r_{ECM}^{max}	μg^3/cm^6 cell day	7.59e9	Olsen et al. (1995)
C_{ECM}	μg/cm^3	1e-3	Olsen et al. (1995)
η	-	2	Olsen et al. (1995)
R_{ECM}	μg/cm^3	3e5	Olsen et al. (1995)
d_{ECM}	cm^3/cell day	7.59e-8	Olsen et al. (1995)
D_{gf}	cm^2/day	5e-2	Olsen et al. (1995)
k_{gf}	cm^3/cell day	7.5e-6	Javierre et al. (2009)
ξ	-	1	Olsen et al. (1995)
C_{gf}	μg/cm^3	1e-2	Olsen et al. (1995)
d_{gf}	day^{-1}	0.462	Javierre et al. 2009)
μ_1	N day/cm^2	200	Javierre et al. (2009)
μ_2	N day/cm^2	200	Javierre et al. (2009)
E	N/cm^2	33.4	Khatyr et al. (2004)
ν	-	0.3	Khatyr et al. (2004)
ζ	cm^3/cell	1e-3	Olsen et al. (1995)

(Continued)

Table 5.19 (Continued)

Parameter	Unit	Value	References
R_T	μg/cm^3	5e2	Olsen et al. (1995)
s	N/cm μg	5e-4	Javierre et al. 2009)
$\boldsymbol{u}^0$	Cm	0	Khaksar (2020)
ρ_{gf}^0	μg/cm^3	1e-2	Olsen et al. (1995)
ρ_{ECM}^0	μg/cm^3	1e5	Olsen et al. (1995)
ρ_{fb}^0	cells/cm^3	1e4	Olsen et al. (1995)
ρ_{mfb}^0	μg/cm^3	0	Khaksar (2020)
ρ_{ECM}^{0d}	μg/cm^3	1e3	Olsen et al. (1995)

Source: Reproduced from Khaksar 2020/University of Tehran.

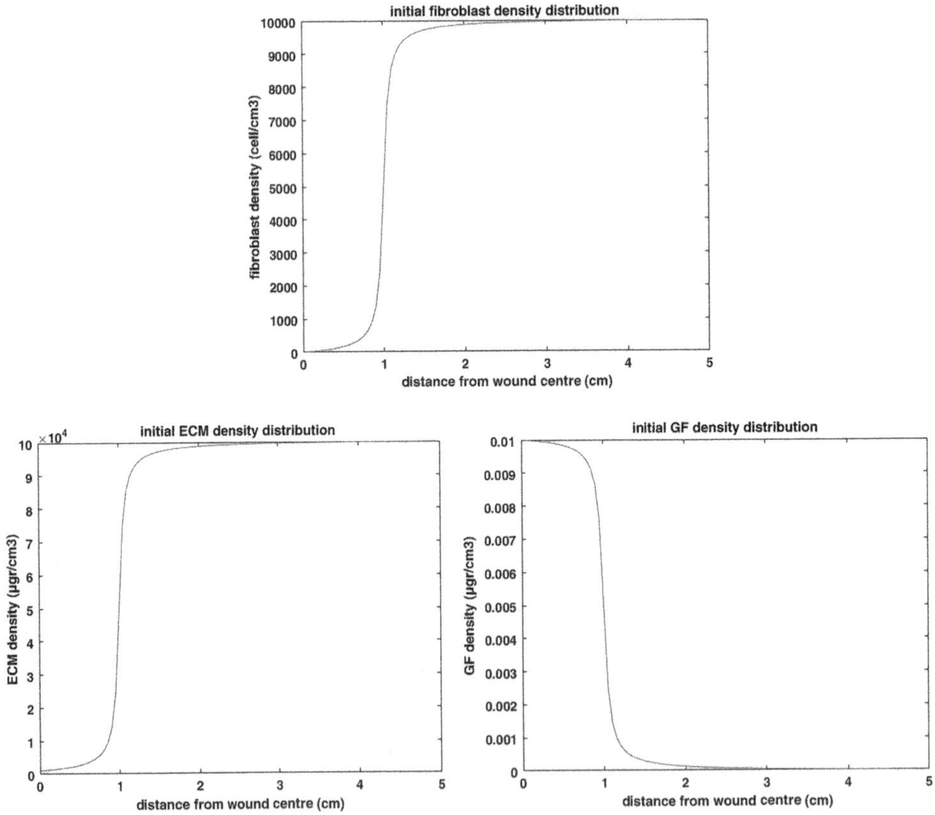

Figure 5.69 Initial distributions of densities of fibroblast, collagen (ECM) and growth factors. *Source:* Reproduced from Khaksar 2020/the University of Tehran.

5.8.4.1.1 *Changes of Fibroblast Density Over Time*

In the first step of tissue healing, fibroblasts are not present at the site of the injury. During the contraction, fibroblasts move from the surroundings into the injured region and gradually grow in numbers.

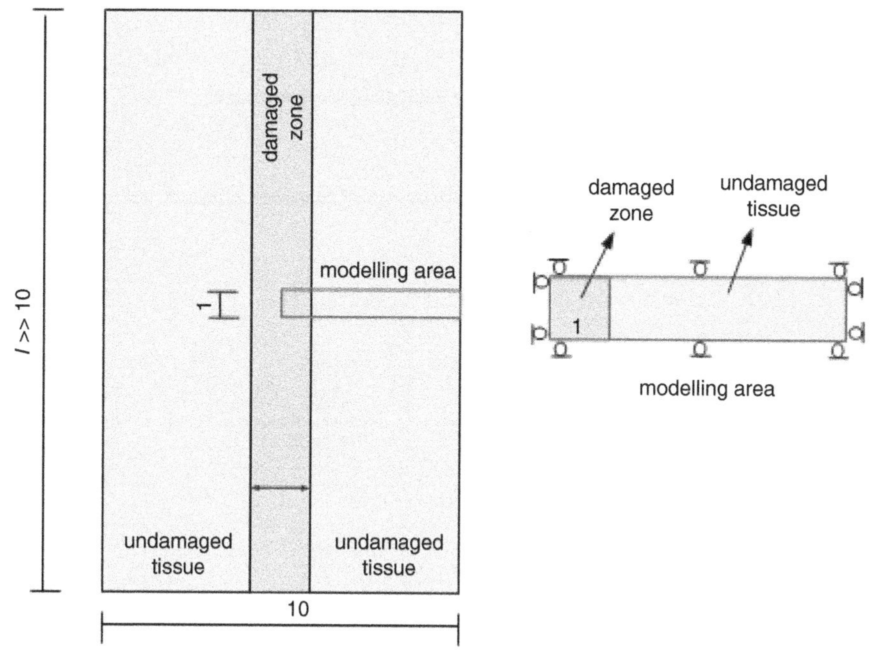

Figure 5.70 A planar strip of wounded skin.

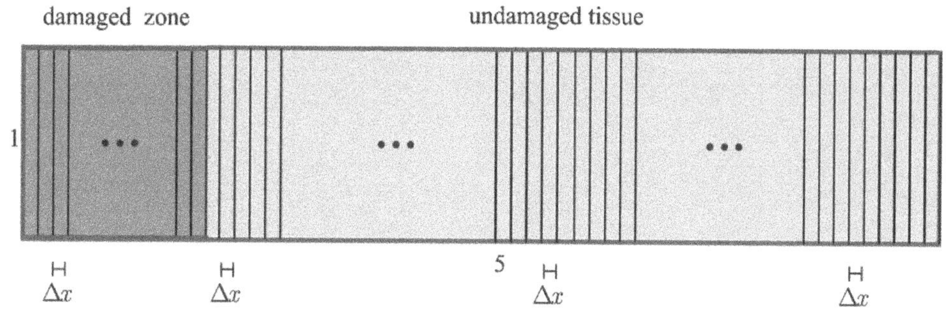

Figure 5.71 The numerical model of the planar strip of wounded skin.

Time variations of the fibroblast density are illustrated in Figure 5.72. Within the first two days of healing, the cells move from the healthy surroundings into the damaged region, which is demonstrated by the decrease in the cell density of the healthy tissue and the increase in density in the damaged area. Afterwards, the cell density in the damaged medium increases, while it does not change in the healthy tissue. Clearly, the process of reproduction of cells in the damaged environment is properly simulated.

The contraction phase eventually stops at about 12 days after the initiation stage, showing a constant fibroblast density. It is noted that at the end of this stage the density of fibroblasts near the wound edge becomes lower than the one of healthy tissue. Moreover, Figure 5.73 compares the present results (at 12 days) with the long-term reference data (Javierre et al. 2009), which shows a similar trend.

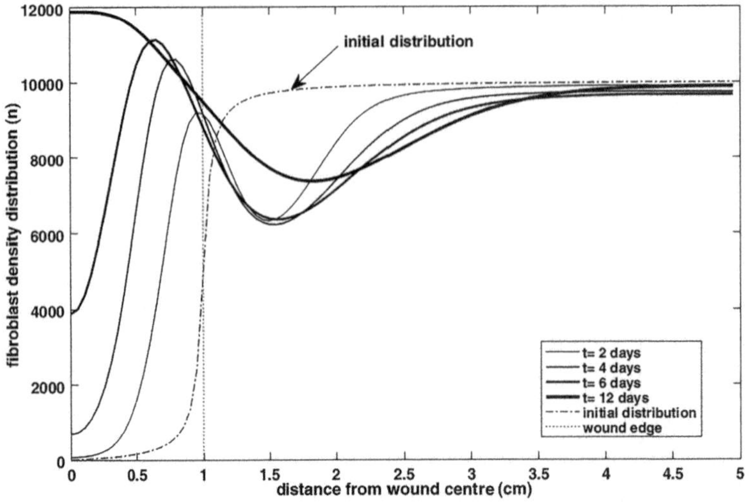

Figure 5.72 Distribution of the fibroblast density over time. *Source:* Adapted from Khaksar and Mohammadi 2020/the Sahand University of Technology.

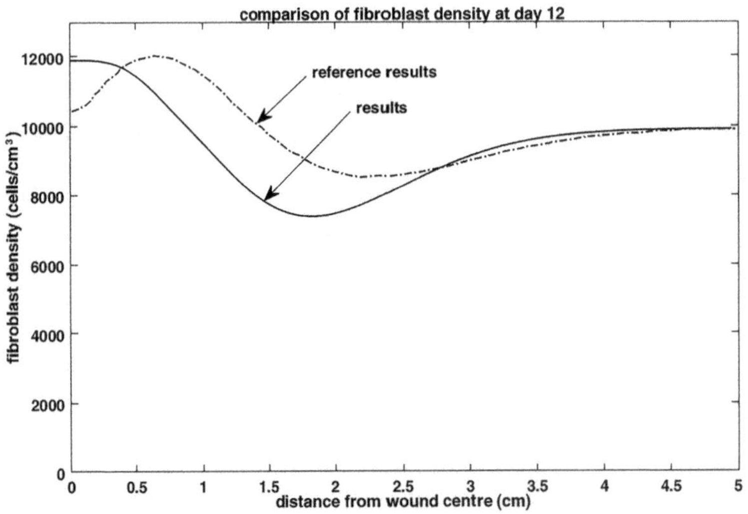

Figure 5.73 Comparison of the distribution of the fibroblast density after 12 days. *Source:* Reproduced from Khaksar 2020/the University of Tehran.

5.8.4.1.2 Changes of Myofibroblast Density Over Time

Myofibroblasts are immobile cells that are responsible for applying force on the edges of the healthy tissue to contract the damaged area. These cells are not present in the tissue before the beginning of contraction and are proliferated by fibroblasts during the contraction stage. Eventually, after the contraction is completed, these cells are removed

Figure 5.74 shows the results of the analysis for the density of myofibroblasts. As expected, myofibroblast density starts to monotonically vary from zero and peaks at about the middle of the process. The peak time corresponds to the maximum contraction rate. This density then decreases until very low densities are achieved, which is again expected from the simulations. These results are also very close to the reference results (Javierre et al. 2009).

5.8.4.1.3 Changes of Collagen Density Over Time

Production of collagen is a very slow process that may take up to a year. As a result, only a small amount of production of collagen is expected in the contraction stage of the wound healing. Therefore, the distribution of collagen density during the contraction phase can mostly be attributed to the tissue movement.

The results of distribution of collagen fibre density in time are presented in Figure 5.75. It is observed that the height of the tissue (indirectly represented by the collagen density) decreases as it moves into the affected area. Moreover, the collagen density increases slightly during the contraction process due to the collagen production.

5.8.4.1.4 Changes in Growth Factor Concentration Over Time

Growth factors are the primary triggers for the progression of the healing process. The density of growth factors at the end of the inflammatory phase is the primary starting factor for the later stages of the healing, such as the contraction.

At the beginning of the healing process, shown in Figure 5.76, the growth factors are highly concentrated in the affected area, but their concentration gradually decreases. At the end of the contraction, the concentration of the growth factors in the affected area reaches a negligible value.

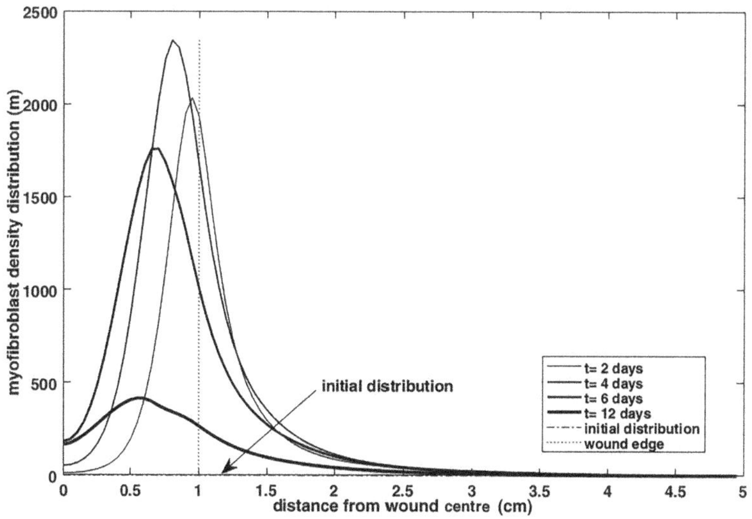

Figure 5.74 Distribution of the myofibroblast density over time. *Source:* Reproduced from Khaksar 2020/the University of Tehran.

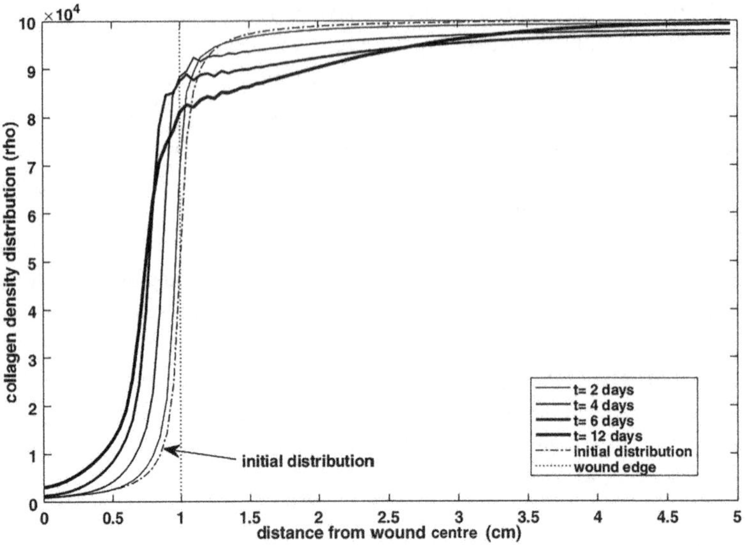

Figure 5.75 Distribution of the collagen density over time. *Source:* Reproduced from Khaksar 2020 /the University of Tehran.

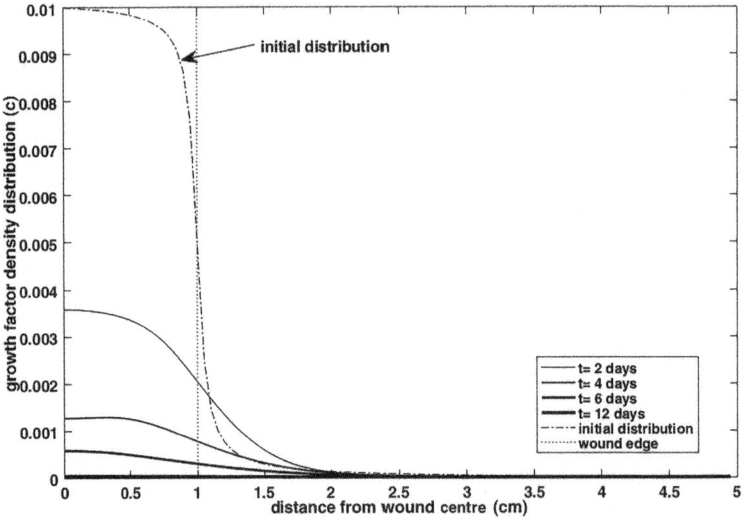

Figure 5.76 Concentration of the growth factor density over time. *Source:* Reproduced from Khaksar 2020/the University of Tehran.

5.8.4.1.5 Displacement Profile Over Time

The main feature of the contraction process of the damaged tissue is the reduction of the size of the damaged area.

According to Figure 5.77, the deformation results clearly indicate the contraction of the damaged region. The highest contraction rate, as expected, occurs in the middle of the healing process. Moreover, the maximum deformation is associated with the wound border.

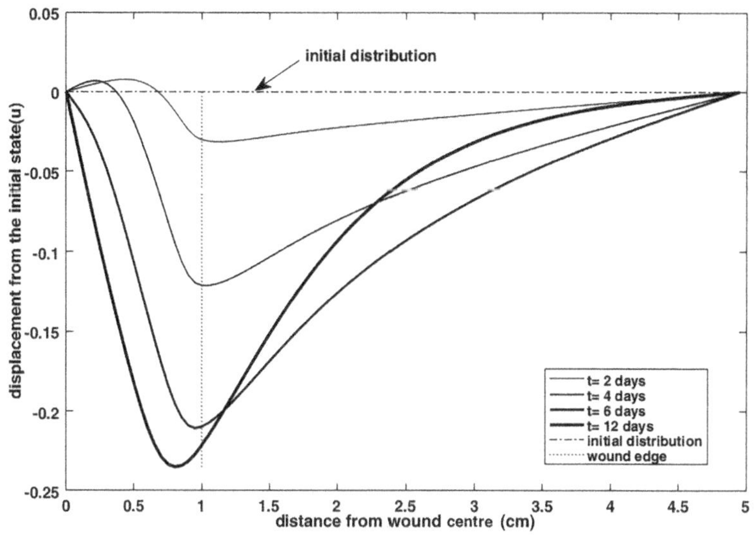

Figure 5.77 Displacement profile over time. *Source:* Reproduced from Khaksar 2020/the University of Tehran.

5.8.4.2 Healing of a Circular Wound

Figure 5.78 shows the geometry of a circular wound and its surrounding healthy skin. Only a quarter of the damaged tissue can be modelled by the finite element mesh, as presented in Figure 5.79.

Figure 5.80 shows the initial prescribed distributions of fibroblast, collagen and growth factors in the circular damaged wound.

Figure 5.81 illustrates the deformed shape of the skin at 2, 7 and 12 days after contraction. The extent of the contraction due to the healing process is clearly observed.

Distributions of the fibroblast density at 2, 7 and 12 days after contraction are presented in Figure 5.82. The density of fibroblast after the healing becomes larger than the healthy tissue, as similarly observed in the strip model.

Figure 5.83 illustrates the distributions of myofibroblast density after 2, 7 and 12 days. The density of myofibroblast is larger near the boundary with the healthy tissue.

Distributions of collagen/ECM density follows a similar pattern as in the strip model, as presented in Figure 5.84 for 2, 7 and 12 days after contraction.

The growth factors are concentrated at the damaged region at the beginning of the healing process. It is gradually reduced by time, as presented in Figure 5.85.

5.8.4.3 Anisotropic Healing of the Prestressed Circular Wound

Existence of the preferred direction of collagen in the tissue requires the principal mechanical characteristics of skin to be anisotropic. Based on the studies of Khatyr et al. (2004), the anisotropic modulus of a sample of skin tissue can be defined by

$$D_{anisotropic} = \begin{bmatrix} D_{11} & D_{12} & 0 \\ D_{21} & D_{22} & 0 \\ 0 & 0 & D_{33} \end{bmatrix} = \begin{bmatrix} 0.657 & 0.132 & 0 \\ 0.132 & 0.13 & 0 \\ 0 & 0 & 0.132 \end{bmatrix} \text{MPa} \tag{5.8.63}$$

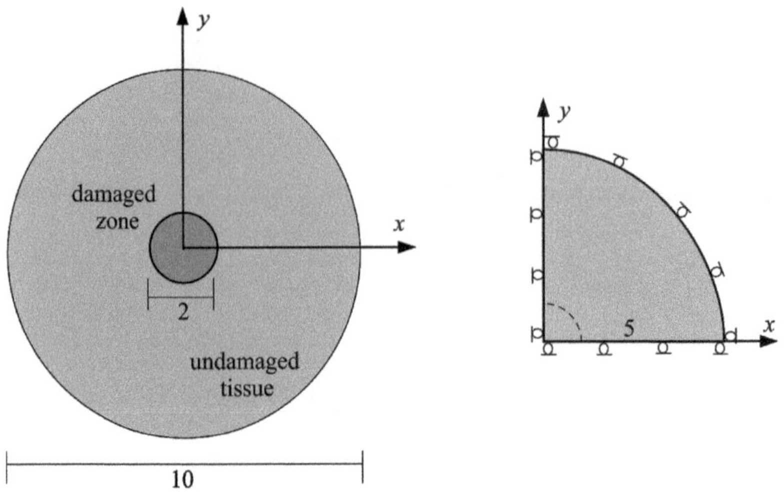

Figure 5.78 A circular damaged skin. *Source:* Reproduced from Khaksar 2020/the University of Tehran.

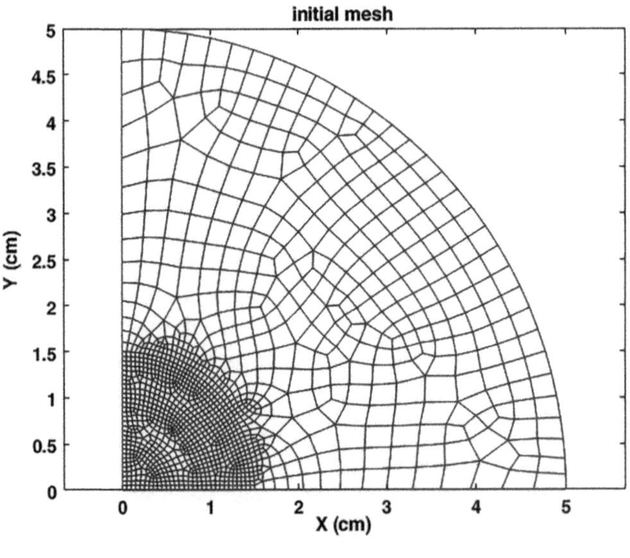

Figure 5.79 The finite element mesh of the circular damaged skin. *Source:* Reproduced from Khaksar 2020/the University of Tehran.

Moreover, the skin is inherently in a prestress state, which can be modelled by the concept of an initial stress $\boldsymbol{\sigma}_{initial}$, as reported by Flynn et al. (2011):

$$\boldsymbol{\sigma}_{initial} = \begin{Bmatrix} \sigma_1 \\ \sigma_2 \\ \sigma_3 \end{Bmatrix} = \begin{Bmatrix} 53.3 \\ 41.4 \\ 0 \end{Bmatrix} \text{ kPa} \tag{5.8.64}$$

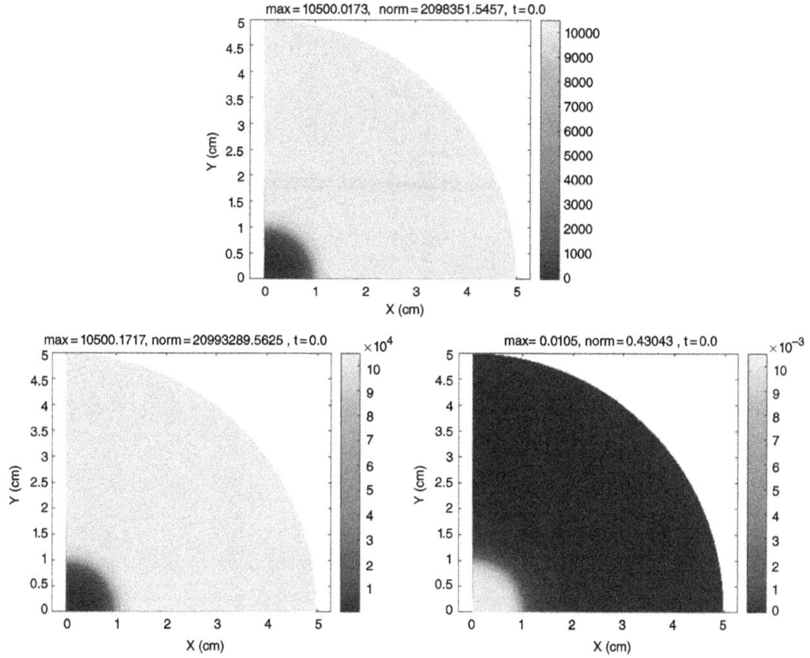

Figure 5.80 Initial density distributions of the fibroblast, collagen and growth factors. *Source:* Reproduced from Khaksar 2020/the University of Tehran.

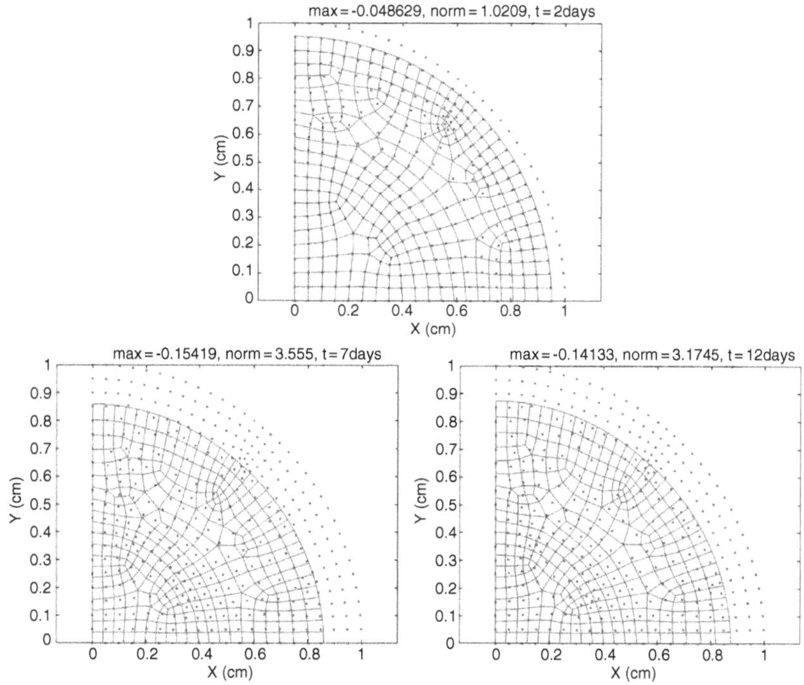

Figure 5.81 Deformed shapes of the skin tissue at 2, 7 and 12 days after contraction. *Source:* Reproduced from Khaksar 2020/the University of Tehran.

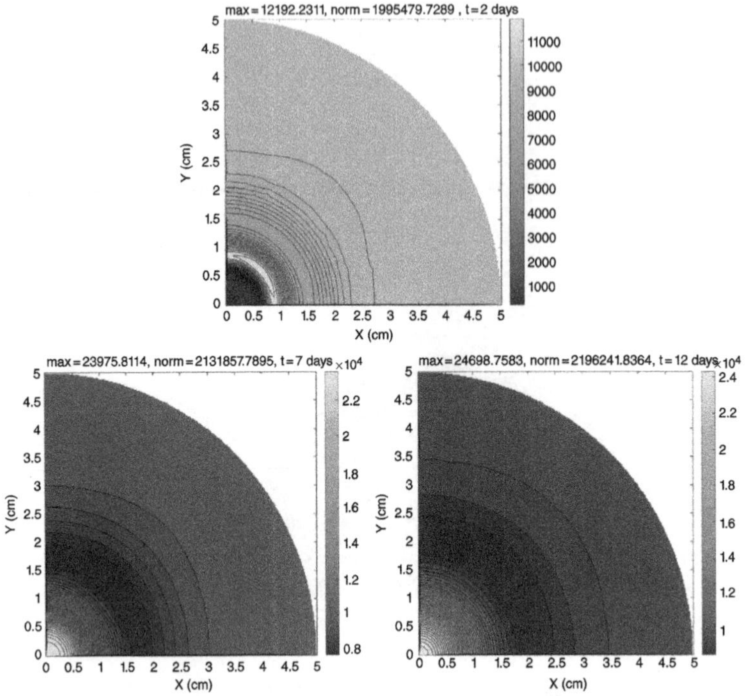

Figure 5.82 Distribution of the fibroblast density at 2, 7 and 12 days after contraction. *Source:* Reproduced from Khaksar 2020/the University of Tehran.

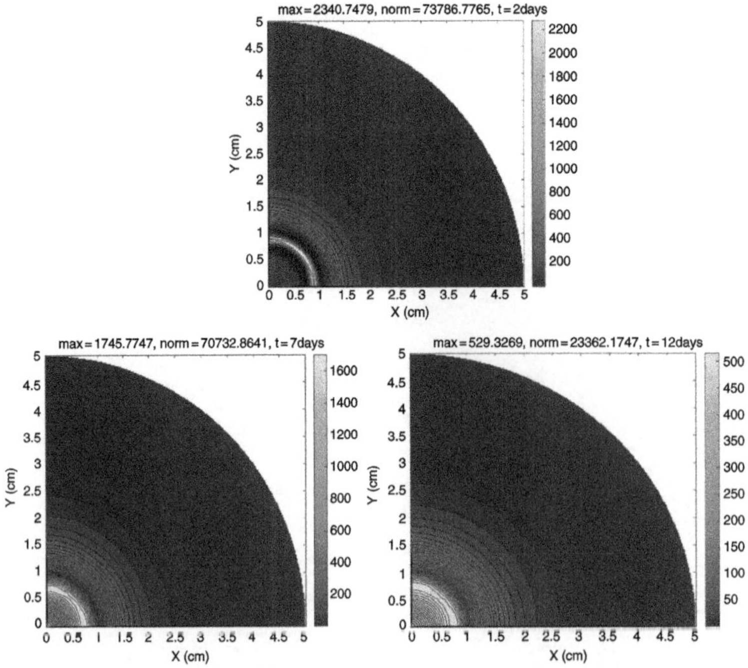

Figure 5.83 Distribution of the myofibroblast density at 2, 7 and 12 days after contraction. *Source:* Reproduced from Khaksar 2020/the University of Tehran.

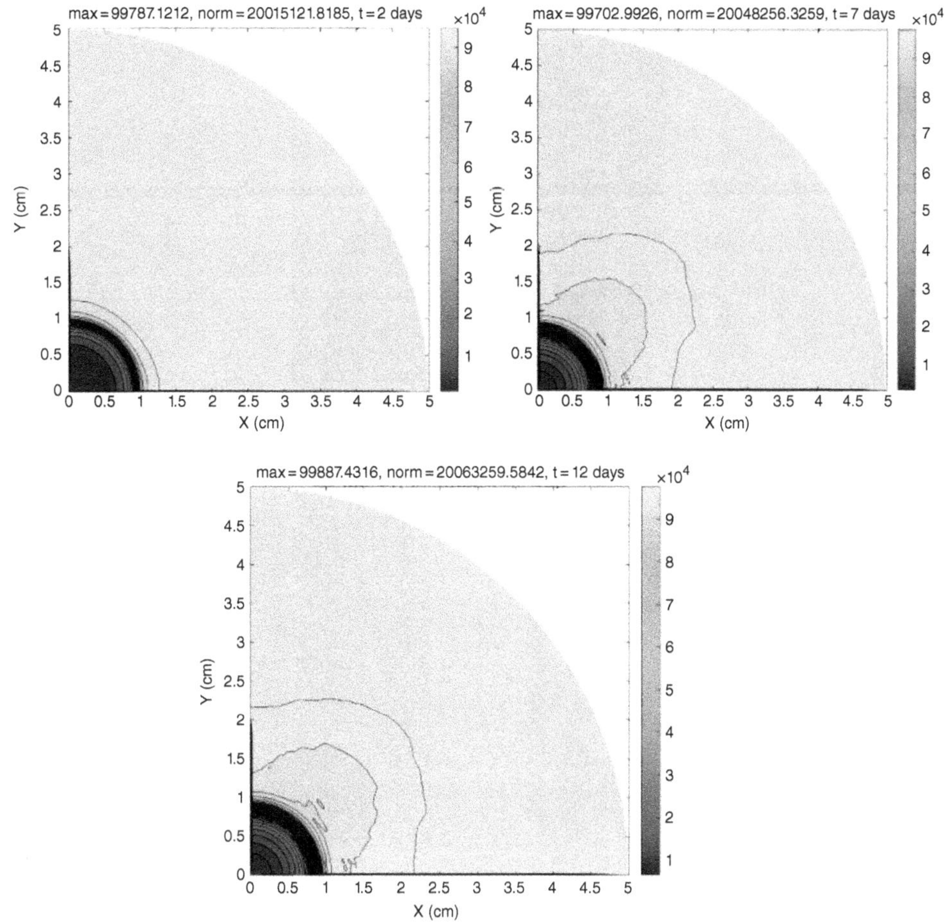

Figure 5.84 Distribution of the collagen density at 2, 7 and 12 days after contraction. *Source:* Reproduced from Khaksar 2020/the University of Tehran.

Taking into account these two modifications, the resultant deformations of the tissue during the contraction process after 2, 7 and 12 days are presented in Figure 5.86. A clear anisotropic deformation is observed after 12 days.

Finally, variation of the normalized area of the damaged tissue in time is compared with the reference results (McGrath and Simon 1983) in Figure 5.87.

5.9 Hierarchical Multiscale Modelling of a Degraded Arterial Wall

5.9.1 Definition of the Problem

Collagen fibres are the main mechanical load-bearing component of soft tissues, including the arterial wall. Any defect, mechanical degradation and even insufficient volume content

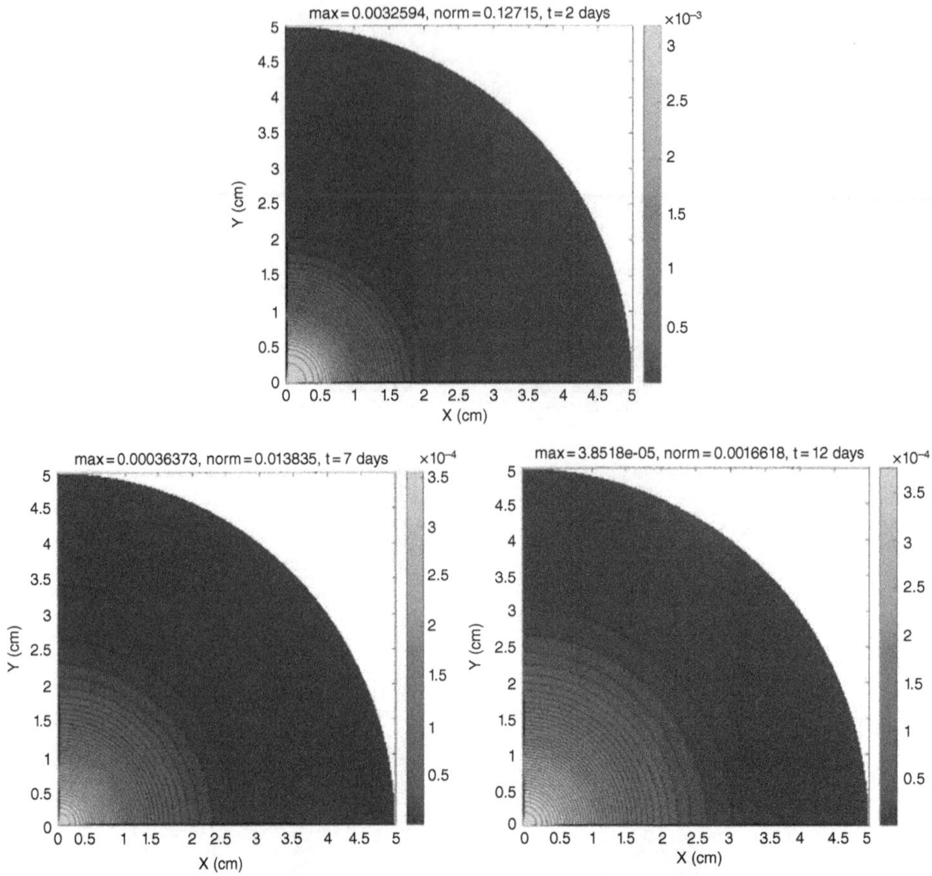

Figure 5.85 Distribution of the growth factor at 2, 7 and 12 days after contraction. *Source:* Reproduced from Khaksar 2020/the University of Tehran.

of the collagen fibre may lead to sever consequences in the functioning of the arterial wall and may ignite its rupture failure in high blood pressures.

In this section, the hierarchical multiscale method is employed to simulate the response of a typical degraded artery. The section follows the original work of Hatefi Ardakani et al. (2022). The source of degradation is assumed to be due to the reduction of the collagen fibre content.

The model of an artery is presented in Figure 5.88. The long cylindrical artery is subjected to a uniform internal pressure. In the mid part of the artery, a local micromechanical degradation zone of reduced collagen fibre content is assumed.

The soft tissue of the artery is composed of adventitia, media and intima layers, as depicted in Figure 5.89. The three layers include two circumferential sets of inclined collagen fibres within the ECM matrix, with orientation angles 2γ of 0°, 14° and 98°, respectively (Holzapfel et al. 2000; Ferrara and Pandolfi 2008). Only the media layer, which is the primary load-bearing constituent of the artery, is considered in this study, as similarly considered by Stylianopoulos and Barocas (2007). In fact, Figure 5.88 only illustrates the model

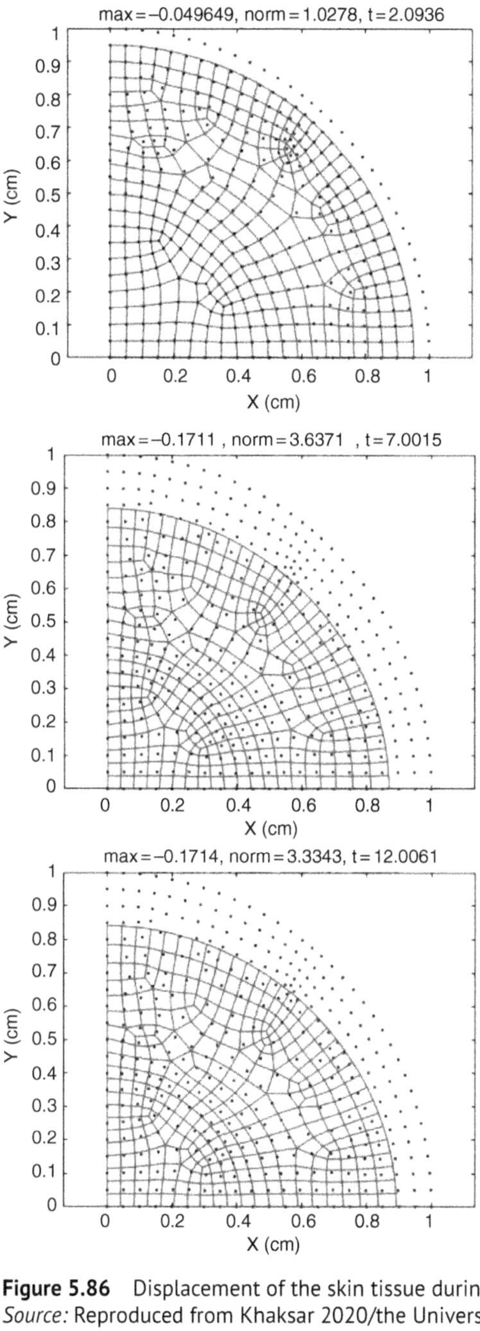

Figure 5.86 Displacement of the skin tissue during the contraction process after 2, 7 and 12 days. *Source:* Reproduced from Khaksar 2020/the University of Tehran.

of media layer with a locally degraded zone. More advanced and comprehensive modelling can also be performed.

In an intact media, the content of the collagen fibre is about 30% (Fung 1993). The pattern of reduction of collagen fibre content is assumed to occur at 10, 5 and 2% from the edges towards the centre of this zone, as shown in Figure 5.88.

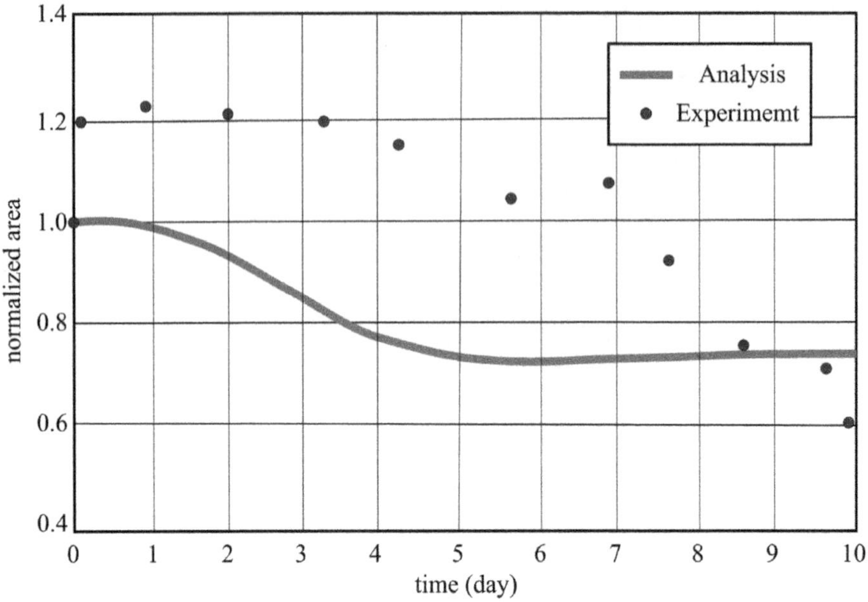

Figure 5.87 Comparison of the area change between the model and experiment. *Source:* Adapted from McGrath and Simon 1983; Khaksar 2020.

The hierarchical multiscale model of the artery includes a macroscale finite element model along with the microscale analysis of RVEs, positioned at each integration point of the degraded zone.

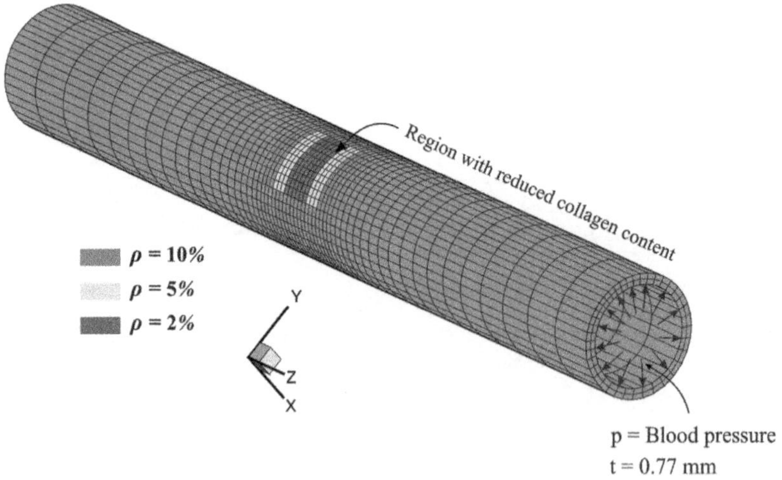

Figure 5.88 Geometry of the simplified model of a degraded artery. *Source:* Adapted from Hatefi Ardakani et al. 2022.

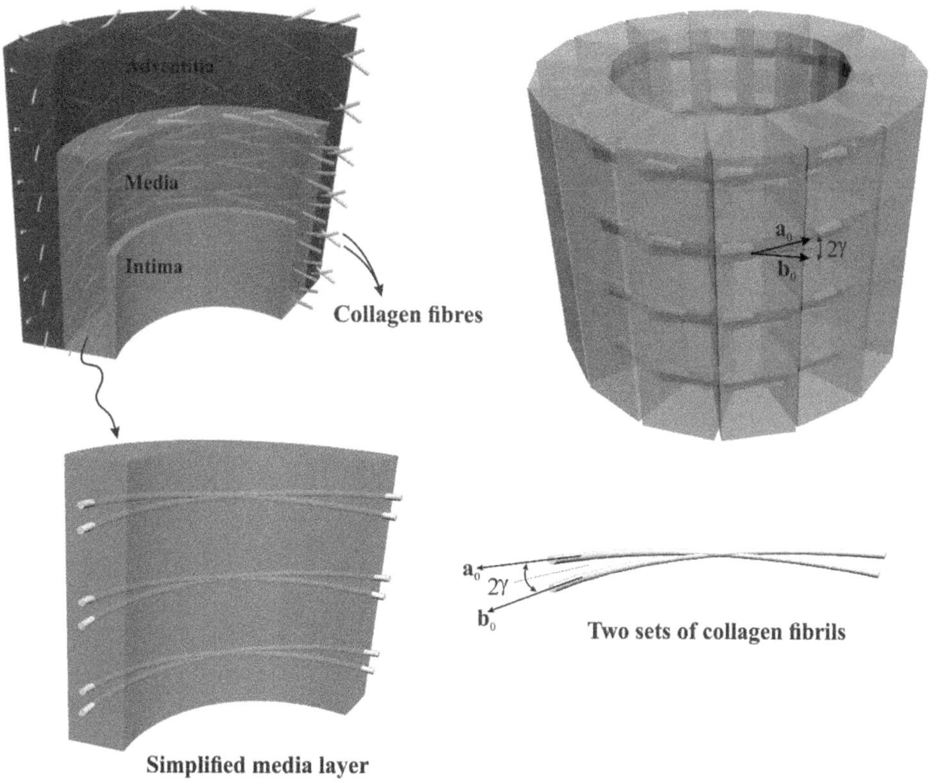

Figure 5.89 Composition of the artery wall and the media layer with two sets of collagen fibres within its ECM matrix. Distribution of two sets of inclined collagen fibres may be viewed as a pseudo helical arrangement. *Source:* Adapted from Hatefi Ardakani et al. 2022.

5.9.2 Multiscale Model

Figure 5.90 illustrates the model of an artery along with two different microscale RVE models for the multiscale analysis. The whole model is made of the media layer and the localized zone represents part of the tissue with the reduced collagen fibre content.

Due to an extremely high run time of a complete multiscale analysis, only a nonlinear large deformation single macroscale framework is employed for the non-degraded regions. The single scale framework is based on a proper hyperelastic model to account for the response of the non-degraded fibrous tissue, without getting into the microscale details. The degraded zone, which is expected to experience high gradients, is analysed by the hierarchical multiscale method based on the microstructure of the media, represented by two different RVEs, as presented in Figure 5.90. The RVEs are designed by the reduced collagen fibre content of the Gauss point of the degraded zone.

RVE I is a simplified layer-wise model that comprises one matrix layer in between the two other layers, each including the two sets of collagen fibres, as presented in Figure 5.91. While this simplified model may not be fully consistent with available constitutive models for fibrous tissues, it may still be adopted to obtain an overall estimation of the response.

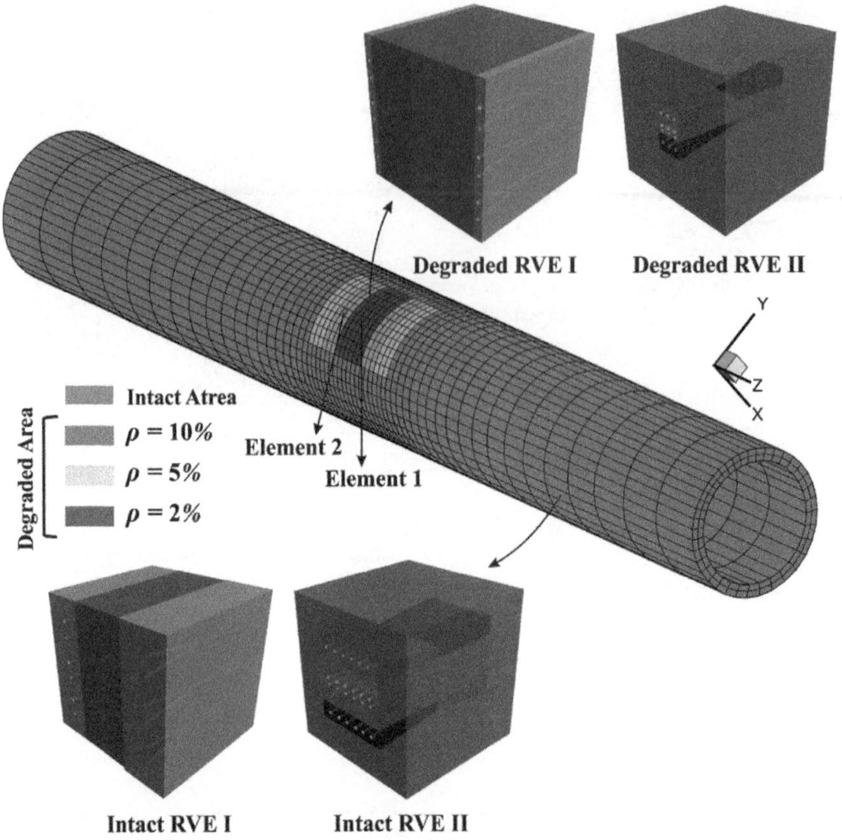

Figure 5.90 Partitioning of the degraded arterial wall with two RVE types for the multiscale analysis of the degraded zone. *Source:* Adapted from Hatefi Ardakani et al. 2022.

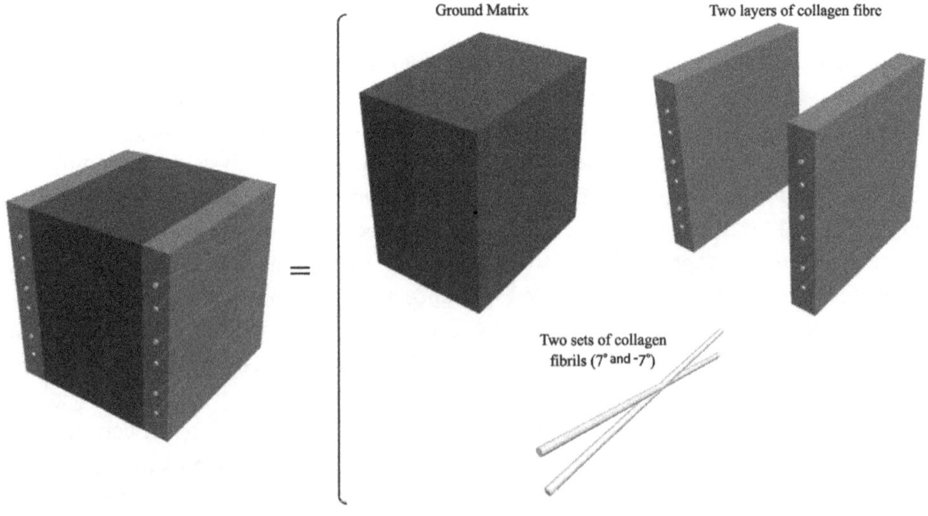

Figure 5.91 Definition of simplified layer-wise RVE I. *Source:* Adapted from Hatefi Ardakani et al. 2022.

RVE II is composed of three different parts of fibrils, which form the collagen fibres, within a ground matrix. The three parts include two independent sets of sole fibrils with the orientations of 7° and -7°, respectively, and a third part that contains the two intersecting sets of fibrils (Figure 5.92). Clearly, RVE II is better suited to most constitutive models to represent the microstructure of fibrous tissues.

5.9.3 Hyperelastic Material Models

Hyperelastic material models are employed for the matrix and fibre sets, as comprehensively presented in Section 5.3.6. The incompressible neo-Hookean model (with μ constant) is adopted for the ECM, and the collagen fibre layers follow the well-developed Holzapfel model (Holzapfel et al. 2000) with two sets of 7° and -7° inclined collagen fibrils (with λ, μ, c_{1f_1}, c_{2f_1} constants).

To begin the solution procedure, the hyperelastic material properties for the homogenized macromodel, and for the ECM and collagen fibres of the micromodel, should be determined.

In macro scale, fortunately, sufficient data are available for the material properties of an artery, as presented in Table 5.20, according to the reference reports of Ferrara and Pandolfi (2008) and Holzapfel et al. (2004).

In contrast, very limited information is available on the properties of the microscale constituents of an artery. Therefore, a major perquisite challenge in micromechanical analysis of RVE is to determine the appropriate material properties of the ECM and collagen fibres.

Hatefi Ardakani et al. (2022) adopted a genetic algorithm (GA) (Holland 1992) to determine the unknown material constants of ECM and fibres for the microscale models of RVE I and RVE II by best fitting their force-stretch predictions with the available macroscale response (Ferrara and Pandolfi 2008).

The GA procedure is initiated by assuming the reference (Ferrara and Pandolfi 2008) value of $\mu = 10.77$ kPa for the shear modulus of the matrix of RVE and the matrix of the

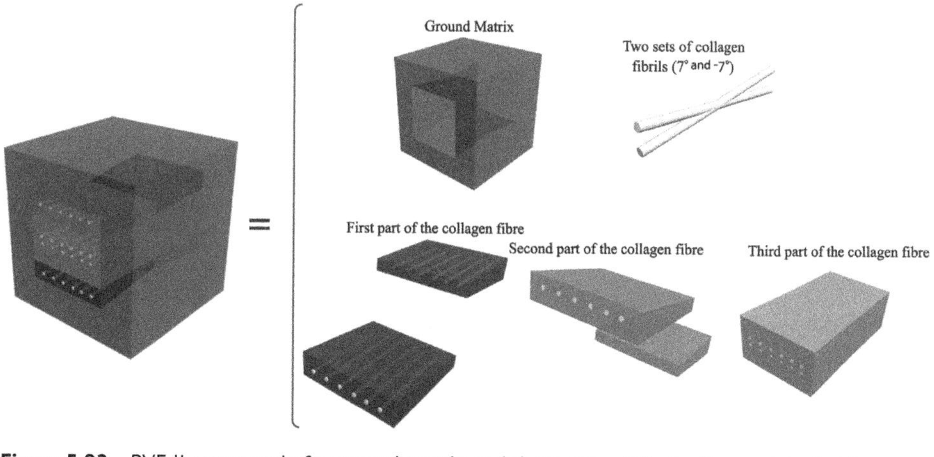

Figure 5.92 RVE II composed of a ground matrix and three sets of fibrils. *Source:* Adapted from Hatefi Ardakani et al. 2022.

Table 5.20 Homogenized properties of the media layer.

Material	Material parameters	Values
Matrix	λ(kPa)	1,667
	μ(kPa)	10.77
Fibre set 1	γ_{f_1} (°)	7
	c_{1f_1}	4.71
	c_{2f_1} (kPa)	483
Fibre set 2	γ_{f_2} (°)	-7
	c_{1f_2}	4.71
	c_{2f_2} (kPa)	483

Source: Adapted from Ferrara and Pandolfi 2008; Holzapfel et al. 2004.

Table 5.21 Material constants predicted by the GA analysis for RVE I and RVE II.

	Matrix	Fibres			
	Neo-Hookean	Holzapfel	Holzapfel		
	Ferrara and Pandolfi (2008)	Ferrara and Pandolfi (2008)	GA prediction		
Material parameters	μ(kPa)	μ(kPa)	λ(kPa)	c_{1f_1}	c_{2f_1} (kPa)
RVE I	10.77	10.77	5175	4.69	16.09
RVE II	10.77	10.77	7664	4.67	22.37

Source: Adapted from Hatefi Ardakani et al. 2022.

collagen fibre part of RVE. Then the minimization procedure is performed for determining the remaining parameters λ, c_{1f_1} and c_{2f_1} of RVE. The final GA predictions of the material constants are presented in Table 5.21.

Figure 5.93 illustrates the way the GA approach can be adopted to determine the micro-scale material constants by fitting their force-stretch predictions with an available macro-scale response.

5.9.4 Computational Framework of the Hierarchical Multiscale Homogenization

The adopted multiscale approach is performed in the large strain regime to accurately account for the hyperelastic response of various soft tissues in macro- and microscales. Similar to the conventional computational homogenization technique, the macromodel is analysed at each increment of the loading. Then, the macroscale deformation gradient is

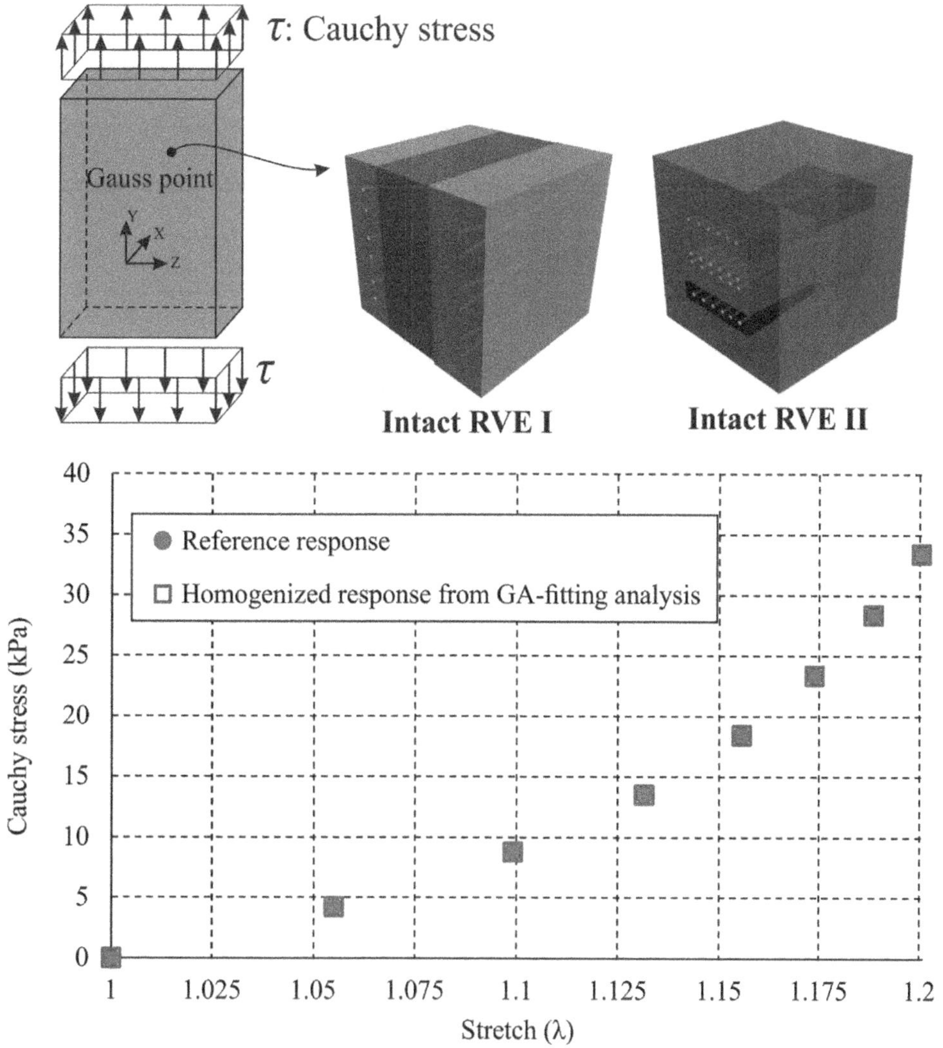

Figure 5.93 GA procedure for determination of material constants for ECM and collagen fibres. *Source:* Adapted from Hatefi Ardakani et al. 2022.

applied on the microstructure and the homogenized stress and material modulus are determined from the microscale solution. A Lagrangian description is employed to solve both scales, as briefly presented in the following sections. Figure 5.94 briefly illustrates the adopted computational homogenization procedure for a biomechanical model in terms of its initial/reference and current configurations.

5.9.4.1 Macroscale

The governing macroscale equation, defined in the initial/reference configuration Ω_0^{M}, can be written in terms of the first Piola–Kirchhoff stress $\boldsymbol{P}^{\mathrm{M}}$ as

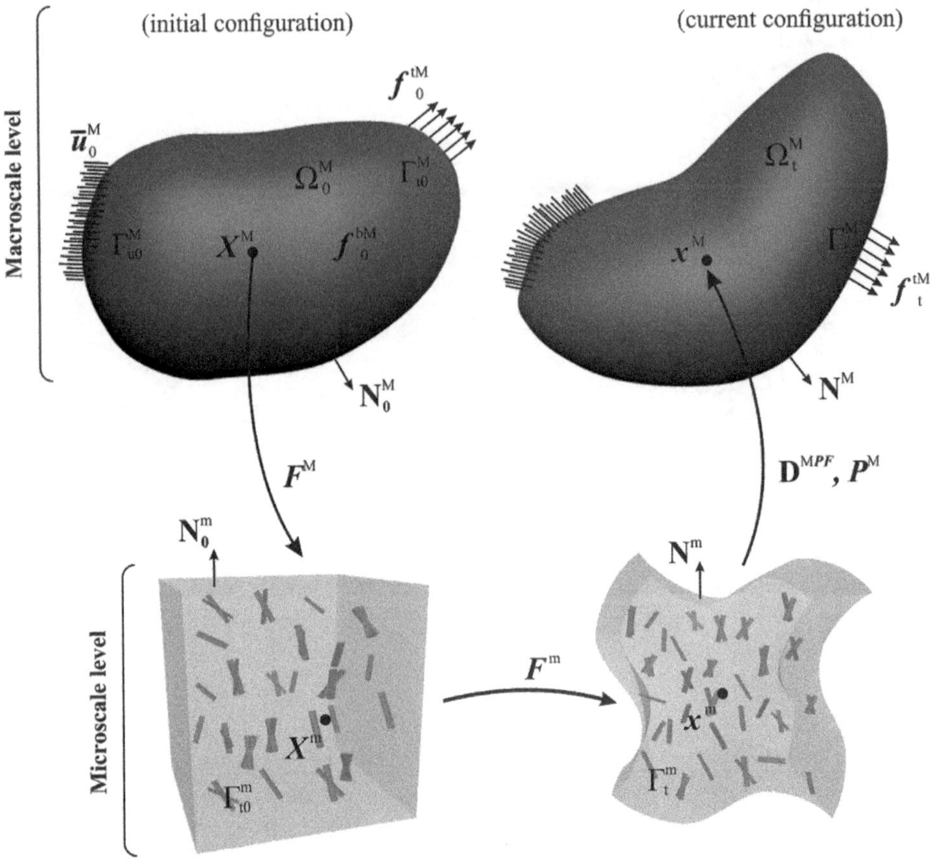

Figure 5.94 Schematic illustration of the adopted computational homogenization. *Source:* Adapted from Hatefi Ardakani et al. 2022.

$$\nabla^{M} \cdot \boldsymbol{P}^{M}(\mathbf{x}^{M}) + \boldsymbol{f}_0^{bM}(\mathbf{x}^{M}) = \mathbf{0},\ \mathbf{x}^{M} \in \Omega_0^{M} \tag{5.9.1}$$

where $\boldsymbol{f}_0^{bM}$ is the body force. Prescribed displacement $\bar{\boldsymbol{u}}_0^{M}$ and traction $\boldsymbol{f}^{tM}$ boundary conditions can be defined as

$$\boldsymbol{P}^{M}\boldsymbol{N}_0^{M} = \boldsymbol{f}^{tM} \quad \text{on } \Gamma_{t0}^{M} \tag{5.9.2}$$

$$\boldsymbol{u}^{M} = \bar{\boldsymbol{u}}_0^{M} \quad \text{on } \Gamma_{u0}^{M} \tag{5.9.3}$$

where Γ_{t0}^{M} and Γ_{u0}^{M} are the natural (traction) and essential (displacement) boundaries, respectively. $\boldsymbol{N}_0^{M}$ represents the unit vector normal to the boundary Γ_{t0}^{M}.

Having known the hyperelastic strain energy function w^{sM}, defined by Equation (5.3.68), the macroscale first Piola–Kirchhoff stress $\boldsymbol{P}^{M}$ and material modulus $\mathbf{D}^{MPF}$ can be derived from

$$\boldsymbol{P}^{\mathrm{M}} = \frac{\partial w^{\mathrm{sM}}}{\partial \boldsymbol{F}^{\mathrm{M}}} \tag{5.9.4}$$

$$\mathbf{D}^{MPF} = \frac{\partial \boldsymbol{P}^{\mathrm{M}}}{\partial \boldsymbol{F}^{\mathrm{M}}} = \frac{\partial^2 w^{\mathrm{sM}}}{\partial \boldsymbol{F}^{\mathrm{M}} \partial \boldsymbol{F}^{\mathrm{M}}} \tag{5.9.5}$$

where

$$\boldsymbol{F}^{\mathrm{M}} = \frac{\mathrm{d}\boldsymbol{x}^{\mathrm{M}}}{\mathrm{d}\boldsymbol{X}^{\mathrm{M}}} \tag{5.9.6}$$

5.9.4.2 Microscale

Similar to the macrolevel, the governing equation for the microscale model Ω_0^m can be written in terms of the first Piola–Kirchhoff stress $\boldsymbol{P}^{\mathrm{m}}$ as

$$\nabla^{\mathrm{m}} \cdot \boldsymbol{P}^{\mathrm{m}}(\mathbf{x}^{\mathrm{m}}) + \boldsymbol{f}^{\mathrm{bm}}(\mathbf{x}^{\mathrm{m}}) = 0, \ \mathbf{x}^{\mathrm{m}} \in \Omega_0^{\mathrm{m}} \tag{5.9.7}$$

where $\boldsymbol{f}_0^{\mathrm{bm}}$ is the body force. The Prescribed displacement $\bar{\boldsymbol{u}}_0^{\mathrm{m}}$ and traction $\boldsymbol{f}^{\mathrm{tm}}$ boundary conditions can be defined as

$$\boldsymbol{P}^{\mathrm{m}} \boldsymbol{N}_0^{\mathrm{m}} = \boldsymbol{f}^{\mathrm{tm}} \ \text{ on } \ \Gamma_{t0}^{\mathrm{m}} \tag{5.9.8}$$

$$\boldsymbol{u}^{\mathrm{m}} = \bar{\boldsymbol{u}}_0^{\mathrm{m}} \ \text{ on } \ \Gamma_{u0}^{\mathrm{m}} \tag{5.9.9}$$

where Γ_{t0}^m and Γ_{u0}^m are the natural (traction) and essential (displacement) boundaries, respectively, and $\boldsymbol{N}_0^{\mathrm{m}}$ is the normal vector on the boundary Γ_{t0}^{m} of the microscale RVE. Note that $\boldsymbol{u}^{\mathrm{m}}$ in the microscale model represents the microdisplacement fluctuation.

The microscale first Piola–Kirchhoff stress $\boldsymbol{P}^{\mathrm{m}}$ and the material modulus $\mathbf{D}^{mPF}$ can be derived from the hyperelastic strain energy function w^{sm} (Saeb et al. 2016):

$$\boldsymbol{P}^{\mathrm{m}} = \frac{\partial w^{\mathrm{sm}}}{\partial \boldsymbol{F}^{\mathrm{m}}} \tag{5.9.10}$$

$$\mathbf{D}^{mPF} = \frac{\partial \boldsymbol{P}^{\mathrm{m}}}{\partial \boldsymbol{F}^{\mathrm{m}}} = \frac{\partial^2 w^{\mathrm{sm}}}{\partial \boldsymbol{F}^{\mathrm{m}} \boldsymbol{F}^{\mathrm{m}}} \tag{5.9.11}$$

where the deformation gradient in the microscale $\boldsymbol{F}^{\mathrm{m}}$ is defined as

$$\boldsymbol{F}^{\mathrm{m}} = \frac{\mathrm{d}\mathbf{x}^{\mathrm{m}}}{\mathrm{d}\mathbf{X}^{\mathrm{m}}} \tag{5.9.12}$$

w^{sm} is set to the neo-Hookean incompressible material model for the ground matrix and the Holzapfel material model for the fibre parts of the RVE.

5.9.4.3 Micro–Macro Transition

According to the computational homogenization technique, the two micro- and macroscale solutions are coupled together by the Hill–Mandel principle in the form of an averaging of microscale solutions to obtain the homogenized macroscale variables at the corresponding macroscale Gauss point (Saeb et al. 2016):

$$\boldsymbol{F}^{\mathrm{M}} = \frac{1}{\Omega_0^{\mathrm{m}}} \int_{\Omega_0^{\mathrm{m}}} \boldsymbol{F}^{\mathrm{m}} \mathrm{d}\Omega \tag{5.9.13}$$

$$\boldsymbol{P}^{\mathrm{M}} = \frac{1}{\Omega_0^{\mathrm{m}}} \int_{\Omega_0^{\mathrm{m}}} \boldsymbol{P}^{\mathrm{m}} \mathrm{d}\Omega \tag{5.9.14}$$

and the corresponding material modulus $\mathbf{D}^{MPF}$ can be defined in terms of the microscale $\mathbf{D}^{mPF}$ as (Miehe et al. 1999; Hatefi Ardakani et al. 2022):

$$\mathbf{D}^{MPF} = \frac{1}{\Omega_0^{\mathrm{m}}} \int_{\Omega_0^{\mathrm{m}}} \mathbf{D}^{\mathrm{m}PF} \, \mathrm{d}\Omega - \left[\frac{1}{\Omega^{\mathrm{m}}} \left(\mathbf{L}^{\mathrm{m}}\right)^T \left(\mathbf{K}^{\mathrm{m}}\right)^{-1} \mathbf{L}^{\mathrm{m}} \right] \tag{5.9.15}$$

with

$$\mathbf{K}^{\mathrm{m}} = \sum_{e=1}^{\mathrm{nelem}} \int_{\Omega_e^{\mathrm{m}}} \mathbf{B}^{\mathrm{T}} \mathbf{D}^{\mathrm{m}PF} \mathbf{B} \, \mathrm{d}\Omega \tag{5.9.16}$$

$$\mathbf{L}^{\mathrm{m}} = \sum_{e=1}^{nelem} \int_{\Omega_e^{\mathrm{m}}} \mathbf{B}^{\mathrm{T}} \mathbf{D}^{\mathrm{m}PF} \, \mathrm{d}\Omega \tag{5.9.17}$$

and $\mathbf{B}$ is the matrix of derivatives of the classical finite element shape functions.

An iterative procedure, such as the Newton–Raphson, is required to solve the governing equations. For details refer to Section 4.2.

5.9.5 Numerical Results

5.9.5.1 Uniaxial Test

Before examining the main multiscale simulation, the results of a uniaxial test, used for the GA best-fitting analysis, is discussed to indicate better the effect of type of RVE. The model is depicted in Figure 5.95 and the material properties are according to Tables 5.20 and 5.21.

In order to provide a better presentation of the overall multiaxial state at the complex microscale structure, the von Mises stress contours on RVE I and RVE II (associated with the specified Gauss point in Figure 5.95) are presented in Figures 5.96 and 5.97, respectively.

Clearly, the parts with collagen fibre largely contribute in the load bearing of the tissue. Moreover, a uniform distribution is observed in the simplified model of RVE I, which contradicts the expectations from the complex microstructure, even though it could fit the global solution by proper selection of material constants through the GA algorithm. In contrast, the non-uniform distribution in RVE II captures well the effects of two sets of

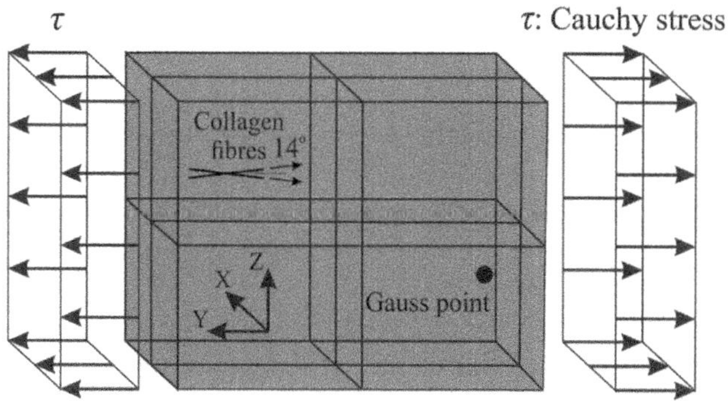

Figure 5.95 The uniaxial tensile test on the fibrous tissue. *Source:* Adapted from Hatefi Ardakani et al. 2022.

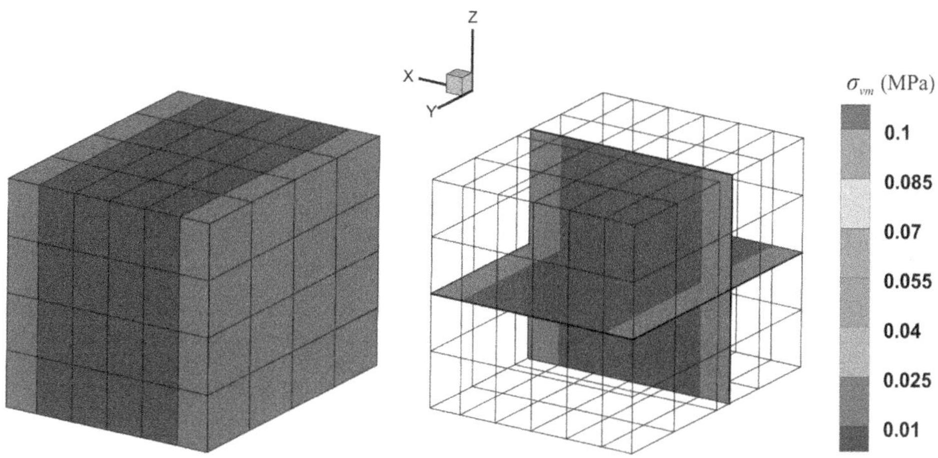

Figure 5.96 The von Mises stress distribution for RVE I (uniaxial tensile test). *Source:* Adapted from Hatefi Ardakani et al. 2022.

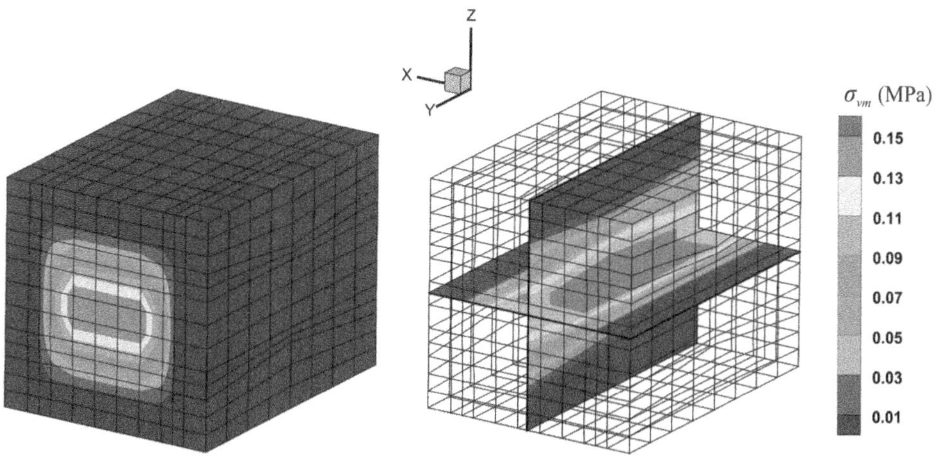

Figure 5.97 The von Mises stress distribution for RVE II (uniaxial tensile test). *Source:* Adapted from Hatefi Ardakani et al. 2022.

fibre distributions. In addition, the stiffer microstructure of RVE II (due to multiple sets of inclined fibres) leads to a higher share of stress for the fibres (maximum 150 kPa compared to 100 kPa for RVE I).

Since the goal is to accurately model the degraded zone (Figure 5.88), the RVEs are examined by different collagen fibre contents of 10, 5 and 2%. Different collagen fibre contents are represented by different thicknesses of the parallel layers of fibres in RVE I and with different cross-sections of each inclined part of the fibre in RVE II. Descriptions of RVEs are schematically presented in Figure 5.98.

The uniaxial tensile force-stretch response of RVEs with different collagen contents, presented in Figure 5.99, shows a significant macro-scale stiffness reduction as the fibre content is reduced. As a result, a damaged artery (with reduced collagen fibre content) is highly susceptible to a high local deformation gradient, high levels of stress concentration and even a potential rupture in high blood pressures.

5.9.5.2 Multiscale Simulation of the Degraded Artery

The simulation model is presented in Figure 5.90 with the details of RVEs in Figures 5.91 and 5.92. The degraded zone is assumed to be represented by different collagen fibre contents of 10, 5 and 2%. For the rest of the model, the collagen fibre content is set to 30%.

The results of the uniaxial test in Figure 5.99 implies that significant differences in deformation are expected in elements or points with different contents of collagen fibres. As a result, a potential local inflation in the degraded soft tissue is expected to occur, an indication of a potential aneurysm. Figure 5.100 illustrates the variation of the maximum

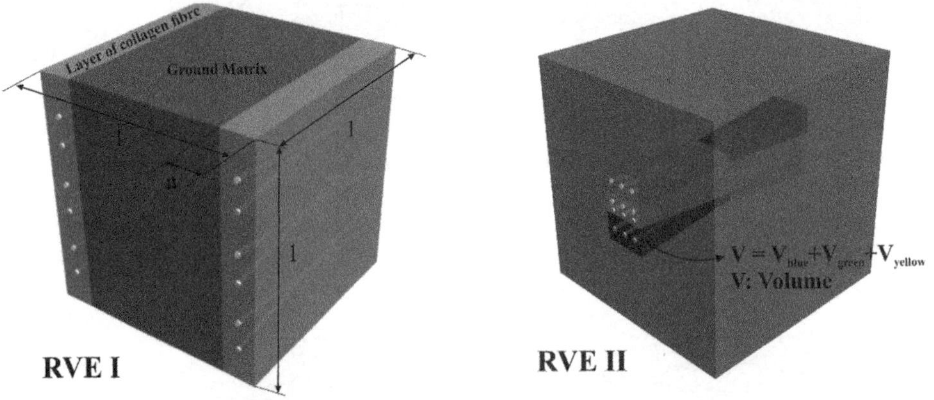

Collagen fibre content (%)	a RVE I	V RVE II
2	0.01	0.02
5	0.025	0.05
10	0.05	0.1

Figure 5.98 Descriptions of RVE I and RVE II for different collagen fibre contents of 2, 5 and 10%. *Source:* Adapted from Hatefi Ardakani et al. 2022.

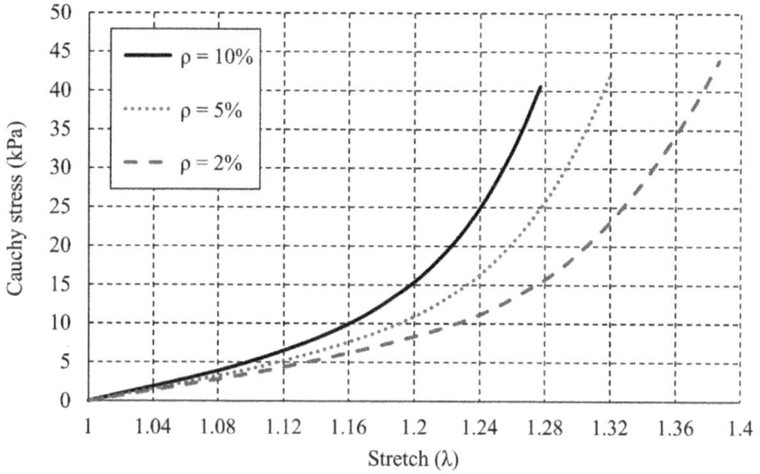

Figure 5.99 Tensile force-stretch responses with different collagen fibre contents. *Source:* Adapted from Hatefi Ardakani et al. 2022.

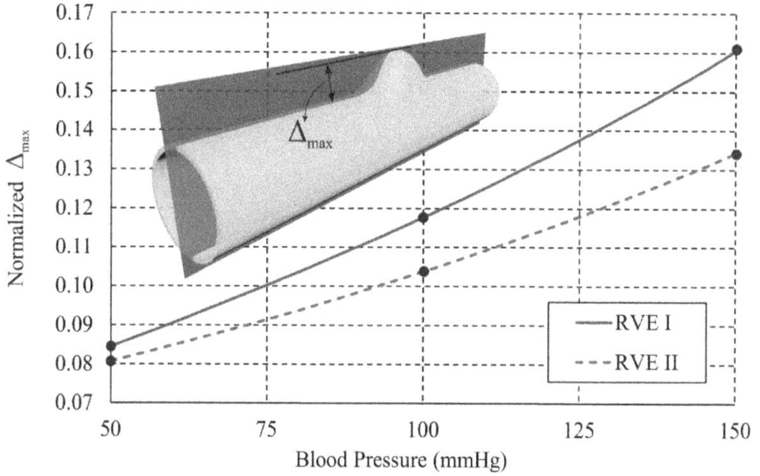

Figure 5.100 Maximum deformation of the degraded artery for different blood pressures. *Source:* Adapted from Hatefi Ardakani et al. 2022.

deformation of the degraded part of the artery with respect to the applied blood pressure. Again, a stiffer response is observed for RVE II.

Figures 5.101 and 5.102 illustrate the von Mises stress distributions on the external surface of the artery for different blood pressures, showing a significant increase in the maximum stress level at the local high deformation gradient within the degraded zone, as similarly discussed by Rodrıguez et al. (2008) and Roy et al. (2014). More or less, similar patterns are observed for both micromechanical RVE models.

Similar patterns can be observed in Figures 5.103 and 5.104 for the internal surface of the artery.

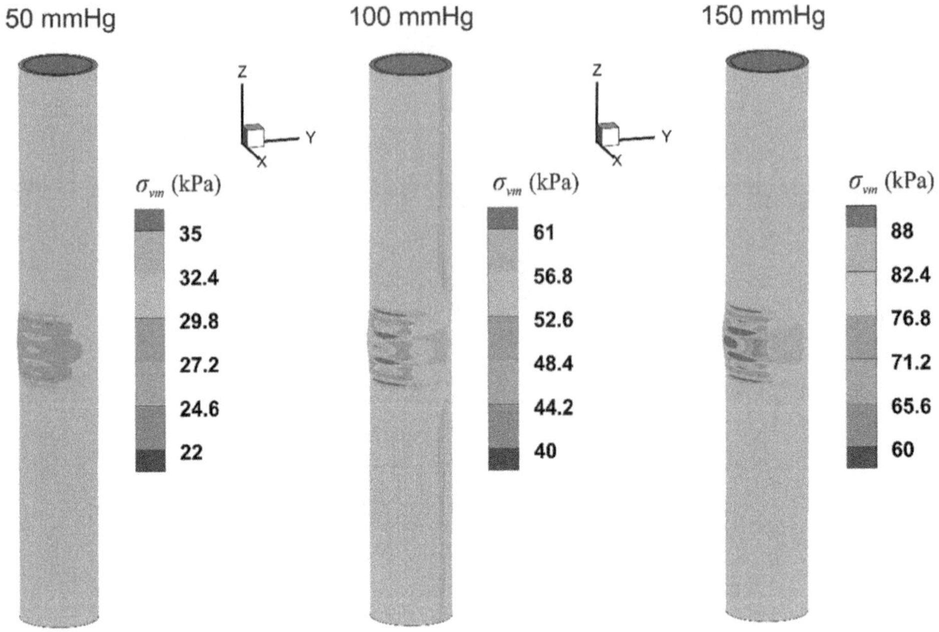

Figure 5.101 Contours of the von Mises stress for RVE I on the external surface of the artery for different blood pressures: 50, 100 and 150 mmHg. *Source:* Adapted from Hatefi Ardakani et al. 2022.

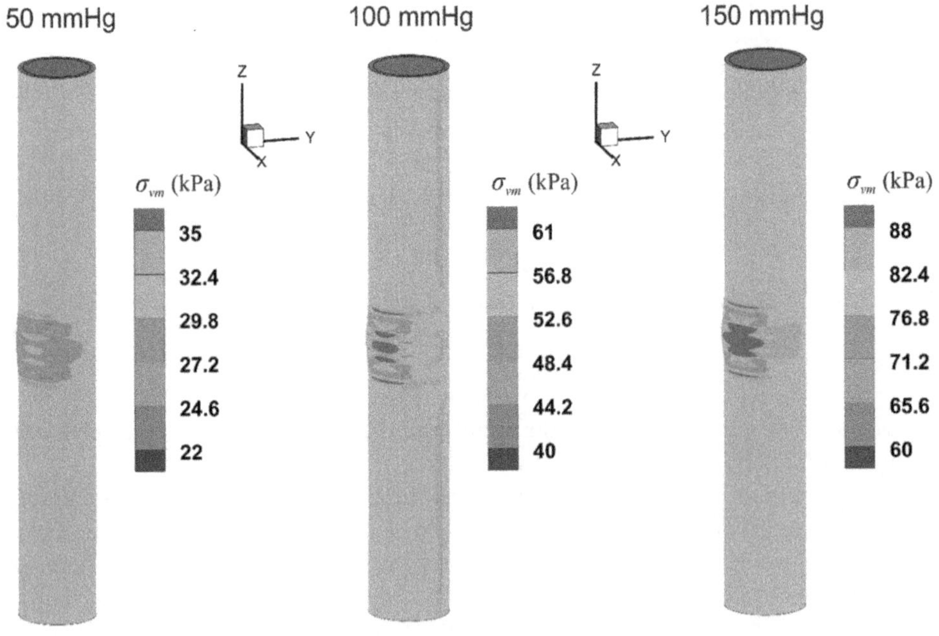

Figure 5.102 Contours of the von Mises stress for RVE II on the external surface of the artery for different blood pressures: 50, 100 and 150 mmHg. *Source:* Adapted from Hatefi Ardakani et al. 2022.

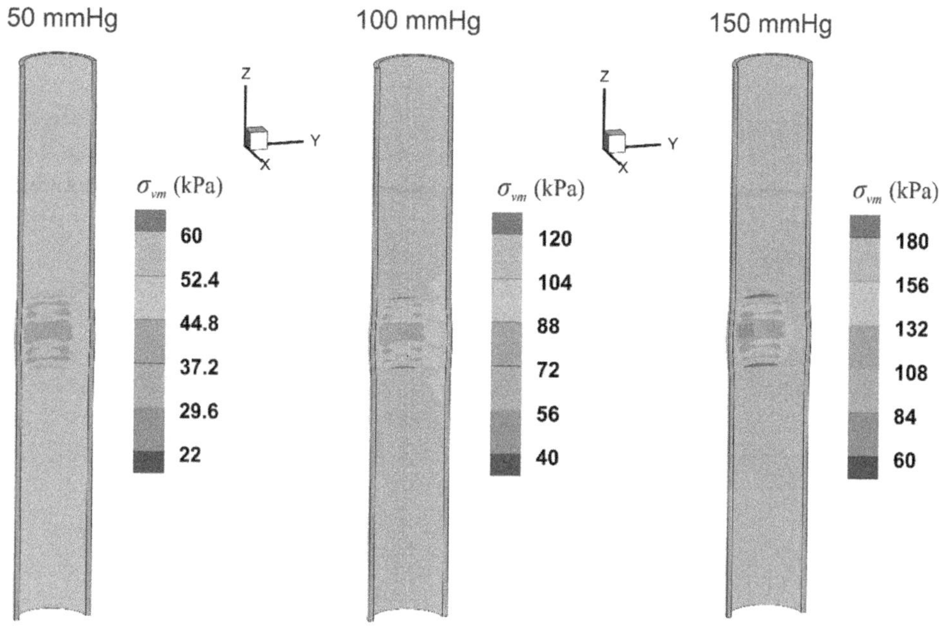

Figure 5.103 Contours of the von Mises stress for RVE I on the internal surface of the artery for different blood pressures: 50, 100 and 150 mmHg. *Source:* Adapted from Hatefi Ardakani et al. 2022.

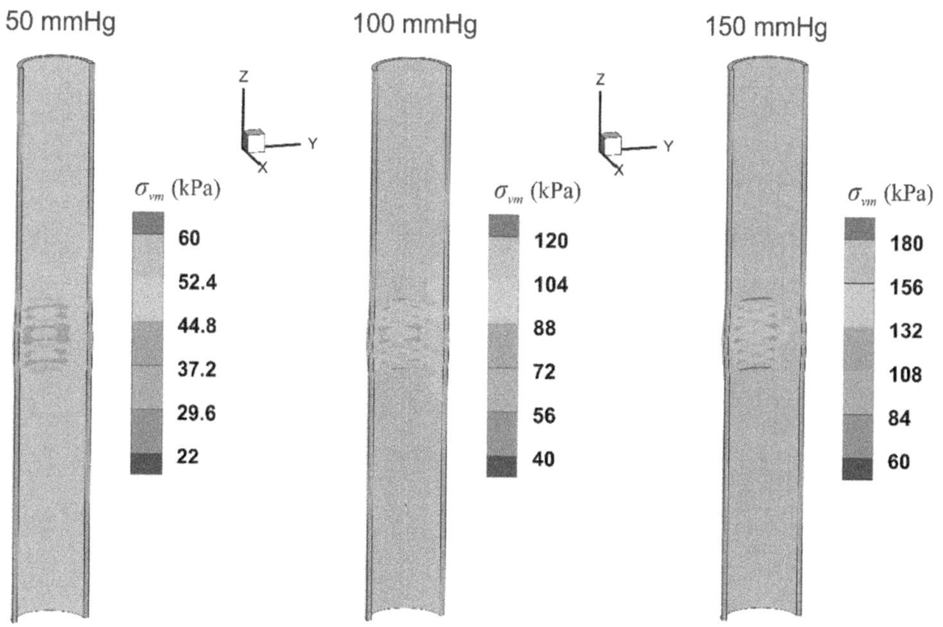

Figure 5.104 Contours of the von Mises stress for RVE II on the internal surface of the artery for different blood pressures: 50, 100 and 150 mmHg. *Source:* Adapted from Hatefi Ardakani et al. 2022.

Table 5.22 compares the maximum values of the von Mises stress on external and internal surfaces of the degraded zone of the artery. The maximum values for both cases of RVE I and RVE II remain almost the same, but with different extents within the local degraded zone, as presented in Figures 5.101 to 5.104. Clearly, the internal surface experiences much larger stress levels, making it more susceptible to failure or subsequent initiation phases of rupture.

In order to examine the effect of the microstructure (collagen fibre content) better, the von Mises stress contours for RVE I and RVE II, associated with two different elements in the degraded zone of the macroscale model (shown in Figure 5.90), are presented in Figures 5.105 and 5.106, respectively.

The final deformation of the artery at two different locations, illustrated in Figure 5.107, shows that a significant reduction in the thickness of the artery wall occurs at the position with the lowest content of the collagen fibre. As a result, a major risk of aneurysm complications and even potential rupture of the artery wall for high blood pressures should be considered.

Table 5.22 Comparison of the maximum von Mises stress in the degraded zone of the artery.

	RVE I		RVE II	
Blood pressure	**Internal surface**	**External surface**	**Internal surface**	**External surface**
50 mmHg	60 kPa	35 kPa	60 kPa	35 kPa
100 mmHg	120 kPa	61 kPa	120 kPa	61 kPa
150 mmHg	180 kPa	88 kPa	180 kPa	88 kPa

Source: Adapted from Hatefi Ardakani et al. 2022.

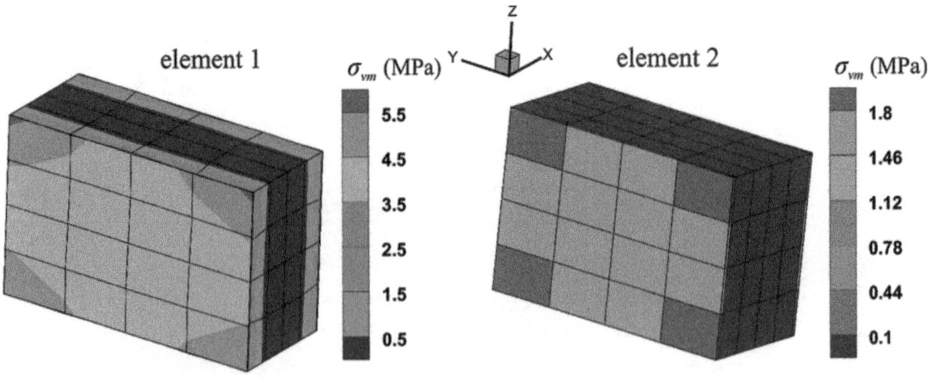

Figure 5.105 Distribution of the von Mises stress at different macro positions (based on RVE I). *Source:* Adapted from Hatefi Ardakani et al. 2022.

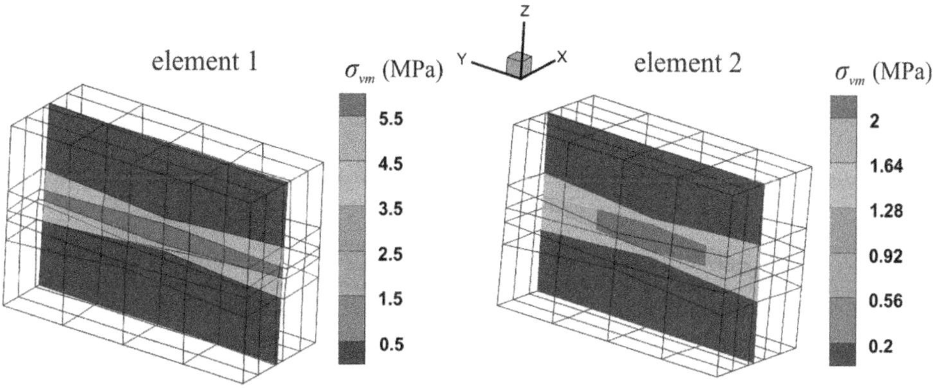

Figure 5.106 Distribution of the von Mises stress at different macro positions (based on RVE II). *Source:* Adapted from Hatefi Ardakani et al. 2022.

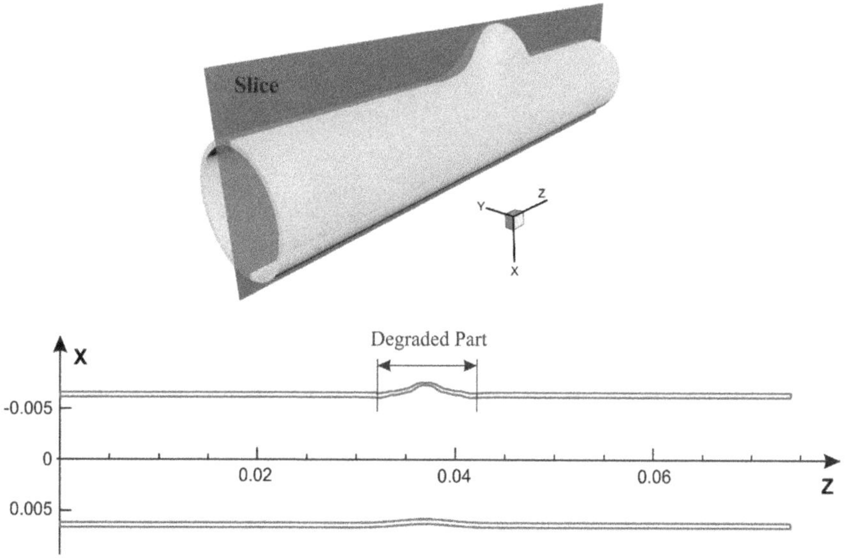

Figure 5.107 Slimming of the degraded zone of the artery wall. *Source:* Adapted from Hatefi Ardakani et al. 2022.

5.10 Multiscale Modelling of the Brain

5.10.1 Introduction

Apart from biomedical causes, the brain may be damaged or malfunction due to daily practices such as running and fall of objects or may be exposed to extreme conditions such as the effects of car accidents on passengers and the blast waves on occupants of an industrial complex. The situation is so important that many countries have developed code of practices and standards to regulate procedures to reduce the potential life consequences of damage on the human brain (traumatic brain injury, TBI). The multiscale biomechanics is an advanced efficient approach to study TBI and other biomechanical responses of the brain across a wide range of scales – from macromodelling to cellular scale.

Figure 5.108 shows illustrations of the brain in three successive scales of organ, tissue and cell level. In an impact loading on the brain, fracture of the skull, excessive acceleration, extreme levels of tensile and shear stress and strains, and internal pressures may lead to disproportionate deformations between the brain and skull and between different parts of the brain tissue, which leads to damage of nerve cells.

5.10.2 Biomechanics of the Brain

Brain tissue is similar to a composite material composed of a matrix and axon fibres, as depicted in Figure 5.109.

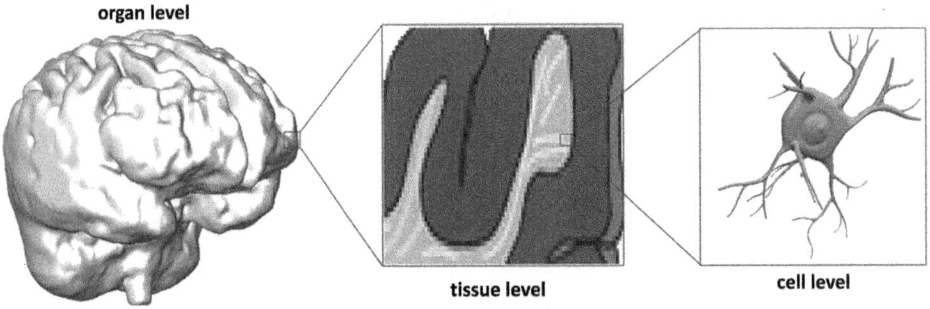

Figure 5.108 Typical structures of organ, tissue and cell levels of the brain.

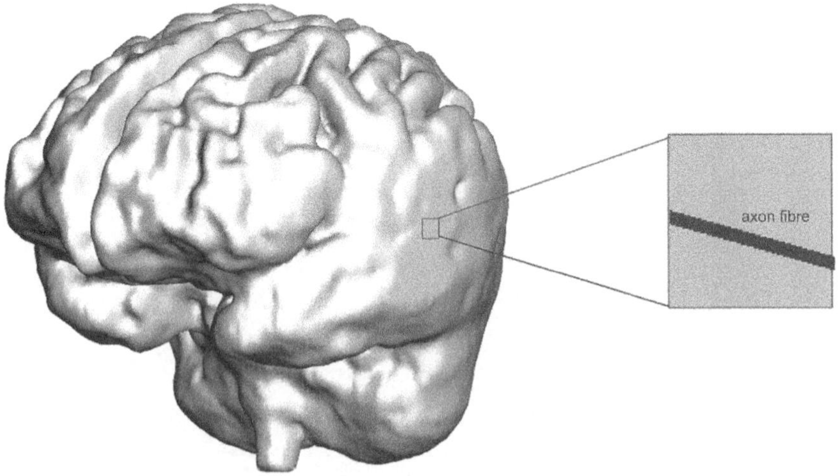

Figure 5.109 Axon fibre in the composite structure of the brain tissue.

While some reports believe that the white matter is highly anisotropic (Pfefferbaum et al. 2000), others have adopted isotropic hyperelastic models, at least in low strains (Arbogast et al. 1997a, 1997b; Donnelly and Medige 1997). Shuck and Advani (1972) experimentally studied the white and grey matter and concluded that they can be assumed to be isotropic in macroscale simulations. Later, Arbogast et al. (1997a) showed that the white matter is stiffer than the grey matter. In contrast, Prange et al. (2000) discussed that in large shear strains, the grey matter shows a larger elastic modulus than the white matter.

Damage to the brain tissue may be local and concentrated, which are usually caused by mechanical effects. Studies on helmet safety, airbags, etc. are largely performed in the car industries in order to reduce the potential damage on the brain to a medically acceptable level. An external mechanical impact is transmitted to the brain through the skin, skull, meninges membrane and cerebrospinal fluid. Many numerical simulations may simplify the modelling procedure by ignoring some of the external layers and applying the equivalent mechanical effects directly on part of the brain.

Alternatively, diffuse damage may occur in various parts of the internal tissues (Silver et al. 2011). Internal damages are usually in the microscale level and are very hard to detect, even with advanced medical instruments.

In both cases, the predicted stress and strain states can be used to determine the level of damage on brain tissue or its microstructures.

A simple damage criterion, called the abbreviated injury scale (AIS), is based on medically observable indices (Khosravi and Ebrahimi 2008; Schmitt et al. 2009) in six stages; from a low damage state to eventual death.

Another criterion is the head injury criterion (HIC), proposed in 1972, which uses a mathematical relation to define the damage index in terms of the applied acceleration $a(t)$ (in terms of g) in a time interval $[t_1, t_2]$:

$$\mathrm{HIC} = \max\left\{ \left[\frac{1}{t_2 - t_1} \int_{t_1}^{t_2} a(t)dt \right]^{2.5} (t_2 - t_1) \right\} \tag{5.10.1}$$

Alternatively, the maximum acceleration criterion correlates the potential level of damage and skull fracture to the maximum applied acceleration (Mertz et al. 1997). Recently, the axon elongation criterion in terms of the strain or its rate has been used for investigating the damage on brain tissue on a micro level (Chatelin et al. 2012; Dagro et al. 2013). It is reported that the axon elongation is the practical cause of axon damage (rather than its reorientation) (Bain and Meaney 2000; Wright et al. 2013; Sahoo et al. 2015).

5.10.3 Multiscale Modelling of the Brain (neo-Hookean Model)

In the multiscale simulation, a hierarchical modelling is performed. The macromodel is analysed subject to the macroscale mechanical loading. Then, the critical part of the brain is further simulated in the microscale model with proper boundary conditions to determine the maximum deformations of axons and to assess the state of the damage criterion.

Accurate modelling of the boundary conditions of the brain subject to a mechanical impact loading is not straight-forward due to the existence of the skull, fluid, etc. Therefore, simplified models with sliding contact and even fixed boundaries have been frequently used (Darvish and Crandall 2002). Here, the simple model of Garimella et al. (2022) with two fixed top and bottom boundaries, which circumvents rotation of the brain, is adopted. The loading is applied in an incremental displacement control quasi-static approach.

5.10.3.1 Creation of Geometry of the Brain

First, MRI scan photos are transformed into a three-dimensional model, as presented in Figure 5.110 (Janfada 2018). The powerful Mimix software can be adopted to generate a

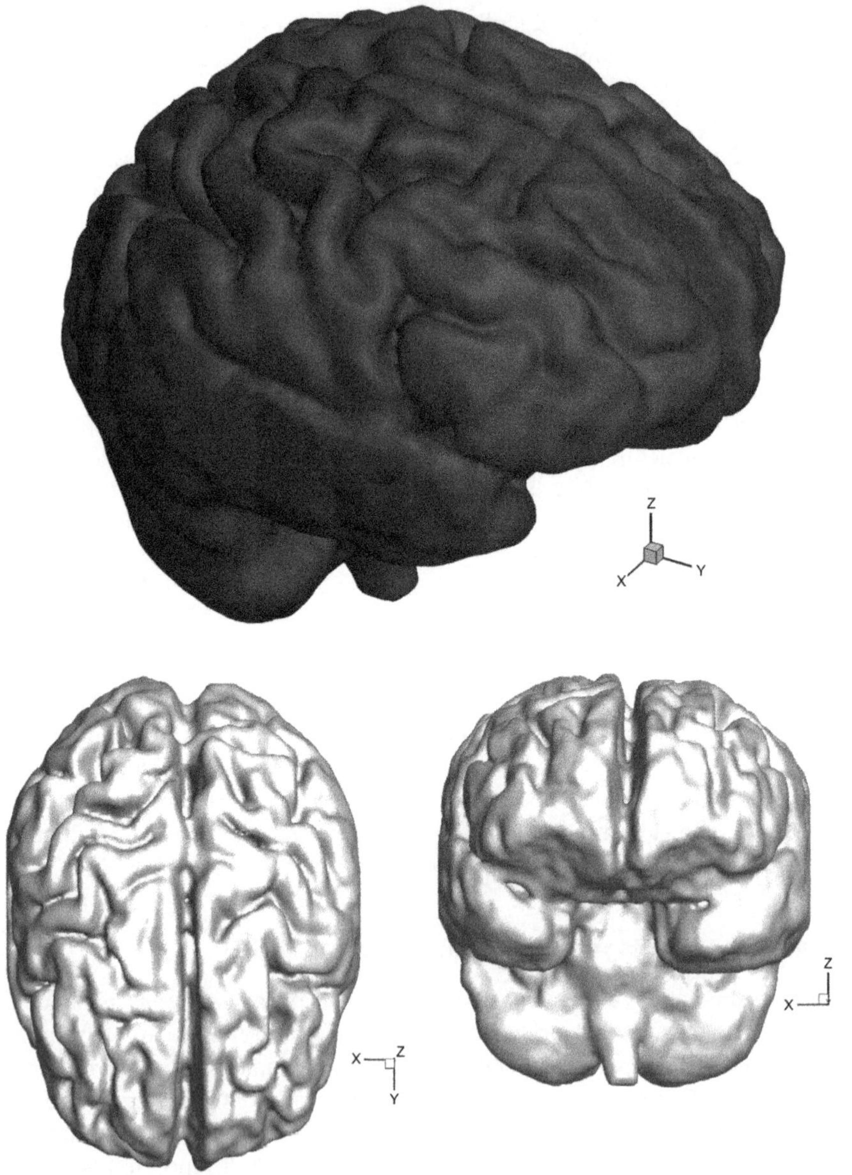

Figure 5.110 Three-dimensional image of the brain. *Source:* Adapted from Janfada 2018; Zolghadr 2022.

three-dimensional geometry of biosystems from digital scans (Baseri et al. 2016). Different tissues can be distinguished from the density of the grey colour according to known tables. The final smoothed geometry can be exported to the finite element software to generate the required mesh and to analyse the model.

5.10.3.2 Macro Model

The three-dimensional finite element model at the macro level is composed of 83,056 nodes and 3,222,924 four-node elements, as illustrated in Figure 5.111, to simulate the whole brain. Moreover, local details of the brain geometry can be accurately modelled by the very fine mesh of finite elements.

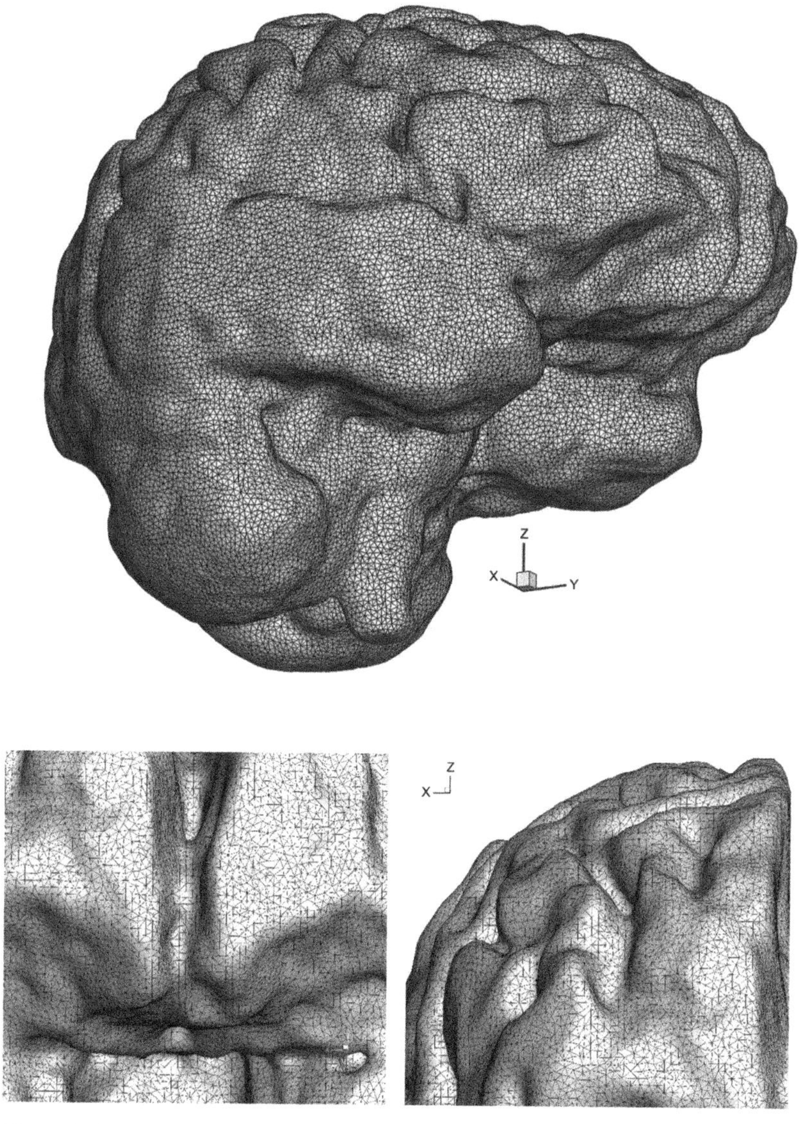

Figure 5.111 Three-dimensional finite element model of the brain. *Source:* Reproduced from Zolghadr 2022/the University of Tehran.

The neo-Hookean hyperelastic model is adopted to represent the constitutive equation of the brain tissue subject to a quasi-static concentrated loading.

The material properties, presented in Table 5.23, are taken from Budday et al. (2017) based on the characteristics of the corpus callosum part of the brain.

The model is subjected to a displacement control loading of 2 mm at the front of the brain. Distribution of the u_y is presented in Figure 5.112 from different viewpoints. More or less similar patterns are expected for other components.

Table 5.23 Material properties of the corpus callosum part of the brain for macroscale analysis.

Bulk modulus K (Pa)	Shear modulus μ (Pa)
15000	290

Source: Adapted from Zolghadr 2022/the University of Tehran.

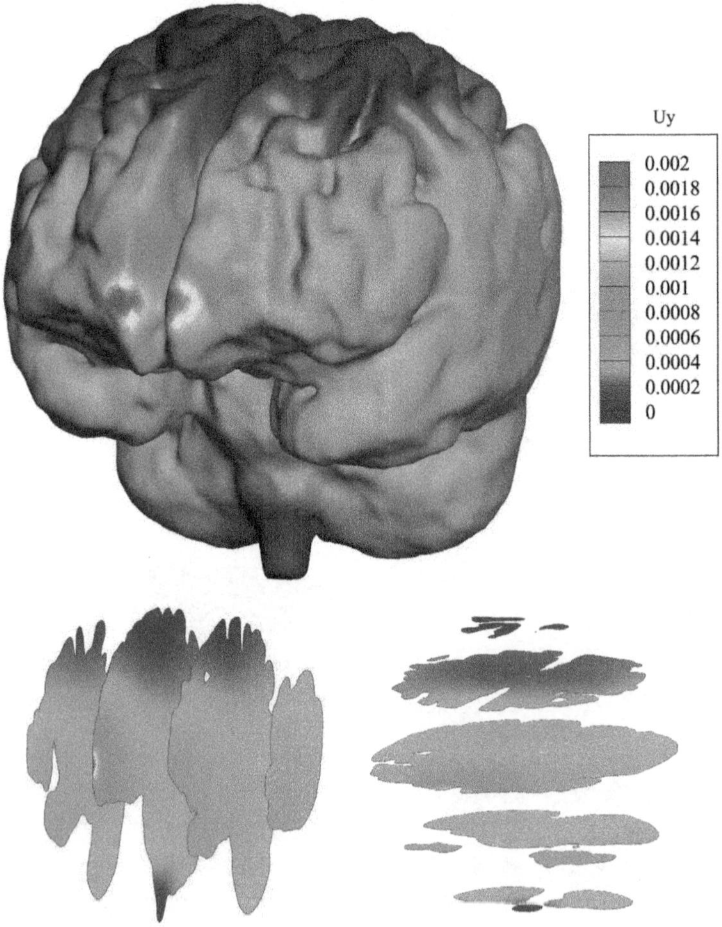

Figure 5.112 Displacement u_y of the brain. *Source:* Reproduced from Zolghadr 2022/the University of Tehran.

Distributions of the von Mises strain and stress components, as presented in Figures 5.113 and 5.114, respectively, show concentrations around the position of loading. The maximum value of 4,979 Pa can be observed for the von Mises stress. The maximum strain of about 46.5% occurs around the position of loading.

5.10.3.3 Micromodelling of the Brain

According to the hierarchical multiscale procedure, the critical points in the macromodel are determined. An RVE on the microscale with a detailed microstructure is then constructed for each critical point and is subjected to the corresponding deformation gradient $\boldsymbol{F}$ of the macromodel.

The selected $1\times1\times1\mu m$ RVE with periodic boundary conditions is modelled by the finite element mesh, as depicted in Figure 5.115. A finer mesh or different types of RVE structures may be considered. The dark grey parts represent the axon fibres, which constitute 53% of the volume of RVE, and the rest contains the extracellular matrix (Karami et al. 2009). Full continuity is assumed in between the axon and matrix in all stages of the analysis.

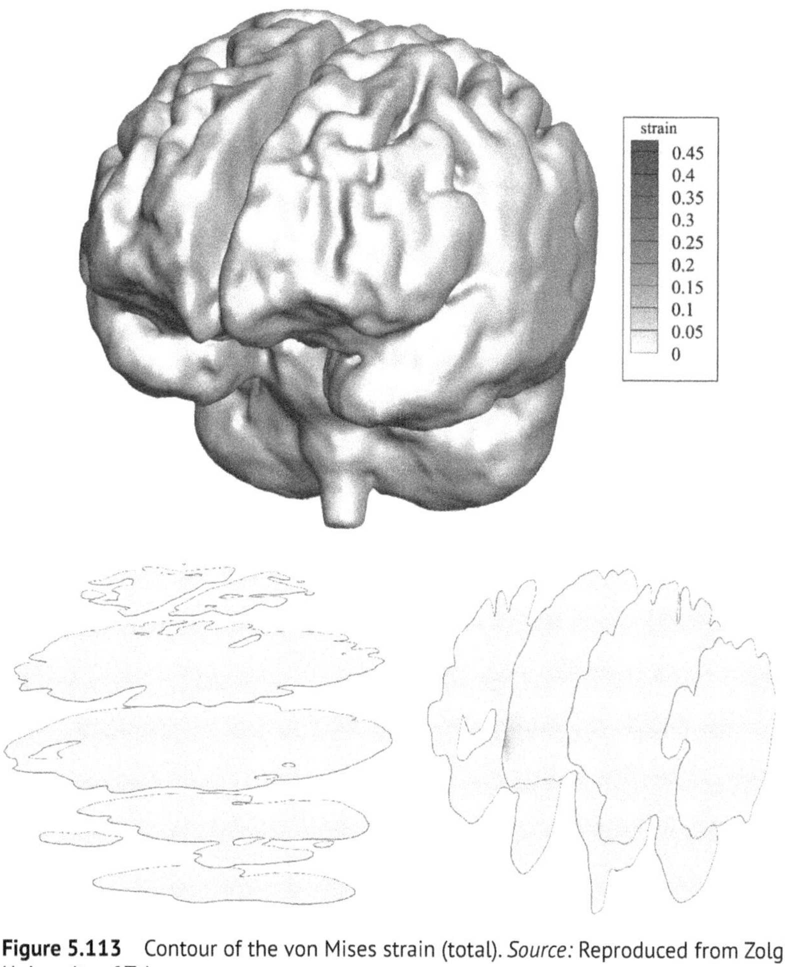

Figure 5.113 Contour of the von Mises strain (total). *Source:* Reproduced from Zolghadr 2022/the University of Tehran.

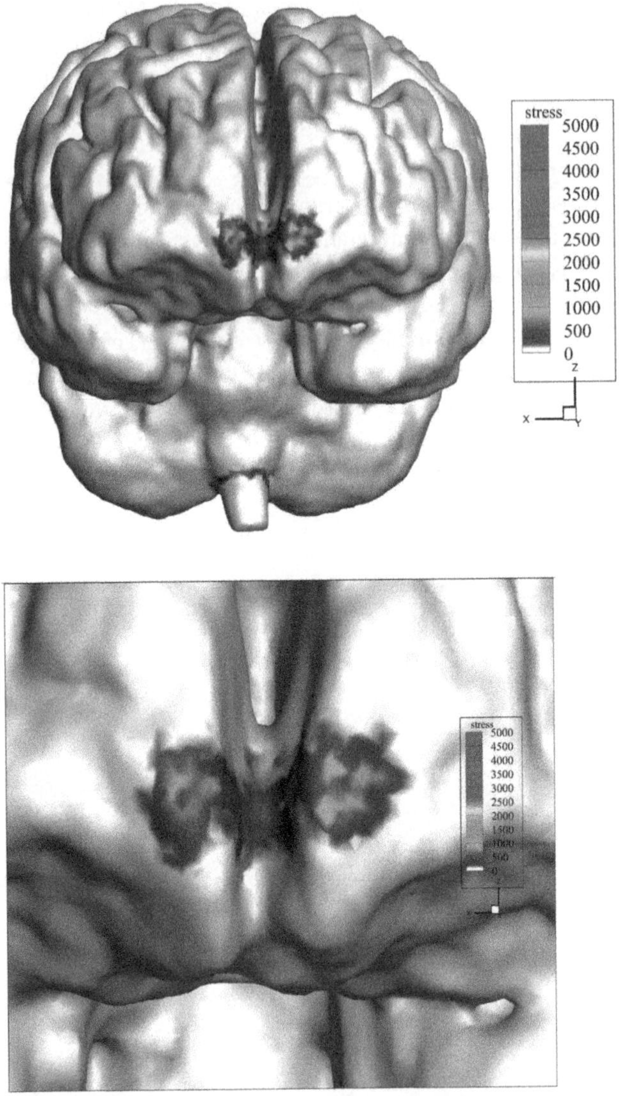

Figure 5.114 Contour of the von Mises stress (total). *Source:* Reproduced from Zolghadr 2022/the University of Tehran.

One difficulty of modelling at the microscale is that the necessary hyperelastic parameters may not be known, due to the fact that the same material properties of the macroscale brain cannot be used for the microscale constituents. Fortunately, results of unidirectional tests are available (Velardi et al. 2006) and a genetic algorithm (GA) can be efficiently followed to obtain the material parameters of the constituents at the microscale in an iterative procedure to achieve the same unidirectional experimental response. The obtained results are presented in Table 5.24.

Based on the micromechanical analysis of RVE, the results of deformations, strain and stress components can be obtained from the finite element model, as presented in Figures 5.116 to 5.118.

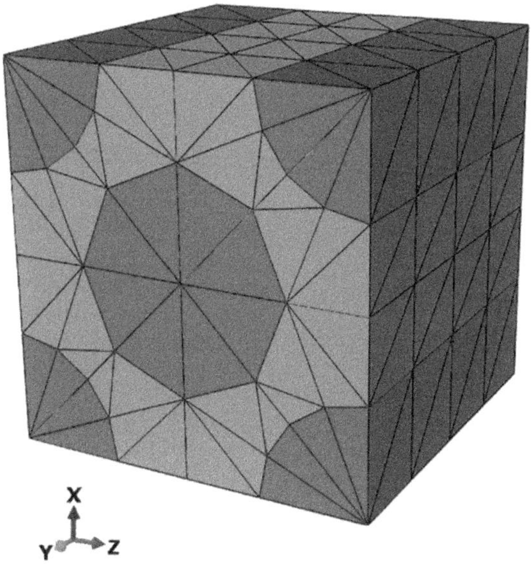

Figure 5.115 Finite element mesh of the micromodel of the brain, including the axons (dark grey) and the ECM (light grey). *Source:* Reproduced from Janfada 2018/the University of Tehran.

Table 5.24 Neo-Hookean material properties of ECM and axon for microscale modelling.

	Bulk modulus K (Pa)	**Shear modulus μ (Pa)**
ECM	4,354	70
Axon	23,625	380

Source: Adapted from Chavoshnejad et al. 2021; Zolghadr 2022.

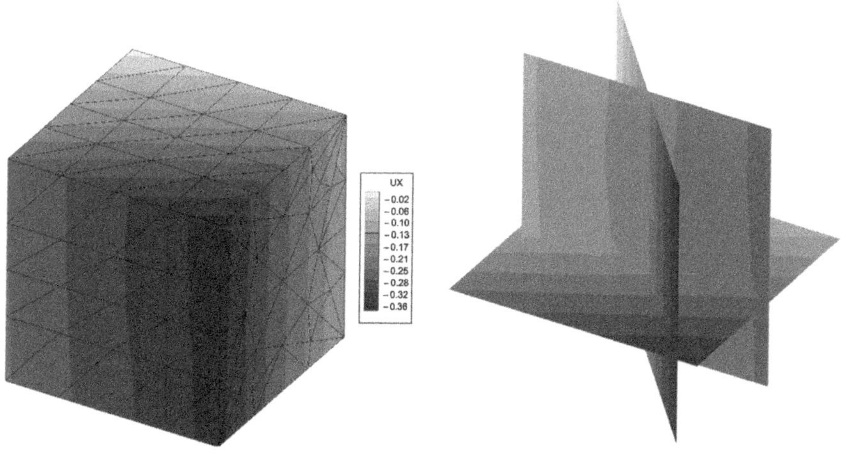

Figure 5.116 Displacement contours on the RVE. *Source:* Reproduced from Zolghadr 2022/the University of Tehran.

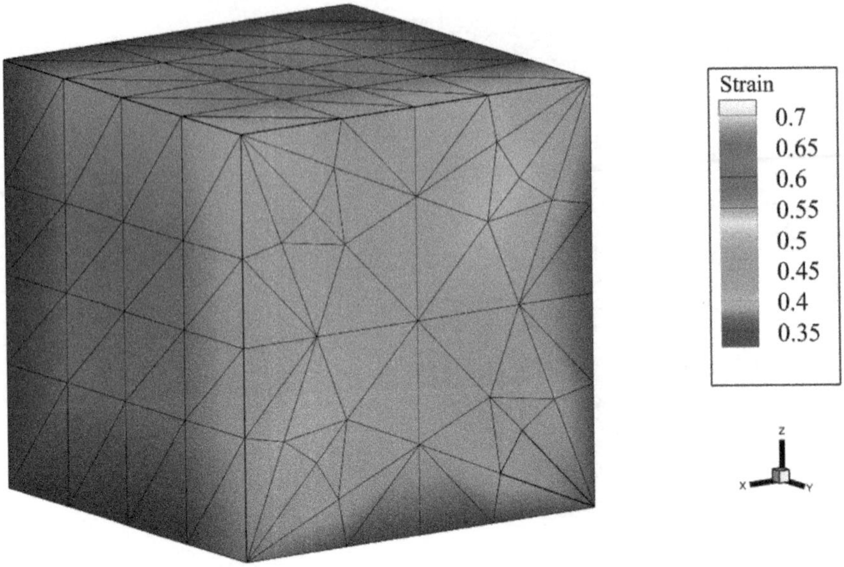

Figure 5.117 Contour of the von Mises strain (total) on the RVE. *Source:* Reproduced from Zolghadr 2022/the University of Tehran.

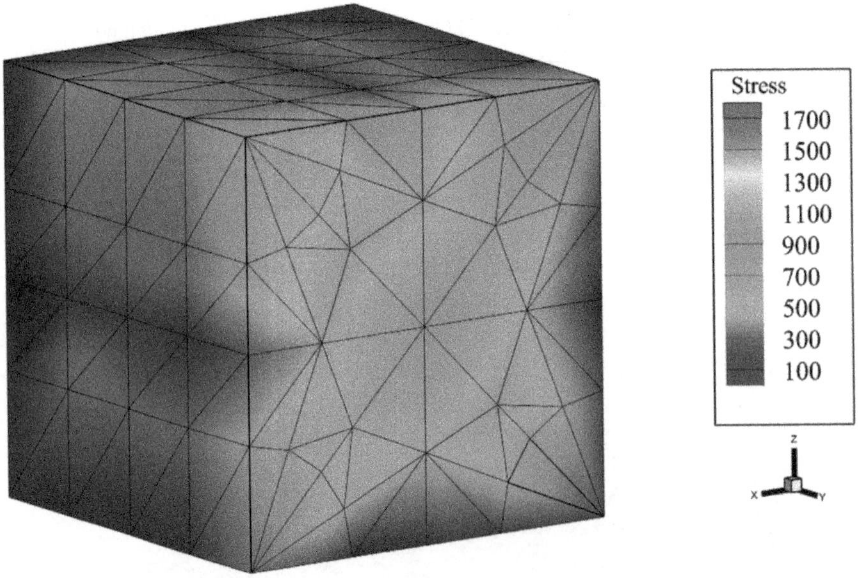

Figure 5.118 Contour of the von Mises stress (total) on the RVE. *Source:* Reproduced from Zolghadr 2022/the University of Tehran.

The von Mises strain varies between 40 and 50% in most parts of the RVE (Figure 5.117), indicating reasonable accuracy of RVE modelling, despite the large difference between the shear modules of macroscale and microscale models. Moreover, the maximum longitudinal strain in the Y direction is about 16.2%. This value can be used to check against the axon damage criterion.

Table 5.25 Mooney–Rivlin material properties of the brain tissue for macroscale modelling.

A_1(Pa)	A_2(Pa)
15	130

Source: Adapted from Chavoshnejad et al. 2021; Zolghadr 2022.

Figure 5.118 shows the contour of the von Mises stress on the RVE. Due to the large difference between the stiffness of axons and the matrix, axons bear most of the loading (between 95 and 1,700 Pa), which increase their potential risk of damage. The maximum stress in the axon is 1,732 Pa.

5.10.3.4 Macromodelling by the Mooney–Rivlin Law

The same model (geometry, loading and the finite element mesh) is now considered using the Mooney–Rivlin hyperelastic constitutive equation. The material properties are defined in Table 5.25.

The macromodel is subjected to 2 mm tensile deformation and the periodic boundary conditions are prescribed.

Contours of the displacement, von Mises strain and von Mises stress are presented in Figures 5.119 to 5.121.

The maximum strain (von Mises) is about 49.6% and occurs around the position of loading, as presented in Figure 5.120.

5.10.3.5 Micromodelling by the Mooney–Rivlin Law

For the micromodel, the same RVE model is analysed subject to the maximum deformation gradient experienced in the macromodel (associated with the von Mises stress of 866 Pa

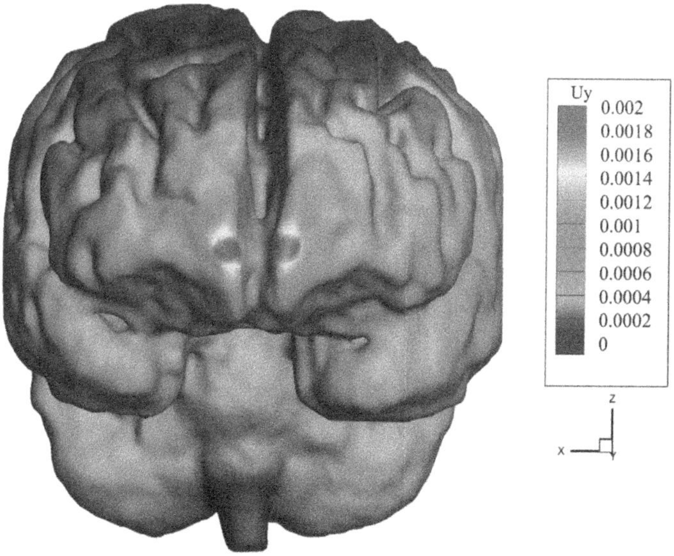

Figure 5.119 Displacement u_y of the brain. *Source:* Reproduced from Zolghadr 2022/the University of Tehran.

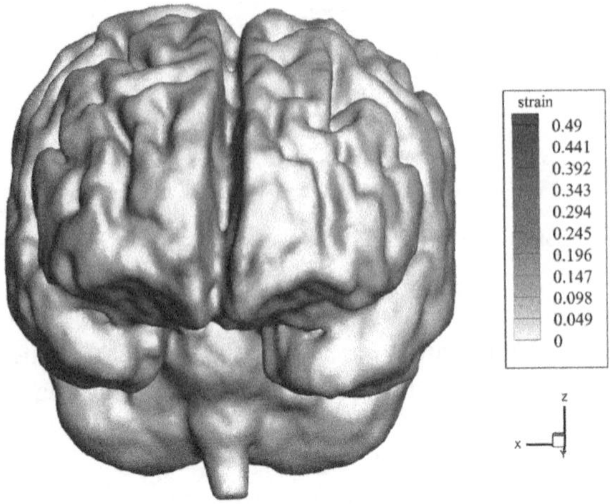

Figure 5.120 Von Mises strain (total) contour on the brain. *Source:* Reproduced from Zolghadr 2022/the University of Tehran.

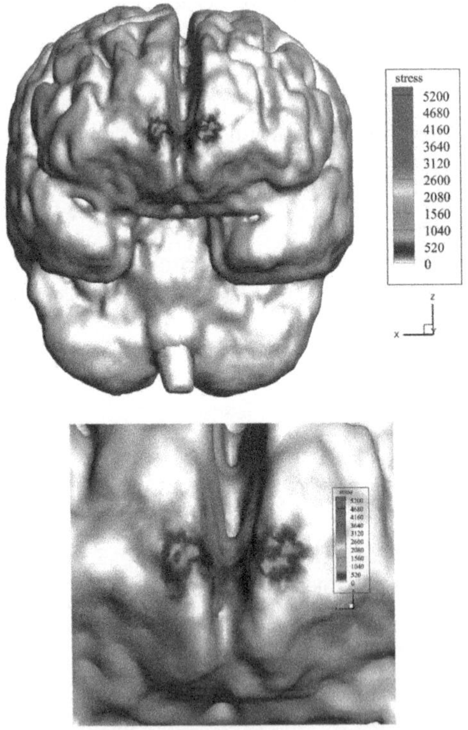

Figure 5.121 Von Mises stress (total) contour on the brain. *Source:* Reproduced from Zolghadr 2022/the University of Tehran.

and von Mises strain of 49%). The Mooney–Rivlin material properties are assumed from (Chavoshnejad et al. 2021), as defined in Table 5.26.

The von Mises strain and stress contours on the RVE are presented in Figures 5.122 and 5.123. The von Mises strains vary between 35 and 75%, but most of the regions are around 40 to 50%.

The von Mises stress varies between 181 and 2,762 Pa with the maximum value related to the axon. Clearly, higher stresses are generated in axons in the micromodel than the brain tissue of the micromodel due to the significant difference of the hyperelastic properties of the brain and axon.

Table 5.26 Mooney–Rivlin material properties of the ECM and axon in microscale modelling.

	A_1(Pa)	A_2(Pa)	κ(Pa)
ECM	60	6	4354
Axon	250	106	23625

Source: Adapted from Chavoshnejad et al. 2021; Zolghadr 2022.

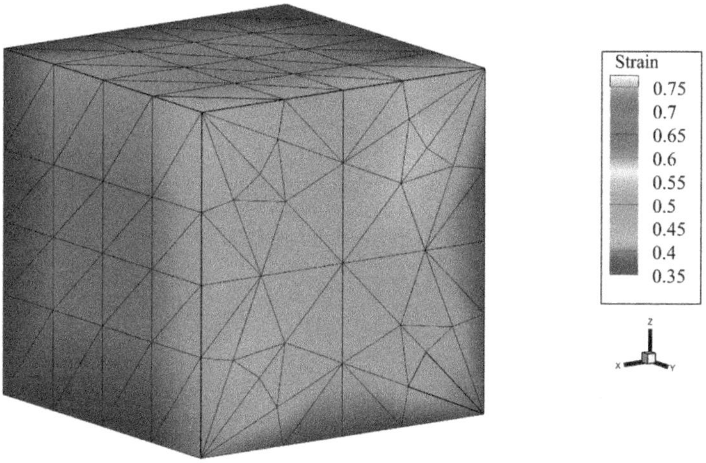

Figure 5.122 Von Mises strain (total) contour on RVE. *Source:* Reproduced from Zolghadr 2022/the University of Tehran.

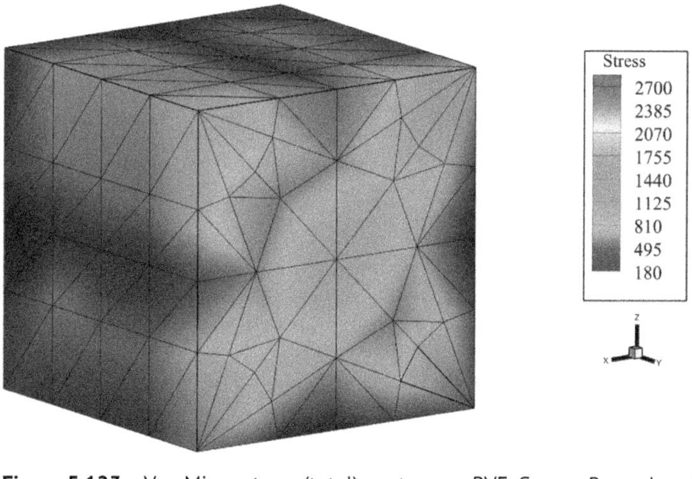

Figure 5.123 Von Mises stress (total) contour on RVE. *Source:* Reproduced from Zolghadr 2022/the University of Tehran.

Table 5.27 compares the results of neo-Hookean and Mooney–Rivlin assumptions. While they are generally in relative agreement, the Mooney–Rivlin constitutive model leads to higher levels of stress and strain at the micro level.

5.10.3.6 Nominal Damage Assessment

Axons that are subjected to maximum tension are most vulnerable to damage. The number of axons with excessive stretch is increased by the increase of loading effects on the microstructure of the tissue, leading to higher brain damage (Bain et al. 2003).

Table 5.28 compares the maximum equivalent Green–Lagrange strain of the present study with some of the available reference results. The Green–Lagrange strain ε_{GL} is obtained from the nominal strain ε_{eng} by

$$\varepsilon_{GL} = \frac{1}{2}\left[(\varepsilon_{eng} + 1)^2 - 1\right] \tag{5.10.2}$$

According to Bain and Meaney (2000), such a level of straining would damage the axons and disrupt the functioning of the nerve system of the brain. Some other references, however, may require higher strains as the onset of damage.

5.10.4 Viscoelastic Modelling of the Brain

Due to time dependency of the response of brain tissue and its constituents, a set of viscoelastic simulations are presented. The viscoelastic constitutive assumption is adopted for both macro (brain) and micro (RVE) simulations.

Table 5.27 Comparison of maximum von Mises strain and stress values for neo-Hookean and Mooney–Rivlin models.

	Neo-Hookean		Mooney–Rivlin	
	von Mises stress (Pa)	von Mises strain (%)	von Mises stress (Pa)	von Mises strain (%)
Macro	4,979	49.6	5,216	49.5
Micro	1,732	70	2,762	77

Source: Reproduced from Zolghadr 2022/the University of Tehran.

Table 5.28 Comparison of Green–Lagrange strains ε_{GL}.

	ε_{GL} (%)
Bain and Meaney (2000)	19.6
Wright et al. (2013)	17.3
Sahoo et al. (2015)	14.7
Present multiscale (neo-Hookean)	16.2

Source: Reproduced from Janfada 2018/the University of Tehran.

5.10.4.1 Macromodel of the Brain

The macromodel is now analysed by the viscoelastic model. The viscoelastic parameters reported by Finan et al. (2012), as presented in Table 5.29, are assumed to be valid for the present model.

The macromodel is subjected to a constant displacement for 5 s. The obtained short-term stress in various parts of the brain varies between 0 to 643 Pa, as presented in Figure 5.124.

After stress relaxation, the long-term von Mises stress is largely reduced, with the maximum value of about 103 Pa, as depicted in Figure 5.125.

After the instantaneous displacement is applied (in 0.001 s), the strain state remains constant for 5 s, with the maximum value of 6%, as illustrated in Figure 5.126.

Table 5.29 Elastic and viscoelastic material parameters for the brainstem.

Elastic modulus E_0 (Pa)	**Lame modulus λ_0 (Pa)**		**Shear modulus μ_0 (Pa)**		**Bulk modulus K_0 (Pa)**	
3751.8	61691		1,259		62,350	
Viscoelastic parameters						
E_∞ (Pa)	E_1 (Pa)	η_1 (Pa.s)	E_2 (Pa)	η_2 (Pa.s)	E_3 (Pa)	η_3 (Pa.s)
408.3	3,751.2	41.0	2,112.8	237.4	742.0	2,556
Alternative viscoelastic parameters						
g_∞	g_1	τ_1 (s)	g_2	τ_2 (s)	g_3	τ_3 (s)
0.108	0.563	0.0194	0.197	0.320	0.130	5.23

Source: Adapted from Finan et al. 2012; Zolghadr 2022.

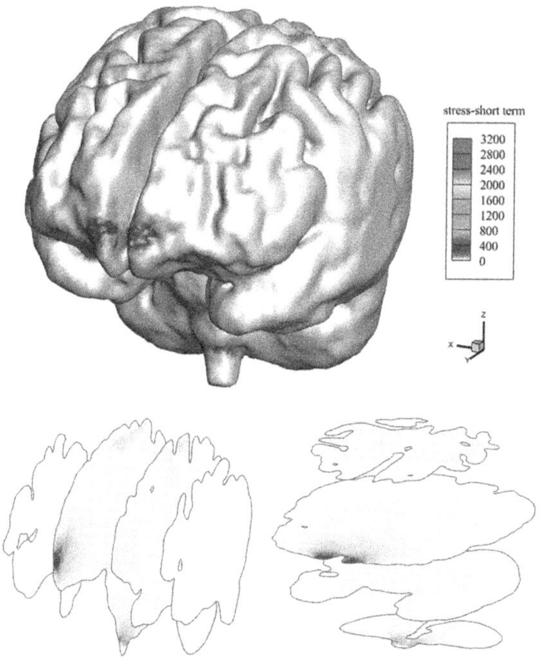

Figure 5.124 Short-term von Mises stress contour on the brain. *Source:* Reproduced from Zolghadr 2022/the University of Tehran.

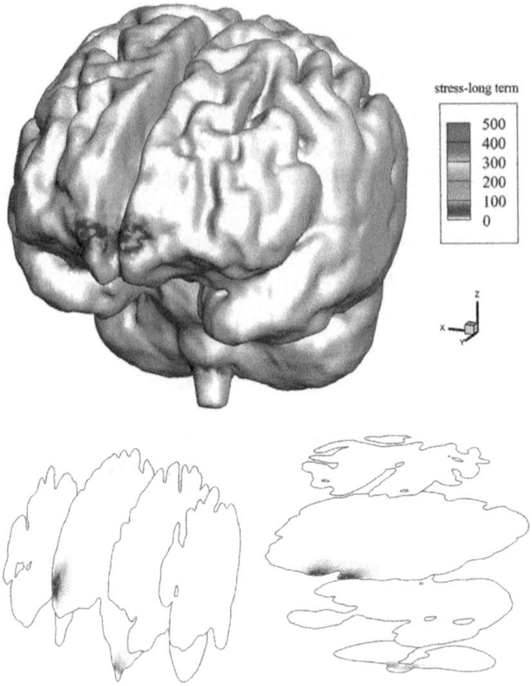

Figure 5.125 Long-term von Mises stress contour on the brain. *Source:* Reproduced from Zolghadr 2022/the University of Tehran.

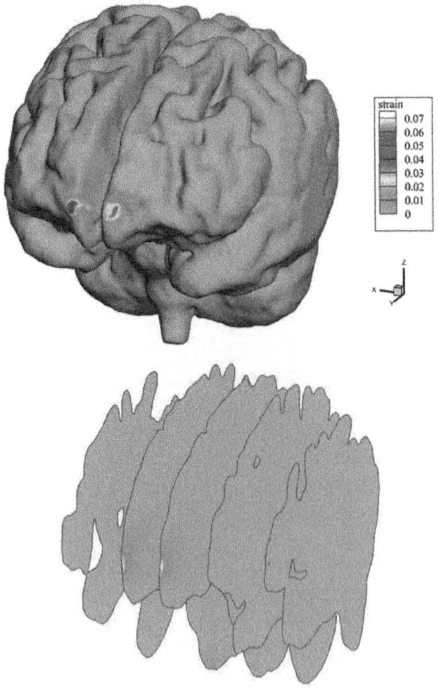

Figure 5.126 Von Mises strain contour on the brain. *Source:* Reproduced from Zolghadr 2022/the University of Tehran.

5.10.4.2 Viscoelastic Modelling of the Microscale Model

An RVE from the brain tissue with dimension of $1\times1\times1\mu m$, previously studied by Javid et al. (2014), is considered. Randomly generated unidirectional axons form 52.7% of the volume fraction and the rest is matrix (Figure 5.127). The finite element mesh is composed of 21,725 elements and 23,920 nodes. The RVEs are constrained by the periodic boundary conditions and are subjected to various instantaneous total strainings in the longitudinal (z) direction, which are kept constant for 5 s. The time-dependent responses of RVEs in a relaxation viscoelastic solution are obtained by the Newton–Raphson nonlinear solution.

The two-element Maxwell model is considered and the material parameters for the axon and matrix are defined in Table 5.30 (Javid et al. 2014).

Figure 5.128 compares the present results with those of Javid et al. (2014), which shows a good agreement.

The second study is to analyse different compositions of RVE for a corpus callosum sample of the brain and to compare the results with the available reference results (Velardi et al. 2006; Karami et al. 2009). The volume fractions of axons are assumed to be 40%, 52.7% and 60% and three different RVEs (Figure 5.129) are analysed subject to a stress relaxation viscoelastic process with 1%, 3% and 5% strains in 5 s (Figure 5.130).

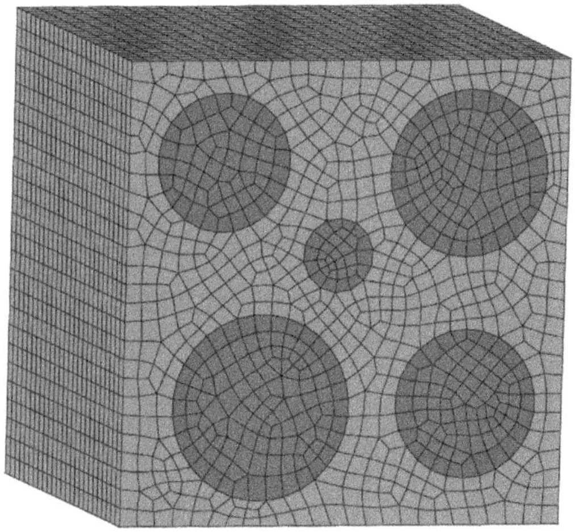

Figure 5.127 RVE with 52.7% volume fraction of axons. *Source:* Adapted from Zolghadr and Mohammadi 2022.

Table 5.30 Viscoelastic material parameters for the axon and ECM of the brainstem.

	Elastic modulus E_0 (Pa)		Lame modulus λ_0 (Pa)	Shear modulus μ_0 (Pa)	Bulk modulus K_0 (Pa)
Axon	12,860		211,456	4,315.4	214,333
ECM	4,290		70,540	1,439.6	71,500
	Viscoelastic parameters				
	E_∞ (Pa)	E_1 (Pa)	η_1 (Pa.s)	E_2 (Pa)	η_2 (Pa.s)
Axon	3,701.1	7,766.1	4,667.2	192.7	694.5
ECM	1,030.2	2,145.0	13.36	1,114.8	1,003.3
	Alternative viscoelastic parameters				
	g_∞	g_1	τ_1 (s)	g_2	τ_2 (s)
Axon	0.2878	0.6039	0.6020	0.1083	0.4987
ECM	0.2401	0.5000	0.0062	0.2599	0.9

Source: Adapted from Javid et al. 2014; Zolghadr and Mohammadi 2022.

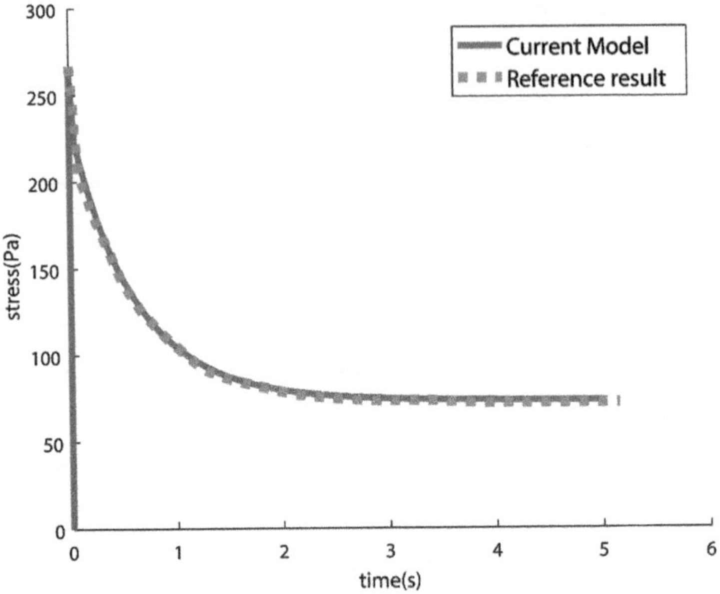

Figure 5.128 Comparison of the present and reference results. *Source:* Adapted from Javid et al. 2014; Zolghadr and Mohammadi 2022.

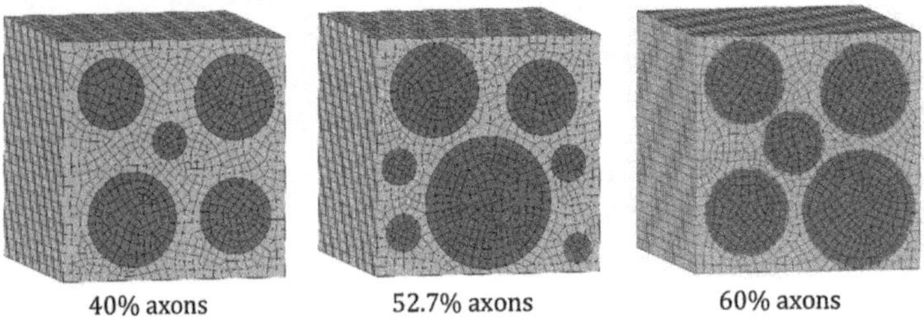

Figure 5.129 Different RVEs with 40%, 52.7% and 60% axons. *Source:* Adapted from Zolghadr and Mohammadi 2022.

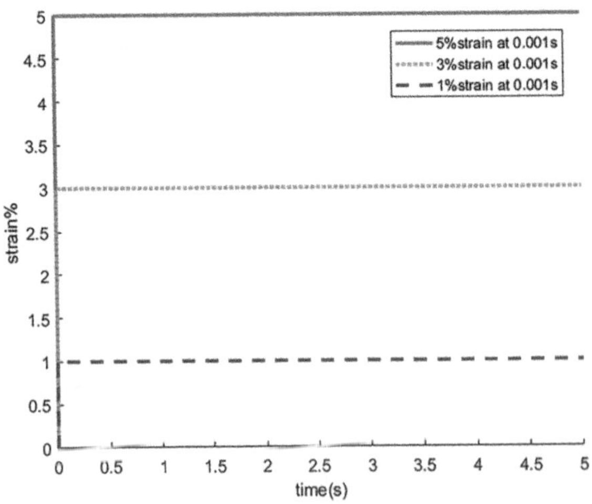

Figure 5.130 Step 1%, 3% and 5% strains (initial peaks at 0.001 s) for 5 s in a stress relaxation process. *Source:* Reproduced from Zolghadr 2022/the University of Tehran.

The long term σ_z stress distribution contours after 5 s are presented for all RVEs in Figures 5.131 to 5.133.

Temporal variations of the relaxation of average stress σ_z of the tissue based on different volumes of axons (different RVEs) are shown in Figure 5.134, which are associated with the three levels of 1%, 3% and 5% instantaneous straining (0.001 s, according to Figure 5.130).

Tables 5.31 and 5.32 compare the short-term and long-term stress σ_z for different RVEs (different volume fractions of axons).

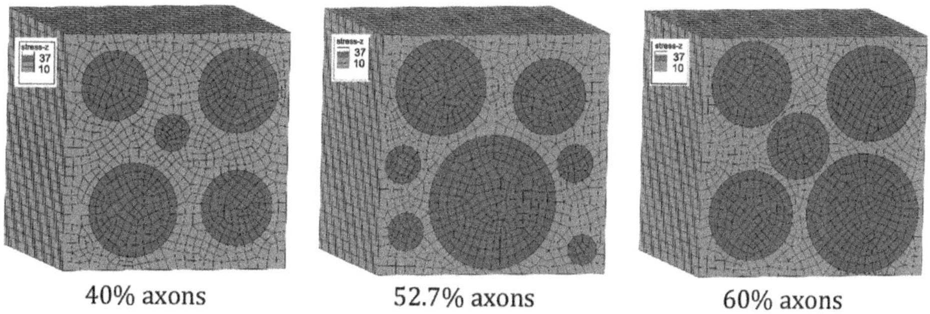

Figure 5.131 Contours of long term σ_z on different RVEs with 40%, 52.7% and 60% axons subject to 1% strain. *Source:* Adapted from Zolghadr and Mohammadi 2022.

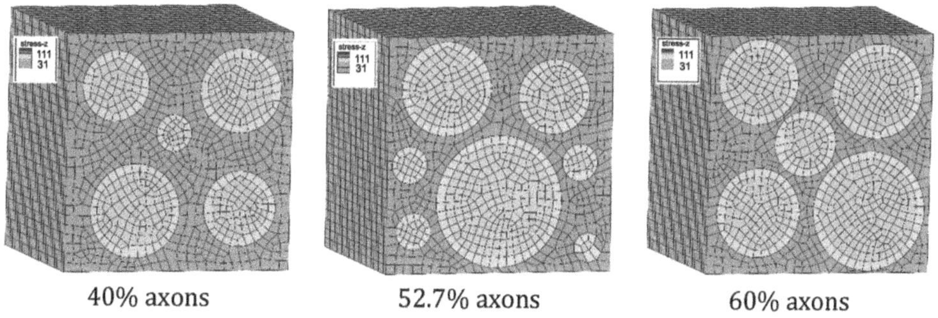

Figure 5.132 Contours of long term σ_z on different RVEs with 40%, 52.7% and 60% axons subject to 3% strain. *Source:* Adapted from Zolghadr and Mohammadi 2022.

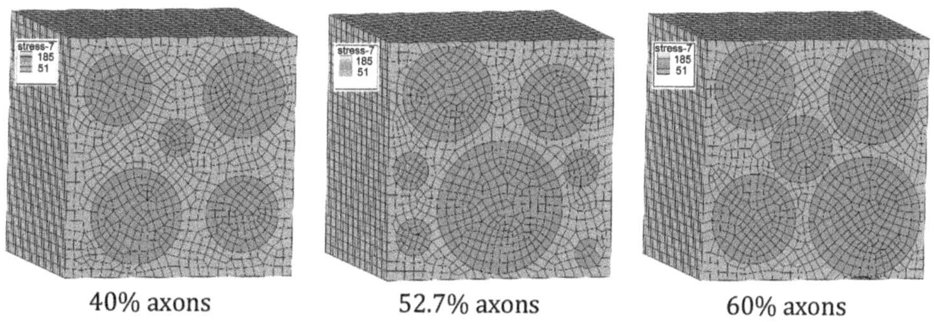

Figure 5.133 Contours of long term σ_z on different RVEs with 40%, 52.7% and 60% axons subject to 5% strain. *Source:* Adapted from Zolghadr and Mohammadi 2022.

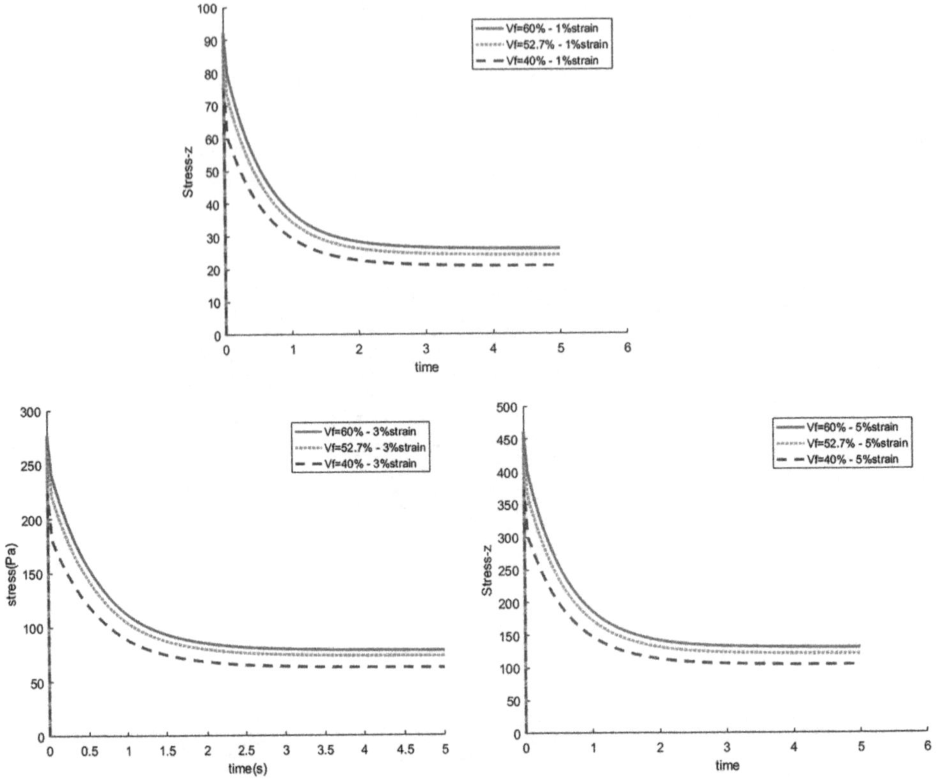

Figure 5.134 Relaxation of stress σ_z due to 1%, 3% and 5% straining. *Source:* Reproduced from Zolghadr 2022/the University of Tehran.

Table 5.31 Short-term and long-term stress σ_z for different RVEs (different volume fraction of axons).

S_{33} (Pa)	Strain	RVE1 (40% axon)	RVE2 (52.7% axon)	RVE3 (60% axon)
Short-term (0.001 s)	1%	74	87	92
	3%	225	260	280
	5%	375	430	461
Long-term (4.5 s)	1%	20	24	26
	3%	58	70	80
	5%	107	120	130

Source: Reproduced from Zolghadr 2022/the University of Tehran.

In order to examine the effect of strain rates, the 3% strain is applied at three different times of 0.001 s, 0.2 s and 0.5 s, as depicted in Figure 5.135. Figure 5.136 shows that the applied strain in 0.05 s generates less short-term stress than the two higher strain rates. By increasing the time of loading in a stress relaxation analysis, the brain tissue experiences lower short-term stress, whereas the long-term stress remains unaltered.

Table 5.32 Long-term stress σ_z in axon and matrix for different RVEs (different volume fraction of axons).

Long-term S_{33} (Pa)	Strain	RVE1 (40% axon)	RVE2 (52.7% axon)	RVE3 (60% axon)
axon	1%	37	37	37
	3%	111	111	111
	5%	185	185	185
matrix	1%	10	10	10
	3%	31	31	31
	5%	51	51	51

Source: Reproduced from Zolghadr 2022/the University of Tehran.

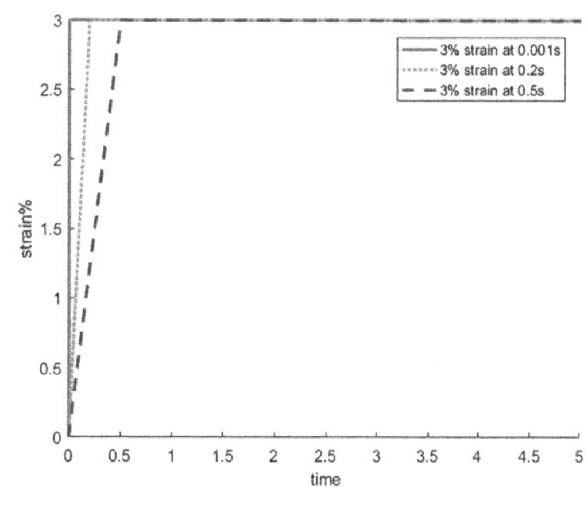

Figure 5.135 Various strain rates for the relaxation analysis. *Source:* Reproduced from Zolghadr 2022/the University of Tehran.

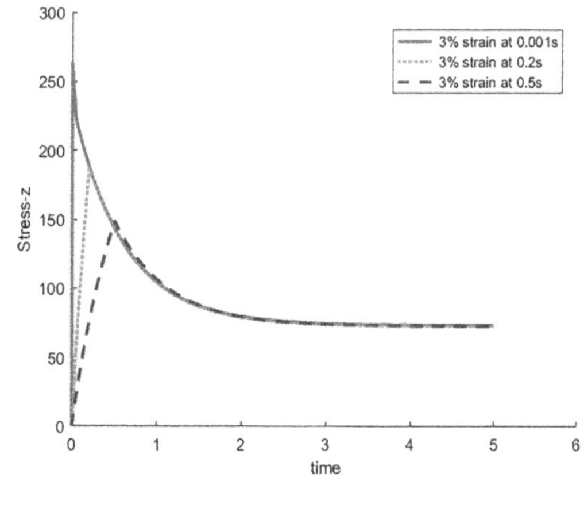

Figure 5.136 The effect of strain rate on variations of the relaxation stress σ_z. *Source:* Reproduced from Zolghadr 2022/the University of Tehran.

6

Biomechanics of Hard Tissues

6.1 Introduction

6.1.1 Hard Tissues

Hard tissues constitute the essential parts of the animal living system. They are composed of relatively light mineralized tissues that provide support and stability for the body or an internal organ (Nawathe et al. 2014), interconnect the flow of force and movement of muscles (Hoo 2011) and may contribute in various biological and metabolic functions as a source of calcium, phosphorus and other minerals (Hamed et al. 2013).

Depending on their roles, they exist in widely different shapes, densities and compositions. For instance, the skull plays a protective role for the brain and the ribcage covers the heart, lungs and many other internal organs. The hip bone helps in stability of the body at a standing situation and is best shaped to facilitate the sitting positions. Moreover, the cavity of long bones provides the appropriate conditions for blood production (Figure 6.1).

6.1.2 Chemical Composition of Bone

A typical bone is composed of two main parts: the compact outer layer of the cortical tissue and the spongy inner part of the trabecular tissue (Hoo 2011).

From a different point of view, a typical bone is composed of an inorganic mineral phase (mainly carbonated hydroxyapatite, and small amounts of citrate, magnesium, fluoride, etc.), an organic phase (collagen, etc.), and cells and water (Hoo 2011; Hamed et al. 2012, 2013, 2015). Such a diverse composition from the soft collagen, stiff minerals, plasticizing water and cells naturally creates a stiff and light component that well serves all the structural and physiological requirements of the bone (Hamed et al. 2013). The basic multicellular unit of bones is composed of two main groups of cells, osteoclasts and osteoblasts, both of which are controlled by osteocytes (Colloca et al. 2014a).

6.1.3 Multiscale Structure of Bone

The hierarchical micromechanical composition of bone is clearly heterogeneous (Colloca et al. 2014b), yet is naturally optimized in geometrical topology, shape, size and density.

Multiscale Biomechanics: Theory and Applications, First Edition. Soheil Mohammadi.

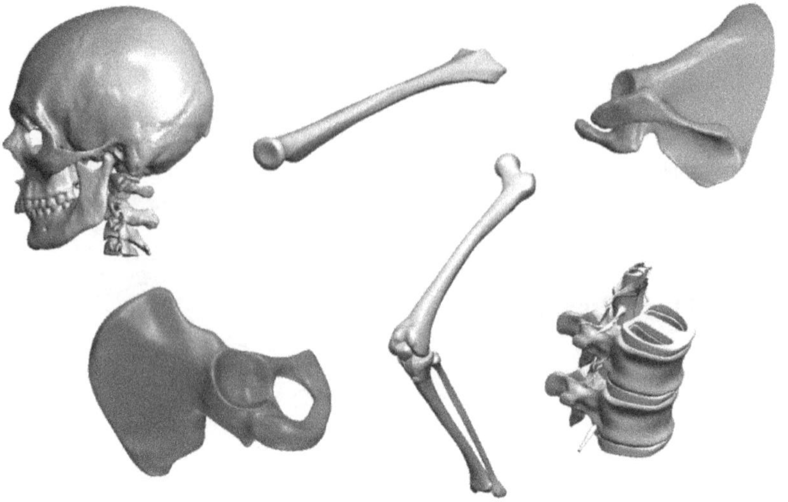

Figure 6.1 Different shapes of hard tissues (some of the designs are by GrabCAD).

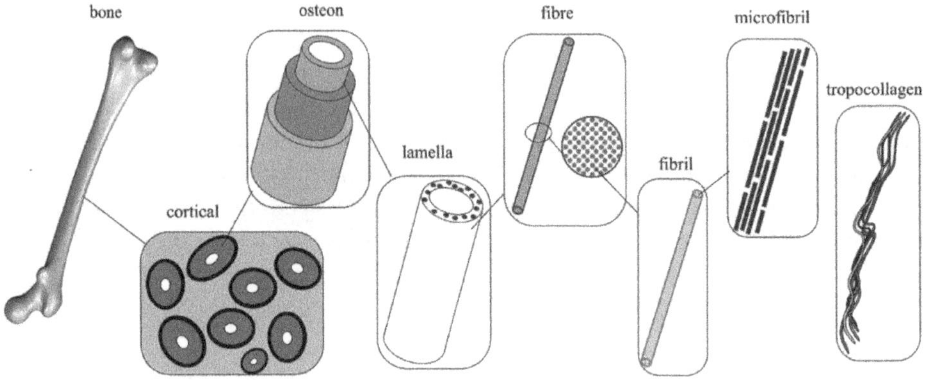

Figure 6.2 Hierarchical structure of the bone. *Source:* Adapted from Podshivalov et al. 2014; Barkaoui et al. 2014.

Both heterogeneous and optimized characteristics exist in various length scales (Nawathe et al. 2014). Such a multiple scale structure, as illustrated in Figure 6.2, is a major challenge for biomechanical analysis of hard tissue problems.

Some studies have proposed several hierarchical levels for specifying the bone structure at various scales. Alternatively, categories based on macro, meso, micro, nano and sub-nano scales have been proposed. Some other studies use a tissue-type description to define the scales (Beniash et al. 2019; Khaterchi et al. 2013; Ural and Vashishth 2014; Lemaire et al. 2013; Barkaoui et al. 2014; Podshivalov et al. 2014; Katsamenis et al. 2015).

Hoo (2011) provided a review on the various levels of microstructure of the bone as:

- Minerals, fibrils, water, etc. (~1–10 nm)
- Mineralized collagen fibrils (~500 nm)
- Fibril arrays (~1 μm)
- Lamellar (~1–10 μm)

- Osteons and single trabecula (~10–500 μm)
- Trabecular and cortical bones (mm– μm)
- Macroscale bone or organ level (cm)

This section briefly describes the material structure of each hierarchical level of bone and its mechanical function (Hoo 2011).

6.1.3.1 Hierarchical Level 1: Minerals, Fibrils, Water and Proteins

At the finest scale, the bone is composed of plate-shape minerals, collagen molecules, forming fibrils and non-collagenous organic proteins. The plate-shape minerals are in the size ranges of 5 to 50 nm and grow along the collagen fibrils (with diameters of about 10–50 nm and length of about 300 nm). The collagen molecules are made of helical protein chains, with diameters of about 1.5 nm and lengths of 300 nm. The diameter of forming fibrils is about 50–100 nm. The non-collagenous organic proteins (NCPs) form a cohesive matrix for attachment of the mineralized fibrils (Katsamenis et al. 2015). While the collagen fibrils provide the main contribution to the strength of the bone from its lowest scales, the ratio of mineral to organic components largely influences the overall macroscale characteristics (Hoo 2011) (Figure 6.3).

6.1.3.2 Hierarchical Level 2: Mineralized Fibrils

The mineralized collagen fibrils are made of mineral hydroxyapatite crystals and collagen molecules (Figure 6.4). The mechanical properties at this level are orthotropic, due to strong directional characteristics caused by stiff fibrils and low strengths of interfacial properties (Hoo 2011).

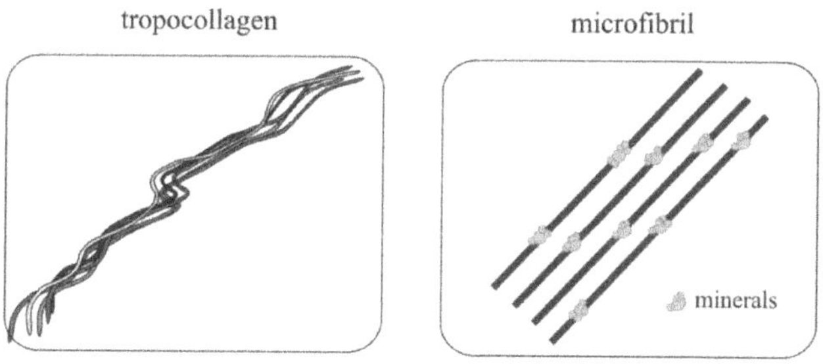

Figure 6.3 Tropocollagen and microfibril.

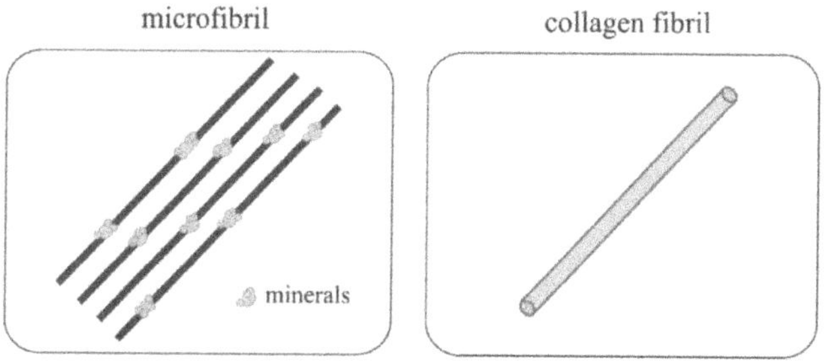

Figure 6.4 Assembly of collagen molecules into fibrils (mineralized collagen fibril).

6.1.3.3 Hierarchical Level 3: Parallel Set of Fibrils

Arrays of collagen fibrils (Figure 6.5) are formed by clusters of parallel mineralized fibrils, providing strong orthotropic mechanical properties (Hoo 2011).

6.1.3.4 Hierarchical Level 4: Lamellae

Collagen fibrils reinforced with some crystals form a lamella, which is a set of closely positioned thin plate-like structures with an open space in between (Podshivalov et al. 2014).

Bone lamellae are composed of collagen, mineral and water. In contrast to a porous spongy structure, which is created by randomly distributed minerals that fill the spaces between the fibrils, a lamellar component is made up of sub-layers of mineralized collagen fibrils, which can be organized in parallel, woven and plywood-like structures, as presented in Figure 6.6 (Hoo 2011).

6.1.3.5 Hierarchical Level 5: Osteon and Single Trabecula

Depending on the orientations of the stacked lamellae, they may form either osteons in the cortical tissue or trabecular rods in the trabecular tissue (Podshivalov et al. 2014; Nawathe et al. 2014).

Osteon is a cylindrical hollow structure formed by wrapping layers of collagen fibres (lamella). Its central hole is an ideal path for the blood vessels and nerve cells (Figure 6.7).

A trabecula is a tiny rod- or plate-shaped tissue with the size order of tens to hundreds of micrometres, found in the trabecular bone tissue. Its structure, which is composed of variably oriented layers of lamellae and cement, determine the anisotropic properties of the porous bone tissue (Podshivalov et al. 2014).

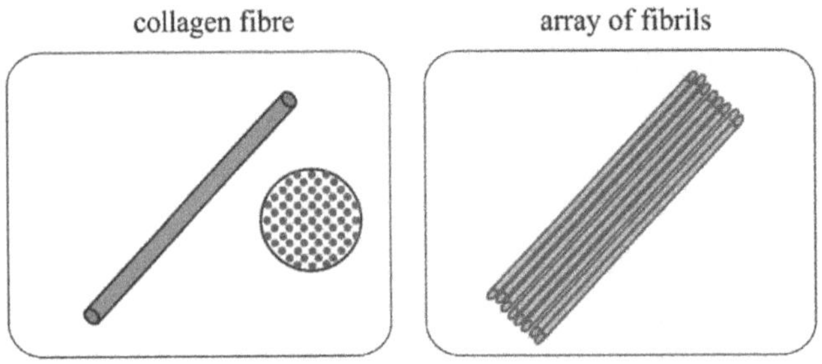

Figure 6.5 Array of collagen fibrils.

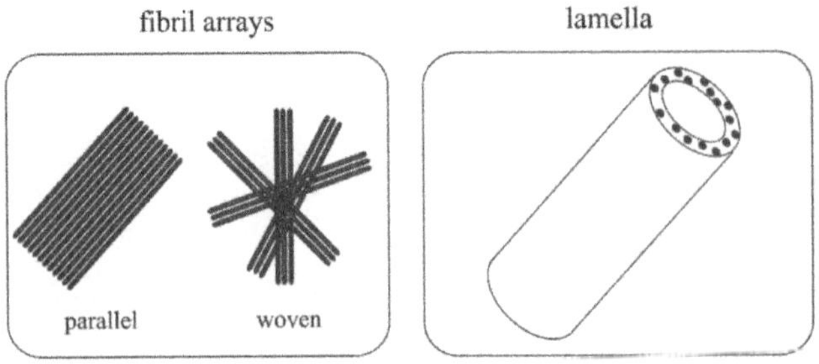

Figure 6.6 Different assembly patterns of fibrils and lamellae.

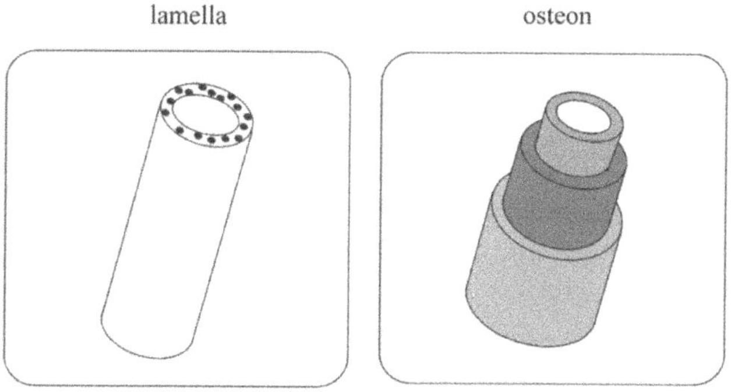

Figure 6.7 Illustrations of lamella and osteon.

6.1.3.6 Hierarchical Level 6: Cortical and Trabecular and Bone Tissues

Lamellae may be distributed in different orientations to form either osteons in cortical bone or trabecular parts of the trabecular bone (Podshivalov et al. 2014; Nawathe et al. 2014) (Figure 6.8).

The cortical tissue is the compact and dense solid part of the bone, which forms the outer shell of the bone. The primary role of the cortical bone is to provide sufficient overall strength for the bone and the skeleton system, to protect the organs and the weaker porous trabecular tissue, and to store and release minerals and chemicals. The mechanical characteristics of the cortical bone is either orthotropic or transversely isotropic (Odgaard et al. 1997; Pahr and Zysset 2014; Yang et al. 1999; Hoo 2011; Podshivalov et al. 2014). It contains microstructural voids and channels for blood vessels and neurovascular systems.

The trabecular bone is a highly spongy or cancellous tissue, usually in the three-dimensional lattice of trabeculae rods and plates, inside the cortical shell of the bone (Schwiedrzik et al. 2016; Podshivalov et al. 2014), as presented in Figure 6.8. The mechanical response of the cellular trabecular bone is heterogeneous and anisotropic and contributes to the overall mechanical response of the bone, despite its relatively low volume content. Moreover, it has a significant part to play in the energy absorption mechanisms (Ramezani et al. 2012; Podshivalov

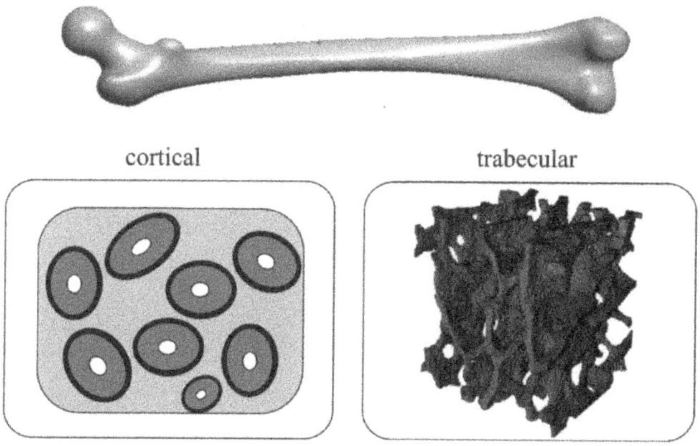

Figure 6.8 Cortical and trabecular bone tissues.

et al. 2014). Cavities or pores of the trabecular bone are used for the production of red blood cells and the storage of minerals (Nawathe et al. 2014).

Both cortical and trabecular tissues are vulnerable to damage due to cyclic loadings and show time-dependent viscous characteristics (Podshivalov et al. 2014).

6.1.3.7 Hierarchical Level 7: Macroscale Bone or Organ Level

The macroscale bones exist in various shapes and compositions, and consist of both cortical and trabecular tissues, as explained in previous sections (Figure 6.9).

An alternative classification is to replace the component-based definitions of levels by the structural size-based descriptions of sub-nano, nano, sub-micro, micro and macro levels as (Podshivalov et al. 2014):

- Minerals, collagen fibrils, water and organic proteins (<100 nm)
- Mineralized collagen fibrils (100 nm–1 μm)
- Lamellae (1–10 μm)
- Osteons and trabeculae (10 to 500 μm)
- Cortical and trabecular bones (> 1 mm)

6.1.4 Bone Remodelling

Despite its solid structure, a bone is continuously adapting its shape and even composition to comply with the natural growth and the very diverse ranges of mechanical and physiological demands over the long period of its functioning. The extent of adaptation depends on the required functioning of the bone. For instance, the skull may be subjected to an impact

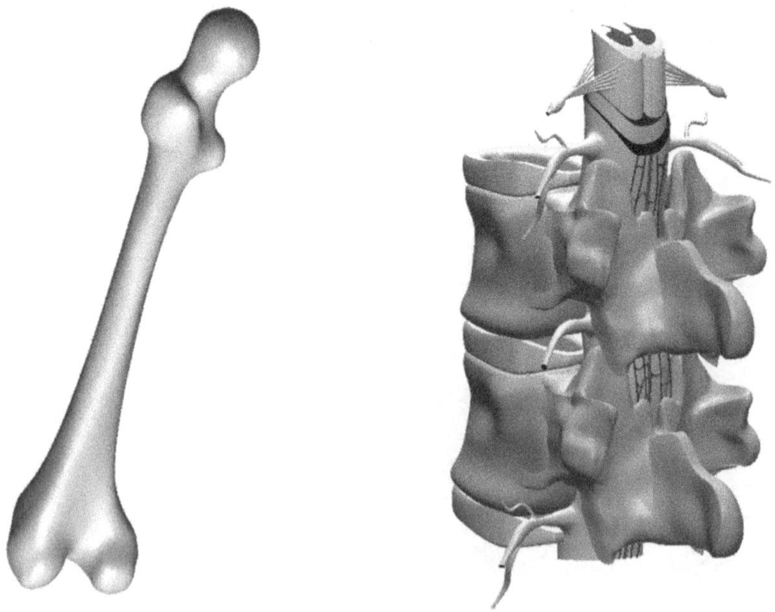

Figure 6.9 Samples of macro-level bones.

(car accident or an object impact), but it is not expected to experience such extreme mechanical loadings frequently. On the other hand, common loadings with lower maximum values are quite natural for the forearm, hip, femur, teeth and other bones (Podshivalov et al. 2014). These bones should be studied for potential micro damages and cracks in cyclic loadings and potential fatigue failures, which may even lead to macroscale damage, failure and crack.

Bones renew their shape, size and density constantly to comply with the growth conditions of the body. Interestingly, bones are naturally capable of healing or repairing a damaged part in a reasonable period of time. The very complex process of healing is based on the microstructure of the bone and the availability and diffusion of necessary cells and growth factors into the damaged area to allow for eventual generation of new cells and tissues (Christen et al. 2015a).

The phenomenon of the bone remodelling represents the continuous process of bone adaptation, where the bone is gradually created in the positions with high loading demands and may reduce in zones with low loading conditions (Christen et al. 2012). This may also justify the need for practicing and sport in maintaining the necessary density and characteristics of different bones.

Osteocytes regulate the adaptive response of osteoclasts and osteoblasts to properly respond to potential damage in the bone and the extent of existing mechanical effects (Nawathe et al. 2014; Christen et al. 2015b). This is achieved by gradually adding or removing the bone tissue accordingly. In a damaged hard tissue, osteoclasts remove the damaged matrix. It will be followed by an increase of growth factors and movement of the cells into the damaged region to stimulate formation of the new matrix and subsequent mineralization (Hoo 2011; Podshivalov et al. 2014).

6.1.5 Contents of the Chapter

After this introductory section on the physiological aspects of the bone, the next section reviews the concepts of fracture and failure modelling of the hard tissues.

Details of simulations of the femur bone fracture at multiple scales are then examined in Section 6.3, covering the microscale modelling of the trabecular bone, the mesoscale XFEM fracture analysis of heterogeneous cortical bone and the macroscale cracking of the femur bone.

The chapter concludes in Section 6.4 with a comprehensive discussion on the healing process of damaged/fractured bones by examining the large set of coupled space–time partial differential equations in terms of the seven independent variables of mesenchymal stem cells, chondrocyte, osteoblast, cartilage matrix and bone matrix densities, and chondrogenic and osteogenic growth factor concentrations.

6.2 Concepts of Fracture Analysis of Hard Tissues

The mechanical properties of bone decrease with time due to natural age effects and physiological processes, different diseases, abnormalities such as osteoporosis, and mechanical causes (Lekadir et al. 2016). The resultant reduction of the toughness of bone, both in its micro- and macrostructures, increases the bone fragility and the potential risk of cracking and fracture, even from small mechanical loads (Hamed et al. 2013; Katsamenis et al.

2015). The medical consequences of bone fracture is very high for a patient and the painful healing process takes a long time and largely affects normal life (Johnell and Kanis 2006; Katsamenis et al. 2015). The number of osteoporotic-related bone fractures are increasingly high (Metzger et al. 2015; Badilatti et al. 2015). As a result, the study of bone fracture is a major topic in biomedical and biomechanical applications (Lekadir et al. 2016).

Figure 6.10 illustrates X-ray scans of a fractured bone and the way a metal system is used to fix the fractured region from relative movements in order to allow for the natural healing process of the damaged area.

6.2.1 Numerical Studies of Bone Fracture

Both single and multiscale models have been widely used to study the fracture strength of bones.

Single-scale simulations are usually performed by the finite element method. For instance, the cohesive zone model (Ural et al. 2013; Ural and Mischinski 2013) can be adopted to examine bone cracking at both micro- and macroscales (Figure 6.11), and the

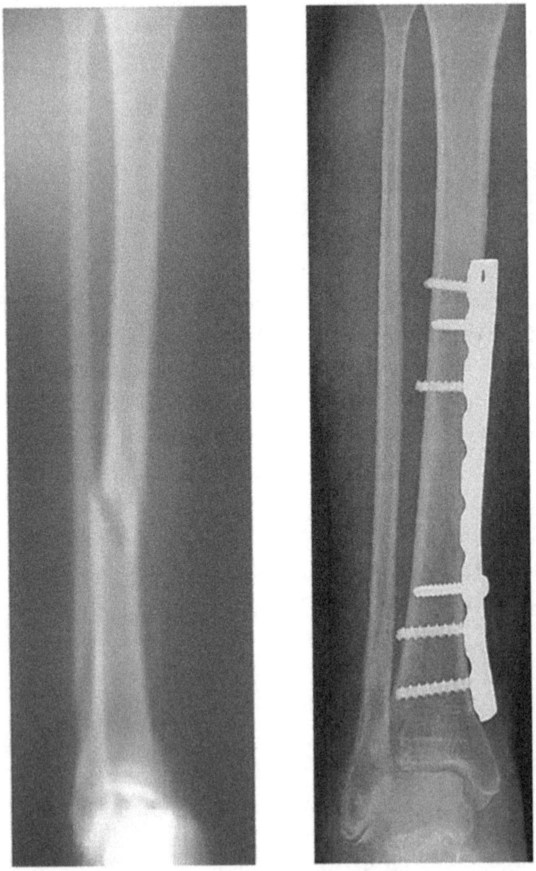

Figure 6.10 X-ray scans of a fractured tibia bone and the metal system for fixing the fractured zone (with permission from the patient, 2017).

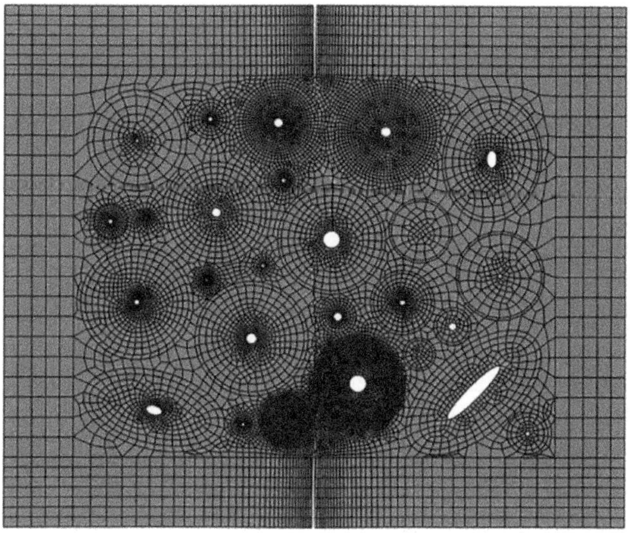

Figure 6.11 A typical finite element mesoscale model from a section of an inhomogeneous bone. *Source:* Reproduced from Torabizadeh and Mohammadi, 2022a; 2022b/University of Tehran.

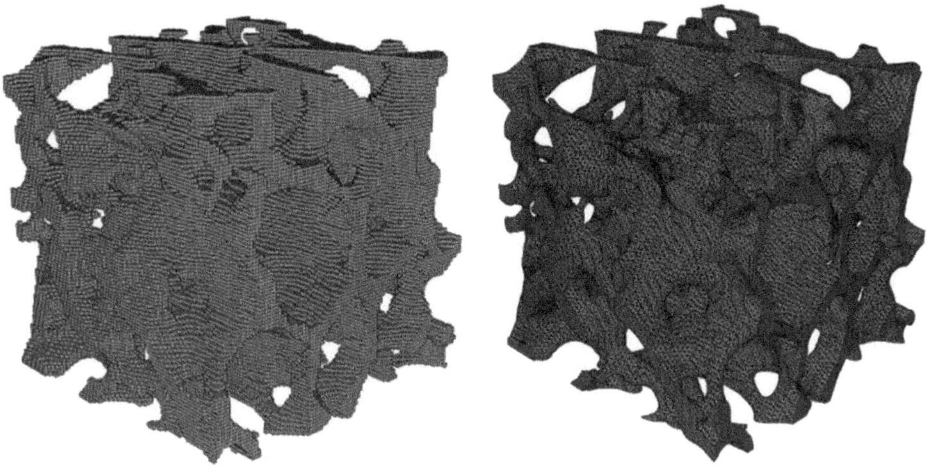

Figure 6.12 Voxel-based hexahedral (left) and tetrahedral (right) finite element models for simulation of bone microstructures. *Source:* Reproduced from Torabizadeh and Mohammadi, 2022a; 2022b/University of Tehran.

continuum damage mechanics is employed to analyse the bone remodelling (Hambli et al. 2012, 2013; Liu et al. 2012; Li et al. 2013; Ali et al. 2014).

Specifically designed finite elements based on the voxel conversion technique to simulate the bone microstructure with solid brick elements (Figure 6.12) are also available. However, the method generates non-smooth surfaces in the model, which may lead to inaccuracies in the microscopic solutions. Better results may be obtained by using tetrahedral finite elements (Figure 6.12). Moreover, alternative models based on the isogeometric technique (Section 3.6), which somehow integrates the modelling and analysis procedures, are available (Verhoosel et al. 2015).

The extended finite element method (XFEM) (Section 3.5) is probably the most efficient method for general stability and propagation analysis of discontinuity problems, including the bone fracture, as typically presented in Figure 6.13 (Mohammadi 2008, 2012). In addition to scientific and research centres, XFEM is now available on major commercial finite element softwares (Gracie et al. 2008; Huynh and Belytschko 2009). Moreover, XFEM has been employed for simulation of biomechanical applications, such as modelling of fracture of the ceramic hip liners (Elkins et al. 2013), impact analysis of the proximal femur (Liu and Li 2010), tensile fracture in the femoral neck and greater trochanter (Morgan and Keaveny 2001) and crack propagation in the trabecular bone (Cook and Zioupos 2009).

Multiscale simulations of bone fractures study the response of different scales (Figure 6.14) to provide a better understanding of the failure or fracture mechanisms at different scales through the use of computational homogenization techniques (Hellmich et al. 2004; Hamed

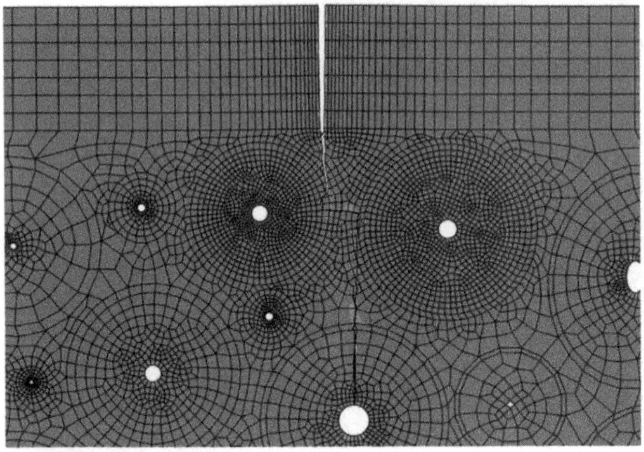

Figure 6.13 Sample XFEM crack propagation simulation of a bone fracture in a heterogeneous bone. *Source:* Reproduced from Torabizadeh and Mohammadi, 2022a; 2022b/University of Tehran.

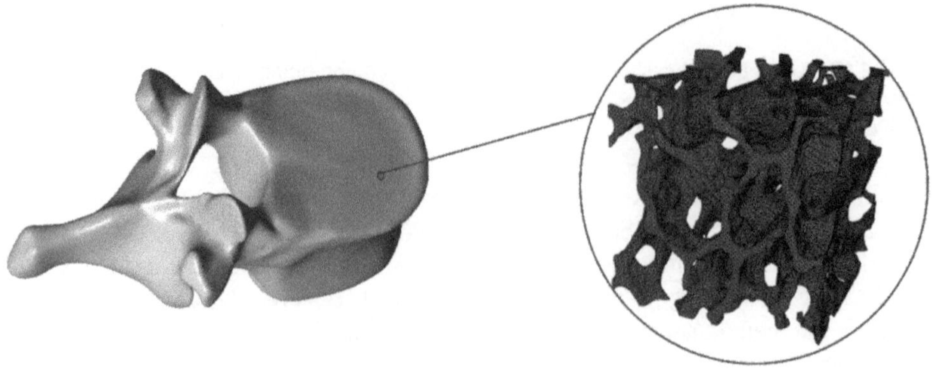

Figure 6.14 Typical macro- and microscale structures of a bone.

et al. 2012; Vaughan et al. 2012; Dall'Ara et al. 2013; Khaterchi et al. 2013), by coupling the micromechanical theories and CT-based FE models (Rahmoun et al. 2014), etc.

6.2.2 Constitutive Response of the Bone

The constitutive model that governs the response of the complex composite structure of bone tissue should be selected based on the scale and the forming tissue. In any case, the material constants and the overall response should be verified and calibrated with the experimental data (Khaterchi et al. 2013). The models include the isotropic or orthotropic laws for the linear elastic part of the response.

Physically, the material structure of the bone requires its mechanical properties to be orthotropic/anisotropic (Martínez-Reina et al. 2011; Hamed et al. 2015; Lekadir et al. 2016). Nevertheless, many studies have adopted the very simplified model of the linear elastic isotropic response for simulations of bone (Sansalone et al. 2010; Hambli et al. 2011; Khaterchi et al. 2013).

6.2.3 Poroelastic Nature of Bone Tissues

A number of biological organs, in general, and bones, in particular, are made from poroelastic tissues. Two different approaches of the microscale description and the macroscale formulation can be adopted to study the mechanical responses of porous bones/tissues.

In the microscale analysis, the porous geometry is constructed as a solid material based on the information obtained from the medical scans. The governing constitutive relation and the solution strategy follow the solid constituent. Most of such porous solid micromodels do not consider the potential fluid flows (such as blood flow) or ignore the existence of nerve systems, as they are not expected to contribute to the overall mechanical properties of the bone. The porous micromodel is used for defining the constitutive relation of the macroscale problem, but it cannot be adopted for deriving the general governing equations on a macro level. A clear example is the modelling of trabecular bone, presented in Figure 6.12.

In the poroelastic theory, however, a macro description with some idea of average properties of the mixed problem of porous solid and fluid flow is adopted. The second part is usually less important in bone fracture problems. For a complete review of the governing equations and numerical issues refer to Section 2.2.6 in Chapter 2.

6.2.4 Plasticity and Damage

There is no general agreement on the post-elastic response model of different hard tissues, as a variety of maximum principal strain (MPS), eccentric von Mises, Drucker–Prager, Mohr–Coulomb and Drucker–Lode plasticity models, among others, have been used to simulate the cortical and trabecular bone tissues (Schwiedrzik et al. 2016). Both isotropic and kinematic hardening rules can be adopted to determine the evolution of the flow/yield surface (Podshivalov et al. 2014). The asymmetry of tension/compression responses should be accounted for when modelling bone tissue.

Damage constitutive laws can be adopted in the form of elastic damage or elastoplastic damage formulations to study the failure of bones and its potential post-failure response

(Charlebois et al. 2010; Podshivalov et al. 2014; Hazrati Marangalou et al. 2015). The elastoplastic damage analyses may be accompanied with large deformations and require various softening formulations for different levels of bone tissue simulations (Chevalley et al. 2013; Lekadir et al. 2016).

For a bone with small elastic strains, the additive decomposition of the Green–Lagrange strain into the elastic and plastic parts can be adopted (Schwiedrzik et al. 2016):

$$\boldsymbol{E} = \boldsymbol{E}^{\mathrm{e}} + \boldsymbol{E}^{\mathrm{p}} \tag{6.2.1}$$

where the second Piola–Kirchhoff stress $\boldsymbol{S}$ is defined as

$$\boldsymbol{S} = \mathbf{D}^{\mathrm{SE}}(\boldsymbol{E} - \boldsymbol{E}^{\mathrm{p}}) \tag{6.2.2}$$

and $\mathbf{D}^{\mathrm{SE}}$ is the isotropic material modulus.

The quadratic Drucker–Prager yielding criterion can be described by (Schwiedrzik et al. 2016)

$$Y(\boldsymbol{S}) = \sqrt{\boldsymbol{S}:\boldsymbol{\mathcal{F}}:\boldsymbol{S}} + \boldsymbol{G}:\boldsymbol{S} - r(\kappa) \tag{6.2.3}$$

with

$$\boldsymbol{\mathcal{F}} = -\zeta_0 \bar{\sigma}_0^2 (\mathbf{I} \otimes \mathbf{I}) + (\zeta_0 + 1)\bar{\sigma}_0^2 (\mathbf{I} \bar{\otimes} \mathbf{I}) \tag{6.2.4}$$

$$\boldsymbol{G} = \frac{1}{2}\left(\frac{1}{\sigma_0^{\mathrm{t}}} - \frac{1}{\sigma_0^{\mathrm{c}}}\right)\mathbf{I} \tag{6.2.5}$$

where $\bar{\sigma}_0$ is defined in terms of the tensile and compressive yield stresses, σ_0^{t} and σ_0^{c}, respectively,

$$\bar{\sigma}_0 = \frac{\sigma_0^{\mathrm{t}} + \sigma_0^{\mathrm{c}}}{2\sigma_0^{\mathrm{t}}\sigma_0^{\mathrm{c}}} \tag{6.2.6}$$

and σ_0^{t} and σ_0^{c} are the tensile and compressive yield stresses, respectively, ζ_0 is an interaction parameter that defines the shape and pressure sensitivity of the criterion in stress space (plastic compressibility or dilatant plasticity) and $\mathbf{I}$ is the second-order unit tensor.

A linear hardening function $r(\kappa)$ is assumed in terms of the accumulated plastic strain κ:

$$\kappa = \int \left|\dot{\boldsymbol{E}}^{p}\right| \mathrm{d}t \tag{6.2.7}$$

$$r(\kappa) = 1 + k_{\mathrm{h}} E_0 \left(\frac{\sigma_0^{\mathrm{t}} + \sigma_0^{\mathrm{c}}}{2\sigma_0^{\mathrm{t}}\sigma_0^{\mathrm{c}}}\right)\kappa \tag{6.2.8}$$

where k_{h} is the linear hardening parameter and E_0 is the Young's modulus of the tissue.

Alternatively, evolution of the quasi-brittle damage variable can be governed by the maximum value of the equivalent strain $\varepsilon_{\mathrm{eq}}$,

$$\kappa = \max(\varepsilon_{\mathrm{eq}}) \tag{6.2.9}$$

where

$$\varepsilon_{eq} = \sqrt{\frac{2}{3}\varepsilon_{ij}\varepsilon_{ij}} \tag{6.2.10}$$

The damage evolution depends on the damage loading function $f(\varepsilon_{eq},\varepsilon_0)$ (Giry et al. 2009):

$$f(\varepsilon_{eq},\varepsilon_0) = \varepsilon_{eq} - \max(\kappa,\varepsilon_0) \tag{6.2.11}$$

where ε_0 is associated with the onset of damage initiation.

In an elastic damage model, proposed by Hambli (2013) and Li et al. (2013), the isotropic elastic model is coupled with the quasi-brittle damage law to describe the progressive initiation and propagation of cracks:

$$\sigma_{ij} = (1-\varphi_d)\mathbf{D}_{ijkl}\varepsilon_{ij} \tag{6.2.12}$$

where $\mathbf{D}_{ijkl}$ is the elasticity tensor and φ_d is the damage variable.

The damage law, which should be calibrated by the experimental results, is expressed as

$$\begin{cases} \varphi_d = 0, & \varepsilon_{eq} \le \varepsilon_0 \\ \varphi_d = D_c\varepsilon_{eq}^n, & \varepsilon_0 < \varepsilon_{eq} < \varepsilon_f \\ \varphi_d = \varphi_{dc}, & \varepsilon_{eq} \ge \varepsilon_f \end{cases} \tag{6.2.13}$$

where D_c, n and ε_c are the critical damage, the damage exponent and the strain associated with the onset of fracture.

For further details, refer to Section 2.2.3.

6.2.5 Hyperelastic Response

Dissimilar to soft tissues, which may experience very large elastic deformations and strains (prior to tearing), hard tissues are expected to remain in very small deformation/strain regimes for their service life. In fact, they are prone to fracture and fatigue failures before reaching a stretch level, which can be considered large.

Nevertheless, a number of studies have adopted hyperelastic formulations to simulate macro- and microscale models. For instance, Levrero-Florencio et al. (2016) adopted the Heckney isotropic hyperelastic model to study the solid phase of the trabecular bone.

For a discussion on available hyperelastic models for hard tissues, the reader may refer to parts of discussions in Section 5.3, which is basically dedicated to the hyperelastic models of soft tissues.

6.3 Simulation of the Femur Bone at Multiple Scales

In this section, details of simulations of the femur bone in different scales are presented. The general description of the multiscale model is presented in Figure 6.15. At the microscale, a detailed microstructure of the bone is considered. It may be a highly porous medium

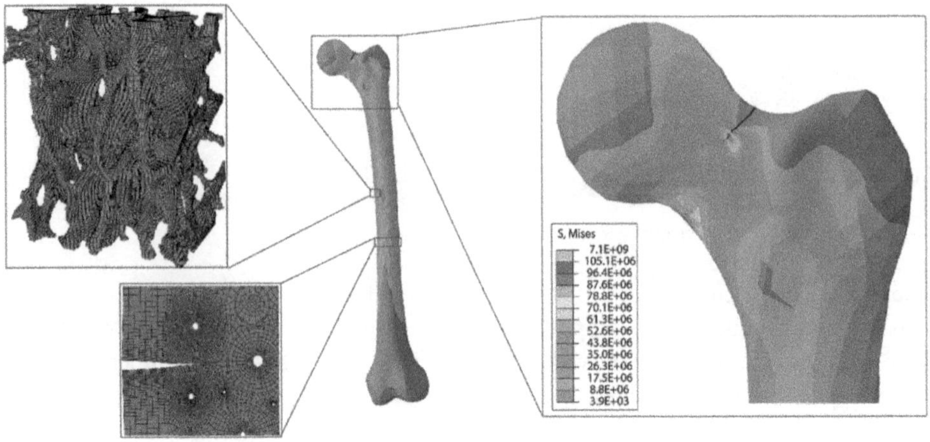

Figure 6.15 Modelling of the femur bone in multiple scales.

of the trabecular bone. The microstructure is obtained from the digital micro scanning. The overall response of the microscale simulation leads to an equivalent stress–strain response of the bone. In a larger mesoscale, a heterogeneous section of the bone can be considered to determine the potential damage and crack propagation responses of the bone. The meso model usually consists of the bone constituents, such as osteon, interstitial matrix, the cement lines and the voids and canals of the nervous system and blood vessels. The macro model is generally related to a main bone, which is analysed subject to an extreme mechanical loading, such as impact, to determine its failure or fracture state. The constitutive material models of different parts of the bone may be obtained from the experimental measurements or from the microscale predictions of the stress–strain response.

6.3.1 Microscale Simulation of the Trabecular Bone

The microstructure of the trabecular bone can be determined by high-resolution imaging techniques such as the micro-computed tomography (μ-CT) (Hamed et al. 2015), where a cubic model of a human trabecular bone is obtained from the segmented CT scans.

The voxel-conversion procedure can be used to generate a micro mesh of finite elements with equally sized brick elements, as presented in Figure 6.16 for a model composed of 220,440 hexahedral elements and 307,550 nodes. Clearly, a rather rough surface is obtained around the voids of the porous medium. Adopting a very fine mesh would reduce the size of the roughness.

A different analysis may be performed on the microscale model. For instance, the model can be considered as an RVE with periodic boundary conditions subject to a macro-based deformation gradient in a concurrent multiscale homogenization technique. In a simplified alternative procedure, the model is subjected to a specific loading condition to determine its overall microscale stress–strain response.

In the present modelling, the bottom edge is constrained and a tensile normal traction is applied on the top edge. The elastic material properties are defined in Table 6.1.

Contour of the von Mises stress is depicted in Figure 6.17, showing an almost uniform distribution except for very localized parts in the porous microstructural solid links.

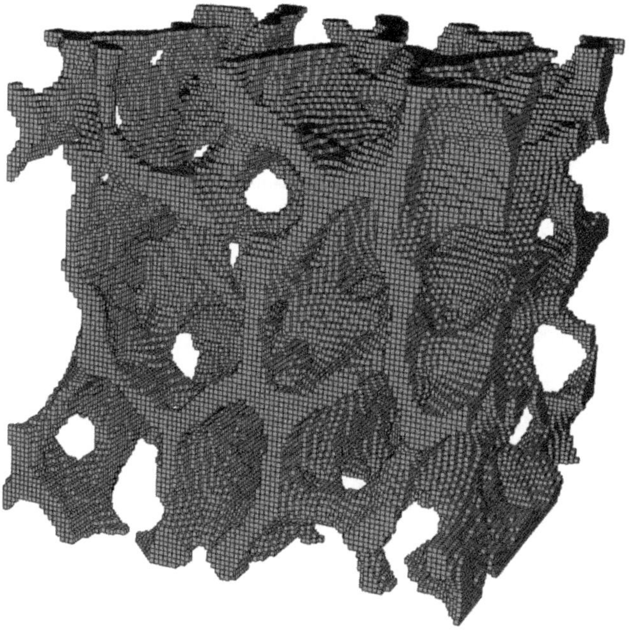

Figure 6.16 Voxel-converted finite element model of the trabecular bone. *Source:* Reproduced from Torabizadeh and Mohammadi, 2022a; 2022b/University of Tehran.

Table 6.1 Elastic constants of the trabecular bone.

Young's modulus	Poisson's ratio
10,000 MPa	0.3

Alternatively, a very smoother finite element mesh can be generated by the four-node tetrahedral finite elements, as depicted in Figure 6.18. This model contains 1,543,480 elements and 376,633 nodes. The same material properties, boundary conditions and loading are assumed.

The resultant von Mises stress contour is presented in Figure 6.19. Except for very tiny regions where thin solid junctions exist in the porous medium, the rest of the model experiences almost a uniform constant stress. The stress concentration factors at the mentioned junctions (shown in Figure 6.20) may reach to 20 and higher, indicating the potential for very localized damage in the porous trabecular bone.

Finally, Figure 6.21 compares the force–displacement responses of the two models. They provide more or less the same linear stiffness coefficient, but the fine tetrahedral mesh is expectedly softer with a lower value of global stiffness for the trabecular bone.

6.3.2 Two-dimensional XFEM Mesoscale Fracture Simulation of the Cortical Bone

In the second part, a micromodel of the cortical bone is analysed by the XFEM cohesive crack propagation technique. The model consists of four phases of osteons, interstitial

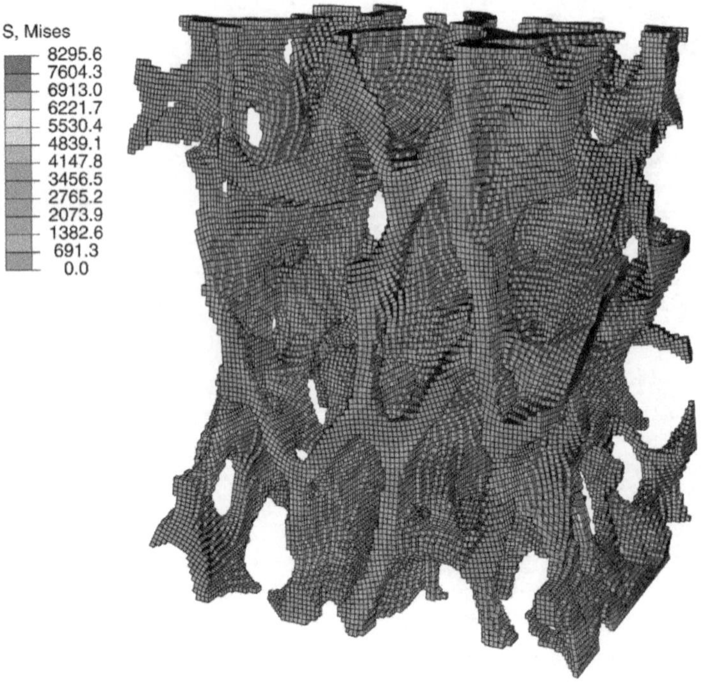

Figure 6.17 Distribution of the von Mises stress on the deformed shape of the voxel-converted finite element microscale model of the trabecular bone. *Source:* Reproduced from Torabizadeh and Mohammadi, 2022a; 2022b/University of Tehran.

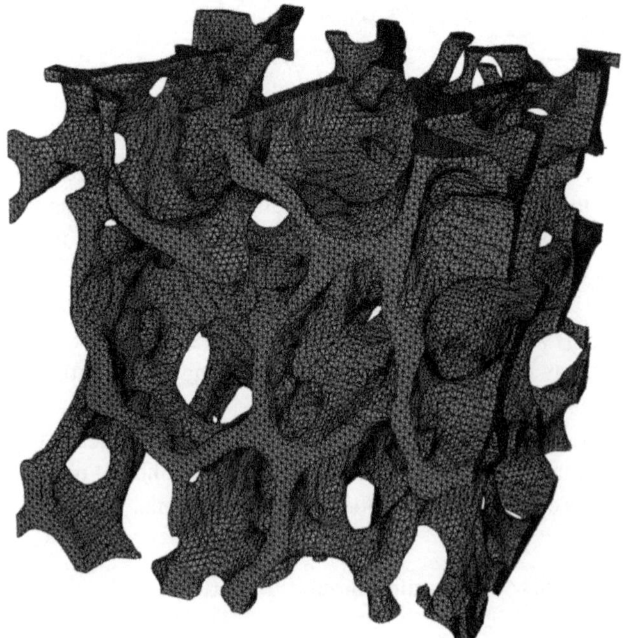

Figure 6.18 Tetrahedral finite element microscale model of the trabecular bone. *Source:* Reproduced from Torabizadeh and Mohammadi, 2022a; 2022b/University of Tehran.

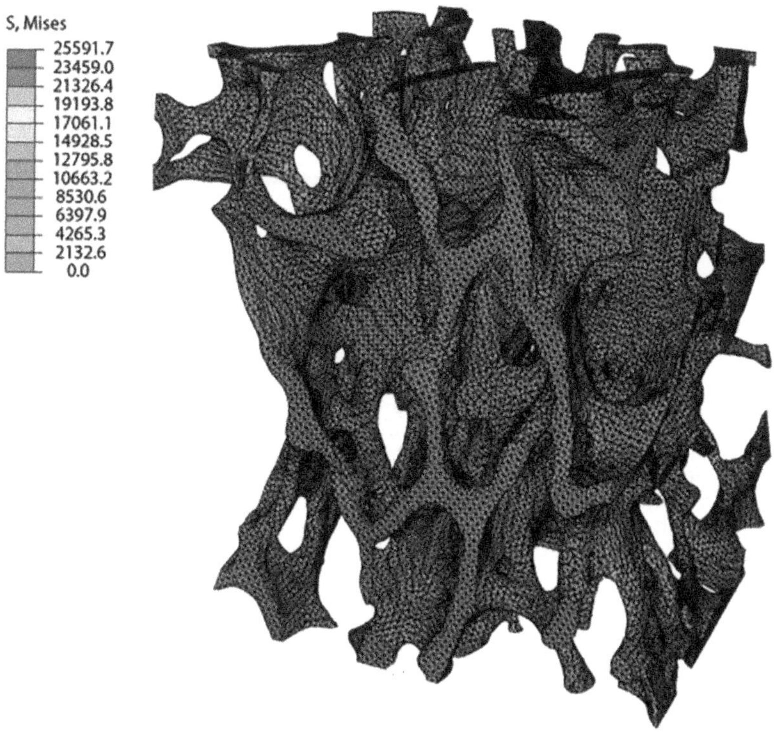

Figure 6.19 Distribution of the von Mises stress on the deformed configuration of the tetrahedral finite element microscale model of the trabecular bone. *Source:* Reproduced from Torabizadeh and Mohammadi, 2022a; 2022b/University of Tehran.

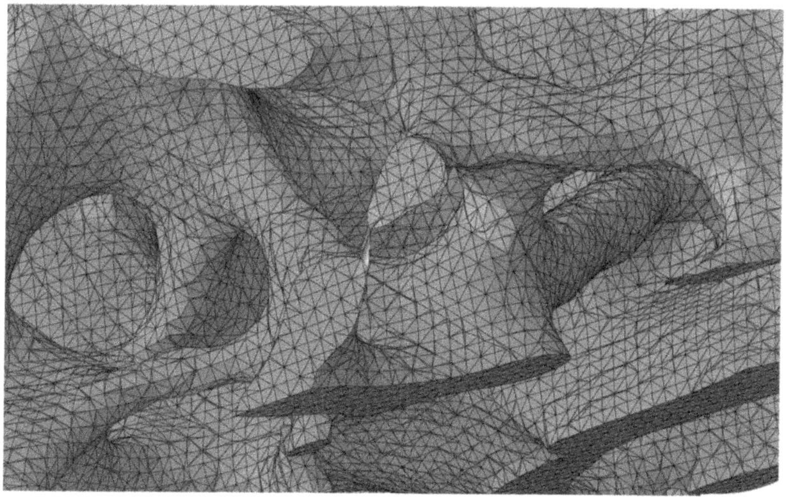

Figure 6.20 Extreme stress concentration on the microscale level of the material structure of the trabecular bone. *Source:* Reproduce from Torabizadeh and Mohammadi, 2022a; 2022b/University of Tehran.

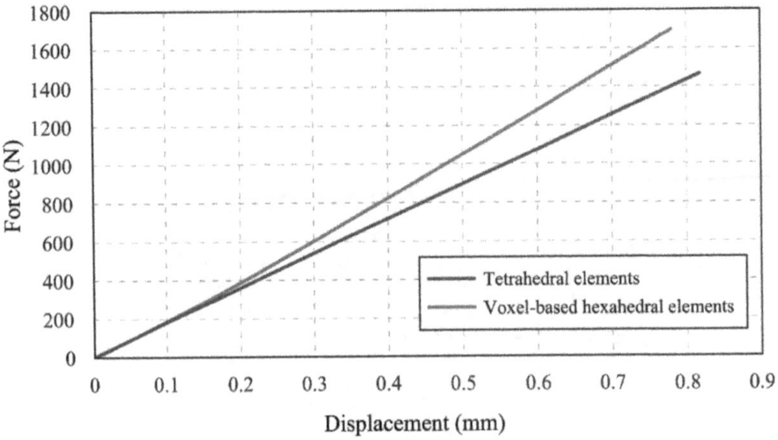

Figure 6.21 Stress–strain response of the voxel-based hexahedral and tetrahedral finite element models of the microscale trabecular bone. *Source:* Reproduced from Torabizadeh and Mohammadi, 2022a; 2022b/University of Iran.

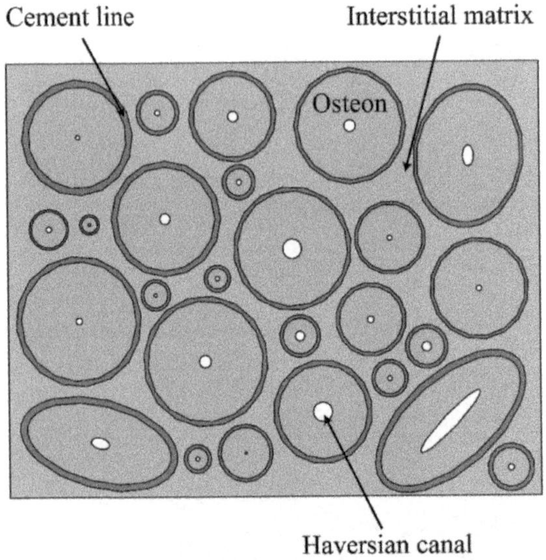

Figure 6.22 A microstructural sample of cortical bone.

bone, cement lines and the Haversian canals, as depicted in Figure 6.22. A similar procedure was previously followed by Li et al. (2013).

The employed material properties, taken from the experimental data of Nalla et al. (2003), are summarized in Table 6.2.

Geometry, boundary conditions and the loading patterns are presented in Figure 6.23. The model includes two initial cracks at the middle part of the specimen in the homogenized material. The aim is to study the crack propagation patterns in the highly heterogeneous model and to determine stress distributions in two cases of normal and shear tractions.

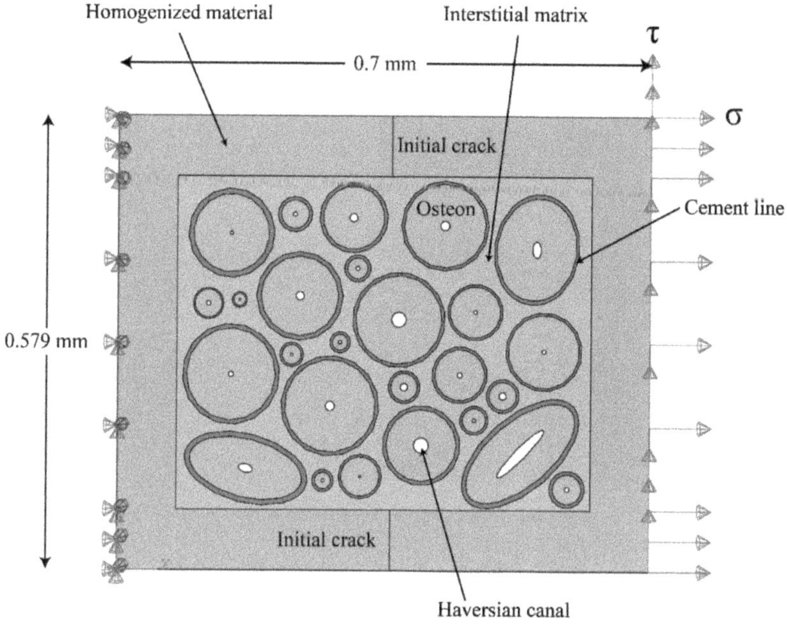

Figure 6.23 Microstructural model of heterogeneous cortical bone. *Source:* Reproduced from Torabizadeh and Mohammadi, 2022a; 2022b/University of Tehran.

Table 6.2 Material properties of the micro XFEM model of cortical bone.

Model	Elastic modulus (MPa)	Poisson's ratio	Yield strain (%)	Fracture initiation strain (%)	Fracture energy release rate (N/mm)
Homogenized material	11,180	0.167	0.6	0.65	2.043
Osteon	12,850	0.170	0.6	0.65	0.860
Interstitial matrix	14,120	0.153	0.6	0.65	0.238
Cement line	9,640	0.490	0.6	0.65	0.146

The cohesive-crack XFEM model is composed of 13,484 quadrilateral finite elements and 13,854 nodes for simulation of all parts of the model, as depicted in Figure 6.24. All phases are assumed to follow a similar strain-based yield criterion and limit. The fracture energy release rates are different for the components of the model.

Contours of the von Mises stress distribution along with the final crack propagation paths for the tensile and shear specimens are presented in Figures 6.25 and 6.26, respectively. The stress concentrations can be observed around the crack tips and in the vicinity of the holes (Haversian canals). The zoomed illustrations show that the crack paths are

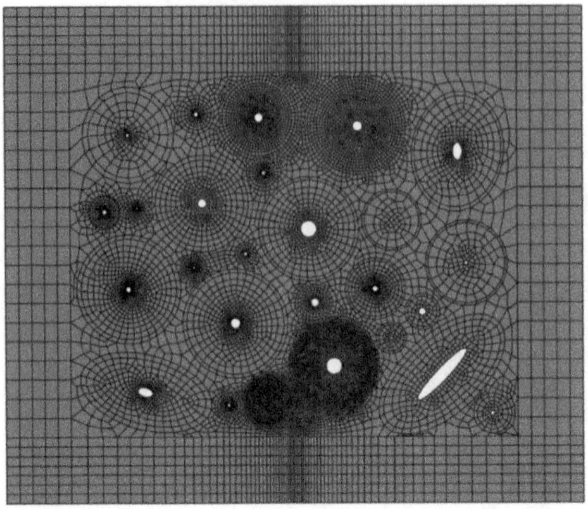

Figure 6.24 XFEM mesh of the cortical bone. *Source:* Reproduced from Torabizadeh and Mohammadi, 2022a; 2022b/University of Tehran.

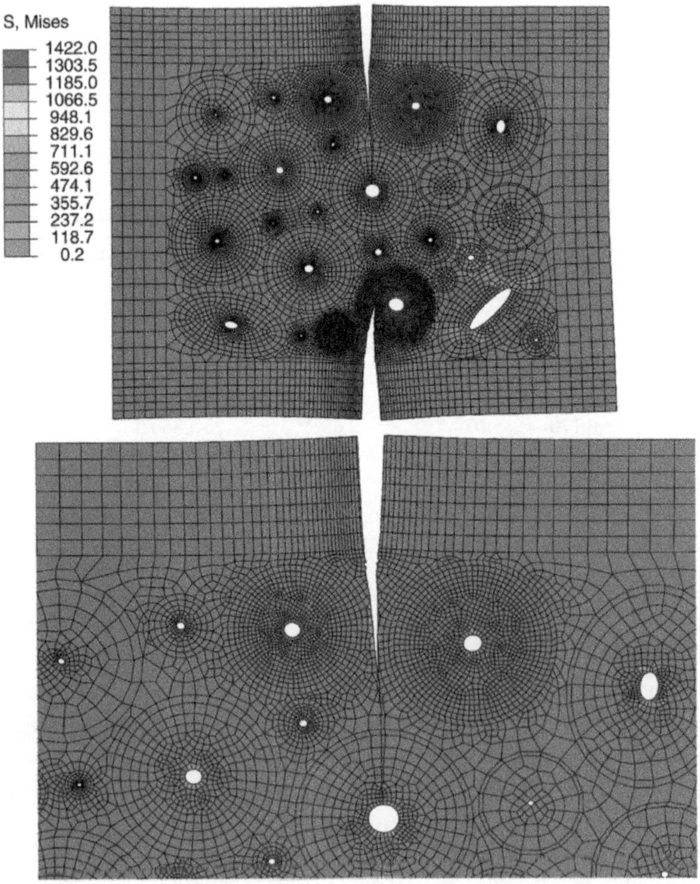

Figure 6.25 Von Mises stress contour and the cracking path of the tensile specimen. *Source:* Reproduced from Torabizadeh and Mohammadi, 2022a; 2022b/University of Tehran.

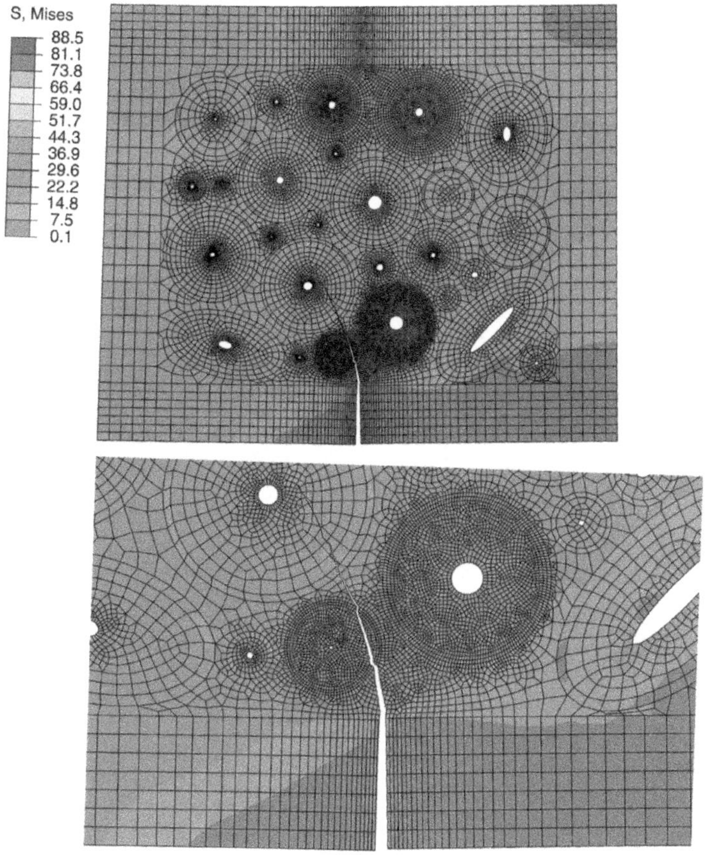

Figure 6.26 Von Mises stress contour and the cracking path of the shear specimen. *Source:* Reproduced from Torabizadeh and Mohammadi, 2022a; 2022b/University of Tehran.

mainly in the interstitial matrix, but they also pass through the cement lines and even the osteons in some stages of the crack propagation.

The force-displacement response of the bone specimen is illustrated in Figure 6.27. It is observed that the stiffness of the model is reduced to almost a quarter of its original value due to the extensive crack propagation.

The force-crack opening displacement of Figure 6.28 shows a similar weakening effect due to the crack propagation. It approaches a horizontal line at the final stages of the loading, indicating the onset of full rupture of the specimen.

6.3.3 Macroscale Simulation of the Femur

The femur bone may be subjected to various extreme loading conditions, resulting in different fracture patterns. Fracture analysis of macro model of femur can be performed by XFEM. The model is constructed by 108,199 six-node tetrahedral elements and 21,725 nodes, as presented in Figure 6.29.

In general, the constitutive model of each part of the bone may be obtained from the experimental measurements, the microscans or from the results of the microscale analysis.

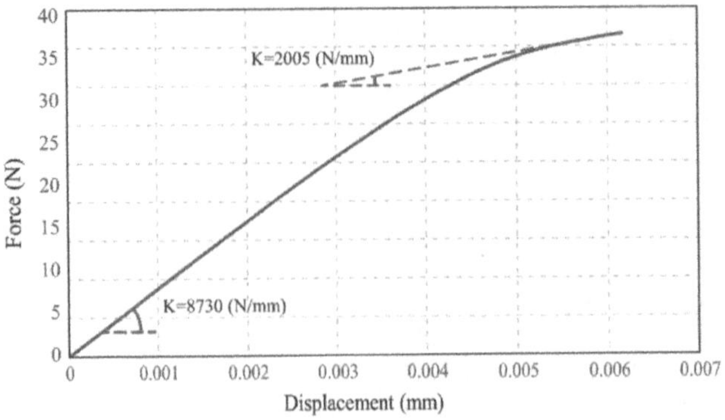

Figure 6.27 Force-displacement response of the cracked bone specimen. *Source:* Reproduced from Torabizadeh and Mohammadi, 2022a; 2022b/University of Tehran.

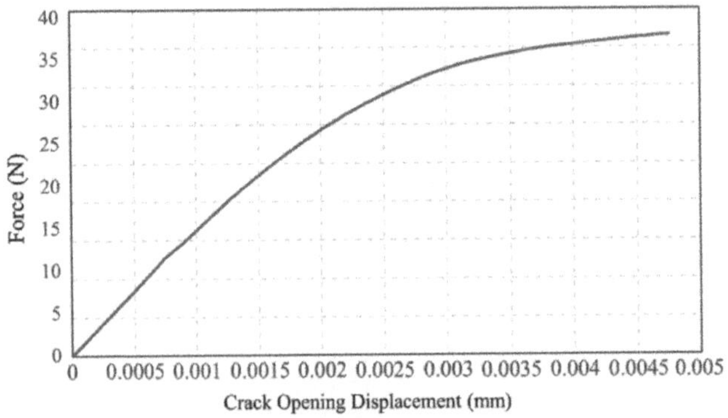

Figure 6.28 Force-crack opening response of the cracked bone specimen. *Source:* Reproduced from Torabizadeh and Mohammadi, 2022a; 2022b/University of Tehran.

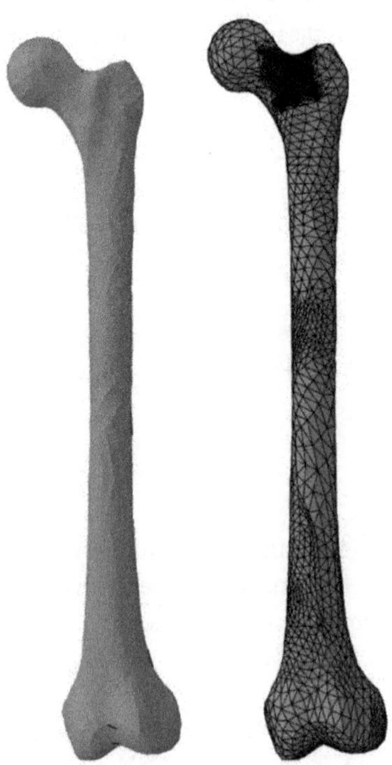

Figure 6.29 Geometry and the finite element model of femur. *Source:* Reproduced from Torabizadeh and Mohammadi, 2022a; 2022b/University of Tehran.

The aim here is to analyse the fracture resistance and crack propagation around the femur head. Similar linear elastic material properties are assigned to different parts of the model, as defined in Table 6.3.

The constrained nodes on the femur head and the loaded nodes are depicted in Figure 6.30.

Figure 6.31 illustrates the distribution of the von Mises stress on the outer surface of the femur. The failed elements and the generated crack appear in the middle part of the bone head.

Table 6.3 Elastic constants of the femur model.

Young's modulus (MPa)	Poisson's ratio	Fracture initiation strain (%)	Fracture energy release rate (N/mm)
11,000	0.167	0.65	2.043

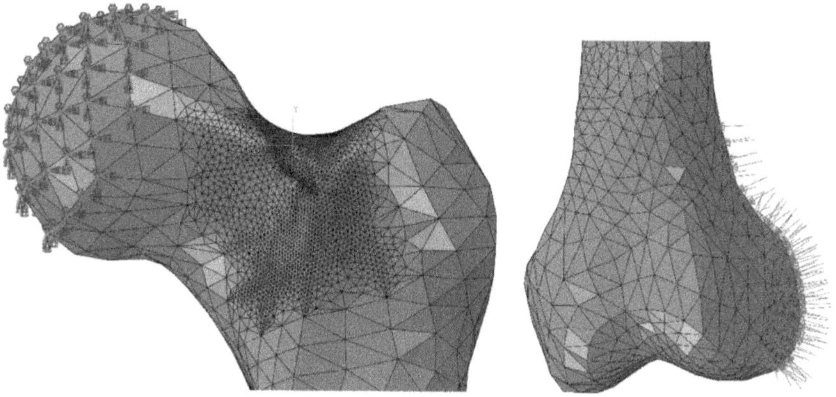

Figure 6.30 Boundary condition and loadings on the femur. *Source:* Reproduced from Torabizadeh and Mohammadi, 2022a; 2022b/University of Tehran.

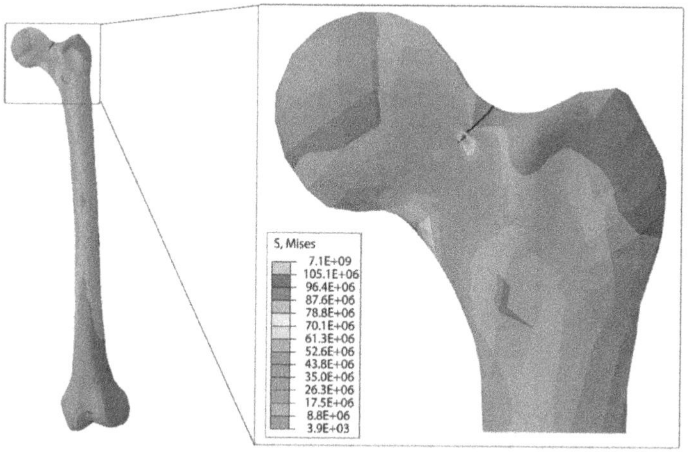

Figure 6.31 Distribution of the von Mises stress on the outer surface of the femur. *Source:* Reproduced from Torabizadeh and Mohammadi, 2022a; 2022b/University of Tehran.

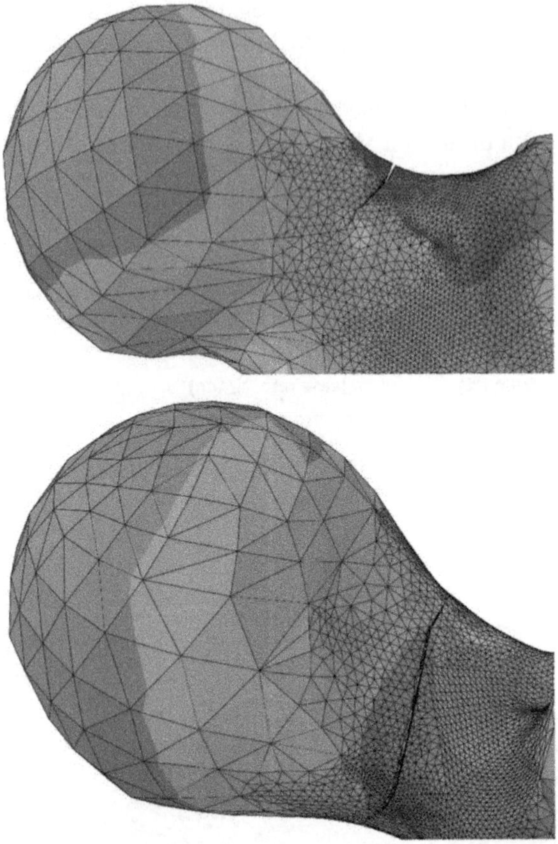

Figure 6.32 Distribution of the von Mises stress on the outer surface of the femur head at different stages of crack propagation. *Source:* Reproduced from Torabizadeh and Mohammadi, 2022a; 2022b/University of Tehran.

Finally, the extent of crack propagations at the femur head at two early and final stages of the loadings are presented in Figure 6.32.

6.4 Healing in Damaged Hard Tissue

6.4.1 Introduction

The study of the healing process of damaged hard biological tissues, such as the analysis of bone fracture healing mechanisms, is an active area of research in biomechanics. Moreover, the way that living materials repair the position of a broken bone can be stimulated to make new prosthetic tissues or even to design materials that are capable of healing themselves or help certain tissues to do so.

The process of bone-fracture healing follows a well-defined sequence of four partially overlapping stages, beginning with an inflammatory phase, followed by two repair phases

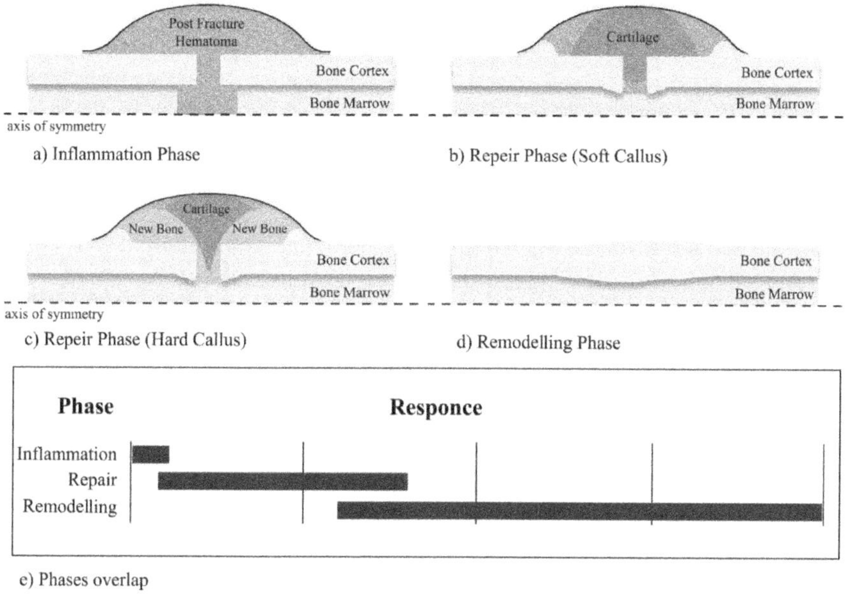

Figure 6.33 Stages of healing of a fractured hard tissue. *Source:* Reproduced from Zamani 2022/ University of Tehran.

of soft and hard callus production, and concluded by a remodelling phase (Pivonka and Dunstan 2012), as depicted in Figure 6.33.

6.4.1.1 Inflammation Phase

After the occurrence of a bone trauma, peripheral, intramedullary and bone marrow cells combine to form a hematoma (Kon et al. 2014; Cho et al. 2002). Early in the phase of inflammation, growth factors are produced (Cho et al. 2002) and blood vessel rupture-induced hypoxic environments release angiogenic factors (Pivonka and Dunstan 2012).

6.4.1.2 Soft Callus Formation

The repair phases (soft and hard callus) require the deposition of new extracellular matrix (ECM) from the mesenchymal stem cells (MSCs) and their differentiation into chondrocytes and osteoblasts, along with the necessary condition of stability of the fractured zone (Strube et al. 2008). MSCs are mostly drawn from the bone marrow, blood vessels, periosteum, surrounding muscle tissue or the circulation (Cuthbert et al. 2013). The rate at which a bone fracture heals is limited by the time it takes to find sufficient MSCs (Kon et al. 2014).

At the middle of the initial callus and near the fracture, MSCs differentiate into cartilage-producing chondrocytes, creating the primary ECM soft cartilage (Wagegg et al. 2012). Then ossification and mineralization of the cartilage callus gradually occur due to the increase of chondrocytes (Bailon-Plaza and Van der Meulen 2001). The mineralized cartilage is degraded by osteoclasts and is replaced by the cancellous bone produced by osteoblasts, resulting in a mechanically more stable callus to provide a sufficiently strong scaffold around the fracture zone (Wehner et al. 2014).

6.4.1.3 Hard Callus Formation

In the third stage of healing, intramembranous and endochondral ossifications continue and a bone bridge is formed. This bone bridge continues to change as the body heals. Prior to endochondral replacement, mature chondrocytes stop proliferation and calcify the extracellular matrix. Chondrocytes undergo apoptosis and blood vessels expand into the vacated area. These capillaries contain osteoprogenitor cells that transform into osteoblasts, which deposit bone on the calcified cartilage matrix. Once osteoblasts have encased themselves in the bone, they undergo apoptosis or differentiate into osteocytes, or bone lining cells. Endochondral ossification continues until the cartilage is fully replaced by bone and a bone bridge fills the fractured space (Bailon-Plaza and Van der Meulen 2001). A necessary condition for the whole procedure is the overall stability of the callus (Morgan et al. 2010; Claes et al. 2002; Einhorn and Gerstenfeld 2014).

6.4.1.4 Remodelling Phase

Cortical bone replaces the cancellous bone that bridges the fracture gap at the final phase of fracture healing. Because of the formation of secondary osteons, the cancellous bone that is formed in the fracture gap is turned into the cortical bone. During the phase of remodelling, vascularization returns to its pre-fracture levels (Pivonka and Dunstan 2012).

6.4.2 Physical Foundation of Bone Tissue Healing

To represent each phase of the process by which a damaged living tissue recovers, one must examine the differential equations that govern the events. The main independent variables, which are functions of space $\mathbf{x}$ and time t, are in the forms of cellular densities (number of cells ml^{-1}), matrix densities (μg ml^{-1}) and growth factor concentrations (μg ml^{-1}). They are:

$\rho_m(\mathbf{x},t)$: MSCs density
$\rho_c(\mathbf{x},t)$: Chondrocyte density
$\rho_b(\mathbf{x},t)$: Osteoblast density
$\rho_{cM}(\mathbf{x},t)$: Cartilage matrix density
$\rho_{bM}(\mathbf{x},t)$: Bone matrix density
$\rho_{cF}(\mathbf{x},t)$: Chondrogenic growth factor concentration
$\rho_{bF}(\mathbf{x},t)$: Osteogenic growth factor concentration

where the sum of cartilage and bone densities equals the total matrix density ρ_{tM}:

$$\rho_{tM}(\mathbf{x},t) = \rho_{cM}(\mathbf{x},t) + \rho_{bM}(\mathbf{x},t) \tag{6.4.1}$$

The formulation is accompanied with the assumption that no mechanical deformation is possible and other external mechanical effects are ignored.

6.4.2.1 Diffusion of Densities of Cells and Growth Factor

Similar to what was previously discussed in Section 5.7.2.2, various factors such as MSCs, chondrocyte, osteoblast and ECM densities affect the balances of concentrations and densities in the healing process. The governing equation for the distribution of each variable in time is coupled and cannot be obtained independently.

The governing diffusion equation for time variations of a specific variable Q can be expressed in terms of the flux $\boldsymbol{J}_{\mathrm{Q}}$ and the rate of change of the variable in the control volume (f_{Q}) (Figure 5.68):

$$\frac{\partial Q}{\partial t} = -\boldsymbol{\nabla} \cdot \boldsymbol{J}_{\mathrm{Q}} + f_{\mathrm{Q}} \tag{6.4.2}$$

The equations governing the diffusion of densities and concentrations are the corrected set of equations by Bailon-Plaza and Van der Meulen (2001).

6.4.2.1.1 Chondrogenic Growth Factor Concentration ρ_{cF}

The diffusion equation for the chondrogenic growth factor concentration ρ_{cF} can be written as (Bailon-Plaza and Van der Meulen 2001)

$$\frac{\partial \rho_{\mathrm{cF}}}{\partial t} = -\boldsymbol{\nabla} \cdot (\mathbf{J}_{\mathrm{rdf}}) + (f_{\mathrm{pc}} + f_{\mathrm{nd}}) \tag{6.4.3}$$

where the three contributing terms are based on the wound healing model of Olsen et al. (1995, 1998):

$$\text{Random dispersal flux} = \boldsymbol{J}_{\mathrm{rdf}} = -D_{\mathrm{cF}} \boldsymbol{\nabla} \rho_{\mathrm{cF}} \tag{6.4.4}$$

$$\frac{\text{Production}}{\text{Consumption}} \text{rate of chemicals} = f_{\mathrm{pc}} = \left(\frac{r_{\mathrm{cF}}^{\max} \rho_{\mathrm{cF}}}{K_{\mathrm{cF}} + \rho_{\mathrm{cF}}}\right) \frac{\rho_{\mathrm{tM}}}{\left(\alpha_{\mathrm{cF}}^{3} + \rho_{\mathrm{tM}} 3\right)} \rho_{\mathrm{c}} \tag{6.4.5}$$

$$\text{Natural decay rate} = f_{\mathrm{nd}} = -d_{\mathrm{cF}} \rho_{\mathrm{cF}} \tag{6.4.6}$$

where D_{cF} is the chondrogenic growth factor diffusion rate, $r_{\mathrm{cF}}^{\max}$ is the chondrogenic growth factor production constant, K_{cF} is the Michaelis–Menten constant for chondrogenic growth factor production, α_{cF} is the saturation level of the total matrix density, d_{cF} is the chondrogenic growth factor deterioration rate, ρ_{tM} is the total matrix density and ρ_{c} is the chondrocyte density.

6.4.2.1.2 Osteogenic Growth Factor Concentration ρ_{bF}

The diffusion equation for the osteogenic growth factor concentration ρ_{bF} can be written as (Bailon-Plaza and Van der Meulen 2001)

$$\frac{\partial \rho_{\mathrm{bF}}}{\partial t} = -\boldsymbol{\nabla} \cdot (\boldsymbol{J}_{\mathrm{rdf}}) + (f_{\mathrm{pc}} + f_{\mathrm{nd}}) \tag{6.4.7}$$

where the three contributing terms are

$$\text{Random dispersal flux} = \boldsymbol{J}_{\mathrm{rdf}} = -D_{\mathrm{bF}} \boldsymbol{\nabla} \rho_{\mathrm{bF}} \tag{6.4.8}$$

$$\frac{\text{Production}}{\text{Consumption}}\text{rate of chemicals} = f_{pc} = \frac{r_{bF}^{max}\ \rho_{bF}}{K_{bF} + \rho_{bF}}\rho_b \tag{6.4.9}$$

$$\text{Natural decay rate} = f_{nd} = -d_{bF}\rho_{bF} \tag{6.4.10}$$

where D_{bF} is the osteogenic growth factor diffusion rate, r_{bF}^{max} is the osteogenic growth factor production constant, K_{bF} is the Michaelis–Menten constant for osteogenic growth factor production, ρ_b is the osteoblast density and d_{bF} is the osteogenic growth factor deterioration rate.

6.4.2.1.3 Cartilage Matrix Density ρ_{cM}

The equation governing the cartilage matrix density ρ_{cM} can be written as

$$\frac{\partial \rho_{cM}}{\partial t} = f_{es} + f_{ed} \tag{6.4.11}$$

where the two contributing terms are (Bailon-Plaza and Van der Meulen 2001):

Synthesis by both MSCs and chondrocyte cells =

$$f_{es} = r_{cM}\left(1 - \frac{\rho_{cM}}{\rho_{cM}^{max}}\right) \times (\rho_m + \rho_c) \tag{6.4.12}$$

$$\text{Matrix degradation by osteoblasts} = f_{ed} = -d_{cM}\rho_b\rho_{cM} \tag{6.4.13}$$

where r_{cM} is the cartilage matrix production rate, ρ_{cM}^{max} is the maximum cartilage matrix density, ρ_m is the MSC density, ρ_c is the chondrocyte density, d_{cM} is the constant of cartilage matrix decay, ρ_{cM} is the cartilage matrix density and ρ_b is the osteoblast density.

6.4.2.1.4 Bone Matrix Density ρ_{bM}

The equation governing the bone matrix density ρ_{bM} can be written as

$$\frac{\partial \rho_{bM}}{\partial t} = f_{es} \tag{6.4.14}$$

where f_{es} is defined as (Bailon-Plaza and Van der Meulen 2001)

$$\text{Synthesis by osteoblast cells} = f_{es} = r_{bM}\left(1 - \frac{\rho_{bM}}{\rho_{bM}^{max}}\right)\rho_b \tag{6.4.15}$$

where r_{bM} is the bone matrix production rate, ρ_{bM}^{max} is the maximum bone matrix density and ρ_b is the osteoblast density.

6.4.2.1.5 Mesenchymal Stem Cell Density ρ_m

The diffusion equation for the MSC density ρ_m can be written as

$$\frac{\partial \rho_m}{\partial t} = -\nabla \cdot (\boldsymbol{J}_{rdf} + \boldsymbol{J}_{cmcf}) + (f_m + f_{dmb} + f_{dmc}) \tag{6.4.16}$$

where the five contributing terms are (Bailon-Plaza and Van der Meulen 2001)

$$\text{Random dispersal flux} = \boldsymbol{J}_{rdf} = -\frac{a_m \rho_{tM}}{\left(b_m^2 + \rho_{tM}^2\right)} \nabla \rho_m \tag{6.4.17}$$

$$\text{Cell} - \text{matrix creep flux} = \boldsymbol{J}_{cmcf} = \frac{c_m}{\left(d_m + \rho_{tM}\right)^2} \rho_m \nabla \rho_{tM} \tag{6.4.18}$$

$$\text{Mitosis} = f_m = \left(\frac{r_m^{init} \rho_{tM}}{K_m^2 + \rho_{tM}^2}\right) \rho_m \left(1 - \frac{\rho_m}{\rho_m^{max}}\right) \tag{6.4.19}$$

$$\text{Differentiation into osteoblast} = f_{dmb} = -\frac{\Phi_{tmb} \rho_{bF}}{\Psi_{tmb} + \rho_{bF}} \rho_m \tag{6.4.20}$$

$$\text{Differentiation into chondrocyte} = f_{dmc} = -\frac{\Phi_{tmc} \rho_{cF}}{\Psi_{tmc} + \rho_{cF}} \rho_m \tag{6.4.21}$$

where a_m is the constant of the diffusion and b_m is the total density of the matrix that gives the fastest diffusion rate and controls the MSC diffusion rate due to the random dispersal flux, ρ_{tM} is the total matrix density, c_m is the constant of the diffusion and d_m is the total density of the matrix that gives the fastest diffusion rate and controls the MSCs diffusion rate due to the cell-matrix creep, r_m^{init} and K_m are the constants that control the MSCs proliferation rate, ρ_m^{max} is the maximal MSCs capacity of callus, Φ_{tmb} and Ψ_{tmb} are the constants of the function that simulates MSCs differentiation into osteoblasts, ρ_{bF} is the osteogenic growth factor concentration, Φ_{tmc} and Ψ_{tmc} are the constants of the function that simulates MSC differentiation into chondrocytes and ρ_{cF} is the chondrogenic growth factor concentration.

6.4.2.1.6 Chondrocyte Density ρ_c

The equation for the chondrocyte density ρ_c can be written as

$$\frac{\partial \rho_c}{\partial t} = f_m + f_{dmc} + f_{dcb} \tag{6.4.22}$$

where the three contributing terms are (Bailon-Plaza and Van der Meulen 2001)

$$\text{Mitosis} = f_m = \left(\frac{r_c^{init} \rho_{tM}}{K_c^2 + \rho_{tM}^2}\right) \rho_c \left(1 - \frac{\rho_c}{\rho_c^{max}}\right) \tag{6.4.23}$$

$$\text{Differentiation from MSCs} = f_{\text{dmc}} = \frac{\Phi_{\text{tmc}}\rho_{\text{cF}}}{\Psi_{\text{tmc}} + \rho_{\text{cF}}}\rho_{\text{m}} \tag{6.4.24}$$

$$\text{Differentiation into osteoblast} = f_{\text{dcb}} = -\frac{\rho_{\text{cM}}^{6}}{\beta^{6} + \rho_{\text{cM}}^{6}} \times \frac{\Phi_{\text{tcb}}\rho_{\text{bF}}}{\Psi_{\text{tcb}} + \rho_{\text{bF}}}\rho_{\text{c}} \tag{6.4.25}$$

where $r_{\text{c}}^{\text{init}}$ and K_{c} are the constants that control the chondrocyte proliferation rate, ρ_{tM} is the total matrix density, $\rho_{\text{c}}^{\text{max}}$ is the maximal chondrocyte capacity of callus, Φ_{tmc} and Ψ_{tmc} are the constants of the function that simulates MSC differentiation into chondrocyte, ρ_{cF} is the chondrogenic growth factor concentration, ρ_{m} is the MSC density, β is endochondral ossification regulator parameter, ρ_{cM} is the cartilage matrix density, Φ_{tcb} and Ψ_{tcb} are the constants of the function that simulates chondrocytes differentiation into osteoblast and ρ_{bF} is the osteogenic growth factor concentration.

6.4.2.1.7 Osteoblast Density ρ_{b}

The equation for the osteoblast density ρ_{b} can be written as

$$\frac{\partial \rho_{\text{b}}}{\partial t} = f_{\text{m}} + f_{\text{dmb}} + f_{\text{dcb}} + f_{\text{d}} \tag{6.4.26}$$

where the four contributing terms are (Bailon-Plaza and Van der Meulen 2001)

$$\text{Mitosis} = f_{\text{m}} = \left(\frac{r_{\text{b}}^{\text{init}}\rho_{\text{tM}}}{K_{\text{b}}^{2} + \rho_{\text{tM}}^{2}}\right)\rho_{\text{b}}\left(1 - \frac{\rho_{\text{b}}}{\rho_{\text{b}}^{\text{max}}}\right) \tag{6.4.27}$$

$$\text{Differentiation from MSCs} = f_{\text{dmb}} = \frac{\Phi_{\text{tmb}}\rho_{\text{bF}}}{\Psi_{\text{tmb}} + \rho_{\text{bF}}}\rho_{\text{m}} \tag{6.4.28}$$

$$\text{Differentiation from chondrocytes} = f_{\text{dcb}} = \frac{\rho_{\text{cM}}^{6}}{\beta^{6} + \rho_{\text{cM}}^{6}} \times \frac{\Phi_{\text{tcb}}\rho_{\text{bF}}}{\Psi_{\text{tcb}} + \rho_{\text{bF}}}\rho_{\text{c}} \tag{6.4.29}$$

$$\text{Death} = f_{\text{d}} = -d_{\text{bd}}\rho_{\text{b}} \tag{6.4.30}$$

where $r_{\text{b}}^{\text{init}}$ and K_{b} are the constants that control the osteoblast proliferation rate, ρ_{tM} is the total matrix density, $\rho_{\text{b}}^{\text{max}}$ is the maximum osteoblast capacity of callus, Φ_{tmb} and Ψ_{tmb} are the constants of the function that simulates MSC differentiation into osteoblast, ρ_{bF} is the osteogenic growth factor concentration, ρ_{m} is the MSC density, β is the endochondral ossification regulator parameter, ρ_{cM} is the cartilage matrix density, Φ_{tcb} and Ψ_{tcb} are the constants of the function that simulates chondrocytes differentiation into osteoblast, ρ_{bF} is the osteogenic growth factor concentration and d_{bd} is the osteoblast death rate.

6.4.2.2 Set of Coupled Equations

The full set of seven coupled differential equations in terms of ρ_{m}, ρ_{c}, ρ_{b}, ρ_{cM}, ρ_{bM}, ρ_{cF} and ρ_{cF} are summarized in the next section.

6.4.2.2.1 MSCs, Chondrocyte and Osteoblast Densities ρ_m, ρ_c and ρ_b

$$\frac{\partial \rho_m}{\partial t} = \left\{ \nabla \cdot \left[\frac{a_m \rho_{tM}}{\left(b_m{}^2 + \rho_{tM}{}^2\right)} \nabla \rho_m - \frac{c_m}{\left(d_m + \rho_{tM}\right)^2} \rho_m \nabla \rho_{tM} \right] \right\} + \left\{ \left[\left(\frac{r_m^{init} \rho_{tM}}{K_m{}^2 + \rho_{tM}{}^2} \right) \rho_m \left(1 - \frac{\rho_m}{\rho_m^{max}} \right) - \frac{\Phi_{tmb} \rho_{bF}}{\Psi_{tmb} + \rho_{bF}} \rho_m - \frac{\Phi_{tmc} \rho_{cF}}{\Psi_{tmc} + \rho_{cF}} \rho_m \right] \right\} \tag{6.4.31}$$

$$\frac{\partial \rho_c}{\partial t} = \left(\frac{r_c^{init} \rho_{tM}}{K_c{}^2 + \rho_{tM}{}^2} \right) \rho_c \left(1 - \frac{\rho_c}{\rho_c^{max}} \right) + \frac{\Phi_{tmc} \rho_{cF}}{\Psi_{tmc} + \rho_{cF}} \rho_m - \frac{\rho_{cM}{}^6}{\beta^6 + \rho_{cM}{}^6} \times \frac{\Phi_{tcb} \rho_{bF}}{\Psi_{tcb} + \rho_{bF}} \rho_c \tag{6.4.32}$$

$$\frac{\partial \rho_b}{\partial t} = \left(\frac{r_b^{init} \rho_{tM}}{K_b{}^2 + \rho_{tM}{}^2} \right) \rho_b \left(1 - \frac{\rho_b}{\rho_b^{max}} \right) + \frac{\Phi_{tmb} \rho_{bF}}{\Psi_{tmb} + \rho_{bF}} \rho_m + \frac{\rho_{cM}{}^6}{\beta^6 + \rho_{cM}{}^6} \times \frac{\Phi_{tcb} \rho_{bF}}{\Psi_{tcb} + \rho_{bF}} \rho_c - d_{bd} \rho_b \tag{6.4.33}$$

where the constants are defined in Table 6.4.

Table 6.4 Constants of the diffusion equations for the densities of MSCs, chondrocyte and osteoblast.

Parameter	Unit	Description
a_m	μg/cm day	Constant of the diffusion coefficient due to random dispersal flux
b_m	μg/cm^3	Total density of the matrix that gives the fastest diffusion rate due to random dispersal flux
c_m	μg/cm day	Constant of the diffusion coefficient due to the cell-matrix creep
d_m	μg/cm^3	Total density of the matrix that give the fastest diffusion rate due to the cell-matrix creep
r_m^{init}	μg/cm^3 day	MSCs proliferation rate at initial matrix density
K_m	μg/cm^3	Matrix density which provides half of maximum MSCs proliferation rate
ρ_m^{max}	cell/cm^3	Maximal MSCs capacity of callus
Φ_{tmb}	day^{-1}	Maximum MSCs into osteoblast differentiation rate
Ψ_{tmb}	μg/cm^3	Matrix density which provides half of Maximum MSCs into osteoblast differentiation rate
Φ_{tmc}	day^{-1}	Maximum MSCs into chondrocyte differentiation rate
Ψ_{tmc}	μg/cm^3	Matrix density which provides half of Maximum MSCs into chondrocyte differentiation rate
r_c^{init}	μg/cm^3 day	Chondrocyte proliferation rate at initial matrix density
K_c	μg/cm^3	Matrix density which provides half of maximum chondrocyte proliferation rate

(Continued)

Table 6.4 (Continued)

Parameter	Unit	Description
ρ_c^{max}	cell/cm^3	Maximal chondrocyte capacity of callus
β	μg/cm^3	Endochondral ossification regulator parameter
Φ_{tcb}	day^{-1}	Maximum chondrocyte into osteoblast differentiation rate
Ψ_{tcb}	μg/cm^3	Matrix density which provides half of maximum chondrocyte into osteoblast differentiation rate
r_b^{init}	μg/cm^3 day	Osteoblast proliferation rate at initial matrix density
K_b	μg/cm^3	Matrix density which provides half of maximum osteoblast proliferation rate
ρ_b^{max}	cell/cm^3	Maximal osteoblast capacity of callus
d_{bd}	day^{-1}	Osteoblast death rate

6.4.2.2.2 Cartilage Matrix Density ρ_{cM} and Bone Matrix Density ρ_{bM}

$$\frac{\partial \rho_{cM}}{\partial t} = r_{cM}\left(1 - \frac{\rho_{cM}}{\rho_{cM}^{max}}\right) \times (\rho_m + \rho_c) - d_{cM}\rho_b\rho_{cM} \tag{6.4.34}$$

$$\frac{\partial \rho_b}{\partial t} = r_{bM}\left(1 - \frac{\rho_{bM}}{\rho_{bM}^{max}}\right)\rho_b \tag{6.4.35}$$

where the constants are defined in Table 6.5.

Table 6.5 Constants of cartilage matrix density ρ_{cM} and bone matrix density ρ_{bM} equations.

Parameter	Unit	Description
r_{cM}	μg/cell day	Cartilage matrix production rate
ρ_{cM}^{max}	μg/cm^3	Maximum cartilage matrix density
d_{cM}	cm^3/cell day	Cartilage matrix decay constant
r_{bM}	μg/cell day	Bone matrix production rate
ρ_{bM}^{max}	μg/cm^3	Maximum bone matrix density

6.4.2.2.3 Chondrogenic Growth Factor Concentration ρ_{cF} and Osteogenic Growth Factor Concentration ρ_{bF}

$$\frac{\partial \rho_{cF}}{\partial t} = \{\nabla \cdot (D_{cF}\nabla \rho_{cF})\} + \left\{\left(\frac{r_{cF}^{max}\rho_{cF}}{K_{cF} + \rho_{cF}}\right)\frac{\rho_{tM}}{\left(\alpha_{cF}^3 + \rho_{tM}{}^3\right)}\rho_c - d_{cF}\rho_{cF}\right\} \tag{6.4.36}$$

Table 6.6 Constants of chondrogenic growth factor concentration ρ_{cF} and osteogenic growth factor concentration ρ_{bF} diffusion equations.

Parameter	Unit	Description
D_{cF}	cm^2/day	Chondrogenic growth factor diffusion rate
r_{cF}^{max}	$\mu g^3/cm^6$ cell day	Chondrogenic growth factor production constant
K_{cF}	$\mu g/cm^3$	Michaelis–Menten constant for chondrogenic growth factor production
α_{cF}	$\mu g/cm^3$	Saturation level of total matrix density
d_{cF}	day^{-1}	Chondrogenic growth factor deterioration rate
D_{bF}	cm^2/day	Osteogenic growth factor diffusion rate
r_{bF}^{max}	μg/cell day	Osteogenic growth factor production constant
K_{bF}	$\mu g/cm^3$	Michaelis–Menten constant for osteogenic growth factor production
d_{bF}	day^{-1}	Osteogenic growth factor deterioration rate

$$\frac{\partial \rho_{bF}}{\partial t} = \{\nabla \cdot (D_{bF} \nabla \rho_{bF})\} + \left\{ \frac{r_{bF}^{max} \rho_{bF}}{K_{bF} + \rho_{bF}} \rho_b - d_{bF} \rho_{bF} \right\} \tag{6.4.37}$$

where the constants are defined in Table 6.6.

6.4.3 Solution Procedure

6.4.3.1 Solution in Time

The generalized trapezoidal rule can be adopted to formulate the time integration procedure of the coupled set of equations (Chawla et al. 1996). Denoting each variable H at the time i by H^i, the recursive relation to compute the rate ΔH^i of variations of H^i in time can be defined by

$$\Delta H^i = \left(\frac{\partial H}{\partial t} \right)^i = \frac{H^i - \left(H^{i-1} + \Delta t (1 - \alpha) V_H^{\ i-1} \right)}{\alpha \Delta t} \tag{6.4.38}$$

where $\alpha = 0.5$ for the trapezoidal rule:

$$\Delta H^i = \left(\frac{\partial H}{\partial t} \right)^i = 2 \times \frac{H^i - H^{i-1}}{\Delta t} - \Delta H^{i-1} \tag{6.4.39}$$

The time marching technique can be simplified by assuming zero initial velocities for all variables in the first timestep. Only the definitions of the initial distributions of the variables, ρ_m, ρ_c, ρ_b, ρ_{cM}, ρ_{bM}, ρ_{cF} and ρ_{bF} are required (defined in Table 6.7).

Table 6.7 Initialization of the main variables.

Parameter	Unit	Description
$\rho_{\mathrm{m}}^{\mathrm{init}}$	cell/cm^3	Initial MSCs density at the onset of healing
$\rho_{\mathrm{c}}^{\mathrm{init}}$	cell/cm^3	Initial chondrocyte density at the onset of healing
$\rho_{\mathrm{b}}^{\mathrm{init}}$	cell/cm^3	Initial osteoblast density at the onset of healing
$\rho_{\mathrm{cM}}^{\mathrm{init}}$	µg/cm^3	Initial cartilage matrix density at the onset of healing
$\rho_{\mathrm{bM}}^{\mathrm{init}}$	µg/cm^3	Initial bone matrix density at the onset of healing
$\rho_{\mathrm{cF}}^{\mathrm{init}}$	µg/cm^3	Initial chondrogenic growth factor concentration at the onset of healing
$\rho_{\mathrm{bF}}^{\mathrm{init}}$	µg/cm^3	Initial osteogenic growth factor concentration at the onset of healing

6.4.3.2 Finite Element Solution (in Each Timestep)

The governing equations are in the general strong form of the diffusion equation:

$$\frac{\partial Q}{\partial t} = -\boldsymbol{\nabla} \cdot \boldsymbol{J}_{\mathrm{Q}} + f_{\mathrm{Q}} \tag{6.4.40}$$

which can be expressed in the following weak form:

$$\int_{\Omega} \frac{\partial Q}{\partial t} v_Q \mathrm{d}\Omega - \int_{\Omega} \boldsymbol{J}_Q \cdot \boldsymbol{\nabla} v_Q \mathrm{d}\Omega = \int_{\Omega} (f_Q v_Q) \mathrm{d}\Omega - \int_{\Gamma Q} (\boldsymbol{J}_Q \cdot \boldsymbol{n} v_Q) \mathrm{d}\Gamma \tag{6.4.41}$$

where v_Q is an arbitrary function (virtual density).

The finite element shape function $\mathbf{N}_Q(\boldsymbol{x})$ can be used to discretize the field variable $Q(\mathbf{x},t)$ to the nodal values $\boldsymbol{Q}(t)$:

$$Q(\mathbf{x},t) = \mathbf{N}_Q(\mathbf{x})\boldsymbol{Q}(t) \tag{6.4.42}$$

The final discretized form of the coupled set of equations at a time t can be written in a form suitable for nonlinear solution procedures:

$$F_Q = F_Q^{\mathrm{int}} - F_Q^{\mathrm{ext}} = 0 \tag{6.4.43}$$

Using the final set of strong form Equations (6.4.31) to (6.4.37) and assuming the domain boundaries are far enough to ignore the boundary terms, the internal force vectors can be expressed as

$$\boldsymbol{F}_{\rho_{\mathrm{m}}}^{\mathrm{int}} = \int_{\Omega} \mathbf{N}_{\mathrm{m}}^{\mathrm{T}} \frac{\partial \rho_{\mathrm{m}}}{\partial t} \mathrm{d}\Omega + \int_{\Omega} \nabla \mathbf{N}_{\mathrm{m}}^{\mathrm{T}} \left[\frac{a_{\mathrm{m}} \rho_{\mathrm{tM}}}{(b_{\mathrm{m}} 2 + \rho_{\mathrm{tM}} 2)} \boldsymbol{\nabla} \rho_{\mathrm{m}} - \frac{c_{\mathrm{m}}}{(d_{\mathrm{m}} + \rho_{\mathrm{tM}})^2} \rho_{\mathrm{m}} \boldsymbol{\nabla} \rho_{\mathrm{tM}} \right] \mathrm{d}\Omega \tag{6.4.44}$$

$$\boldsymbol{F}_{\rho_c}^{\text{int}} = \int_\Omega \boldsymbol{N}_c^{\text{T}} \frac{\partial \rho_c}{\partial t} d\Omega \tag{6.4.45}$$

$$\boldsymbol{F}_{\rho_b}^{\text{int}} = \int_\Omega \boldsymbol{N}_b^{\text{T}} \frac{\partial \rho_b}{\partial t} d\Omega \tag{6.4.46}$$

$$\boldsymbol{F}_{\rho_{cM}}^{\text{int}} = \int_\Omega \boldsymbol{N}_{cM}^{\text{T}} \frac{\partial \rho_{cM}}{\partial t} \mathrm{d}\Omega \tag{6.4.47}$$

$$\boldsymbol{F}_{\rho_{bM}}^{\text{int}} = \int_\Omega \boldsymbol{N}_{bM}^{\text{T}} \frac{\partial \rho_{bM}}{\partial t} \mathrm{d}\Omega \tag{6.4.48}$$

$$\boldsymbol{F}_{\rho_{cF}}^{\text{int}} = \int_\Omega \boldsymbol{N}_{cF}^{T} \frac{\partial \rho_{cF}}{\partial t} \mathrm{d}\Omega + \int_\Omega \nabla \boldsymbol{N}_{cF}^{T} \left[D_{cF} \boldsymbol{\nabla} \rho_{cF} \right] \mathrm{d}\Omega \tag{6.4.49}$$

$$\boldsymbol{F}_{\rho_{bF}}^{\text{int}} = \int_\Omega \boldsymbol{N}_{bF}^{\text{T}} \frac{\partial \rho_{bF}}{\partial t} \mathrm{d}\Omega + \nabla \boldsymbol{N}_{bF}^{\text{T}} \left[D_{bF} \boldsymbol{\nabla} \rho_{bF} \right] \mathrm{d}\Omega \tag{6.4.50}$$

and the external force vectors:

$$\boldsymbol{F}_{\rho_m}^{\text{ext}} = \int_\Omega \boldsymbol{N}_m^{\text{T}} \left[\left(\frac{r_m^{\text{init}} \rho_{tM}}{K_m{}^2 + \rho_{tM}{}^2} \right) \rho_m \left(1 - \frac{\rho_m}{\rho_m^{\max}} \right) - \frac{\Phi_{\text{tmb}} \rho_{bF}}{\Psi_{\text{tmb}} + \rho_{bF}} \rho_m - \frac{\Phi_{\text{tmc}} \rho_{cF}}{\Psi_{\text{tmc}} + \rho_{cF}} \rho_m \right] \mathrm{d}\Omega \tag{6.4.51}$$

$$\boldsymbol{F}_{\rho_c}^{\text{ext}} = \int_\Omega \boldsymbol{N}_c^{\text{T}} \left[\left(\frac{r_c^{\text{init}} \rho_{tM}}{K_c{}^2 + \rho_{tM}{}^2} \right) \rho_c \left(1 - \frac{\rho_c}{\rho_c^{\max}} \right) + \frac{\Phi_{\text{tmc}} \rho_{cF}}{\Psi_{\text{tmc}} + \rho_{cF}} \rho_m - \frac{\rho_{cM}{}^6}{\beta^6 + \rho_{cM}{}^6} \times \frac{\Phi_{\text{tcb}} \rho_{bF}}{\Psi_{\text{tcb}} + \rho_{bF}} \rho_c \right] \mathrm{d}\Omega \tag{6.4.52}$$

$$\boldsymbol{F}_{\rho_b}^{\text{ext}} = \int_\Omega \boldsymbol{N}_b^{\text{T}} \left[\left(\frac{r_b^{\text{init}} \rho_{tM}}{K_b{}^2 + \rho_{tM}{}^2} \right) \rho_b \left(1 - \frac{\rho_b}{\rho_b^{\max}} \right) + \frac{\Phi_{\text{tmb}} \rho_{bF}}{\Psi_{\text{tmb}} + \rho_{bF}} \rho_m + \frac{\rho_{cM}{}^6}{\beta^6 + \rho_{cM}{}^6} \times \frac{\Phi_{\text{tcb}} \rho_{bF}}{\Psi_{\text{tcb}} + \rho_{bF}} \rho_c - \mathrm{d}_{bd} \rho_b \right] \mathrm{d}\Omega \tag{6.4.53}$$

$$\boldsymbol{F}_{\rho_{cM}}^{\text{ext}} = \int_\Omega \boldsymbol{N}_{cM}^{\text{T}} \left[r_{cM} \left(1 - \frac{\rho_{cM}}{\rho_{cM}^{\max}} \right) \times (\rho_m + \rho_c) - \mathrm{d}_{cM} \rho_b \rho_{cM} \right] \mathrm{d}\Omega \tag{6.4.54}$$

$$\boldsymbol{F}_{\rho_{bM}}^{\text{ext}} = \int_\Omega \boldsymbol{N}_{bM}^{\text{T}} \left[r_{bM} \left(1 - \frac{\rho_{bM}}{\rho_{bM}^{\max}} \right) \rho_b \right] \mathrm{d}\Omega \tag{6.4.55}$$

$$\boldsymbol{F}_{\rho_{cF}}^{\text{ext}} = \int_\Omega \boldsymbol{N}_{cF}^{\text{T}} \left[\left(\frac{r_{cF}^{\max} \rho_{cF}}{K_{cF} + \rho_{cF}} \right) \frac{\rho_{tM}}{\left(\alpha_{cF}^3 + \rho_{tM} 3 \right)} \rho_c - \mathrm{d}_{cF} \rho_{cF} \right] \mathrm{d}\Omega \tag{6.4.56}$$

$$\boldsymbol{F}_{\rho_{bF}}^{\text{ext}} = \int_\Omega \boldsymbol{N}_{bF}^{\text{T}} \left[\frac{r_{bF}^{\max} \rho_{bF}}{K_{bF} + \rho_{bF}} \rho_b - \mathrm{d}_{bF} \rho_{bF} \right] \mathrm{d}\Omega \tag{6.4.57}$$

The final discretized nonlinear set of equations becomes

$$\begin{Bmatrix} -\boldsymbol{F}_{\rho_\mathrm{m}} \\ -\boldsymbol{F}_{\rho_\mathrm{c}} \\ -\boldsymbol{F}_{\rho_\mathrm{b}} \\ -\boldsymbol{F}_{\rho_\mathrm{cM}} \\ -\boldsymbol{F}_{\rho_\mathrm{bM}} \\ -\boldsymbol{F}_{\rho_\mathrm{cF}} \\ -\boldsymbol{F}_{\rho_\mathrm{bF}} \end{Bmatrix} = \begin{Bmatrix} \boldsymbol{F}^{\mathrm{int}}_{\rho_\mathrm{m}} \\ \boldsymbol{F}^{\mathrm{int}}_{\rho_\mathrm{c}} \\ \boldsymbol{F}^{\mathrm{int}}_{\rho_\mathrm{b}} \\ \boldsymbol{F}^{\mathrm{int}}_{\rho_\mathrm{cM}} \\ \boldsymbol{F}^{\mathrm{int}}_{\rho_\mathrm{bM}} \\ \boldsymbol{F}^{\mathrm{int}}_{\rho_\mathrm{cF}} \\ \boldsymbol{F}^{\mathrm{int}}_{\rho_\mathrm{bF}} \end{Bmatrix} - \begin{Bmatrix} \boldsymbol{F}^{\mathrm{ext}}_{\rho_\mathrm{m}} \\ \boldsymbol{F}^{\mathrm{ext}}_{\rho_\mathrm{c}} \\ \boldsymbol{F}^{\mathrm{ext}}_{\rho_\mathrm{b}} \\ \boldsymbol{F}^{\mathrm{ext}}_{\rho_\mathrm{cM}} \\ \boldsymbol{F}^{\mathrm{ext}}_{\rho_\mathrm{bM}} \\ \boldsymbol{F}^{\mathrm{ext}}_{\rho_\mathrm{cF}} \\ \boldsymbol{F}^{\mathrm{ext}}_{\rho_\mathrm{bF}} \end{Bmatrix} = \begin{Bmatrix} \mathbf{0} \\ \mathbf{0} \\ \mathbf{0} \\ \mathbf{0} \\ \mathbf{0} \\ \mathbf{0} \\ \mathbf{0} \end{Bmatrix} \tag{6.4.58}$$

which should be solved by a proper nonlinear solver.

6.4.3.3 Nonlinear Solution

The iterative Newton–Raphson nonlinear solution for a general form of nonlinear equation $F(x)=0$ can be expressed as (at iteration number n):

$$\frac{dF(x_n)}{dx}\{x_{n+1}-x_n\}=-F(x_n) \tag{6.4.59}$$

The iterative solution continues until a specific convergence criterion is met.

Adopting the same procedure, the coupled set of Equations (6.4.58) can be written as

$$\begin{bmatrix}
\frac{\partial \boldsymbol{F}_{\rho_\mathrm{m}}}{\partial \rho_\mathrm{m}} & \frac{\partial \boldsymbol{F}_{\rho_\mathrm{m}}}{\partial \rho_\mathrm{c}} & \frac{\partial \boldsymbol{F}_{\rho_\mathrm{m}}}{\partial \rho_\mathrm{b}} & \frac{\partial \boldsymbol{F}_{\rho_\mathrm{m}}}{\partial \rho_\mathrm{cM}} & \frac{\partial \boldsymbol{F}_{\rho_\mathrm{m}}}{\partial \rho_\mathrm{bM}} & \frac{\partial \boldsymbol{F}_{\rho_\mathrm{m}}}{\partial \rho_\mathrm{cF}} & \frac{\partial \boldsymbol{F}_{\rho_\mathrm{m}}}{\partial \rho_\mathrm{bF}} \\
\frac{\partial \boldsymbol{F}_{\rho_\mathrm{c}}}{\partial \rho_\mathrm{m}} & \frac{\partial \boldsymbol{F}_{\rho_\mathrm{c}}}{\partial \rho_\mathrm{c}} & \frac{\partial \boldsymbol{F}_{\rho_\mathrm{c}}}{\partial \rho_\mathrm{b}} & \frac{\partial \boldsymbol{F}_{\rho_\mathrm{c}}}{\partial \rho_\mathrm{cM}} & \frac{\partial \boldsymbol{F}_{\rho_\mathrm{c}}}{\partial \rho_\mathrm{bM}} & \frac{\partial \boldsymbol{F}_{\rho_\mathrm{c}}}{\partial \rho_\mathrm{cF}} & \frac{\partial \boldsymbol{F}_{\rho_\mathrm{c}}}{\partial \rho_\mathrm{bF}} \\
\frac{\partial \boldsymbol{F}_{\rho_\mathrm{b}}}{\partial \rho_\mathrm{m}} & \frac{\partial \boldsymbol{F}_{\rho_\mathrm{b}}}{\partial \rho_\mathrm{c}} & \frac{\partial \boldsymbol{F}_{\rho_\mathrm{b}}}{\partial \rho_\mathrm{b}} & \frac{\partial \boldsymbol{F}_{\rho_\mathrm{b}}}{\partial \rho_\mathrm{cM}} & \frac{\partial \boldsymbol{F}_{\rho_\mathrm{b}}}{\partial \rho_\mathrm{bM}} & \frac{\partial \boldsymbol{F}_{\rho_\mathrm{b}}}{\partial \rho_\mathrm{cF}} & \frac{\partial \boldsymbol{F}_{\rho_\mathrm{b}}}{\partial \rho_\mathrm{bF}} \\
\frac{\partial \boldsymbol{F}_{\rho_\mathrm{cM}}}{\partial \rho_\mathrm{m}} & \frac{\partial \boldsymbol{F}_{\rho_\mathrm{cM}}}{\partial \rho_\mathrm{c}} & \frac{\partial \boldsymbol{F}_{\rho_\mathrm{cM}}}{\partial \rho_\mathrm{b}} & \frac{\partial \boldsymbol{F}_{\rho_\mathrm{cM}}}{\partial \rho_\mathrm{cM}} & \frac{\partial \boldsymbol{F}_{\rho_\mathrm{cM}}}{\partial \rho_\mathrm{bM}} & \frac{\partial \boldsymbol{F}_{\rho_\mathrm{cM}}}{\partial \rho_\mathrm{cF}} & \frac{\partial \boldsymbol{F}_{\rho_\mathrm{cM}}}{\partial \rho_\mathrm{bF}} \\
\frac{\partial \boldsymbol{F}_{\rho_\mathrm{bM}}}{\partial \rho_\mathrm{m}} & \frac{\partial \boldsymbol{F}_{\rho_\mathrm{bM}}}{\partial \rho_\mathrm{c}} & \frac{\partial \boldsymbol{F}_{\rho_\mathrm{bM}}}{\partial \rho_\mathrm{b}} & \frac{\partial \boldsymbol{F}_{\rho_\mathrm{bM}}}{\partial \rho_\mathrm{cM}} & \frac{\partial \boldsymbol{F}_{\rho_\mathrm{bM}}}{\partial \rho_\mathrm{bM}} & \frac{\partial \boldsymbol{F}_{\rho_\mathrm{bM}}}{\partial \rho_\mathrm{cF}} & \frac{\partial \boldsymbol{F}_{\rho_\mathrm{bM}}}{\partial \rho_\mathrm{bF}} \\
\frac{\partial \boldsymbol{F}_{\rho_\mathrm{cF}}}{\partial \rho_\mathrm{m}} & \frac{\partial \boldsymbol{F}_{\rho_\mathrm{cF}}}{\partial \rho_\mathrm{c}} & \frac{\partial \boldsymbol{F}_{\rho_\mathrm{cF}}}{\partial \rho_\mathrm{b}} & \frac{\partial \boldsymbol{F}_{\rho_\mathrm{cF}}}{\partial \rho_\mathrm{cM}} & \frac{\partial \boldsymbol{F}_{\rho_\mathrm{cF}}}{\partial \rho_\mathrm{bM}} & \frac{\partial \boldsymbol{F}_{\rho_\mathrm{cF}}}{\partial \rho_\mathrm{cF}} & \frac{\partial \boldsymbol{F}_{\rho_\mathrm{cF}}}{\partial \rho_\mathrm{bF}} \\
\frac{\partial \boldsymbol{F}_{\rho_\mathrm{bF}}}{\partial \rho_\mathrm{m}} & \frac{\partial \boldsymbol{F}_{\rho_\mathrm{bF}}}{\partial \rho_\mathrm{c}} & \frac{\partial \boldsymbol{F}_{\rho_\mathrm{bF}}}{\partial \rho_\mathrm{b}} & \frac{\partial \boldsymbol{F}_{\rho_\mathrm{bF}}}{\partial \rho_\mathrm{cM}} & \frac{\partial \boldsymbol{F}_{\rho_\mathrm{bF}}}{\partial \rho_\mathrm{bM}} & \frac{\partial \boldsymbol{F}_{\rho_\mathrm{bF}}}{\partial \rho_\mathrm{cF}} & \frac{\partial \boldsymbol{F}_{\rho_\mathrm{bF}}}{\partial \rho_\mathrm{bF}}
\end{bmatrix}
\begin{Bmatrix} \delta_{\rho_\mathrm{m}} \\ \delta_{\rho_\mathrm{c}} \\ \delta_{\rho_\mathrm{b}} \\ \delta_{\rho_\mathrm{cM}} \\ \delta_{\rho_\mathrm{bM}} \\ \delta_{\rho_\mathrm{cF}} \\ \delta_{\rho_\mathrm{bF}} \end{Bmatrix} = \begin{Bmatrix} -\boldsymbol{F}_{\rho_\mathrm{m}} \\ -\boldsymbol{F}_{\rho_\mathrm{c}} \\ -\boldsymbol{F}_{\rho_\mathrm{b}} \\ -\boldsymbol{F}_{\rho_\mathrm{cM}} \\ -\boldsymbol{F}_{\rho_\mathrm{bM}} \\ -\boldsymbol{F}_{\rho_\mathrm{cF}} \\ -\boldsymbol{F}_{\rho_\mathrm{bF}} \end{Bmatrix} \tag{6.4.60}$$

6.4.4 Numerical Analysis

Healing of a bone fracture often involves the production of irregular and asymmetric bone callus at the fracture site. The solution region may be reduced and simplified to a quarter

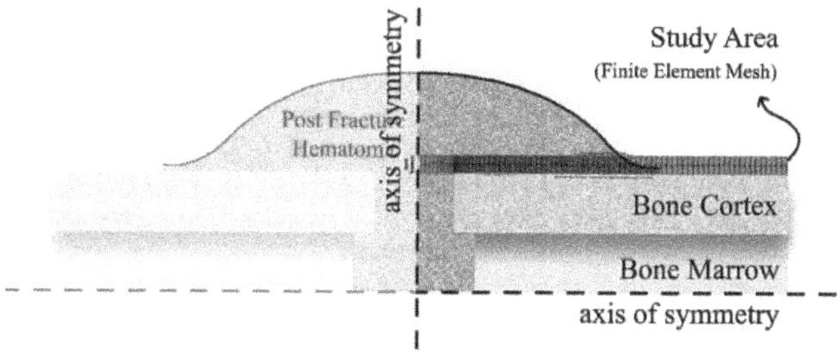

Figure 6.34 A strip of fracture callus. *Source:* Reproduced from Zamani 2022/University of Tehran.

of a two-dimensional domain if axial and longitudinal symmetries can be assumed for the callus produced around a long and cylindrical bone. Only a simple case of a strip callus is considered here. The simulation investigates a conceptual one-dimensional region of broken bone callus, as depicted in Figure 6.34.

In this simplified numerical model, it is sufficient to examine a small portion of the content in the vicinity and in line with the bone, as illustrated in Figure 6.35. The size of the unbroken tissue is considered to be nine times the size of the fracture, which is adequate to avoid the effects of distant boundary conditions.

The material constants for the governing healing equations are defined in Table 6.8.

The initial distributions of the MSC density, the chodrogenic growth factor concentration and the osteogenic growth factor concentration are presented in Figure 6.36.

6.4.4.1 Temporal Changes of the MSC Density

MSCs are commonly predicted to migrate into the fracture callus and diffuse there. Moreover, when MSCs are exposed to high amounts of chondrogenic and osteogenic growth factors, they differentiate into chondrocytes and osteoblasts, respectively.

The area around the bone periosteum has the highest density of MSCs based on the initial conditions (Figure 6.36). The osteogenic growth factors are also dominant in this area. As a results, differentiation of MSCs into osteoblasts is anticipated. On the other hand, due to the initial presence of chondrogenic growth factors near the fracture site, the MSCs are differentiated into chondrocytes near the fracture zone.

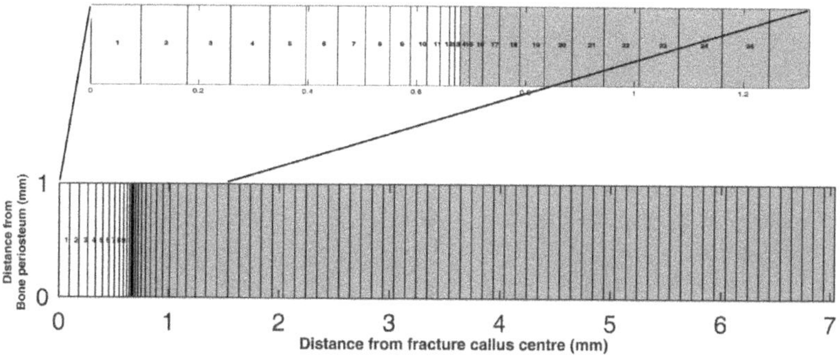

Figure 6.35 The numerical model of the strip of fracture callus. *Source:* Reproduced from Zamani 2022/University of Tehran.

Table 6.8 Material constants for numerical simulations.

Parameter	Value	Unit
a_{m}	171.5	μg/cm day
b_{m}	25,000	$\mu g/cm^3$
c_{m}	41.65	μg/cm day
d_{m}	50,000	$\mu g/cm^3$
$r_{\mathrm{m}}^{\mathrm{init}}$	101,000	$\mu g/cm^3$ day
K_{m}	10,000	$\mu g/cm^3$
$\rho_{\mathrm{m}}^{\mathrm{max}}$	1,000,000	$cell/cm^3$
Φ_{tmb}	10	day^{-1}
Ψ_{tmb}	0.01	$\mu g/cm^3$
Φ_{tmc}	5	day^{-1}
Ψ_{tmc}	0.01	$\mu g/cm^3$
$r_{\mathrm{c}}^{\mathrm{init}}$	101,000	$\mu g/cm^3$ day
K_{c}	10,000	$\mu g/cm^3$
$\rho_{\mathrm{c}}^{\mathrm{max}}$	1,000,000	$cell/cm^3$
β	150,000	$\mu g/cm^3$
Φ_{tcb}	1,000	day^{-1}
Ψ_{tcb}	0.01	$\mu g/cm^3$
$r_{\mathrm{b}}^{\mathrm{init}}$	20,200	$\mu g/cm^3$ day
K_{b}	10,000	$\mu g/cm^3$
$\rho_{\mathrm{b}}^{\mathrm{max}}$	1,000,000	$cell/cm^3$
d_{bd}	0.1	day^{-1}
r_{cM}	0.02	μg/cell day
$\rho_{\mathrm{cM}}^{\mathrm{max}}$	100,000	$\mu g/cm^3$
d_{cM}	0.000002	cm^3/cell day
r_{bM}	0.2	μg/cell day
$\rho_{\mathrm{bM}}^{\mathrm{max}}$	100,000	$\mu g/cm^3$
D_{cF}	0.00062208	cm^2/day
$r_{\mathrm{cF}}^{\mathrm{max}}$	0.5	$\mu g^3/cm^6$ cell day
K_{cF}	0.1	$\mu g/cm^3$
α_{cF}	10,000	$\mu g/cm^3$
d_{cF}	0.35	day^{-1}
D_{bF}	0.00062208	cm^2/day
$r_{\mathrm{bF}}^{\mathrm{max}}$	0.0000005	μg/cell day
K_{bF}	0.1	$\mu g/cm^3$
d_{bF}	0.25	day^{-1}

(Continued)

Table 6.8 (Continued)

Parameter	Value	Unit
ρ_{m}^{init}	1,000,000	cell/cm^3
ρ_{c}^{init}	0	cell/cm^3
ρ_{b}^{init}	0	cell/cm^3
ρ_{cM}^{init}	10,000	µg/cm^3
ρ_{bM}^{init}	0	µg/cm^3
ρ_{M}^{init}	10,000	µg/cm^3
ρ_{cF}^{init}	2	µg/cm^3
ρ_{bF}^{init}	2	µg/cm^3

Source: Adapted from Zamani, 2022; Bailon-Plaza and Van der Meulen, 2001; Geris et al., 2008; Carlier et al., 2015.

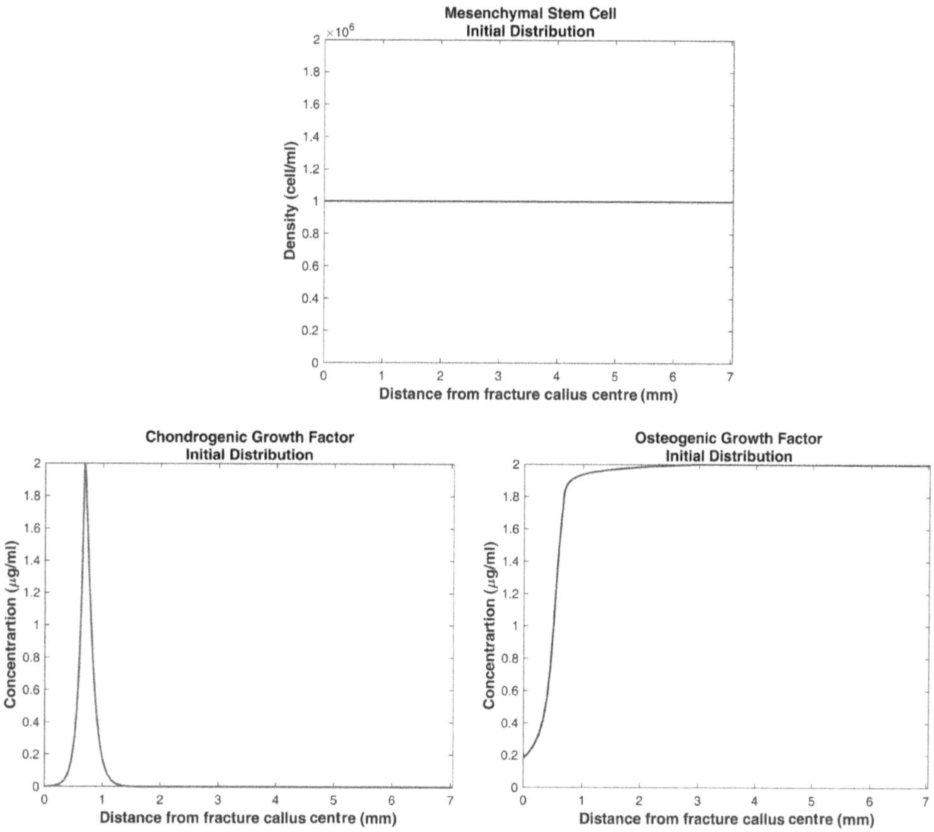

Figure 6.36 Initial distributions of MSC density, chodrogenic growth factor concentration and osteogenic growth factor concentration. *Source:* Reproduced from Zamani 2022/University of Tehran.

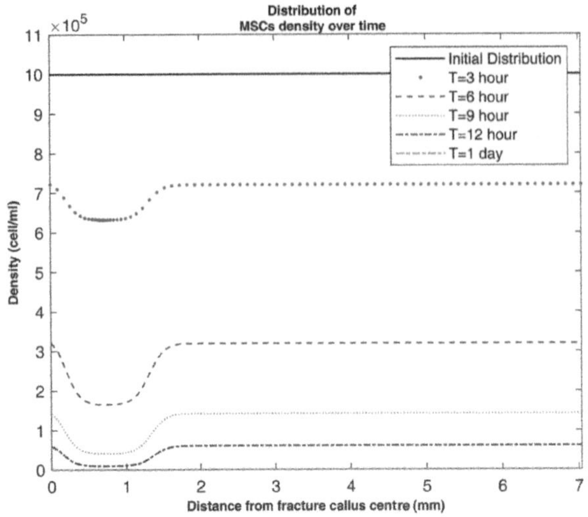

Figure 6.37 Distribution of the MSC density over time. *Source:* Reproduced from Zamani 2022/University of Tehran.

According to Figure 6.37, MSC density is reduced around the bone periosteum and the site of the fracture in the early stages of fracture healing. This reduction can be interpreted as a result of the differentiation of MSCs into osteoblasts. Moreover, the density will decline further around the centre of the fracture callus due to the large portion of MSCs being differentiated into chondrocytes.

6.4.4.2 Temporal Changes of the Chondrocyte Density

Initially, no chondrocytes exist in the fracture callus. If there is a minimal amount of matrix surrounding them, they will grow rapidly and increase in density. In addition, chondrocyte proliferation will cease if they are exposed to an environment with a dense matrix. If the cartilage matrix is dense enough, the cells will then differentiate into osteoblasts.

The initial conditions assume that MSCs and chondrogenic growth factors are present at the fracture site. Consequently, MSCs differentiate into chondrocytes in the vicinity of the fractured bone, as illustrated in Figure 6.38 by the increased chondrocyte density. Due to the presence of chondrocytes in this zone and the minimal amount of matrix provided by the hematoma, chondrocytes begin to proliferate until they reach the point of density saturation.

6.4.4.3 Temporal Changes of the Osteoblast Density

Similar to the case of chondrocytes, the fracture callus initially contains no osteoblasts, which are differentiated from MSCs and chondrocytes. Moreover, in the absence of a sufficient surrounding matrix, they proliferate and become denser, whereas osteoblast proliferation ceases in extremely dense matrices. The osteoblasts are eventually captured by the matrix and transformed into osteocytes, or a fraction of them are eliminated by osteoclasts. In a rich osteogenic growth factor environment along the bone periosteum, more osteoblasts are differentiated from the existing MSCs and then proliferate until the surrounding matrix becomes dense. In contrast, by increasing the cartilage matrix density near the fracture site, more chondrocytes transform into osteoblasts until the environmental matrix is saturated.

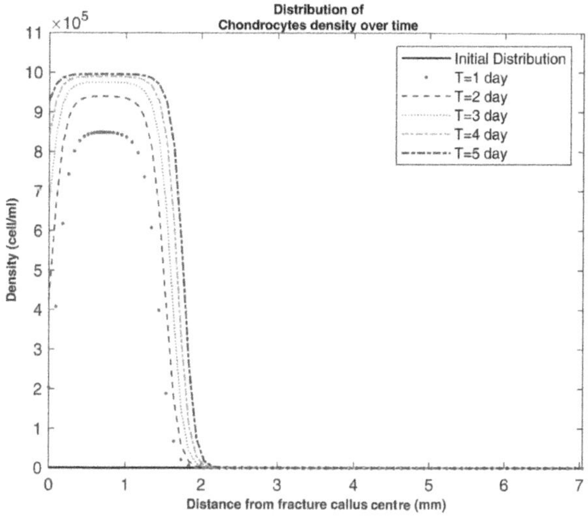

Figure 6.38 Distribution of the chondrocyte density over the time. *Source:* Reproduced from Zamani 2022/University of Tehran.

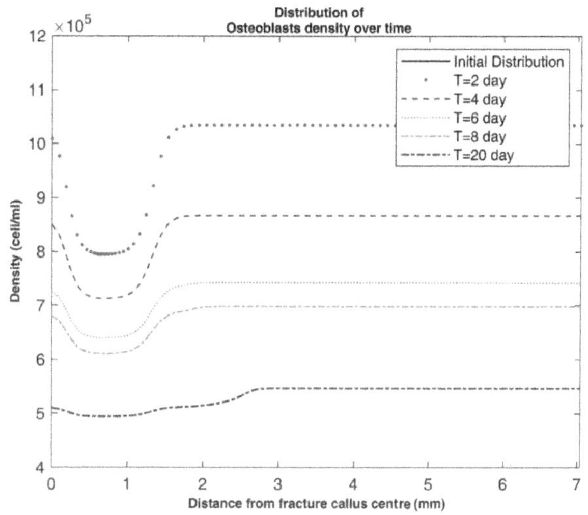

Figure 6.39 Distribution of the osteoblast density over time. *Source:* Reproduced from Zamani 2022/University of Tehran.

According to Figure 6.39, the density of osteoblasts is initially increased in the vicinity of periosteum bone by the differentiation of chondrocytes into osteoblasts due to endochondral ossification, which is then followed by cell proliferation until the matrix becomes saturated. Due to osteoblast cell death, the osteoblast density decreases following the cessation of osteoblast proliferation.

6.4.4.4 Temporal Changes of the Cartilage Matrix Density

Despite the fact that only a small amount of cartilage matrix is produced during the inflammatory phase, it is essential to the healing process. It is known that the cartilage production decreases as the matrix density increases.

In the early stages of bone fracture healing, mesenchymal stem cells differentiate into the osteoblast and chondrocyte cells. The first set decomposes the primary cartilage matrix around the periosteum and the second set increases the matrix density at the fracture site.

As the amount of cartilage matrix in the investigated area (near the periosteum of the bone) increases over the time, the rate of matrix synthesis decreases and the rate of matrix degradation by osteoblasts increases. Therefore, the density of cartilage matrix in this region cannot be maximized, as depicted in Figure 6.40.

6.4.4.5 Temporal Changes of the Bone Matrix Density

As discussed for the cartilage matrix, osteoblasts produce bone matrix at a rate proportional to the matrix density of their environment. It is noted that the effects of osteoclasts degrading bone matrix are not included in the present model and the fracture healing is followed only up to the stage of remodelling (in which the cancellous bone is replaced by the cortical bone).

Figure 6.41 shows that the changes of bone matrix density begin on the first day of the healing process, as the osteoblasts differentiate and start to synthetize the bone matrix. Increased bone matrix density is inversely proportional to the spread of osteoblast cells. As a result, it rises first in the vicinity of the periosteum and then gradually at the fracture site, known as the intramembranous and the endochondral ossification processes, respectively. Finally, the ossification stage is completed by making the bone matrix as dense as the maximum attainable level.

6.4.4.6 Temporal Changes of the Chondrogenic Growth Factor Concentration

Chondrogenic growth factors are diffused into the domain according to their concentration gradients. The existing chondrocytes are capable of producing these factors until their concentrations reach the saturation level.

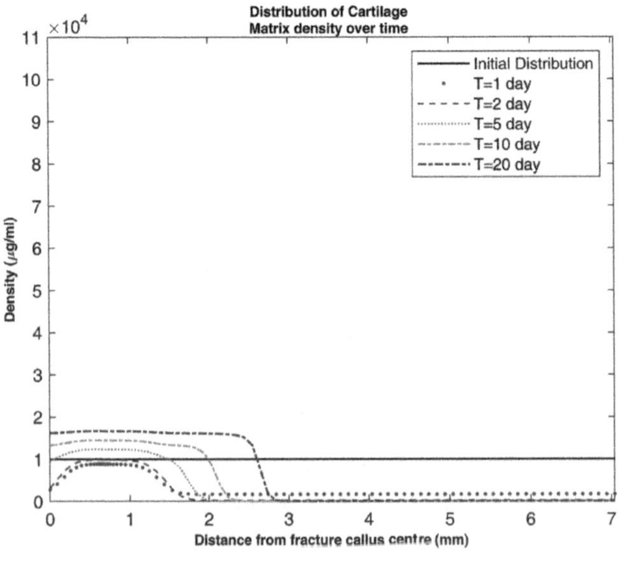

Figure 6.40 Distribution of the cartilage matrix density over time. *Source:* Reproduced from Zamani 2022/University of Tehran.

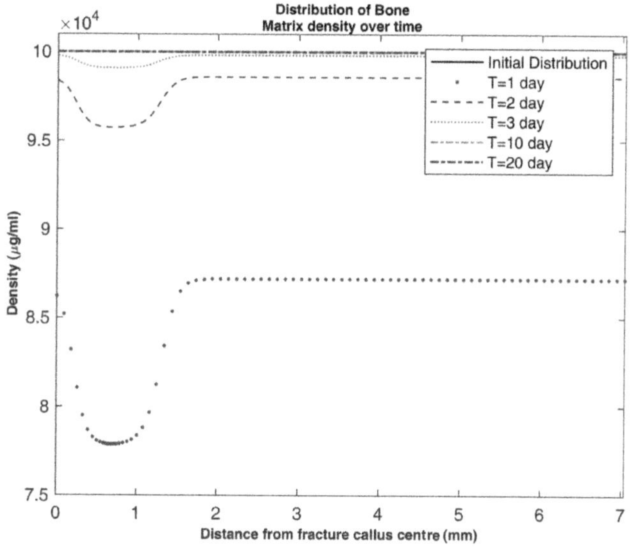

Figure 6.41 Distribution of the bone matrix density over time. *Source:* Reproduced from Zamani 2022/University of Tehran.

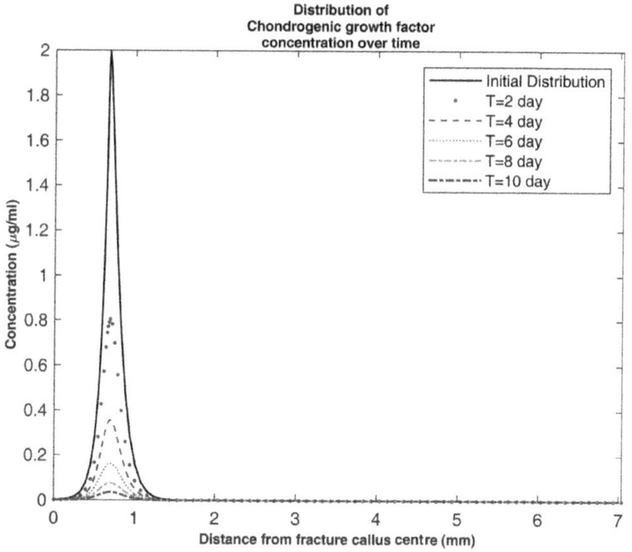

Figure 6.42 Distribution of the chondrogenic growth factor concentration over time. *Source:* Reproduced from Zamani 2022/University of Iran.

The fractured bone cortex releases a significant amount of chondrogenic growth factors into the surrounding environment during the initial stages of the healing process. As time passes and chondrocytes at the fracture site become present and begin to produce a minimal matrix, the production of chondrogenic growth factors accelerates. These factors are consumed and eliminated due to their short half-lives. Variations of the chondrogenic growth factor concentration over time are presented in Figure 6.42.

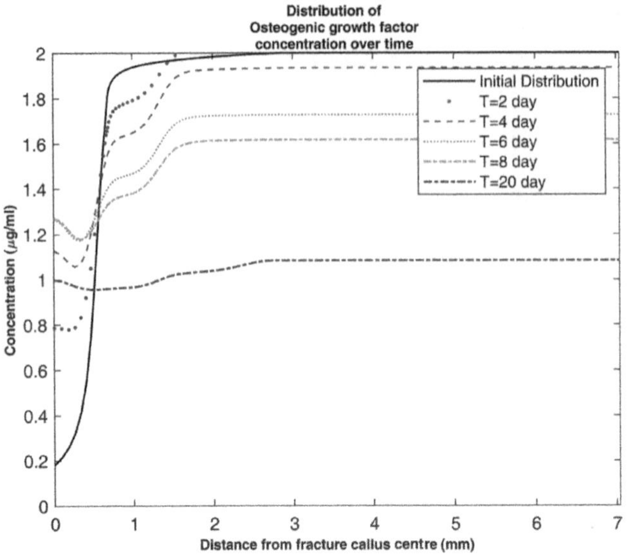

Figure 6.43 Distribution of the osteogenic growth factor concentration over time. *Source:* Reproduced from Zamani 2022/University of Iran.

6.4.4.7 Temporal Changes of the Osteogenic Growth Factor Concentration

According to Figure 6.43, the initial distribution of osteogenic growth factor concentration is diffused towards the centre of the fracture callus and the concentration of osteogenic growth factor in the damaged area is increased due to osteoblast synthesis. This concentration factor is eventually decreased due to degradation and consumption of the osteogenic growth factors by other present organisms.

7

Supplementary Topics

7.1 Introduction

The main goal of the present chapter is to open up some other biomechanical topics that may not be specifically categorized as a soft or hard tissue. Only some preliminary remarks are provided on the way the computational modelling (either single or multiscale analysis) may be performed without going into the details of numerical solutions and results.

Some of the subjects are related to the routine daily practice of medical teams, whereas some others are associated with the new technologies (either medical or computational) evolved in the recent decade or so.

Five different types of problems are briefly examined and reviewed in this chapter. First, the shape memory alloy (SMA) stenting procedure in an artery is considered. It involves the phase transformation properties of shape memory alloys, which characterizes its super-elasticity and shape memory effects. The basic concept is to deploy the folded SMA stent in the proper position, and then unfold it to the predefined configuration (based on the needs of the patient/artery) by available stimulation techniques.

The next section describes the way a multiscale computational approach may be adopted to determine the vitreous pressure on the optical nerve head, which plays a significant role in glaucoma and a number of other eye-related diseases.

It is followed by the fluid–solid interaction modelling of the aorta, with an eye on future FSI particle-based modellings of the formation of plagues inside the veins and around the installed stents.

Then, the novel idea of a drug delivery system by the help of shape memory polymers (SMPs) is examined. Specially designed SMPs with pre-injected or positioned drug substances may be designed to deform to a folded position at the end of the manufacturing process. The polymer may then be stimulated inside the body to deform to an unfolded configuration to allow for the drug substances to separate and reach the target cells/tissues.

The concluding section is dedicated to a brief introduction to the computational technology of artificial intelligence (AI) and deep learning (DL), which has revolutionized the science and technology and everyday practices of human life. The basics concepts of physics-informed AI and DL are briefly described and its potential application in biomechanical problems are noted.

Multiscale Biomechanics: Theory and Applications, First Edition. Soheil Mohammadi.

7.2 Shape Memory Alloy (SMA) Stenting of an Artery

7.2.1 Stenting Procedures

Cardiovascular diseases are among the main causes of death worldwide. One of the frequent types of these diseases is the blockage of an artery by blood clots or various types of plagues, which must be examined regularly and should be removed, when necessary.

Plagues on blocked coronary arteries are usually removed or reduced in size by angioplasty (PTCA) by inserting a small balloon into the blocked region and inflating it. The balloon is deflated to examine the level of compression of the plague and its size reduction. The procedure may be repeated several times to reach the required reduction of the plague and widening of the artery.

The artery opened by a balloon may be kept in a stationary condition by deployment of a stent (wire tube). It may also be directly used to remove or reduce the blockage of the artery and to create a rather free passage for blood flow, as schematically presented in Figure 7.1.

The stenting procedure is a delicate medical procedure which should be performed with extreme care. Any additional pressure or misplacement may lead to permanent damage or tearing of the artery with life-threatening consequences. It may also lead to a further obstacle inside the vascular system. Therefore, it should be properly designed, precisely installed and safely removed if needed.

Several types of stents are available from different manufacturers with a variety of specifications, depending on the needs of patients. One class of stent, which is made from shape memory alloys (SMAs), is activated by the existing conditions inside the vein. Another class of stent needs some other devices, such as small balloons, for installation and/or activation.

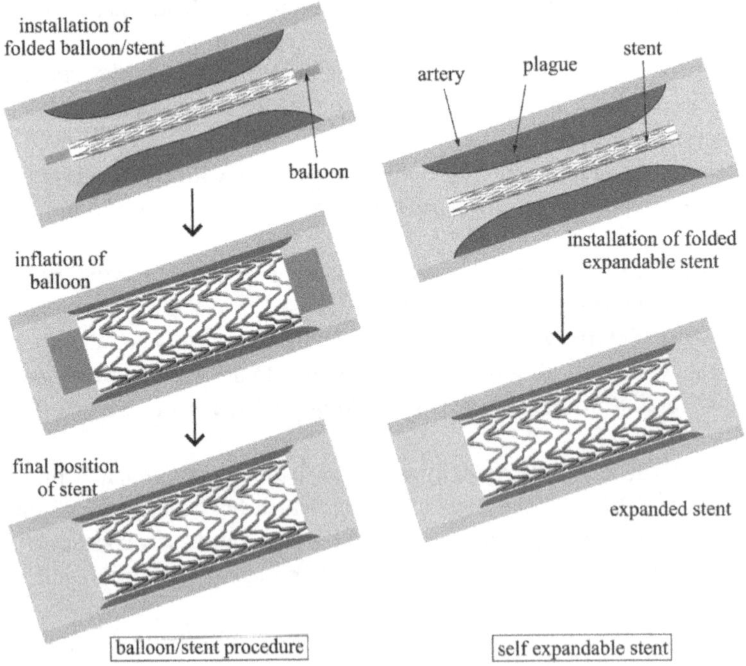

Figure 7.1 Typical procedures of stent installation.

SMAs have been largely used in aerospace (McDonald Schetky 1991; Bil et al. 2013), composite structures (Furuya 1996), robotics (Kheirikhah et al. 2010) and biomechanics (Van Humbeeck 1999). Nitinol (nickle-titanium) SMA alloys have been used in dental implants and related appliances (Andreasen and Hilleman 1971; Kauffman and Mayo 1997) and orthopaedic industries (Morgan 2004), among others.

Stents are usually made in different forms, such as helical wires or wire grids (Stoeckel et al. 2002). Self-expandable stents are usually made in a helical form, while the balloon-stent sets are practised in both forms. Stents that are expanded by balloons are usually made from nickel–cobalt alloys, while low-carbon steels and nitinol can be used for self-expandable stents (Stoeckel et al. 2002; Mani et al. 2007). The SMA stents are assumed to open and extend without the need of balloon pressure.

Analysis of the stenting procedure requires a complicated set of numerical simulations of the shape memory stent, the balloon, the soft tissue of the vein wall, and the contact interactions among them. The first attempts of such simulations were made by Teo et al. (2000) and Dumoulin and Cochelin (2000) to simulate a single stent and the artery. Similar attempts, followed by Migliavacca et al. (2002), David Chua et al. (2002) and Boivin et al. (2002), were based on imposing a uniform internal pressure on the stent, which was in contact with the artery. Some models ignored the complicated composite structure of the artery wall and its deformation.

David Chua et al. (2003) proposed the modelling of a balloon-stent system using a contact interaction algorithm. In 2005, Lally et al. (2005) presented the stent–vein model. They used a linear elastic model for the stent and a hyperelastic model for the artery wall. The same methodology was followed by Wu et al. (2007) and Capelli et al. (2009). More recently, Nolan and Lally (2019) studied the re-blockage of the artery after stenting using a damage mechanics approach embedded within the finite element model.

Numerical simulations are expected to provide an insight into the levels of stress and strain inside the soft tissue of the vein wall in different stages of stenting, as well as the mechanical state of the stent itself. As a result, such a comprehensive set of data may provide useful information for decision making of cardiovascular experts to perform the necessary procedure for any specific patient. In the following, a number of important aspects of numerical modelling of the stenting procedure are briefly reviewed.

7.2.2 SMA Constitutive Equations

A physical feature of the nitinol shape memory alloy is that while it is in the so-called martensite phase (M) microstructure in low temperatures, it changes to the austenite phase microstructure in higher temperatures. In addition, the cycles of stress loading/unloading may lead to various phase transformations. These important characteristics allow for SMAs to show two important responses of the shape memory effect (SME) and superelasticity. These effects have found extensive interest in high-tech aerospace and biomechanical applications, as well as in control devices in civil and mechanical engineering systems. Figure 7.2 shows a triple axes view of the stress–strain–temperature SME and superelastic responses of SMA in various mechanical and thermal loading and unloading paths.

Modelling of SMAs has been comprehensively studied in the past two decades (Hatefi Ardakani et al. 2015, Hashemi Yazdi et al. 2015; Afshar et al. 2015a, 2015b). The response of SMA is not only based on the stress and strain states but is largely affected by the thermal

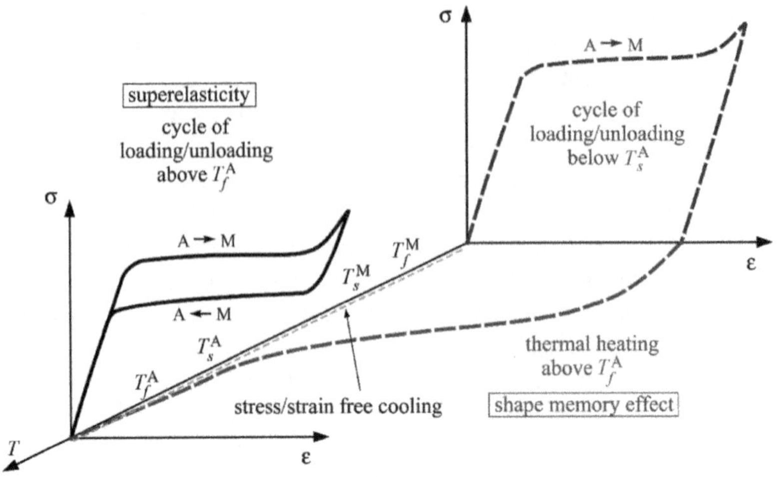

Figure 7.2 Nitinol SMA response. *Source:* Adapted from Karimi et al. 2019; Zhou et al. 2019.

effects, as the phase change of SMA is governed by the thermodynamic principles (Fatemi et al. 2017; Karimi et al. 2019).

The SMA stent is assumed to follow the constitutive model proposed by Boyd and Lagoudas (1996). In this constitutive model, the total strain ε is computed from the sum of thermoelastic and transformation strains, ε^{th} and ε^{t}, respectively (Lagodas 2008; Fatemi et al. 2017):

$$\varepsilon = \varepsilon^{\text{th}} + \varepsilon^{\text{t}} \tag{7.2.1}$$

The effective material properties $\chi = \mathbf{C}, \alpha, c, s_0, u_0$ in terms of the martensitic volume fraction ξ and the corresponding values at the austenitic A and martensitic M phases can be computed from (Lagodas 2008)

$$\chi(\xi) = \chi^{\text{A}} + \xi\Delta\chi = \chi^{\text{A}} + \xi\left(\chi^{\text{M}} - \chi^{\text{A}}\right),\ \chi = \mathbf{C}, \alpha, c, s_0, u_0 \tag{7.2.2}$$

where $\mathbf{C}$ and α are the effective compliance and the thermal expansion tensors, respectively, and c, s_0 and u_0 represent the effective specific heat, entropy and internal energy, respectively.

The rates of transformation strain $\dot{\varepsilon}^t$ and the martensitic volume fraction $\dot{\xi}$ during the forward ($\dot{\xi} > 0$) and reverse ($\dot{\xi} < 0$) martensite transformations are related by the transformation tensor $\mathbf{\Lambda}$ (Lagodas 2008)

$$\dot{\varepsilon}^t = \mathbf{\Lambda}\dot{\xi} \tag{7.2.3}$$

$$\mathbf{\Lambda} = \begin{cases} \dfrac{3}{2}\varepsilon^{\max} \dfrac{\boldsymbol{\sigma}^{\text{dev}}}{\sqrt{\dfrac{3}{2}\left\|\boldsymbol{\sigma}^{\text{dev}}\right\|^2}}, & \dot{\xi} > 0 \\ \varepsilon^{\max} \dfrac{\boldsymbol{\varepsilon}^{\text{tr}-R}}{\sqrt{\dfrac{2}{3}\left\|\boldsymbol{\varepsilon}^{\text{tr}-R}\right\|^2}}, & \dot{\xi} < 0 \end{cases} \tag{7.2.4}$$

where $\boldsymbol{\sigma}^{\text{dev}}$ is the deviatoric stress tensor, $\varepsilon^{\max}$ represents the maximum transformation strain, and $\boldsymbol{\varepsilon}^{\text{tr}-R}$ is the transformation strain at the reversal point (based on $\xi = 1, 0$ for the martensite and austenite reversal points, respectively).

Forward ($\dot{\xi} > 0$), reverse ($\dot{\xi} < 0$) and no phase transformation ($\dot{\xi} = 0$) processes can be defined by the transformation function $\boldsymbol{\Phi}$ (Lagodas 2008; Hatefi Ardakani et al. 2015),

$$\boldsymbol{\Phi} = \begin{cases} \Pi - \mathrm{Y} & (\mathrm{A} \rightarrow \mathrm{M}) \\ -\Pi - \mathrm{Y} & (\mathrm{M} \rightarrow \mathrm{A}) \end{cases} \tag{7.2.5}$$

with (Lagodas 2008; Afshar et al. 2015a, 2015b)

$$\begin{aligned} \Pi(\boldsymbol{\sigma}, T, \xi) = {} & \boldsymbol{\sigma} : \boldsymbol{\Lambda} + \frac{1}{2}\boldsymbol{\sigma} : \Delta\mathbf{C} : \boldsymbol{\sigma} + \boldsymbol{\sigma} : \boldsymbol{\alpha}(T - T_0) - \rho\Delta c\left[\left(T - T_0\right) - T\ln\left(\frac{T}{T_0}\right)\right] \\ & + \rho\Delta s_0 T - \rho\Delta u_0 - \frac{\partial h(\xi)}{\partial \xi} \end{aligned} \tag{7.2.6}$$

$$\mathrm{Y} = \frac{1}{4}\rho\Delta s_0\left(T_f^{\mathrm{M}} + T_s^{\mathrm{M}} - T_f^{\mathrm{A}} - T_s^{\mathrm{A}}\right) \tag{7.2.7}$$

where T is the temperature, T_0 is the reference temperature and h is the hardening function (Boyd and Lagoudas 1996; Lagodas 2008):

$$h(\xi) = \begin{cases} -\frac{1}{2}\rho\Delta s_0\left(T_s^{\mathrm{M}} - T_f^{\mathrm{M}}\right)\xi^2 + \left(\mu_1 + \mu_2\right)\xi, & \dot{\xi} > 0 \\ -\frac{1}{2}\rho\Delta s_0\left(T_s^{\mathrm{A}} - T_f^{\mathrm{A}}\right)\xi^2 + \left(\mu_1 - \mu_2\right)\xi, & \dot{\xi} < 0 \end{cases} \tag{7.2.8}$$

with parameters μ_1 and μ_2:

$$\mu_1 = \frac{1}{2}\rho\Delta s_0\left(T_s^{\mathrm{M}} + T_f^{\mathrm{A}}\right) - \rho\Delta u_0 \tag{7.2.9}$$

$$\mu_2 = \frac{1}{2}\rho\Delta s_0\left(T_s^{\mathrm{A}} + T_s^{\mathrm{M}} - T_f^{\mathrm{A}} - T_f^{\mathrm{M}}\right) - \rho\Delta u_0 \tag{7.2.10}$$

7.2.3 Contact Mechanics

Contact mechanisms should be adopted to account for interaction of the inflating balloon and the stent, as well as the expanding SMA and the soft tissue of an artery. Due to very low stiffness of the balloon and relatively low stiffness of the soft tissue, large deformations are expected to occur on contact. For accurate results, consistent contact formulations have to be employed (see Mohammadi 2003).

7.2.4 Modelling of Stenting

In this section, a brief presentation is provided on a number of main steps of modelling of an expanding balloon–stent system or a self-expanding SMA stent. Intentionally, only

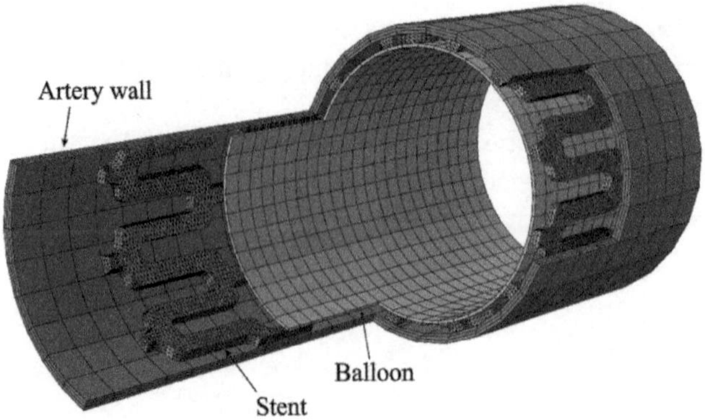

Figure 7.3 Typical finite element modelling of the balloon, the stent and the artery wall. *Source:* Reproduced from Torabizadeh and Mohammadi 2022a; 2022b/University of Tehran.

qualitative results are presented. Figure 7.3 shows the typical finite element modelling of all parts of the problem. No plague is considered in this sample presentation.

Material properties of different components of the model can be adopted from the available literature (for instance, Schiavone and Zhao 2015, etc.). The balloon is assumed to be a linear hyperelastic material, usually based on the Mooney–Rivlin constitutive law (Section 5.3.4).

Due to the availability of various stents, the specifications of stents are usually obtained from the manufacturers. Sample specifications can be found in the works of Conti et al. (2009) and Pelton et al. (2000).

Selection of the proper constitutive equation for the response of the artery wall or its forming layers is important to obtain accurate results. Section 5.3 has comprehensively discussed the available hyperelastic models for the soft tissues. Moreover, the multiscale modelling of a degraded artery has been examined in Section 5.9. The same constitutive models and material properties can be used for modelling the soft tissue of the artery in a stenting procedure.

The well-developed anisotropic Holzapfel model (Holzapfel et al. 2000) with two sets of inclined fibres (with λ, μ, c_{1f}, c_{2f} constants) (Section 5.3.6) can be adopted to simulate various layers of the artery (intima, media and adventia) (Schiavone and Zhao 2015):

$$w^s(J_1, J_f, J) = \frac{1}{2}\mu(J_1 - 3) + \frac{c_{2f}}{2c_{1f}}\left[e^{c_{1f}(J_f-1)^2} - 1\right] + \frac{1}{2}\lambda\left(\frac{J^2-1}{2} - \ln J\right) \tag{7.2.11}$$

In a simplified approach, average values for a homogenized layer for the whole artery can be adopted.

7.2.5 Basics of Modelling

In a simple case, typically presented in Figure 7.4, the stress-free stent and the balloon are assumed to be in contact. Then, the balloon is radially expanded, pushing the stent to further expand until it reaches the artery wall and begins to push into it.

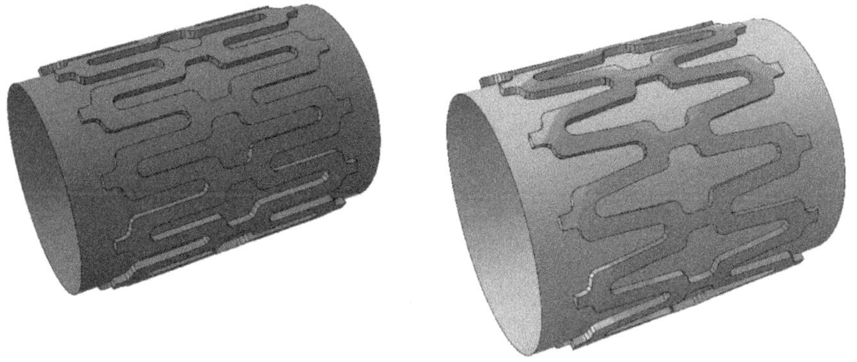

Figure 7.4 Deformed configurations of the balloon/stent system before and after balloon inflation.

According to Figure 7.5, the von Mises stress distribution is changed from a free stress state at the beginning of the analysis to a distribution with localized zones around the corners of the stent.

At this stage, the modelling should continue with the inclusion of the soft tissue as the expanded stent is expected to reach a contact position with the artery wall. The modelling may consider application of the blood pressure.

The self-expanding nitinol SMA stent can be similarly simulated. First, the stent is assumed to be deployed in the proper place with respect to the artery wall. The expansion process is then activated, leading to the contact and deformation of the stent and soft tissue, as typically presented in Figure 7.6. In a practical procedure, certain levels of resultant stress and deformation are required to allow for the removal of the obstacles inside the vein. In principle, it is possible to model the obstacle if sufficient mechanical properties of the obstacle material are known.

Patterns of stress concentration are expected to be similar to the previous case, as they are localized around the corners of the stent, which is in agreement with the available reference reports (David Chua et al. 2004; Schiavone and Zhao 2015).

Due to the stress concentrations, phase transformation zones may be formed, as typically presented in Figure 7.7.

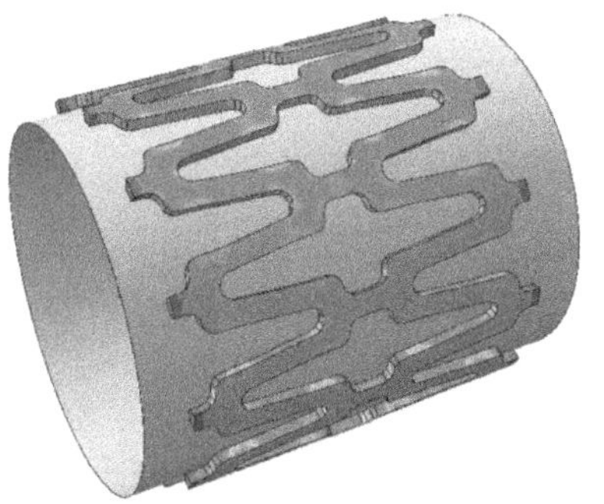

Figure 7.5 Typical distribution of the von Mises stress on the expanded stent.

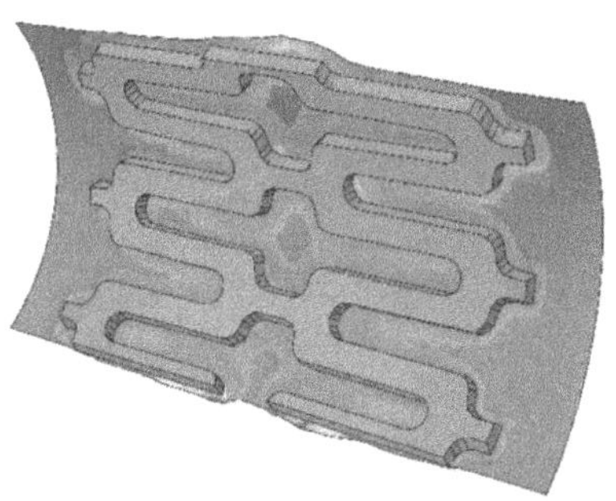

Figure 7.6 Deformed configurations of the expanded stent and the artery wall.

Figure 7.7 Zones of phase transformation in the SMA stent.

For a comprehensive discussion on the subject, see Schiavone and Zhao (2015).

7.3 Multiscale Modelling of the Eye

Multiscale analysis of the human eye allows for the study of a number of phenomena in the optical system. For instance, it is possible to determine a reliable estimate for the vitreous pressure on the internal tissues, and the optical nerve head (ONH) in particular, as it is known to be a significant factor in a number of eye-related diseases such as the primary open angle glaucoma (POAG) and its common form of normal tension glaucoma (NTG).

Figure 7.8 defines the major parts of the human eye. The first step of the simulation is to create a model with all possible details of the main parts, as typically presented in Figure 7.9.

The material properties for different parts of the model can be obtained from the available literature (for instance, Uchio et al. 1999).

A fine finite element mesh, presented in Figure 7.10, should then be employed to discretize the model with all the necessary details.

Simulation is based on multiscale modelling of the eye (organ level) to determine, ultimately, the pressure on the microstructure of the optical nerve head (ONH) (Figure 7.11).

The macroscale solution requires a conventional fluid–structure interaction (FSI) theory, which considers the fluid and solid components of the model and their interaction through the FSI mechanisms (see Section 2.4). Most of the commercial finite element

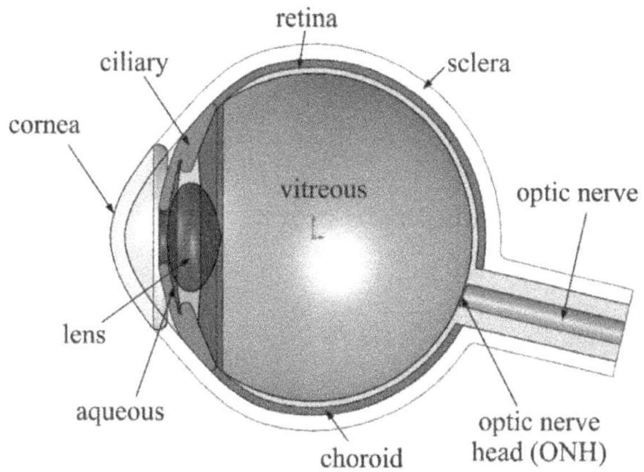

Figure 7.8 Illustrations of the geometry and parts of the model for a human eye. *Source:* Reproduced from Shahi, 2013/University of Tehran.

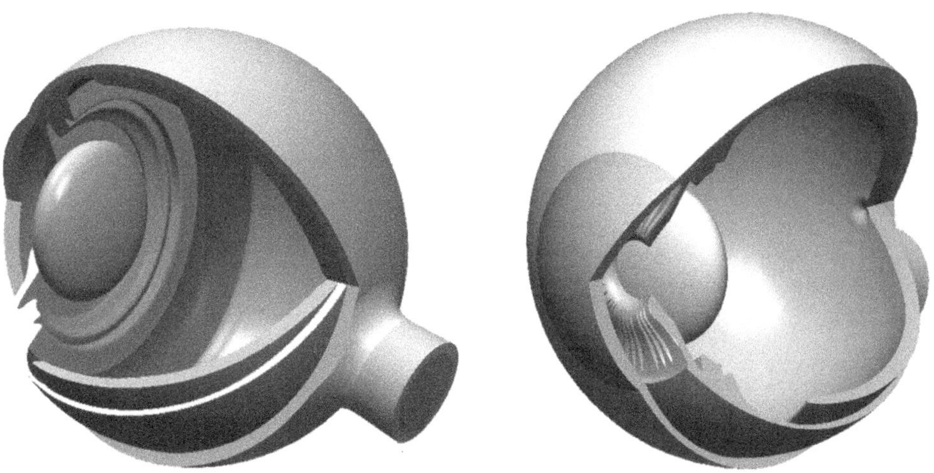

Figure 7.9 Illustrations of the geometry and the main parts of the model of a human eye. *Source:* Reproduced from Shahi, 2013/University of Tehran.

packages have reliable and efficient FSI solvers. The vitreous fluid is assumed to fully fill the inside cavity with the initial nominal intraocular pressure (IOP).

The common results of the macroscale FSI analysis can be qualitatively expressed in terms of the strain and stress distributions. The maximum horizontal strain occurs along the back edge of the outer (cornea-sclera) shell, just above the ONH, whereas the critical point for the vertical strain is at the top and bottom edges of the cornea-sclera shell.

On the other hand, the principle horizontal stress is localized around the optic nerve disk/head, but the maximum vertical stress is generated at the top and bottom edges of the

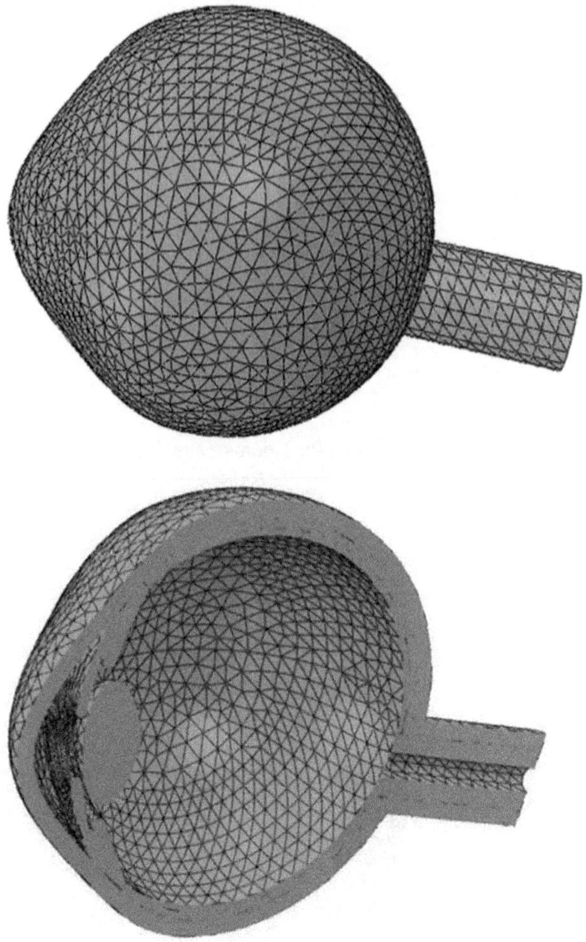

Figure 7.10 Finite element mesh of the eye model. *Source:* Reproduced from Shahi, 2013/University of Tehran).

sclera shell. Noticeably, the maximum overall effective von Mises stress occurs at the position of the lens and ONH. Due to a comparatively large stiffness of the lens, the critical pressure point of the eye is at the position of ONH, which increases the risk of damage to the optical nerve cells.

The next stage of the multiscale analysis of the present problem is to transform the deformation and stress states of the critical point at the ONH into a microscale model to determine a more accurate insight into the state of the optic nerve cells. For the microscale analysis (see Figure 7.11), a computational homogenization procedure (Section 4.2) can be followed to correlate the state of the microstructure with the pressure and other states of the macro model of the ONH to determine a fully consistent macro–micro solution.

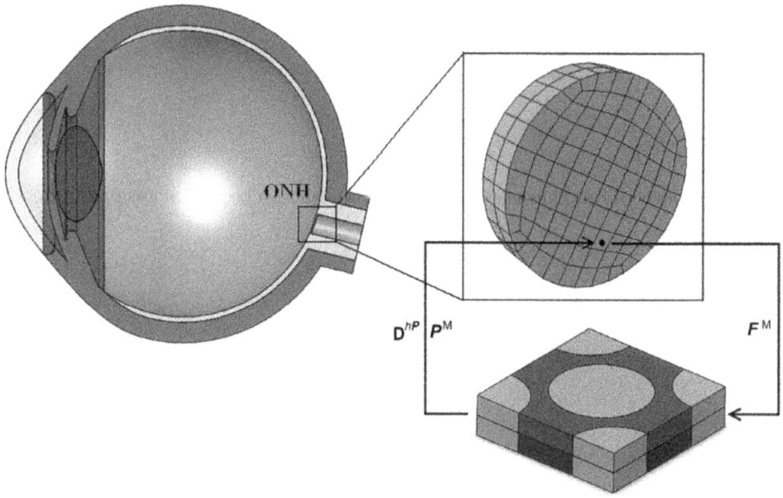

Figure 7.11 Multiscale analysis of the eye to determine pressure on the ONH cell level. *Source:* Reproduced from Shahi, 2013/University of Tehran.

7.4 Pulsatile Blood Flow in the Aorta

7.4.1 Description of the Problem

This brief section reviews a fluid–solid interaction biomechanical problem. Only a macro-scale model is described here, but the model can be extended to multiscale solutions to investigate the effects of pulsatile blood flow on microstructural layers of the aorta, similar to Section 5.9, which discussed the arterial aneurysm.

The problem is to simulate the response of the aorta subject to the effect of a single blood flow pulse (Figure 7.12). While the advanced ALE technique (Section 2.4) can be adopted, the present fluid–solid interaction biomechanical problem can be analysed by the rather simpler semi-coupled (one-way) interaction method. According to this simplified approach, first a CFD analysis is performed and the corresponding velocity and pressure fields are computed. A mechanical analysis is then conducted to determine the deformation and stress fields of the solid (soft tissue of the artery).

Specifications for the fluid (blood) and the soft solid (aorta) are well documented and can be found in the literature (for instance, refer to Vasava et al. 2008). The blood is assumed to be a viscous fluid, and the aorta may follow a simple neo-Hookean hyperelastic model, for simplified studies, or is assumed to be according to the complete form of the anisotropic fibrous tissues of Section 5.3.6. If damage in the soft tissue is important, the model of Section 5.3.8 can be adopted.

Figure 7.13 illustrates the sample contour of velocity of blood inside the deformable soft tissue of the aorta. Clearly, proper flux boundary conditions are required for the inlet cross-sections of the vein branches to allow for accurate CFD analysis.

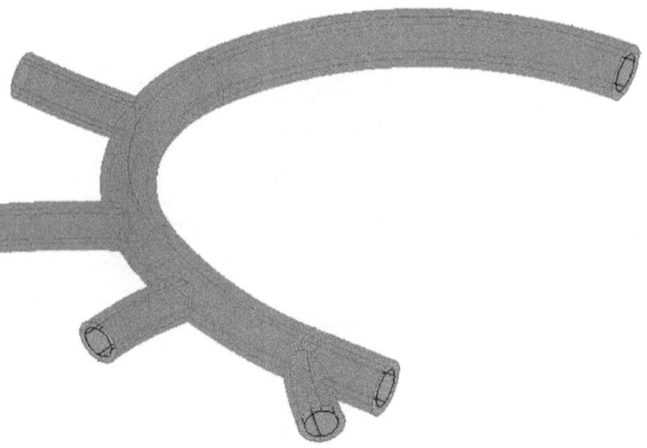

Figure 7.12 Geometric model of the aorta. *Source:* Reproduced from Shahi, 2013/University of Tehran.

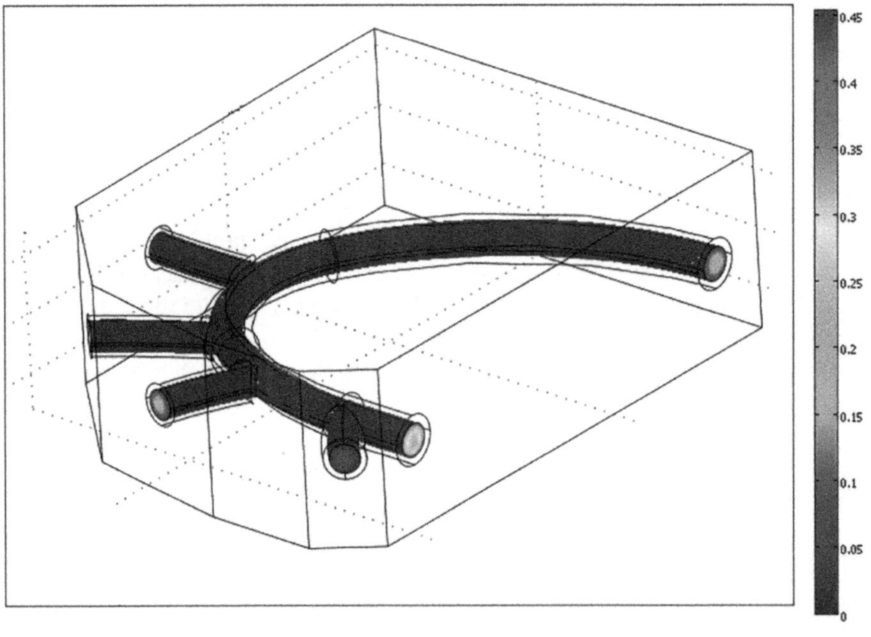

Figure 7.13 Contour of the velocity field. *Source:* Reproduced from Shahi, 2013/University of Tehran.

Distribution of the normal stress/pressure on the aorta wall is depicted in Figure 7.14, which clearly shows the stress/pressure concentration regions mainly around the vein branching corners.

Such an FSI analysis may be upgraded to include floating particles in the blood and their potential sedimentation to form plaques inside the vein. Perhaps the meshless smoothed particle hydrodynamics (SPH) methodology (Section 3.7) is better suited to solve such a complicated problem.

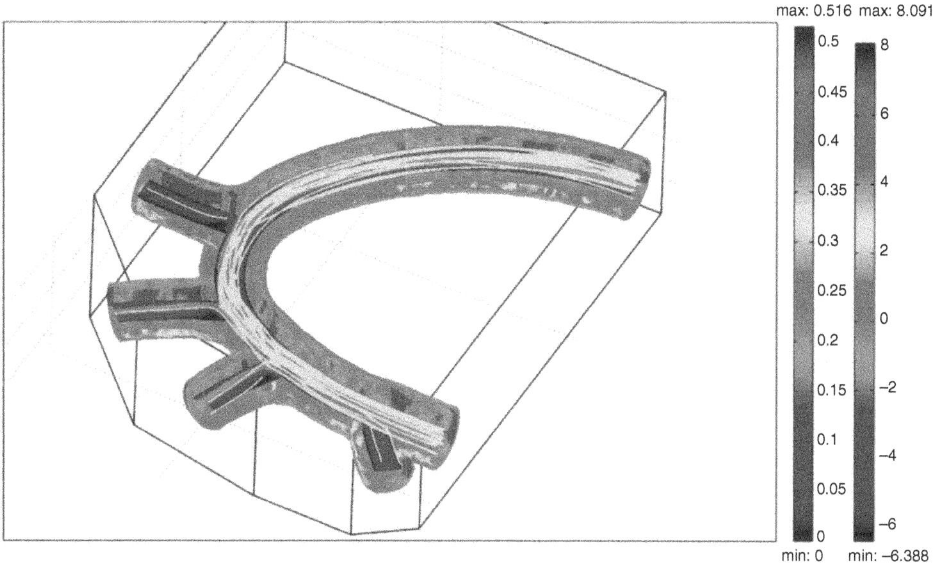

Figure 7.14 Distribution of normal stress in the aorta. *Source:* Reproduced from Shahi, 2013/University of Tehran.

Moreover, the pulsatile blood flow in an artery with an installed stent may provide better understanding of the mechanism of plaque formations around a solid object (stent). In this case, the ALE technique for fluid–solid interactions may be required to ensure the accuracy of the results for a flow that contains moving and deformable solid objects.

7.5 Shape Memory Polymer Drug Delivery System

The idea to improve the drug efficacy through the control of drug release and drug delivery mechanisms has received considerable interest in recent years (Zhao et al. 2019; Baniasadi et al. 2021). Significant developments on drug stability, drug absorption and drug targeting accuracy are expected by the design of more advanced drug delivery devices.

The idea of a drug delivery system by the help of shape memory polymers (SMPs) is briefly reviewed here. Specifically designed nanosize SMPs with pre-injected or positioned drug substances may be designed to deform to a folded position at the end of the manufacturing process to hold the drug inside. The polymer may then be stimulated inside the body to deform to an unfolded configuration to allow the drug substances to separate and reach the target cells/tissues (Figure 7.15).

The biocompatible properties of SMPs have made them one of the high-potential candidates for drug delivery purposes. For example, Li et al. (2012) found that employing high-intensity focused ultrasound with different high frequencies can activate SMPs to accomplish various predefined configurations. The possibility of programming multiple temporary shapes for SMPs facilitates the staged capture of arbitrary configurations and release of a drug in specific locations of human organs based on the exposure time,

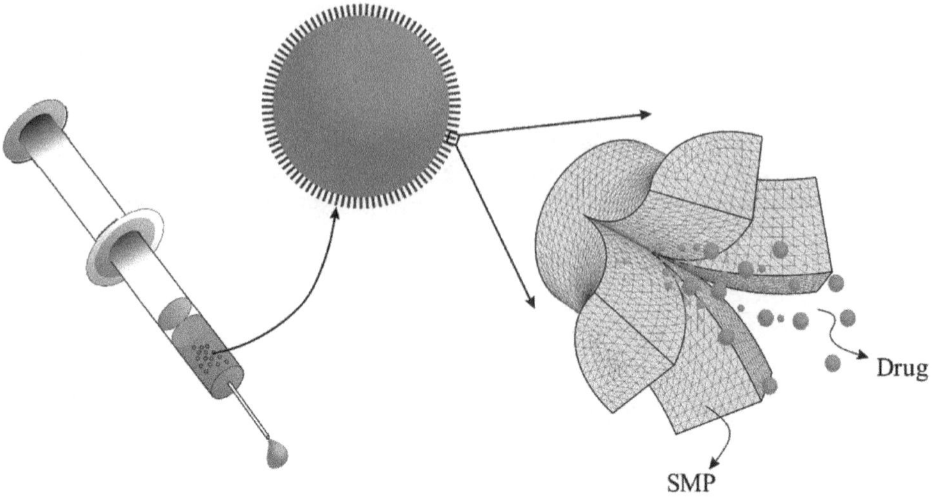

Figure 7.15 The concept of an SMP drug delivery system. *Source:* Reproduced from Foyouzat and Mohammadi, 2022/University of Tehran.

frequency, etc. Similarly, Wischke et al. (2009) used the shape memory property of SMPs to study their capability in controlling the drug release.

The idealized process starts with embedding the drug in an SMP capsule at its first step of the thermomechanical cycle (Figure 7.16). The SMP capsule then undergoes a pure mechanical loading to reach a fully folded configuration to function as a safe container for the drug. Afterwards, while the deformation is kept fixed, the SMP is cooled down to a lower temperature of $T < T_g$ to stabilize its temporary shape after the unloading process.

The final step of the thermomechanical cycle is accomplished when this drug is transferred into an organ. At this stage, the SMP is activated by an increasing temperature or other stimulations, leading to an unfolded configuration and a controlled drug release at specified locations (Figure 7.17).

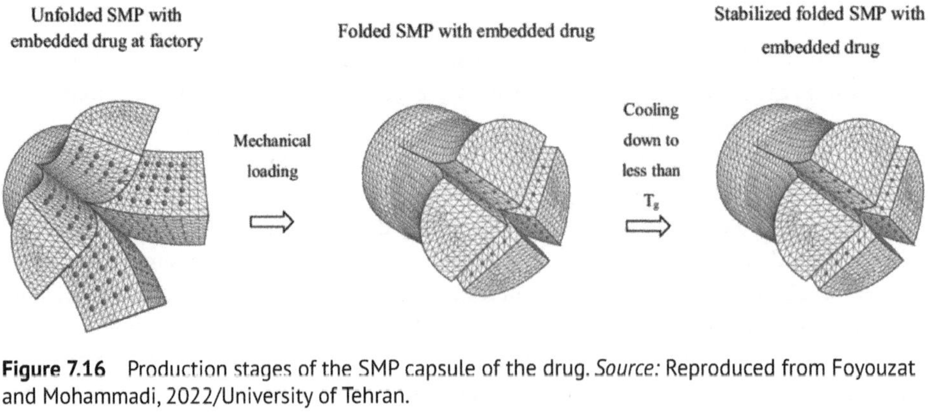

Figure 7.16 Production stages of the SMP capsule of the drug. *Source:* Reproduced from Foyouzat and Mohammadi, 2022/University of Tehran.

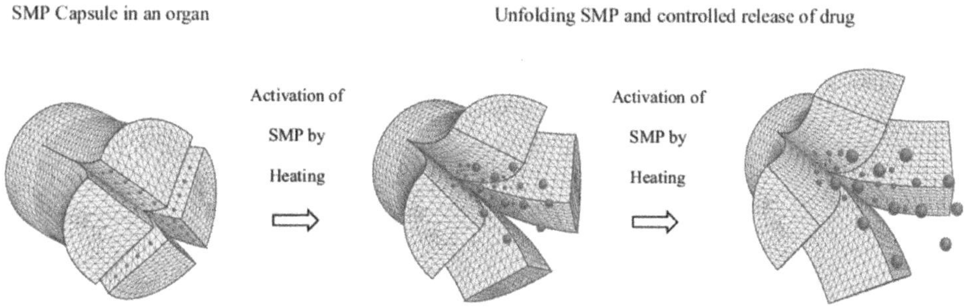

Figure 7.17 Drug delivery concept of an SMP capsule. *Source:* Reproduced from Foyouzat and Mohammadi, 2022.

From a macroscopic point of view, two classes of constitutive models can characterize the SMP behaviour, namely viscoelastic (Tobushi et al. 1997; Balogun and Mo 2014; Gu et al. 2019; Baniasadi et al. 2020) and phase transition (Liu et al. 2006b; Chen and Lagoudas, 2008a, 2008b; Baghani et al. 2013; Boatti et al. 2016; Foyouzat et al. 2021) models. The thermo-viscoelastic constitutive model developed by Balogun and Mo (2014) is based on the original model of Zener–Maxwell, as typically shown in the Figure 7.18.

The governing formulation of SMP behaviour can be briefly described by (Balogun and Mo 2016)

$$\varepsilon = \varepsilon_{\mathrm{t}} + \varepsilon_{\mathrm{T}} + \varepsilon_{\mathrm{is}} \tag{7.5.1}$$

$$\boldsymbol{\sigma}_{\mathrm{t}} = \boldsymbol{\sigma}_{\mathrm{m}} + \boldsymbol{\sigma}_{\mathrm{e}} \tag{7.5.2}$$

where ε is the total strain, ε_{T} refers to the thermal strain and $\varepsilon_{\mathrm{is}}$ is the irrecoverable strain. Moreover, $\boldsymbol{\sigma}_{\mathrm{m}}$ and $\boldsymbol{\sigma}_{\mathrm{e}}$ are the stress tensors in the Maxwell and elastic arms, respectively. The compatibility equation can then be written as

$$\varepsilon_{\mathrm{t}} = \varepsilon_{\mathrm{e}} = \varepsilon_{\mathrm{m}} + \varepsilon_{\mu} \tag{7.5.3}$$

where ε_{e}, ε_{m} and ε_{μ} are the elastic, the Maxwell and the dashpot strains, respectively. The corresponding time derivatives for the material behaviour are expressed as

$$\dot{\varepsilon}_{\mathrm{t}} = \dot{\varepsilon}_{\mathrm{m}} + \dot{\varepsilon}_{\mu} \tag{7.5.4}$$

$$\dot{\boldsymbol{\sigma}}_{\mathrm{t}} = \dot{\boldsymbol{\sigma}}_{\mathrm{m}} + \dot{\boldsymbol{\sigma}}_{\mathrm{e}} \tag{7.5.5}$$

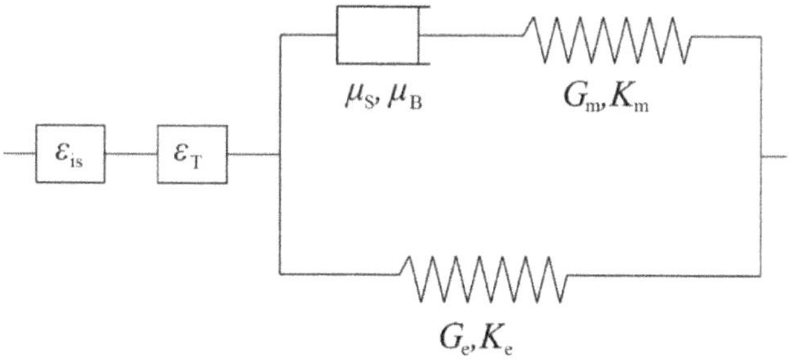

Figure 7.18 The thermo-viscoelastic constitutive model. *Source:* Adapted from Balogun and Mo, 2014.

Considering the stress–strain relationship for each arm as

$$\boldsymbol{\sigma}_{\rm e} = \mathbf{C}_{\rm e}\boldsymbol{\varepsilon}_{\rm e} \tag{7.5.6}$$

$$\boldsymbol{\sigma}_{\rm m} = \mathbf{C}_{\rm m}\boldsymbol{\varepsilon}_{\rm m} \tag{7.5.7}$$

$$\boldsymbol{\sigma}_{\mu} = \mathbf{C}_{\mu}\boldsymbol{\varepsilon}_{\mu} \tag{7.5.8}$$

the final governing equation for the material can be written as (Balogun and Mo 2014)

$$\begin{aligned}&\left(1+\frac{G_{\rm e}}{G_{\rm m}}\right)(\dot{\varepsilon}-\dot{\varepsilon}_{\rm is})+\left(\frac{K_{\rm e}}{K_{\rm m}}-\frac{G_{\rm e}}{G_{\rm m}}\right)\frac{1}{3}{\rm tr}(\dot{\varepsilon}-\dot{\varepsilon}_{\rm is})\mathbf{1}+\frac{G_{\rm e}}{\mu_{\rm s}}\varepsilon-\frac{G_{\rm e}}{3\mu_{\rm s}}{\rm tr}(\varepsilon)\mathbf{1}-\alpha\dot{T}\left(1+\frac{K_{\rm e}}{K_{\rm m}}\right)\mathbf{1}\\&=\left(\frac{1}{3K_{\rm m}}-\frac{1}{2G_{\rm m}}\right)\frac{1}{3}{\rm tr}(\dot{\sigma})\mathbf{1}+\frac{1}{2G_{\rm m}}\dot{\sigma}-\frac{1}{6\mu_{\rm s}}{\rm tr}(\sigma)\mathbf{1}+\frac{1}{2\mu_{\rm s}}\sigma\end{aligned} \tag{7.5.9}$$

The required material properties for the glass and rubber phases of the SMP include the elastic modules E_g, E_r, the viscoelastic parameters μ_g, μ_r and λ_g, λ_r and the minimum, glass and maximum temperatures, T_{min}, T_g and T_{max}, respectively.

Figure 7.19 shows the typical variation of the capsule gate closing ratio (defined as the gate closing Δ to the radius of the capsule) with the normalized temperature in various stages of the production of the SMP capsule of the drug and the delivery process. Clearly, it adapts the openings of the capsule to keep or deliver the drug in the production line and inside the body, respectively.

Finally, samples of the von Mises stress contours of the SMP product in different stages of the production and delivery of the drug are presented in Figure 7.20.

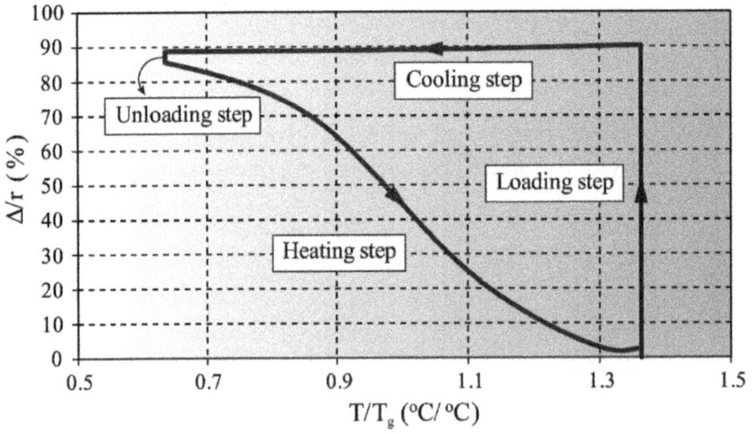

Figure 7.19 Variation of the capsule gate closing ratio with the temperature in various stages of the production of the SMP capsule. *Source:* Reproduced from Foyouzat and Mohammadi, 2022/ University of Tehran.

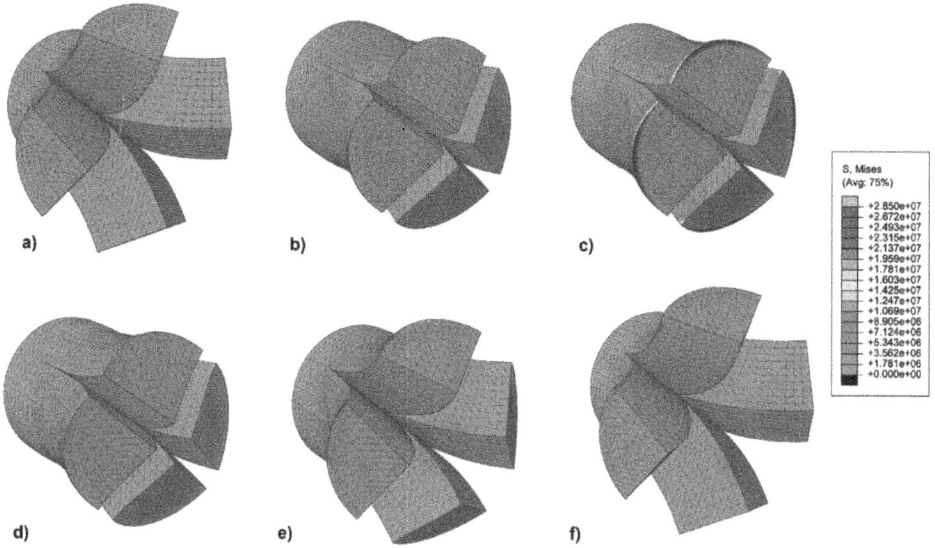

Figure 7.20 Variation of the von Mises stress contours of the SMP at different stages of the production and delivery of the drug: (a) initial configuration, (b) pure mechanical loading at $T = T_{max}$, (c) cooling the material to $T = T_{min}$, (d) pure mechanical unloading at $T = T_{min}$, (e) heating the material to $T = T_g$ and (f) heating the material to $T = T_{max}$ (permanent shape). *Source:* Reproduced from Foyouzat and Mohammadi, 2022/University of Tehran.

7.6 Artificial Intelligence in Biomechanics

This section is dedicated to a brief introduction to the application of artificial intelligence and deep learning in biomechanical problems. A short description of artificial intelligence and machine learning is presented and the concept of a physics-informed neural network is addressed. The same concepts can be employed in primitive multiscale modelling of nanoscale problems. The methods are being investigated for multiscale biomechanical problems, which are expected to be publicly reported in the near future.

Artificial intelligence is closely related to data science (see Figure 7.21). An extremely huge amount of data is produced, collected and stored every second worldwide. The data may be collected from social networks, banking transactions, maps and GPSs, surveys, online gamings, phone calls, surveillance cameras, experimental tests, computer simulations, etc. Such a huge amount of valuable data can well be used for training of machine learning methods through data mining techniques (Schmidt 2022).

7.6.1 Artificial Intelligence and Machine Learning

Artificial intelligence (AI) has evolved from scientific communities and expanded to practically every aspect of human life. In public view, AI is considered as a computer program capable of doing everything automatically. It is occasionally regarded as the ultimate power in the hand of humans, and at the same time it is feared by some others to be the beginning of the takeover of earth by intelligent robots.

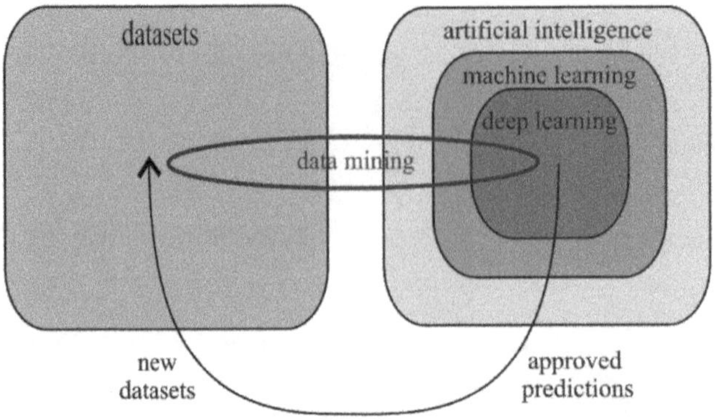

Figure 7.21 A simple illustration of data science and artificial intelligence.

Since the beginning of the introduction and development of AI in more than half a century ago in pure scientific communities, it has become part of our daily life in the recent decade. It is accompanied with the big data created every second worldwide and the availability of high-performance computing capabilities over the clouds. AI combines big sets of data and iteratively processes them to learn from the existing patterns/features and to predict the best solutions.

In contrast to public interpretation of AI, machine learning (ML) can be regarded as the expert part of AI, which is related to the development of efficient AI programming techniques. Sometimes ML is regarded as the science for AI computers to perform and solve every problem without being programmed for any specific problem.

ML techniques can be categorized into three types of supervised learning, unsupervised learning and reinforcement learning (Genc 2019).

In supervised learning, each data is accompanied with a known output label, according to some advance knowledge of the so-called supervisor. The available data can be used to select and set the necessary tagging. The ML procedure will then learn to determine the pattern and decide on the tagging of new data. The approach includes two main parts of the regression (linear or logistic) for prediction of continuous fields and the decision tree classification for determining the appropriate group/class for a certain data.

In unsupervised learning, no pre-set output tagging is available for the dataset and a hidden pattern is sought. Clearly, unsupervised learning is a more difficult approach than supervised learning. Unsupervised learning is developed in two types of clustering (based on grouping of similar data) and association (based on some rules or observations associated with the data).

The third approach, called reinforcement learning (RL), is based on a trial-and-error training by the reward/penalty procedure. The approach requires the response of an observing agent and continues to reach higher rewards. RL can also be regarded as a type of deep learning.

For a complete review of machine learning subjects, refer to Schmidt (2022) and other text books on AI and ML.

7.6.2 Deep Learning

Deep learning (DL) is a class of ML techniques inspired by the functionality of brain cells and is based on the multilayered artificial neural networks (ANN) to accomplish two functionalities of receiving input and applying an activation function based on a user set threshold. It has been successfully implemented in applications for daily routine practices and is capable of extracting the necessary features from a dataset and learn from them without any known pre-set specific rules for the dataset (Genc 2019).

In ANN, each input node is associated with a numerical value and each neural connection is assigned a real number of weight (w). The weighted sum includes the input nodes and weights of the connections and acts as a threshold measure for the transfer/activation function. The output node is associated with the function of the weighted sum of the input nodes. ANN iteratively runs to optimize the values of weight w, which is a very expensive computational task.

Deep learning techniques can be classified into three main models of convolutional neural networks, recurrent neural networks and generative adversarial networks (Genc 2019).

The convolutional neural networks (CNN) technique is employed to classify images through a series of filtered convolution layers. It begins by random weight numbers in the forward pass to predict the output labels of an existing data set with known labels. Comparison allows for evaluation of the loss function and a backward pass with the updated weight numbers to reduce the loss. The procedure continues until an acceptable low level of loss is achieved. The trained approach is then adopted (tested) for different image data sets.

The second approach, the recursive (recurrent) neural networks (RNN), generates a memory by saving previous input sources within the calculations. This characteristic is an advantage when handling sequential processes.

The advanced generative adversarial networks (GAN) approach allows for the neural networks to create data sets. In the starting phase, a random data set is generated, which leads to a random (fake) output. Then the discriminator module gradually improves the output.

7.6.3 Physics-Informed Neural Networks (PINNs)

The concept of physics-informed neural networks (PINNs) has recently been used for solution of different partial differential equations (PDEs) that govern engineering problems and physical phenomena. PINN is expected to overcome the difficulties of multi-objective optimization encountered in non-convex problems by the use of sequential training and adaptive weighting (Amini et al. 2022).

Beginning with a governing PDE of a boundary value problem based on the unknown field variables $\boldsymbol{u}(\mathbf{x},\boldsymbol{t})$ (deformation, temperature, etc.):

$$\mathcal{A}u(\mathbf{x},t) = f(\mathbf{x},t) \text{ in } \Omega \tag{7.6.1}$$

where $\mathcal{A}$ is a partial differential operator. The PDE Equation (7.6.1) is accompanied by the following boundary and initial conditions:

$$u(\mathbf{x}_\Gamma,t) = \bar{u}(\mathbf{x}_\Gamma,t) \text{ on } \Gamma \tag{7.6.2}$$

$$u(\mathbf{x},t_0) = u_0(\mathbf{x},t_0) \text{ in } \Omega \tag{7.6.3}$$

PINN uses an l-layer fully connected neural network $\aleph^l$ to express the field variable $u(\mathbf{x},t)$ with the approximate u^h (Raissi et al. 2019):

$$u^h(\mathbf{x},t) = \aleph^l(\mathbf{x},t) \circ \aleph^{l-1}(\mathbf{x},t) \circ \cdots \circ \aleph^l(\mathbf{x},t) \tag{7.6.4}$$

where $\aleph^l$ is a nonlinear transformation of inputs $\mathbf{x}^{h,l-1}$ to outputs $\mathbf{y}^{h,l}$ of layer l,

$$\mathbf{y}^{h,l} = \aleph^l\left(\mathbf{x}^{h,l-1}\right) = \sigma^l\left(W^l \cdot \mathbf{x}^{h,l-1} + b^l\right) \tag{7.6.5}$$

where σ^l is the activation function and W^l and b^l represent the set of weight and bias network parameters, respectively. Linear, sigmoid and hyperbolic functions are the common choices for the activation function.

Equation (7.6.5) can be re-written in the following form, which is more appropriate for the optimization description:

$$u^h(\mathbf{x},t) = \mathfrak{N}^u(\mathbf{x},t;\chi) \tag{7.6.6}$$

where χ is the set of all parameters of the network. Equation (7.6.6) allows built-in analytical differentiation in many deep learning frameworks (Abadi et al. 2016; Baydin et al. 2018).

In physics-based learning techniques, the goal of DL is to find the set of output solutions that simultaneously matches a given dataset and satisfies the PDEs. This goal can be achieved by the multiobjective optimization techniques. Accordingly, the loss function $\mathcal{L}(\mathbf{x},t;\chi)$ can be expressed as

$$\begin{aligned}\mathcal{L}(\mathbf{x},t;\chi) = \lambda_1 \left[\!\left[Au^h(\mathbf{x},t) - f(\mathbf{x},t) \right]\!\right]_\Omega + \lambda_2 \left[\!\left[u^h(\mathbf{x}_\Gamma,t) - \bar{u}(\mathbf{x}_\Gamma,t) \right]\!\right]_\Gamma \\ + \lambda_3 \left[\!\left[u^h(\mathbf{x},t_0) - u_0(\mathbf{x},t_0) \right]\!\right]_\Omega\end{aligned} \tag{7.6.7}$$

where λ_i are the penalty terms to impose the constraints and $[\![\,]\!]$ is a proper norm, usually the mean-squared error (MSE). This goal can be achieved by the multiobjective optimization techniques. The final solution $u(\mathbf{x},t)$ is obtained by minimization of the loss function $\mathcal{L}(\mathbf{x},t,\chi)$ on the set of $(\mathbf{X},U)$ points, leading to the optimized ANN parameters χ^{opt} (Nocedal and Wright 2006; Haghighat et al. 2021; Wang et al. 2022),

$$\chi^{\text{opt}} : \min \mathcal{L}(\mathbf{X},U;\chi) \tag{7.6.8}$$

The solution procedure of PINN can be summarized as

1) Sample $(\mathbf{X},U)$.
2) Initialize randomly the set of network parameters $\chi^n, n=1$.
3) Loop while $\varepsilon^n > \bar{\varepsilon}$.
   3.1 Optimize $\mathcal{L}(\mathbf{x},t;\chi)$ for χ over $(\mathbf{X},U)$ to obtain χ^n.
   3.2 Evaluate $u^{h,n}$ from χ^n.
   3.3 Check for convergence: $\varepsilon^n = \dfrac{\left\|\chi^n - \chi^{n-1}\right\|}{\left\|\chi^n\right\|}$.
   3.4 $n = n+1$.
4) End Loop.

7.6.4 Biomechanical Applications of Artificial Intelligence

The human neuromusculoskeletal system is susceptible to damages, diseases and disorders that may disrupt its functioning, with a significant effect on the human life. Low level or internal damages may be difficult to diagnose. Artificial intelligence and deep learning techniques may conceptually provide the set of predictions and decision-making strategies based on the conditions of each specific patient.

There are significant potentials for AI in biomechanical applications, such as development of intelligent diagnostic tools for assessing the conditions of different biosystems, providing better healthcare and clinical managements, automatic extraction of data that are relevant to identification of a specific disease or required for a specific medical treatment, anticipation of the effects of new drags, prediction of the widespread epidemic of new viruses, optimization of the shape, size and compositions of new synthesized tissues and implants, analysis of potential psychological responses, production of optimized data acquisition systems and measurement devices, design of personalized sport trainings, and many more to come.

Conceptually, the corresponding governing Equations (7.6.1) to (7.6.3) should be defined for each biomechanical problem. The general procedure of Section 7.6.3 can then basically be followed to develop a deep learning solution for that particular biomechanical problem. In practice, however, development and implementation of the solution procedure is not easy due to the fact that the governing formulation of the biomechanical problem is usually very complex and may even involve multiple scales. Therefore, there is still a long way to go to reach practical physically informed machine learning solutions for general biomechanical applications – certainly for multiscale biomechanical problems.

In practice, other data-only driven machine learning techniques are frequently employed. As a sample application, Shahi et al. (2022) have recently developed a machine learning approach for prediction of the complex signals during cardiac arrhythmias. They tested both the experimental and synthetic datasets and reported improved predictions.

References

Abadi, M., Barham, P., Chen, J., et al. (2016). TensorFlow: a system for large-scale machine learning. In: *12th USENIX Symposium on Operating Systems Design and Implementation (OSDI 16)*, 265–283.

Abolfathi, N., Nik, A., Sotoudeh Chafi, M., et al. (2009). A micromechanical procedure for modelling the anisotropic mechanical properties of brain white matter. *Computer Methods in Biomechanics and Biomedical Engineering* 12 (3): 249–262.

Abraham, F.F., Brodbeck, D., Rudge, W.E., et al. (1998). *Ab initio* dynamics of rapid fracture. *Modelling and Simulation in Materials Science and Engineering* 6: 639–670.

ADINA (2010) ADINA, theory and modeling guide - ADINA CFD & FSI.

Afshar, A., Daneshyar, A., and Mohammadi, S. (2015a). XFEM analysis of fiber bridging in mixed-mode crack propagation in composites. *Composite Structures* 125: 314–327.

Afshar, A., Hatefi Ardakani, S., Hashemi, S., and Mohammadi, S. (2015b). Numerical analysis of crack tip fields in interface fracture of SMA/elastic bi-materials. *International Journal of Fracture* 195: 39–52.

Afshar, A., Hatefi, S., and Mohammadi, S. (2016). Transient analysis of stationary interface cracks in orthotropic bi-materials using oscillatory crack tip enrichments and the interaction integral method. *Composite Structures* 142: 200–214.

Afshar, A., Hatefi, S., and Mohammadi, S. (2018). Stable discontinuous space–time analysis of dynamic interface crack growth in orthotropic bi-materials using oscillatory crack tip enrichment functions. *International Journal of Mechanical Sciences* 140: 557–580.

Ahmadian, H., Hatefi, S., and Mohammadi, S. (2015). Strain-rate sensitivity of unstable localized phase transformation phenomenon in shape memory alloys using a non-local model. *International Journal of Solids and Structures* 63: 167–183.

Alastrué, V., Rodríguez, J., Calvo, B., and Doblaré, M. (2007). Structural damage models for fibrous biological soft tissues. *International Journal of Solids and Structures* 44: 5894–5911.

Ali, A.A., Cristofolini, L., Schileo, E., et al. (2014). Specimen-specific modeling of hip fracture pattern and repair. *Journal of Biomechanics* 47: 536–543.

Alizadeh, O. (2019). *Enriched Multiscale Methods*. Ph.D. Thesis, School of Civil Engineering, University of Tehran, Iran.

Alizadeh, O. and Mohammadi, S. (2019). The variable node multiscale approach: coupling the atomistic and continuum scales. *Computational Materials Science* 160: 256–274.

Multiscale Biomechanics: Theory and Applications, First Edition. Soheil Mohammadi.

Alizadeh, O. and Mohammadi, S. (2022) The enriched multiscale method; a novel concurrent multiscale method based on atomistic enrichment term. Submitted for publication.

Allen, M.P. (2004). Introduction to molecular dynamics simulation. *Computational Soft Matter: From Synthetic Polymers to Proteins* 23: 1–28.

Amini, D., Haghighat, E., and Juanes, R. (2022). Physics-informed neural network solution of thermo-hydro-mechanical (THM) processes in porous media, *ArXiv*, 2203.01514.

Anderson, T. (1995). *Fracture Mechanics: Fundamentals and Applications*. USA: CRC Press.

Andreasen, G.F. and Hilleman, T.B. (1971). An evaluation of 55 cobalt substituted Nitinol wire for use in orthodontics. *Journal of American Dental Association* 82 (6): 1373–1375.

Arbogast, K.B., Prange, M.T., Meaney, D.F., and Margulies, S.S. (1997a). Properties of cerebral gray and white matter undergoing large deformation. In: *Symposium Proceedings of the Center for Disease Control, Wayne State University*, 33–39.

Arbogast, K.B., Thibault, K.L., Pinheiro, B.S., et al. (1997b). A high-frequency shear device for testing soft biological tissues. *Journal of Biomechanics* 30 (7): 757–759.

Arruda, E.M. and Boyce, M.C. (1993). A three-dimensional constitutive model for the large stretch behavior of rubber elastic materials. *Journal of the Mechanics and Physics of Solids* 41: 389–412.

Asadpoure, A. and Mohammadi, S. (2007). A new approach to simulate the crack with the extended finite element method in orthotropic media. *International Journal for Numerical Methods in Engineering* 69: 2150–2172.

Asadpoure, A., Mohammadi, S., and Vafai, A. (2006). Modeling crack in orthotropic media using a coupled finite element and partition of unity methods. *Finite Elements in Analysis and Design* 42 (13): 1165–1175.

Asadpoure, A., Mohammadi, S., and Vafai, A. (2007). Crack analysis in orthotropic media using the extended finite element method. *Thin Walled Structures* 44 (9): 1031–1038.

Atluri, S.N. (1982). Path-independent integrals in finite elasticity and inelasticity, with body forces, inertia and arbitrary crack-face conditions. *Engineering Fracture Mechanics* 16 (3): 341–364.

Atluri, S.N. and Shen, S. (2002). *The Meshless Local Petrov–Galerkin (MLPG) Method*. USA: Tech Science Press.

Atluri, S.N. and Zhu, T. (1998). A new meshless local Petrov-Galerkin (MLPG) approach in computational mechanics. *Computational Mechanics* 22: 117–127.

Babuska, I. and Miller, A. (1984). The post-processing approach to finite element method – part II. *International Journal for Numerical Methods in Engineering* 20: 1111–1129.

Badilatti, S.D., Christen, P., Levchuk, A., et al. (2015). Large-scale microstructural simulation of load-adaptive bone remodeling in whole human vertebrae. *Biomechanics and Modeling in Mechanobiology* 15 (1): 83–95.

Baghani, M., Naghdabadi, R., and Arghavani, J. (2013). A large deformation framework for shape memory polymers: constitutive modeling and finite element implementation. *Journal of Intelligent Material Systems and Structures* 24 (1): 21–32.

Bailon-Plaza, A. and Van der Meulen, M.C.H. (2001). A mathematical framework to study the effects of growth factor influences on fracture healing. *Journal of Theoretical Biology* 212: 191–209.

Bain, A.C. and Meaney, D.F. (2000). Tissue-level thresholds for axonal damage in an experimental model of central nervous system white matter injury. *Journal of Biomechanical Engineering* 122 (6): 615–622.

Bain, A.C., Shreiber, D.I., and Meaney, D.F. (2003). Modeling of microstructural kinematics during simple elongation of central nervous system tissue. *Journal of Biomechanical Engineering* 225 (6): 798–804.

Balogun, O. and Mo, C. (2014). Shape memory polymers: three-dimensional isotropic modeling. *Smart Materials and Structures* 23 (4): 045008.

Balogun, O.A. and Mo, C. (2016). Three-dimensional 3hermos-mechanical viscoelastic model for shape memory polymers with binding factor. *Journal of Intelligent Material Systems and Structures* 27 (14): 1908–1916.

Balzani, D., Brinkhues, S., and Holzapfel, G.A. (2012). Constitutive framework for the modeling of damage in collagenous soft tissues with application to arterial walls. *Computer Methods in Applied Mechanics and Engineering* 213: 139–151.

Balzani, D., Neff, P., Schröder, J., and Holzapfel, G.A. (2006). A polyconvex framework for soft biological tissues. Adjustment to experimental data. *International Journal of Solids and Structures* 43: 6052–6070.

Baniasadi, M., Foyouzat, A., and Baghani, M. (2020). Multiple shape memory effect for smart helical springs with variable stiffness over time and temperature. *International Journal of Mechanical Sciences* 182: 105742.

Baniasadi, M., Yarali, E., Foyouzat, A., and Baghani, M. (2021). Crack self-healing of 4hermos-responsive shape memory polymers with application to control valves, filtration, and drug delivery capsule. *European Journal of Mechanics-A/Solids* 85: 104093.

Barkaoui, A., Chamekh, A., Merzouki, T., et al. (2014). Multiscale approach including microfibril scale to assess elastic constants of cortical bone based on neural network computation and homogenization method. *International Journal for Numerical Methods in Biomedical Engineering* 30: 318–338.

Baseri, A., Kiani Borojeni, K., and Abdi, H. (2016). *Modeling of Biological Tissues by Mimics* (in Persian). Parsia Press, Tehran, Iran.

Bathe, K.-J. (1982). *Finite Element Procedures in Engineering Analysis*. Prentice-Hall, Inc., New Jersey.

Bathe, K.-J. (2006). *Finite Element Procedures*. Published by the author, taken from the original 1982 book published by Prentice-Hall, Pearson Education, Inc., USA.

Baydin, A.G., Pearlmutter, B.A., Radul, A.A., and Siskind, J.M. (2018). Automatic differentiation in machine learning: a survey. *Journal of Machine Learning Research* 18: 1–23.

Bayesteh, H. (2018) *Multiscale Failure Analysis of Heterogeneous Media*. Ph.D. Thesis, School of Civil Engineering, University of Tehran.

Bayesteh, H. and Mohammadi, S. (2011). XFEM fracture analysis of shells: the effect of crack tip enrichments. *Computational Materials Science* 50: 2793–2813.

Bayesteh, H. and Mohammadi, S. (2013). Fracture analysis of orthotropic functionally graded materials by XFEM. *Journal of Composites, Part B*. 44: 8–25.

Bayesteh, H. and Mohammadi, S. (2017). Micro-based enriched multiscale homogenization method for analysis of heterogeneous materials. *International Journal of Solids and Structures* 125: 22–42.

Bazant, Z.P. and Jirasek, M. (2002). Nonlocal integral formulations of plasticity and damage: survey of progress. *Journal of Engineering Mechanics* 128 (11): 1119–1149.

Bechet, E., Scherzer, M., and Kuna, M. (2009). Application of the X-FEM to the fracture of piezoelectric materials. *International Journal for Numerical Methods in Engineering* 77: 1535–1565.

Bell, E., Ivarsson, B., and Merrill, C. (1979). Production of a tissue-like structure by contraction of collagen lattices by human fibroblasts of different proliferative potential in vitro. *Proceedings of the National Academy of Sciences* 76 (3): 1274–1278.

Belytschko, T. and Black, T. (1999). Elastic crack growth in finite elements with minimal remeshing. *International Journal of Fracture Mechanics* 45: 601–620.

Belytschko, T. and Chen, H. (2004). Singular enrichment finite element method for elastodynamic crack propagation. *International Journal of Computational Methods* 1 (1): 1–15.

Belytschko, T., Krongauz, Y., Organ, D., et al. (1996). Meshless methods: an overview and recent developments. *Computer Methods in Applied Mechanics and Engineering* 139 (1–4): 3–47.

Belytschko, T., Liu, W.K., and Moran, B. (2000). *Nonlinear Finite Elements for Continua and Structures*, 1e. Wiley.

Belytschko, T., Liu, W.K., Moran, B., and Elkhodary, K.I. (2014). *Nonlinear Finite Elements for Continua and Structures*, 2e. Wiley.

Belytschko, T., Lu, Y.Y., and Gu, L. (1994). Element-free Galerkin methods. *International Journal for Numerical Methods in Engineering* 37 (2): 229–256.

Beniash, E., Stifler, C.A., Sun, C.-Y. et al. (2019). The hidden structure of human enamel. *Nature Communications* 10: 4383.

Benson, D.J., Bazilevs, Y., De Luycker, E. et al. (2010). A generalized finite element formulation for arbitrary basis functions: from isogeometric analysis to XFEM. *International Journal for Numerical Methods in Engineering* 8 (6): 765–785.

Bentz (2005). *CEMHYD3D: A Three-Dimensional Cement Hydration and Microstructure Development Modeling Package*, Version 3.0. National Institute of Standards and Technology NISTIR 7232.

Berger, L. (2015). *A Low Order Finite Element Method for Poroelasticity with Applications to Lung Modelling*. Ph.D. Thesis, Keble College, University of Oxford.

Bil, C., Massey, K., and Abdullah, E.J. (2013). Wing morphing control with shape memory alloy actuators. *Journal of Intelligent Material Systems and Structures* 24 (7): 879–898.

Billiar, K. and Sacks, M. (2000). Biaxial mechanical properties of the natural and glutaraldehyde treated aortic valve cusp – Part I: experimental results. *Transactions–American Society of Mechanical Engineers Journal of Biomechanical Engineering* 122 (1): 23–30.

Bishnoi, S. and Scrivener, K.L. (2009). μic: a new platform for modelling the hydration of cements. *Cement and Concrete Research* 39 (4) 266–274.

Bitaraf, M. and Mohammadi, S. (2008). Analysis of chloride diffusion in concrete structures for prediction of initiation time of corrosion using a new meshless approach. *Construction & Building Materials* 22 (4) 546–556.

Bitaraf, M. and Mohammadi, S. (2010). Large deflection analysis of flexible plates by the finite point method. *Thin-Walled Structures* 48, 200–214.

Blazek, J. (2001). *Computational Fluid Dynamics: Principles and Applications*. Elsevier.

Boatti, E., Scalet, G., and Auricchio, F. (2016). A three-dimensional finite-strain phenomenological model for shape-memory polymers: formulation, numerical simulations, and comparison with experimental data. *International Journal of Plasticity* 83: 153–177.

Boivin, M., Boyer, J.-C., Thollet, G., et al. (2002). Mechanical properties of coronary stents determined by using finite element analysis. *Journal of Biomechanics* 34 (8): 1065–1075.

Bolotin, V.V. (2019). *Mechanics of Fatigue*. Blackwell.

Bonet, J. and Wood, R.D. (1997). *Nonlinear Continuum Mechanics for Finite Element Analysis*, 1e. USA: Cambridge University Press.

Bonet, J. and Wood, R.D. (2008). *Nonlinear Continuum Mechanics for Finite Element Analysis*, 2e. USA: Cambridge University Press.

Boroomand, B., Najjar, M., and Onate, E. (2009). The generalized finite point method. *Computational Mechanics* 44: 173–190.

Boroomand, B., Tabatabaei, A.A., and Onate, E. (2005). Simple modifications for stabilization of the finite point method. *International Journal for Numerical Methods in Engineering* 63: 351–379.

Boroomand, B. and Zienkiewicz, O.C.Z. (1997). Recovery by equilibrium (REP). *International Journal for Numerical Methods in Engineering* 40: 137–164.

Bosco, E., Kouznetsova, V.G., Coenen, E.W., et al. (2014). A multiscale framework for localizing microstructures towards the onset of macroscopic discontinuity. *Computational Mechanics* 54: 299–319.

Bouchbinder, E., Livne, A., and Finebergm, J. (2009). The 1/*r* singularity in weakly nonlinear fracture mechanics. *Journal of the Mechanics and Physics of Solids* 57: 1568–1577.

Bouchbinder, E., Livne, A., and Finebergm, J. (2010). Weakly nonlinear fracture mechanics: experiments and theory. *International Journal of Fracture* 162: 3–20.

Bower, A.F. (2012) http://solidmechanics.org/text/Chapter3_2/Chapter3_2.htm (date of access November 2022).

Boyd, J.G. and Lagoudas, D.C. (1996). A thermodynamic constitutive model for the shape memory alloy materials. Part I. The monolithic shape memory alloy. *Int. J. Plasticity* 12 (6): 805–842.

Budday, S., Sommer, G., Birkl, C., et al. (2017). Mechanical characterization of human brain tissue. *Acta Biomaterialia* 48: 319–340.

Calvo, B., Pena, E., Martinez, M., and Doblaré, M. (2007). An uncoupled directional damage model for fibred biological soft tissues. Formulation and computational aspects. *International Journal for Numerical Methods in Engineering* 69: 2036–2057.

Capaldi, F.M., Boyce, M.C., and Rutledge, G.C. (2004). Molecular response of a glassy polymer to active deformation. *Polymer* 45 (4): 1391–1399.

Capelli, C., Gervaso, F., Petrini, L., et al. (2009). Assessment of tissue prolapse after balloon-expandable stenting: influence of stent cell geometry. *Medical Engineering and Physics* 31 (4): 441–447.

Car, R. and Parrinello, M. (1985). Unified approach for molecular dynamics and density-functional theory. *Physics Review Letter* 55: 2471.

Car, R. and Parrinello, M. (1987). The unified approach to density functional and molecular dynamics in real space. *Solid State Communications* 62 (6): 403–405.

Carlier, A., Geris, L., van Gastek, N., et al. (2015). Oxygen as a critical determinant of bone fracture healing – a multiscale model. *Journal of Theoretical Biology* 365: 247–264.

Carmody, C., Burriesci, G., Howard, I., and Patterson, E. (2006). An approach to the simulation of fluid–structure interaction in the aortic valve. *Journal of Biomechanics* 39 (1): 158–169.

Cataloglu, A., Clark, R.E., and Gould, P.L. (1977). Stress analysis of aortic valve leaflets with smoothed geometrical data. *Journal of Biomechanics* 10 (3): 153–158.

Chandra, N. and Namilae, S. (2006). Tensile and compressive behavior of carbon nanotubes: effect of functionalization and topological defects. *Mechanics of Advanced Materials and Structures* 13: 115–127.

Charlebois, M., Jirásek, M., and Zysset, P.K. (2010). A nonlocal constitutive model for trabecular bone softening in compression. *Biomechanics and Modeling in Mechanobiology* 9: 597–611.

Chatelin, S., Vappou, J., Roth, S., et al. (2012). Towards child versus adult brain mechanical properties. *Journal of the Mechanical Behaviour of Biomedical Materials* 6: 166–173.

Chavoshnejad, P., German, G.K., and Razavi, M.J. (2021). Hyperelastic material properties of axonal fibers in brain white matter. *Brain Multiphysics* 2: 100035.

Chawla, M.M., Al-Zanaidi, M.A., and Evans, D.J. (1996). A class of generalized trapezoidal formulas for the numerical integration of. *International Journal of Computer Mathematics* 62 (1–2): 131–142.

Chen, J.K. and Beraun, J.E. (2000). A generalized smoothed particle hydrodynamics method for nonlinear dynamic problems. *Computer Methods in Applied Mechanics and Engineering* 190: 225–239.

Chen, Y.C. and Lagoudas, D.C. (2008a). A constitutive theory for shape memory polymers. Part I: large deformations. *Journal of the Mechanics and Physics of Solids* 56 (5): 1752–1765.

Chen, Y.C. and Lagoudas, D.C. (2008b). A constitutive theory for shape memory polymers. Part II: a linearized model for small deformations. *Journal of the Mechanics and Physics of Solids* 56 (5): 1766–1778.

Chessa, J. and Belytschko, T. (2004). Arbitrary discontinuities in space–time finite elements by level sets and X-FEM. *International Journal for Numerical Methods in Engineering* 61: 2595–2614.

Chessa, J. and Belytschko, T. (2006). A local space–time discontinuous finite element method. *Computer Methods in Applied Mechanics and Engineering* 195: 1325–1343.

Chevalley, T., Bonjour, J.P., van Rietbergen, B., et al. (2013). Fracture history of healthy premenopausal women is associated with a reduction of cortical microstructural components at the distal radius. *Bone* 55: 377–383.

Cho, T.-J., Gerstenfeld, L., and Einhorn, T.A. (2002). Differential temporal expression of members of the transforming growth factor B superfamily during murine fracture healing. *Journal of Bone and Mineral Research* 17: 3.

Christen, P., Ito, K., Dos Santos, A.A., et al (2012). Validation of a bone loading estimation algorithm for patient-specific bone remodelling simulations. *Journal of Biomechanics* 46: 941–948.

Christen, P., Ito, K., Galis, F., and van Rietbergen, B. (2015b). Determination of hip-joint loading patterns of living and extinct mammals using an inverse Wolff's law approach. *Biomech Model Mechanobiol* 14 (2): 427–432.

Christen, P., Ito, K., and van Rietbergen, B. (2015a). A potential mechanism for allometric trabecular bone scaling in terrestrial mammals. *Journal of Anatomy* 226: 236–243.

Claes, L., Eckert-Hubner, K., and Augat, P. (2002). The effect of mechanical stability on local vascularization and tissue differentiation in callus healing. *Journal of Orthopaedic Research* 20: 1099–1105.

Clark, R.A.F. (1989). Wound repair. *Current Opinion in Cell Biology* 1 (5): 1000–1008.

Clayton, J.D. (2010). *Nonlinear Mechanics of Crystals*. Springer Science & Business Media.

Colloca, M., Blanchard, R., Hellmich, C., et al. (2014a). A multiscale analytical approach for bone remodeling simulations: linking scales from collagen to trabeculae. *Bone* 64: 303–313.

Colloca, M., Ito, K., and van Rietbergan, B. (2014b). An analytical approach to investigate the evolution of bone volume fraction in bone remodeling simulation at the tissue and cell level. *Journal of Biomechanical Engineering* 136: 031004–1.

Comi, C., Mariani, S., and Perego, U. (2007). An extended FE strategy for transition from continuum damage to mode I cohesive crack propagation. *International Journal for Numerical and Analytical Methods in Geomechanics* 31: 213–238.

Conti, M., De Beule, M., Mortier, P., et al. (2009). Nitinol embolic protection filters: design investigation by finite element analysis. *Journal of Materials Engineering and Performance* 18: 787–792.

Cook, R.B. and Zioupos, P. (2009). The fracture toughness of cancellous bone. *Journal of Biomechanics* 18 (42): 2054–2060.

Cornell, W.D., Cieplak, P., Bayly, C.I., et al. (1995). A second generation force field for the simulation of proteins, nucleic acids, and organic molecules. *Journal of the American Chemical Society* 117 (19): 5179–5197.

Cottrell, A.H. (1961) Theoretical aspects of radiation damage and brittle fracture in steel pressure vessels. *Iron and Steel Institute Special Report*, No. 69, Iron and Steel Institute, London, UK. pp. 281–296.

Cottrell, J.A., Hughes, T.J.R., and Bazilevs, Y. (2009). *Isogeometric Analysis: Towards Integration of CAD and FEA*. John Wiley & Sons Ltd.

Criado, F.J. (2011). Aortic dissection: a 250-year perspective. *Texas Heart Institute Journal* 38: 694.

Cuthbert, R.J., Churchman, S.M., Tan, H.B., et al. (2013). Induced periosteum a complex cellular scaffold for the treatment of large bone defects. *Bone* 57: 484–492.

Dagro, A.M., MoKee, P.J., Kraft, R.H., et al. (2013) A preliminary investigation of traumatically induced axonal injury in a three-dimensional (3-D) finite element model (FEM) of the human head during blast-loading, Army Research Lab Aberdeen Proving Ground MD Weapons and Materials Research Directorate, ADA588181.

Dall'Ara, E., Luisier, B., Schmidt, R., et al. (2013). A nonlinear QCT-based finite element model validation study for the human femur tested in two configurations in vitro. *Bone* 52: 27–38.

Daneshyar, A.R. and Mohammadi, S. (2011). Simulation of strong tangential discontinuity for XFEM shear band evolution. In: *International Conference on Extended Finite Element Methods – XFEM 2011* (eds. S. Bordas, B. Karihaloo, and P. Kerfriden). Cardiff, United Kingdom.

Darvish, K.K. and Crandall, J.R. (2002) Influence of brain material properties and boundary conditions on brain response during dynamic loading. In: *Proceedings of the 2002 International IRCOBI Conference on the Biomechanics of Impacts*, Munich, Germany pp. 339–350.

David Chua, S.N., Mac Donald, B.J., and Hashmi, M.S.J. (2002). Finite-element simulation of stent expansion. *Journal of Materials Processing Technology* 120 (1–3): 335–340.

David Chua, S.N., Mac Donald, B.J., and Hashmi, M.S.J. (2004). Effects of varying slotted tube (stent) geometry on its expansion behaviour using finite element method. *Journal of Materials Processing Technology* 55: 1764–1771.

Detournay, E. and Cheng, A.H.-D. (1993). 5 – Fundamentals of poroelasticity. In: *Analysis and Design Methods* (ed. C. Fairhurst), Pergamon. pp. 113–171.

Dolbow, J.E. (1999) *An Extended Finite Element Method with Discontinuous Enrichment for Applied Mechanics*. Ph.D. Dissertation, Theoretical and Applied Mechanics, Northwestern University, USA.

Dolbow, J., Moës, N., and Belytschko, T. (2000a). Discontinuous enrichment in finite elements with a partition of unity method. *Finite Elements in Analysis and Design* 36: 235–260.

Dolbow, J., Moës, N., and Belytschko, T. (2000b). Modeling fracture in Mindlin–Reissner plates with the extended finite element method. *International Journal of Solids and Structures* 37: 7161–7183.

Dolbow, J., Moës, N., and Belytschko, T. (2000c). An extended finite element method for modeling crack growth with frictional contact. *Finite Elements in Analysis and Design* 36 (3): 235–260.

Donea, J., Huerta, A., Ponthot, J.-P., and Rodrıguez-Ferran, A. (2004). Arbitrary Lagrangian–Eulerian methods, Chapter 14. In: *Encyclopedia of Computational Mechanics*, 1: Fundamentals (eds. E. Stein, R. de Borst and T.R.J. Hughes). John Wiley & Sons, Ltd.

Donnelly, B.R. and Medige, J. (1997). Shear properties of human brain tissue. *Journal of Biomechanical Engineering* 119 (4): 423–432.

Dumoulin, C. and Cochelin, B. (2000). Mechanical behaviour modelling of balloon-expandable stents. *Journal of Biomechanics* 33 (11): 1461–1470.

Dunne, F. and Petrinic, N. (2004). *Introduction to Computational Plasticity*. Oxford University Press.

Dupuy, L.M., Tadmor, E.B., Miller, R.E., and Phillips, R. (2005). Finite-temperature quasicontinuum: molecular dynamics without all the atoms. *Physical Review Letters* 95: 060202.

Ebrahimi, S.H. (2012) *Development of the Extended Finite Element Method in Modelling Frictional Contact Problems*. Ph.D. Thesis, University of Tehran, Iran.

Ebrahimi, S.H., Mohammadi, S., and Mahmoudzadeh Kani, I. (2013). A local PUFEM modeling of stress singularity in sliding contact with minimal enrichment for direct evaluation of generalized stress intensity factors. *Engineering Fracture Mechanics* 105: 16–40.

Edholm, O. (2014) Molecular Dynamics. Lecture Notes in Computational Physics.

Eftekhari, M. (2015) *Application of Multiscale Method for Analysis of Cyclic Behavior of Concrete Structures Reinforced by Defected Carbon Nanotubes*. Ph.D. Thesis, Faculty of Engineering, Islamic Azad University (in Persian).

Eftekhari, M., Ardakani, S.H., and Mohammadi, S. (2014). An XFEM multiscale approach for fracture analysis of carbon nanotube reinforced concrete. *Theoretical and Applied Fracture Mechanics* 72: 64–75.

Eftekhari, M. and Mohammadi, S. (2015) Application of multiscale analysis to investigate the effect of defects of carbon nano tubes in cyclic response of reinforced concrete specimen. Iran Nano Technology Initiative Council, HRDC, 2015 (in Persian).

Eftekhari, M. and Mohammadi, S. (2016a). Multiscale dynamic fracture behavior of the carbon nanotube reinforced concrete under impact loading. *International Journal of Impact Engineering* 87: 55–64.

Eftekhari, M. and Mohammadi, S. (2016b). Molecular dynamics simulation of the nonlinear behavior of the CNT-reinforced calcium silicate hydrate (C–S–H) composite. *Composites: Part A* 82: 78–87.

Eftekhari, M., Mohammadi, S., and Khanmohammadi, M. (2018). A hierarchical nano to macro multiscale analysis of monotonic behavior of concrete columns made of CNT-reinforced cement composite. *Construction and Building Materials* 175: 134–143.

Eftekhari, M., Mohammadi, S., and Khoei, A.R. (2013). Effect of defects on the local shell buckling and post-buckling behavior of single and multi-walled carbon nanotubes. *Computational Materials Science* 79: 736–744.

Ehret, A.E. and Itskov, M. (2009). Modeling of anisotropic softening phenomena: application to soft biological tissues. *International Journal of Plasticity* 25: 901–919.

Einhorn, T.A. and Gerstenfeld, L.C. (2014). Fracture healing: mechanisms and interventions. *Nature Reviews, Rheumatology* 11 (1): 45–54.

Elguedj, T., Gravouil, A., and Combescure, A. (2006). Appropriate extended functions for X-FEM simulation of plastic fracture mechanics. *Computer Methods in Applied Mechanics and Engineering* 195: 501–515.

Elkins, M.R., Moseley, A.M., Sherrington, C., et al. (2013). Growth in the physiotherapy evidence database (PEDro) and use of the PEDro scale. *British Journal of Sports Medicine* 47 (4): 188–189.

Emamzadeh, S.S., Ahmadi, M.T., Mohammadi, S., and Biglarkhani, M. (2015). Dynamic adaptive finite element analysis of acoustic wave propagation due to underwater explosion for fluid–structure interaction problems. *Journal of Marine and Structural Applications* 14 (3): 302–315.

Emdadi, A., Kansa, E.J., Libre, N.A., et al. (2008). Stable PDE solution methods for large multiquadric shape parameters. *Computer Modeling in Engineering Sciences* 25 (1): 23–42.

Erdogan, F. and Sih, G.C. (1963). On the crack extension in plates under plane loading and transverse shear. *Journal of Basic Engineering* 85: 519–527.

Eshelby, J.D. (1956). The continuum theory of lattice defects. In: *Solid State Physics*, 3 (ed. F. Seitz and D. Turnbull), New York: Academic Press. 79–141.

Eshelby, J.D. (1974). Calculation of energy release rate. In: *Prospect of Fracture Mechanics* (eds. G.C. Sih, H.C. Van Elst and D. Brock), Nordhoff, UK. 69–84.

Esna Ashari, S. and Mohammadi, S. (2009). XFEM delamination analysis of composite laminates by new orthotropic enrichment functions. In: *Proceedings of 1st International Conference on Extended Finite Element Method – Recent Developments and Applications (XFEM2009)*. Aachen, Germany.

Esna Ashari, S. and Mohammadi, S. (2010). Modeling delamination in composite laminates using XFEM by new orthotropic enrichment functions. In: *WCCM/APCOM 2010; IOP Conference Series: Materials Science and Engineering* Sydney, Australia: IOP Publishing. 10, 012240, 1–8.

Esna Ashari, S. and Mohammadi, S. (2011a). Delamination analysis of composites by new orthotropic bimaterial extended finite element method. *International Journal for Numerical Methods in Engineering* 86 (13): 1507–1543.

Esna Ashari, S. and Mohammadi, S. (2011b). Debonding propagation analysis of FRP reinforced beams by the extended finite element method. In: *International Conference on Extended Finite Element Methods – XFEM 2011* (eds. S. Bordas, B. Karihaloo and P. Kerfriden). Cardiff, United Kingdom.

Esna Ashari, S. and Mohammadi, S. (2012). Fracture analysis of FRP reinforced beams by orthotropic XFEM. *Journal of Composite Materials – Part B* 46: 1357–1389.

Eymard, R., Gallouet, T., and Herbin, R. (2019). Finite volume methods. In: *An update appeared in Handbook of Numerical Analysis*, 7 (eds. P.G. Ciarlet and J.L. Lions), 713–1020.

Fallah, N.A., Bailey, C., Cross, M., and Taylor, G.A. (2000). Comparison of finite element and finite volume methods application in geometrically nonlinear stress analysis. *Applied Mathematical Modelling* 24: 439–455.

Fatemi, P., Hatefi, S., Bayesteh, H., and Mohammadi, S. (2017). 3D hierarchical multiscale analysis of heterogeneous SMA based materials. *International Journal of Solids and Structures* 118-119: 24–40.

Fathi, F. (2015) *Multiscale Simulation of Soft Biomechanical Systems*, M.Sc. Thesis, School of Civil Engineering, University of Tehran, Tehran, Iran.

Fathi, F., Ardakani, S.H., Dehaghani, P.F., and Mohammadi, S. (2017). A finite strain integral-type anisotropic damage model for fiber-reinforced materials: application in soft biological tissues. *Computer Methods in Applied Mechanics and Engineering* 322: 262–295.

Fathi, F. and Mohammadi, S. (2015). Multi-scale analysis of the soft biological nano-composites: the role of fibril. In: *3rd International Conference on Nanotechnology (ICN2015)*, 27–28 August 2015, Istanbul, Turkey.

Ferrara, A. and Pandolfi, A. (2008). Numerical modelling of fracture in human arteries. *Computer Methods in Biomechanics and Biomedical Engineering* 11: 553–567.

Ferrara, A. and Pandolfi, A. (2010). A numerical study of arterial media dissection processes. *International Journal of Fracture* 166: 21–33.

Finan, J.D., Pearson, E.M., and Morrison, B. (2012). Viscoelastic properties of the rat brain in the horizontal plane. In: *2012 IRCOBI Conference Proceedings – International Research Council on the Biomechanics of Injury*, 474–485.

Fish, J. (2006). Bridging the scales in nano engineering and science. *Journal of Nanoparticle Research* 8: 577–594.

Fish, J. (2009). *Multiscale Methods: Bridging the Scales in Science and Engineering*. Oxford Scholarship Online.

Flynn, C., Taberner, A., and Nielsen, P. (2011). Modeling the mechanical response of in vivo human skin under a rich set of deformations. *Annals of Biomedical Engineering* 39 (7): 1935–1946.

Forouzan-sepehr, S. and Mohammadi, S. (2010). A fast mesh-free Galerkin method for the analysis of steady-state heat transfer. *Journal of Aerospace Science and Technology* 6 (1): 13–24.

Foyouzat, A., Bayesteh, H., and Mohammadi, S. (2021). Phase evolution based thermomechanical crack closure mechanism of shape memory polymers. *Mechanics of Materials* 160: 103998.

Foyouzat, A. and Mohammadi, S. (2022). *Bio-compatible Shape Memory Polymers in Drug Delivery Systems*. Report HPC2022-06, High Performance Computing Lab, School of Civil Engineering, University of Tehran, Iran.

Franceschini, G. (2006). *The Mechanics of Human Brain Tissue*. Ph.D., University of Trento.

Fung, Y.C. (1993). *Biomechanics: Mechanical Properties of Living Tissues*, 2e. USA: Springer.

Furuya, Y. (1996). Design and material evaluation of shape memory composites. *Journal of Intelligent Material Systems and Structures* 7 (3): 321–330.

Gadalaa, M.S., Movahhedya, M.R., and Wang, J. (2002). On the mesh motion for ALE modeling of metal forming processes. *Finite Elements in Analysis and Design* 38: 435–459.

Galletti, P.M. and Mora, C.T. (1995). Cardiopulmonary bypass: the historical foundation, the future promise. In: *Cardiopulmonary Bypass* (eds. C.T. Mora, R.A. Guyton, D.C. Finlayson, and R.L. Rigatti). New York, USA: Springer.

Garcia-sanchez, F., Zhang, C., and Saez, A. (2008). A two-dimensional time-dependent boundary element method for dynamic crack problems in anisotropic solids. *Engineering Fracture Mechanics* 75 (6): 1412–1430.

Garimella, H.T., Kraft, R.H., and Przekwas, A.J. (2022). Do blast induced skull flexures result in axonal deformation? *PLoS ONE* 13 (3): e0190881.

Gaspar, F., Rodrigo, C., Hu, X., et al. (2018) Stabilized finite element discretizations for poroelasticity, In: *International Conference on Numerical Methods and Applications*, August 20–24, 2018, Borovets (Bulgaria).

Gasser, T.C. and Holzapfel, G.A. (2006). Modeling the propagation of arterial dissection. *European Journal of Mechanics – A/Solids* 25: 617–633.

Gasser, T.C., Ogden, R.W., and Holzapfel, G.A. (2006). Hyperelastic modelling of arterial layers with distributed collagen fibre orientations. *Journal of the Royal Society Interface* 3: 15–35.

Gee, R.H. and Boyd, R.H. (1998). The role of the torsional potential in relaxation dynamics: a molecular dynamics study of polyethylene. *Computational and Theoretical Polymer Science* 8: 93–98.

Genc, O. (2019) Notes on artificial intelligence, machine learning and deep learning for curious people. Available online.

Geris, L., Gerisch, A., Vander Sloten, J., et al. (2008). Angiogenesis in bone fracture healing: a bioregulatory model. *Journal of Theoretical Biology* 251: 137–158.

Ghorashi, S.S., Valizadeh, N., and Mohammadi, S. (2012). Extended isogeometric analysis (XIGA) for analysis of stationary and propagating crack. *International Journal for Numerical Methods in Engineering* 89 (9): 1069–1101.

Ghorashi, S.S., Valizadeh, N., Mohammadi, S., and Rabzuk, T. (2015). T-spline based XIGA for fracture analysis of orthotropic media. *Computers and Structures* 147: 138–146.

Ghosh, K., Pan, Z., Guan, E., et al. (2007). Cell adaptation to a physiologically relevant ECM mimic with different viscoelastic properties. *Biomaterials* 28 (4): 671–679.

Gingold, R.A. and Monagan, J.J. (1977). Smoothed particle hydrodynamics: theory and application to non-spherical stars. *Monthly Notices of the Royal Astronomical Society* 181: 375–389.

Gingold, R.A. and Monagan, J.J. (1983). Shock simulation by the particle method SPH. *Journal of Computational Physics* 52 (2): 374–389.

Giry, C., Bottoni, M., Dufour, F., et al. (2009). Endommagement et fissuration du beton arme: passage continu – discret. In: *19 eme Congres Franc¸ais de Mecanique*. Marseille.

Goh, S.M., Charalambides, M.N., and Williams, J.G. (2004). Determination of the constitutive constants of non-linear viscoelastic materials. *Mechanics of Time-Dependent Materials* 8: 225–268.

Goudarzi, M. and Mohammadi, S. (2014). Weak discontinuity in porous media: an enriched EFG method for fully coupled layered porous media. *International Journal for Numerical and Analytical Methods in Geomechanics* 38: 1792–1822.

Goudarzi, M. and Mohammadi, S. (2015). Analysis of cohesive cracking in saturated porous media using an extrinsically enriched EFG method. *Computers and Geotecnics* 63: 183–198.

Goodarzi, M., Mohammadi, S., and Jafari, A. (2011). Numerical study of induced gas pressure on controlled blasting. In: *6th National Congress of Civil Engineering (6NCCE)*, Semnan, Iran (In Persian).

Goodarzi, M., Mohammadi, S., and Jafari, A. (2015). Numerical analysis of rock fracturing by gas pressure using the extended finite element method. *Petroleum Science* 12: 304–315.

Gracie, R., Ventura, G., and Belytschko, T. (2007). A new fast finite element method for dislocations based on interior discontinuities. *International Journal for Numerical Methods in Engineering* 69: 423–441.

Gracie, R., Wang, H., and Belytschko, T. (2008). Blending in the extended finite element method by discontinuous Galerkin and assumed strain methods. *International Journal for Numerical Methods in Engineering* 74 (11): 1645–1669.

Grande, K.J., Cochran, R.P., Reinhall, P.G., and Kunzelman, K.S. (1998). Stress variations in the human aortic root and valve: the role of anatomic asymmetry. *Annals of Biomedical Engineering* 26 (4): 534–545.

Griffith, A.A. (1921). The phenomena of rupture and flow in solids. *Philosophical Transactions of the Royal Society A: Mathematical, Physical and Engineering* 221: 163–197.

Griffith, A.A. (1924). The theory of rupture. *Proceedings of International Congress on Applied Mechanics*, Delft, pp. 55–62.

Griffiths, C.E.M., Barker, J., Bleiker, T. et al. (2016). *Rook's Textbook of Dermatology*. John Wiley & Sons Ltd.

Grotendorst, J., Attig, N., Blugel, S., and Marx, D. (2009). Multiscale simulation methods in molecular sciences. Winter School, 2–6 March 2009, Forschungszentrum Julich, Germany.

Gtoudos, E. (1993). *Fracture Mechanics: An Introduction*. The Netherlands: Kluwer Academic Press.

Gu, J., Leng, J., Sun, H., et al. (2019). Thermomechanical constitutive modeling of fiber reinforced shape memory polymer composites based on thermodynamics with internal state variables. *Mechanics of Materials* 130: 9–19.

Gültekin, O., Dal, H., and Holzapfel, G.A. (2016). A phase-field approach to model fracture of arterial walls: theory and finite element analysis. *Computer Methods in Applied Mechanics and Engineering* 312: 542–566.

Guo, Z., Peng, X., and Moran, B. (2006). A composites-based hyperelastic constitutive model for soft tissue with application to the human annulus fibrosus. *Journal of the Mechanics and Physics of Solids* 54: 1952–1971.

Haghighat, E., Raissi, M., Moure, A., et al. (2021). A physics-informed deep learning framework for inversion and surrogate modelling in solid mechanics. *Computer Methods in Applied Mechanics and Engineering* 379: 113741.

Hambli, R. (2013). A quasi-brittle continuum damage finite element model of the human proximal femur based on element deletion. *Medical & Biological Engineering & Computing* 51: 219–231.

Hambli, R., Benhamou, C.-L., Jennane, R., et al. (2013). Combined finite element model of human proximal femur behaviour considering remodeling and fracture. *IRBM* 34: 191–195.

Hambli, R., Bettamer, A., and Allaoui, S. (2012). Finite element prediction of proximal femur fracture pattern based on orthotropic behaviour law coupled to quasi-brittle damage. *Medical Engineering & Physics* 34: 202–210.

Hambli, R., Katerchi, H., and Benhamou, C.-L. (2011). Multiscale methodology for bone remodelling simulation using coupled finite element and neural network computation. *Biomechanics and Modeling in Mechanobiology* 10: 133–145.

Hamed, E., Jasiuk, I., Yoo, A., et al. (2012). Multi-scale modelling of elastic moduli of trabecular bone. *Journal of Royal Society Interface* 9: 1654–1673.

Hamed, E., Ma, D., and Keten, S. (2015). Multiple PEG chains attached onto the surface of a helix bundle: conformations and implications. *ACS Biomaterials Science & Engineering* 1 (2): 79–84.

Hamed, E., Xu, T., and Keten, S. (2013). Poly(ethylene glycol) conjugation stabilizes the secondary structure of α-helices by reducing peptide solvent accessible surface area. *Biomacromolecules* 14: 4053–4060.

Hanwell, M.D., Curtis, D.E., Lonie, D.C., et al. (2012). Avogadro: an advanced semantic chemical editor, visualization, and analysis platform. *Journal of Cheminformatics* 4: 17.

Hao, X., Qiang, H., and Xiaohu, Y. (2008). Buckling of defective single-walled and double-walled carbon nanotubes under axial compression by molecular dynamics simulation. *Composites Science and Technology* 68 (7–8): 1809–1814.

Harper, G., Liu, J., Tavener, S., and Wang, Z. (2018). A two-field finite element software for poroelasticity on quadrilateral meshes. In: *International Conference on Computational Science, ICSS 2018*, 76–88.

Harris, A.K., Stopak, D., and Wild, P. (1981). Fibroblast traction as a mechanism for collagen morphogenesis. *Nature* 290 (5803): 249–251.

Harvey, W. (1628). Exercitatio anatomica de motu cordis et sanguinis in animalibus (an anatomical disquisition on the motion of the heart and blood in living beings). Frankfurt.

Hashemi Yazdi, S.S., Ahmadian, H., and Mohammadi, S. (2015). An extended thermo-mechanically coupled algorithm for simulation of superelasticity and shape memory effect in shape memory alloys. *Frontiers of Structural and Civil Engineering* 9 (4): 466–477.

Hassani, B. and Hinton, E. (1998a). A review of homogenization and topology optimization I – homogenization theory for media with periodic structure. *Computers and Structures* 69 (6): 707–717.

Hassani, B. and Hinton, E. (1998b). A review of homogenization and topology optimization II—analytical and numerical solution of homogenization equations. *Computers and Structures* 69 (6): 719–738.

Hassani, B. and Hinton, E. (1999). *Homogenization and Topology Optimization: Theory, Practice, and Software*. Springer.

Hatefi Ardakani, S., Ahmadian, H., and Mohammadi, S. (2015). Thermo-mechanically coupled fracture analysis of shape memory alloys using the extended finite element method. *Smart Materials and Structures* 24 (2015): 045031.

Hatefi Ardakani, S., Fatemi Dehghani, P., Moslemzadeh, H., and Mohammadi, S. (2022). 3D large strain hierarchical multiscale analysis of soft fiber-reinforced tissues: application to a degraded arterial wall. *Engineering Computations* 39 (6): 2108–2143.

Hatefi, S. and Mohammadi, S. (2012) An XFEM model for transition of micro damage mechanics to macro crack analysis, University of Tehran, Report: 8102051.

Hatefi, S., Moslemzadeh, H., and Mohammadi, S. (2019). Delamination analysis in bimaterials consisting of shape memory alloy and elastoplastic layers. *Composite Structures* 225: 111149.

Hazrati Marangalou, J., Ito, K., and van Rietbergen, B. (2015). A novel approach to estimate trabecular bone anisotropy from stress tensors. *Biomech. Model Mechanobiol.* 14: 39–48.

Hellmich, C., Barthélémy, J.-F., and Dormieux, L. (2004). Mineral–collagen interactions in elasticity of bone ultrastructure – a continuum micromechanics approach. *European Journal of Mechanics A/Solids* 23: 783–810.

Holland, J.H. (1992). *Adaptation in Natural and Artificial Systems: An Introductory Analysis with Applications to Biology, Control, and Artificial Intelligence*. Cambridge, Massachusetts: MIT Press.

Holzapfel, G. and Fereidoonnezhad, B. (2017). Modeling of damage in soft biological tissues, Chapter 5. In: *Biomechanics of Living Organs*, Elsevier. 101–123.

Holzapfel, G.A. (2000). *Nonlinear Solid Mechanics, A Continuum Approach for Engineering*. John Wiley & Sons Ltd.

Holzapfel, G.A. (2002). Nonlinear solid mechanics: a continuum approach for engineering science. *Meccanica* 37 (4): 489–490.

Holzapfel, G.A. and Gasser, T.C. (2001). A viscoelastic model for fiber-reinforced composites at finite strains: continuum basis, computational aspects and applications. *Computer Methods in Applied Mechanics and Engineering* 190: 4379–4403.

Holzapfel, G.A., Gasser, T.C., and Ogden, R.W. (2000). A new constitutive framework for arterial wall mechanics and a comparative study of material models. *Journal of Elasticity* 61: 1–48.

Holzapfel, G.A. and Ogden, R.W. (2009). Constitutive modelling of passive myocardium: a structurally based framework for material characterization. *Philosophical Transactions of The Royal Society A, Mathematical, Physical and Engineering Sciences* 367 (1902): 3445–3475.

Holzapfel, G.A. and Ogden, R.W. (2010). Constitutive modelling of arteries. *Proceedings of the Royal Society A, Mathematical, Physical and Engineering Sciences* 466 (2118): 1551–1597.

Holzapfel, G.A., Sommer, G., and Regitnig, P. (2004). Anisotropic mechanical properties of tissue components in human atherosclerotic plaques. *Journal of Biomechanical Engineering* 126: 657–665.

Holzapfel, G.A., Stadler, M., and Gasser, T.C. (2005). Changes in the mechanical environment of stenotic arteries during interaction with stents: computational assessment of parametric stent designs. *Journal of Biomechanical Engineering* 127: 166–180.

Holzapfel, G.A., Stadler, M., and Schulze-Bauer, C.A. (2002). A layer-specific three-dimensional model for the simulation of balloon angioplasty using magnetic resonance imaging and mechanical testing. *Annals of Biomedical Engineering* 30: 753–767.

Hoo, R.P. (2011). *Multi-Scale Mechanics of Bone*. Ph.D. Thesis, Materials Science and Engineering, School of Materials, Science and Engineering, Faculty of Science, University of New South Wales, Australia.

Hossain, D., Tschopp, M.A., Ward, D.K., et al. (2010). Molecular dynamics simulations of deformation mechanisms of amorphous polyethylene. *Polymer* 51: 6071–6083.

Hu, Y. and Suo, Z. (2012). Viscoelsticity and poroelasticity in elastomeric gels. *Acta Mechanical Solida Sinical* 25 (5): 441–458.

Huang, H.Y.S. (2005). *Micromechanical Simulations of Heart Valve Tissues*. University of Pittsburgh, PA, USA.

Hughes, T.J.R., Cottrell, J.A., and Bazilevs, Y. (2005). Isogeometric analysis: CAD, finite elements, NURBS, exact geometry and mesh refinement. *Computer Methods in Applied Mechanics and Engineering* 194: 4135–4195.

Hughes, T.J.R., Feijóo, G.R., Luca, M., and Quincy, J.-B. (1998). The variational multiscale method – a paradigm for computational mechanics. *Computer Methods in Applied Mechanics and Engineering* 166 (1–2): 3–24.

Hutchinson, J. (1968). Singular behavior at the end of a tensile crack tip in a power-law hardening material. *Journal of Mechanics and Physics of Solids* 16: 13–31.

Huynh, D.B.P. and Belytschko, T. (2009). The extended finite element method for fracture in composite materials. *International Journal for Numerical Methods in Engineering* 77 (2): 214–239.

Inglis, C.E. (1913). Stresses in a plate due to the presence of cracks and sharp corners. *Transactions of Institute of Naval Architects* 55: 219–241.

Irwin, G.R. (1948). Fracture dynamics fracturing of metals. *American Society for Metals, Cleaveland* 19-9: 147–166.

Irwin, G.R. (1957). Analysis of stresses and strains near the end of a crack transversing a plate. *Journal of Applied Mechanics, Transactions ASME* 24: 361–364.

Irwin, G.R. (1961). Plastic zone near a crack tip and fracture toughness. In: *Proceedings of the 7th Sagamore Conference*, IV, New York, USA 63–76.

Janfada, M. (2018). *Multiscale Modeling of Brain Failure Due to Impulsive Loading*. M.Sc. Thesis, School of Civil Engineering, University of Tehran, Iran.

Javid, S., Rezaei, A., and Karami, G. (2014). A micromechanical procedure for viscoelastic characterization of the axons and ECM of the brainstem. *Journal of the Mechanical Behavior of Biomedical Materials* 1 (30): 290–299.

Javierre, E., Moreo, P., Doblaré, M., and García-Aznar, J.M. (2009). Numerical modeling of a mechano-chemical theory for wound contraction analysis. *International Journal of Solids and Structures* 46 (20): 3597–3606.

Jirásek, M. (2004). Non-local damage mechanics with application to concrete. *Revue Française de Génie Civil* 8: 683–707.

Jirasek, M. (2007). Nonlocal damage mechanics. *Revue Européenne de Génie Civil* 11 (7–8): 993–1021.

Johnell, O. and Kanis, J.A. (2006). An estimate of the worldwide prevalence and disability associated with osteoporotic fractures. *Osteoporosis International* 17: 1726–1733.

Ju, J. (1989). On energy-based coupled elastoplastic damage theories: constitutive modeling and computational aspects. *International Journal of Solids and Structures* 25: 803–833.

Kachanov, L. (1958). Time of the rupture process under creep conditions. *Izvestia Akademii Nauk SSSR, Otdelenie Technicheskikh Nauk* 8: 26–1.

Kaliske, M. and Rothert, H. (1997). Formulation and implementation of three-dimensional viscoelasticity at small and finite strains. *Computational Mechanics* 19 (3): 228–239.

Kanninen, M. (1984) Application of fracture mechanics to fiber composite materials and adhesive joint, a review. *Third International Conference on Numerical Methods in Fracture Mechanics*, Swansea, UK.

Kansa, E.J. (1990). Multiquadrics – A scattered data approximation scheme with applications to computational fluid-dynamics – II solutions to parabolic, hyperbolic and elliptic partial differential equations. *Computers and Mathematics with Applications* 19: 147–161.

Karami, G., Grundman, N., Abolfathi, N., et al. (2009). A micromechanical hyperelastic modeling of brain white matter under large deformation. *Journal of the Mechanical Behavior of Biomedical Materials* 2 (3): 243–254.

Karimi, M., Bayesteh, H., and Mohammadi, S. (2019). An adapting cohesive approach for crack-healing analysis in SMA fibers-reinforced composites. *Computer Methods in Applied Mechanics and Engineering* 349: 550–575.

Karlsson, A. and Backlund, J. (1978). J-integral at loaded crack surfaces. *International Journal of Fracture* 14 (6): R311–R314.

Katsamenis, O.L., Jenkins, T., and Thurner, P.J. (2015). Toughness and damage susceptibility in human cortical bone is proportional to mechanical inhomogeneity at the osteonal-level. *Bone* 76: 158–168.

Kattan, P.I. and Voyiadjis, G.Z. (2001). *Damage Mechanics with Finite Elements*. Springer.

Kauffman, G.B. and Mayo, I. (1997). The story of nitinol: the serendipitous discovery of the memory metal and its applications. *The Chemical Educator* 2 (2): 1–21.

Keyhani, A., Goudarzi, M., Mohammadi, S., and Roumina, R. (2015). XFEM-dislocation dynamics multi-scale modeling of plasticity and fracture. *Computationals Materials Science* 104: 98–107.

Khaksar, K. (2020). Simulation of healing process in damaged domains with application in soft biomechanical systems. M.Sc. Thesis, School of Civil Engineering, University of Tehran, Iran.

Khaksar, K. and Mohammadi, S. (2020). Simulation of healing in bio-composites: a case study of human skin. In: *The 7th International Conference on Composites: Characterization, Fabrication and Application (CCFA-7)*, December 2020, Sahand University of Technology, Tabriz, Iran.

Khaterchi, H., Chamekh, A., and Belhadjsalah, H. (2013). Multi-scale modelling of orthotropic properties of trabecular bone in nanoscale. In: *Design and Modeling of Mechanical Systems, LNME* (eds. M. Haddar, L. Romdhane, J. Louati and A. Ben Amara), Berlin and Heidelberg: Springer-Verlag. 557–566.

Khatyr, F., Imberdis, C., Vescovo, P., et al. (2004). Model of the viscoelastic behaviour of skin in vivo and study of anisotropy. *Skin Research and Technology* 10 (2): 96–103.

Khazal, H., Bayesteh, H., Mohammadi, S., et al. (2016). An extended element free Galerkin method for fracture analysis of anisotropic functionally graded materials. *Mechanics of Advanced Materials and Structures* 23 (5): 513–528.

Kheirikhah, M.M., Rabiee, S., and Edalat, M.E. (2010). A review of shape memory alloy actuators in robotics. *Robocup2010: Robot Soccer World Cup* XIV: 206–217.

Khosravi, A. and Ebrahimi, H. (2008). To evaluate the outcomes of patients with trauma admitted to the Imam Hossein Hospital, Shahrood, using the trauma and injury severity score (TRISS). *Iranian Journal of Epidemiology* 4 (2): 35–41.

Kim, H., Lu, J., Sacks, M.S., and Chandran, K.B. (2008). Dynamic simulation of bioprosthetic heart valves using a stress resultant shell model. *Annals of Biomedical Engineering* 36 (2): 262–275.

Kim, H.S. (2009). *Nonlinear Multi-scale Anisotropic Material and Structural Models for Prosthetic and Native Aortic Heart Valves*. Georgia Institute of Technology, GA, USA.

Kircsh, G. (1898). Infinite plate containing a circular hole (Die theorie der elastizitat und die bedurfnisse der festigkeitslehre). *Zeitschrift der Vereines Deutscher Ingenieure* 42: 797–807.

Koch, T., Reddy, B., Zilla, P., and Franz, T. (2010). Aortic valve leaflet mechanical properties facilitate diastolic valve function. *Computer Methods in Biomechanics and Biomedical Engineering* 13 (2): 225–234.

Kohlhoff, S., Gumbsch, P., and Fischmeister, H.F. (1991). Crack propagation in b.c.c. crystals studied with a combined finite-element and atomistic model. *Philosophical Magazine A* 64 (4): 851–878.

Kolosov, G.V. (1909). On an application of complex function theory to a plane problem of the mathematical theory of elasticity. *Infinite Plate Containing an Elliptical Hole*, Yuriev.

Komorniczak, M.M. (2012). *Skin Layers – Wikimedia Commons*. Available online: https://commons.wikimedia.org/wiki/File:Skin_layers.svg.

Kon, T., Cho, T.-J., Aizawa, T., et al. (2014). Expression of osteoprotegerin, receptor activator of NF-kB ligand (osteoprotegerin ligand) and related proinflammatory cytokines during fracture healing. *Journal of Bone and Mineral Research* 16 (6): 1004–1014.

Kordestani, S.S. (2019). *Chapter 3 – Wound healing process*. In: *Atlas of Wound Healing*, 11–22. Elsevier.

Kouznetsova, V.G. (2002) *Computational Homogenization for the Multi-scale Analysis of Multi-phase Materials*. Ph.D. Thesis, Eindhoven University of Technology.

Kouznetsova, V.G., Geers, M.G.D., and Brekelmans, W.A.M. (2004). Multi-scale second-order computational homogenization of multi-phase materials: a nested finite element solution strategy. *Computer Methods in Applied Mechanics and Engineering* 193: 5525–5550.

Kugo, H., Zaima, N., Tanaka, H., et al. (2017). Pathological analysis of the ruptured vascular wall of hypoperfusion-induced abdominal aortic aneurysm animal model. *Journal of Oleo Science* 66 (5): 499–506.

Kumpel, H.-J. (2004) *Theory of Poroelasticity in 10 Lessons*, Lecture Notes, TU Clausthal, Institute of Geophysics, WT 2003/04.

Laborde, P., Pommier, J., Renard, Y., and Salaün, M. (2005). High-order extended finite element method for cracked domains. *International Journal for Numerical Methods in Engineering* 64: 354–381.

Lagodas, D.C. (2008). *Shape Memory Alloys: Modeling and Engineering Applications*. Springer.

Lai, W.M., Rubin, D., and Krempl, E. (2010). *Introduction to Continuum Mechanics*. Elsevier.

Lally, C., Dolan, F., and Prendergast, P.J. (2005). Cardiovascular stent design and vessel stresses: a finite element analysis. *Journal of Biomechanics* 38 (8): 1574–1581.

Langer, K. (1861). Zur Anatomie und Physiologie der Haut. Über die Spaltbarkeit der Cutis. *Sitzungsbericht der Mathematisch-naturwissenschaftlichen Classe der Wiener Kaiserlichen Academie der Wissenschaften Abt*, 44. (On the anatomy and physiology of the skin). *British Journal of Plastic Surgery* 31 (1): 3–8. 1978.

Langer, R. and Vacanti, J.P. (1993). Tissue engineering. *Science* 5110 (260): 920–926.

Lecampion, B. (2009). An extended finite element method for hydraulic fracture problems. *Communications in Numerical Methods in Engineering* 25 (2): 121–133.

Legay, A., Chessa, J., and Belytschko, T. (2006). An Eulerian–Lagrangian method for fluid-structure interaction based on level sets. *Computer Methods in Applied Mechanics and Engineering* 195 (17–18): 2070–2087.

Lekadir, K., Noble, C., Hazrati-Marangalou, J., et al. (2016). Patient-specific biomechanical modeling of bone strength using statistically-derived fabric tensors. *Annals of Biomedical Engineering* 44 (1): 234–246.

Lekhnitskii, S.G. (1968). *Anisotropic Plates*. New York, USA: Gordon and Breach Science Publishers.

Lemaire, T., Kaiser, J., Naili, S., and Sansalone, V. (2013). Textural versus electrostatic exclusion-enrichment effects in the effective chemical transport within the cortical bone: A numerical investigation. *International Journal for Numerical Methods in Biomedical Engineering* 29: 1223–1242.

Lemaitre, J. (1984). How to use damage mechanics. *Nuclear Engineering* 80 (2): 23–245.

Lemaitre, J. (1996). *A Course on Damage Mechanics*, 2e. Germany.

Lemaitre, J. and Desmorat, R. (2005). *Engineering Damage Mechanics*. Netherlands: Springer.

LeVeque, R.J. (2007). *Finite Difference Methods for Ordinary and Partial Differential Equations*. Siam.

Levrero-Florencio, F., Margetts, L., Sales, E., et al. (2016). Evaluating the macroscopic yield behaviour of trabecular bone using a nonlinear homogenization approach. *Journal of the Mechanical Behaviour of Biomedical Materials* 61: 384–396.

Li, C. and Borja, R.I. (2005) *Finite Element Formulation of Poro-elasticity Suitable for Large Deformation Dynamic Analysis*, Report No. 147, The John A. Blume Earthquake Engineering Center, Department of Civil and Environmental Engineering, Stanford University.

Li, F.Z., Shih, C.F., and Needleman, A. (1985). A comparison of methods for calculating energy release rates. *Engineering Fracture Mechanics* 21 (2): 405–421.

Li, G., Fei, G., Xia, H., et al. (2012). Spatial and temporal control of shape memory polymers and simultaneous drug release using high intensity focused ultrasound. *Journal of Materials Chemistry* 22 (16): 7692–7696.

Li, S., Abdel-Wahab, A., Demirci, E., and Silberschmidt, V.V. (2013). Fracture process in cortical bone: X-FEM analysis of microstructured models. *International Journal of Fracture* 184: 43–55.

Libre, N.A., Emdadi, A., Kansa, E., et al. (2009). A stabilized RBF collocation scheme for Neumann type boundary value problems. *Computer Moldeling in Engineering and Sciences* 24 (1): 61–80.

Liu, G.R. (2003). *Mesh Free Methods: Moving Beyond the Finite Element Method*. C.R.C. Press.

Liu, G.R. and Gu, Y.T. (2005). *An Introduction to Meshfree Methods and Their Programming*. Springer.

Liu, G.R. and Liu, M.B. (2003). *Smoothed Particle Hydrodynamics, A Meshfree Particle Method*. World Scientific.

Liu, H. and Li, B. (2010). p53 control of bone remodeling. *Journal of Cellular Biochemistry* 111 (3): 2529–2534.

Liu, W.K., Jun, S., and Zhang, Y.F. (1995). Reproducing kernel particle methods. *International Journal for Numerical Methods in Engineering* 20 (8-9): 1081–1106.

Liu, W.K., Park, H.S., Qian, D., et al. (2006a). Bridging scale methods for nanomechanics and materials. *Computer Methods in Applied Mechanics and Engineering* 195: 1407–1421.

Liu, Y., Gall, K., Dunn, M.L., et al. (2006b). Thermomechanics of shape memory polymers: uniaxial experiments and constitutive modeling. *International Journal of Plasticity* 22 (2): 279–313.

Liu, X., Qin, X., and Du, -Z.-Z. (2012). Bone fracture analysis using the extended finite element method (XFEM) with Abaqus, In: *Proceedings LiuBONFA2010*.

Lucy, L.B. (1974). An iterative technique for the rectification of observed distributions. *The Astronomical Journal* 79 (6): 745–754.

Lukkassen, D. and Milton, G.W. (2008). On hierarchical structures and reiterated homogenization. *Function Spaces, Interpolation Theory and Related Topics. Proceedings of the International Conference in Honour of Jaak Peetre on his 65th Birthday. Lund, Sweden, August 17–22, 2000* (ed. M. Cwikel, M. Englis, A. Kufner, et al.), Berlin, New York: De Gruyter. 355–368.

Maceri, F., Marino, M., and Vairo, G. (2010). A unified multiscale mechanical model for soft collagenous tissues with regular fiber arrangement. *Journal of Biomechanics* 43: 355–363.

Malekmohammadi, H. (2012). Interesting Facts about the Human Brain (in Persian).

Mani, G., Feldman, M.D., Patel, D., and Agrawal, C.M. (2007). Coronary stents: a materials perspective. *Biomaterials* 28 (9): 1689–1710.

Marques, S.P.C. and Creus, G.J. (2012). *Computational Viscoelasticity*. Springer.

Martínez, L., Andrade, R., Birgin, E.G., and Martínez, J.M. (2009). PACKMOL: a package for building initial configurations for molecular dynamics simulations. *Journal of Computational Chemistry* 30 (13): 2157–2164.

Martínez-Reina, J., Dominguez, J., and Garcia-Aznar, J.M. (2011). Effect of porosity and mineral content on the elastic constants of cortical bone: a multiscale approach. *Biomechanics and Modeling in Mechanobiology* 10: 309–322.

Martins, P., Pena, E., Natal Jorge, R.M., et al. (2012). Mechanical characterization and constitutive modelling of damage process in rectus sheath. *Journal of the Mechanical Behaviour of Biomedical Materials* 8: 111–122.

Mayo, S.L., Olafson, B.D., and Goddard, W.A., III (1990). DREIDING: A generic force field for molecular simulations. *The Journal of Physics Chemistry* 94: 8897–8909.

McDonald Schetky, L. (1991). Shape memory alloy applications in space systems. *Materials & Design* 12 (1): 29–32.

McGrath, J.A. and Uitto, J. (2010). Anatomy and organization of human skin. In: Chapter 3, *Rook's Textbook of Dermatology*, 1, 8e (eds. T. Burns, S. Breathnach, N. Cox, and C. Griffiths). John Wiley & Sons Ltd.

McGrath, M.H. and Simon, R.H. (1983). Wound geometry and the kinetics of wound contraction. *Plastic and Reconstructive Surgery* 72 (1): 66–72.

Meguid, S.A. (1989). *Engineering Fracture Mechanics*. UK: Elsevier Applied Science.

Melenk, J.M. and Babuska, I. (1996). The partition of unity finite element method: basic theory and applications. *Seminar fur Angewandte Mathematik, Eidgenossische Technische Hochschule*, Research Report No. 96-01, January, CH-8092 Zurich, Switzerland.

Mertz, H.J., Prasad, P., and Irwin, A.L. (1997), Injury risk curves for children and adults in frontal and rear collisions. In: *41st Stapp Car Crash Conference*, 973318.

Merxhani, A. (2016). An introduction to linear poroelasticity. *Archive of Physics, Geophysic.* arXiv, 1607.04274, 1–38.

Metzger, T.A., Kriepl, T.C., Vaughan, T.J., et al. (2015). The in situ mechanics of trabecular bone marrow: the potential for mechanobiological response. *Journal of Biomechanical Engineering* 137 (1): 10.115.

Micoulaut, M. (2013) *Basics for Molecular Simulations, Atomic Modeling of Glass*. Lecture Notes LECTURE4 MD BASICS, Université Pierre et Marie Curie.

Miehe, C. and Koch, A. (2002). Computational micro-to-macro transitions of discretized microstructures undergoing small strains. *Archive of Applied Mechanics* 72: 300–317.

Miehe, C., Schotte, J., and Schroder, J. (1999). Computational micro–macro transitions and overall moduli in the analysis of polycrystals at large strains. *Computational Materials Science* 16: 372–382.

Miehe, C., Welschinger, F., and Hofacker, M. (2010). Thermodynamically consistent phase_ field models of fracture: Variational principles and multi-field FE implementations. *International Journal for Numerical Methods in Engineering* 83: 1273–1311.

Migliavacca, F., Petrini, L., Colombo, M., et al. (2002). Mechanical behavior of coronary stents investigated through the finite element method. *Journal of Biomechanics* 35 (6): 803–811.

Miller, R., Tadmor, E.B., Phillips, R., and Ortiz, M. (1998). Quasicontinuum simulation of fracture at the atomic scale. *Modelling and Simulation in Materials Science and Engineering* 6: 607–638.

Milton, G.W. (2002). *The Theory of Composites*. Cambridge University Press.
Minamel (2018). *The Differences of White and Grey Matters of the Brain*. Minamel Medical Co. (in Persian).
Moës, N. and Belytschko, T. (2002a). Extended finite element method for cohesive crack growth. *Engineering Fracture Mechanics* 69: 813–833.
Moës, N., Dolbow, J., and Belytschko, T. (1999). A finite element method for crack growth without remeshing. *International Journal for Numerical Methods in Engineering* 46: 131–150.
Mohammadi, S. (2003). *Discontinuum Mechanics Using Finite and Discrete Elements*. UK: WIT Press.
Mohammadi, S. (2008). *Extended Finite Element Method for Fracture Analysis of Structures*. UK: Wiley/Blackwell.
Mohammadi, S. (2012a). *XFEM Fracture Analysis of Composites*. John Wiley & Sons Ltd.
Mohammadi, S. (2012b) *Multiscale Analysis*, Lecture notes, University of Tehran, Tehran, Iran.
Mohammadi, S. and Malekafzali, S. (2011) *Application of Disclocation Dynamics for Analysis of New Orthotropic Materials in Nano-Scale*, Report HRDC-22139, Iran Nano Technology Initiative Council.
Moreo, P., García-Aznar, J.M., and Doblaré, M. (2008). Modeling mechanosensing and its effect on the migration and proliferation of adherent cells. *Acta Biomaterialia* 4 (3): 613–621.
Morgan, E.F. and Keaveny, T.M. (2001). Dependence of yield strain of human trabecular bone on an atomic site. *Journal of Biomechanics* 34 (5): 569–577.
Morgan, E.F., Salisbury Palomares, K.T., Gleason, R.E., et al. (2010). Correlations between local strains and tissue phenotypes in an experimental model of skeletal healing. *Journal of Biomechanics* 43 (12): 2418–2424.
Morgan, N.B. (2004). Medical shape memory alloy applications – the market and its products. *Materials Science and Engineering: A* 378 (1–2): 16–23.
Moslemzadeh, H., Alizadeh, O., and Mohammadi, S. (2019). Quasicontinuum multiscale modeling of the effect of rough surface on nanoindentation behavior. *Meccanica* 54 (3): 411–427.
Moslemzadeh, H. and Mohammadi, S. (2022) An atomistic entropy based finite element multiscale method for modeling amorphous materials. *International Journal of Solids and Structures* 256: 111983.
Motamedi, D. and Mohammadi, S. (2010a). Dynamic analysis of fixed cracks in composites by the extended finite element method. *Engineering Fracture Mechanics* 77: 3373–3393.
Motamedi, D. and Mohammadi, S. (2010b). Dynamic crack propagation analysis of orthotropic media by the extended finite element method. *International Journal of Fracture* 161: 21–39.
Motamedi, D. and Mohammadi, S. (2012). Fracture analysis of composites by time independent moving-crack orthotropic XFEM. *International Journal of Mechanical Sciences* 54: 20–37.
Moukalled, F., Mangani, L., and Darwish, M. (2015). *The Finite Volume in Computational Fluid Dynamics*. Springer.
Nakayama, Y. and Boucher, R.F. (1999). *Introduction to Fluid Mechanics*. Butterworth-Heinemann.
Nalla, R.K., Immbeni, V., Kinney, J.H., et al. (2003). *In vitro* fatigue behavior of human dentin with implications for life prediction. *Journal of Biomedical Materials Research* 66A (1): 10–20.

Nawathe, S.M., Akhlaghpour, H., Bouxsein, M.L., and Keaveny, T.M. (2014). Microstructural failure mechanisms in the human proximal femur for sideways fall loading. *Journal of Bone and Mineral Research* 29 (2): 507–515.

Nepf, H. (2008). *Transport Processes in the Environment*. MIT OpenCourseWare.

Neto, E.D.S., Peric, D., and Owen, D.R.J. (2008). *Computational Methods for Plasticity – Theory and Applications*. UK: John Wiley & Sons Ltd.

Nocedal, J. and Wright, S. (2006). *Numerical Optimization*. Springer Science & Business Media.

Nolan, D.R. and Lally, C. (2019). An investigation of damage mechanisms in mechanobiological models of in-stent restenosis. *Journal of Computational Science* 24: 132–142.

Norouzi, E., Moslemzadeh, H., and Mohammadi, S. (2019). Maximum entropy based finite element analysis of porous media. *Frontiers of Structural and Civil Engineering* 13 (2): 364–379.

Odell, G.M., Oster, G., Alberch, P., and Burnside, B. (1981). The mechanical basis of morphogenesis. *Developmental Biology* 85 (2): 446–462.

Odgaard, A., Kabel, J., van Rietbergen, B., et al. (1997). Fabric and elastic principal directions of cancellous bone are closely related. *Journal of Biomechanics* 30 (5): 487–495.

Olsen, L., Maini, P.K., and Sherratt, J.A. (1998). Spatially varying equilibria of mechanical models: application to dermal wound contraction. *Mathematical Biosciences* 147 (1): 113–129.

Olsen, L., Sherratt, J.A., and Maini, P.K. (1995). A mechanochemical model for adult dermal wound contraction and the permanence of the contracted tissue displacement profile. *Journal of Theoretical Biology* 177 (2): 113–128.

Onate, E., Sacco, C., and Idelsohn, S. (1999) A finite point method for incompressible flow problems. Research Reports of the International Centre for Numerical Methods in Engineering (CIMNE).

Orowan, E. (1948). Fracture and strength of solids. *Reports on Progress in Physics* XII: 185–232.

Osher, S. and Sethian, J.A. (1988). Fronts propagating with curvature-dependent speed: algorithms based on Hamilton–Jacobi formulations. *Journal of Computational Physics* 79 (1): 12–49.

Ostad, H. and Mohammdi, S. (2009). Analysis of shock wave reflection from fixed and moving boundaries using a stabilized particle method. *Particuology* 7: 373–383.

Ostad, H. and Mohammdi, S. (2010). Unsteady fluid–solid interaction by a kernel based particle method. *International Journal for Numerical Methods in Biomedical Engineering* 26: 1596–1603.

Ostad, H. and Mohammdi, S. (2012). A stabilized particle method for large deformation dynamic analysis of structures. *International Journal of Structural Stability and Dynamics* 12 (4): 1250026.

Ostad-Hossein, H. and Mohammadi, S. (2008). A field smoothing stabilization of particle methods in elastodynamics. *Finite Elements in Analysis and Design* 44: 564–579.

Oster, G.F., Murray, J.D., and Harris, A.K. (1983). Mechanical aspects of mesenchymal morphogenesis. *Journal of Embryology and Experimental Morphology* 78: 83–125.

Owen, D.R.J. and Fawkes, A. (1983). *Engineering Fracture Mechanics: Numerical Methods and Applications*. Swansea, UK: Pineridge Press Ltd.

Owen, D.R.J. and Hinton, E. (1980). *Finite Elements in Plasticity: Theory and Practice*. Pineridge Press.

Pahr, D.H. and Zysset, P.K. (2014). A comparison of enhanced continuum FE with micro FE models of human vertebral bodies. *Journal of Bone and Mineral Research* 29 (2): 507–515. 2014.

Parchei, M., Mohammadi, S., and Zafarani, H. (2011). Two-dimensional dynamic extended finite element method for simulation of seismic fault rupture. In: *International Conference on Extended Finite Element Methods – XFEM 2011* (eds. S. Bordas, B. Karihaloo and P. Kerfriden). Cardiff, United Kingdom.

Paris, P. and Erdogan, F. (1963). A critical analysis of crack propagation laws. *Journal of Basic Engineering* 85 (4): 528–533.

Paris, P.C., Gomez, M.P., and Anderson, W.E. (1961). A rational analytic theory of fatigue. *The Trend in Engineering* 13 (1): 9–14.

Pelton, A.R., DiCello, J., and Miyazaki, S. (2000). Optimisation of processing and properties of medical grade Nitinol wire. *Minimally Invasive Therapy & Allied Technologies* 9: 107–118.

Peña, E. (2011a). A rate dependent directional damage model for fibred materials: application to soft biological tissues. *Computational Mechanics* 48: 407–420.

Peña, E. (2011b). Damage functions of the internal variables for soft biological fibred tissues. *Mechanics Research Communications* 38: 610–615.

Pericevic, I., Lally, C., Toner, T., and Kelly, D.J. (2009). The influence of plaque composition on underlying arterial wall stress during stent expansion: the case for lesion-specific stents. *Medical Engineering & Physics* 31: 428–433.

Petrenko, R. and Meller, J. (2010). *Molecular Dynamics*. John Wiley & Sons Ltd.

Pfefferbaum, A., Sullivan, E.V., Hedehus, M., et al. (2000). Age-related decline in brain white matter anisotropy measured with spatially corrected echo-planar diffusion tensor imaging. *Magnetic Resonance in Medicine* 44 (2): 259–268.

Piegl, L. and Tiller, W. (1997). *The NURBS book (Monographs in Visual Communication)*, 2e. Springer-Verlag.

Pierce, G.F., Vande Berg, J., Rudolph, R., et al. (1991) Platelet-derived growth factor-BB and transforming growth factor beta1 selectively modulate glycosaminoglycans, collagen, and myofibroblasts in excisional wounds. *American Journal of Pathology*, 138 (3): 629–646.

Pivonka, P. and Dunstan, C.R. (2012). Role of mathematical modeling in bone fracture healing. *BoneKEy Reports* 1: 221.

Plimpton, S. (1995). Fast parallel algorithms for short-range molecular dynamics. *Journal of Computational Physics* 117 (1): 1–19.

Podshivalov, L., Fischer, A., and Bar-Yoseph, P.Z. (2014). On the road to personalized medicine: multiscale computational modeling of bone tissue. *Archives of Computational Methods in Engineering* 21: 399–479.

Pokluda, J. and Sandera, P. (2010). *Micromechanisms of Fracture and Fatigue in a Multi-scale Context*. Springer.

Powers, J.M. (2019) Lecture Notes on Gas Dynamics. Department of Aerospace and Mechanical Engineering, University of Notre Dame, France.

Prange, M.T., Meaney, D.F., and Margulies, S.S. (2000). Defining brain mechanical properties: effects of region, direction, and species. *Stapp Car Crash Journal* 44: 205–213.

Rafii-Tabar, H. and Mansoori, G.A. (2004). Interatomic potential models for nanostructures. *Encyclopedia of Nanoscience and Nanotechnology* 4: 231–248.

Rahmoun, J., Auperrin, A., Delille, R., et al. (2014). Characterization and micromechanical modeling of the human cranial bone elastic properties. *Mechanics Research Communications* 60: 7–14.

Raissi, M., Perdikaris, P., and Karniadakis, G.E. (2019). Physics-informed neural networks: A deep learning frame-work for solving forward and inverse problems involving nonlinear partial differential equations. *Journal of Computational Physics* 378: 686–707.

Ramezani, H., El-Hraiech, A., Jeong, J., and Benhamou, C.L. (2012). Size effect method application for modeling of human cancellous bone using geometrically exact Cosserat elasticity. *Computer Methods in Applied Mechanics and Engineering* 237–240 (2012): 227–243.

Rashetnia, R. and Mohammadi, S. (2015). Finite strain fracture analysis using the extended finite element method with a new set of enrichment functions. *International Journal for Numerical Methods in Engineering* 102: 1316–1351.

Rice, J.R. and Rosengren, G.F. (1968). Plane strain deformation near a crack tip in a power-law hardening material. *Journal of Mechanics and Physics of Solids* 16: 1–12.

Ritchie, R.O. and Knott, J.F. (1973). Mechanisms of fatigue crack growth in low alloy steel. *Acta Merallurgica* 21 (5): 639–648.

Rodrıguez, J.F., Ruiz, C., Doblare, M., and Holzapfel, G.A. (2008). Mechanical stresses in abdominal aortic aneurysms: influence of diameter, asymmetry, and material anisotropy. *Journal of Biomechanical Engineering* 130: 021023.

Roh, H.Y. and Cho, M. (2004). The application of geometrically exact shell elements to B-spline surfaces. *Computer Methods in Applied Mechanics and Engineering* 193: 2261–2299.

Roy, D., Holzapfel, G.A., Kauffmann, C., and Soulez, G. (2014). Finite element analysis of abdominal aortic aneurysms: geometrical and structural reconstruction with application of an anisotropic material model. *The IMA Journal of Applied Mathematics* 79: 1011–1026.

Rugonyi, S. and Bathe, K.J. (2001). On finite element analysis of fluid flows fully coupled with structural interactions. *Computer Modeling in Engineering and Sciences* 2 (2): 195–212.

Sacks, M.S. and Yoganathan, A.P. (2007). Heart valve function: a biomechanical perspective. *Philosophical Transactions of the Royal Society B: Biological Sciences* 362 (1484): 1369–1391.

Sadd, M.H. (2005). *Elasticity: Theory, Applications, and Numeric*. Elsevier.

Sadeghirad, A. and Mohammadi, S. (2007). Equilibrium on line method (ELM) for imposition of Neumann boundary conditions in the finite point method (FPM). *International Journal for Numerical Methods in Engineering* 69 (1): 60–86.

Saeb, S., Steinmann, P., and Javili, A. (2016). Aspects of computational homogenization at finite deformations: a unifying review from Reuss' to Voigt's bound. *Applied Mechanics Reviews* 68: 050801.

Sahoo, D., Deck, C., and Willinger, R. (2015). Axonal strain as brain injury predictor based on real-world head trauma simulations. In: *2015 IRCOBI Conference Proceedings – International Research Council on the Biomechanics of Injury*, 186–197.

Sakakura, K., Nakano, M., Otsuka, F., et al. (2013). Pathophysiology of atherosclerosis plaque progression. *Heart, Lung and Circulation* (Edited by Birkedal-Hansen, H., Moore, W.G., Bodden, M.K., Windsor, L.J., Birkedal-Hansen, B., DeCarlo, A. and Engler, J.A.). 22 (6): 399–411.

Sanchez, F. and Zhang, L. (2010). Interaction energies, structure, and dynamics at functionalized graphitic structure–liquid phase interfaces in an aqueous calcium sulfate solution by molecular dynamics simulation. *Carbon* 48 (4): 1210–1223.

Sánchez, P.J., Blanco, P.J., Huespe, A.E., and Feijóo, R.A. (2013). Failure-oriented multi-scale variational formulation: micro-structures with nucleation and evolution of softening bands. *Computer Methods in Applied Mechanics and Engineering* 257: 221–247.

Sansalone, V., Naili, S., Bousson, V., et al. (2010). Determination of the heterogeneous anisotropic elastic properties of human femoral bone: from nanoscopic to organs scale. *Journal of Biomechanics* 43: 1857–1863.

Saouma, V.E. (2000) *Fracture Mechanics*. Lecture notes CVEN-6831, University of Colorado, USA.

Saouma, V.E., Ayari, M., and Leavell, D. (1987). Mixed mode crack propagation in homogeneous anisotropic solids. *Engineering Fracture Mechanics* 27 (2): 171–184.

Schiavone, A. and Zhao, L.G. (2015). A study of balloon type, system constraint and artery constitutive model used in finite element simulation of stent deployment. *Mechanics of Advanced Materials and Modern Processes* 1 (1): 10.1186.

Schlick, T. (2002). *Molecular Modeling and Simulation: An Interdisciplinary Guide*. Springer.

Schmidt, M. (2022). *100 Lectures on Machine Learning*. Available online.

Schmitt, K.-U., Niederer, P., Muser, M., and Walz, F. (2009). *Trauma Biomechanics: Accidental Injury in Traffic and Sports*. Springer Science & Business Media.

Schofield, J. and van Zon, R. (2008). *Topics in Statistical Mechanics: The Foundations of Molecular Simulation*. Lecture Notes, CHM1464H Fall 2008, University of Toronto.

Schwiedrzik, J., Gross, T., Bina, M., et al. (2016). Experimental validation of a nonlinear μFE model based on cohesive-frictional plasticity for trabecular bone. *International Journal for Numerical Methods in Biomedical Engineering* 32 (4): e02739.

Shahi, S. (2013) *Multiscale Simulation of Biomechanical Systems*, M.Sc. Thesis, School of Civil Engineering, University of Tehran, Tehran, Iran.

Shahi, S., Fenton, F.H., and Cherry, E.M. (2022). A machine-learning approach for long-term prediction of experimental cardiac action potential time series using an autoencoder and echo state networks. *Chaos* 32: 063117.

Shahi, S. and Mohammadi, S. (2013). A multiscale finite element simulation of human aortic heart valve. *Applied Mechanics and Materials* 367: 275–279.

Sharma, K., Biu, T.Q., Zhang, C., and Bhargava, R.R. (2012). Analysis of subinterface crack in piezoelectric bimaterials with the extended finite element method. *Engineering Fracture Mechanics* 104: 114–139.

Shenoy, V.B., Miller, R., Tadmor, E.B., et al. (1998). Quasicontinuum models of interfacial structure and deformation. *Physical Review Letters* 80 (4): 742–745.

Shenoy, V.B., Miller, R., Tadmor, E.B., et al. (1999). An adaptive finite element approach to atomic-scale mechanics – the quasicontinuum method. *Journal of the Mechanics and Physics of Solids* 47 (3): 611–642.

Sherratt, J.A., Sage, E.H., and Murray, J.D. (1993) Chemical control of eukaryotic cell movement: a new model. *Journal of Theoretical Biology* 162 (1): 23–40.

Shih, C., de Lorenzi, H., and German, M. (1976). Crack extension modelling with singular quadratic isoparametric elements. *International Journal of Fracture* 12: 647–651.

Shilkrot, L.E., Miller, R.E., and Curtin, W.A. (2002). Coupled atomistic and discrete dislocation plasticity. *Physical Review Letters* 89 (2): 025501.

Shuck, L.Z. and Advani, S.H. (1972). Rheilogical response of human brain tissue in shear. *Journal of Basic Engineering* 94 (4): 905–911.

Sih, G.C., Paris, P., and Irwin, G. (1965). On cracks in rectilinearly anisotropic bodies. *International Journal of Fracture Mechanics* 1 (3): 189–203.

Silver, J.M., McAllister, T.W., and Yudofsky, S.C. (2011). *Textbook of Traumatic Brain Injury*. American Psychiatric Publication.

Simo, J. (1987). On a fully three-dimensional finite-strain viscoelastic damage model: formulation and computational aspects. *Computer Methods in Applied Mechanics and Engineering* 60: 153–173.

Simo, J. and Ju, J. (1987). Strain and stress-based continuum damage models – I. Formulation. *International Journal of Solids and Structures* 23: 821–840.

Simo, J.C. and Pister, K.S. (1984). Remarks on rate constitutive equations for a finite deformation problem: computational implications. *Computer Methods in Applied Mechanics and Engineering* 46: 201–215.

Singer, A.J. and Clark, R.A.F. (1999). Cutaneous wound healing. *New England Journal of Medicine* 341 (10): 738–746.

Sommer, G., Gasser, T.C., Regitnig, P., et al. (2008). Dissection properties of the human aortic media: an experimental study. *Journal of Biomechanical Engineering* 130: 021007.

Sowinskil, D.R., McGarry, M.D.J., Van Houten, E.E.V., et al. (2021). Poroelasticity as a model of soft tissue structure: hydraulic permeability reconstruction for magnetic resonance elastography in silico. *Frontiers in Physics* 8: 617582.

Stadelmann, W.K., Digenis, A.G., and Tobin, G.R. (1998). Physiology and healing dynamics of chronic cutaneous wounds. *The American Journal of Surgery* 176 (2): 26–38.

Stein, E., de Borst, R., and Hughes, T.J.R. (eds.) (2007). *Encyclopedia of Computational Mechanics*. John Wiley & sons Ltd.

Stillinger, F.H. and Weber, T.A. (1983). Dynamics of structural transitions in liquids. *Physical Review A* 28 (4): 2408–2416.

Stoeckel, D., Bonsignore, C., and Duda, S. (2002). A survey of stent designs. *Minimally Invasive Therapy & Allied Technologies* 11 (4): 137–147.

Strachan, A. (2017). *Lectures on Molecular Dynamics Simulations of Materials*. School of Materials Engineering, Purdue University.

Strube, P., Sentuerk, U., Riha, T., et al. (2008). Influence of age and mechanical stability on bone defect healing: age reverses mechanical effects. *Bone* 42: 758–764.

Stukowski, A. (2009). Visualization and analysis of atomistic simulation data with OVITO – the open visualization tool. *Modelling and Simulation in Materials Science and Engineering* 18: 015012.

Stylianopoulos, T. and Barocas, V.H. (2007). Multiscale, structure-based modeling for the elastic mechanical behavior of arterial walls. *Journal of Biomechanical Engineering* 129: 611–618.

Stynes, M. and Stynes, D. (2010). *Convection-Diffusion Problems, An Introduction to Their Analysis and Numerical Solution*. American Mathematical Society, Rhode Island.

Sukumar, N., Huang, Z., Prevost, J.H., and Suo, Z. (2004). Partition of unity enrichment for bimaterial interface cracks. *International Journal for Numerical Methods in Engineering* 59: 1075–1102.

Suresh, S. (1998). *Fatigue of Materials*, 2e. Cambridge University Press.

Swope, W.C. and Anderson, H.C. (1984) A molecular dynamics method for calculating the solubility of gases in liquids and the hydrophobic hydration of inert-gas atoms in aqueous solution. *The Journal of Physical Chemistry A*, 88 (26): 6548–6556.

Tadmor, E.B. (1996) *The Quasicontinuum Method*. Ph.D. Thesis, Division of Engineering, Brown University.

Tadmor, E.B., Miller, R., and Phillips, R. (1999). Nanoindentation and incipient plasticity. *Journal of Materials Research* 14 (6): 2233–2250.

Tadmor, E.B. and Miller, R.E. (2011). *Modeling Materials; Continuum, Atomistic and Multiscale Techniques*. Cambridge University Press.

Tadmor, E.B., Miller, R.E., and Elliot, R.S. (2012). *Continuum Mechanics and Thermodynamics*. Cambridge University Press.

Tadmor, E.B., Ortiz, M., and Phillips, R. (1996). Quasicontinuum analysis of defects in solids. *Philosophical Magazine A* 73 (6): 1529–1563.

Tadmor, E.B., Waghmare, U.V., Smith, G.S. and Kaxiras, E. (2002) Polarization switching in $PbTiO_3$: an ab initio finite element simulation. *Acta Materials* 50: 2989–3002.

Tang, H., Buehler, M.J., and Moran, B. (2009). A constitutive model of soft tissue: from nanoscale collagen to tissue continuum. *Annals of Biomedical Engineering* 37: 1117–1130.

Teo, E.C., Yuan, Q., and Yeo, J.H. (2000). Design optimization of coronary stent using finite element analysis. *Asaio Journal* 46 (2): 201.

Tersoff, J. (1988). New empirical approach for the structure and energy of covalent systems. *Physical Review B* 37 (12): 6991–7000.

Thomas, J.W. (1995). *Numerical Partial Differential Equations: Finite Difference Methods*. Springer.

Thomas, V.S. (2010). *A Comparative Study for the Effect of Tissue Anisotropy on the Behavior of a Single Cardiac Pressure Cycle for a Symmetric Tri-leaflet Valve*. The University of Akron.

Tobushi, H., Hashimoto, T., Hayashi, S., and Yamada, E. (1997). Thermomechanical constitutive modeling in shape memory polymer of polyurethane series. *Journal of Intelligent Material Systems and Structures* 8 (8): 711–718.

Torabizadeh, A.R. and Mohammadi, S. (2022a). *Analysis of SMA Stenting*. Report HPC2022-03, High Performance Computing Lab, School of Civil Engineering, University of Tehran, Iran.

Torabizadeh, A.R. and Mohammadi, S. (2022b). *Multiscale Analysis of Bone Fracture*. Report HPC2022-04, High Performance Computing Lab, School of Civil Engineering, University of Tehran, Iran.

Tortora, G.J. and Derrickson, B.H. (2017). *Tortora's Principles of Anatomy and Physiology*. John Wiley & Sons Ltd.

Tranquillo, R.T. and Murray, J.D. (1992). Continuum model of fibroblast-driven wound contraction: inflammation-mediation. *Journal of Theoretical Biology* 158 (2): 135–172.

Uchio, E., Ohno, S., Kudoh, J., et al. (1999). Simulation model of an eyeball based on finite element analysis on a supercomputer. *British Journal of Ophthalmology* 83: 1106–1111.

Ural, A., Bruno, P., Zhou, B., et al. (2013). A new fracture assessment approach coupling HR-pQCT imaging and fracture mechanics-based finite element modeling. *Journal of Biomechanics* 46: 1305–1311.

Ural, A. and Mischinski, S. (2013). Multiscale modeling of bone fracture using cohesive finite elements. *Engineering Fracture Mechanics* 103: 141–152.

Ural, A. and Vashishth, D. (2014). Hierarchical perspective of bone toughness – from molecules to fracture. *International Materials Reviews* 59 (5): 245–263.

Van Humbeeck, J. (1999). Non-medical applications of shape memory alloys. *Materials Science and Engineering. A* 273: 134–148.

Vasava, P., Jalali, P., and Dabagh, M. (2008). Pulsatile blood flow simulations in aortic arch: effects of blood pressure and the geometry of arch on wall shear stress. In: *ECIFMBE 2008, IFMBE Proceedings 22* (eds. J. Vander Sloten, P. Verdonck, M. Nyssen and J. Haueisen), Springer. 1926–1929.

Vaughan, T.J., McCarthy, C.T., and McNamara, L.M. (2012). A three-scale finite element investigation into the effects of tissue mineralisation and lamellar organisation in human cortical and trabecular bone. *Journal of the Mechanical Behaviour of Biomedical Materials* 12: 50–62.

Velardi, F., Fraternali, F., and Angelillo, M. (2006). Anisotropic constitutive equations and experimental tensile behavior of brain tissue. *Biomechanics and Modeling in Mechanobiology* 5 (1): 53–61.

Verhoosel, C.V., Scott, M.A., Hughes, T.J.R., and de Borst, R. (2011a). An isogeometric analysis approach to gradient damage models. *International Journal for Numerical Methods in Engineering* 86 (1): 115–134.

Verhoosel, C.V., Scott, M.A., de Borst, R., and Hughes, T.J.R. (2011b). An isogeometric approach to cohesive zone modeling. *International Journal for Numerical Methods in Engineering* 87 (1–5): 336–360.

Verhoosel, C.V., van Zwieten, G.J., van Rietbergen, B., and de Borst, R. (2015). Image-based goal-oriented adaptive isogeometric analysis with application to the micro-mechanical modelling of trabecular bone. *Computer Methods in Applied Mechanics and Engineering* 284: 138–164.

Violeau, D. (2012). *Fluid Mechanics and the SPH Method*. Oxford University Press.

Vokhshoori, M. and Mohammadi, S. (2022) *Multiscale Analysis of Polymer Chains*. Report HPC2022-02, High Performance Computing Lab, School of Civil Engineering, University of Tehran, Iran.

Voyajilis, G.Z. and Kattan, P.I. (2005). *Damage Mechanics*. CRC Press, Taylors & Francis.

Voyiadis, G.Z. and Woelke, P. (2008). *Elastoplastic and Damage Analysis of Plates and Shells*. Springer.

Voyiadjis, G.Z. and Kattan, P.I. (1996). *Advances in Damage Mechanics*. UK: Elsevier.

Waffenschmidt, T., Polindara, C., and Menzel, A. (2014). A gradient-enhanced continuum damage model for fibre-reinforced materials at finite strains. *Proceedings in Applied Methematics and Mechanics* 14 (1): 153–154.

Wagegg, M., Gaber, T., Lohanatha, F.L., et al. (2012). Hypoxia promotes osteogenesis but suppresses adipogenesis of human mesenchymal stromal cells in a hypoxia-inducible factor-1 dependent manner. *PLoS ONE* 7 (9): e46483.

Wagner, G.J. and Liu, W.K. (2003). Coupling of atomistic and continuum simulations using a bridging scale decomposition. *Journal of Computational Physics* 190 (1): 249–274.

Wahnstrom, G. (2018) *Molecular Dynamics*, Lecture Notes, Goteborg.

Waite, L. and Fine, J. (2007). *Applied Biofluid Mechanics*. New York, USA: McGraw-Hill.

Wang, D. and Xuan, J. (2010). An improved NURBS-based isogeometric analysis with enhanced treatment of essential boundary conditions. *Computer Methods in Applied Mechanics and Engineering* 199: 2425–2436.

Wang, H.F. (2000). *Theory of Linear Poroelasticity with Applications to Geomechanics and Hydrology*. Princeton.

Wang, J.G. and Liu, G.R. (2002). On the optimal shape parameters of radial basis functions used for 2-D meshlesss methods. *Computer Methods in Applied Mechanics and Engineering* 191: 2611–2630.

Wang, R. and Kumpel, H.-J. (2003). Poroelasticity: efficient modelling of strongly coupled, slow deformation processes in a multilayered half-space. *Geophysics* 68 (2): 705–717.

Wang, S., Yu, X., and Perdikaris, P. (2022). When and why PINNs fail to train: A neural tangent kernel perspective. *Journal of Computational Physics* 449: 110768.

Wang, X. and Hong, W. (2012). A visco-poroelastic theory for polymeric gels. *Proceedings of the Royal Society, A: Mathematical, Physical and Engineering Sciences* 468: 3824–241.

Wehner, T., Steiner, M., Ignatius, A., and Claes, L. (2014). Prediction of the time course of callus stiffness as a function of mechanical parameters in experimental rat fracture healing studies – a numerical study. *PLoS ONE* 9 (12): e115695.

Weinberg, E.J. and Kaazempur Mofrad, M.R. (2007a). Transient, three-dimensional, multiscale simulations of the human aortic valve. *Cardiovascular Engineering* 7 (4): 140–155.

Weinberg, E.J. and Kaazempur Mofrad, M.R. (2007b). A finite shell element for heart mitral valve leaflet mechanics, with large deformations and 3D constitutive material model. *Journal of Biomechanics* 40 (3): 705–711.

Weinberg, E.J. and Kaazempur Mofrad, M.R. (2008). A multiscale computational comparison of the bicuspid and tricuspid aortic valves in relation to calcific aortic stenosis. *Journal of Biomechanics* 41 (16): 3482–3487.

Weiss, J.A. (1995) *A Constitutive Model and Finite Element Representation for Transversely Isotropic Soft Tissues*. Ph.D. Thesis, Department of Bioengineering, University of Utah.

Wells, A. (1963) *Application of Fracture Mechanics At and Beyond General Yielding*. Report M13, British Welding Research Association, UK.

Wells, A.A. (1955) *The Condition of Fast Fracture in Aluminium Alloys with Particular Reference to Comet Failures*. British Welding Research Association Report.

Wenik, J.M. and Meguid, S. (2011). Multiscale modeling of the nonlinear response of nano-reinforced polymers. *Acta Mechanica* 217 (1): 1–16.

Westergaard, H. (1939). Bearing pressures and cracks. *ASME Journal of Applied Mechanics* 6: 49–53.

Wijn, P.F.F., Brakkee, A.J.M., Kuiper, J.P., and Vendrik, A.J.H. (1981). The alinear viscoelastic properties of human skin in vivo related to sex and age. *Bioengineering and the Skin (eds. R. Marks and P.A. Payne)*, Netherlands: Springer. 135–145.

Williams, M.L. (1952). Stress singularities resulting from various boundary conditions in angular corners of plates in extension. *Journal of Applied Mechanics, Transactions ASME* 19: 526–528.

Wischke, C., Neffe, A.T., Steuer, S., and Lendlein, A. (2009). Evaluation of a degradable shape-memory polymer network as matrix for controlled drug release. *Journal of Controlled Release* 138 (3): 243–250.

Wright, R.M., Post, A., Hoshizaki, B., and Ramesh, K.T. (2013). A multiscale computational approach to estimating axonal damage under inertial loading of the head. *Journal of Neurotrauma* 30 (2): 102–118.

Wu, W., Wang, W.-Q., Yang, D.-Z., and Qi, M. (2007). Stent expansion in curved vessel and their interactions: a finite element analysis. *Journal of Biomechanics* 40 (11): 2580–2585.

Xiao, J., Liu, B., Huang, Y., et al. (2007). Collapse and stability of single – and multi-wall carbo nanotubes. *Nanotechnology* 18 (39): 395703.

Xiao, S. and Belytschko, T. (2004). A bridging domain method for coupling continua with molecular dynamics. *Computer Methods in Applied Mechanics and Engineering* 193 (17–20): 1645–1669.

Yang, G., Kabel, J., van Rietbergen, B. et al. (1999). The anisotropic Hooke's law for cancellous bone and wood. *Journal of Elasticity* 53: 125–146.

Zamani, M. (2022) *Numerical Simulation of Healing Process in Hard Biological Tissues*. M.Sc. Thesis, School of Civil Engineering, University of Tehran, Iran.

Zhao, W., Liu, L., Zhang, F., et al. (2019). Shape memory polymers and their composites in biomedical applications. *Materials Science and Engineering: C* 97: 864–883.

Zhou, B., Kang, Z., Wang, Z., and Xue, S. (2019). Finite element method on shape memory alloy structure and its applications. *Chinese Journal of Mechanical Engineering* 32: 84.

Zhu, H. (2018). *Graphene; Fabrication, Characterizations, Properties and Applications*. Elsevier Inc.

Zienkiewicz, O.C., Taylor, R.L., and Zhu, J.Z. (2005). *The Finite Element Method*, 6e. USA: Elsevier.

Zigmond, S.H. (1981). Consequences of chemotactic peptide receptor modulation for leukocyte orientation. *Journal of Cell Biology* 88 (3): 644–647.

Zigmond, S.H., Sullivan, S.J., and Lauffenburger, D.A. (1982). Kinetic analysis of chemotactic peptide receptor modulation. *Journal of Cell Biology* 92 (1): 34–43.

Zohdi, T.I., Oden, J.T., and Rodin, G.J. (1996). Hierarchical modeling of heterogeneous bodies. *Computer Methods in Applied Mechanics and Engineering* 138: 273–298.

Zolghadr, S. (2022) *Multiscale Analysis of Brain Damage Due to Quasi Static Loading*. M.Sc. Thesis, School of Civil Engineering, University of Tehran, Iran.

Zolghadr, S. and Mohammadi, S. (2022). Viscoelastic response of brain tissue in quasi-static loadings. In: *30th Annual Conference of ISME*.

Index

Multiscale Biomechanics: Theory and Applications, First Edition. Soheil Mohammadi.

c

d

e

f

g

h

n

o

p

t

y

z

μ